SYMBOL	DESCRIPTION

Symbol	Description
R	(1) A random variable denoting the number of successes obtained from several trials of a Bernoulli process (2) The test statistic for the matched-pairs sign test for comparing two populations; R is the number of positive sign differences for the paired difference between two sample observations
R_a	Number of runs of type a obtained in a sample; used as the test statistic for the number-of-runs test
$R_{Y \cdot 12},$ $R_{Y \cdot 123}$	Sample multiple correlation coefficient; the square root of the coefficient of multiple determination
$R^2_{Y \cdot 12},$ $R^2_{Y \cdot 123}$	Sample coefficient of multiple determination; represents the proportion of variation in the dependent variable Y that can be explained by the multiple regression equation of Y on independent variables X_1, X_2 (and X_3)
r	(1) One of several possible values of the random variable R (2) The sample correlation coefficient; expresses the strength and direction of the relationship between variables X and Y (3) Number of rows in an analysis-of-variance layout
r^2	Sample coefficient of determination; represents the proportion of the variation in the dependent variable Y that can be explained by linear regression on the independent variable X
r_s	Spearman rank correlation coefficient
r^2_{Y1}	Sample coefficient of determination for regression of the dependent variable Y on the independent variable X_1; used in multiple correlation analysis
$r^2_{Y \cdot 12},$ $r^2_{Y3 \cdot 12}$	Coefficients of partial determination for multiple regression of the dependent variable Y on the independent variables X_1, X_2 (and X_3); measures the proportional reduction in previously unexplained variation in Y by adding X_2 or X_3 to the multiple regression analysis while incorporating the effects of the other independent variables
ρ (rho)	Population correlation coefficient; estimated by r
ρ^2	Population coefficient of determination; estimated by r^2
S	Number of successes in a population; used in conjunction with the hypergeometric distribution
$S_1, S_2, \ldots$	Strategies in a decision structure
S_t	Seasonal component of the classical time series model
$S_{Y \cdot 12},$ $S_{Y \cdot 123}$	Standard error of the estimate about the multiple regression plane
SSA	Sum of squares between columns for a two-factor analysis of variance; A represents one of the factors about which inferences are to be made
SSB	Sum of squares between rows for a two-factor analysis of variance when: (1) B represents the blocking variable (2) B represents one factors about which inferences are to be made
$SSCOL$	Sum of squares between columns for a Latin-square design in a three-factor analysis of variance
SSE	Error sum of squares: (1) (within columns) for a one-factor analysis of variance (2) (residual) for a two-factor or a three-factor analysis of variance
$SSROW$	Sum of squares between rows for a Latin-square design in a three-factor analysis of variance
SST	Sum of squares: (1) (between columns) for the treatments in a one-factor analysis of variance (2) (between letters) in a three-factor analysis of variance using a Latin-square design
s	Sample standard deviation
s^2	Sample variance
s^2_A, s^2_B	Variances of samples from populations A and B; used in two-sample hypothesis tests
s_X	Sample standard deviation of the independent variable X; used in correlation analysis
s_Y	Sample standard deviation of the dependent variable Y; used in correlation analysis
$s_{Y \cdot X}$	Standard error of the estimate about the regression line; $s^2_{Y \cdot X}$ is an unbiased estimator of $\sigma^2_{Y \cdot X}$
s_d	The estimator of the standard error σ_d for the difference between: (1) two sample means (2) two sample proportions
$s_{d\text{-paired}}$	Sample standard deviation of matched-pairs differences
$s_{d\text{-small}}$	Sample standard deviation of the difference in means of independent samples
$\sum$	Summation sign
σ (sigma)	Population standard deviation
σ^2	Population variance
$\sigma(X)$	Standard deviation of random variable X
$\sigma^2(X)$	Variance of random variable X
σ^2_A, σ^2_B	Variance of populations A and B compared by two-sample inferences
σ_P	Standard error (also called the standard deviation) of the sample proportion P
$\sigma_{\bar{X}}$	Standard error (also called the standard deviation) of the sample proportion $\bar{X}$
$\sigma_{Y \cdot X}$	Standard deviation for the conditional probability distribution of the depen-

STATISTICS FOR MODERN BUSINESS DECISIONS

Third Edition

STATISTICS FOR MODERN BUSINESS DECISIONS

Third Edition

LAWRENCE L. LAPIN
San José State University

Harcourt Brace Jovanovich, Inc.
New York   San Diego   Chicago   San Francisco   Atlanta
London   Sydney   Toronto

Preface

In writing this introductory statistics book for students of business and economics, my overriding goal has been to enliven statistics, to make it more interesting and relevant and easier to learn. It is no secret that today's student too often finds statistics boring and irrelevant and more difficult than necessary.

To illustrate that statistics is neither boring nor irrelevant, this book treats it as essentially a decision-making tool and includes many modern concepts and applications. All topics are introduced by carefully chosen examples that illustrate more than technical mathematical concepts. The stage is set, the motivation is provided, and the rationale is given as each new concept is presented. Accordingly, the importance of inferences is highlighted when a statistics professor compares the effectiveness of a new textbook with the text she currently uses. Testing for the equality of several proportions, an ad agency wishes to determine whether there are any differences in terms of reader recall among three different kinds of magazine advertisements. The fact that sampling is just one source of potential error is emphasized by detailing some of the blunders committed in taking the U.S. Decennial Census. Many of the examples and exercises are clearly related to present-day issues, such as health, conservation, and the environment. Several examples of each major area of business and economics—such as accounting, marketing, finance, production, forecasting, and consumer behavior—serve to richly illustrate the applications of statistics.

A course in high school algebra is the only background required. And although this book should prove more accessible than most, the relevant nature of its presentation has not been achieved at the cost of avoiding reputedly difficult material. Probability, hypothesis testing, and other more difficult topics receive generous explanation, often from several viewpoints. The reader is encouraged to rely more on intuition than on rote memory; less than· the customary emphasis is placed on the mechanical and computational aspects of statistics. The computer's role in statistical analysis in managing the more onerous calculations is highlighted throughout the book. Purely mathematical symbology has been minimized; for example, instead of the Boolean cup and bowl notation, the italicized *and* and *or* are used in describing probability concepts.

This third edition features many improvements over the second. The organization has been streamlined, and a smoother topical transition has been achieved. Over 100 new problems have been added, and many of the original ones have been updated or modified. Most sections within chapters have their own problem sets, permitting the student to relate the questions easily to the concepts just covered. This arrangement also gives the instructor flexibility in picking topics within a chapter. As an added feature, chapters now end with review exercises, allowing the student to gain experience in determining which procedures and concepts to apply. All problems are graded, so that each set begins with easy exercises and increases in difficulty. As a further improvement, much of the statistical jargon and notation has been simplified. All of the third edition's changes should make the book easier to use and to teach from. *Users of the second edition will find a detailed synopsis of the book's changes in the Instructor's Manual.*

The book has been thoroughly class tested in a variety of circumstances and courses in many colleges and universities. The experience of hundreds of instructors has been drawn upon in writing this third edition.

An examination of the table of contents reveals that there is much to choose from in this book. The chapters have been constructed to make it easy for the instructor to design a course to fit individual needs. Classical statistics has not been mixed with statistical decision theory—this book may be used with either emphasis. Many modern texts supplant classical statistics; both the old and the new are available here. Overall, the presentation is familiar except for the inclusion of several topics omitted from most texts. Chapter 14 considers two-sample inferences. Chapters 15 and 16 discuss chi-square applications and analysis of variance. Chapter 17 includes several probability distributions—the hypergeometric, Poisson, exponential, and uniform. These may be conveniently omitted or incorporated in a course without loss of continuity. This third edition includes probability tables for both the binomial and Poisson distributions. Chapter 18 contains some nonparametric statistics most useful for business applications. The Bayesian decision-making procedures of Chapters 19–24 emphasize decision trees; much of the symbology and terminology of decision theory is avoided to allow a simpler and more pragmatic presentation. The three chapters on so-called Bayesian methods have been used primarily to extend statistics to areas where classical procedures have proved inadequate in analyzing decisions under uncertainty.

A glossary of statistical symbols is provided on the endpapers for easy reference. Abbreviated answers to all even-numbered exercises are included in the back of the book. Complete solutions to all exercises are available in the Instructor's Manual, along with nearly 200 additional exercises and more than 500 examination questions and their solutions. The Instructor's Manual also provides teaching suggestions and hints on structuring courses. A Study Guide containing over 200 solved problems is also available for student use as a workbook.

I am greatly indebted to the many people who have assisted me in preparing this book. Special thanks go to my colleagues whose comments were invaluable in setting the tone and to the reviewers of this edition: Frank Alt, University of Maryland; Joel Gibbons, Illinois Institute of Technology (Chicago); and Jim Willis, Louisiana State University. I am deeply grateful to my students who over many years of debugging the book helped identify the problems a reader might face.

Lawrence L. Lapin

Contents

1 Introduction   1

1-1   The Meaning of Statistics   1

1-2   The Role of Statistics   3

1-3   Descriptive and Inferential Statistics   3

1-4   Deductive and Inductive Statistics   4

1-5   Statistical Error   5

2 Describing Statistical Data   7

2-1   The Population and the Sample   7

Elementary Units
The Frame

Qualitative and Quantitative Populations
Exercises

2-2   The Frequency Distribution   11

Finding a Meaningful Pattern for the Data
Graphical Displays: The Histogram
Graphical Displays: Frequency Polygons and Curves
Descriptive Analysis
Constructing a Frequency Distribution: Width and Number of Class
    Intervals
Qualitative Populations: The Frequency Distribution in Cryptanalysis
Exercises

**2-3   Relative and Cumulative Frequency
Distributions   26**

Relative Frequency Distributions
Cumulative Frequency Distributions
Exercises

2-4   Common Forms of the Frequency Distribution   32

Exercises

Review Exercises   37

3 *Summary Descriptive Measures*   *39*

3-1   Statistics, Parameters, and Decision Making   40

3-2   The Arithmetic Mean   41

Symbolic Expressions for Calculating the Arithmetic Mean
Calculating the Mean from Grouped Data
Exercises

3-3   The Median and the Mode   45

The Median
The Median Contrasted with the Mean
The Mode
Positional Comparison of Measures: Skewed Distributions
Bimodal Distributions
Exercises

3-4   Measuring Variability   53

Importance of Variability
Fractiles, Percentiles, and Distance Measures of Dispersion
Measures of Average Deviation
The Variance
The Standard Deviation
Shortcut Calculations of the Variance and the Standard Deviation
The Meaning of the Standard Deviation
Exercises

3-5   The Proportion   64

Exercises

Review Exercises   66

4 *The Statistical Sampling Study*   71

4-1   The Need for Samples   71

The Economic Advantages of Using a Sample
The Time Factor
The Very Large Population
Partly Inaccessible Populations
The Destructive Nature of the Observation
Accuracy and Sampling
Exercises

4-2   Designing and Conducting a Sampling Study   75

The Importance of Planning
Data Collection
Data Analysis and Conclusions

4-3   Bias and Error in Sampling   78

Sampling Error
Sampling Bias
Nonsampling Error
Exercises

4-4   Selecting the Sample   83

The Convenience Sample
The Judgment Sample

The Random Sample
Sample Selection Using Random Numbers
Types of Random Samples
Exercises

4-5   **Selecting Statistical Procedures   91**

Illustration: Evaluating a Beverage Container Law

Review Exercises   94

5   *Probability*   95

5-1   **Fundamental Concepts   96**

The Event
Basic Definitions of Probability
Certain and Impossible Events
Alternative Expressions of Probability
Exercises

5-2   **Events and Their Relationships   102**

Events Having Components: Union and Intersection
Relationships Between Events
The Law of Large Numbers
Exercises

5-3   **Probabilities for Compound Events   108**

The Addition Law
Events That Are Mutually Exclusive and Collectively Exhaustive
Application to Complementary Events
The Multiplication Law for Independent Events
Finding Joint Probabilities for Dependent Events
Exercises

5-4   **Conditional Probability and the Joint
Probability Table   116**

Conditional Probability
The Joint Probability Table and Marginal Probabilities
Computing Conditional Probability
Testing for Independence by Comparing Probabilities
Exercises

5-5   **The General Multiplication Law, Probability, Trees, and Sampling   123**

The General Multiplication Law
The Matching-Birthdays Problem
Random Sampling
The Probability Tree Diagram
Sampling with and Without Replacement
Exercises

5-6   **Common Errors in Applying the Laws of Probability   132**

Review Exercises   134

5-7   **Optional Topic: Counting Techniques   136**

Underlying Principle of Multiplication
Number of Ways of Sequencing Objects: The Factorial
The Number of Permutations
The Number of Combinations
Optional Exercises

5-8   **Optional Topic: Revising Probabilities Using Bayes' Theorem   146**

Bayes' Theorem
Crooked Die Cube Illustration
Posterior Probability as a Conditional Probability
Posterior Probabilities in Jury Selection
Optional Exercises

6   *Probability Distributions, Expected Value, and Sampling*   *153*

6-1   **Discrete Probability Distributions   154**

The Random Variable
The Probability Distribution of a Random Variable
Discrete and Continuous Random Variables
Exercises

6-2   **Expected Value and Variance   158**

The Expected Value of a Random Variable
The Variance of a Random Variable
Expected Value and Sampling
Exercises

6-3   **The Sampling Distribution of the Mean   163**

Illustration: GMAT Scores
Expected Value of the Sample Mean
Sampling with and Without Replacement
Standard Error of the Sample Mean
Exercises

6-4   **Binomial Probabilities:
The Sampling Distribution of the Proportion   170**

A Coin-Tossing Illustration
The Bernoulli Process
The Number of Combinations
The Binomial Formula
The Proportion of Successes
The Binomial Distribution Family
The Cumulative Probability Distribution
Using Binomial Probability Tables
The Sampling Distribution of the Proportion
Exercises

6-5   **Continuous Probability Distributions   188**

Smoothed Curve Approximation
Probability Density Function
Cumulative Probability Distribution
The Expected Value and the Variance

Review Exercises   193

7 *The Normal Distribution*   *195*

7-1   **Characteristics of the Normal Distribution   195**

7-2   **Finding Areas Under the Normal Curve   200**

Using the Normal Curve Table
Cumulative Probabilities and Percentiles

The Standard Normal Random Variable
Concluding Remarks
Exercises

7-3   **Sampling Distribution of the Sample Mean
for a Normal Population   211**

The Role Played by the Standard Error of $\bar{X}$
Exercises

7-4   **Sampling Distribution of $\bar{X}$ When the Population
Is Not Normal   217**

Effect of Sample Size n on the Sampling Distribution of $\bar{X}$
The Central Limit Theorem
Exercises

7-5   **Sampling Distribution of P and the
Normal Approximation   224**

Advantages of Approximating the Binomial Distribution
Normal Distribution as an Approximation
An Application of the Normal Approximation to Acceptance Sampling
Exercises

Review Exercises   232

7-6   **Optional Topic: Sample Distribution of $\bar{X}$ When
the Population Is Small   234**

Finite Population Correction Factor
Optional Exercises

8 *Statistical Estimation*   **237**

8-1   **Estimators and Estimates   237**

The Estimation Process
Choosing an Estimator
Criteria for Statistics Used as Estimators
Commonly Used Estimators
Exercises

8-2   Interval Estimates of the Mean Using a Large Sample   243

Confidence Interval Estimate of the Mean When σ Is Known
The Meaning of the Interval Estimate
Confidence Interval Estimate of the Mean When σ Is Unknown
Illustration: Estimating Mean Bill Payment Times
Features Desired in a Confidence Interval
Exercises

8-3   Interval Estimates of μ Using Small Sample Sizes   250

The Student t Distribution
Constructing the Confidence Interval
Illustration: Estimating Mean Lost Production Time
Exercises

8-4   Interval Estimates of the Population Proportion   257

Exercises

8-5   Determining the Required Sample Size   259

Error and Reliability
Steps for Finding the Required Sample Size to Estimate the Mean
Three Influences of Sample Size
Steps for Finding the Required Sample Size to Estimate the Proportion
Work-Sampling Applications
Reliability Versus Confidence and Tolerable Error Versus Precision
Exercises

Review Exercises   271

9 Hypothesis Testing   275

9-1   Basic Concepts of Hypothesis Testing   275

Illustration: Testing a Dietary Supplement for Approval
The Null and Alternative Hypotheses
Making the Decision
Finding the Error Probabilities
Interpreting the Sample Results
Choosing the Decision Rule
Formulating Hypotheses
Exercises

9-2   **Large-Sample Tests Using the Mean**   **284**

Further Concepts: Acceptance Sampling
Finding the Acceptance and Rejection Regions
Testing with z Instead of $\bar{X}$
Upper-Tailed and Lower-Tailed Tests
Illustration: The Trend in Human Growth
Summary of Hypothesis-Testing Steps
Type II Error Considerations
Exercises

9-3   **Two-Sided Hypothesis Tests Using the Mean**   **293**

The Testing Procedure: A Quality-Control Illustration
Explanation of Hypothesis Testing in Terms of Confidence Intervals
Exercises

9-4   **Tests of the Mean Using Small Samples**   **296**

Illustration: Artificial Blood
An Assumption of t Test
Exercises

9-5   **Hypothesis Tests Using the Proportion**   **301**

Illustration: Testing a Theory on Right-Handedness
Exercises

9-6   **Selecting the Test**   **304**

Some Important Questions
The Power Curve
Efficiency and Power

9-7   **Limitations of Hypothesis-Testing Procedures**   **307**

Review Exercises   **308**

10   *Regression and Correlation*   *311*

10-1   **Regression Analysis**   **312**

The Scatter Diagram
The Data and the Regression Equation
Some Characteristics of the Regression Line

Fitting the Data by a Straight Line
Regression Curves
Exercises

10-2   The Method of Least Squares   323

Rational for Least Squares
Finding the Regression Equation
Illustration of the Method
Meaning and Use of the Regression Line
Measuring Variability of Results
Exercises

10-3   Assumptions and Properties of Linear Regression Analysis   332

Assumptions of Linear Regression Analysis
Estimating the True Regression Equation

10-4   Predictions and Statistical Inferences Using the Regression Line   335

Making Predictions with the Regression Equation
Prediction Intervals for the Conditional Mean
Prediction Intervals for an Individual Value of Y Given X
Dangers of Extrapolation in Regression Analysis
Inferences Regarding the Slope of the Regression Line
Exercises

10-5   Correlation Analysis   343

Measuring Degree of Association Between X and Y
The Coefficient of Determination
The Correlation Coefficient
Correlation Explained Without Regression
Correlation and Causality
Exercises

10-6   Common Pitfalls and Limitations of Regression and Correlation Analysis   352

Relevancy of Past Data
Cause-and-Effect Relationships

10-7   Curvilinear Regression Analysis   354

Review Exercises   355

10-8 **Optional Topic: Inferences Regarding Regression Coefficients** 357

Confidence Interval Estimate of B
Testing Hypotheses Regarding σ^2
Using B to Make Statistical Inferences About ρ Rho
Optional Exercises

11 *Multiple Regression and Correlation* 361

11-1 **Linear Multiple Regression Involving Three Variables** 362

Regression in Three Dimensions

11-2 **Multiple Regression Using the Method of Least Squares** 364

A Supermarket Illustration
Advantages of Multiple Regression
The Standard Error of the Estimate
Assumptions of Multiple Regression
Prediction Intervals in Multiple Regression
Inferences Regarding Regression Coefficients
Exercises

11-3 **Regression with Many Variables: Computer Applications** 375

Using the Computer in Multiple Regression Analysis
Stepwise Multiple Regression
Exercises

11-4 **Dummy Variable Techniques** 381

Using a Dummy Variable
Using a Dummy Variable with Time Series
Exercises

11-5 **Multiple Correlation** 390

Sample Coefficient of Multiple Determination
Partial Correlation
Exercises

Review Exercises 396

12 Time-Series Analysis and Forecasting   399

12-1   **The Time Series and its Components**   399

12-2   **The Classical Time-Series Model**   401

Exercises

12-3   **Analysis of Secular Trend**   408

Describing Trend
Determining Linear Trend Using Least Squares
Modifying Trend for Periods Shorter Than One Year
Nonlinear Trend: Exponential Trend Curve
Exercises

12-4   **Forecasting Using Seasonal Indexes**   422

Ratio-to-Moving-Average Method
Monthly Data
Deseasonalized Data
Making the Forecast
Exercises

12-5   **Identifying Cycles and Irregular Fluctuation**   432

12-6   **Exponential Smoothing**   432

Single-Parameter Exponential Smoothing
Two-Parameter Exponential Smoothing
Further Exponential-Smoothing Procedures
Exercises

Review Exercises   436

13 Index Numbers   439

13-1   **Price Indexes**   439

13-2   **Aggregate Price Indexes**   441

Exercises

13-3   **Price Relatives Indexes**   447

Exercises

13-4   **The Consumer Price Index**   450

13-5   **Deflating Time Series Using Index Numbers**   453

Exercises

13-6   **Quantity Indexes**   455

Exercises

13-7   **Shifting and Splicing Index-Number Series**   457

Shifting the Base
Splicing Index-Number Time Series
Exercises

14   *Inferences Using Two Samples*   461

14-1   **Confidence Intervals for the Difference Between Means Using Large Samples**   462

Independent Samples
Illustration: Comparing the Effectiveness of Two Textbooks
Matched-Pairs Samples
Matched Pairs Compared to Independent Samples
Exercises

14-2   **Hypothesis Tests for Comparing Two Means Using Large Samples**   472

Independent Samples: Testing Exhaust Emissions
Matched-Pairs Samples: Smoking and Heart Disease
The Efficiency of Matched-Pairs Testing
Exercises

14-3   **Inferences for Two Means Using Small Samples**   478

Estimating the Difference Between Means: Independent Samples
Estimating the Difference Between Means: Matched-Pairs Samples
Comparing Means Using Independent Samples: Effect of Coffee on Sleep

Comparing Means Using Matched-Pairs Samples
Exercises

14-4   **Hypothesis Tests for Comparing Proportions**   **486**

Illustration: Television Programming
Exercises

14-5   **Further Remarks**   **490**

Review Exercises   **491**

15 Chi-Square Applications   **495**

15-1   **Testing for Independence and the Chi-Square Distribution**   **496**

Independence Between Qualitative Variables
Contingency Tables and Expected Frequencies
The Chi-Square Statistic
The Chi-Square Distribution
The Hypothesis-Testing Steps
Special Considerations
Exercises

15-2   **Testing for Equality of Several Proportions**   **510**

Illustration: Remembering Advertisements
Exercises

15-3   **Other Chi-Square Applications**   **514**

Review Exercises   **514**

15-4   **Optional Topic: Inferences Regarding the Population Variance**   **516**

Probabilities for the s^2 Variable
Confidence Interval Estimate of σ^2
Testing Hypotheses Regarding σ^2
Illustration: Variability in Waiting Time
Normal Approximation for Large Sample Sizes
Optional Exercises

16 *Analysis of Variance* 525

16-1 Analysis of Variance and the *F* Distribution 526

Testing for Equality of Means
Summarizing the Data
Using Variability to Identify Differences
The Basis for Comparison
The ANOVA Table
The *F* Statistic
The *F* Distribution
Testing the Null Hypothesis
Additional Comments
Exercises

16-2 Estimating Treatment Means and Differences 541

Confidence Intervals for Treatment Population Means
Comparing Treatment Means Using Differences
Multiple Comparisons
Exercises

16-3 Two-Factor Analysis of Variance 544

The Randomized Block Design
Sample Mean Calculations
The Two-Factor ANOVA Table
A New Source of Explained Variation
Testing the Null Hypothesis for One Factor
Testing the Null Hypothesis for Both Factors: Evaluating Swimming-Pool
    Chemicals
Interactions
Exercises

16-4 Latin Squares and Other ANOVA Designs 556

Important ANOVA Considerations
The Latin-Square Design: Comparing Training Methods
Exercises

Review Exercises 563

16-5 Optional Topic: Analysis of Variance for Multiple Regression Results 565

17 Further Probability Distributions and the Goodness-of-Fit Test   569

17-1   The Hypergeometric Distribution   570

Finding Probabilities When Sampling Without Replacement
Expressing the Hypergeometric Distribution
Illustrations with Poker Hands
Hypergeometric Distribution for the Proportion of Successes
Exercises

17-2   The Poisson Distribution   577

The Poisson Process
Expressing the Poisson Distribution
Assumptions of a Poisson Process
Practical Limitations of the Poisson Distribution
Using Poisson Probability Tables
Applications Not Involving Time
Exercises

17-3   The Uniform Distribution   585

Exercises

17-4   The Exponential Distribution   589

Finding Interevent Probabilities
Applications of the Exponential Distribution
Exercises

17-5   Testing for Goodness of Fit   597

Importance of Knowing the Distribution
Conducting the Test: The Distribution of SAT Scores
The Chi-Square Test Statistic
The Hypothesis-Testing Steps
Testing with Unknown Population Parameters
Illustration: Poisson Probability Distribution for the Number of
    Typesetters' Errors
Exercises

Review Exercises   612

18 *Nonparametric Statistics* 615

18-1   The Need for Nonparametric Statistics   616

Advantages of Nonparametric Statistics
Disadvantages of Nonparametric Statistics

18-2   Comparing Two Populations Using Independent Samples: The Wilcoxon Rank-Sum Test   617

Description of the Test: Comparing Two Fertilizers
Application to One-Sided Tests: Evaluating Compensation Plans
The Problem of Ties
Comparison of the Wilcoxon Test and Student t Tests
Exercises

18-3   The Mann-Whitney *U* Test   625

18-4   Comparing Two Populations Using Matched Pairs: The Sign Test   626

Description of the Test: Comparing Two Razor Blades
Illustration: Evaluating a New Reading Program
The Normal Approximation
A Comparison of the Sign Test and the Student t Test
Exercises

18-5   The Wilcoxon Signed-Rank Test for Matched Pairs   635

Description of the Test: Comparing Reading Programs
A Comparison of the Wilcoxon Signed-Rank Test and the Sign Test
Exercises

18-6   Testing for Randomness: The Number-of-Runs Test   639

The Need to Test for Randomness
The Number-of-Runs Test
Testing the 1970 Draft Lottery for Randomness
The Meaning of the Draft Lottery Results
Other Tests for Randomness
Exercises

18-7 **The Rank Correlation Coefficient**   649

Usefulness of Rank Correlation
Exercises

18-8 **One-Factor Analysis of Variance: The Kruskal-Wallis Test**   652

Alternative to the F Test
Description of the Test: College Major and Programming Attitude
Exercises

Review Exercises   655

18-9 **Optional Topic: Testing for Goodness of Fit—The Kolmogorov-Smirnov One-Sample Test**   656

An Alternative to the Chi-Square Test
Description of the Test: Aircraft Delay Times
Finding the Decision Rule and the Critical Value
Investigating the Type II Error
Comparison with the Chi-Square Test
Optional Exercises

19 *Basic Concepts of Decision Making*   665

19-1 **Certainty and Uncertainty in Decision Making**   666

19-2 **Elements of Decisions**   666

The Decision Table
The Decision Tree Diagram

19-3 **Ranking the Alternatives and the Payoff Table**   669

Objectives and Payoff Values
Tippi-Toes: A Case Illustration
Exercises

19-4 **Reducing the Number of Alternatives: Inadmissible Acts**   674

Exercise

19-5 **Maximizing Expected Payoff: The Bayes Decision Rule** **675**

Exercises

19-6 **Decision Tree Analysis** **676**

Ponderosa Record Company: A Case Illustration
The Decision Tree Diagram
Determining the Payoffs
Assigning Event Probabilities
Backward Induction
Additional Remarks
Exercises

Review Exercises **687**

20 *Elements of Decision Theory* **689**

20-1 **Decision Criteria** **690**

The Maximin Payoff Criterion
The Maximum Likelihood Criterion
The Criterion of Insufficient Reason
The Preferred Criterion: The Bayes Decision Rule

20-2 **The Bayes Decision Rule and Utility** **695**

Exercises

20-3 **Decision Making Using Strategies** **698**

The Cannery Inspector: A Case Illustration
Extensive and Normal Form Analysis
Exercises

20-4 **Opportunity Loss and the Expected Value of Perfect Information** **705**

Opportunity Loss
The Bayes Decision Rule and Opportunity Loss
The Expected Value of Perfect Information

EVPI and Opportunity Loss
Exercises

Review Exercises   711

21 *Utility: Accounting for Attitude*   713

21-1   Attitudes, Preferences, and Utility   714

The Decision to Buy Insurance

21-2   Numerical Utility Values   716

The Saint Petersburg Paradox
The Validity of Logarithmic Values
Outcomes Without a Natural Payoff Measure
Exercises

21-3   The Assumptions of Utility Theory   719

Preference Ranking
Transivity of Preference
The Assumption of Continuity
The Assumptions of Substitutability
Increasing Performance

21-4   Assigning Utility Values   723

Gambles and Expected Utility
The Reference Lottery
Obtaining Utility Values
Attitude Versus Judgment
Utility and the Bayes Decision Rule
Exercises

**21-5   The Utility for Money and Attitudes
Toward Risk   730**

Applying the Utility Function in Decision Analysis
Utility and the Decision to Buy Insurance
Attitudes Toward Risk and the Shape of the Utility Curve
Exercises

Review Exercises   739

22 Subjective Probability: Accounting for Judgment   *743*

22-1   **Probabilities Obtained from History**   **743**

22-2   **Subjective Probabilities**   **744**

Betting Odds
Substituting a Hypothetical Lottery for the Real Gamble
Subjective Probability Distributions
Exercises

22-3   **Determining a Normal Curve Judgmentally**   **747**

Finding the Mean
Finding the Standard Deviation
Exercises

22-4   **The Judgmental Probability Distribution: The Interview Method**   **751**

Common Shapes of Subjective Probability Curves
Approximating the Subjective Probability Distribution
Exercises

22-5   **Finding a Probability Distribution from Actual/Forecast Ratios**   **761**

Exercises

22-6   **Additional Remarks**   **763**

Review Exercises   **763**

23 Bayesian Analysis of Decisions Using Experimental Information   *765*

23-1   **Probability Analysis for Decision Making**   **766**

The Oil Wildcatter: A Case Illustration
Using Probability Trees
Exercises

23-2   Posterior Analysis   772

Decision Tree Analysis
Obviously Nonoptimal Strategies
Exercises

23-3   The Decision to Experiment: Preposterior Analysis   776

The Role of EVPI
Exercises

Review Exercises   780

24 *Bayesian Analysis of Decisions Using Sample Information*   783

24-1   Decision Making with the Proportion   784

Charles Stereo: An Acceptance-Sampling Illustration
Structuring the Decision
Determining the Event Probabilities
The Decision Rule
Exercises

24-2   Decision Making with the Mean   792

A Computer Memory Device Decision
Decision Making with Sample Information
Exercises

24-3   Deciding About the Sample: Preposterior Analysis   798

Determining the Optimal Sample Size
The Role of Perfect Information (EVPI)
The Expected Value of Sample Information
The Expected Net Gain of Sampling
Exercises

24-4   Decision Theory and Traditional Statistics   806

Hypothesis-Testing Concepts Reviewed
Contrasting the Two Approaches
Exercises

Review Exercises   811

Selected References   815

Appendix   819

Answers to Even-Numbered Exercises   851

Index   865

STATISTICS FOR MODERN BUSINESS DECISIONS

Third Edition

The true foundation of theology is to ascertain the character of God. It is by the aid of Statistics that law in the social sphere can be ascertained and codified, and certain aspects of the character of God thereby revealed. The study of statistics is thus a religious service.

FLORENCE NIGHTINGALE

Introduction

There is perhaps no other subject quite like statistics. Everyone—from such remote people as the Bushmen in Africa's Kalahari Desert to space scientists who plan grand tours of the planets—uses statistics. Surprisingly, the universal application of statistics is not what most people envision. The word *statistics* commonly brings to mind masses of numbers, graphs, and tables. These do play a role in statistics, but a limited one. The Kalahari Bushmen cannot count very high, know little about mathematics, draw no graphs, and assemble no tables, but they do use statistics.

THE MEANING OF STATISTICS 1-1

One common thread linking scientist and Bushman is that both *must make decisions in the face of uncertainty*. A successful Bushman locates water in a manner that puzzles some geologists. He sucks through a reed stuck in the sand in what his experience tells him is a likely spot, as if he were drinking a thick milkshake from a straw. Sometimes no water is found, but often enough the Bushman does find a "drink hole." His very survival depends on his skill at locating drink holes—on using his experience to cope with uncertainty.

Similarly, scientists depend on experience to determine an optimal space-vehicle configuration that will withstand the rigors of an environment far more hostile than earth's atmosphere and that will survive long enough to fly past Jupiter and eventually to even more distant planets. If the communications system fails, the mission itself will be a failure. The system must be super-reliable—designed so that its chances for survival are high. Millions of hours of human experience can contribute to design decisions. Even so, there are no guarantees that the chosen design will function as intended. Again, decisions must be made in the face of uncertainty.

Both the Bushman's and the space scientist's mechanisms for making choices utilize limited experience, and in this respect they are not substantially different. The remaining element that makes their mechanisms statistical is that *both use numerical evidence.* To evaluate the reliability of communications equipment, components must be tested under stress to determine the number of hours they can function before failing. The scientist's evidence is obviously numerical. But what about the Bushman's? He knows that one set of conditions favors the presence of water more than another. Every wet hole reinforces his knowledge of factors that are positive to water, and each dry hole strengthens his awareness of the negative factors. Thus, the Bushman's numerical evidence is the *frequency* of successful drink holes he finds under various prevailing conditions. Even though the presence of water is a qualitative factor, frequencies of occurrence are themselves numerical.

This book describes some of the more useful methods and procedures that can be applied to numerical evidence to facilitate decision making in the face of uncertainty. Certain basic principles are explained within a theoretical framework to enable you to achieve a greater appreciation and understanding of how and why they work.

A concise statement summarizing the subject of this book is the

DEFINITION
Statistics is a body of methods and theory that is applied to numerical evidence when making decisions in the face of uncertainty.

This definition treats statistics as a separate academic field, just like physics, history, or biology. As a discipline, statistics has advanced rapidly during the twentieth century to become recognized as a branch of mathematics. Partly due to the relative newness of statistics as a subject of study, the word *statistics* has acquired several meanings.

In its earlier and most common usage, statistics means a collection of numerical facts or data. In sports, for example, batting averages, rushing yardage, and games won or lost are all statistics. Closing prices from the New York Stock Exchange, figures showing sources of income and expenditures for the federal budget, and distances between major cities are also statistics. Your reported income on last year's tax return is a statistic. Examples of this usage are the *vital statistics*—figures on births, deaths, marriages, and divorces—and the contents of publications listing population and economic data, such as the *Statistical Abstract of the United States.* The main distinction between our definition of statistic and this usage of the word is its form: singular versus plural. Statistics (singular) *is* a subject of study; statistics (plural) *are* numerical facts.

THE ROLE OF STATISTICS   1-2

Our definition of statistics is particularly appropriate for readers who are primarily interested in applications to business and economics, where there is a high incidence of decisions made under uncertainty. Managerial decisions involving both numerical data and uncertainty are required daily, on matters that range from ordering raw material for products for which demand is uncertain to hiring personnel whose performance cannot be predicted. Statistics is used to answer long-range planning questions, such as when and where to locate facilities to handle future sales of uncertain levels. Statistics can be helpful in formulating strategic policies that affect a firm's survival, such as new-product development, pricing, and financing. Economists must choose among alternative policies whose outcomes are determined partly by chance in order to rescue the economy from unemployment and inflation and to sustain economic growth. They are called upon by government, business, foundations, and unions to forecast future economic conditions. In doing this they must not only decide what types of data are appropriate, but they must also choose from a dazzling array of techniques. Economists rely on statistical techniques in building, verifying, or implementing economic models.

The social sciences, especially psychology and sociology, also make heavy use of statistics, and so do the physical and life sciences. Geneticists are highly dependent on statistics. Although this book emphasizes business and economic applications, whereas other statistics books may be oriented to scientific fields, the statistical elements are identical. The differences are primarily in the examples chosen and the particular techniques emphasized.

DESCRIPTIVE AND INFERENTIAL STATISTICS   1-3

The emphasis on the decision-making aspects of statistics is a recent one. In its early years, the study of statistics largely consisted of methodology for summarizing or describing numerical data. This area of study has become known as *descriptive statistics* because it is concerned largely with summary calculations and graphical displays. These methods are in contrast to the modern statistical approach in which generalizations are made about the whole (called the *population*) by investigating a portion (referred to as the *sample*).

Thus, the average income of *all* families in the United States can be estimated from figures obtained from a *few hundred* families. Such a prediction or estimate is an example of an inference. The stu ly of how inferences are made from numerical data is called *inferential statistics*.

Our formal definition for the field of statistics applies to both its descriptive and its inferential forms. This book emphasizes inferential statistics because it can play such a dynamic role in decision making. But first we will focus on descriptive statistics in Chapters 2 and 3. We begin by describing the population. Later, we will study how generalizations about a population can be made by means of a sample.

Inferential statistics acknowled₁ es that the potential for error exists in making generalizations from a sample. The seasonal controversy over decisions made by te evision networks to drop c ·tain programs illustrates some of the principles ₁nd problems involved. The ₁nformation on which these decisions are based is obtained from a sample of a few hundred TV viewers who are believed to be a representative microcosm of the viewing public at large. Only those shows indicated by the sample to be most popular are allowed to continue. Although articulate segments of the public complain that their tastes are not represented, the agency responsible for the collection of sample data has demonstrated that its audience is selected in accordance with accepted scientific sampling procedures. This still does not guarantee that correct decisions will be made,   ₁ samples may deviate from the underlying population. Such discrepancie₁ ₁re called *sampling errors*.

The sampling agency knows that its sampling errors are likely to be insignificant because it has selected its sample *randomly*. Using probability theory, the likelihood of making large sampling errors is known to be small. Probability theory measures the chance that an untypical sample will be selected from a population whose characteristics are known. Inferential statistics is based on probability theory, extending its concepts to the measurement of the chance of making erroneous generalizations, even when the characteristics of the population are unknown or uncertain.

Because of the important role it plays in statistical inference, we will discuss probability in Chapter 5, after we examine descriptive statistics for populations having *known* properties. This will prepare us for developing statistical inferences about populations whose characteristics are *unknown*.

1-4   DEDUCTIVE AND INDUCTIVE STATISTICS

There is another dichotomy encountered in statistics. Deduction ascribes properties to the specific, starting with the general. For example, probability tells us that if a person is chosen by lottery from a group containing nine men and one women, then the odds against picking the women are 9 to 1. We deduce that in about 90% of such samples the person will be male. The use of probability to determine the chance of obtaining a particular kind of sample result is known as *deductive statistics*.

In Chapters 5, 6, and 7, we will learn how to apply deductive techniques when we know everything regarding the population in advance and we are concerned with studying the characteristics of the possible samples that may arise from *known* populations. In much of the remainder of the book, we will reverse direction and use *inductive statistics*. Induction involves drawing general conclusions from the specific. In statistics this means that inferences about populations are drawn from samples. The sample is all that is known; we must determine uncertain characteristics of the population from the incomplete information available.

Inductive and deductive statistics are completely complementary. We must study how samples are generated before we can learn to generalize from a sample. Deductive statistics is not wholly satisfactory in performing much statistical analysis because the pragmatic aspects of statistics are largely inductive. But we cannot understand inductive statistics unless we study deductive statistics first.

STATISTICAL ERROR   1-5

Statistics is characterized partly as art and partly as science. It is an art because we must rely heavily on experience and judgment in choosing from the vast number of procedures available for analyzing sample information. But statistics embodies, to various degrees, all elements of the scientific method—most notably the element of *error*. Because statistics concerns uncertainty, there is always a chance of making erroneous inferences. Statistical procedures are available both to control and to measure the risks of reaching erroneous conclusions. To illustrate this point, again let's consider television-viewer sampling.

Those who disagree with the network choices also base their arguments on a sample—the opinions of their own friends and acquaintances. The problem here is that such a sample is *biased* in favor of persons who have similar tastes, education, and social experience. The validity of opinion surveyor's claims is supported by a random selection from the public at large in which everyone has an equal chance of representation. As we will see in Chapter 8, the survey agency chooses a sample large enough to minimize the chance of error.

The survey agency cannot be legitimately criticized, unless it can be shown that its samples were not random and were therefore biased in favor of specific groups. Even when large samples are taken, bias can be introduced by statistical errors due to improper procedures. We will discuss some of the more serious sources of bias in Chapter 4.

No study is less alluring or more dry and tedious than statistics, unless the mind and imagination are set to work.

WILLIAM PLAYFAIR (1801)

Describing Statistical Data

T he arrangement and display of statistical data are important elements of descriptive statistics. Raw numbers alone do not provide any underlying pattern from which conclusions can be drawn. For example, we might believe that doctors generally have high incomes and that younger people usually earn less than older people. But even if we had access to all the tax returns on file with the Internal Revenue Service, without arranging, sifting, and sorting the original income data, we would not be able to determine if doctors earn more than dentists, if some lawyers earn more than most doctors, or if greater experience in a particular occupation increases salary. Only by organizing the data can we gain information to use in career planning.

Such descriptive statistics sets the stage for everything else.

THE POPULATION AND THE SAMPLE   2-1

The basic element of statistics is a single data point, which we call an *obervation* because it represents what we actually see. An observation may be a physical measurement (weight or height), an answer to a question (yes or no), or a classification (defective or nondefective). Observations relevant to a particular decision constitute a *population*. Stated more formally, we have the following

DEFINITION
A statistical population is the collection of all possible observations of a specified characteristic of interest.

Note that a statistical population exists whether all, some, or none of the possible observations are actually made. It may be real (such as the height of 30-year-old males in Schenectady in June 1982) or hypothetical (such as the longevity of laboratory rats fed a special diet that is not yet widely used). Because it consists of all possible observations, a population is often referred to synonymously as a *universe*.

In contrast to the statistical population, which consists of all possible observations, the *sample* contains only some of the observations. We make the following

DEFINITION
A sample is a collection of observations representing only a portion of the population.

Whether a collection of observations is a population or a sample depends on the purpose of the analysis. For example, a medical association may wish to compare physicians' fees to the costs of other kinds of hospital services. Here, one population (physicians' fees) would be compared to another population (the cost of medical services). Study objectives determine whether the populations would comprise fees and costs for the entire nation, for a state, or for a county.

If the population of service costs covered all patients released during a particular calendar year, then the charges for all patients released on a *single day* would be a sample. The costs for patients at a *single hospital* during the entire year would also be a sample of the same population.

If the medical association is concerned only with fees and charges for patients discharged on March 1, those data would constitute populations rather than samples. Similarly, if fees at different hospitals were to be compared, then the costs at one hospital would be a separate population.

Elementary Units

A statistical population consists of *observations* of some characteristic of interest associated with the individuals concerned, *not* the individual items or persons themselves. A company may need to know the ages of its employees in order to analyze proposed changes in its retirement program. To find these, the employee records would be searched and the ages determined from dates of birth. Each number determined constitutes an observation of the age charac-

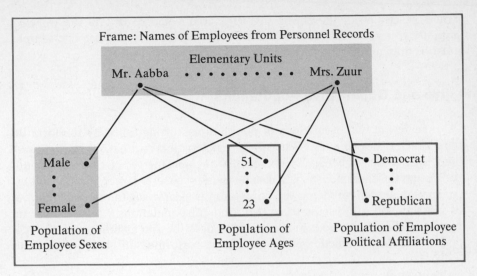

FIGURE 2-1

Illustration of how several different populations may be obtained from the same elementary units.

tertistic. The entire collection of numbers so obtained is the population; the employees themselves are referred to as the *elementary units* of the population.

A limitless variety of different populations may be obtained from the same elementary units, depending on the characteristic of interest. This principle is illustrated in Figure 2-1. Thus, a company could have different populations for political affiliation, sex, martial status, years of education, height, weight, eye color, job classification, and years on present job, where each population is composed of observed characteristics of the same elementary units—the employees.

The Frame

How do we define the elementary units of a population? The employer can define the employees as those persons on the payroll, so that the names of all employees may be obtained from payroll records. Such a source of elementary units is called a *frame*.

The frame is important in statistical studies because it helps to define the population. Suppose that a sample of voter preference toward candidates is taken to predict the outcome of an election. The sample should represent the population of votes to be cast for a specific office. The population of interest is referred to as a *target population*. In selecting the sample of voters, the only frame available to the opinion surveyor is the roll of registered voters. The elementary units listed in this frame may differ from those of the target population, because some people who are registered will not vote. The votes that would be cast if all those who are registered voted constitute the *working population*, because its frame is the only one currently available. The sample must be drawn from this population.

Qualitative and Quantitative Populations

There are two basic kinds of populations, distinguished by the form of the characteristic of interest. When the characteristic is one that can be expressed numerically, such as height, weight, cost, or income, the population is *quantitative*. When the characteristic is non-numerical, such as sex, race, martial status, occupation, or college major, the population is *qualitative*. In our presentation of statistics, we will discuss these two kinds of populations separately but in parallel. Different methods are required to describe and summarize each type of population: Arithmetic operations can be performed on numbers, so that,

TABLE 2-1

Illustration of Possible Observations
for Various Quantitative and Qualitative Populations

	Quantitative Population Observations		
Elementary Unit	Characteristic of Interest	Unit of Measurement	Possible Variate
person	age	years	21.3 yr
microcircuit	defective solder joints	number	5
tire	remaining tread	millimeters	10 mm
account balance	amount	dollars	$5,233.46
employer	female employees	percent	35%
common stock	earnings per share	dollars	$3.49
keypuncher	errors	proportion	.02
light bulb	lifetime	hours	581 hr
can of food	weight of contents	ounces	15.3 oz

	Qualitative Population Observations	
Elementary Unit	Characteristic of Interest	Possible Attributes
person	sex	male, female
security	type	bond, common stock, preferred stock
building	exterior materials	brick, wood, aluminum
employee	experience	applicable, not applicable
television	quality	defective, nondefective
firm	legal status	corporation, partnership, proprietorship
patient	condition	satisfactory, critical
student	residence	on-campus, off-campus

for example, we can calculate the average height of a collection of 30-year-old men; other procedures must be employed to summarize a qualitative population.

We refer to a particular observation of a qualitative characteristic as an *attribute*. Thus, for the characteristic of marital status, we might observe any one of the following attributes: single, married, divorced, or widowed. An observation of a quantitative characteristic such as income will be a particular numerical value or *variate*, such as $11,928.23, $21,234.15, or $1,095.61. Table 2-1 presents examples of observations from various qualitative and quantitative populations.

EXERCISES

2-1 Consider the number of days employees lost due to illness during August 1982. Give an example of a goal for which we may consider these data to be a population, to be a sample.

2-2 For each of the following problem situations, (1) provide an example of an elementary unit; (2) provide an exmple of a characteristic of interest; and (3) state whether the population would be qualitative or quantitative.

(a) An aircraft manufacturer's investment plans will be affected by the fate of a congressional appropriation bill containing funds for a new fighter plane. The company wishes to conduct a survey to facilitate planning.

(b) A politician plans to study the voter response to pending legislation to determine her platform.

(c) The California Public Utilities Commission is seeking operating cost data that will help it establish new rates for the San Francisco–Los Angeles airline routes.

2-3 For each of the following situations, discuss whether the suggested frame would be suitable.

(a) An insurance company is using home-theft claims processed in the past year as its frame to determine the number of thefts from its policyholders of items valued below $50.

(b) A public health official uses doctors' records of persons diagnosed as suffering from flu as his frame to study the effects of a recent flu epidemic.

(c) A stock exchange official studying the investment attitudes of owners of listed securities uses the active accounts of member brokerage firms as her frame.

THE FREQUENCY DISTRIBUTION   2-2

Finding a Meaningful Pattern for the Data

The ages of a sample of 100 statistics students are given in Table 2-2. If we wish to describe this sample, how should we proceed? The values in Table 2-2 have not been summarized or rearranged in a meaningful manner, so we

TABLE 2-2

Ages of a Sample of 100 Statistics Students

				Age (years)					
20.9	33.4	18.7	24.2	22.1	18.9	21.9	20.5	21.9	37.3
57.2	25.3	24.6	29.0	26.3	19.1	48.7	23.5	23.1	28.6
21.3	22.4	22.3	20.0	30.3	31.7	34.3	28.5	36.1	32.6
33.7	19.6	18.7	24.3	27.1	20.7	22.2	19.2	26.5	27.4
22.8	51.3	44.4	22.9	20.6	32.8	27.3	23.5	23.8	22.4
18.1	23.9	20.8	41.5	20.4	21.3	19.3	24.2	22.3	23.1
22.9	21.3	29.7	25.6	33.7	24.2	24.5	21.2	21.5	25.8
21.5	21.5	27.0	19.9	29.2	25.3	26.4	22.7	27.9	22.0
23.3	28.1	24.8	19.6	23.7	26.3	30.1	29.7	24.8	24.7
23.5	22.9	26.0	25.2	23.6	21.0	30.9	21.7	28.3	22.1

refer to them as *raw data*. We might begin by grouping the ages in a meaningful way.

A convenient way to accomplish this is to group the ages into categories of two calendar years, beginning with age 18. Each age will then fall into one of the categories 18.0–under 20.0, 20.0–under 22.0, and so on. We refer to such groupings of values as *class intervals*. Now if we summarize the raw data simply by counting the number of ages in each class interval, we should be able to identify some properties of the sample.

First we list the class intervals in increasing sequence, as in the first column of Table 2-3. Then, reading down the list of raw data for each successive age, we place a tally mark beside the corresponding class interval and a check mark beside the number on the original list so that it will not be counted twice. When the tally is complete, we count the marks to determine the total number of observations in each class interval. The final result indicates the *frequency* with which ages occur in each class interval and is called a *sample frequency distribution*. In Table 2-3, we find that 10 of the 100 statistics students fall into the interval 18.0–under 20.0. The number 10 can be referred to as the *class frequency* for the first class interval.

Each class interval has two limits. For the first interval, the *lower class limit* is 18.0 and the *upper class limit* is "under 20.0." (We use the word *under* to distinguish this limit from 20.0 exactly, which serves as the lower limit of the second class interval.) No matter how precisely the raw data are measured, this designation prevents ambiguity in assigning an observation to a particular interval. For example, the age 19.99 years is represented by the first class, whereas 20.01 years would fall into the second class (20–under 22.0) The *width* of a class interval is found by subtracting its lower limit from the lower

TABLE 2-3
Sample Frequency Distribution of Student Ages

Age	Tally	Number of Persons (Class Frequency)
18.0−under 20.0	⊬⊬⊬ ⊬⊬⊬	10
20.0−under 22.0	⊬⊬⊬ ⊬⊬⊬ ⊬⊬⊬ ///	18
22.0−under 24.0	⊬⊬⊬ ⊬⊬⊬ ⊬⊬⊬ ⊬⊬⊬ ///	23
24.0−under 26.0	⊬⊬⊬ ⊬⊬⊬ ////	14
26.0−under 28.0	⊬⊬⊬ ⊬⊬⊬	10
28.0−under 30.0	⊬⊬⊬ ///	8
30.0−under 32.0	////	4
32.0−under 34.0	⊬⊬⊬	5
34.0−under 36.0	/	1
36.0−under 38.0	//	2
38.0−under 58.0	⊬⊬⊬	5
	Total	100

limit of the succeeding class interval. For example, the width of the first class interval is $20.0 - 18.0 = 2.0$. All intervals except the last class interval have the same width. As we will see, beginning and ending class intervals must sometimes be treated differently.

This frequency distribution of student ages applies to sample data. If available, the raw data for the entire population of student ages could be similarly arranged. In this case, a table of frequencies would constitute a *population frequency distribution*. Ordinarily, only sample results are available, but the present discussion applies to raw data taken from either a sample or a population.

Graphical Displays: The Histogram

A visual display can be a very useful starting point in describing a frequency distribution. Figure 2-2 graphically portrays the frequency distribution of student ages. Age is represented on the horizontal axis, which is divided into class intervals of two years in width. The classes are represented by bars of varying heights, corresponding to the class frequency (the number of observations in each interval). Thus, the vertical axis represents frequency. Such a graphical portrayal of a frequency distribution is called a *histogram*.

Since 23 of the ages are 22.0−under 24.0, the corresponding bar is of height 23. Only one person is between 34.0 and 36.0 years old, so the height of that bar is 1. Neighboring bars touch, emphasizing that age is a continuous scale.

Note that the highest class interval, 38.0–under 58.0, is 20 years wide—ten times as wide as the standard intervals. With only five ages falling between 38.0 and 58.0, there is, on the average, one-half an observation every two years in this interval. This bar therefore has a height of $\frac{1}{2}$. Treating the two-year width as a standard unit, *the area inside any bar* (its height times its width in standard units) *equals the number of observations in the corresponding class interval.* Thus, the area inside the first bar is $10 \times 1 = 10$, and the area inside the last bar is

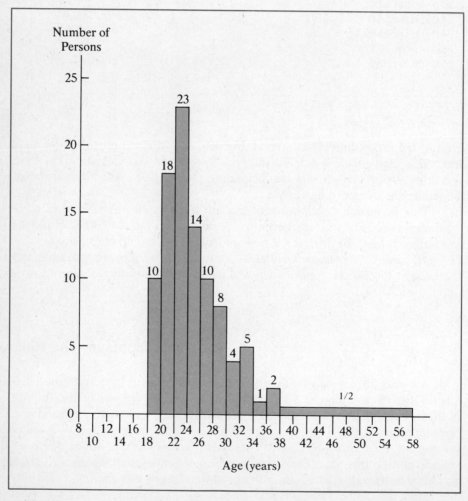

FIGURE 2-2
Histogram for the frequency distribution of student ages.

one-half the number of standard two-year class intervals, or $\frac{1}{2} \times 10 = 5$, which is the number of observations in that category.

Graphical Displays: Frequency Polygons and Curves

An alternative graphical portrayal of a frequency distribution is the *frequency polygon*, shown in Figure 2-3 for the same sample of student ages. In Figure 2-3, each class interval is represented by a dot positioned above its

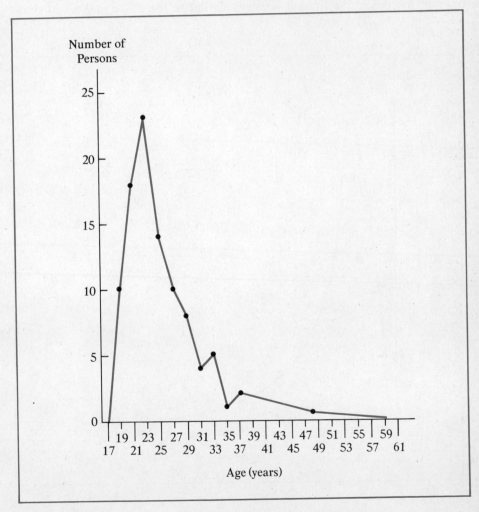

FIGURE 2-3
Frequency polygon for sample of student ages.

midpoint at a height equal to the class frequency. Each midpoint is typical of all values in the interval and is defined by the average of the interval's class limits. Thus, the midpoint of the first interval is $(18.0 + 20.0)/2 = 19.0$; the midpoint of the second interval is 21.0; and so on. The midpoint of the last interval is $(38.0 + 58.0)/2 = 48.0$. The dots are connected by line segments to make the graph more readable. To complete the frequency polygon, line segments are drawn from the first and last dots to the horizontal axis at points one-half the width of a standard class interval below the lowest and above the highest class intervals, respectively (in this case, touching the axis at 17 and 59).

When the number of observations is large, the shape of the polygon for a sample frequency distribution becomes similar to the shape that might result

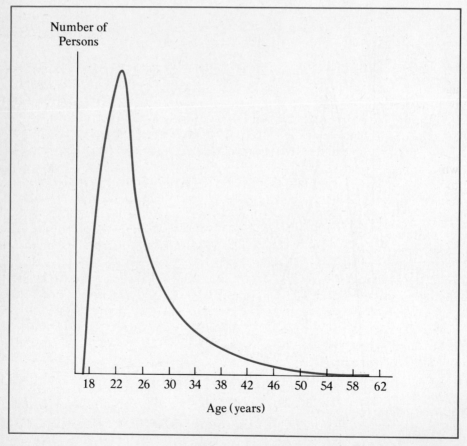

FIGURE 2-4
Suggested shape of a smoothed frequency curve for the entire population.

from raw data representing the entire population. But since data for the entire population are not ordinarily available, the population frequency distribution is usually portrayed in terms of a *frequency curve* like the one shown in Figure 2-4. The basic shape of a population's frequency curve is usually suggested by the histogram or frequency polygon originally constructed from the sample data. Because populations are usually quite large in relation to the number of observations in a sample, the frequency curve would resemble a frequency polygon or a histogram with many class intervals of tiny widths. If these intervals were individually plotted, the entire collection of population data would present a less jagged graph than one obtained from the sample data alone, and this graph would be almost totally smooth (like the curve in Figure 2-4). Later in this chapter, we will examine some forms of the frequency distribution that are commonly encountered for populations.

Descriptive Analysis

We have managed to translate the confusion of our raw data on student ages into a pattern based on frequencies. The frequency distribution tells us two things: It shows how the observations cluster around a central value, and it illustrates the degree of dispersion or difference between observations.

We know that no student is younger than 18 and that ages below 28 are most typical. Of these, the most common age is somewhere between 22 and 24, which (from general information obtained from the registrar's office) we know to be higher than usual for the student who enters college right after high school and graduates at about age 22. The students in the sample are generally older, but we can rule out the possibility that they might be graduate students because of the substantial number of younger persons. It is possible that the population could be made up of night students and that the older persons work on their degrees on a part-time basis while holding full-time jobs.

The appropriateness of this conclusion may be substantiated by the incidence of persons over 30, who are most likely to have financial burdens and thus not be full-time students. Although predominantly young, the sample is peppered with persons approaching or exceeding the age of 40. Five students are relatively so much older that they have been put into a special group—the "over 38s." Their ages (41.5, 44.4, 48.7, 51.3, and 57.2) are spread so thinly along the age scale that it is more convenient to lump these more extreme ages into a single special category.

This descriptive analysis provides us with an image of the student sample that is not immediately available from the raw data. The entire description is based on frequencies of occurrence—the heart of all statistical analysis. As we will see in Chapter 5, frequency is the basis for probability theory and as such plays a fundamental role in all procedures of statistical inference.

Constructing a Frequency Distribution:
Width and Number of Class Intervals

A frequency distribution converts raw data into a meaningful pattern for statistical analysis. In doing this, detail must be sacrificed to insight. Because it lumps individual variates into class intervals, the frequency distribution cannot tell us how the observations within a certain category differ from each other. Only the raw data can provide this information. Thus, some less important information is lost to gain more useful information.

How can the raw data best be condensed into class intervals? This question leads to two others: What range of values should be included in a single category? How many intervals should be used?

The class intervals should be of equal width to facilitate comparisons between classes and provide simpler calculations of the more concise summary values to be discussed later in this chapter. As we noted about the distribution of the age data in Table 2-3, it is sometimes better to lump the extreme values into a single category. This becomes quite evident during the construction of a frequency distribution for individual annual earnings. Most persons earn incomes of less than $50,000, and a few earn between $50,000 and $100,000. But the rare, extremely well-to-do have substantially higher incomes. A frequency distribution graph with equal intervals of income width of, say, $5,000 would be hard to draw and difficult to read. The resulting histogram would be analogous to a few mountain peaks (representing frequencies for the lower income levels) sitting to the left of many successively smaller anthills (representing the number of persons in each progressively rarer, higher income category). This difficulty can be avoided by grouping the higher incomes into a single class interval of, say, "$100,000 or more."

We cannot really know how wide the class intervals ought to be without knowing how many intervals are to be used. No totally accepted rule tells us this. It is generally recommended that between 5 and 20 class intervals be used, but the number is really a matter of personal judgment. When all intervals are to be the same width, the following rule may be applied to find the required

CLASS INTERVAL WIDTH

$$\text{Width} = \frac{\text{Largest value} - \text{Smallest value}}{\text{Number of class intervals}}$$

We illustrate this for the 100 observations of fuel consumption in miles per gallon (mpg) for a fleet of cars listed in Table 2-4. Suppose we wish to construct five class intervals of equal width. We find the required width by taking the difference between the largest and smallest values, $23.9 - 14.1 = 9.8$, and dividing by 5 to obtain 1.96 mpg. To simplify our task, we can round off

TABLE 2-4

Fuel Consumption (miles per gallon) Achieved
by 100 Large-Sized Cars (raw data)

19.0	20.8	22.0	22.7	20.0	18.9	16.6	16.8	20.8	14.7
15.1	21.8	21.1	21.5	21.1	15.5	19.3	15.1	20.6	16.8
18.2	20.5	15.3	16.2	16.3	22.8	22.7	21.9	22.5	17.1
19.1	21.6	19.0	18.3	18.6	22.1	17.5	22.9	21.7	18.7
21.9	20.2	14.5	14.1	22.9	20.2	17.3	22.6	19.3	21.7
21.5	22.6	18.7	19.2	22.8	21.6	21.7	20.5	22.7	20.4
18.8	15.1	16.5	20.5	19.1	17.4	19.7	19.2	16.4	21.9
14.3	19.2	19.7	17.1	21.4	21.9	21.7	19.2	23.9	19.6
20.9	18.5	20.2	18.2	20.2	22.4	20.4	21.6	21.3	22.4
20.5	18.1	20.7	21.3	16.9	20.3	23.9	18.8	21.1	21.9

1.96 to 2.0 for the width and 14.1 to 14.0 for the lower limit of our first class interval. The frequency distribution is provided in Table 2-5 and graphed in Figure 2-5(a).

Figure 2-5 also contains histograms for the same fuel consumption data using 10 class intervals (b) and 20 class intervals (c). Notice that the histogram profiles become progressively "lumpier" as the number of classes increases. Both the 5-interval and 10-interval graphs provide concise summaries of the data, but a pronounced "sawtooth" effect occurs with 20 intervals. The poorness of the 20-interval summary reflects the large number of intervals in relation to the size of the sample. One way to decide how many class intervals to use would be to try several—plot a histogram for each and select the one that provides the most logical explanation of the underlying population pattern. A large number of class intervals will provide more data, but too many will produce meaningless oscillations.

TABLE 2-5

Frequency Distribution of Fuel Consumption
of 100 Large-Sized Cars Using Five
Class Intervals

Miles per Gallon	Number of Cars
14.0–under 16.0	9
16.0–under 18.0	13
18.0–under 20.0	24
20.0–under 22.0	38
22.0–under 24.0	16
Total	100

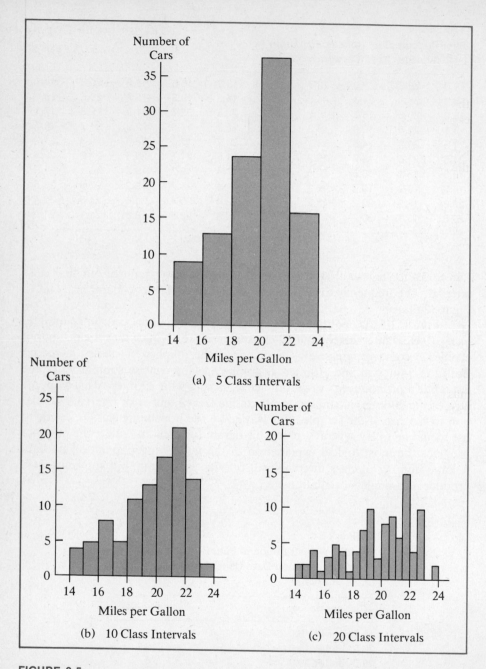

FIGURE 2-5

Histograms for fuel consumption of 100 large-sized cars, using three different sets of class intervals.

Qualitative Populations:
The Frequency Distribution in Cryptanalysis

A vivid illustration of how frequency of occurrence can be used to reduce the confusion of raw data is provided by *cryptanalysis*. This also serves to explain frequency distributions for qualitative populations.

Cryptanalysis is the breaking of secret codes or ciphers. In its simplest form, a *cipher* is a message in which another alphabet has been substituted for the ordinary one. Consider the following message:

AOYNS	YIRXJ	AJRRS	OOYIR	YGDYP
MQQCY	CYMOQ	JAPQM	QDPQD	RMGMI
MGXPD	PDQMG	GRVPS	PQJRO	YMQYJ
OKYOA	OJHRJ	IASPD	JIHYM	IDIBA
SGGXM	OOMIB	YKISH	UYOPR	MIQYG
GSPMP	QJOXM	IKCYG	LSPRC	JJPYH
YQCJK	PMIKL	OJRYK	SOYPA	JOBYI
YOMGD	ZDIBM	UJSQL	JLSGM	QDJIP

To decipher this message, we must find the counterparts to the letters in the original message. Cryptanalysis begins with the frequency distribution of letters in the cipher message (see Table 2-6). The ciphertext letters are the elementary units from a qualitative population of alphabet symbols. Here a count has been made of the number of times each letter occurs. Each letter is a separate category, just like a quantitative class interval.

TABLE 2-6
Frequency Distribution of Letters in the Cipher Message

Letter	Tally	Number of Letters	Letter	Tally	Number of Letters
A	⁣卌 //	7	N	/	1
B	////	4	O	卌 卌 卌 /	16
C	卌	5	P	卌 卌 卌 /	16
D	卌 卌	10	Q	卌 卌 ////	14
E		0	R	卌 卌 /	11
F		0	S	卌 卌 /	11
G	卌 卌 //	12	T		0
H	////	4	U	//	2
I	卌 卌 卌	15	V	/	1
J	卌 卌 卌 //	17	W		0
K	卌 /	6	X	////	4
L	////	4	Y	卌 卌 卌 卌 /	21
M	卌 卌 卌 ///	18	Z	/	1

Total 200

TABLE 2-7

Frequency Distribution of 200 Letters of Ordinary English
Language Text

Letter	Frequency	Letter	Frequency	Letter	Frequency
a	16	j	1	s	12
b	3	k	1	t	18
c	6	l	7	u	6
d	8	m	6	v	2
e	26	n	14	w	3
f	4	o	16	x	1
g	3	p	4	y	4
h	12	q	$\frac{1}{2}$	z	$\frac{1}{2}$
i	13	r	13	Total	200

SOURCE: David Kahn, *The Codebreakers: The Story of Secret Writing* (New York: Macmillan, 1967), p. 100. Copyright © 1967 by David Kahn.

The cryptanalyst knows the frequency with which each letter occurs in the ordinary English language. This is provided in Table 2-7, where the letters are shown in lower case to distinguish them from the ciphertext alphabet. The frequency distribution of ordinary English letters is graphed as a histogram in Figure 2-6. When the observations are attributes rather than variates, the bars do not touch each other, reflecting the fact that each category is discrete. Sometimes vertical spikes like the ones in Figure 2-6 are used instead of bars to emphasize that the observations do not range along a continuous scale.

The frequency distributions for samples from qualitative populations, where the characteristic of interest takes the form of various attributes (such as male or female, defective or nondefective, or professional categories such as doctor, lawyer, engineer, or educator), would be represented similarly.

The letters occurring in ordinary English may be arranged in sequence of decreasing frequency:

English letters	e	t	a	o	n	i	r	s	h	d	l	u	c	m	p	f	y	w	g	b	v	j	k	x	q	z
Frequency	26	18	16	16	14	13	13	12	12	8	7	6	6	6	4	4	4	3	3	3	2	1	1	1	$\frac{1}{2}$	$\frac{1}{2}$

| | *High* | | *Medium* | | *Low* |

A similar arrangement of the ciphertext letters may be made:

Ciphertext	Y	M	J	P	O	I	Q	G	S	R	D	A	K	C	H	B	L	X	U	N	V	Z	E	F	T	W
Frequency	21	18	17	16	16	15	14	12	11	11	10	7	6	5	4	4	4	4	2	1	1	1	0	0	0	0

It would be highly unusual for this particular message to provide an identical match—letter for letter—strictly on the basis of frequency. But the

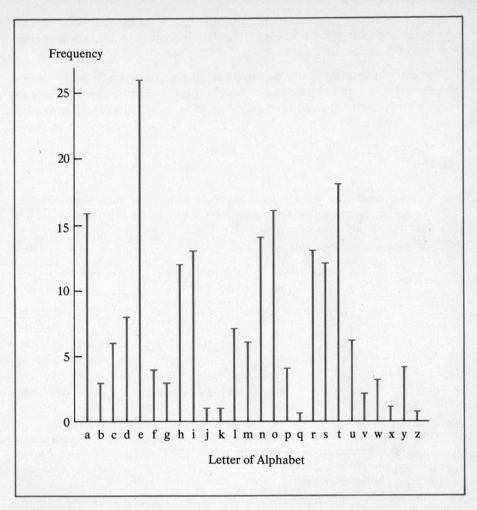

FIGURE 2-6
Qualitative frequency distribution for 200 letters of ordinary English language text.

frequencies can help us to find the more likely possibilities. By trial and error (for instance, substituting e for Y or M and t for one of the other high-frequency letters), portions of the original words can be found. This is a slow process at first, but as each letter is identified, it becomes easier to match the remaining ones.

The complete key for our message is

Ciphertext: M U R K Y A B C D E F G H I J L N O P Q S T V W X Z

Original text: a b c d e f g h i j k l m n o p q r s t u v w x y z

The cipher is constructed by using the word MURKY for the first five original-text letters and listing the remaining ciphertext letters in alphabetical sequence. The text of the full message reads

Frequency of occurrence lies at the heart of statistical analysis. It allows us to create order from confusion. Meaningfully arranged, numbers can tell us a story and help us choose methods and procedures for generalizing about populations.

EXERCISES

2-4 A statistics instructor is compiling a frequency distribution of examination scores. The lowest score is 61 and the highest is 98. Calculate the class interval width if eight intervals are desired.

2-5 Using the family income data:

$ 3,145	$15,879	$ 6,914	$ 4,572	$11,374
12,764	9,061	8,245	10,563	8,164
6,395	8,758	17,270	10,755	10,465
7,415	9,637	9,361	11,606	7,836
13,517	7,645	9,757	9,537	23,957
8,020	8,346	12,848	8,438	6,347
21,333	9,280	7,538	7,414	11,707
9,144	7,424	25,639	10,274	4,683
5,089	6,904	9,182	12,193	12,472
8,494	6,032	16,012	9,282	3,331

(a) Construct a table for the frequency distribution that has six class intervals of equal width. Round the width to the nearest $1,000, and use $3,000 as the lower class limit of the first interval.

(b) Plot a histogram of the frequency distribution you constructed in (a).

2-6 Using the family income data given in Exercise 2-5:

(a) Construct a table for the frequency distribution that has 12 class intervals of width $2,000, with $3,000 as the lower limit of the first interval.

(b) Plot a frequency polygon of the frequency distribution you constructed in (a).

(c) Do you think a more accurate data summary could be obtained by using a fewer or a greater number of class intervals? Discuss the reasons for your choice.

2-7 A brewery has two divisions, one in the East and the other in the West. Separate frequency distributions for the hourly wages of workers in the East and West are provided below.

(a) Plot frequency polygons for each division separately but on the same graph.

(b) Combine the East and West data into a single table for the all-company frequency distribution.

(c) Plot a frequency polygon for the data obtained in part (b).

(d) Comparing the graphs in part (a) to that in part (c), which presentation do you think provides more meaningful managerial information? Explain.

East		West	
Hourly Wage	*Number of Workers*	*Hourly Wage*	*Number of Workers*
$ 5–under 6	150	$ 5–under 6	0
6–under 7	300	6–under 7	0
7–under 8	150	7–under 8	0
8–under 9	100	8–under 9	0
9–under 10	50	9–under 10	50
10–under 11	50	10–under 11	150
11–under 12	50	11–under 12	150
12–under 13	50	12–under 13	200
13–under 14	50	13–under 14	100

2-8 Using a separate category for each alphabetical character and treating upper- and lower-case letters as equal, construct a table for the frequency distribution of the first 200 letters of Abraham Lincoln's Gettysburg Address:

> Four-score and seven years ago our fathers brought forth on this continent a new nation, conceived in liberty, and dedicated to the proposition that all men are created equal.
>   Now we are engaged in a great civil war, testing whether that nation, or . . . .

Compare your distribution with the one in Table 2-7. Do you think that Lincoln used "ordinary" English? Explain.

2-9 A sample of 100 engineering students has been categorized by sex, marital status, and major. The number of persons in each category is summarized below.

Sex	Marital Status	Major	Number of Students
male	married	electrical	14
male	married	mechanical	16
male	married	civil	10
male	single	electrical	5
male	single	mechanical	19
male	single	civil	4
female	married	electrical	5
female	married	mechanical	12
female	married	civil	0
female	single	electrical	5
female	single	mechanical	8
female	single	civil	2
		Total	100

Construct a table for the sample frequency distribution characterized by (a) sex, (b) marital status, and (c) major.

2-10 The following frequency distribution of hospital-patient ages has too many classes. Construct a table for the frequency distribution with half as many intervals, each twice as wide.

Class Interval (years)	Frequency	Class Interval (years)	Frequency
10–under 15	5	45–under 50	21
15–under 20	13	50–under 55	39
20–under 25	25	55–under 60	28
25–under 30	32	60–under 65	59
30–under 35	33	65–under 70	37
35–under 40	27	70–under 75	11
40–under 45	32	75–under 80	1

2-11 Comment on the appropriateness of each of the following class interval widths for a frequency distribution of rates of pay for 100 selected hourly workers in a metropolitan area: $0.05, $0.25, $0.50, $1.00, $2.00.

2-12 Criticize each of the following designations for class intervals of monthly household electricity bills:

(a)	(b)	(c)
$10–15	$11–22	$10–under 15
15–20	21–32	16–under 20
20–25	31–42	21–under 25
etc.	etc.	etc.

2-3   RELATIVE AND CUMULATIVE FREQUENCY DISTRIBUTIONS

Two useful extensions of the basic frequency distribution are the *relative* and the *cumulative* frequency distributions.

Relative Frequency Distributions

A *relative frequency* is the ratio of the number of observations in a particular category to the total number of observations. Relative frequency may be determined for both quantitative and qualitative data and is a convenient basis for the comparison of similar groups of different size. For example, consider comparing family incomes in sparsely populated Nevada with those in California, the most populous state. It would not be very meaningful to say that only

TABLE 2-8

Calculation of Relative Frequency Distribution of Age
of Department Store Accounts Receivable

(1) Age (days)	(2) Number of Accounts (Frequency)	(3) Relative Frequency
0–under 30	532	.330
30–under 60	317	.196
60–under 90	285	.176
90–under 120	176	.109
120–under 150	158	.098
150–under 180	147	.091
Totals	1,615	1.000

6,000 families in Nevada have achieved a particular level of income and that
250,000 families in California receive the same income. But if we say that 1%
of the families in Nevada and only .8% of the families in California have achieved
this income level, we can make a realistic comparison between family incomes
in the two states. We must compare the *proportions* (percentage)—not the
total frequencies—of families who have achieved this income level in each
state.

The relative frequencies for a population are calculated by dividing the
number of observations in each category by the total number of observations.
Once the frequency distribution itself is obtained, the relative frequency
computations are a matter of simple arithmetic. This is illustrated in Table 2-8
for each class interval of data on the age of department store accounts receivable (accounts 180 days or older are written off as uncollectible). The relative
frequency of accounts less than 30 days old (the class interval of 0–under 30)
is 532/1,615 = .330, which means that 33% of all accounts are less than 30 days
old. The other relative frequencies are also calculated by dividing the class
frequency in column (2) by the total number of accounts. The sum of all the
relative frequencies must always be 1. Columns (1) and (3) constitute the *relative
frequency distribution* for these data. The histogram for this distribution is
shown in Figure 2-7. A histogram constructed using relative frequencies will
have the same shape as one constructed using original tallies, or absolute frequencies. Only the scale on the vertical axis will change.

Analysis of relative frequencies can sometimes be useful when analysis of
absolute frequencies is not. A department store's credit sales will fluctuate
considerably over the seasons, but the same seasonal patterns will generally
persist from year to year. If the credit worthiness of the store's customers
remains about the same as it has been in the past, then the relative frequency

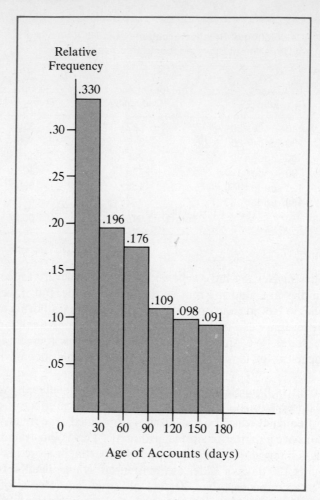

FIGURE 2-7
Relative frequency histogram for age of department store accounts receivable.

distributions for the ages of accounts receivable obtained during the same calendar month in two successive years should not be noticeably different when growth is steady. A substantial difference in relative frequencies may be due to erratic growth or to a change in the quality of credit customers or in collection procedures. Absolute frequencies cannot isolate the effect of sales growth from the effect of changes in credit procedures.

Relative frequency may also be used to compare two qualitative populations. Table 2-9 shows the frequency distribution by sex of students on two campuses. The number of men competing for dates in the large university is 6,000, whereas the number in the small college is only 400. The relative frequency

TABLE 2-9

Frequency Distributions of Students by Sex

Large University		Small College	
Sex	Frequency	Sex	Frequency
male	6,000	male	400
female	4,000	female	100
Totals	10,000		500

TABLE 2-10

Relative Frequency Distributions of Students by Sex

Large University		Small College	
Sex	Frequency	Sex	Frequency
male	.60	male	.80
female	.40	female	.20
Totals	1.00		1.00

distributions are provided in Table 2-10. In terms of relative frequency, there is less competition in the university, where the proportion of men is smaller (.60 in the university versus .80 at the college).

Cumulative Frequency Distributions

A *cumulative frequency*, used only when the observations are numerical, is the sum of the frequencies for successively higher class intervals. For example, we might find that 8,439 men in a population of 10,000 are less than 6 feet tall; stated differently, 84.39% of the population is shorter than 6 feet. Cumulative frequencies can provide useful descriptions of a population, especially when they are expressed relatively as percentages or proportions. We can judge our chances of being accepted by a graduate school, for example, partly by the percentage of persons who obtained lower scores on the admissions examination. (Cumulative frequency is relevant to some probability concepts we will encounter in Chapter 6.)

The cumulative frequencies for a population are determined by adding the frequency for each class interval to the frequencies for preceding intervals. Table 2-11 illustrates this for a sample of 100 annual unit sales figures for television sets sold by retail stores. The cumulative frequency for sales under

TABLE 2-11

Calculation of Cumulative Frequencies for Unit Television Set Sales

(1) Units Sold	(2) Number of Stores (Frequency)	(3) Cumulative Frequency
100–under 200	0	0
200–under 300	5	0 + 5 = 5
300–under 400	39	5 + 39 = 44
400–under 500	31	44 + 31 = 75
500–under 600	16	75 + 16 = 91
600–under 700	6	91 + 6 = 97
700–under 800	3	97 + 3 = 100

300 units is found by adding the frequency of sales at or above 100 (but under 200) to the number of stores having sales at or above 200 (but under 300), or 0 + 5 = 5. The cumulative frequency for a class interval represents the total number of stores having sales levels falling in or below that interval. Columns (1) and (3) of Table 2-11 constitute the *cumulative frequency distribution* for the television sales data.

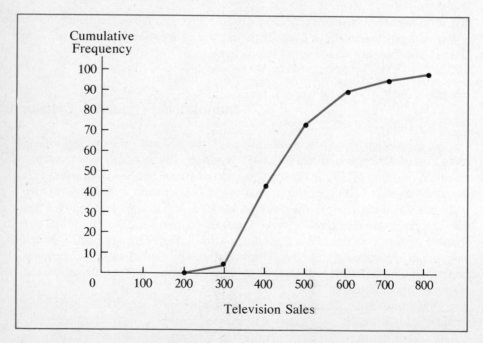

FIGURE 2-8

Ogive for cumulative frequency distribution of television set sales.

The cumulative frequency distribution for the television sales data is shown graphically in Figure 2-8. The ordinate (vertical height) of a plotted point represents the cumulative frequency for values less than the abscissa (the horizontal scale value). Such a curve is called an *ogive*.

Sometimes it is desirable to calculate cumulative relative frequencies. These are found by adding successive relative frequencies. The largest possible cumulative relative frequency would be 1.0.

EXERCISES

2-13 The frequency distribution of times between arrivals of injured patients at a hospital emergency room is as follows:

Time (minutes)	Number of Patients
0.0–under 3.0	210
3.0–under 6.0	130
6.0–under 9.0	75
9.0–under 12.0	40
12.0–under 15.0	20
15.0–under 18.0	15
18.0–under 21.0	10

(a) Make a table for the relative frequency distribution. Then plot the resulting distribution as a histogram.

(b) Make a table for the cumulative frequency distribution. Then plot an ogive for this distribution.

2-14 The following is the frequency distribution of consecutive days absent during 1982 by employees in an automobile assembly plant who are classified as active:

Consecutive Days Absent	Number of Employees
0–under 5	3,100
5–under 10	810
10–under 15	510
15–under 20	320
20–under 25	120
25–under 30	30
30–under 60	110

(Employees absent a total of 60 or more consecutive days are reclassified as inactive. Each employee is counted just once, being placed in the category corresponding to his or her longest absence.)

(a) Make a table for the relative frequency distribution.

(b) Make a table for the cumulative *relative* frequency distribution.

(c) Plot an ogive using the data obtained in part (b).

2-15 The president of an insurance company orders occasional spot checks of the degree of managerial discipline in five regional offices. One factor considered is the relative frequency of tardy and absent employees. On a particular day he ordered an audit of payroll records, which resulted in the following frequency distributions:

| | Number of Employees by Office | | | | |
	A	B	C	D	E
Tardy	60	60	10	56	2
Absent	60	30	20	24	15
On job and on time	1,080	510	270	720	233
Totals	1,200	600	300	800	250

(a) Determine the relative frequency distributions for each office.
(b) Using office A as a standard, which regions have an excessively high number of tardy employees?
(c) Compared to office A, which regions have an unusually large number of absent employees? Do you think that the managers of these regions are necessarily "softer" than the manager in region A? Explain.

2-16 The cumulative relative frequency distribution for the number of shares owned by each of the 1,000 shareholders in a small nonpublic corporation is provided below.

Number of Shares	Cumulative Proportion of Shareholders
0–under 5,000	0.36
5,000–under 10,000	0.63
10,000–under 15,000	0.81
15,000–under 20,000	0.90
20,000–under 25,000	0.97
25,000–under 30,000	1.00

(a) Make a table for the relative frequency distribution.
(b) Using your answer to part (a), determine the original frequency distribution.
(c) Using your answer to part (b), determine the cumulative frequency distribution.

2-4   COMMON FORMS OF THE FREQUENCY DISTRIBUTION

Statistical methodology has been developed to analyze samples taken from populations having various general forms. By placing populations in various categories, we can increase our analytical powers considerably by using

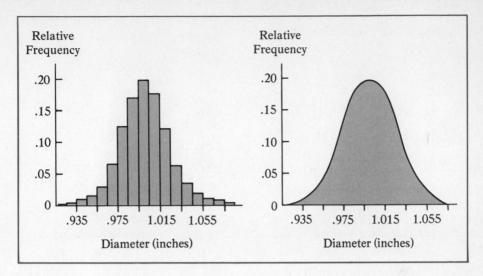

FIGURE 2-9
Sample histogram (left) and population frequency curve (right) for steel rod diameters.

techniques that apply to all populations in the same *distribution family*. All quantitative populations have some form of distribution. Most populations can be classified into a few well-known distributions, although the mathematical equations describing them may be complex. The shape of the sample histogram usually indicates the form of the population distribution.

In the following figures, some of the more common general shapes of frequency curves appear beside representative sample histograms. Figure 2-9 represents the relative frequency distribution of the diameters of a production batch of 1-inch-thick steel reinforcing rods. Note that the histogram is fairly symmetrical about the interval .995–1.005 inches, with frequency dropping for the next higher and lower intervals. The histogram bars on the left and right become progressively shorter, tapering off at about .91 and 1.09 inches, respectively. The smoothed frequency curve beside the histogram is bell-shaped and belongs to a class of populations having the *normal distribution*. We will have more to say about normal curves in Chapter 7. A large number of populations have frequency distributions in this category, including many physical measurements.

Figure 2-10 shows the histogram for the television sales data given in Table 2-11. Here the data are *skewed* to the right, because a few stores had sales levels that were quite high. Beside the histogram is a frequency curve in which the right "tail" is longer than the left. This general shape corresponds to a class of skewed distributions in which there are a few small observations

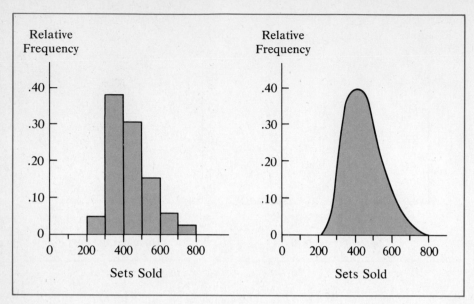

FIGURE 2-10
Sample histogram (left) and population frequency curve (right) for television sales.

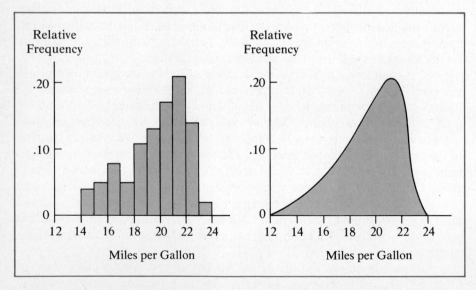

FIGURE 2-11
Sample histogram (left) and population frequency curve (right) for gasoline mileages.

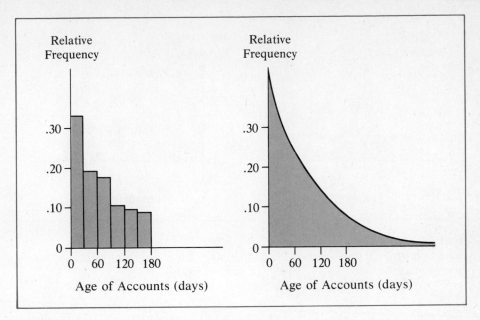

FIGURE 2-12
Sample histogram (left) and population frequency curve (right) for the age of department store accounts receivable.

and in which proportionately more large observations fall over a wide range of values.

Another type of skewed distribution is provided in Figure 2-11, where the gasoline mileage frequency curve has a longer left tail. In this case, the smaller values are widely spread from 14 through 18 miles per gallon, and the rarer large values fall into the single class 23.0–under 24.0.

The distribution given in Table 2-8 for the age of department store accounts receivable is plotted in Figure 2-12. Here, the smoothed frequency curve has the general shape of a reversed letter J. This type of distribution is sometimes called the *exponential distribution*. Such a frequency curve approximates a great many populations in which the observations involve items that exhibit changes in status over time. Exponential distributions have been used to characterize equipment lifetimes until failure. They are also used to describe the time between successive arrivals by cars at a toll booth or by patients in a hospital emergency room and are therefore useful in the analysis of waiting-line or queuing situations.

The left portion of Figure 2-13 is the histogram for the lengths of pieces of scrap roofing lumber from a construction job. Any piece shorter than 20 inches (the width between studs) cannot be used. Because the lumber arrives from the mill in various lengths, a piece of scrap is likely to be any length

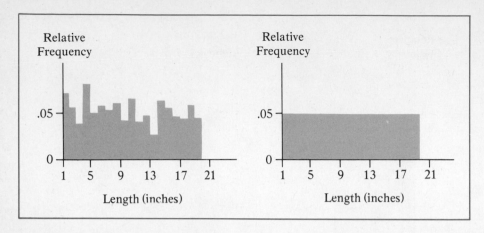

FIGURE 2-13
Sample histogram (left) and population frequency curve (right) for lengths of scrap pieces of lumber.

between 1 and 20 inches. (Scrap shorter than 1 inch is not counted.) The histogram is therefore approximated by a rectangle. Because no particular width is favored, such a population is a member of the *uniform distribution* family.

Later we will encounter other, less common shapes, and we will discuss how one particular shape—the bimodal—can help identify nonhomogeneous population influences. Classifying the various shapes of frequency distributions enables us to obtain better population descriptions and to be selective in choosing statistical techniques for analysis.

EXERCISES

2-17 For each of the following populations for the *ages* of persons, sketch an appropriate shape for the frequency distribution and explain the reasons for your choice.
(a) Science-fiction novel readers.
(b) Persons who do not work full-time.
(c) Persons who have false teeth.
(d) Children in a grammar school.

2-18 For each of the following populations, sketch an appropriate shape for the frequency distribution and explain the reasons for your choice.
(a) The times required by a pharmacist to fill new prescriptions.
(b) The last two digits in numbers assigned to telephones in a metropolitan area.
(c) The number of persons employed by manufacturing firms.
(d) The duration of stay of patients hospitalized with nonchronic diseases.

REVIEW EXERCISES

2-19 The following gasoline mileage data for a taxicab fleet are incomplete:

Miles per Gallon	Number of Cars	Relative Frequency	Cumulative Frequency
6–under 8	—	—	—
8–under 10	23	—	29
10–under 12	—	.34	—
12–under 14	17	.17	—
14–under 16	—	—	92
16–under 18	—	—	—
Totals	100	1.00	

(a) Determine the missing values.

(b) Plot the cumulative frequency ogive for this population.

2-20 A sample of students has been selected at a university, and the students' ages have the frequency distribution given below. Construct a table providing the relative and cumulative frequency distributions.

Age (years)	Number of Students
20–under 25	18
25–under 30	23
30–under 35	15
35–under 40	13
40–under 45	7
45–under 50	6
50–under 55	5
55–under 60	5
60–under 65	8
Total	100

2-21 Consider the following family income data (in thousands of dollars) from a sample of residents in a large apartment complex.

17	25	32	41	43
31	28	27	39	36
25	19	21	28	26
30	32	26	27	34
21	24	20	25	31

(a) Construct a frequency distribution table using six class intervals of width 5 thousand dollars. Use a lower limit of 15 thousand dollars for your first class interval.

(b) Determine the relative frequencies for each class from part (a).

(c) Determine the cumulative frequencies for each class from part (a).

2-22 A group of 350 persons has been categorized by sex, marital status, and occupation. The number of persons in each category combination follows:

Sex	Marital Status	Occupation	Number of Persons
male	married	blue collar	75
male	married	white collar	37
male	married	professional	12
male	single	blue collar	55
male	single	white collar	38
male	single	professional	3
female	married	blue collar	13
female	married	white collar	32
female	married	professional	2
female	single	blue collar	12
female	single	white collar	66
female	single	professional	5
		Total	350

Construct a table for the frequency distributions for the populations characterized by (a) sex; (b) marital status; (c) occupation.

3

Summary Descriptive Measures

We have seen how the frequency distribution arranges raw data into a meaningful pattern. Now we are ready to investigate further ways to summarize statistical data. Knowledge of the frequency distribution alone is not sufficient to answer many statistical questions. For example, in evaluating a new drug, researchers must establish whether it improves patient recovery. The new drug must be compared in some way to the drug in current use. Sample data can provide recovery times for the two treatments, but a direct comparison of histograms would be cumbersome. Instead, the *average* recovery times for the two drugs would more clearly establish whether the new treatment is better. If the new drug significantly speeds up the average patient's recovery, it should replace the old drug.

This chapter considers a variety of summary measures. Each is a number precisely measuring various properties of the observations. There are two major classes of summary numerical values for quantitative data. One measures *central tendency* or *location*, a value around which the observations tend to cluster and which typifies their magnitude. The *arithmetic mean* is one of the more commonly used measures of central tendency. Another broad category of numbers provides measures of *dispersion* or *variability* among the observation values. These measures indicate how observed values differ from each other. Conceptually, the simplest of these is the *range*, which expresses the difference between the largest and smallest observations. A useful summary measure for qualitative data is the *proportion*, which indicates how frequently a particular attribute is observed.

3-1   STATISTICS, PARAMETERS, AND DECISION MAKING

Summary data measures may correspond either to populations or to samples. A summary measure based on population data is called a *population parameter* and expresses a particular property of the entire collection of potential observations. Ordinarily, not all of the population will be observed, and usually only a sample is taken. A summary measure obtained from sample data is referred to as a *statistic*. A statistic is a number conveying a property of only those data that are actually observed. Corresponding to every population parameter is an analogous sample statistic that can be computed in a similar manner. For instance, the arithmetic mean may be calculated for both a population and a sample. Usually the population's mean is not known but may be estimated from the sample's arithmetic mean.

Population parameters serve a role in decision making not provided by the frequency distribution alone. For example, a production manager must periodically develop schedules for personnel and equipment based on some reasonable expectation of performance. Thus, the manager may find that average past productivity levels provide the best indicators of future performance. A frequency distribution for the number of parts produced daily by the assembly department would provide less specific information for a monthly plan than the average daily output. Similarly, a mail-order distributor can use sales to readers of a magazine advertisement in evaluating the effectiveness of magazine advertising. A frequency distribution of dollar sales per reader may be less useful for this purpose than the average sales per reader.

When a population is qualitative, the parameter used is the proportion of observations having a particular attribute. Thus, a federal agency investigating sex discrimination in employment may be interested in the proportion of females employed in various job categories. If this proportion is judged too low in certain jobs, pressures may be applied to remedy the situation. In quality control, the proportion of defective items often serves as a guideline for accepting or rejecting a production batch or for initiating remedial action.

Populations may be easily compared using parameter values. A young man choosing a career might consider the typical earnings levels achieved by members of various professions. He may find that 50% of all the doctors in his state earn above $85,000, a *median* income level of this population, whereas the median income for lawyers is only $35,000. This information may influence him to prefer a medical career to a legal one. As another example, a sales manager wishes to spend her last advertising dollars in a medium providing the greatest difference between average sales per 1,000 viewers, listeners, or readers and cost per 1,000. She would thus compare television and radio commercial possibilities to various newspaper, magazine, and billboard advertisements. If, for the same cost per 1,000 readers, an advertisement in magazine A resulted in increased average sales of only $20 per 1,000, whereas magazine B increased sales on the average by $25 per 1,000, magazine B would be more desirable than A.

When generalizing about a population from a sample, inferences about the value of a parameter are the most common. The median income of all physicians is estimated from a limited sample of doctors. The average test score of all persons who will be taking an aptitude test is projected from the scores of the few persons who initially take the test. The proportion of defectives in a shipment, found from detailed testing of a sample, often determines whether the shipment is accepted or rejected. The procedures of statistical inference are largely concerned with estimating the value of a particular population parameter or with testing to determine whether an assumed value holds.

THE ARITHMETIC MEAN   3-2

The arithmetic mean is the most commonly encountered and best understood of the measures of central tendency. Consider the cash balances (in thousands of dollars) of five manufacturing firms: 101.3, 34.5, 17.6, 83.4, and 52.7. The mean 57.9 is calculated by adding these values and dividing the sum by the number of firms (5):

$$\frac{101.3 + 34.5 + 17.6 + 83.4 + 52.7}{5} = \frac{289.5}{5} = 57.9$$

If the five cash balances represent only a sample from a population of observations of all medium-sized companies in the nation, then the result of the calculation is a *sample mean* of 57.9 thousand dollars. If, instead, they are the entire set of observations for the population of cash balances at the foundries in Marlborough County, then 57.9 thousand dollars is the *population mean*.

Symbolic Expressions for Calculating the Arithmetic Mean

Since the arithmetic mean is calculated in the same way for any group of raw data, it is convenient to express this calculation symbolically. Because we may not yet know the value of a particular observation, we may also have to assign a special symbol to each observation. Traditionally, the letter X is used to represent an observed value. To distinguish each observation, we use the numbers 1, 2, 3, . . . as *subscripts*. Thus, X_1 represents the first observation value; X_2, the second; X_3, the third; and so on. The symbol X_1 is referred to as "X sub 1." Another advantage of expressing observation values symbolically is that algebraic expressions can be used to state precisely how each statistical measure is calculated, so that the same procedure can be applied to any set of data.

The sample mean is represented by the special symbol $\bar{X}$, called "X bar." To calculate the sample mean, we divide the sum of the values by the number of observations made (the *sample size*), which is represented by the letter n. The following formula is used to calculate the

SAMPLE MEAN

$$\bar{X} = \frac{X_1 + X_2 + \cdots + X_n}{n}$$

In our previous illustration, the sample size was $n = 5$, and the observed values were $X_1 = 101.3$, $X_2 = 34.5$, $X_3 = 17.6$, $X_4 = 83.4$, and $X_5 = 52.7$.

A similar calculation applies for the *population mean*, which is traditionally represented by μ (the lowercase Greek *mu*). μ is the equivalent of "m," the first letter in mean. (Several other population parameters are also represented by Greek letters.) When all the observation values from the population are available, μ is found in the same way $\bar{X}$ is. Generally, we do not compute μ directly, because the entire population is not usually observed. Ordinarily, the value of μ remains unknown; only $\bar{X}$ is calculated, and this value serves as an estimate of μ.

Sometimes it is convenient to use an even more concise formula for the sample mean. Just as we use a plus sign $(+)$ to add two quantities together, we can represent the sum of several values by a *summation sign*, which takes the form of $\sum$ (the uppercase Greek *sigma*). The sample mean can then be expressed as

$$\bar{X} = \frac{\sum X}{n}.$$

Calculating the Mean from Grouped Data

When the number of observations is quite large, calculating the mean can be tedious and filled with potential error. If the raw data are to be analyzed by computer, then the procedure we have just learned for finding the sample mean should be used. Otherwise, a shortcut method may be used.

Generally, the first step in analyzing or describing a collection of raw data is to construct a frequency distribution by arranging the raw data into groups or classes according to size. It is usually possible to obtain a good approximation to the sample mean using only these summary data. This shortcut procedure not only saves time and effort, but it may be the only way to obtain a mean.

TABLE 3-1

Calculation of the Mean Gasoline Mileage
Using Shortcut with Grouped Data

Gasoline Mileage (miles per gallon)	Number of Cars f	Class Interval Midpoint X	fX
14.0–under 16.0	9	15	135
16.0–under 18.0	13	17	221
18.0–under 20.0	24	19	456
20.0–under 22.0	38	21	798
22.0–under 24.0	16	23	368
Totals	100		1,978

$$\bar{X} = \frac{\sum fX}{n} = \frac{1,978}{100} = 19.78 \text{ miles per gallon}$$

Sometimes published data are already grouped into classes and the raw data are unavailable.

Table 3-1 shows the sample mean calculation using grouped data for the frequency distribution for gasoline mileages. Every class interval is represented by its midpoint. The first of these, 15, is found by averaging the limits of the first class interval: $(14.0 + 16.0)/2 = 15$. The remaining midpoints can then be easily found by adding one class interval width (here, 2 miles per gallon) to the preceding class midpoint. Each midpoint is multiplied by its respective class frequency. The sum of these products is then divided by the sample size n. Representing successive midpoints by X's and class frequencies by f's, the procedure is summarized by the following expression for calculating the

SAMPLE MEAN USING GROUPED DATA

$$\bar{X} = \frac{\sum fX}{n}$$

The resulting value is a *weighted average*. Each class midpoint X is weighted by the respective frequency f. The sum of the weighted products is then divided by n, which equals the sum of the frequencies or weights. We will encounter weighted averages again in Chapter 6 when we discuss special topics in probability.

The value 19.78 found in Table 3-1 is only an *approximation* of the sample mean and can be expected to differ from the value calculated directly from the raw data. This is because all gasoline mileages in a particular class interval are represented by a single number. The raw gasoline mileage data are provided in Table 2-4 (page 19). The arithmetic mean of these data, calculated without grouping, is 19.718 miles per gallon. This differs only slightly from the value calculated for $\bar{X}$ using the shortcut method. For practical purposes, the approximation is usually close enough.

EXERCISES

3-1 The following numbers of persons were hired by a sample of firms during a certain month:

$$14, \ 6, \ 12, \ 19, \ 2, \ 35, \ 5, \ 4, \ 3, \ 7, \ 5, \ 8$$

Calculate the sample mean number of hirings.

3-2 A data-processing manager has purchased remote-terminal processing time to run special jobs at two different computer "utilities." She wishes to sign a long-term contract with the firm whose computer causes the least delay on the average. During trial periods with each firm, the numbers of minutes of delayed processing per week are

CompuQuick	210, 15, 47, 93, 104
Dial-a-Pute	18, 341, 523, 25, 19, 293, 115, 203

Assuming that trial experience is representative of future performance, which firm should receive the business? Substantiate your answer with appropriate calculations.

3-3 The sales manager of Kleen Janitorial Supplies allows sales representatives to give special introductory prices to new customers. He wishes to weed out those salespersons who take undue advantage of specials, "milking" them for commissions and bringing in little new continuing business. Investigating the files for the past ten months, he has determined for each of the five sales representatives the following sample percentages of specials buyers who placed second orders:

Percentage of Specials Buyers Retained

A	B	C	D	E
36	9	11	33	18
43	16	5	17	17
49	21	6	45	23
18	14	14	29	6
17	33	25	17	31
32	8	12	27	42
24	19	11	61	19
19	17	9	47	13
28	26	28	35	26
36	11	14	14	33

(a) Find the mean percentage of specials buyers retained for each salesperson.

(b) Using your answers to part (a), determine the mean percentage of such customers retained for the sales force as a whole.

(c) Assuming that the sales manager will take remedial action against those salespersons who produce lower than average retention, which salespersons should be singled out for "milking" specials?

3-4 Using the following frequency distribution for the times to failure of a certain kind of fuse, calculate the sample mean.

Time to Failure (hundreds of hours)	Number of Fuses
0–under 20	500
20–under 40	250
40–under 60	125
60–under 80	61
80–under 100	15
100–under 120	6

3-5 Consider the percentages of specials buyers retained by all salespersons in Exercise 3-3 as a single sample.

(a) Construct a frequency distribution using six classes of equal width, starting with 5.0–under 15.0.

(b) Calculate the sample mean using the grouped data of the frequency distribution obtained in part (a).

(c) Calculate the sample mean using ungrouped data. By how much is this above or below your answer to part (b)?

THE MEDIAN AND THE MODE   **3-3**

The Median

After the mean, the most common measure of central tendency is the *median*. Like the mean, the median provides a numerical value that is typical of quantitative sample observations. The *sample median*, denoted by m, is the central observation when all the data are arranged in increasing sequence. For the heights 66, 68, 69, 73, and 74 inches, the median is $m = 69$ inches, the central value. If a person 70 inches tall were added to the initial group, then the median would be obtained by averaging the two central values from 66, 68, 69, 70, 73, 74. Thus, the sample median would be $m = (69 + 70)/2 = 69.5$ inches.

In general, the **median is the value above or below which lies an equal number of observations.** We find the median from the raw data by listing the observations in sequence from lowest to highest and then selecting the central value if there is an odd number of values or averaging the two central values if there is an even number of observations.

When the data have already been grouped into a frequency distribution, the median will fall somewhere in the first class interval above which fewer than half the observations lie. (Its approximate position in this case can be found by using an interpolation procedure.)

The counterpart population parameter to m is the *population median*, denoted by M. This median may be found in the same way from the entire collection of observed values. Ordinarily, we directly calculate a value for m only, and this value then serves as the basis for estimating the value of M, as $\bar{X}$ does for μ.

The Median Contrasted with the Mean

In different senses, the arithmetic mean and the median are both averages. The mean is the arithmetic average of variates, and the median is the average of position. When we use the term *average*, we speak of the arithmetic mean.

A mean can be algebraically manipulated; the combined mean of two populations can be calculated from the individual means. Due to these convenient mathematical properties, many statistical techniques employing the mean have been developed. The median is not as well suited to mathematical operations. For example, the median of a combined population cannot be obtained from the separate component population medians. Fewer statistical techniques employ the median due to the mathematical difficulties associated with it.

On the other hand, the mean is influenced by extreme values to a much greater degree than the median. Consider the net worth levels of your close friends. Suppose that you meet one of the world's few billionaires and include him in your circle of friends. The addition of this person would distort the mean level of wealth so greatly that "on the average" all of your friends would be multimillionaires—hardly a meaningful summary. But the median would not be significantly influenced by the billionaire. The median is more democratic, giving each elementary unit only one "vote" in establishing the central location.

Generally, the median provides a better measure of central tendency than the mean when there are some extremely large or small observations.

The Mode

A third measure that may be used to describe central tendency is the *mode*—the *most frequently occurring* value. Consider the simple illustration provided by the collection of five observations: 2, 3, 4, 4, and 7. The mode is 4, because 4 occurs most often.

The interpretation of the statistical mode is analogous to that of the fashion mode. A person dressing in the current style is "in the mode." But a current fashion can be a poor description of what most people are wearing, because a variety of styles is worn by the general public. In statistics, the mode only tells us which single value occurs most often; it may therefore represent a minority of the observations.

As a basis for decisions, the mode can be insidiously undemocratic. For example, most shoe stores stock only the most popular sizes, so the mode is their deciding parameter. One reason for this is that the turnover on unusual sizes is so low that they are unprofitable. When such sizes are stocked, there is ordinarily little choice of color or style (and what there is will usually be on the conservative side). A significant number of people who wear unpopular sizes find it difficult to buy shoes in stores and may have to resort to mail-order purchasing. Not being of modal size, they are forced to be hopelessly out of mode in the sense of fashion as well.

When the data are grouped into classes, the mode is represented by the midpoint of the interval having the greatest class frequency. We refer to this group as the *modal class*. When the frequency distribution is portrayed as a smoothed curve like the one in Figure 3-1, the mode corresponds to the possible observation value lying beneath the highest point on the frequency curve—the location of maximum clustering. The mode can therefore serve as a basis for comparing the typicalness of the other measures of central tendency.

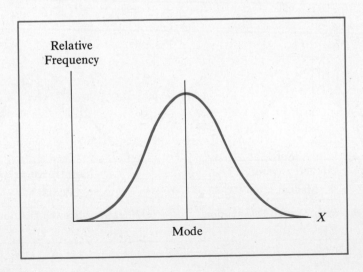

FIGURE 3-1
Illustration of the location of the mode.

Positional Comparison of Measures: Skewed Distributions

When the population has a *symmetrical* frequency curve like the one in Figure 3-2(a), the mean, median, and mode coincide. When the population is not symmetrical, the mean and median will lie to the same side of the mode. The frequency curve in Figure 3-2(b) has a tail tapering off to the right. Such an asymmetrical frequency distribution is said to be *skewed to the right*, meaning that the population values cluster around a relatively low value, although there are some extremely large observation values. In this case, the mean lies to

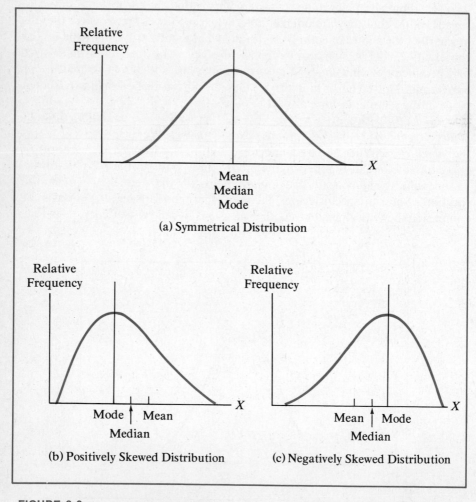

FIGURE 3-2
Positional comparison of center measures for symmetrical and skewed frequency distributions.

the right of the mode, reflecting the influence of the larger values in raising the arithmetic average. The median, which is less sensitive to extremes, must lie to the left of the mean. A rightward skewed distribution will also have a mode that is smaller than the median. The median will therefore lie somewhere between the mode and the mean.

Because a rightward skewed population will always have a median smaller than its mean, subtracting the median from the mean results in a positive difference. For this reason, such a frequency distribution is said to be *positively skewed*. This expression also reflects the fact that the long tail of the distribution tapers off in a positive direction from the center of the population.

Generally, a characteristic having a lower limit but no theoretical upper boundary will result in positively skewed populations. This is true of the annual wages earned by construction workers in the United States. A lower limit is established by the legal minimum wage. Some blue-collar aristocrats, such as operating engineers (who drive heavy equipment), plumbers, electricians, and steeplejacks, earn more than many doctors when their overtime premiums are considered. In total numbers, however, the high earners are relatively sparse, creating a long, thin, rightward tail in the frequency curve of this population. Economic data often have positively skewed frequency distributions.

Figure 3-2(c) shows a frequency curve that is *skewed to the left*—the direction toward which its long tail points. Because the extremes are relatively small values, the mean lies below the mode. Again, the median will lie somewhere in between the mean and the mode. Subtracting the median from the mean (a smaller value) results in a negative difference, so the frequency curve is said to be *negatively skewed*.

Negatively skewed distributions result when the observed values have an upper limit and no significant lower boundary. For example, the ages of viewers who watch a television program on which "old favorites" are played would be a negatively skewed population. Aimed at a predominantly middle-aged and elderly audience, with commercials touting denture deodorants, laxatives, and home health-care remedies, the show would not be that appealing to younger viewers. The few young people who might watch it would bring the mean viewing age below what would be considered typical, so that the *median* age would more truly represent a typical population value in this case.

The median is the most realistic measure of location for data having skewed distributions.

Bimodal Distributions

The mode has been defined as the most frequently occurring population value. But if two or more values occur with equal or nearly equal frequency, then there are two or more modes. A population that has two modes is said to be *bimodal*.

The presence of more than one mode has a special significance in statistical analysis: It indicates potential trouble. Comparing or drawing conclusions about bimodal populations can be dangerous because they usually arise when some nonhomogeneous factor is present in the population. Figure 3-3(a) presents the frequency curve of a population of heights of adult patients admitted to a large hospital during the past year. The curve is bimodal, with the two modes occurring at 5′3″ and 5′10″. In this case, the nonhomogeneous factor is sex: The two modes result from the fact that an equal number of male and female patients have been measured. It would be more meaningful to separate

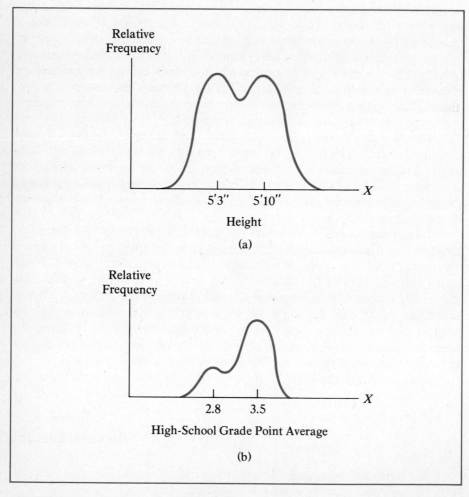

FIGURE 3-3
Illustrations of bimodal frequency distributions.

the population into male and female populations and analyze the two more homogeneous populations separately.

To illustrate the potential difficulty, suppose we wish to compare patient heights in the year before the new maternity ward opened to those in the current year. The proportion of female patients would be higher in the second year, so the overall mean height would be less. The patients would not have become shorter, as a cursory analysis might indicate. By separately comparing the two female populations, the mean heights would not noticeably change over the two-year period.

Figure 3-3(b) shows another population having a double-humped frequency curve. This curve represents the high-school grade point average of students at a university. Humps in frequency occur at 2.8 and 3.5. Although the mode is 3.5, we would still classify this as a bimodal population, because a nonhomogeneous factor influences the grades of the high-school students. A relatively small group of disadvantaged students is to be given remedial work. Although these remedial students are comparatively ill-prepared now, their *college* grade point averages are not expected to differ substantially from those of the other students.

EXERCISES

3-6 The numbers of major automobile accidents occurring each month during the past year in a certain city were

$$0, \ 1, \ 3, \ 4, \ 5, \ 2, \ 2, \ 6, \ 7, \ 2, \ 0, \ 1$$

Letting these data represent sample data for accidents in cities of similar size, calculate the (a) sample mean, (b) median, and (c) mode.

3-7 The sample frequency distribution for the lifetimes of a particular stereo cartridge follows:

Lifetime (hours)	Number of Cartridges
0.0–under 50.0	5
50.0–under 100.0	16
100.0–under 150.0	117
150.0–under 200.0	236
200.0–under 250.0	331
250.0–under 300.0	78
300.0–under 350.0	27
350.0–under 400.0	·8

49·99 – 99·99 → 50.0–under 100.0
99·99 – 149·99 100.0–under 150.0

(a) Calculate the sample mean.
(b) Find the sample mode.

MIDPOINT OF MODAL CLASS

E) FIND TILL 38TH percentile
F) FIND TILL STANDARD DEVIATION
4) STATE HOW WE DIST. IS SKEWED
C) FIND MEDIAN
D) FIND VALUE OF 3RD QUARTILE
DO NOT DO

(c) If the value of the sample median is $m = 205$, does the mean lie above or below the median? Does this indicate that the distribution is positively or negatively skewed?

3-8 Using the following family installment-debt data, determine the sample median.

4) INTER QUARTILE RANGE

$2,032	$ 232	$ 493	$5,555
597	4,893	4,432	4,444
203	796	978	329
97	852	1,427	972
3,333	1,712	2,121	438
1,212	1,940	5,067	705
5,769	1,843	4,337	3,976
2,347	3,525	5,213	3,034
2,137	3,414	4,896	5,035
2,049	4,327	2,172	4,222

3-9 The instructor in a statistics course gives four exams of equal weight. From the scores on these, she will determine a single central score value for each student. Suppose that she lets each student decide *in advance of the first test* whether his or her particular grade will be determined by a mean or by a median test score.
(a) Would you request the mean or the median? Why?
(b) For each of the following hypothetical sets of test scores, indicate whether the mean or the median would provide the greatest central value.

(1)	(2)	(3)	(4)
95	80	80	60
60	50	75	80
75	75	65	90
65	70	60	90

3-10 For each of the following situations, indicate a possible source of nonhomogeneity in the data, and discuss why the population should or should not be split to better serve the purposes of the statistical study.
(a) Six months ago, a new driver safety program was initiated by a state's highway patrol. For the past year, the governor's office has maintained records of the number of weekly accidents. A study is being made to evaluate highway construction standards.
(b) An appliance manufacturer wishes to make a powerful home vacuum cleaner. To be most effective, the cleaner must be heavier than the normal weight of such an appliance. A representative group of employees, including men and women, is used to obtain data on how much weight a person can carry up a flight of stairs without excessive fatigue. These data will be used to set a maximum vacuum cleaner weight.
(c) An automobile manufacturer has obtained data on the total man-hours required to assemble each car at two identically equipped plants of the same size. Each plant produces the models of cars ordered by dealers in its geographical region. These data will be used to establish production standards for each car model.

MEASURING VARIABILITY   3-4

Importance of Variability

Measuring variability is just as important as finding central tendency. The role of variability is illustrated in Figure 3-4, where the frequency curves for the net earnings of dairies and ranches are compared. These populations, reflecting both good and bad years, were determined in such a way that the median incomes were the same. In each case, the populations are positively skewed. Losses occurred for a portion of both dairies and ranches, but the losses incurred by some ranchers were more frequent and were often larger than those of the least profitable dairy. On the other hand, many ranches were more profitable than the best dairies. Ranch earnings exhibit *greater variability* than dairy incomes. Although both forms of agriculture involve cattle and must meet comparable feed requirements, dairy farmers are blessed with stable milk prices (often due to regulation), whereas ranchers must sell their beef at the volatile market price. The greater variability in income makes ranching a riskier venture than owning a dairy, even though the earnings of both enterprises have identical central tendencies (equal medians). Here, statistical variability might help the new agricultural college graduate to choose between ranching or dairy farming.

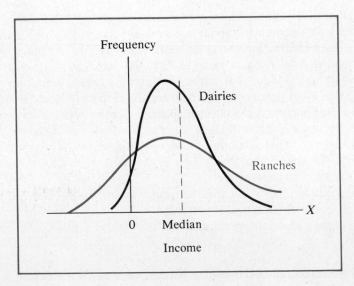

FIGURE 3-4
Frequency curves for the earnings of dairies and ranches.

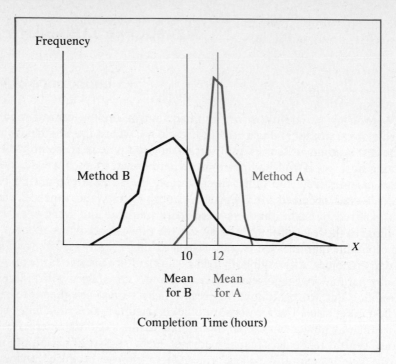

FIGURE 3-5
Frequency distributions for the completion times of two chemical processes.

Another illustration of the importance of variability in statistical analysis is provided by a decision to modify a chemical process. Figure 3-5 shows the frequency distributions for the completion times of two chemical processes being tested. Method A has a mean of 12 hours, whereas method B takes an average of 10 hours. Method B shows substantially greater variability than A. Assuming that both processes cost the same on an hourly basis, B appears to be cheaper. But process A would make planning easier because its completion times are less varied. If the final product is perishable or must be made to customer order, so that it cannot be stored in quantity for very long, then A might be the superior method. *In a great many populations, reduction in variability is itself an improvement.*

Variability or dispersion may be measured in two basic ways: in terms of *distances* between particular observation values, or in terms of the *average deviations* of individual observations about the central value.

Fractiles, Percentiles, and Distance Measures of Dispersion

Distance measures of dispersion are popular when the only purpose is to describe a collection of data. The most common distance measure is the *range*, which is obtained by subtracting the smallest observation from the largest. For example, suppose that the five students in a college accounting honors program have the following IQs: 111, 118, 126, 137, 148. The range of these values is $148 - 111 = 37$. The figure 37 represents the total spread in these observations. The range provides a concise summary of the total variation in a sample or a population, but its major disadvantage as a useful measure is that it ignores all except the two most extreme observations. These two numbers may be untypical values, even among the higher and lower observations.

Other distance measures ignore the most extreme observations. These are *interfractile ranges*, which express the differences between two values called fractiles. A *fractile* is a point below which some specified proportion of the value lies. The median is the .50 fractile because half of all observations lie

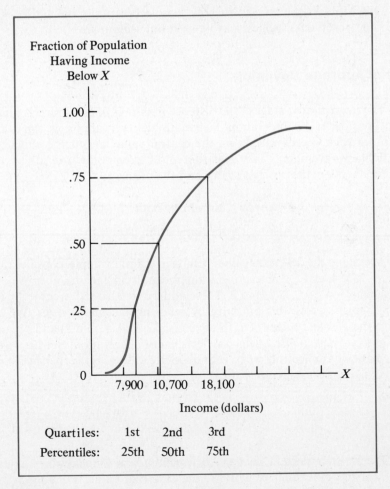

FIGURE 3-6
Cumulative relative frequency distribution for family income.

below this value. When expressed as a percentage, the analogous measure is called a *percentile*; the median is the 50th percentile. The most commonly used fractiles or percentiles are the *quartiles*. These divide the observations into four groups, each of which contains about 25% of the observations. The .25 fractile, or 25th percentile, is the *first quartile*. This is the value below which 25% of the observations lie. The median is the *second quartile*, and the 75th percentile is the *third quartile*.

The most common, dispersion measure based on fractiles is the *interquartile range*. This is the difference between the third and the first quartiles, or *the middle 50% of the observation values*. Figure 3-6 shows an ogive for family income. The fractile values may be read from this graph. The first quartile is $7,900 and the third quartile is $18,100, so that the interquartile range is $18,100 - $7,900 = $10,200.

Other interfractile ranges, such as those representing the middle 90% or 99% of the observation values, may be used but seldom are.

Measures of Average Deviation

The main disadvantage of the distance measures of dispersion is that they do not consider every observation. To include all observations, we can calculate how much each one deviates from the central value and then combine these deviations by averaging. The deviation most commonly considered is the difference between the observed value and the mean:

$$X - \bar{X} \quad \text{for sample data}$$

$$X - \mu \quad \text{for population data}$$

Averaging the deviation values results in 0. For example, consider the five values 1, 2, 3, 4, and 5. Subtracting 3 (the mean) from each number, we obtain deviations of -2, -1, 0, 1, and 2. These add up to 0, so the average deviation is 0. We might avoid this difficulty by ignoring the minus signs. We could average the *absolute values* of each deviation (2, 1, 0, 1, and 2) and obtain the *mean absolute deviation* 1.2, which reflects every observation. But because it is mathematically difficult to work with absolute values, two other measures of dispersion are more commonly used.

The Variance

The most important measure of variability is found by averaging the *squares* of the individual deviations; the resulting value is the *mean of the squared deviations*. Performing this calculation on the IQs for the five accounting students cited earlier, the mean value is

$$\frac{111 + 118 + 126 + 137 + 148}{5} = 128$$

The deviations from the mean value are

$$111 - 128 = -17$$
$$118 - 128 = -10$$
$$126 - 128 = -2$$
$$137 - 128 = 9$$
$$148 - 128 = 20$$

and the mean of the squared deviations is

$$\frac{(-17)^2 + (-10)^2 + (-2)^2 + (9)^2 + (20)^2}{5} = \frac{289 + 100 + 4 + 81 + 400}{5}$$

$$= \frac{874}{5} = 174.8$$

The measure of variability obtained in this manner is referred to as the *variance*. It may be calculated using data from either the sample or the entire population. When all the population data are available, this procedure provides the *population variance*, denoted by σ^2. (The symbol σ is the lowercase Greek *sigma*, and σ^2 is called "sigma squared.") The expression used to calculate the population variance is

$$\sigma^2 = \frac{\sum(X - \mu)^2}{N}$$

Here the sum of the squared deviations from the population mean μ is found, and—because every possible observation is made—this sum is divided by the *population size*, denoted by the letter N. In our example here, the *entire population* consists of $N = 5$ observations.

In practice, a value for σ^2 cannot be computed because most populations are so large that usually only a sample is taken. For sample data, variability is measured by the *sample variance*, denoted by s^2. From the following analogous expression *using sample size n in place of N and sample mean $\bar{X}$ in place of μ*, we calculate the

SAMPLE VARIANCE

$$s^2 = \frac{\sum(X - \bar{X})^2}{n - 1}$$

Here, the squared deviations from $\bar{X}$ are averaged using only the n *sample observations*. The expression for s^2 is slightly different in form from that for σ^2, with $n - 1$ as the divisor instead of the complete sample size n. For reasons explained in Chapter 8, using $n - 1$ for the divisor makes the resulting s^2 value a more accurate estimate of the usually unknown value for σ^2.

Two practical difficulties are associated with the use of the variance. First, the variance is usually a very large number compared to the number of observations themselves. Thus, if the observations are largely in the thousands, the variance is often in the millions. Second, the variance is not expressed in the same units as the observations. In the previous example, the variance is 174.8 *squared* IQ points. This is because the deviations, measured in IQ points, have all been squared. The variance for heights, originally measured in feet, will therefore be expressed in square feet. The variance for gasoline mileages will be in squared miles per gallon.

In spite of these difficulties, the mathematical properties of the variance make it extremely important in statistical theory. Furthermore, the difficulties may be overcome simply by working with the square root of the variance, called the *standard deviation*. (Appendix Table A on page 820 provides square roots for numbers between 0 and 100.)

The Standard Deviation

The square root of the variance in IQ levels is

$$\sqrt{174.8} = 13.22$$

which yields the standard deviation in IQ level. The standard deviation is expressed in the same units as the observations themselves; the value 13.22 is a point on the same numerical scale.

The *sample standard deviation* is represented by the letter s without its exponent 2. The following expression may be used to calculate the

SAMPLE STANDARD DEVIATION

$$s = \sqrt{\frac{\sum (X - \bar{X})^2}{n - 1}}$$

Note that both sides of the equation simply contain the square root of the expression for the sample variance. Likewise, for the population parameters, the *population standard deviation* is denoted by σ, and σ is equal to $\sqrt{\sigma^2}$. Both the variance and the standard deviation provide the same information; one can always be obtained from the other. The standard deviation is a practical

TABLE 3-2

Calculation of Sample Standard Deviation for Heights

Height (inches) X	Deviation $X - \bar{X}$	Squared Deviation $(X - \bar{X})^2$
66	$66 - 70 = -4$	16
73	$73 - 70 = 3$	9
68	$68 - 70 = -2$	4
69	$69 - 70 = -1$	1
74	$74 - 70 = 4$	16
Totals 350	0	46

$$\bar{X} = \frac{\sum X}{n} = \frac{350}{5} = 70 \text{ inches}$$

$$s = \sqrt{\frac{\sum(X - \bar{X})^2}{n - 1}} = \sqrt{\frac{46}{5 - 1}}$$

$$= \sqrt{\frac{46}{4}} = \sqrt{11.5} = 3.391 \text{ inches}$$

descriptive measure of dispersion, whereas the variance is generally used in developing statistical theory.

The sample standard deviation is computed in Table 3-2 for $n = 5$ for heights. (These five observations are a sample from a very large population.)

Shortcut Calculations of the Variance and the Standard Deviation

The sample standard deviation may be calculated from individual values with a mathematically equivalent equation that is usually simpler to use.

SHORTCUT CALCULATION

FOR SAMPLE VARIANCE

$$s^2 = \frac{\sum X^2 - n\bar{X}^2}{n - 1}$$

FOR SAMPLE STANDARD DEVIATION

$$s = \sqrt{\frac{\sum X^2 - n\bar{X}^2}{n - 1}}$$

The use of this formula for the standard deviation is illustrated in Table 3-3. The shortcut procedure is recommended whenever s is computed by hand or with a calculator.

TABLE 3-3

Calculation of Sample Standard
Deviation Using Shortcut Procedure
with Ungrouped Height Data

Height (inches) X	X^2
66	4,356
73	5,329
68	4,624
69	4,761
74	5,476
Totals 350	24,546

$$\bar{X} = \frac{\sum X}{n} = \frac{350}{5} = 70 \text{ inches}$$

$$s = \sqrt{\frac{\sum X^2 - n\bar{X}^2}{n-1}} = \sqrt{\frac{24,546 - 5(70)^2}{5-1}}$$

$$= \sqrt{11.5} = 3.391 \text{ inches}$$

Just as we calculated the mean from the frequency distribution in Section 3-2, we can use grouped data to calculate the sample variance and the standard deviation:

$$s^2 = \frac{\sum f(X - \bar{X})^2}{n-1} \qquad s = \sqrt{\frac{\sum f(X - \bar{X})^2}{n-1}}$$

However, we would usually follow this shortcut procedure:

GROUPED DATA CALCULATION

FOR SAMPLE VARIANCE

FOR SAMPLE STANDARD DEVIATION

$$s^2 = \frac{\sum fX^2 - n\bar{X}^2}{n-1} \qquad s = \sqrt{\frac{\sum fX^2 - n\bar{X}^2}{n-1}}$$

The grouped data computations for the standard deviation are illustrated in Table 3-4, using the earlier gasoline mileages. We find that $s = 2.34$ miles per gallon.

TABLE 3-4

Calculation of Sample Standard Deviation with Grouped Gasoline Mileage Data

(1)	(2)	(3)	(4)	(5)	(6)
		Class			
Gasoline	Number	Interval			
Mileage	of Cars	Midpoint			
(miles per gallon)	f	X	fX	X²	fX²
14.0–under 16.0	9	15	135	225	2,025
16.0–under 18.0	13	17	221	289	3,757
18.0–under 20.0	24	19	456	361	8,664
20.0–under 22.0	38	21	798	441	16,758
22.0–under 24.0	16	23	368	529	8,464
Totals	100		1,978		39,668

$$\bar{X} = \frac{\sum fX}{n} = \frac{1,978}{100} = 19.78 \text{ miles per gallon}$$

$$s = \sqrt{\frac{\sum fX^2 - n\bar{X}^2}{n-1}} = \sqrt{\frac{39,668 - 100(19.78)^2}{99}}$$

$$= \sqrt{5.4865} = 2.34 \text{ miles per gallon}$$

As in the case of calculating the mean from grouped data, only an *approximate* value for the standard deviation can be obtained using this procedure. For most purposes, the approximate value is close enough to the true value that would be calculated directly from the raw data.

The Meaning of the Standard Deviation

The standard deviation is a parameter that, when combined with statistical techniques, provides a great deal of information. When the population has a special frequency distribution called the normal curve, we can find the percentage of observations falling within distances of one, two, or three standard deviations from the mean. About 68% of all observations lie within the region $\mu \pm 1\sigma$. For example, suppose a group of men have a mean height of $\mu = 5'9''$ and a standard deviation of $\sigma = 3''$. If these heights constitute a normal distribution, then 68% of all men will be between $\mu - \sigma = 5'6''$ and $\mu + \sigma = 6'0''$ tall. Furthermore, about 95.5% of the population will lie within $\mu \pm 2\sigma$, and 99.7% will fall within $\mu \pm 3\sigma$.

The normal curve is described mathematically in terms of only two parameters, μ and σ. Thus, for populations characterized by this curve, we can construct a close representation of the entire frequency distribution simply by knowing the mean and the standard deviation. We will discuss this more thoroughly in Chapter 7.

A theoretical result called *Chebyshev's Theorem*, named for the mathematician who proposed it, indicates that the standard deviation plays a key role in any population.

CHEBYSHEV'S THEOREM

The proportion of observations falling within k standard deviations of the mean is at least $1 - 1/k^2$.

This theorem says that regardless of the characteristics of the population, the following holds:

Standard Deviations k	Minimum Proportion of Observations Within $\mu \pm k\sigma$
1	$1 - 1/1^2 = 0$
2	$1 - 1/2^2 = 0.75$
3	$1 - 1/3^2 = 0.89$

Since Chebyshev's Theorem applies for any population, it is too general to be of much practical use. Usually a more precise determination of how many observations lie within k standard deviations of the mean can be found when the form of a population's frequency distribution is known, as we have shown for the normal curve. The practical significance of Chebyshev's Theorem is that it tells us that a great deal of information is imparted by the population standard deviation.

EXERCISES

3-11 The following sample gasoline sales data (in thousands of gallons) have been obtained for a city's service stations:

$$35, \ 47, \ 57, \ 16, \ 12, \ 33, \ 38$$

(a) Calculate the sample range.
(b) Calculate the sample variance and the sample standard deviation.

3-12 The following data have been obtained to represent the price-earnings ratios of stocks listed on the New York Stock Exchange:

$$25, \ 16, \ 50, \ 19, \ 42, \ 37$$

Calculate the sample variance and the sample standard deviation.

3-13 For each of the following decision situations involving populations, discuss why a central population value may not be a wholly adequate summary measure by itself.

(a) The population of temperatures actually achieved in each room of a building with standard temperature control settings is being used to determine whether the system should be modified.

(b) Balancing an automobile assembly line requires that sufficient personnel and equipment be positioned at each work station so that no one station will be excessively idle. The population of task-completion times at each station is being used to help determine how this may be done.

(c) To plan for facilities expansion, a hospital administrator requires data on the convalescent times of surgical patients.

3-14 Referring to the fuse-failure data in Exercise 3-4, calculate the sample standard deviation.

3-15 The Educational Testing Service administers the Graduate Management Admissions Test (GMAT). Following is the cumulative relative frequency distribution for the base period:

Test Score	Cumulative Relative Frequency
200–below 250	.01
250–below 300	.03
300–below 350	.09
350–below 400	.19
400–below 450	.36
450–below 500	.55
500–below 550	.74
550–below 600	.88
600–below 650	.96
650–below 700	.99
700–below 750	1.00

(a) On graph paper, plot the ogive for these data.

(b) Assuming that intermediate values may be read to a good approximation from your graph, determine the .25, .50, and .75 fractiles for GMAT scores.

(c) Use your answers to (b) to calculate the interquartile range for GMAT scores.

3-16 The following frequency distribution has been obtained for a sample of $n = 100$ university-student grade point averages (GPA).

GPA	Number of Students
0.0–under 1.0	1
1.0–under 2.0	7
2.0–under 3.0	58
3.0–under 4.0	34

Using grouped data calculations, determine the sample mean and the sample standard deviation.

3-17 In describing the extent of quality control in her company, the president of a rubber company states that the mean weight of a particular tire is 40 pounds, with a standard deviation of 1 pound. She adds that about 68% of all tires weigh between 39 and 41 pounds, whereas almost all tires weigh between 37 and 43 pounds. State the assumptions on which these statements rest.

3-5   THE PROPORTION

In describing a qualitative population, the key measure of interest is the *proportion* of observations that fall into a particular category. Like the measures we have already discussed, the population parameter is a separate entity from the sample statistic. The population parameter is referred to as the *population proportion* and is denoted by π (the lowercase Greek *pi*).* The analogous sample statistic is the *sample proportion*, which is represented by P. The following ratio is used to calculate the

SAMPLE PROPORTION

$$P = \frac{\text{Number of observations in category}}{\text{Sample size}}$$

A proportion may assume various values between 0 and 1, according to the relative frequency with which the particular attribute occurs. Like the other population parameters, π is ordinarily unknown and may be estimated from the sample results. The statistical procedures for estimating π from P are analogous to those used to estimate a population mean from a sample. But the differences between these procedures are substantial enough to require parallel statistical development throughout this book.

The proportion is important in many kinds of statistical analysis. For example, it is often used as the basis for taking remedial action. An unusually high proportion of sales returns will be singled out as a managerial problem. A machine that produces a large proportion of oversized or undersized items must be adjusted or repaired. The level of impurities in a drug might be expressed as a proportion; if this figure is too high, the drug cannot be used. In many elections for public office in the United States, the winner is the candidate

---

* Statistics has its own special notation, and here π is not the 3.1416 used in geometry to express the ratio of the circumference to the diameter of a circle. Just as μ and σ, the Greek equivalents of *m* and *s*, are the first letters in the words *mean* and *standard deviation*, π, the Greek p, is the first letter in the word *proportion*.

who receives a plurality—that is, the person who receives the highest proportion of votes.

If a sample of $n = 500$ persons contains 200 men and 300 women, then the sample proportion of women is $P = 300/500 = .60$. If a machine has produced a sample of $n = 100$ parts, 5 of which are defective, then the sample proportion of defective parts would be $P = 5/100 = .05$. This same machine might actually produce defective items at a consistently higher rate than experienced in the sample, so that the population of defective parts could be a different value, such as $\pi = .06$. The proportion may also be expressed as a percentage. Thus, 60% of the first sample are women, and 5% of the items in the second sample are defective.

The proportion is the only measure available for qualitative data. It indicates relatively how many observations fall into a particular category. (When the observations are attributes, such as male or female, central tendency or variability have no meaning.) The type of question answered by P or π is *how many* rather than *how much*.

Sometimes we may wish to express the proportion of observations having a collection of attributes. For example, Table 2-7 (page 22) presents the frequency distribution of the letters in a representative sample of 200 letters of ordinary English text. We may wish to find the proportion of vowels. We find that "a" occurs 16 times and that "e" occurs 26 times; the frequencies for "i," "o," and "u" are 13, 16, and 6, respectively. Thus, in 200 letters, vowels occur $16 + 26 + 13 + 16 + 6 = 77$ times. The sample proportion of vowels is therefore $P = 77/200 = .385$. These five letters, taken together, occur 38.5% of the time in the sample of ordinary English text.

The proportion is not limited to qualitative data. We may also use it to represent the relative frequency of a quantitative category. For example, the median is the .50 fractile; thus, the proportion of observations falling below the sample median is $P = .50$. Likewise, the sample proportion of items falling below the third quartile is $P = .75$.

EXERCISES

3-18 A government inspector is investigating companies to determine whether they exhibit pronounced discrimination against employees based on sex. The data on the number of men and women in management positions in five firms within the same industry are:

			Firm		
	A	B	C	D	E
Men	2,342	532	849	1,137	975
Women	156	115	57	145	139

Calculate the proportion of women managers in each firm. Assuming the inspector will more thoroughly study the firm with the lowest proportion of women managers, which firm would be selected?

3-19 Referring to the personal debt data in Exercise 3-8 (page 52), determine the sample proportion of families whose indebtedness lies below (a) $1,000; (b) $1,500; (c) $3,000; (d) $5,000.

3-20 A quality-control inspector must reject incoming shipments if the proportion of inspected items found to be defective exceeds .05. For each of the following four shipments, (a) determine the sample proportion of defectives found, and (b) state whether the inspector would accept or reject the shipment.

	Shipment			
	(1)	(2)	(3)	(4)
Items inspected	100	500	600	1,000
Items defective	7	25	10	39

3-21 In each of the following situations, discuss whether a mean, a proportion, or both would be an appropriate parameter on which to base decisions.

(a) A garment manufacturer wishes to ship dresses of the highest quality. Many things can cause a dress to be defective, including incorrect sizing, improper seams, creases, and missing stitches.

(b) The Federal Trade Commission requires that the weights of ingredients in packaged goods be indicated on the label. A soap manufacturer wishes to comply. Underweight production batches (populations) are reprocessed rather than shipped. For the sake of efficiency, a large number of packages (elementary units) are weighed simultaneously, and the weight of the packaging material is subtracted to obtain the weight of the ingredients.

(c) A drug maker is testing a new food supplement believed to reduce levels of anemia. The supplement is not expected to work on all patients, but when it does, the extent to which it will reduce anemia by increasing the red corpuscle count in the blood should be measurable.

REVIEW EXERCISES

3-22 A statistics instructor needs to analyze the grade point data of a random sample of ten statistics students. The possible scores range downward from 4 points for "A" to 0 points for "F." The following data have been obtained from departmental records:

$$
\begin{array}{ccccc}
3 & 2 & 4 & 1 & 0 \\
3 & 2 & 2 & 2 & 1
\end{array}
$$

(a) Determine the sample range in grade points.
(b) Calculate the sample mean.

(c) Determine the sample variance for grade points.

(d) Determine the sample standard deviation in grade points.

3-23 The proportions of weed seed found in boxes of lawn seed mixtures have the following cumulative relative frequency distribution:

Proportion of Weed Seed	Cumulative Relative Frequency
.00–under .01	.05
.01–under .02	.12
.02–under .03	.25
.03–under .04	.37
.04–under .05	.50
.05–under .06	.67
.06–under .07	.75
.07–under .08	.84
.08–under .09	.89
.09–under .10	.90
.10–under .11	.95
.11–under .12	.98
.12–under .13	1.00

For the proportion of weed seed, determine:

(a) the .90 fractile   (d) the third quartile

(b) the 37th percentile   (e) the interquartile range

(c) the first quartile   (f) the median

3-24 The mean lifetimes for cartons of 100-watt, long-life lightbulbs have been established to be normally distributed, with mean 1,500 hours and standard deviation 100 hours. Find the upper and lower bounds for the following limits, and indicate the percentage of all cartons for which the mean lifetimes fall within these limits:

(a) $\mu \pm 1\sigma$ (b) $\mu \pm 2\sigma$ (c) $\mu \pm 3\sigma$

3-25 All entering freshmen at a particular college must take a series of aptitude tests. The following percentiles have been determined:

Score	Percentile
450	25th
573	50th
615	75th
729	90th
738	95th
752	99th

(a) Find the first, second, and third quartiles.

(b) Determine the median score.

(c) Compute the interquartile range.

3-26 The following frequency distribution has been obtained for sample data related to the viable shelf life of a certain brand of bread:

Hours to Deterioration	Number of Loaves
90–under 110	23
110–under 130	37
130–under 150	26
150–under 170	14

Calculate the sample mean and the sample variance from these grouped data.

3-27 The IQ scores achieved by students in a particular school have the following cumulative relative frequency distribution:

IQ Score	Cumulative Proportion of Students
70–less than 80	.02
80–less than 90	.12
90–less than 100	.24
100–less than 110	.46
110–less than 120	.65
120–less than 130	.79
130–less than 140	.88
140–less than 150	.97
150–less than 160	1.00

(a) Plot the ogive for these data.
(b) Read the following values from your graph. (The figures obtained will be only estimates of the actual values.)
   (1) the 50th percentile
   (2) the .75 fractile
   (3) the first quartile
   (4) the median
   (5) the 90th percentile
(c) Use your answers from (b) to compute the interquartile range.

3-28 (a) Calculate the sample mean and the sample variance for the following ungrouped sample data regarding examination scores:

84	77	67	94	90
81	56	89	77	88
74	76	28	80	58
66	77	89	81	78
77	72	94	93	79
93				

(b) The sample data in (a) have been grouped into the following frequency distribution:

Class Interval	f
25–under 35	1
35–under 45	0
45–under 55	0
55–under 65	2
65–under 75	4
75–under 85	11
85–under 95	8
Total	26

Calculate the sample mean and the sample variance using these grouped data.

(c) Calculating population parameters using grouped data provides only approximations of the population parameters that would be obtained directly from the ungrouped raw data. Find the amount of error in this approximation procedure for the sample mean by subtracting your answer in (b) from the value you obtained in (a).

The Statistical Sampling Study

4

Now that we have covered the basic elements of descriptive statistics, we are ready to consider *statistical sampling*. To draw conclusions about populations from samples, we must use inferential statistics, which enables us to determine a population's characteristics by directly observing only a portion, or sample, of the population.

We obtain a sample rather than a complete enumeration (a census) of the population for many reasons. Obviously, it is cheaper to observe a part rather than the whole, but we should prepare ourselves to cope with the dangers of using samples. In this chapter, we will investigate various kinds of sampling procedures. Some are better than others, but all may yield samples that are inaccurate and unreliable. We will learn how to minimize these dangers, but some potential error is the price we must pay for the convenience and savings that samples provide.

THE NEED FOR SAMPLES 4-1

There would be no need for statistical theory if a census, rather than a sample, was always used to obtain information about populations. But a census may not be practical and is almost never economical.

71

Sample information must be relied on for most applications. There are six main reasons for sampling in lieu of the census: economy, timeliness, the large size of many populations, inaccessibility of some of the population, destructiveness of the observation, and accuracy.

The Economic Advantages of Using a Sample

Obviously, taking a sample—directly observing only a portion of the population—requires fewer resources than a census. For example, consider a consumer survey to determine the reactions of all the owners of a popular automobile to several proposed colors for next year's model. A questionnaire is to be printed and mailed to all owners. Imagine the clerical chore of simply addressing the envelopes, and the cost of the postage alone could easily run into five figures. The bookkeeping problems in tabulating the replies would be overwhelming.

Perhaps the greatest difficulty would be to ensure a 100% response. A high proportion of the questionnaires will simply be ignored along with the plethora of other "junk" mail. Most persons can be reached by telephone, telegram, or personal visit. Invariably, however, an irascible few will slam their doors even in the face of the company president, and some will have to be bribed with a new car to respond. A great many car owners will have moved and will be hard to locate, necessitating the services of a detective agency. Some owners will die without responding before the survey is completed. The company could have more difficulty taking a census of its owner population than the FBI has arresting its ten most wanted men.

As unrealistic as this example is, it does illustrate the very high cost of a census. For the type of information desired, the car manufacturer would be wise to interview a small portion of the car owners. Rarely does a circumstance require a census of the population, and even more rarely does one justify the expense.

The Time Factor

A sample may provide an investigator with needed information quickly. Often speed is of paramount importance, as in a political polling, where the goal is to determine from sample evidence voter preferences toward candidates for public office. The voting public as a whole is extremely fickle; polls have indicated that voters fluctuate in preference right up to the time of the election. If a poll is to be used to gauge public opinion, it must be very current; the opinions must be obtained, tabulated, and published within a very short period of time. Even if it were physically possible to use a census in such an opinion survey, its results would not be valid due to significant shifts in public opinion over the time required to conduct it.

Similarly, managerial decisions based on a population's characteristics must frequently be made very quickly. In a highly competitive environment there may not be enough time to wait for a 100% response before introducing a new product or changing service patterns. Since a census would not be feasible, sample evidence must be relied on.

The Very Large Population

Many populations about which inferences must be made are quite large. For example, consider the population of high-school seniors in the United States—a group numbering about 3,000,000. The plans of these soon-to-be graduates will directly affect the status of universities and colleges, the military, and prospective employers—all of whom must have specific knowledge about the students' plans in order to make compatible decisions to absorb them during the coming year. But the sheer size of the population of postgraduate plans will probably make it physically impossible to conduct a census. Sample evidence may be the only available way to obtain information from high-school seniors.

Partly Inaccessible Populations

Some populations contain elementary units so difficult to observe that they are in a sense inaccessible. One example is the population of crashed aircraft: Planes that crash and sink in the deep ocean are inaccessible and cannot be directly studied to determine the physical causes of their failure. Similar conditions hold in determining consumer attitudes. Not all of the users of a product can be queried. Some may be in prisons, nursing homes, or hospitals and have limited contact with the outside world. Others may be insulated from harassment; consider the limited accessibility to the President of the United States or the Pope.

Whenever some elementary units are inaccessible, sampling must be used to provide the desired information about the population. These illustrations demonstrate physical inaccessibility, but an elementary unit's accessibility may be limited for economic reasons alone: Those observations deemed too costly to be made are, in a real sense, also inaccessible.

The Destructive Nature of the Observation

Sometimes the very act of observing the desired characteristic of the elementary unit destroys it for the use intended. Classical examples of this occur in quality control: To test a fuse to determine whether it it defective, it must be destroyed; to obtain a census of the quality of a shipment of fuses, all of them must be destroyed. This negates any purpose served by quality-control testing. Clearly, a *sample* of fuses must be tested to assess the quality of the shipment.

Accuracy and Sampling

A sample may be more accurate than a census. A sloppily conducted census can provide less reliable information than a carefully obtained sample. Indeed, the 1950 U.S. Census provides us with an excellent example.

EXAMPLE: INDIANS AND TEEN-AGE WIDOWS

Two statisticians noticed a very unusual anomaly in the change of two categories between 1940 and 1950.* An abnormally large number of 14-year-old widows were reported in the 1950 figures—20 times as many as in 1940. Indian divorcees, who had also been extremely rare, had increased by a like magnitude. Unable to justify the changes by any sociological trends, the statisticians analyzed the data-gathering and data-processing procedures of the census. They concluded that the figures could have resulted from the erroneous reversal of the entries on a few of the punched cards used to process the data. In this instance, a census seemed to provide far from reliable or credible results.

A large utility company had problems of a somewhat different nature with punched cards.

EXAMPLE: WORK SAMPLING

The procedure for allocating a data center's keypunching costs was based on a census compiled from detailed logs maintained by keypunch operators. These records were not very accurate, and record keeping reduced operator productivity. A company statistician found that operators spent 10% of their time maintaining the logs and proposed that management use *work sampling*. At random time intervals, a bell would ring and all operators would stop work and fill out a simplified form. Keypunching costs were allocated to the various departments according to the percentage of forms returned, which was accurate since heavier users would accumulate more forms. Operator productivity increased, and everyone was pleased with the greater accuracy.

---

*Ansley J. Coale and Frederick F. Stephan, "The Case of the Indians and Teen-age Widows," *Journal of the American Statistical Association*, **LVII** (June 1962), 388–437.

EXERCISES

4-1 In each of the following situations, indicate whether a sample or a census would be more appropriate. Explain your choice.

(a) A manufacturer of automobile spark plugs knows that due to variations in processing, a certain percentage of the plugs will be defective. A thorough test of a spark plug destroys it, but the production output must be tested to determine whether remedial action is necessary.

(b) NASA must determine the quality of the components of a manned space vehicle.

(c) A hospital administrator attempting to improve patient services wishes to determine the attitudes of persons treated.

(d) A medical researcher wants to determine the possible harmful side effects from a chemical that eases the pain of arthritis.

4-2 For each of the following reasons, give an example of a situation for which a census would be less desirable than a sample. In each case, explain why.

(a) economy   (c) size of population   (e) accuracy

(b) timeliness   (d) inaccessibility   (f) destructive observations

---

DESIGNING AND CONDUCTING   4-2
A SAMPLING STUDY

Much of this book is concerned with how to collect samples so that meaningful conclusions can be drawn about an entire population. Figure 4-1 shows the major stages of a sampling study. The first stage—*planning*—requires careful attention so that the sampling results will achieve the best impact. In the second stage—*data collection*, where the plans are followed—a major goal is to ensure that observations are free from bias. The last phase—*data analysis and conclusions*—is the area to which a major portion of this book will be devoted.

The Importance of Planning

Planning is the most important step in a statistical study. Inadequate planning can lead to needless expense when actual data collection begins. Poor planning may result in the eventual invalidation of the entire study. The outcome of the study may even prove strongly counterproductive. Several extreme cases of poor planning will be given in this chapter.

Planning begins with the identification of a population that will achieve the study's goals. This is followed by the selection of an observation procedure that might require a questionnaire, which would then have to be designed, or an observation-measuring instrument. The latter is a broad category that includes mechanical devices (to record physical data, such as dimensions and weight),

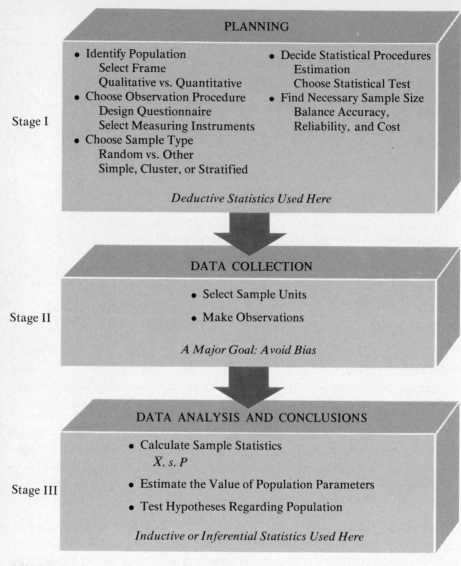

FIGURE 4-1
Major stages in a statistical sampling study.

various tests (for intelligence, aptitude, personality, or knowledge), sources of data (for example, family income could be obtained from tax records, from employers, or by directly questioning a family member), and survey techniques (written reply, telephone response, or personal interview).

Of equal importance in planning is the choice of the sample type itself. Alternative kinds of samples range from the most "scientific" random sample, like the one used in the major public opinion polls, to the convenience sample taken in campus straw votes, where practically no attention is paid to how representative the sample might be. The random sample is the most important, because statistical theory applies to it alone. As we will see, there are a variety of ways to select a sample.

If proper and valid conclusions are to be drawn from the sample data, planning must include a presampling choice of the statistical procedure to be used later in analyzing the results. Choosing an analytical technique *after* sifting through the results may lead to the most insidious kind of bias—the innocent or intentional selection of a tool that strongly supports the desired conclusion.

An important planning consideration in any sampling study is determining how many observations to make. In general, the larger its size, the more closely a random sample is expected to represent its population. But where do we draw the line? Isn't a sample ten times as large ten times as good? The first question will be answered in Chapter 8. To the second, the answer is *no*. As we will see, selection of sample size is largely an economic matter; scarce resources (funds available for gathering data) must be traded off against the accuracy and reliability of the result. A point of diminishing returns is soon reached where an extra dollar buys very little in increased sample quality.

Both choosing a statistical procedure and selecting an appropriate sample size involve a great amount of deductive statistics, since both of these planning aspects require us to look at how samples are generated from "textbook" populations whose characteristics are fully known.

Data Collection

The second phase of the statistical sampling study—obtaining the raw data—usually requires the greatest amount of time and effort. Adequate controls must be provided to prevent observations from becoming biased.

Potential problem areas associated with collecting and managing the data are described in detail in subsequent sections of this chapter. Such problems include observational biases that result when the sample units are measured, queried, or investigated. A poorly asked question and an improperly operated device both lead to incorrect responses. Another source of bias arises because some parts of the population are inaccessible or difficult to observe. Additional problems may appear after the data are collected: The inevitable ambiguities and contradictions must be dealt with and the results processed so that statistical analyses can be made.

Data Analysis and Conclusions

This last stage of a statistical sampling study is extremely important. Here, many of the descriptive tools we discussed earlier are used to communicate the results. Our emphasis, however, is placed on drawing conclusions from the necessarily incomplete information available from any sample. This is the point at which statistical inferences are made. The lion's share of statistical theory is directly related to or in support of making generalizations or *inferences* about populations from samples. Inferences generally fall into the three broad categories of estimation, hypothesis testing, and association. The goals of the study and the resouces available dictate the statistical inference and technique that will be used.

4-3   BIAS AND ERROR IN SAMPLING

The manner in which the observed units are chosen significantly affects the adequacy of the sample. Here we will examine some of the pitfalls commonly encountered in sampling and learn how to avoid many of them.

A sample is expected to mirror the population from which it comes. However, there is no guarantee that any sample will be precisely representative of the population; chance may dictate that a disproportionate number of untypical observations will be made. In practice, it is rarely known when a sample is unrepresentative and should be discarded, because some population characteristics are unknown (or else no sample would be needed in the first place). Steps must be taken during the planning of the study to minimize the chance that the sample will be untypical.

Sampling Error

What can make a sample unrepresentative of its population? One of the most frequent causes is *sampling error*, for which we provide the following

DEFINITION
Sampling error comprises the differences between the sample and the population that are due solely to the particular elementary units that happen to have been selected.

To illustrate sampling error, suppose that a sample of 100 American men are measured and are all found to be taller than seven feet. An obvious conclusion would be that most American males are taller than seven feet. Of course, this is absurd; most people do not know a man who is seven feet tall and might not even be sure one existed if a few basketball players were not that tall. Yet it is possible for a statistician to be unfortunate enough to obtain such a highly unrepresentative sample. However, this is highly unlikely; nature has distributed the rare seven-footers widely among the population.

More dangerous is the less obvious sampling error against which nature offers little protection. It is not difficult to envision a sample in which the average height is overstated by only an inch or two rather than a foot. It is the insidious, unobvious error that is our major concern.

There are two basic causes for sampling error. One is chance: Bad luck may result in untypical choices. Unusual elementary units do exist, and there is always a possibility that an abnormally large number of them will be chosen. The main protection against this type of error is to use a large enough sample. Another cause of sampling error—sampling bias—is not as easy to remedy.

Sampling Bias

The size of the sample has little to do with the effects of *sampling bias*. In the present context, we provide the

DEFINITION
Sampling bias is a tendency to favor the selection of elementary units that have particular characteristics.

Sampling bias is usually the result of a poor sampling plan. Consider the following classical example from political polling.

EXAMPLE: THE *LITERARY DIGEST* SAID ROOSEVELT WOULD LOSE

A poll was conducted by the now defunct *Literary Digest* in 1936, when Franklin D. Roosevelt was running against Republican Alfred M. Landon for the Presidency. Based on several million responses obtained from a sample of the voting public, the *Digest* conclude that Landon would win by a record margin. Exactly the opposite happened: Roosevelt won one of the most one-sided victories in American history. The poll's erroneous results were later attributed to sampling bias. The *Digest* had selected its sample from magazine subscription listings and telephone directories. During the Great Depression, both sources contained a disproportionate number of prosperous persons who favored the laissez-faire Republican platform. Because a large proportion of the disgruntled majority, who strongly favored Roosevelt, had no telephones and did not subscribe to magazines, these voters were not adequately represented in the sample.

The avoidance of sampling bias is a major concern of statisticians. A means of selecting the elementary units must be designed to avoid the more obvious forms of bias. In practice, it is difficult (if not impossible) to eliminate all forms of bias. Most notable is the *bias of nonresponse*, when—for whatever reason—some elementary units have no chance of appearing in the sample. The cost considerations involved in making sample observations almost guarantee that this particular form of bias will be present to some degree in any sample.

An excellent example is provided by consumer surveys involving food products. It is relatively cheap to obtain opinions about a company's product from persons who remain at home during the day. But a large proportion of adult residents in many areas are employed and are at home only in the evenings and on weekends. If the sampling survey is conducted during normal working hours, these consumers who make food-buying decisions will not be represented. A company wishing to determine whether to expand its line of prepared frozen foods may obtain survey results indicating that there is not a great demand for such products. But working people have less leisure time and are often enthusiastic users of such items. A sample that neglected their opinions could lead to an erroneous marketing decision.

Nonsampling Error

The other main cause of unrepresentative samples is *nonsampling error*. This type of error can occur whether a census or a sample is being used. Like sampling error, nonsampling error may either be produced by participants in the statistical study or be an innocent byproduct of the sampling plans and procedures.

DEFINITION

A nonsampling error is an error that results solely from the manner in which the observations are made.

The simplest example of nonsampling error is inaccurate physical measurement due to malfunctioning instruments or poor procedures. Consider the observation of human weights. If persons are asked to state their own weights, no two answers will be of equal reliability. The people will have weighed themselves on different scales in various states of poor calibration. An individual's weight fluctuates diurnally by several pounds, so that the time of weighing will affect the answer. The scale reading will also vary with the person's state of undress. Responses will not be of comparable validity unless all persons are weighed under the same circumstances.

Biased observations due to inaccurate measurement can be innocent but devastating. A French astronomer once proposed a new theory based on spectroscopic measurements of light emitted by a particular star. When his colleagues discovered that the measuring instrument had been contaminated by cigarette smoke, they rejected his findings.

In surveys of personal characteristics, unintended errors may result from (1) the manner in which the response is elicited, (2) the idiosyncrasies of the persons surveyed, (3) the purpose of the study, or (4) the personal biases of the interviewer or survey writer.

No two interviewers are alike, and the same person may provide different answers to different interviewers. The manner in which a question is formulated can also result in inaccurate responses. Individuals tend to provide false answers to particular questions. A good example of this is a person's age. Some people will lie about their age. If asked their age in years, most people will give their chronological age at their previous birthday, even if it is more than 11 months past. Both problems can be alleviated by asking the date of birth. It would require quick arithmetic to give a false date, and a date of birth is much more accurate than a person's age on his or her last birthday.

Respondents might also give incorrect answers to impress the interviewer. This is especially prevalent when survey questions are based on a knowledge of current events or on intellectual accomplishments. Many people have said that they "know" about famous people who are fictitious or that they have read books that were never written. This type of error is the most difficult to prevent, because it results from outright deceit on the part of the respondee. It is important to acknowledge that certain psychological factors induce incorrect responses, and great care must be taken to design a study that minimizes their effect.

Knowing why a study is being conducted may create incorrect responses. A classic example is the answer to the question: What is your income? If a government agency is asking, a different figure may be provided than the respondent would give on an application for a home mortgage. A teacher's union seeking to justify a wage increase may receive a different response from a member than would the brother-in-law of the member. The union would probably not be given (nor would it want to know) the member's full income, incorporating earnings from moonlighting and summer employment, whereas the brother-in-law would probably be informed about the income derived from these sources as well as any successful investments. One way to guard against such bias is to camouflage the study's goals; an independent opinion survey firm might even be employed to keep secret the objective of the investigation. Another remedy is to make the questions very specific, allowing little room for personal interpretation. For example, "Where are you employed?" could be followed by "What is your salary? and "Do you have any extra jobs?" A sequence of such questions may produce more accurate information.

Finally, it should be noted that the personal prejudices of either the designer of the study or the data collector may tend to *induce bias*. A person should not design a questionnaire related to a subject he or she feels strongly about. Questions may be slanted in such a way that a particular response will be obtained even though it is inaccurate. An example of this is a preference survey sponsored by the makers of brand A, prefaced by the question "In your opinion, which of the following factors that make brand A a superior-quality product is the most important?" This could be followed by a question asking for a preference ranking of a list of products, including brand A. (Most blatantly, brand A may be the first name on the list.) The preliminary question encourages the respondent to think about the good features of brand A, creating an atmosphere that will make the later comparison unfair. Another instance of induced error may occur in an experiment in which a technical specialist selects the elementary units to be treated with a new product. For example, when doctors favoring a new drug decide which patients are going to receive it, they may select (consciously or not) those they feel have the best chance of benefiting from its administration. An agronomist may apply a new fertilizer to certain key plots, knowing that they will provide more favorable yields than others. To protect against induced bias, an individual trained in statistics should have a measure of control over the design and implementation of the entire statistical study, and someone who is aware of these pitfalls should serve in an auditing capacity.

EXERCISES

4-3 A government official must delegate the responsibility for determining employee attitudes toward a training program conducted by the personnel department. What are the dangers in asking the personnel manager to handle this matter? Suggest a procedure that would overcome these obstacles.

4-4 People who filled out the "long" form for the 1970 U.S. Census were asked to answer questions about their employment. One question was "How many hours did you work last week?" Successive questions relating to the same work week asked about type of employer, duties, and so on. The public was instructed to provide the data by April 1, 1970. This was the Wednesday right after Easter, and most of the nation's educators had been on a holiday for the entire week in question. Comment on the possible errors that may have resulted from this set of questions. How could the questionnaire have been improved to avoid these errors?

4-5 A drugstore has recently experienced a significant decline in sales due to the presence of some new discount stores in the city. A questionnaire asked customers who visited the store during one week to compare its prices, convenience, and level of service with the discount stores. What do you think about the manner in which the information was obtained? What sampling procedure could you recommend?

4-6 A questionnaire prepared for a study on consumer buying habits contains the question "What is the value of your present automobile(s)?" What errors may result from this question? Suggest a better way to obtain the desired response.

4-7 Many personality profiles require that an individual provide answers to multiple-choice questions. The answers are then compiled, and various trait ratings are found. Comment on the pitfalls of this procedure. Suggest how some of them may be alleviated.

4-8 Testing a new drug therapy sometimes involves two patient groups. The patients in one group receive the medication under investigation. The others may receive no medication at all or may be given an innocuous substance called a *placebo* (often just a sugar pill) that looks like the drug being tested. Why is the placebo used?

4-9 A well-known television rating agency provides networks and advertisers with estimates of the audience size for all nationally televised programs. These estimates are based on detailed logs, maintained by persons in the households surveyed, that indicate which stations were watched during each period of the day. The agency has a difficult time recruiting households for its sample, because it is bothersome and inconvenient to keep the logs. Independent observers have estimated that an average of 50 families must be approached before one agrees to join the sample. Comment on the pitfalls that may be encountered when the sample results are taken as representative of the nation's television viewing habits. Can you suggest any ways to minimize these pitfalls? Do you think that the sample observations obtained are in danger of being untypical? Explain.

SELECTING THE SAMPLE 4-4

In Section 4-3, we examined the most common problems associated with statistical studies. The desirability of a sampling procedure depends on both its vulnerability to error and its cost. However, economy and reliability are competing ends, because to reduce error often requires an increased expenditure of resources.

Of the two types of statistical error, only sampling error can be controlled by exercising care in determining the method for choosing the sample. We have established that sampling error may be due to either bias or chance. The chance component exists no matter how carefully the selection procedures are implemented, and the only way to minimize chance sampling errors is to select a sufficiently large sample. Sampling bias, on the other hand, may be minimized by the judicious choice of a sampling procedure.

There are three primary kinds of samples: the convenience sample, the judgment sample, and the random sample. They differ in the manner in which their elementary units are chosen. Each sample will be described in turn, and its desirable and undesirable features will be discussed.

The Convenience Sample

A convenience sample results when the more convenient elementary units are chosen from a population for observation. Convenience samples are the cheapest to obtain. Their major drawback is the extent to which they may be permeated with sampling bias, which tends to make them highly unreliable. An important decision made on the basis of a convenience sample is in great danger of being wrong. The letters received by a member of Congress constitute a convenience sample of constituent attitudes. The legislator does not have to expend any time, money, or effort to receive mail, but the opinions voiced there will rarely be indicative of the attitudes of the entire constituency. Persons who have special interests to protect are far more likely to write than the typical voter.

The bias of nonresponse, when present, usually results in a convenience sample. This type of bias can be expected when the telephone—a very convenient device—is used to obtain responses, because the incidence of nonresponse may be relatively high. Many calls will be unanswered, and some people will have unlisted numbers or no telephones at all.

Despite the obvious disadvantages of convenience samples, they are sometimes suitable, depending on the purpose of the study. If only approximate information about a population is needed, then a convenience sample may be adequate. Or funds for a study may be so limited that only a convenience sample is feasible. But in such cases, the obvious drawbacks must be acknowledged. On the other hand, a convenience sample may result in an insignificant source of bias. For example, when there is no reason to believe that nonrespondents would provide information that would differ on the whole from the data already available, there may be little danger in using a telephone survey; but we can never be sure.

The Judgment Sample

A judgment sample is obtained according to the discretion of someone who is familiar with the relevant characteristics of the population. The elementary units are selected judgmentally when the population is highly heterogeneous, when the sample is to be quite small, or when special skill is required to ensure a representative collection of observations.

EXAMPLE: CONSUMER PRICE INDEX

The Consumer Price Index of the U.S. Department of Labor is partly a judgment sample. The items to be included in establishing the index are chosen on the basis of judgment in order to measure a dollar's purchasing power. Thus, only items or commodities that are used by most people are included; for example, television sets are included, but tape recorders are not. Over time, certain items become obsolete and must be replaced by others; for instance, refrigerators have supplanted iceboxes. Judgment must be employed in deciding what will be replaced and what will be substituted. Judgment is also used in establishing weights for each item; toothpaste should only be weighted in proportion to its position in a typical family's total expenditure—and the typical expenditure is itself a matter of judgment. Finally, judgment is employed in selecting the cities to be represented in the surveys; these, too, must be assigned weights.

The judgment sample is obviously prone to bias. Its adequacy is limited by the ability of the individual who selects the sample to perceive differences. All of the dangers mentioned in Section 4-3 in connection with induced bias apply to the judgment sample. Clearly, there are instances in which this type of sample is preferred. Indeed, the Consumer Price Index would be unworkable if it were not partly based on judgment samples.

The Random Sample

Perhaps the most important type of sample is the random sample. A random sample allows a known probability that each elementary unit will be chosen. For this reason, it is sometimes referred to as a *probability sample*.

In its simplest form, a random sample is selected in the manner of a raffle. For example, a random sample of ten symphony conductors from a population of 100 could be obtained by lottery. The name of each symphony conductor would be written on a slip of paper and placed into a capsule. All 100 capsules would then be placed in a box and thoroughly mixed. An impartial party would select the sample by drawing ten capsules from the box.

In actual practice, a lottery can be physically cumbersome and even its randomness can be questioned.

EXAMPLE: THE 1970 DRAFT LOTTERY FIASCO

The 1970 U.S. draft lottery determined priorities of selection by date of birth. A physical method similar to a raffle was used, but the results appeared to have a pattern. December birthdates were assigned a disproportionate amount of early numbers, and January birthdates were assigned predominantly later numbers. Investigation showed that the undesirable outcome resulted from the fact that the capsules had not been mixed at all. Since the December capsules were the last to be placed in the hopper, many December dates were drawn first. Because persons with these birth dates were almost certain to be drafted, considerable controversy arose and lawsuits were filed to invalidate the entire lottery.

Sample Selection Using Random Numbers

A more acceptable way to obtain a random sample is to use *random numbers*—digits generated by a lottery process that allows for the equal probability that each possible number will appear next. For example, in a list of five-digit random numbers, each value between 00000 and 99999 has the same chance of appearing at each location. Appendix Table B is a list of five-digit random numbers that was generated by an electromechanical process at the

Rand Corporation. These particular numbers have passed many tests and have been certified by the scientific community. If the Selective Service had used such a table of random numbers to select its birthdates, it would have been guilt-free in the eyes of everyone except those who were chosen first.

There is no guarantee that a list of random numbers will not exhibit some sort of pattern. For example, 8 may follow 7 more frequently than any other value. But in a long list, any kind of pattern is highly unlikely (8 should follow 7 about 10% of the time).

To use random numbers to select our sample of ten conductors, first we must assign a number to each member of the population. In this case, we can use two digits; the assigned numbers will therefore range from 00 through 99. It makes no difference how these numbers are assigned, but the person making the assignments should not know which location of the random number table is being used. In Table 4-1, the 100 conductors have been arranged in alphabetical order and assigned the numbers 00 through 99 in sequence.

We will choose our sample of ten conductors by reading down the first column of the random number table until we obtain ten different values. Note

TABLE 4-1
Alphabetical Listing of a Population of 100 Symphony Conductors

00. Abbado	25. Golschmann	50. Mehta	75. Santini
01. André	26. Hannikainen	51. Mitropoulos	76. Sargent
02. Anosov	27. Hollingsworth	52. Monteux	77. Scherchen
03. Ansermet	28. Horenstein	53. Morel	78. Schippers
04. Argenta	29. Horvat	54. Mravinsky	79. Schmidt-Isserstedt
05. Barbirolli	30. Jacquillat	55. Newman	80. Sejna
06. Beecham	31. Jorda	56. Ormandy	81. Serafin
07. Bernstein	32. Karajan	57. Ozawa	82. Silvestri
08. Black	33. Kempe	58. Patanè	83. Skrowaczewski
09. Bloomfield	34. Kertesz	59. Pedrotti	84. Slatkin
10. Bonynge	35. Klemperer	60. Perlea	85. Smetáček
11. Boult	36. Kletzki	61. Prêtre	86. Solti
12. Cantelli	37. Klima	62. Previn	87. Stein
13. Cluytens	38. Kondrashin	63. Previtali	88. Steinberg
14. Dorati	39. Kostelanetz	64. Prohaska	89. Stokowski
15. Dragon	40. Koussevitzky	65. Reiner	90. Svetlanov
16. Erede	41. Krips	66. Reinhardt	91. Swarowsky
17. Ferencsik	42. Kubelik	67. Rekai	92. Szell
18. Fiedler	43. Lane	68. Rignold	93. Toscanini
19. Fistoulari	44. Leinsdorf	69. Ristenpart	94. Van Otterloo
20. Fricsay	45. Maag	70. Rodzinski	95. Van Remoortel
21. Frühbeck de Burgos	46. Maazel	71. Rosenthal	96. Vogel
22. Furtwängler	47. Mackerras	72. Rowicki	97. Von Matacic
23. Gamba	48. Markevitch	73. Rozhdestvensky	98. Walter
24. Giulini	49. Martin	74. Sanderling	99. Watanabee

that we only require two-digit numbers but that the table entries contain five digits, so we will use only the first two digits of each entry and ignore the extra digits. If a value is obtained more than once before all ten conductors have been chosen, then that number will be skipped and the next number on the list will be used instead. The values obtained by starting at the top of the first column of Appendix Table B (page 821) and reading down are listed here, with the corresponding conductors who now comprise our sample:

12651	Cantelli	74146	Sanderling
81769	Serafin	90759	Svetlanov
36737	Kletzki	55683	Newman
82861	Silvestri	79686	Schmidt-Isserstedt
21325	Frühbeck de Burgos	70333	Rodzinski

It does not matter how the random number table is read (from left to right, right to left, top to bottom, bottom to top, or diagonally), but it is important not to read the same number location more than once and not to look ahead before deciding in which direction to proceed. If the numbers are larger than required, the beginning or ending digits can be ignored. If a number on the list has no counterpart in the population (which may happen when the population is smaller than the possible number of random numbers), it should be skipped.

Unlike judgment and convenience samples, random samples are free from sampling bias. No particular elementary units are favored. This still does not guarantee that a random sample will be extremely representative of the population from which it came, because the chance effect may cause it not to be. But it will be possible to assess the reliability of a sample in terms of probability, which can be done only with random samples. When sampling bias is present, an objective basis for measuring its effect does not exist. Therefore, *only the random sample has a theoretical basis for the quantitative evaluation of its quality.* In later chapters, we will see how probability is used for such evaluations.

Types of Random Samples

The way we chose our sample of ten conductors is only one type of random sampling scheme. At times, it may be desirable to modify the selection procedure without altering the essential features of the random sample. The type of sampling we just discussed is a *simple random sample*, for which we provide the following

DEFINITION

A simple random sample is obtained by choosing elementary units in such a way that each unit in the population has an equal chance of being selected.

A simple random sample is free from sampling bias. However, using a random number table to choose the elementary units can be cumbersome. If the sample is to be collected by a person untrained in statistics, then instructions may be misinterpreted and selections may be made improperly. Instead of using a list of random numbers, data collection can be simplified by selecting, say, every 10th or 100th unit after the first unit has been chosen randomly. Such a procedure is called a *systematic random sample*, for which we have the following

DEFINITION

A systematic random sample is obtained by selecting one unit on a random basis and choosing additional elementary units at evenly spaced intervals until the desired number of units is obtained.

EXAMPLE: SAMPLING WITH TELEPHONE NUMBERS

A telephone company is conducting a billing study to justify a new rate structure. Data are to be collected by taking a random sample of individual telephone bills from cities and metropolitan areas classified according to population. Billing data are stored by telephone number. A company statistical analyst has designed a procedure for each regional office to employ in collecting the samples.

In one city, 10,000 telephones share the 825 prefix. The telephone numbers range from 825-0000 through 825-9999. A sample size of 200 is required. Dividing the number of telephones by 200, the interval width of 50 is obtained. Thus, the bill for every 50th telephone number will be incorporated into the sample. The analyst chooses a number between 0 and 49 at random. Suppose she chooses 37. Her instructions to the regional accounting offices specify that the billing figures for every 50th telephone number, starting with 825-0037, be forwarded to her office. It then becomes a simple manual task for a clerk to pull every 50th telephone bill from the files, and the analyst will receive the figures for 825-0037, 825-0087, 825-0137, 825-0187, and so on.

Sometimes we must ensure that various subgroups of a population are represented in the sample. (In statistics, these subgroups are referred to as *strata*.) To do this, we must collect a sample that contains the same proportions of each subgroup as the population itself does. Thus, we obtain a stratified sample, for which we provide the

DEFINITION

A stratified sample is obtained by independently selecting a separate simple random sample from each population stratum.

Although it depends on the purpose of the study, stratification is usually required if certain nonhomogeneities are present in the population. There are other reasons why a stratified sample may be desirable. It may be more convenient to collect samples from separate regions. Or there may be some prevailing reason to divide the population, such as to compare strata.

EXAMPLE: PROBLEM IN PREDICTING ELECTIONS

Voter preference polls have been taken prior to each presidential election. Due to the expense involved, the pollsters draw a random sample of voters from the entire United States. Sampling is not conducted independently within each state.

The electoral college determines who will be President. The membership of the college has historically voted as a block by state, and all of a state's votes have gone to the candidate winning in that state. American history shows that the popular-vote leader can still lose the election in the electoral college. The winner must only achieve small margins in several large states. Thus, a candidate might win a majority of electoral votes while losing the national plurality. This has happened twice since the Civil War: In 1876, Rutherford B. Hayes beat Samuel J. Tilden, who received a greater number of popular votes; in 1888, Benjamin Harrison beat the larger popular-vote getter, Grover Cleveland. The situation is further distorted by the fact that electoral votes are rationed to each state in accordance with its representation in *both* houses of Congress. Thus, Alaska, with about one hundredth the voting population of California, has about one-fourteenth as many electoral votes, making each Alaskan voter about seven times as powerful as a Californian.

A sampling scheme designed to predict the election outcome must treat each state as a separate stratum to adequately cope with the distortion of the electoral college. The sample results must be separately compiled for each state, and the winner of each state must be determined. The results must then be combined in the same proportion as the electoral votes to present a distortion-free prediction.

Another form of random sampling is used prevalently in consumer surveying, where it is more economical to interview several persons in the same neighborhood. For example, contrast the time, effort, and expense of collecting 1,000 responses in a city of 100,000 by simple random sampling with the relative ease of completely canvassing a few selected neighborhoods. In the former case, the elementary units are scattered throughout the city; in the latter, the elementary units live next door to each other. Random sampling would therefore require considerable travel between interviews, whereas the door-to-door neighborhood canvass would minimize travel time. Because the interviewer is normally paid by the hour, the simple random sample would be much more expensive.

Neighborhood groupings of people are referred to as *clusters*. Populations may be divided into clusters according to a criterion such as geographical proximity, creating groups that are easy to observe in their entirety. Most often, the clusters that are to be observed are chosen randomly, and so we can state the

DEFINITION
A cluster sample is obtained by selecting clusters from the population on the basis of simple random sampling. The sample comprises a census of each random cluster selected.

The only justification for using cluster sampling is that it is economical. Cluster sampling is highly susceptible to sampling bias. In neighborhood surveys, similar responses may be obtained from the entire cluster, as people of similar age, family size, income, and ethnic and educational backgrounds tend to live in the same vicinity. Thus, there is a high risk that persons with certain backgrounds or preferences will not be represented in the sample because the clusters in which they are predominant are not chosen; conversely, other groups may be unduly represented.

At this point, it might be helpful to emphasize the similarities and differences between cluster and stratified samples. Both separate the population into groups, but the basis for distinction is usually the homogeneity within strata versus the accessibility of clusters. In stratified samples, all groups are represented, whereas the cluster sample comprises only a fraction of the groups in a population. In stratified sampling, a sample is taken from within each stratum; in cluster sampling, a census is conducted within each group. The goal of stratified sampling is to eliminate certain forms of bias; the goal of cluster sampling it to do the job cheaply, which enhances the chance of bias.

Systematic, stratified, and cluster sampling are approximations to simple random sampling. To the extent that the statistical theory applicable to simple random sampling is applied to samples obtained via these other procedures, there may be erroneous conclusions. Throughout much of the remainder of this book, techniques will be developed primarily for analyzing simple random samples. The more complex sampling schemes require more complicated methodology.

EXERCISES

4-10 What type of sample does each of the following represent?
  (a) Telephone callers on a radio talk show.
  (b) The 30 stocks constituting the Dow-Jones Industrial Average.
  (c) The record of a gambler's individual wins and losses from a day of placing bets of $1 on black in roulette.
  (d) The oranges purchased by a homemaker from the produce section of a market.
4-11 In the following situations, would you recommend the use of judgment, convenience, or random sampling, or some combination of these? Explain the reasons for your choice in each case.

(a) A salewoman for a large paper manufacturer is preparing a kit of product samples to carry on her next road trip to printing concerns.
(b) A teacher asks his students for suggestions as to how to improve curriculum. He plans to use their suggestions as a basis for a questionnaire concerning curriculum preferences.
(c) A city government wants to compare the wages of its clerical workers with the wages of persons holding comparable jobs in private industry.
(d) A utility company plans to purchase a fleet of standard-sized cars from one of four different manufacturers. The criterion for selection will be economy of operation.
(e) A newspaper editor selects letters from the day's mail to print on the editorial page. (Answer from the editor's point of view.)

4-12 An oil company wishes to obtain a random sample of its customers to estimate the relative proportion of purchases made from other oil companies. An accountant on the operations staff has recommended that a random sample of credit-card customers be selected. An independent survey firm would be retained to obtain detailed records of the following month's purchases by brand and quantity for each sample unit. Evaluate this procedure.

4-13 A medical research foundation wishes to obtain a random sample of expectant mothers to investigate postnatal developments in their newborn babies. Various metropolitan areas are to be represented. Suggest a procedure that might be used to select the sample in a particular city.

4-14 A certain U.S. Senator is reputed to follow the majority opinion of his constituents on important pending legislation. Over the years, he has accumulated a panel of 20 persons whom his secretary contacts before a major Senate vote. Panel members are replaced only when they die or change their state of residence. The Senator's votes usually coincide with the majority of this panel. Comment on this policy.

4-15 A random sample must be obtained in each of the following situations. Indicate whether a cluster, stratified, or combined sampling procedure should be used. Explain the reasons for your choice in each case.
(a) A large retail food chain wishes to determine its customers' attitudes toward trading stamps. Because the funds for the survey are limited, the sample replies are to be obtained by interviewing customers on the store premises.
(b) An agronomist wants to take a sample to compare the effectiveness of two fertilizers on a variety of crops under a range of climate and soil conditions.
(c) An airline is taking a sample of its first-class and tourist-class passengers to assess the quality of its in-flight meals. The passengers' ratings must be taken directly after they have eaten their meals.

SELECTING STATISTICAL PROCEDURES   4-5

A complete examination of statistical procedures is made in several later chapters. All of these procedures make generalizations about a population when only sample data are available. We call this process *inductive* or *inferential statistics.* There are several types of inference, and the procedures to be used depend on the goals of the study.

Statistical procedures should be determined during the planning stage. There are many ways to do this. The simplest procedure is to estimate a population's parameter, such as it mean, standard deviation, or proportion. The most complicated procedures test some hypothesis regarding the population's characteristics. Ordinarily, several testing schemes are employed. In all tests, the decision is indicated by the sample results. Many tests actually compare two or more populations and require that a separate sample be obtained from each.

In the fields of education, psychology, and medicine, the two-sample testing procedure is a popular way to decide which of two approaches is better. Often the status quo is compared to a proposed improvement. For example, a new method of teaching reading might be evaluated in terms of comprehension test scores. The population of scores achieved by students in the current reading program could be compared to the test scores achieved by another group of readers exposed to the new program. Because the new program might or might not be adopted, the associated population would not exist yet. Nevertheless, a sample of readers in the new program would represent this potentially large group, making it a target population. The actual comparison would be made between separate samples from both populations.

The readers in the sample from the current program are referred to as the *control group*, because they are not subjected to anything new. The readers in the second sample constitute the *experimental group*, because these students are exposed to a new procedure on a trial basis primarily to evaluate its merit. Control and experimental groups are common in many testing situations, such as in evaluating a new drug or medical procedure, selecting a more effective fertilizer, or checking new safety devices.

The simplest way to conduct a two-sample test is to select the two groups *independently*. The readers in the control and experimental groups would be chosen using separate lists of random numbers. (Ordinarily, both samples would be derived from a common frame of applicable students, because the experimental group represents only a target population.) One drawback to independent sampling is that any differences in the reading comprehension scores of the two groups could be attributed to causes other than the particular reading programs, such as previous level of reading or home environment. The danger that some outside influence might camouflage the actual differences between the two reading programs could be minimized by applying a stratified sampling scheme. Then two separate groups would be selected each from the same categories of student knowledge and background. If large enough, each group should provide a representative cross section of the student body.

A common method for choosing the control and experimental groups is to pair each elementary unit with one that closely matches it in terms of potential extraneous influences. One member of each pair would be assigned to the experimental group; the other member, to the control group. This is called *matched-pairs sampling*. The procedure is epitomized by studies of identical twins, who are genetically the same and who normally share the same environment as well. In our survey, one twin would be assigned to the existing reading program and the other would participate in the new program. Thus, any differences in reading scores between sample groups could be confidently attributed to differences in the reading programs and not to other causes. For practical purposes, twins are rare and are not normally used in such tests, but unrelated individuals might still be paired. For example, a new drug for heart patients could be tested with pairs matched in terms of age, sex, weight, martial status, smoking habits, medical histories, and so on, thereby producing "near twins." Identical twins who were reared apart have been extensively studied in an attempt to isolate the effects of heredity and environment on intelligence.

Illustration: Evaluating a Beverage Container Law

The experiences of one state during the ecology renaissance of the early 1970s illustrates the pitfalls of inadequate planning when conducting a statistical study. A new law was about to become effective requiring that all soft drink and beer containers be returnable, in order to eliminate the unsightly refuse problem caused by pop-top cans. Bottles were to be used exclusively. To guarantee that the empty bottles would be returned, a deposit was required by the seller, as it had been throughout the United States 20 years earlier.

To assess the effectiveness of the beverage container law, a study was made to determine how it affected accumulations of roadside litter. A random sample of 30 one-mile stretches of highway was chosen, cleared of all beer and soda cans, and left to the vagaries of normal public neglect and mistreatment for three months before the new law became effective. The 30 miles of highway served as the control group, representing state residents' littering habits when beverage containers were not dictated by law. At the end of three months, the litter was cleared and the cans were counted. For the initial three months under the new law, the same 30 miles—now the experimental group—were again left to accumulate beverage containers (which would only be bottles this time). The accumulations of the two groups were then compared.

This study shows how the same elementary units can sometimes be used in both the control and the experimental groups. In this case, each one-mile segment after the law became effective was matched with itself in a preceding time period. In this way, traffic conditions and local idiosyncrasies were alike and could not be responsible for any differences between the two groups. However, using the same units in both groups is valid only if no bias is induced on the second set of observations by the first set—and this is the point at which the study fell apart. When the beverage litter was measured the second time, very little was found. But all other kinds of litter were practically absent, too, and adjacent stretches of highway were abnormally clean. At first blush, environmental officials were ecstatic at the apparent widespread impact of their new law. But they became embarrassed when it was determined that the pattern prevailed only near the test tracts and not throughout the state.

Who absconded with the litter and why? Perhaps environmentalists—well aware that the new law would be temporary unless it proved effective—tried to help the study along by cleaning up the test stretches. Maybe when the state cleaned the one-mile parcels, local residents were shamed into picking up their own litter.

On the advice of a statistician, officials discarded the second set of test results and matched the original 30 one-mile stretches of highway with 30 similar ones that had not been cleaned. The new areas were checked only once, and only the returnable bottles were counted (the old cans did not have to be).

REVIEW EXERCISES

4-16 Starting in the fifth column of the random number table (page 821), select a simple random sample of ten symphony conductors from those listed in Table 4-1. Use only the first two digits of each random number from the table. List the names selected.

4-17 List the names for a systematic random sample of ten symphony conductors chosen from Table 4-1. Assume that the starting random number selects Sir John Barbirolli as the first name.

4-18 Suppose that each successive ten names in Table 4-1 provides a stratum. Redesignate the names in each stratum, starting with 0 and ending with 9, maintaining the original alphabetical sequence. Then select a stratified random sample that contains one conductor from each group. Assume that the following random numbers apply in the respective strata:

$$6, \ 5, \ 5, \ 2, \ 1, \ 9, \ 8, \ 7, \ 7, \ 0$$

List the names selected.

4-19 Treat each successive five names in Table 4-1 as a cluster. A random cluster sample of two clusters from the 20 is to be selected. After assigning a number (1 through 20) to each successive cluster, use the random numbers 18 and 4 to select the sample. List the names you obtained.

Let us imagine . . . a person just brought forth into this world. . . . The Sun would, probably, be the first object that would engage his attention; but after losing it the first night, he would be entirely ignorant whether he should ever see it again. . . . But let him see a second appearance or one return of the Sun, and an expectation would be raised in him of a second return, and he might know that there was an odds of 3 to 1 for some probability of this. This odds would increase, as before represented, with the number of returns. . . . But no finite number of returns would be sufficient to produce absolute or physical certainty.

THE REVEREND THOMAS BAYES (1763)

Probability

Probability plays a special role in all our lives, because we use it to measure uncertainty. We are continually faced with decisions leading to uncertain outcomes, and we rely on probability to help us make our choice. Think of the planned outdoor activities, such as picnics or boating, you canceled because the chance of bad weather seemed too high. Remember those nights before examinations when you decided not to study some topics because they were not likely to be on the test. In business, probability is a pivotal factor in most significant decisions. A department store buyer will order heavily in a new style that is believed likely to sell well. A company will launch a new product when the chance of its success seems high enough to outweigh the possibility of losses due to failure. A new college graduate is hired when the probability for satisfactory performance is judged sufficiently high.

A probability is a numerical value that measures the uncertainty that a particular event will occur. The probability of an event ordinarily represents *the proportion of times under identical circumstances that the event can be expected to occur*. Such a long-run frequency of occurrence is referred to as an *objective probability*. In tossing a fair coin, the probability is 1/2 for getting a head. This can be verified by tossing the coin many times—heads will appear about half the time. However, a probability value is often subjective, set solely on the basis of personal judgment. *Subjective probabilities* are used for events having no meaningful long-run frequency of occurrence. For example, an oil wildcatter may express his uncertainty about the presence of oil beneath a possible drilling site in terms of a probability value such as 1/2. Only one attempt will be made at drilling on that site; since no two sites are identical, there are no other situations like the present one from which the frequency of oil strikes can be determined.

More than 300 years ago, several famous mathematicians initially studied probability scientifically in connection with gambling problems. The theory of probability has since evolved into one of the most elegant and useful branches of mathematics. Today devices ordinarily associated with gambling, such as dice and playing cards, are still useful in illustrating how to find probabilities.

5-1   FUNDAMENTAL CONCEPTS

The Event

Uncertain outcomes are called *events*. A preliminary step in finding an event's probability is to identify all possible outcomes. A complete listing of events is called the *sample space*. Only one event on this master list will occur.

The sample space for tossing a coin contains only two events—head and tail. This may be conveniently expressed in set notation as

Sample space = {head, tail}

Head and tail are the *elements* of the sample space. Only one of these events will occur. We rule out such possibilities as the coin landing on its edge or being lost by specifying that they are not legitimate outcomes. In such instances, the toss is incomplete and must be repeated until either a head or a tail occurs.

Now consider the characteristics of a card drawn from a shuffled deck of 52 ordinary playing cards. Here we have

Sample space = {ace of spades, deuce of spades, . . . , king of diamonds}

It is convenient to represent the sample space pictorially. In Figure 5-1, each point in the figure represents an element of the sample space—in this case, the drawing of a particular card. We refer to the elements of the sample space as *elementary events*. Elementary events are the simplest and most basic outcomes considered.

To extend these concepts, consider the outcomes when three different coins—a penny, a nickel, and a dime—are each tossed. Either a head or a tail will occur for each coin. In the sample space (Figure 5-2), the elementary events are ordered triples; for example, (H, T, H) represents the outcome "head for the penny, tail for the nickel, head for the dime."

SUIT

DENOMINATION	Spades (black)	Hearts (red)	Clubs (black)	Diamonds (red)
King	♠ K •	♥ K •	♣ K •	♦ K •
Queen	♠ Q •	♥ Q •	♣ Q •	♦ Q •
Jack	♠ J •	♥ J •	♣ J •	♦ J •
10	♠ 10 •	♥ 10 •	♣ 10 •	♦ 10 •
9	♠ 9 •	♥ 9 •	♣ 9 •	♦ 9 •
8	♠ 8 •	♥ 8 •	♣ 8 •	♦ 8 •
7	♠ 7 •	♥ 7 •	♣ 7 •	♦ 7 •
6	♠ 6 •	♥ 6 •	♣ 6 •	♦ 6 •
5	♠ 5 •	♥ 5 •	♣ 5 •	♦ 5 •
4	♠ 4 •	♥ 4 •	♣ 4 •	♦ 4 •
3	♠ 3 •	♥ 3 •	♣ 3 •	♦ 3 •
Deuce	♠ 2 •	♥ 2 •	♣ 2 •	♦ 2 •
Ace	♠ A •	♥ A •	♣ A •	♦ A •

SAMPLE SPACE

FIGURE 5-1

Sample space describing the card selected randomly from a fully shuffled deck of 52 ordinary playing cards.

We are usually interested in more complex outcomes than elementary events. The outcome "all coins show the same side" is a *composite event* that occurs whenever either of the elementary events (H, H, H) or (T, T, T) results. We denote the possible ways this event may occur by

$$\text{All coins show same side} = \{(H, H, H), (T, T, T)\}$$

Such a partial listing is an *event set*. Note that an event set is a *subset* of the sample space: All its elements also belong to the sample space. An event set for an elementary event, such as the king of spades, contains a single element:

$$\text{King of spades} = \{\text{king of spades}\}$$

Many composite events are possible for most uncertain situations. Figure 5-3 illustrates a few event sets associated with drawing one card from a deck of 52 ordinary playing cards. Each set is pictured as a grouping of the applicable elementary events.

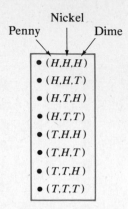

SAMPLE SPACE

FIGURE 5-2
Sample space for the tossing of three coins.

Basic Definitions of Probability

Classically, the probability for an event is the relative frequency with which it occurs when identical circumstances are repeated a large number of times. The probability for some events, such as obtaining a head in a coin toss, can be determined by reasoning alone. Knowing that a coin has two sides, assuming that it is evenly balanced, and presuming that it is tossed fairly, it is reasonable to expect a head to occur in about half of any number of tosses. Thus, we deduce that

$$\Pr[\text{head}] = \frac{1}{2}$$

Similarly, we can reason that the probability for drawing the ace of spades from a shuffled deck of cards is 1/52, since each card should appear with the same frequency in repeated shufflings. Each possible card is *equally likely*.

PROBABILITY WHEN ELEMENTARY EVENTS ARE EQUALLY LIKELY
When the sample space consists of elementary events that are equally likely, then

$$\Pr[\text{event}] = \frac{\text{Number of elementary events in the event set}}{\text{Total number of equally likely elementary events}}$$

Because there is a single way to achieve a head (the size of its event set is 1), this expression tells us that $\Pr[\text{head}] = 1/2$, where the 2 represents the

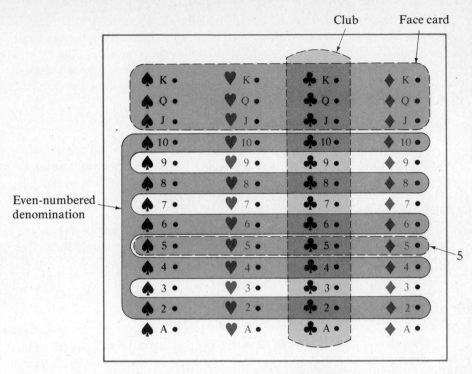

SAMPLE SPACE

FIGURE 5-3
Composite events for drawing a card from a fully shuffled deck of 52 ordinary playing cards.

size of the sample space for one coin toss. The same procedure applies to composite events. For example, the event "club" occurs whenever the drawn card is any one of the 13 elementary events shown in the third column of Figure 5-3. Since there are exactly 52 possible elementary events (cards),

$$Pr[\text{club}] = \frac{13}{52} = \frac{1}{4}$$

We can use this definition to find the probabilities for the other composite events shown in Figure 5-3:

$$Pr[\text{face card}] = \frac{12}{52} = \frac{3}{13}$$

$$Pr[5] = \frac{4}{52} = \frac{1}{13}$$

$$Pr[\text{even-numbered denomination}] = \frac{20}{52} = \frac{5}{13}$$

If the elementary events are not all equally likely, probabilities must be estimated after repeated experimentation. For example, if a die is shaved until it is asymmetrical, it becomes more likely to roll onto some sides than others. Logic cannot tell us what the probabilities are for rolling the faces of a shaved die. These probabilities must be determined by the results actually obtained from many tosses of the die.

Subjective Probability

Not all events have probabilities that can be obtained in the above fashion. For example, the event "the next U.S. President will be a Democrat" cannot be based on past frequencies alone, since conditions differ from election to election. It makes no sense to divide the number of times a Democrat has won in the past by the number of elections held. Rather, a subjective probability value must be used. Such a number is subjective because people might disagree on the proper value. Chapter 22 describes various procedures for determining subjective probabilities.

Certain and Impossible Events

A probability will always be between 0 and 1, inclusively. Two important observations follow. First, an event that is certain to occur will result every time with a frequency of 1. Thus

$$\Pr[\text{certain event}] = 1$$

The event "the next President of the United States will be at least 35 years old" is a certain event and has a probability of 1, because this requirement is specified in the Constitution. The event "food prices will rise, fall, or remain unchanged" is also certain and has a probability of 1.

At the other extreme, an impossible event's frequency will always be 0, because the event will never occur. Thus

$$\Pr[\text{impossible event}] = 0$$

For example, suppose that we draw a card from a shuffled pinochle deck. A pinochle deck contains only 48 cards, all above 8 in denomination. The sample space is therefore not the same as it is for a standard deck of 52 cards, because there are no 8s, 7s, and so on. Thus, the event set of "drawing an 8 from a pinochle deck" has *no* elementary events; it is *empty*. We state this as

$$\text{Drawing an 8 from a pinochle deck} = \{\ \ \}$$

The empty braces indicate that the event set is empty. Since an 8 is impossible,

$$\Pr[\text{drawing an 8 from a pinochle deck}] = 0$$

Alternative Expressions of Probability

In addition to ratios, probabilities can be expressed as percentages, odds, or chances. These other forms are not inconsistent with our definition, because they can be translated into the basic fraction or ratio form. For example, the probability for obtaining the head side of a coin can be expressed in the following forms:

50% probability for a head. [Divide the percent by 100 to obtain the fraction 1/2.]

50–50 chance for a head. [An even chance; a probability of $50/(50 + 50) = 1/2$.]

1 to 1 odds that a head will occur. [To express odds as a fraction, add the two numbers (in this case, $1 + 1 = 2$) and place the result in the denominator; then place the first number in the numerator. For 1 to 1 odds, the result is 1/2. As another example, the odds for drawing a queen are 4 to 48 or 1 to 12, so that $\Pr[\text{queen}] = 4/(4 + 48) = 1/(1 + 12) = 1/13$.]

EXERCISES

5-1 In a special study an accountant found the following history for 850 receivables: (a) 119 were paid early, (b) 340 were settled on time, (c) 221 were paid late, and (d) 170 were uncollectible. Assuming that this experience is representative of the future, estimate the probabilities that a particular account receivable will fall into each of the four categories.

5-2 The number of panelists in each category of a consumer testing panel are provided below.

	Family Income			
Occupation	Low	Medium	High	Total
Homemaker	8	26	6	40
Blue-collar Worker	16	40	14	70
White-collar Worker	6	62	12	80
Professional	0	2	8	10
Total	30	130	40	200

One person is selected at random.

(a) Find the probability that the selected person is a (1) homemaker, (2) blue-collar worker, (3) white-collar worker, (4) professional.

(b) Find the probability that the selected person's family income is (1) low, (2) medium, (3) high.

(c) Find the probability that the selected person is a

    (1) white-collar worker with high family income.

    (2) homemaker with low family income.

    (3) professional with medium family income.

5-3 A pair of six-sided dice, one red and one green, are tossed. The sides of each die cube exhibit the values 1 through 6.

    (a) List the elementary events in the sample space.

    (b) List the possible sums of the two showing face values.

    (c) For each answer to (b), indicate the elementary events that correspond to each possible sum value. Then determine the probability for each sum.

5-4 From your answers to Exercise 5-3, list the elementary events and then determine the probabilities for each of the following composite events:

    (a) The sum value lies between 5 and 7, inclusively.

    (b) The sum value lies between 8 and 12, inclusively.

    (c) The sum value lies outside the range 4 through 9.

    (d) The red and green dice have the same value.

    (e) The red die shows a greater value than the green die.

    (f) The red die shows a greater value than the green die, and the sum value is greater than or equal to 9.

5-5 Determine the probabilities for each event:

    (a) A woman chosen randomly from a group of ten women is a doctor, if the group contains two doctors.

    (b) A single raffle ticket will be drawn out of 10,000 tickets.

    (c) A head will be obtained on one toss each of a dime and a penny. (First determine the sample space.)

    (d) A number greater than 2 will be the value of the showing face from the toss of a six-sided die.

5-2   EVENTS AND THEIR RELATIONSHIPS

In computing probabilities, it is often helpful to examine the relationships between events. Several types of relationships are important. First we will consider how events may be combined.

Events Having Components: Union and Intersection

Some outcomes may be explained by the occurrence of more than one event. For example, obtaining at least two heads when tossing three coins is equivalent to obtaining exactly two heads *or* exactly three heads. Drawing the

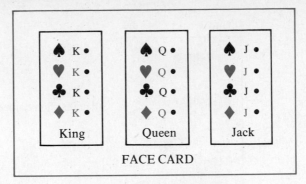

Face Card $=$ King *or* Queen *or* Jack

FIGURE 5-4
Portrayal of an event comprising the union of three component events.

ace of spades can be stated more precisely as drawing a card that is both an ace *and* a spade. The logical connectives *or* and *and*, when used to combine events, are important to probability.

Union of Events

An outcome that occurs whenever any one of several, more specific events happens is called the *union* of those events. This is expressed as

$$A \ or \ B$$

which represents the outcome when event A occurs singly, event B occurs singly, or both A and B occur together. This relationship allows us to simplify many probability computations by first looking at A and B separately. Figure 5-4 illustrates how the union of the events "king," "queen," and "jack" provides the event "face card."

Intersection and Joint Events

An outcome that arises only when both A and B occur is referred to as the *intersection* of events A and B. This is expressed as

$$A \ and \ B$$

We sometimes refer A and B as a *joint event*. Figure 5-5 illustrates this concept for three coin tosses, where the event "all heads" is protrayed as the intersection of the events "dime is a head" and "all coins show same side."

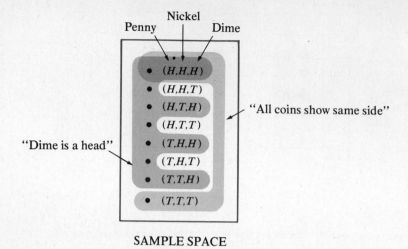

SAMPLE SPACE

FIGURE 5-5
Portrayal of the intersection of two events.

Relationships Between Events

Mutually Exclusive Events

Several events are *mutually exclusive* if the occurrence of any one event automatically rules out the occurrence of the other events. Another way to state this is to say that the joint occurrence of the events is an impossible event. Consider the two events *A* and *B* diagramed in Figure 5-6. Such representations are called *Venn diagrams*, after their originator John Venn.

EXAMPLE: SAN FRANCISCO RAINFALL

Next year's annual rainfall for San Francisco is an uncertain quantity that may range from as low as 10 inches or less to as high as 40 inches or more. Any level of precipitation between these extremes is a possible event. But because there will be only one rainfall level, all the possibilities are mutually exclusive events.

EXAMPLE: BANKRUPTCY AND PROFIT

Consider the events "bankruptcy" and "profit" that may result from the current year's operation of a firm. These events are *not* mutually exclusive, since it is possible for a business to make a profit and yet also be forced into bankruptcy by the claims of impatient creditors.

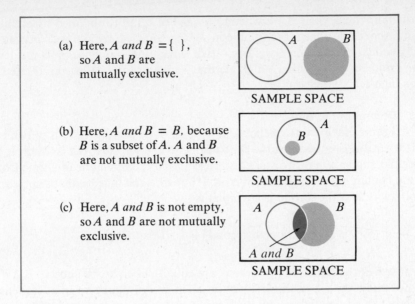

(a) Here, *A and B* = { },
so *A* and *B* are
mutually exclusive.

SAMPLE SPACE

(b) Here, *A and B* = *B*, because
B is a subset of *A*. *A* and *B*
are not mutually exclusive.

SAMPLE SPACE

(c) Here, *A and B* is not empty,
so *A* and *B* are not mutually
exclusive.

A and B

SAMPLE SPACE

FIGURE 5-6
Venn diagrams showing the cases for the intersection of two events.

Collectively Exhaustive Events

A collection of events is *collectively exhaustive* if at least one of those events is bound to occur. For example, consider the following events describing the characteristics of a randomly chosen playing card:

red suit, black suit, spade

One of these events is certain to occur, because taken together they comprise all possibilities. They are not mutually exclusive, because spades are also black cards. A collectively exhaustive grouping of events may be redundant.

Independent Events

Two events are *independent* when the probability for one is not affected by the occurrence of the other. Consider the outcomes of two successive coin tosses. Assuming that both tosses are fair, we can conclude that obtaining a head on the second toss is independent of obtaining a head on the first toss, because the probability for a head is 1/2 whether a head or a tail is obtained first. But we cannot conclude that a randomly chosen person's education and income are independent events, since we know that education influences income levels. The probability of earning a high income is greater for a college graduate than for a high-school dropout.

Dependent Events

Events that are not independent are said to be *dependent*. This only means that when one event occurs, the probability for the second event is affected. For example, suppose that two students are randomly selected one at a time from a group containing 6 women and 4 men. If a man is the first student selected, the probability that the second is a woman is

$$Pr[\text{woman second}] = \frac{6}{9} \quad (\text{man first})$$

because 6 of the remaining 9 persons are equally likely to be chosen. But if we do not know the sex of the first person, then any of the 10 original students is equally likely to be the second pick. Since 6 of these are women,

$$Pr[\text{woman second}] = \frac{6}{10} \quad (\text{first person's sex unknown})$$

We see that the probability for "woman second" differs depending on whether "man first" or "first person's sex unknown" has occurred. The events "woman second" and "man first" are dependent.

The characteristics or variate values of successive units selected during sampling are usually dependent events. Independence and dependence are important considerations in deductive statistics. These concepts apply generally to decision making under uncertainty whenever predictive information is used to revise probabilities. Consider petroleum exploration, where a seismic survey is often used to help find oil. Obviously, the probability of oil is higher when a favorable seismic survey results than otherwise. The events "oil" and "favorable seismic" are dependent events.

The Law of Large Numbers

Another example illustrates a very important property that exists when uncertain situations are repeated and the successive outcomes are independent.

Suppose a fair coin is tossed 20 times, and a head appears every time. The probability for this happening is less than one-millionth. Does this mean that 20 extra tail outcomes must occur in a long series of future tosses to balance the long-run frequency of occurrence ratio and to coincide with "head" having a probability of 1/2? Also, does this increase the probability for obtaining a tail on the next toss?

The answer to both questions is *no*. First, there are no guarantees that any particular sequence of results will exhibit heads even close to one-half the time. This presents no contradiction, because there is a nonzero probability that any number of heads (say, 10,000) will occur in sequence. All that can be stated is that it is not very likely that there will be many more (or less) heads than half the number of tosses. A law of probability, the *law of large numbers*, in essence states that the probability for the result deviating significantly from that indicated by the theoretical long-run frequency of occurrence becomes smaller as the number of repetitions (tosses) increases. Thus, if the coin was tossed several thousand more times, the effect of the 20 "extra" heads on the resulting frequency would barely be noticeable.

If the coin-tossing process is fair (that is, if the process is not biased in favor of heads or tails), then we can infer that the probability for obtaining a head should be 1/2. Also, if the process is fair, then a head would be no more likely to follow a head than a tail would; the events of succeeding and preceding tosses are independent. The probability for obtaining a tail should not increase or decrease if it is accepted that fairness is present. After obtaining 20 heads in a row, human reasoning may lead to the inference that the tossing mechanism is unfair. However, this seemingly aberrant result does not constitute proof of unfairness.

EXERCISES

5-6 For each of the following situations, indicate whether or not the listed events are collectively exhaustive. If they are not, explain why.

(a) A box contains three red, three blue, and three white objects. Two objects are drawn from the box. Following are some possible color combinations for the pair: both red; both blue; red and white; both white; blue and white.

(b) A new job applicant is (1) either male (*M*) or female (*F*); (2) either a college graduate (*G*) or not (not *G*); and (3) either under 30 (*U*) or 30 or over (*O*). The following events describing the attributes of the applicant are of interest:

M, not *G*, and *O*	*F*, *G*, and *O*	*F*, *G*, and *U*
M, *G*, and *U*	*F*, not *G*, and *U*	*M*, *G*, and *O*
M, not *G*, and *U*		

(c) A record disc is to be inspected to determine its quality in terms of (1) whether it has a high (*H*) or a low (*L*) electric charge; (2) whether it is scratched (*S*) or not scratched (not *S*); and (3) whether it is warped (*W*) or flat (*F*). The joint events are

H, *S*, and *W*	*H*, not *S*, and *F*	*L*, not *S*, and *W*
H, *S*, and *F*	*L*, *S*, and *W*	*L*, not *S*, and *F*
H, not *S*, and *W*	*L*, *S*, and *F*	

5-7 For each of the following situations, indicate whether or not the events are mutually exclusive. If they are not, explain why.

(a) The toss of a six-sided die: even-valued result; 1-face; 2-face; 5-face.

(b) Thermometers are inspected and rejected if any of the following is found: poor calibration; inability to withstand extreme temperatures without breaking; not within specified size tolerances.

(c) A manager will reject a job applicant for any of the following reasons: lack of relevant experience; slovenly appearance; too young; too old.

5-8 A group of 100 business students are classified as follows:

	Lower Division (L)	Upper Division (U)	Total
Woman (W)	10	20	30
Man (M)	40	30	70
Total	50	50	100

(a) Find the probability that a particular student is a man.

(b) Considering only the lower-division students, find the probability that a particular student is a man.

(c) Are your answers to (a) and (b) the same? Are the events "man" and "lower-division student" independent?

5-3   PROBABILITIES FOR COMPOUND EVENTS

Any event that has two or more components is a *compound event*. There are two forms of compound events:

$$A \text{ or } B \quad \text{(union)}$$

$$A \text{ and } B \quad \text{(intersection or joint event)}$$

The probability for a compound event (like any other event) is its long-run frequency of occurrence, which can be found by using the basic concepts already described. But we can often take shortcuts in determining these probabilities.

Two basic laws of probability allow us to find $\Pr[A \text{ or } B]$ or $\Pr[A \text{ and } B]$ by first obtaining the values $\Pr[A]$ and $\Pr[B]$ for the component events.

The Addition Law

We may use the *addition law* to obtain the probability for a compound event of the form *A or B*.

ADDITION LAW FOR MUTUALLY EXCLUSIVE EVENTS

$$Pr[A \text{ or } B] = Pr[A] + Pr[B]$$

To illustrate, suppose that the number of customers arriving at a barbershop during the first 10 minutes it is open can be any value between 0 and 4, with the following probabilities:

Number of Persons	Probability
0	.1
1	.2
2	.3
3	.3
4	.1

Applying the addition law gives us the probability that exactly 2 or exactly 3 persons arrive:

$$Pr[2 \text{ or } 3] = Pr[2] + Pr[3]$$
$$= .3 + .3 = .6$$

The addition law applies for any number of components. For example, the probability that at least 1 person arrives can be expressed as

$$Pr[\text{at least } 1] = Pr[1 \text{ or } 2 \text{ or } 3 \text{ or } 4]$$
$$= Pr[1] + Pr[2] + Pr[3] + Pr[4]$$
$$= .2 + .3 + .3 + .1 = .9$$

The above addition law applies only when the compound events are mutually exclusive. If the components are not mutually exclusive, $Pr[A \text{ or } B]$ may instead be found by using the basic probability definition.

For example, let's find the probability of dealing ourselves a playing card that is "ace *or* heart." From Figure 5-1, we easily find

$$Pr[\text{ace}] = \frac{4}{52}$$

$$Pr[\text{heart}] = \frac{13}{52}$$

But we cannot add 4/52 and 13/52 to obtain $Pr[\text{ace } or \text{ heart}]$, because ace and heart are not mutually exclusive events: The ace of hearts belongs to both event sets and would be counted twice. From Figure 5-7, we see that there are 16

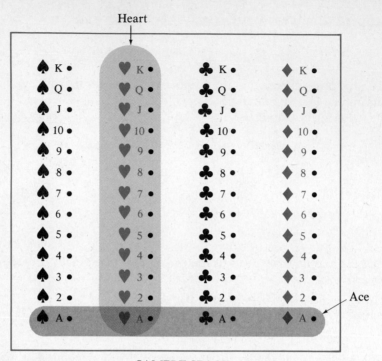

SAMPLE SPACE

FIGURE 5-7
Illustration for calculating the probability for "ace *or* heart."

cards in the event set "ace *or* heart," so the correct calculation is

$$\text{Pr}[\text{ace } or \text{ heart}] = \frac{\text{Number of "ace } or \text{ heart" cards}}{\text{Total number of cards in deck}} = \frac{16}{52}$$

More general forms of the addition law have been developed for situations where events are not mutually exclusive,* although when the elementary events

---

* When A and B are not mutually exclusive, we would have to use the following *general addition law*:

$$\text{Pr}[A \text{ or } B] = \text{Pr}[A] + \text{Pr}[B] - \text{Pr}[A \text{ and } B]$$

which requires that the joint probability, $\text{Pr}[A \text{ and } B]$, be known in advance. If we apply this law to playing cards, we obtain

$$\text{Pr}[\text{ace } or \text{ heart}] = \text{Pr}[\text{ace}] + \text{Pr}[\text{heart}] - \text{Pr}[\text{ace } and \text{ heart}]$$
$$= \frac{4}{52} + \frac{13}{52} - \frac{1}{52} = \frac{16}{52}$$

are equally likely, it is usually simpler to "count and divide," as we have done above.

Events That Are Mutually Exclusive and Collectively Exhaustive

The following property of the addition law can be used *whenever the component events are both mutually exclusive and collectively exhaustive*:

$$\Pr[A \ or \ B \ or \ C] = \Pr[A] + \Pr[B] + \Pr[C] = 1$$

Although this expression involves three events, a greater or lesser number of event components may be involved in some situations. The probabilities of these components sum to 1 because the two properties guarantee that the complex outcome *A or B or C* is *certain*.

Returning to our barbershop customer arrivals, we see that the five values representing the number of persons arriving exhibit the property of mutual exclusiveness and are also collectively exhaustive. (One value is bound to occur, and it is impossible for five or more customers to arrive in the first ten minutes.) Thus

$$\Pr[0 \ or \ 1 \ or \ 2 \ or \ 3 \ or \ 4] = \Pr[0] + \Pr[1] + \Pr[2] + \Pr[3] + \Pr[4]$$
$$= .1 + .2 + .3 + .3 + .1 = 1$$

Viewed another way,

$$\Pr[0 \ or \ \text{at least } 1] = \Pr[0] + \Pr[\text{at least } 1]$$
$$= .1 + .9 = 1$$

The outcomes "0" and "at least 1" have a special relationship: They are opposites, or *complementary events*. This can be a useful feature in finding probabilities.

Application to Complementary Events

An event and its complement are collectively exhaustive, because either one or the other must occur. For example, a person chosen at random must be either male or female (which may be expressed as not male). Since these events are obviously mutually exclusive, the addition law provides for any event A that

$$\text{Pr}[A \textit{ or not } A] = \text{Pr}[A] + \text{Pr}[\text{not } A] = 1$$

From this it follows that

$$\text{Pr}[A] = 1 - \text{Pr}[\text{not } A]$$

This principle is useful when the probability value for the event "not A" is easier to find than the probability value for A itself.

Now suppose we want to determine a faster way to compute the probability that at least one person arrives at the barbershop. "At least 1" means "some," and *the opposite of some is none.* The complementary event is therefore "0" customers, and

$$\text{Pr}[\text{at least } 1] = 1 - \text{Pr}[0]$$

$$= 1 - .1 = .9$$

The Multiplication Law for Independent Events

We may use the *multiplication law* to obtain the probability for a compound event of the form *A and B.*

MULTIPLICATION LAW FOR INDEPENDENT EVENTS

$$\text{Pr}[A \textit{ and } B] = \text{Pr}[A] \times \text{Pr}[B]$$

We can use the multiplication law to find the probability that a playing card taken from a shuffled deck is "ace *and* heart":

$$\text{Pr}[\text{ace } \textit{and } \text{heart}] = \text{Pr}[\text{ace}] \times \text{Pr}[\text{heart}]$$

$$= \frac{1}{13} \times \frac{1}{4} = \frac{1}{52}$$

Of course, it is easier to arrive at the answer more directly: There is only one ace of hearts in a deck of 52 cards, so the answer must be 1/52. However, the multiplication law is vital when $\text{Pr}[A]$ and $\text{Pr}[B]$ are known but the elements in the event set *A and B* cannot be counted or the size of the sample space is unknown. For example, we may know that 60% of the adults in Blahsburg are married and that 40% are college graduates. The multiplication law provides us with the probability that a particular person is both married and a college graduate:

$$Pr[\text{married } and \text{ graduate}] = Pr[\text{married}] \times Pr[\text{graduate}]$$
$$= .60 \times .40 = .24$$

The multiplication law extends to any number of independent events. For example, suppose that just as many men as women read a particular magazine, that married readers are just as predominant among males as females, and that proportionately there are just as many Democrats in each sex and marital status combination. The sex, marital status, and political affiliation of a randomly selected reader yield combinations of three independent events. Thus, to compute the probability for selecting a married woman who is a Democrat, we can use the multiplication law for independent events:

$$Pr[\text{woman } and \text{ married } and \text{ Democrat}]$$
$$= Pr[\text{woman}] \times Pr[\text{married}] \times Pr[\text{Democrat}]$$

Finding Joint Probabilities for Dependent Events

The preceding multiplication law applies only when the component events are independent, so that $Pr[A]$ must have the same value whether or not B occurs. When the components are dependent, $Pr[A \text{ and } B]$ can usually be found by using the basic probability definition.

To illustrate, the number of men and women belonging to various schools at a certain university are as follows:

School	Men	Women	Total
Education	154	236	390
Engineering	337	88	425
Business	924	629	1,553
Humanities	1,185	1,441	2,226
Science	758	422	1,180
Total	3,358	2,816	6,174

For a randomly chosen student, the following probabilities apply:

$$Pr[\text{man}] = 3{,}358/6{,}174 = .544$$
$$Pr[\text{engineering}] = 425/6{,}174 = .069$$

It would be incorrect to find the probability that the student is both a man and in engineering by multiplying .544 and .069. This is because men occur more frequently in engineering than in the entire student body; it follows that the events "man" and "students in engineering" are *dependent*. Therefore, the desired probability must be found by using the basic definition:

$$\text{Pr[man } and \text{ engineering]} = \frac{\text{Number of men engineers}}{\text{Total number of students}} = \frac{337}{6,174} = .055$$

Later in the chapter we will discuss a more general multiplication law that can be applied even when events are dependent.

EXERCISES

5-9   An antique-car parts supplier has determined the following probabilities for the number of annual orders for Locomobile fuel pumps:

Number of Orders	Probability
0	.3
1	.2
2	.1
3	.1
4	.1
5	.1
6	.1
7 or more	.0
Total	1.0

Find the probability that there will be
(a) less than 4 orders.
(b) between 2 and 6 orders, inclusively.
(c) at least 1 order.
(d) between 2 and 4 orders, inclusively.
(e) at the most 2 orders.

5-10   Refer to the data for the consumer testing panel in Exercise 5-2 and to your answers to that problem. One person is selected randomly.
(a) Use the addition law to find the probability that the panelist is
    (1) a homemaker *or* a blue-collar worker.
    (2) a blue-collar worker *or* a professional.
    (3) of low income *or* medium income.
(b) Some of the occupation and income events are independent (but not all). Use the multiplication law to find the probability that the selected panelist is
    (1) a homemaker *and* of medium family income.
    (2) a blue-collar worker *and* of high family income.

5-11   Refer again to the data in Exercise 5-2; one person is selected at random. Because the respective component events are dependent, the multiplication law cannot be used to find the probabilities for the joint events listed below. Therefore, find the joint probabilities for the following using the basic definition of probability for equally likely events.

(a) professional *and* high income
(b) white-collar worker *and* medium income
(c) blue-collar worker *and* low income
(d) professional *and* low income
(e) blue-collar worker *and* medium income
(f) professional *and* medium income
(g) white-collar worker *and* low income
(h) white-collar worker *and* high income
(i) homemaker *and* low income
(j) homemaker *and* high income

5-12 The statistics students at Adams College are 60% male. Exactly 20% of the men and 20% of the women are married. Use the multiplication law to find the probability that a randomly chosen student is

(a) man *and* married.   (c) woman *and* married.
(b) man *and* unmarried.   (d) woman *and* unmarried.

5-13 One card is dealt from a shuffled deck of 52 ordinary playing cards.

(a) Find the probability that the card will be each of the following:
   (1) a ten   (3) a king   (5) a club
   (2) a face card (4) red   (6) a diamond

(b) Using your answers to (a), apply the addition law to find the probability for each of the following compound events:
   (1) ten *or* face   (2) king *or* ten   (3) club *or* diamond   (4) red *or* club

(c) Why can't the addition law be used to determine the probabilities for the following compound events? Use the basic definition to find the probability for each event.
   (1) king *or* diamond   (2) face *or* club   (3) ten *or* red   (4) face *or* red

5-14 An employment agency specializing in clerical and secretarial help classifies candidates in terms of primary skills and years of experience. The skills are bookkeeping, switchboard, and stenography. (We will assume that no candidate is proficient in more than one of these.) Experience categories are less than one year, one to three years, and more than three years. There are 100 persons currently on file, and their skills and experience are summarized in the following table.

| | Skill | | | |
Experience	Bookkeeping	Switchboard	Stenography	Total
Less than One Year	15	5	30	50
One to Three Years	5	10	5	20
More than Three Years	5	15	10	30
Total	25	30	45	100

One person's file is chosen at random. Find the probability that the selected person falls into each of the following categories:

(a) bookeeping
(b) less than one years's experience
(c) switchboard
(d) one to three years' experience
(e) stenography
(f) more than three years' experience

5-15 (*Continuation of Exercise 5-14*):

(a) Assuming the selected person has less than one year of experience, what is the probability that his or her skill is bookkeeping?

(b) Does the bookkeeping probability in (a) differ from the one you found in Exercise 5-14? Are bookkeeping and less than one year's experience statistically independent events?

(c) Find Pr[bookkeeping *and* less than one year's experience]. Can the multiplication law given in this chapter be used to determine this probability?

5-16 (*Continuation of Exercise 5-14*):

(a) Use the addition law to determine the probabilities for the following component events:

(1) bookkeeping *or* switchboard

(2) stenography *or* switchboard

(3) less than one year's experience *or* more than three years' experience

(b) Use the basic definition to determine the probabilities for the following component events:

(1) bookkeeping *or* less than one year's experience

(2) switchboard *or* more than three years' experience

(3) stenography *or* one to three years' experience

5-4 CONDITIONAL PROBABILITY AND THE JOINT PROBABILITY TABLE

We have seen uncertain situations that result in the joint occurrence of two or more events. This may happen because several stages are involved, as in tossing a coin more than once, or it may be due to the simultaneous occurrence of several events, such as selecting a person at random who will be either male or female, married or single, and over or under 21 years old. We now describe a convenient way to organize the probability information when several events are involved. We will also consider new questions that arise when events are dependent. In particular, how do the conditions imposed by the occurrence of some events affect the probabilities of other events?

Conditional Probability

Conditional probabilities are probability values obtained under the stipulation that some events have occurred or will occur.

We can illustrate conditional probability by again considering the outcomes from drawing a playing card. Suppose someone draws a card without letting you see it, but in a brief glimpse you see that it must be a face card. The deck has only 12 face cards. What is the probability that the card is a king? Although the deck contains 4 kings, our answer is not 4/52, since our surreptitiously gained information indicates that some of the 52 cards are impossible. The sample space has been restricted to the 12 face cards, and the only remaining uncertainty is which of these 12 cards has been removed. In a sense, there is a new "sample space" having the 12 face cards as elementary events. Using basic concepts, we determine that the probability for a king is 4/12 = 1/3. Thus, we may state that the conditional probability for a king *given* face card is

$$\Pr[\text{king}|\text{face card}] = \frac{4}{12} = \frac{1}{3}$$

where the vertical bar stands for *given*.

The principle underlying conditional probability is extensively used in our daily decision making. For example, when the sky is heavily overcast, enhancing the chance of rain, you carry an umbrella. Conditional probability is also relied on in making business decisions. Insurance companies, for example, use it to determine their rates.

EXAMPLE: LIFE INSURANCE PREMIUMS

Life insurance companies charge a sizable premium for covering the lives of steeplejacks, miners, divers, and other members of occupational groups who are subject to greater hazards than most people. The mortality tables on which insurance rates are based indicate that such persons have a shorter life expectancy than the population as a whole; in essence, their probability for dying in any given year is higher. Information about an insurance applicant's occupation affects this probability: The likelihood of the event "untimely death" is affected by the occurrence of the event "applicant is a steeplejack."

The Joint Probability Table and Marginal Probabilities

We now illustrate how probability information may be organized when there are several simultaneous events. Consider the following situation.

The credit applicants of a department store are classified in terms of home ownership and job tenure. Suppose that one application is chosen by lottery from 200, which are grouped into four categories in Table 5-1. The letters O, R, L, and M will be used to simplify the following discussion. Using the probability

TABLE 5-1

Number of Credit Applicants by Category

	On Present Job 2 Years or Less (L)	On Present Job More than 2 Years (M)	Total
Owns Home (O)	20	40	60
Rents Home (R)	80	60	140
Total	100	100	200

definition for equally likely events, we can determine the following probability values:

$$\Pr[O \text{ and } L] = \frac{20}{200} = .10 \qquad \Pr[R \text{ and } L] = \frac{80}{200} = .40$$

$$\Pr[O \text{ and } M] = \frac{40}{200} = .20 \qquad \Pr[R \text{ and } M] = \frac{60}{200} = .30$$

The marginal totals may be used to determine the probabilities that the applicant has the respective attributes. For instance, the probability that the applicant owns a home is $\Pr[O] = 60/200 = .30$. In a like manner, we can find $\Pr[R]$, $\Pr[L]$, and $\Pr[M]$. Since only the numbers in the margins of Table 5-1 are needed to compute these probabilities, they are sometimes called *marginal probabilities*.

Construction of the *joint probability table* shown in Table 5-2 may be helpful. The joint events represented by each cell in a joint probability table are mutually exclusive. Thus, all the joint probability values in a particular row (or column) can be added together to obtain the respective marginal probability. For example, the event O has two mutually exclusive components: O and L, O and M. We determine the probability that the applicant is a home-

TABLE 5-2

Joint Probability Table for a Randomly Selected Applicant

	On Present Job 2 Years or Less (L)	On Present Job More than 2 Years (M)	Marginal Probability
Owns Home (O)	.10	.20	.30
Rents Home (R)	.40	.30	.70
Marginal Probability	.50	.50	1.00

owner by

$$Pr[O] = Pr[(O \text{ and } L) \text{ or } (O \text{ and } M)]$$
$$= Pr[O \text{ and } L] + Pr[O \text{ and } M]$$
$$= .10 + .20 = .30$$

Computing Conditional Probabilities

We are sometimes able to compute conditional probabilities by using the joint probability of two events, when it is known, and the probability for the given event.

PROPERTY OF CONDITIONAL PROBABILITY

$$Pr[A|B] = \frac{Pr[A \text{ and } B]}{Pr[B]}$$

Applying this property to the credit applicant illustration, we may compute the conditional probability that the applicant owns a home given job tenure of more than two years:

$$Pr[O|M] = \frac{Pr[O \text{ and } M]}{Pr[M]} = \frac{.20}{.50} = .40$$

The above relationship cannot always be used to compute the conditional probability. Unless values for both $Pr[A \text{ and } B]$ and $Pr[B]$ are known, it is of no use. When either of these probabilities is unknown, $Pr[A|B]$ can usually be found more directly. For example, suppose we know only that in a group of persons there are 30 doctors, and that 10 of the doctors are married. Then for a randomly chosen person it follows that $Pr[\text{married}|\text{doctor}] = 10/30 = 1/3$ (only the 30 doctors are considered). We cannot first find the values for $Pr[\text{married and doctor}]$ and $Pr[\text{doctor}]$, because nobody told us the total number of people involved.

Often in finding probabilities there is a choice of procedure. We may calculate $Pr[O|M]$ in the credit applicant illustration more simply by counting persons and dividing. Using the data from Table 5-1, we have

$$Pr[O|M] = \frac{40}{40 + 60} = \frac{40}{100} = .40$$

Figure 5-8 shows how the same result could be obtained directly from a diagram of the sample space by noting that the condition of being on the job more than 2 years limits the outcomes to a new, smaller sample space of size 100.

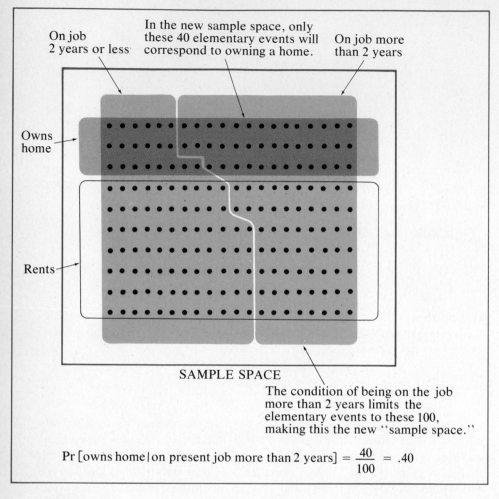

On job
2 years or less

In the new sample space, only
these 40 elementary events will
correspond to owning a home.

On job more
than 2 years

Owns
home

Rents

SAMPLE SPACE

The condition of being on the job
more than 2 years limits the
elementary events to these 100,
making this the new "sample space."

$$\text{Pr}[\text{owns home} \mid \text{on present job more than 2 years}] = \frac{40}{100} = .40$$

FIGURE 5-8

Portrayal of the concept of conditional probability.

Testing for Independence by Comparing Probabilities

Recall that two events are *independent* if the occurrence of one does not affect the probability for the other. We may formally state this by the

DEFINITION
Two events A and B are statistically independent if the chance of one is unaffected by the occurrence of the other; that is, if

$$\text{Pr}[A \mid B] = \text{Pr}[A]$$

Whenever the above equality holds, so must the following:

$$Pr[B|A] = Pr[B]$$

Independence, or the lack of it, may therefore be established by comparing probability values. Consider the following pairs of events:

king versus face card   (playing cards)

O versus M (credit applicants)

To establish whether the events in each pair are independent, we first get *unconditional probabilities* for one event in each pair:

$$Pr[king] = 4/52 = 1/13$$
$$Pr[O] = .30$$

We then obtain the conditional probabilities given the second event in each pair (which were found earlier):

$$Pr[king|face\ card] = 1/3$$
$$Pr[O|M] = .40$$

Comparing the respective conditional and unconditional probabilities, we see that their values differ:

$$Pr[king|face\ card] = 1/3 \neq 1/13 = Pr[king]$$
$$Pr[O|M] = .40 \neq .30 = Pr[O]$$

Thus, in each case the occurrence of the second event affects the probability for the first, and we conclude that king and face card are *not* statistically independent. They are *dependent events*. Likewise, O and M are dependent.

On the other hand, if the unconditional probability for one event is equal to that event's conditional probability given a second event, then the two events are independent. Consider two playing card events:

king versus red

Considering only the 26 red cards, there are but two red kings, so that

$$Pr[king|red] = 2/26 = 1/13$$

Since this is the same value as the unconditional king probability, we have

$$Pr[king|red] = 1/13 = Pr[king]$$

which establishes independence between the events "king" and "red."

EXERCISES

5-17 Consider the following data from the occupation and family income of the members of a consumer testing panel:

	Family Income			
Occupation	Low	Medium	High	Total
Homemaker	8	26	6	40
Blue-Collar Worker	16	40	14	70
White-Collar Worker	6	62	12	80
Professional				10
Total	30	130	40	200

(a) Construct a joint probability table for the occupation and family income events.
(b) Read from your joint probability table the marginal probabilities for the following events:
    (1) low income   (2) homemaker   (3) professional   (4) high income
(c) Read from your joint probability table the joint probabilities for the following joint events:
    (1) white-collar worker *and* low income
    (2) medium income *and* professional
    (3) high income *and* homemaker
    (4) blue-collar worker *and* medium income

5-18 Use the data given in Exercise 5-17 to find the conditional probability that a selected panelist is
(a) a white-collar worker *given* low family income.
(b) from a high-income family *given* a professional occupation.
(c) from a medium-income family *given* a blue-collar occupation.
(d) a homemaker *given* high family income.
(e) a professional *given* low family income.

5-19 The following data apply to the undergraduate students in a business statistics class.

	Class				
Major	Freshman	Sophomore	Junior	Senior	Total
Accounting	6	4	2	0	12
Finance	0	2	4	2	8
Marketing	2	10	6	2	20
Total	8	16	12	4	40

(a) Find the missing numbers in the above table.
(b) Construct the joint probability table.

(c) For one student chosen at random, find the following conditional probabilities:
   (1) freshman *given* accounting   (3) marketing *given* senior
   (2) finance *given* junior   (4) senior *given* accounting

5-20 In computing conditional probabilities, be careful not to confuse the uncertain event with the given event. Referring to the credit applicant data in Table 5-1, find the following conditional probabilities for a randomly selected applicant.
   (a) $\Pr[R|M]$ (c) $\Pr[O|M]$ (e) $\Pr[O|L]$
   (b) $\Pr[M|R]$ (d) $\Pr[M|O]$ (f) $\Pr[L|O]$

5-21 Refer to Exercise 5-19 and to your answers to that problem. Establish the following for a randomly selected student.
   (a) The events "senior" and "marketing" are independent.
   (b) The events "accounting" and "sophomore" are dependent.

5-22 Use the data given in Exercise 5-17 to establish the following for a randomly selected panelist.
   (a) The events "homemaker" and "medium income" are independent.
   (b) The events "professional" and "low income" are dependent.
   (c) The events "blue-collar worker" and "high income" are independent.
   (d) The events "white-collar worker" and "low income" are dependent.

5-23 A new family with two children of different ages has moved into the neighborhood. Suppose that it is equally likely that either child will be a boy or girl. Hence the following situations are equally likely:

Youngest	Oldest	
boy	boy	(B, B)
boy	girl	(B, G)
girl	boy	(G, B)
girl	girl	(G, G)

   (a) Find Pr[at least one girl].
   (b) If you know there is at least one girl, what is the conditional probability that the family has exactly one boy?
   (c) Given that at least one child is a girl, what is the conditional probability that there are two girls?

THE GENERAL MULTIPLICATION LAW,   5-5
PROBABILITY TREES, AND SAMPLING

The multiplication law given earlier applies only when events are independent. Independence between events, however, is the exception. In this section we describe a more general version of the multiplication law that considerably expands our capability of calculating probabilities. Such calculations are often used when the probability information is arranged in the form of *probability trees.* These are useful displays for evaluating all kinds of decisions involving uncertainty and can help us make deductive evaluations of the sampling process.

The General Multiplication Law

This new multiplication law involves both unconditional and conditional probabilities and may be applied even when events are not independent.

GENERAL MULTIPLICATION LAW

$$\Pr[A \text{ and } B] = \Pr[A] \times \Pr[B|A]$$

and

$$\Pr[A \text{ and } B] = \Pr[B] \times \Pr[A|B]$$

We will continue with the credit applicant illustration to show how the general multiplication law can be applied. Recall that $\Pr[M] = .50$ and also that $\Pr[O|M] = .40$. Applying the general multiplication law, we have

$$\Pr[M \text{ and } O] = \Pr[M] \times \Pr[O|M] = .50(40) = .20$$

This is the same joint probability value found earlier.

The general multiplication law is not always needed. But there are situations where only conditional and marginal probabilities are available, and the joint probabilities may be obtained only by using the general multiplication law.

EXAMPLE: AUTOMOBILE ACCIDENTS

A highway commissioner has found that half of all fatal automobile accidents in his state may be blamed on drunken drivers. Only 4 in 1,000 reported accidents have proved fatal, and 10% of all accidents in the state are attributable to drunken drivers. The commissioner wishes to summarize this information in a joint probability table relating to future accidents.

Assuming that the present pattern prevails, the probability that a reported accident happens to be fatal is

$$\Pr[F] = 4/1,000 = .004$$

whereas the probability that a drunken driver causes the accident (fatal or not) is

$$\Pr[D] = .10$$

and the conditional probability that a drunken driver causes the accident given that it is fatal is

$$\Pr[D|F] = .50$$

The multiplication law provides the joint probability that a fatal accident is caused by a drunken driver:

$$\Pr[F \text{ and } D] = \Pr[F] \times \Pr[D|F]$$

$$= .004(.50) = .002$$

TABLE 5-3
Joint Probability Table for Cause and Kind
of Automobile Accident

	Fatal (F)	Nonfatal (not F)	Marginal Probability
Drunken Driver (D)	.002	.098	.100
Other Cause (O)	.002	.898	.900
Marginal Probability	.004	.996	1.000

The complete joint probability table is provided in Table 5-3, where only the black numbers were given directly. The marginal probabilities shown in color were found by subtracting the given probabilities from 1. The remaining joint probabilities were obtained by using the fact that row and column values must sum to the marginal totals.

When we wish to find the joint probability for more than two events, the multiplication law may be extended:

$$\Pr\left[A \text{ and } B \text{ and } C\right] = \Pr\left[A\right] \times \Pr\left[B|A\right] \times \Pr\left[C|A \text{ and } B\right]$$

The Matching-Birthdays Problem

We may use the general multiplication law to determine the probability that there is at least one matching birthday (day and month) among a group of persons. For simplicity, it will be assumed that each day in the year is equally likely to be a person's birthday; February 29 will be combined with March 1.

It will be simplest to find the probability of the complementary event—no matches—by using the multiplication law. Envision each member of a group of size n being asked in succession to state his or her birthday. We conveniently define our events as follows:

A_i = the ith person queried does not share a birthday with the previous
$\quad\quad i - 1$ persons.

Then $Pr[\text{no match}] = Pr[A_1 \text{ and } A_2 \text{ and } \ldots \text{ and } A_{n-1} \text{ and } A_n]$, which is the probability of the event that no person shares a birthday with the preceding persons. Since there is no previous person for the first to match with, the event A_1 is certain, so that

$$Pr[A_1] = 1 = \frac{365}{365}$$

Also, we obtain

$$Pr[A_2 | A_1] = \frac{364}{365}$$

because there are 364 days for which the second person may not match birthdays with the first. Continuing in this same manner, we finally obtain

$$Pr[A_n | A_1 \text{ and } A_2 \text{ and } \ldots \text{ and } A_{n-1}] = \frac{365 - n + 1}{365}$$

since the nth person cannot have a birthday on the previously cited $n - 1$ dates, leaving only $365 - (n - 1)$ days allowable for his or her birthday. Therefore, we may apply the multiplication law to obtain:

$$Pr[\text{no matches}] = \frac{365}{365} \times \frac{364}{365} \times \frac{363}{365} \times \cdots \times \frac{365 - n + 1}{365}$$

The probability for at least one match may be found from this:

$$Pr[\text{at least one match}] = 1 - Pr[\text{no matches}]$$

One interesting issue is finding the size of the group for which the proability exceeds 1/2 that there is at least one match. Knowing this number, you can amaze your less knowledgeable friends and perhaps win a few bets. The "magical" group size turns out to be 23.

Why such a low group size? If it were physically possible to list all the ways (triples, sextuples, septuples, and so forth) in which 23 birthdays can match (there are several million, and for each of them a tremendous number of date possibilities), an intuitive appreciation as to why could be attained. Table 5-4

TABLE 5-4

Probabilities for at Least One Matching Birthday

Group Size	Pr[no matches]	Pr[at least one match]
3	.992	.008
7	.943	.057
10	.883	.117
15	.748	.252
23	.493	.507
40	.109	.891
50	.030	.970
60	.006	.994

shows the matching probabilities for several group sizes. Note that for groups above 60 there is almost certain to be at least one match.

Random Sampling

The selection of a random sample may be considered to have several stages. At each stage an elementary unit is removed from the population, with all remaining units having an equal chance of being chosen at the next stage.

Consider an illustration from manufacturing quality control. Suppose that 5% of all items made by a machine prove to be defective (but this value is unknown to management). The incidence of defectives is sufficiently erratic that the attributes of successive items may be assumed independent. The machinery is readjusted if the number of defectives found in a sample is excessive. Suppose that three successive items are examined for defects. The multiplication law for independent events can be used to calculate probabilities for all possible sample results. It will be helpful to list all elementary events and find the probability for each one, and this can be done using a special graphical display.

The Probability Tree Diagram

The sampling process can be conveniently explained in terms of a *probability tree diagram* like the one in Figure 5-9. The outcome of each successive sample observation is represented by a branch. The probabilities of .05 for a defective item and .95 for a good item appear beside the appropriate branch. The probabilities for the events in any one branching point, or fork, must always sum to 1.

To distinguish the outcomes for each item, we use the subscripts 1, 2, and 3. For instance, D_1 means that the first item will be defective; G_2 means

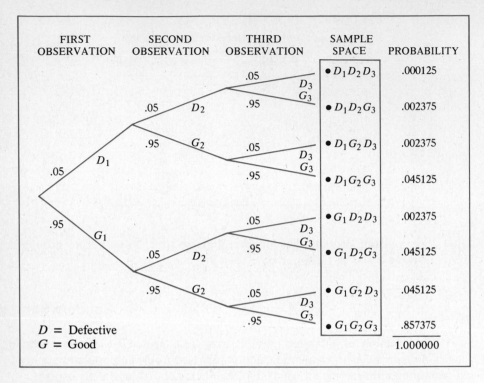

FIGURE 5-9
Probability tree diagram for selecting a quality-control sample of three items from a production process.

that the second item will be good. There are eight paths through the tree, each representing a different sample outcome. Each path leads to a different elementary event, so that the eight end positions provide the sample space for the final sample results.

The probabilities for each elementary event may be found by multiplying the probability values for the branches on the path leading to that particular event. For example, the multiplication law for independent events can be applied to find the probability that all three items are defective:

$$\Pr[D_1 \ and \ D_2 \ and \ D_3] = \Pr[D_1] \times \Pr[D_2] \times \Pr[D_3]$$

$$= (.05)(.05)(.05)$$

$$= .000125$$

The other outcome probabilities shown in Figure 5-9 were determined in the same way.

Sampling With and Without Replacement

Let us slightly modify our example. Suppose that a population of 100 items, of which 5% are defective, is made on the same machine. A three-item sample is randomly selected and set aside, and the quality of each item is determined.

The probability tree diagram of this experiment is provided in Figure 5-10. The results of each selection are *dependent*, because with each selection the composition of the remaining items changes. The proportion of defective items remaining increases or decreases, depending on the quality of the prior selection. Thus, if the first item proves to be defective, so that D_1 occurs, then there are only 4 defectives left out of 99 remaining items, and the probability that the second item is defective D_2 is 4/99. But if G_1 is the first event, then D_2 has a probability of 5/99, because any one of five remaining defective items could be chosen. These values are *conditional probabilities*, because D_2 has a different chance of occurring in each case. The general multiplication law allows us to

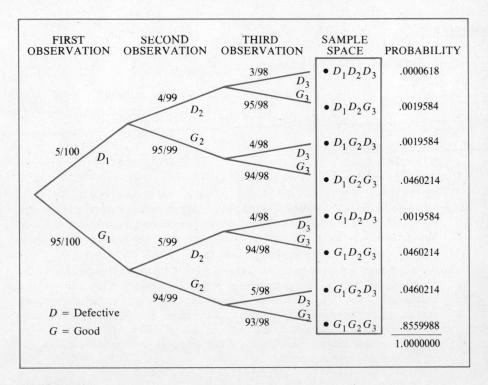

FIGURE 5-10

Probability tree diagram for selecting a quality-control sample of three items without replacement from a population of size 100.

find the probability for getting all defectives:

$$\Pr[D_1 \text{ and } D_2 \text{ and } D_3] = \Pr[D_1] \times \Pr[D_2|D_1] \times \Pr[D_3|D_1 \text{ and } D_2]$$

$$= \left(\frac{5}{100}\right)\left(\frac{4}{99}\right)\left(\frac{3}{98}\right)$$

$$= .0000618$$

Note that this result differs from the one obtained earlier.

When the population units are set aside after each selection, the sampling is done *without replacement*. If, instead, each inspected item were replaced in the group and again allowed the same chance of being chosen as any of the other items, then the outcomes would have identical probabilities for each observation, as in Figure 5-9. Such a procedure is called *sampling with replacement*. Although intuitively wasteful, sampling with replacement makes for simpler probability calculations. More will be said about this in later chapters.

EXERCISES

5-24 The following probability values are given:

$$\Pr[A_2|A_1] = .3 \qquad \Pr[A_1] = .6 \qquad \Pr[A_2] = .4$$
$$\Pr[A_3|A_2] = .4 \qquad \Pr[A_3|A_1 \text{ and } A_2] = .3$$

Use the general multiplication law to find
(a) $\Pr[A_1 \text{ and } A_2]$.
(b) $\Pr[A_2 \text{ and } A_3]$.
(c) $\Pr[A_1 \text{ and } A_2 \text{ and } A_3]$.

5-25 *Matching-birthmonths problem.* For simplicity, suppose that all months of the year are equally likely to be a person's birthmonth. Several people are comparing their birthmonths. Find the probability for at least one matching birthmonth when this group is made up of (a) two people; (b) three people; (c) five people.

5-26 An employment screening test is being evaluated by a company. Historically, 80% of all persons hired by the company have proved to be satisfactory (S); the rest have been unsatisfactory (U). During this evaluation, the screening examination scores are not used to make the final hiring decision. Althougher, 16% of the applicants score low (L) on the test; the remaining scores are high (H). However, only 10% of the satisfactory employees receive a low score. One employee is selected at random.
(a) For that employee, determine:
   (1) $\Pr[S]$ .88 (2) $\Pr[L]$ .16 (3) $\Pr[L|S]$ .10
(b) Construct a joint probability table for the performance quality and the test score of the selected employee.

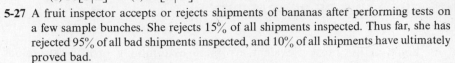

(c) Use your answer to (b) to find the following probabilities:
    (1) $\Pr[S|L]$ (2) $\Pr[L|U]$

5-27 A fruit inspector accepts or rejects shipments of bananas after performing tests on a few sample bunches. She rejects 15% of all shipments inspected. Thus far, she has rejected 95% of all bad shipments inspected, and 10% of all shipments have ultimately proved bad.

(a) Using the above experience as a basis, find the values for the probabilities regarding the outcome of any particular shipment handled by this particular inspector:
    $\Pr[\text{reject}]$ $\Pr[\text{bad}]$ $\Pr[\text{reject}|\text{bad}]$

(b) Find $\Pr[\text{reject } and \text{ bad}]$, using the general multiplication law with the appropriate values found in part (a).

(c) Construct a joint probability table showing the joint probabilities and the marginal probabilities for the inspector's actions (accept or reject) and the quality (good or bad) of the banana shipment.

5-28 Of the ball bearings made by a certain process, 10% are overweight. A random sample of three items is selected. Construct a probability tree diagram for this situation. Find each of the following probabilities describing the final results of the sampling procedure:

(a) No overweight items are selected.

(b) All the items selected are overweight.

(c) Exactly one overweight ball bearing is selected.

5-29 Repeat Exercise 5-28 if the successively weighed items are not replaced.

5-30 An experiment is conducted using three boxes, each containing a mixture of 10 red (R) and white (W) marbles. The three boxes have the following compositions:

Box A	Box B	Box C
6R	4R	7R
4W	6W	3W

Two marbles are selected at random. The first is selected from Box A. If it is red, the second marble is to be taken from Box B, and otherwise Box C. We let R_1 and W_1 represent the color of the first marble and R_2 and W_2 the color of the second.

(a) Construct a probability tree diagram for this experiment.

(b) Read from the branches of your tree the following probabilities:
    (1) $\Pr[R_1]$ (3) $\Pr[R_2|R_1]$ (5) $\Pr[W_2|R_1]$
    (2) $\Pr[W_1]$ (4) $\Pr[R_2|W_1]$ (6) $\Pr[W_2|W_1]$

(c) Apply the general multiplication law to your answers to part (b) to determine the following joint probabilities:
    (1) $\Pr[R_1 \text{ and } R_2]$ (3) $\Pr[W_1 \text{ and } R_2]$
    (2) $\Pr[R_1 \text{ and } W_2]$ (4) $\Pr[W_1 \text{ and } W_2]$

(d) Use the addition law to determine the probabilities for these events:
    (1) One red and one white marble will be chosen.
    (2) Either two red or two white marbles will be chosen.

5-6   COMMON ERRORS IN APPLYING THE LAWS OF PROBABILITY

Some of the most prevalent errors in determining probability values result from the improper use of the laws of probability. Four common mistakes are listed here.

1. Using the addition law to find the probability for the union of several events when they are *not* mutually exclusive.

EXAMPLE: NATURAL DISASTERS

Casualty insurance underwriters have established these probabilities for a city experiencing one of the following natural disasters in the next decade:

Tornado	.5
Flood	.3
Earthquake	.4

We cannot say that the probability for suffering one of these acts is .5 + .3 + .4 = 1.2. Two or more of these disasters may occur over a ten-year period, and some may occur more than once.

2. Using the addition law when the multiplication law should be used, and vice versa. Remember that *or* signifies addition and that *and* signifies multiplication.

   For example, the probability for drawing a red face card is the same as the probability for the event "red *and* face card." Recall that

$$\Pr[\text{red}] = \frac{26}{52} \quad \text{and} \quad \Pr[\text{face card}] = \frac{12}{52}$$

If we add these values to find the joint probability, we obtain a meaningless result:

$$\frac{26}{52} + \frac{12}{52} = \frac{38}{52}$$

Since "red" and "face card" are independent events,

$$\Pr[\text{red } and \text{ face card}] = \frac{26}{52} \times \frac{12}{52} = \frac{6}{52}$$

3. Using the multiplication law for independent component events when the events are dependent.
4. Improperly identifying the complement of an event. For example, the complement of "none" is "some," which may be expressed as "one or more" or "at least one."

The following example, which actually happened, dramatically illustrates how ludicrous results may be obtained by applying the probability laws incorrectly.*

EXAMPLE: WRONGLY CONVICTED BY PROBABILITY

An elderly woman was mugged in the suburb of a large city. A couple was convicted of the crime, although the evidence was largely circumstantial. The multiplication law was used to demonstrate the extremely low probability that a couple fitting the description of the muggers could have committed the crime. The events (the characteristics witnesses ascribed to the couple) are listed below, along with their assumed probabilities:

Characteristic Event	Assumed Probability
Drives yellow car	1/10
Interracial couple	1/1,000
Blond girl	1/4
Girl wears hair in ponytail	1/10
Man bearded	1/10
Man black	1/3

These values were multiplied to obtain the probability that any specific couple, chosen at random from the city's population, would have all six characteristics:

$$\frac{1}{10} \times \frac{1}{1,000} \times \frac{1}{4} \times \frac{1}{10} \times \frac{1}{10} \times \frac{1}{3} = \frac{1}{12,000,000}$$

Since the defendants exhibited all six characteristics and the jury was mystified by the overwhelming strength of the probability argument, they were convicted.

* For a detailed discussion, see "Trial by Mathematics," *Time* (April 26, 1968), p. 41.

The Supreme Court of the state heard the appeal of one of the defendants. The defense attorneys, after obtaining some good advice on probability theory, attacked the prosecution's analysis on two points: (1) the rather dubiously assumed probability values for the events, and (2) the invalid assumption of independence implicit in using the multiplication law in this manner (examples: the proportion of black men having beards may be greater than the proportion of the population as a whole having beards; "interracial couple" and "man black" are not independent events). The judge accepted the arguments of the defense and noted that the trial evidence was misleading on another score: A high probability that the defendants were the *only* such couple should have been determined to demonstrate a strong case. Using the prosecution's original figures and its assumptions of independence, it can be demonstrated that the probability was large that at least one other couple in the area had the same characteristics.

REVIEW EXERCISES

5-31 There are ten applicants for the position of personnel director. Some characteristics of the candidates are provided in the following table:

Name	Age	College Graduate	Marital Status	Previous Experience
Mr. Braun	28	no	single	yes
Mrs. Charles	37	no	married	yes
Mr. Feeley	42	no	single	no
Mr. Gordon	53	yes	single	yes
Ms. Kish	28	yes	married	no
Mr. Lambert	35	no	married	yes
Mr. Minsky	45	yes	married	no
Miss Olivera	33	yes	single	yes
Mr. Snyder	39	no	single	yes
Mr. Wasserman	35	no	married	yes

The file of one applicant is chosen at random. Indicate the elementary events of the event set and the probability for each of the following events:
(a) college graduate
(b) older than 35
(c) same age as another applicant
(d) no previous experience
(e) married

5-32 Referring to the data given in Exercise 5-31, use the basic definition to find the probabilities for the following compound events:
(a) college graduate *and* older than 35
(b) single *and* experienced
(c) college graduate *or* older than 35
(d) man *or* older than 35

5-33 A quality-control inspector accepts only 5% of all bad items and rejects only 1% of all good items. Overall production quality of items is such that only 90% are good.
(a) Using the above percentages as probabilities for the next item inspected, find the following:
(1) $\Pr[\text{accept}|\text{bad}]$ (2) $\Pr[\text{reject}|\text{good}]$ (3) $\Pr[\text{good}]$

(b) Determine the values missing in the following joint probability table:

Quality	Inspector Action		Marginal Probability
	Accept	Reject	
Good			
Bad			
Marginal Probability			

(c) What is the probability that the inspector will accept or reject the next item incorrectly?

5-34 The following probability data are given:

$$\Pr[A] = 1/3 \qquad \Pr[B] = 1/2 \qquad \Pr[C] = 1/4$$

$$\Pr[A \ and \ B] = 1/8 \qquad \Pr[B|C] = 1/3 \qquad \Pr[C|A] = 1/5$$

$$\Pr[A|B \ and \ C] = 1/2 \qquad \Pr[A \ or \ B \ or \ C] = 17/20$$

(a) Determine the following probabilities:
   (1) $\Pr[A \ and \ C]$ (6) $\Pr[C|B]$
   (2) $\Pr[B \ and \ C]$ (7) $\Pr[B \ and \ C|A]$
   (3) $\Pr[A \ and \ B \ and \ C]$ (8) $\Pr[A \ or \ B]$
   (4) $\Pr[A|B]$ (9) $\Pr[A \ or \ C]$
   (5) $\Pr[A|C]$ (10) $\Pr[B \ or \ C]$
(b) Answer the following:
   (1) Are the events in any letter pair mutually exclusive?
   (2) Do A, B, and C form a collectively exhaustive collection of events?
   (3) Which event pairs are *not* independent?

5-35 A pinochle deck consists of 48 cards, all above 8 in denomination. The deck includes aces. A pinochle deck can be compiled from two decks of ordinary playing cards by setting aside the denominations 2 through 8. Suppose that this is done with one regular "Bee" deck and one standard "Bicycle" deck.

(a) Draw a sketch of the sample space for one card taken from the shuffled deck. Use a dot to represent each card in the pinochle deck, and make your sketch similar to the one on page 97. (*Hint:* Use two columns for each suit.)
(b) On your sketch, circle the group of elementary events in the following event sets, and identify each set beside its corresponding area: king, club, face card, and nine. Then determine the probabilities for these events.

5-36 A train detection device for an automated rail network is 99.9% reliable. That is, the device detects a stalled train between stations 99.9% of the time, and it indicates that no train is present 99.9% of the time when there is not one. The probability that any particular departing train will stall before reaching the next station is .005.

(a) Find the probability that a hazardous situation will arise and a stalled train will go undetected.

(b) How many such situations does the law of large numbers imply that we can expect in one million station departures?

5-37 Applicants to a business school are 70% male. Regardless of their sex, 40% are married. Use the multiplication law to find the probability for each of the following characteristics for a particular applicant:
(a) man *and* married   (c) man *and* unmarried
(b) woman *and* married   (d) woman *and* unmarried

5-38 The following probabilities apply to the number of patients waiting in a hospital emergency room at any specific time:

Number Waiting	Probability
0	.34
1	.28
2	.13
3	.10
4	.08
5 or more	.07

Find the probabilities for the following compound events:
(a) 2 or fewer waiting   (c) at least one patient waiting
(b) 4 or more waiting

5-39 A fair coin is tossed three times. Construct the probability tree diagram for this situation. (The events of interest are "head" and "tail.")

5-40 A box contains four marbles, each of a different color: red (R), yellow (Y), white (W), and blue (B). Three marbles are selected randomly without replacement from the box, one at a time, and set aside.
(a) Construct a three-stage probability tree diagram for this situation. Each fork must contain a branch for every possible color. (Remember, there is one less marble possible at each successive stage.) Label each branch with the corresponding color letter (subscripted with a 1, 2, or 3, depending on whether it is the first, second, or third marble drawn) and the probability value for that branch.
(b) Determine the joint probability corresponding to each end position.
(c) Applying the addition law to the values obtained in (b), find the probability for each of the following composite events:
(1) The red marble is selected before the yellow marble.
(2) The blue marble is not selected.
(3) Both the white and the yellow marbles are chosen.
(4) The red marble is selected on the first draw and the white marble on the last draw.

5-7   OPTIONAL TOPIC: COUNTING TECHNIQUES

Sample spaces can be so large that we cannot list all the elementary events. For example, there are about 400 trillion trillion ways to sequence the 26 letters of the alphabet. Probabilities for events involving a huge number of possibilities may be calculated using the techniques presented here. These methods enable us to take shortcuts by counting possibilities in large multiples.

Underlying Principle of Multiplication

Car manufacturing provides a good example of how we might take shortcuts. Each order received by an automobile assembly plant is quite complex; there are so many requested options that it is extremely rare for two independently ordered cars to be just alike. In counting possibilities, it is convenient to view the car-ordering process in several stages. First is the specification of color, then the size of the engine, followed by body style (such as sedan or convertible), and so on. If we know the number of choices in each stage, we can find the total number of possibilities by *multiplying* those numbers together.

EXAMPLE: POSSIBLE WARDROBE SELECTIONS

Consider the number of possible ways that a man can choose what to wear. Suppose that he has 5 suits, 8 shirts, 3 belts, 15 pairs of socks, 6 pairs of shoes, and 20 ties. We may break this situation into stages, as shown in Figure 5-11. The total number

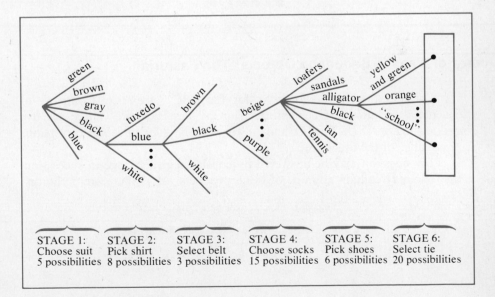

STAGE 1:   STAGE 2:   STAGE 3:   STAGE 4:   STAGE 5:   STAGE 6:
Choose suit   Pick shirt   Select belt   Choose socks   Pick shoes   Select tie
5 possibilities 8 possibilities 3 possibilities 15 possibilities 6 possibilities 20 possibilities

FIGURE 5-11
Diagram of possible wardrobe selections.

of possibilities is found by multiplying together the number of choices at each stage:

$$5 \times 8 \times 3 \times 15 \times 6 \times 20 = 216,000$$

A complete listing of all the possibilities is not necessary in arriving at a total count. (And such a list would require all the space in a large book.)

EXAMPLE: NUMBER OF STRAIGHTS IN POKER

The five top cards are removed from a deck of 52 ordinary playing cards. We will eventually determine the probability that these cards are in a straight denominational sequence. The sample space comprises all possible equally likely five-card results, and we want to know how many of these are straights. We could list the possibilities, but instead we can arrive at the total by multiplying the number of possibilities for each card and for the denominations making up a straight.

Consider the lowest card. It can be a deuce, a three, or any other successive value up to a 10 (aces counting as high). There are *nine* denominational possibilities for the lowest card. For any such straight there are four suit possibilities for each card, so that a 2-3-4-5-6 straight could occur in any of

$$4 \times 4 \times 4 \times 4 \times 4 = 4^5, \text{ or } 1,024$$

different ways. It follows that the total number of five-card straights is

$$9 \times 4^5 = 9,216$$

Number of Ways to Sequence Objects: The Factorial

Consider the number of ways for the ABC Typewriter Company to sequence descriptions of new products in a sales brochure. There are five products to appear (*A*, *B*, *C*, *D*, and *E*). How many different sequence versions are possible for the brochure?

The question may be answered by considering each page in succession. Page one may contain any one of the 5 products—thus there are 5 choices;

Possible Choices of A, B, C, D, and E

Page One	Page Two	Choices for Page Three
A	B	C, D, E
A	C	B, D, E
A	D	B, C, E
A	E	B, C, D
B	A	C, D, E
⋮	⋮	⋮

choose one of them. This will then leave 4 products available for page two—4 choices; again, one of these is to be selected. For the first two pages there have been 20 possible choices, partially listed above (you may verify to your satisfaction that there are 20 by completing the list).

Suppose that product A is chosen for page one, product B for page two. This leaves products C, D, and E to be assigned to pages three through five. The following 6 choices are possible:

Page Three	Page Four	Page Fire
C	D	E
C	E	D
D	C	E
D	E	C
E	C	D
E	D	C

A similar list of equal length may be prepared for any of the other 19 assignments to the first two pages; all such lists would contain 6 entries. For the brochure there are thus 20 × 6 or 120 different product-to-page sequences possible. This figure could also have been obtained from the following multiplication:

5	×	4	×	3	×	2	×	1
number of choices for first page		number of choices for second page		number of choices for third page		number of choices for fourth page		number of choices for fifth page

A multiplication of this type is called a *factorial product*, or simply a factorial, and is denoted by an exclamation point placed after the highest number. Thus, for 5 factorial,

$$5! = 5 \times 4 \times 3 \times 2 \times 1 = 120$$

In general, n factorial ($n!$) may be determined by the

FACTORIAL PRODUCT

$$n! = n \times (n-1) \times (n-2) \times \cdots \times 2 \times 1$$

We define

$$1! = 1$$

$$0! = 1$$

The fact that 0! is 1 is usually a perplexing notion. If you think of 0! as representing the number of different sequences of assigning no (0) products to pages in a brochure with no (0) pages, you may observe that there is only one way of doing this (that is, there is only one way to sequence nothing).

Factorial values literally become astronomically large for modest values of n. For instance,

$$10! = 3,628,800$$

$$20! = \text{approximately 2.4 billion billion}$$

$$100! = \text{approximately 9 followed by 157 zeros}$$

The Number of Permutations

It is sometimes useful to determine the number of ways in which one may select a number of objects from a given group, considering the *order of selection*. Suppose we have n objects from which r are to be removed. The number of ways in which it is possible to select the r objects may be determined from the following product:

$$\underbrace{n}_{\substack{\text{number of} \\ \text{choices for} \\ \text{first object}}} \times \underbrace{(n-1)}_{\substack{\text{number of} \\ \text{choices for} \\ \text{second object}}} \times \underbrace{(n-2)}_{\substack{\text{number of} \\ \text{choices for} \\ \text{third object}}} \times \cdots \times \underbrace{(n-r+1)}_{\substack{\text{number of} \\ \text{choices for} \\ r\text{th object}}}$$

In the sales brochure example, $r = 5$ new-product descriptions were selected one at a time from $n = 5$ for inclusion in the brochure. There, each possibility differed only as to *sequence*, so that $n = r$. But when r is smaller than n (say we took three new-product descriptions for the brochure), each possible group of selections would also differ with respect to *which particular objects were chosen*. Each such possibility is called a *permutation*.

If we multiply the above product by $(n - r)!/(n - r)!$, it can be expressed more simply:

$$\frac{n(n-1)\cdots(n-r+1)(n-r)!}{(n-r)!} = \frac{n!}{(n-r)!}$$

This enables us to make the following statement:

NUMBER OF PERMUTATIONS
The number of possible permutations of r objects from a collection of size n, denoted by P_r^n, is

$$P_r^n = \frac{n!}{(n-r)!}$$

For the number of permutations of $r = 3$ items out of $n = 5$ in our brochure example,

$$P_3^5 = \frac{5!}{(5-3)!} = \frac{5!}{2!} = 5 \times 4 \times 3 = 60$$

When all objects in the collection are taken,

$$P_n^n = \frac{n!}{(n-n)!} = \frac{n!}{0!} = n!$$

so that in this case the number of permutations is the same as the number of ways to sequence n objects.

The Number of Combinations

The number of permutations P_r^n considers not only which r objects are selected but also their sequence of arrangement. Sometimes we do not care about the order of selection and merely wish to determine the number of possible *combinations* of objects.

Suppose that due to a budgetary constraint, ABC Typewriter Company must restrict the number of pages in its sales brochure to three of the five products. How many combinations of 3 objects out of 5 are there?

Each combination may be listed, observing that there are 10:

ABC	*ABE*	*ACD*	*BCD*	*BDE*
ABD	*ACE*	*ADE*	*BCE*	*CDE*

Imagine choosing the items one at a time. We have 5 choices for the first product. This leaves 4 choices for the second. Only 3 products remain for our final choice. The following multiplication thus describes the number of choices:

$$5 \times 4 \times 3$$

But the result of the above product is 60, not 10. The same situations have been accounted for several times each. For instance, the combination ABC has been accounted for in all of the following sequences:

ABC   *BAC*   *CAB*   *ACB*   *BCA*   *CBA*

These differ only in the order in which the items appear, as the same 3 products are involved in all cases. However, the order in which an item is chosen is not now of interest, only whether it is selected. The redundant accounting may be corrected by dividing the earlier product by the number of sequences in which the 3 items ultimately chosen could have been selected: 3!, or $3 \times 2 \times 1 = 6$:

$$\frac{5 \times 4 \times 3}{3!}$$

The fraction in the above example may be transformed into a version that will later prove useful in analyzing other situations. Multiplying the numerator and denominator by 2! (thereby leaving the ratio's value unchanged) yields

$$\frac{(5 \times 4 \times 3) \times 2!}{3! \times 2!}$$

which may be written as

$$\frac{5 \times 4 \times 3 \times 2 \times 1}{3! \times 2!} = \frac{5!}{3! \times 2!}$$

Another representation is

$$\frac{5!}{3! \times (5 - 3)!}$$

From this final representation may be inferred an easily generalized form for the

NUMBER OF COMBINATIONS
The number of combinations of r objects taken from n objects, denoted by C_r^n, may be determined from

$$C_r^n = \frac{n!}{r!(n - r)!}$$

Note that this expression is similar to the one for the number of permutations. However, permutations are distinguished by order, whereas combinations are not. Since there are $r!$ ways to order (or sequence) r objects, the number of combinations is the number of permutations divided by $r!$, so that

$$C_r^n = \frac{P_r^n}{r!}$$

TABLE 5-5
Number of Combinations
of r Objects Taken from n

n	r	$C_r^n = \dfrac{n!}{r!(n-r)!}$
6	3	20
8	5	56
12	7	792
20	5	15,504
100	50	$\approx$ 100,000 trillion trillion

As the expression for the number of combinations is composed of factorials, the numbers may be quite large. Table 5-5 provides some examples of combination sizes.

___

EXAMPLE: NUMBER OF POKER HANDS

A poker hand consists of five cards dealt from a deck of 52 playing cards. The number of combinations of $r = 5$ cards taken from $n = 52$ cards is

$$C_5^{52} = \frac{52!}{5!(52-5)!}$$

In evaluating this type of fraction, do not try to calculate the denominator and numerator separately, because the numbers become far too large to handle easily and the amount of work required is unwarranted. Instead, use the following procedure.

First simplify the above fraction by observing that $(52 - 5)! = 47!$, and then express $52!$ as a product of terms ending with $47!$:

$$\frac{52!}{5!\,47!} = \frac{52 \times 51 \times 50 \times 49 \times 48 \times 47!}{5!\,47!}$$

Cancel the $47!$ terms and factor $5!$:

$$\frac{52 \times 51 \times 50 \times 49 \times 48}{5 \times 4 \times 3 \times 2 \times 1}$$

Further cancellation yields

$$\frac{52 \times 51 \times \overset{10}{\cancel{50}} \times 49 \times \overset{\overset{4}{\cancel{12}}}{\cancel{48}}}{\cancel{5} \times \cancel{4} \times \cancel{3} \times \cancel{2} \times 1} = 52 \times 51 \times 5 \times 49 \times 4 = 2,598,960$$

The resulting figure is the number of poker hand combinations.

To find the probability for getting a straight, we can divide our earlier result for the number of straights by the above figure. We obtain

$$\text{Pr[straight]} = \frac{9 \times 4^5}{\dfrac{52!}{5!(52-5)!}} = \frac{9{,}216}{2{,}598{,}960} \approx \frac{1}{282}$$

After a large number of attempts at drawing five cards from a deck, you would, on the average, expect to encounter a straight only once in 282 tries. That is, the odds against drawing a five-card straight are about 281 to 1.

OPTIONAL EXERCISES

5-41 A General Motors plant manager once remarked, "I've never seen two cars exactly alike come off our assembly line." This is because the number of possible options is so large and the likelihood that two persons will order exactly the same car is so very small.

As an example, consider the abbreviated illustration below, where various options for a hypothetical car model are considered.

Option	Number of Possibilities
Body type (four-door, two-door)	4
Exterior color	8
Interior color	6
Engine size	4
Transmission	3
Suspension	2
Power accessory combinations	4
Sound system combinations	10
Air conditioning	2
Tire type/size	10
Cosmetic combinations	20
Window types	3
Automatic auxiliary features	6

How many possible different cars might be made?

5-42 Calculate the following:

(a) $6!$ (b) $\dfrac{14!}{3!11!}$ (c) $8!$ (d) $\dfrac{7!}{4!3!} \times \dfrac{11!}{7!2!}$ (e) $\dfrac{37!}{4!(37-4)!}$

(*Hint* for (e): Express the numerator as a product of numbers and 33!; then cancel.)

5-43 The following famous nursery rhyme contains a counting problem.

As I was going to St. Ives,
I met a man with seven wives.

Every wife had seven sacks,
  Every sack had seven cats,
Every cat had seven kits,
  Kits, cats, sacks, and wives,
    How many were going to St. Ives?

Find (1) the number of cats and (2) the number of kits encountered on the journey.

5-44 A cafeteria has 3 types of salad (carrot, tossed, and bean); 4 entrees (ham, chop suey, meat loaf, and tacos); 3 vegetables (corn, peas, and stewed tomatoes); 6 drinks (cola, iced tea, hot tea, coffee, lemonade, and orange juice); and 4 desserts (ice cream, apple cobbler, applesauce, and yogurt). For two dollars, you may have 1 salad, 2 different entrees, 2 different vegetables, 1 drink, and 2 different desserts.
   (a) How many ways are there to obtain 2 different entrees? (*Hint:* Ignore the order of selection.)
   (b) How many ways are there to obtain 2 different vegetables?
   (c) How many ways are there to obtain 2 different desserts?
   (d) How many combinations of two-dollar dinners are possible when the full quota of items is selected and when there are no duplications? (*Hint:* Do not attempt to diagram the outcomes.)

5-45 A traveling salesman must visit ten cities in one trip. In how many sequences may he make his stops?

5-46 If the top 13 cards are drawn from a deck of 52 ordinary playing cards that has been fully shuffled, they are the equivalent of a bridge hand. How many possible hands are there when the order in which they are obtained does not matter? How many hands are possible when the order does matter? (Answers in terms of factorials are sufficient.)

5-47 Determine the number of possible occurrences for each of the following situations:
   (a) A coin is tossed ten times.
   (b) Six dice are tossed simultaneously.
   (c) One die is tossed six times.
   (d) Six objects are selected from ten distinguishable items (order of selection is not considered).

5-48 Seating assignments are being made for the guests at an awards banquet. There are 20 guests (ten couples) to be seated along a table with ten chairs on each side. A name card is to appear at every place setting.
   (a) In how many different ways is it possible to place the name cards?
   (b) If no man is to sit aside or across from another man, how many arrangements are possible?
   (c) If each couple is to sit side by side, how many arrangements are possible when (1) the sexes are alternated, and (2) members of the same sex may sit beside each other?

5-49 Have you ever wondered why a royal flush beats four of a kind in straight five-card poker? Recall that a royal flush is an A, K, Q, J, 10 straight of cards in the same suit, whereas four of a kind consists of four cards of the same denomination plus any fifth card.
   Determine the number of five-card hands possible (ignoring the order in which a hand is filled) for obtaining a royal flush and for obtaining four of a kind. Since the rarer hand wins, which one is the best?

5-8   OPTIONAL TOPIC: REVISING PROBABILITIES USING BAYES' THEOREM

In this section, we will introduce a procedure whereby probabilities can be revised when new information is obtained. Revising probabilities is a familiar concept. For example, think of how many times you have left home in the morning with no raincoat, only to look up and notice a menacing cloud cover that sends you back for some protection in case it rains. On first charging outdoors, you judged the probability for rain to be small. But the presence of clouds caused you to revise this probability significantly upward.

We need to be able to revise probabilities so that we can make better use of experimental information. We can accomplish this by applying a fundamental principle that follows immediately from the laws of probability we developed earlier in this chapter. This principle is referred to as *Bayes' Theorem*, which is named after the Reverend Thomas Bayes, who proposed in the eighteenth century that probabilities be revised in accordance with empirical findings.

Most information is not conclusive. Any empirical test can camouflage the truth. For instance, some potentially good students will perform poorly on college entrance examinations, and some poor students will do well. A good example of how such information may be unreliable is illustrated by the geologist's seismic test. A seismic survey can deny the presence of oil in a field where it is already being produced and can confirm the presence of oil under a site already proved dry. Still, such imperfect findings can be valuable. An unfavorable test result can increase the chance of rejecting a poor prospect (college applicant or drilling lease), and a favorable result can enhance the likelihood of selecting a good one.

The information obtained will affect the probabilities for the events that determine the consequences of each act. We can revise the probabilities for these events upward or downward, depending on the evidence we obtain. Thus, the geologist increases the probability for finding oil after obtaining a favorable seismic survey analysis and decreases this probability after obtaining an unfavorable survey.

Bayes' Theorem

Consider a situation for which two uncertain events "E" and "not E" are possible. Suppose that $\Pr[E]$ and $\Pr[\text{not } E]$ have been obtained. These are referred to as *prior probabilities*, because they represent the chances of the events occurring that have been determined *before* the results of the empirical investigation have been obtained. The investigation itself results in several possible outcomes, each statistically dependent on E. For any particular result, denoted by the letter R, the conditional probabilities $\Pr[R|E]$ and $\Pr[R|\text{not } E]$ are often available. The result itself serves to revise the probabilities for "E" and "not E" upward or downward. The values obtained are called *posterior probabilities*, since they apply *after* the outcome has been determined.

The posterior probability values are actually conditional probabilities of the form $\Pr[E|R]$, $\Pr[\text{not } E|R]$ that may be found according to

BAYES' THEOREM
The posterior probability for event E for a particular result R of an empirical investigation can be found from

$$\Pr[E|R] = \frac{\Pr[E]\Pr[R|E]}{\Pr[E]\Pr[R|E] + \Pr[\text{not } E]\Pr[R|\text{not } E]}$$

The principle underlying Bayes' Theorem is best explained in terms of the following example.

Crooked Die Cube Illustration

A box contains four fair dice and one crooked die containing a leaded weight that makes the 6-face appear on two-thirds of all tosses. You are asked to select one die at random and toss it. If the crooked die cannot be distinguished from the fair dice and the result of your toss is a 6-face, what is the probability that you tossed the crooked die?

The events in question are

$$C = \text{crooked die}$$

$$\text{not } C = \text{fair die}$$

The empirical investigation here is the toss itself, so

$$R = \text{6-face}$$

Since 1 out of 5 dice is crooked, the prior probabilities for the type of die tossed are

$$\Pr[C] = 1/5$$

$$\Pr[\text{not } C] = 4/5$$

You know that when the crooked die is tossed,

$$\Pr[R|C] = 2/3$$

If the fair die is tossed, you know that one of the equally likely sides has a 6-face, and

$$\Pr[R \,|\, \text{not } C] = 1/6$$

The posterior probability that the die you tossed is crooked is therefore

$$\Pr[C \,|\, R] = \frac{\Pr[C]\Pr[R \,|\, C]}{\Pr[C]\Pr[R \,|\, C] + \Pr[\text{not } C]\Pr[R \,|\, \text{not } C]}$$

$$= \frac{(1/5)(2/3)}{(1/5)(2/3) + (4/5)(1/6)}$$

$$= \frac{2/15}{4/15} = \frac{1}{2}$$

We can see that the probability for tossing the crooked die must be revised upward from the prior value of 1/5 (which applied when you had no information) to the posterior value of 1/2 now that you know the toss resulted in a 6-face.

Posterior Probability as a Conditional Probability

Although it has a special interpretation, a posterior probability is merely a conditional probability when some relevant result is given, and it can be found in the same manner:

$$\begin{matrix} \text{Posterior} \\ \text{probability} \\ \text{of event} \end{matrix} = \Pr[\text{event} \,|\, \text{result}] = \frac{\Pr[\text{event } and \text{ result}]}{\Pr[\text{result}]}$$

A straightforward procedure for calculating an event's posterior probability is first to find the joint probability that the event will occur with the given result and then to divide by the probability for that result. This is exactly what Bayes' Theorem accomplishes. The numerator is the joint probability found from applying the general multiplication law. The denominator is the probability for obtaining the particular empirical result and is basically the sum of all the joint probabilities for the potential outcomes that might yield that result. In practice, Bayes' Theorem can be cumbersome to use, and the data may be provided in such a way that the posterior probabilities can be found more directly.

For example, if a statistics class contains just as many men as women, the prior probability that an examination paper chosen at random will belong to a man (M) is 1/2. Now suppose that after the exams were graded, 20% of the papers received a mark of "C or better" (C) *and* were written by men; a total of 60% of the exams were scored "C or better." If a randomly selected test sheet was graded "C," we have sufficient information to calculate the poster probability that it was written by a man:

$$\Pr[M \mid C] = \frac{\Pr[M \text{ and } C]}{\Pr[C]} = \frac{.20}{.60} = \frac{1}{3}$$

Here, the information about the grade given to the paper causes us to revise downward the probability that it was written by a man.

Typically, the probability values needed to make such a simple calculation are not immediately available. When we use evidence or empirical results to revise probabilities, our knowledge of the various events involved is usually structured so that some preliminary work is required to obtain the necessary probability values. It may be helpful to construct a joint probability table to accomplish this.

Posterior Probabilities in Jury Selection

A noted lawyer specializing in defending corporate clients in personal-injury suits thinks that a sympathetic jury is half the battle. A jury panel can be winnowed down to a largely sympathetic group by preemptory challenges. Since a potential juror's leaning in a particular case is usually concealed during the selection interview, superficial characteristics must be relied on in accepting or rejecting jury candidates. The lawyer has found that mature and stable persons, those who have "made it on their own," are most likely to be sympathetic to the defendant and that younger jurors tend to have a "social worker attitude" and to favor the plaintiff. From post-trial talks with jurors over the years, the lawyer has determined that 65% of the sympathetic jurors have been older persons. A special bar study has shown that only 30% of all jury panel members in that county are sympathetic to the defendant in a personal-injury suit.

Eleven jurors have been chosen in a negligence suit resulting from an elevator accident, and both lawyers have exhausted their challenges. Six of the potential jurors are younger and four are older. Based on alphabetical sequence, one of the ten will be the last juror.

The joint probability table concerning this last juror appears in Table 5-6. The probability values obtained directly from the data provided here are shown in black type. The numbers in color were obtained from these data after first noting that

$$\Pr[U] = 1 - \Pr[S] = .70$$

The fact that we have been given the conditional probability

$$\Pr[O \mid S] = .65$$

TABLE 5-6
Joint Probability Table Used to Illustrate Posterior Probability Calculation

	Older (O)	Younger (Y)	Marginal Probability
Sympathetic (S)	.195	.105	.300
Unsympathetic (U)	.205	.495	.700
Marginal Probability	.400	.600	1.000

enables us to find the joint probability for obtaining an older, more sympathetic juror:

$$\Pr[S \text{ and } O] = \Pr[S]\Pr[O|S] = .30(.65) = .195$$

The remaining values can be obtained from knowing this value and from the fact that the joint probabilities must sum to the respective marginal values.

The prior probability that a sympathetic juror will be obtained is only .30 (which is also the marginal probability of this event). If an older juror is chosen, the posterior probability that the juror will also be sympathetic is

$$\Pr[S|O] = \frac{\Pr[S \text{ and } O]}{\Pr[O]} = \frac{.195}{.40} = .4875$$

This example shows how we can calculate posterior probabilities by relying on basic concepts, instead of using the complicated expression of Bayes' Theorem. Nevertheless, as Table 5-6 shows, essentially the same steps are required in either case.

OPTIONAL EXERCISES

5-50 A new movie "Star Struck" has a prior probability for success of .20. Ruth Grist is going to review the film. She has liked 70% of all the successful films and has disliked 80% of all the unsuccessful films she has reviewed. Find the posterior probability that "Star Struck" will be a success if (a) Grist likes it; (b) Grist dislikes it.

5-51 A local-television weather reporter makes a daily forecast indicating the probability that it will rain tomorrow. On one particular evening, she announces an 80% chance of rain (E) the next day. The manager of the city golf courses has established a policy that he will water the greens only if the probability for rain is less than 90%. Using the local TV forecast as his prior probability, the manager also relies on his mother-in-law's rheumatism: Historically, she gets a "rain pain" (R) on 90% of all days that are followed by rain, but she also gets a pain on 20% of the days that are not followed by rain. The following probabilities therefore apply:

$$\Pr[E] = .80 \qquad \Pr[R|E] = .90 \qquad \Pr[R|\text{not } E] = .20$$

(a) Assuming that the golf-course manager's mother-in-law is currently receiving pain signals, find the posterior probability that it will rain tomorrow. Should the manager water the greens?

(b) If the manager's mother-in-law feels just fine, what is the posterior probability for rain tomorrow?

5-52 From a given response to a question, a marketing researcher wishes to determine whether a randomly selected person will choose BriDent when next purchasing toothpaste. The question is designed to reveal whether the selected person recalls the name BriDent—an event we will denote by R. Previous testing has established that 99% of the people who bought BriDent previously recalled the name and that only 10% of the people who did not buy BriDent recalled this particular brand name. Since BriDent has cornered 30% of the toothpaste market, the researcher chooses .30 as the prior probability that the person selected will buy BriDent. Denoting this event by B gives us the following probabilities:

$$\Pr[B] = .30 \qquad \Pr[R|B] = .99 \qquad \Pr[R|\text{not } B] = .10$$

(a) If the person who is selected remembers BriDent, what is the posterior probability that BriDent will be purchased next?

(b) If the person who is selected does *not* remember BriDent, what is the posterior probability that BriDent will be purchased next?

5-53 An oil wildcatter has assigned a probability of .50 to striking oil on her property. She orders a seismic survey that has proved to be only 80% reliable in the past: Given oil, it predicts favorably 80% of the time; given no oil, it augurs unfavorably with a frequency of .8.

(a) Given a favorable seismic result, what is the probability for oil?

(b) Given an unfavorable seismic result, what is the probability for oil?

5-54 An employment screening test is being evaluated for possible inclusion in clerical-services hiring decisions. Presently, only 50% of the persons hired for these positions perform satisfactorily. The test itself has been evaluated by outside consultants, who have given it an upside reliability of 90% (90% of all satisfactory employees will pass the test) and a downside reliability of only 80% (80% of all unsatisfactory employees will fail the test). One clerical applicant (acceptable in all other screening activities) is chosen at random. Find the following probabilities:

(a) The prior probability for satisfactory on-the-job performance if the applicant is hired.

(b) The posterior probability for satisfactory on-the-job performance if the applicant passes the screening test.

(c) The posterior probability for satisfactory on-the-job performance if the applicant is hired after failing the screening test.

The curve described by a simple molecule of air vapor is regulated in a manner just as certain as the planetary orbits: the only difference between them is that which comes from our ignorance.

MARQUIS DE LAPLACE (1820)

6

Probability Distributions, Expected Value, and Sampling

This chapter focuses on *deductive statistics:* finding probabilities for obtaining particular sample results when the population values are known. As an illustration, we will determine the possible sample outcomes for a population of Graduate Management Admissions Test scores. Studying how random samples are generated and learning what to expect when working with *known* populations will lay the essential groundwork for evaluating *unknown* populations.

The average test score, or sample mean, is an important value to use in establishing admissions policies. When its value is uncertain (the case *before* the sample is taken), the sample mean is referred to as a *random variable*. The values of a random variable are treated as uncertain events that occur with a set of probabilities called a *probability distribution*.

We can summarize the information in a probability distribution in terms of central tendency and variability. The "average" figure for a random variable is referred to as its *expected value*. We use expected values to compare probability distributions. In sampling, they enable us to analyze the kinds of results that we might obtain.

6-1   DISCRETE PROBABILITY DISTRIBUTIONS

The Random Variable

Let us consider a manager who must decide which projects to include in next year's budget. One project may be to expand the capacity of the plant by purchasing new equipment. Various measures, such as cost or profit, may be used to compare the attractiveness of this purchase with competing projects. One common gauge is the investment's *rate of return.**

The actual rate of return achieved will depend on the particular levels and timing of cash receipts and expenditures attributable to the new equipment. Except for the initial outlay, all of these lie in the future and are therefore *uncertain.*

There are many possible rate-of-return outcomes from investing in the equipment. Each outcome consists of many groupings of potential cash flows, several of which may yield the same rate of return. A rate-of-return figure may be calculated for each, using the principles of compound interest. Since it may assume any one of a number of possible values, we may view the rate of return as a *variable.*

DEFINITION
A random variable is a numerical quantity whose value is determined by chance.

Thus, the rate of return to be achieved from the proposed equipment is a random variable.

A random variable must assume numerical values. For example, the possible outcomes of a coin toss (head and tail) are non-numerical. We can associate a random variable with coin tossing only by assigning numbers to these outcomes, such as in a wager where a head results in winning $1 and a tail results in losing $1. In this case, the winnings (not the side showing) would be the random variable, with possible values of $1 and −$1. An expected value of the random variable can be obtained by averaging. We can determine an average level of winnings from the coin toss, but there is no way we can average the head and the tail, because they are non-numerical.

---

* Rate of return is similar to compound interest on savings. A rate of return of 6% on an investment means that the same profit could be achieved by placing equivalent funds, depositing all proceeds, in a savings account paying 6% compound interest.

TABLE 6-1
Probability Distribution for Rates of Return

Possible Rate of Return	Probability
10%	.05
11	.10
12	.15
13	.17
14	.12
15	.08
16	.09
17	.06
18	.05
19	.05
20	.04
21	.04
Total	1.00

The Probability Distribution of a Random Variable

Random variables are fundamental to statistical theory. The relationship between a random variable's values and their probabilities is summarized by the *probability distribution.*

DEFINITION
The probability distribution provides a probability for each possible value of a random variable.

Many probability distributions can be expressed in terms of a table like Table 6-1. There, the probabilities sum to 1, because the values are mutually exclusive and collectively exhaustive.

Not all probability distributions are expressed in a table. Sometimes they are described by an algebraic formula, because—as the following example demonstrates—there are too many possibilities to list in a table.

EXAMPLE: A RAFFLE

A raffle is to be conducted with 10,000 tickets, numbered consecutively from 1 through 10,000. One ticket is to be selected from a barrel after thorough mixing. Because each ticket is equally likely to be selected, the probability that a particular ticket will be chosen is 1/10,000. If we define a random variable for the raffle as the number of the selected ticket, to be denoted by X, we can shorten the statement "The number of the selected ticket will be 7,777" to read

$$X = 7,777$$

Thus, we can express the probability for this event as

$$\Pr[X = 7,777] = \frac{1}{10,000}$$

Because there are 10,000 possible outcomes, all having the same probability, we can express the probability distribution more concisely by using a statement of the form

$$\Pr[X = x] = \frac{1}{10,000} \quad \text{for values of } x = 1, 2, \ldots, 10,000$$

which is equivalent to duplicating the previous probability expression 10,000 times, once for each ticket number. The lowercase x is a "dummy" variable, representing one particular ticket number.

It is important to keep in mind that the capital letter X represents the number of the ticket that *will be* selected and has no determined value until the raffle is completed. The lowercase letter x is a stand-in for *each* ticket number to avoid 10,000 repetitive statements. We will find that this shorthand representation can be of great value.

Discrete and Continuous Random Variables

Special difficulties arise when random variables do not assume a discrete number of possible values.

EXAMPLE: NO BALL BEARING IS EXACT

For use in large motors, precision 1-inch-diameter ball bearings must be machined to within a tolerance of .01 of an inch. Each bearing is assumed to be so nearly perfectly round that no measuring device can perceive otherwise.

Each bearing sold has been inspected at uniform temperature by using a pair of "go–no go" gauges (stands with holes through which a bearing being tested is dropped). The diameter of one hole is 1.01 inches, the other is .99 inch. Each bearing must be small enough to fall through the large gauge but not small enough to fall through the small gauge.

To determine how many of the 1 million bearings produced annually are precisely 1 inch in diameter, a more accurate measuring system must be established.

A pair of gauges could determine if bearings were within a hundred-thousandth (.00001) of an inch of being exactly 1 inch in diameter, but only about 1,000 ball bearings would pass this test (assuming nearly equal frequency for all values between .99 and 1.01 inches). Bearings that passed this second stage of testing could be measured again with gauge blocks accurate to within a millionth of an inch. About 100 bearings would pass the test. Optical tests would filter out the majority of these bearings, leaving a handful to be measured even more accurately in a fourth stage. But how many would pass the fourth test? Would greater precision of measurement eliminate all of the bearings?

We cannot be certain whether one or more of the bearings would be precisely 1 inch in diameter, as the standard inch is presently defined and within our capability to measure it. But we can conclude that such ball bearings would be extremely rare. Consider the random selection of one bearing. Taking the diameter of the ball bearing as our random variable, we can conclude that the probability it will be exactly 1 inch is so small as to be considered zero for all practical purposes.

The random variable illustrated in this example can assume any value on a continuous scale. Such a random variable is therefore given the

DEFINITION
A continuous random variable is a variable that may assume any numerical value on a continuous scale.

In contrast to the continuous random variable, another class of random variables, called *discrete random variables*, may assume a finite or countable number of numerical values.

The dichotomy of discrete versus continuous random variables is important in probability theory because different mathematical procedures are used to describe the probability distributions of each. However, continuous and discrete random variables share many properties, and the more significant ones attributable to both can be explored using discrete examples.

EXERCISES

6-1 Describe a circumstance that would create a random variable from
  (a) tomorrow's weather.
  (b) the sex of an unborn child.
  (c) the winner of the World Series.

6-2 The point spread from tossing two fair dice is the difference between the number of dots showing on the top faces of the cubes. Determine the probability distribution for this random variable. (You may refer to your solution to Exercise 5-3 on page 102 to obtain the probabilities for the possible dice sums.)

6-3 From the following probability distributions for the receipts and expenses of a charity carnival, determine the probability distribution for the net proceeds (receipts − expenses). Levels for receipts are independent of expenditures.

Receipts	Probability	Expenses	Probability
$30,000	1/3	$30,000	1/3
40,000	1/3	40,000	1/3
50,000	1/3	50,000	1/3
	1		1

6-4 "Craps" is a favorite gambling game in which a pair of six-sided dice are tossed. One way to place a bet is to "play the field," where the bettor places a bet with a complicated payoff, depending on which faces of the dice show. If a "field" number (defined by a sum value of 2 through 4 or 9 through 12) occurs, the player wins. If the roll of dice yields any other total, the player loses. A field gamble is further complicated by varying payoffs: 1 to 1 on all field numbers except 2 or 12; 2 to 1 on a 2; and 3 to 1 on a 12. A winning bettor keeps the original bet and is also paid winnings. A losing bettor forfeits the original bet.

(a) For a bet of $1, find the probability distribution for W, the gambler's net winnings from one field bet in craps. (Refer to your solution to Exercise 5-3 to obtain the probabilities for the possible dice sums.)

(b) Solve (a) for a bet of $2.

6-2   EXPECTED VALUE AND VARIANCE

The Expected Value of a Random Variable

A probability distribution is similar to the frequency distribution of a quantitative population because both provide a long-run frequency for each possible outcome. In Chapter 3, we saw that the population mean is a desirable summary for comparing populations or for making decisions about a population. It is also useful to find the average of the random variable values to be achieved from repeated circumstances. Because the outcomes are in the future, the average result is called an *expected value*.

In Chapter 3, we learned that an average can be calculated in several different ways. The computation that we will use to determine the expected value is analogous to the computation used to find the mean of grouped data.

DEFINITION
The expected value of a discrete random variable X, denoted by $\mu(X)$, is the weighted average of that variable's possible values, where the respective probabilities are used as weights.

TABLE 6-2
Expected Value Calculation for Die Toss Outcome

(1) Possible Value x	(2) Probability Pr[X = x]	(3) Weighted Value (1) × (2) = x Pr[X = x]
1	1/6	1/6
2	1/6	2/6
3	1/6	3/6
4	1/6	4/6
5	1/6	5/6
6	1/6	6/6
	6/6 = 1	$\mu(X) = 21/6 = 3.5$

To illustrate how to calculate the expected value of a random variable, let X represent the number of dots on the face showing after a six-sided die is tossed. The probability distribution for X is provided in the first two columns of Table 6-2. Multiplying each possible result by its corresponding probability produces a weighted value. Summing these results gives the expected value $\mu(X) = 3.5$. On the average, the number of dots obtained for the showing die faces in a large number of tosses will be 3.5. Because it is the average value achieved by the random variable X, we sometimes refer to $\mu(X)$ as the *mean of* X.

The expected value has many uses. In a gambling game, it tells us what our long-run average losses per play will be. Sophisticated gamblers know that slot machines pay poorly in relation to the actual odds and that the average loss per play is less in roulette or dice games. A mathematician, Edward Thorp, caused quite a stir in the early 1960s when he demonstrated that various betting strategies in playing the card game of blackjack result in positive expected winnings.*

The Variance of a Random Variable

Just as the expected value of a random variable is analogous to the weighted mean, the variability of random variables may be measured in much the same way as the variability in a population or a sample. The measure we will consider is the variance. Like the mean, the variance of a random variable represents the same thing that it does for the population: the average of the squared deviations from the mean or expected value.

* *See* Edward Thorp, *Beat the Dealer*, Revised Edition (New York: Random House, 1966). Unlike other gambling games, blackjack allows bets to be placed when the odds are in a player's favor. This is because the card deck may not be reshuffled after each stage of play. By significantly raising bets at these times, a player will make a profit on the average.

TABLE 6-3

Variance Calculation for Die Toss

(1) Possible Values of X x	(2) Probability Pr[X = x]	(3) Deviation $x - \mu(X)$	(4) Squared Deviation $[x - \mu(X)]^2$	(5) Weighted Value [(2) × (4)] $[x - \mu(X)]^2 \Pr[X = x]$
1	1/6 = .167	−2.5	6.25	1.044
2	.167	−1.5	2.25	.376
3	.167	−.5	.25	.042
4	.167	.5	.25	.042
5	.167	1.5	2.25	.376
6	.167	2.5	6.25	1.044
				$\sigma^2(X) = 2.924$

DEFINITION

The variance of a random variable X, denoted by $\sigma^2(X)$, is the average of the squared deviations from the expected value calculated using probability weights.

As an example, again consider the roll of a six-sided die. Table 6-3 shows how the variance $\sigma^2(X) = 2.924$ is computed for the number of dots on the top face.

The *standard deviation* of a random variable X, which we will abbreviate $\sigma(X)$, will have the same meaning as that of the population. It is found by taking the square root of the variance.

TABLE 6-4

Frequency Distribution for the Number of Children in 1,000 Families in a Town

(1) Number of Children X	(2) Number of Families f	(3) fX	(4) X^2	(5) fX^2
0	63	0	0	0
1	185	185	1	185
2	223	446	4	892
3	207	621	9	1,863
4	153	612	16	2,448
5	87	435	25	2,175
6	53	318	36	1,908
7	21	147	49	1.029
8	5	40	64	320
9	1	9	81	81
10	2	20	100	200
Totals	1,000	2,833		11,101

Expected Value and Sampling

RELATION
The value of an elementary unit chosen randomly from a quantitative population is a random variable. Its expected value equals the population mean and its variance equals the population variance.

To illustrate this fact, consider the frequency distribution for the number of children in 1,000 families, presented in Table 6-4. The mean number of children μ and the variance σ^2 may be calculated:

$$\mu = \frac{\sum fX}{N} = \frac{2,833}{1,000} = 2.833$$

$$\sigma^2 = \frac{\sum fX^2 - N\mu^2}{N} = \frac{11,101 - 1,000(2.833)^2}{1,000} = 3.075$$

(We use N rather than $n - 1$ as the divisor for σ^2 because the data pertain to the entire *population*.)

One family is chosen at random. The probability distribution for the number of children in this family may be obtained from the population frequency distribution in Table 6-4. For example, to find the probability that $X = 2$, we divide the number of two-child families by 1,000: $\Pr[X = 2] = 223/1,000 = .223$. The probability distribution is provided in Table 6-5, where

TABLE 6-5
Probability Distribution for the Number of Children in a Randomly Selected Family

(1) Possible Number of Children x	(2) Pr[X = x]	(3) x Pr[X = x]	(4) x − μ(X)	(5) [x − μ(X)]²	(6) [x − μ(X)]² Pr[X = x]
0	.063	.000	−2.833	8.026	.506
1	.185	.185	−1.833	3.360	.621
2	.223	.446	−.833	.694	.155
3	.207	.621	.167	.028	.006
4	.153	.612	1.167	1.362	.208
5	.087	.435	2.167	4.696	.409
6	.053	.318	3.167	10.030	.531
7	.021	.147	4.167	17.364	.365
8	.005	.040	5.167	26.698	.133
9	.001	.009	6.167	38.032	.038
10	.002	.020	7.167	51.366	.103
Totals	1.000	μ(X) = 2.833			σ²(X) = 3.075

$\mu(X)$ and $\sigma^2(X)$ are calculated. Note how similar these calculations are to those used earlier to find the population parameters. We use the probabilities in the same manner as class frequencies, and they have corresponding values. We find that $\mu(X) = 2.833$ and $\sigma^2(X) = 3.075$; these are the same values as the corresponding population parameters, so that $\mu(X) = \mu$ and $\sigma^2(X) = \sigma^2$.

EXERCISES

6-5 A coin is tossed twice. For each tail, you must forfeit $1, and for each head you will receive $2. Determine the probability distribution for your net winnings W. (*Hint:* It will help to determine which outcomes—for example, $H_1 T_2$—apply to each possible amount.) Then calculate $\mu(W)$.

6-6 In roulette there are 38 equally likely slots that a ball may drop into at random: 18 are red, 18 are black, and 2 are green. Players win an amount equal to their bet if the ball falls into a slot of their chosen color, but they lose their wagers otherwise. Determine the probability distribution for the winnings W of a player who bets $1 on red. Calculate $\mu(W)$ and explain the meaning of your answer.

6-7 Following is the probability distribution for the number of requests a credit bureau receives daily for credit verifications:

Number of Requests	Probability
0	.1
1	.2
2	.3
3	.2
4	.1
5	.1

Find the expected daily number of verification requests.

6-8 The following probability distribution applies for the number of customer arrivals during any given minute at a supermarket.

Number of Arrivals	Probability
0	.1
1	.2
2	.3
3	.3
4	.1

Find the expected value and variance of the number of arrivals.

6-9 In investigating the impact of diet on the litter size of a particular breed of rabbit, the following probability distributions apply:

Standard Diet			Special Diet	
Litter Size	*Probability*		*Litter Size*	*Probability*
5	.1		5	.0
6	.2		6	.1
7	.2		7	.2
8	.3		8	.3
9	.1		9	.3
10	.1		10	.1

(a) Find the expected litter sizes using each diet.

(b) Assume the diet that is expected to produce the greatest number of offspring per year will be chosen. Which diet will be used if the expected number of litters per year is 4 using standard diet and 3.5 using the special diet? (*Hint:* Assume that the number of litters per year is unaffected by the sizes of the individual litters.)

6-10 An investor who wishes to buy a stock to be held for one year in anticipation of capital gain has narrowed her choice to High-Volatility Engineering or Stability Power. Both stocks currently sell for $100 per share and yield $5 dividends. The following probability distributions for next year's price have been judgmentally assessed for each stock, where S_1 = selling price of High-Volatility Engineering and S_2 = selling price of Stability Power:

High-Volatility Engineering			Stability Power	
s	$Pr[S_1 = s]$		*s*	$Pr[S_2 = s]$
$ 25	.05		$ 95	.10
50	.07		100	.25
75	.10		105	.50
100	.05		110	.15
125	.10			1.00
150	.15			
175	.12			
200	.10			
225	.12			
250	.14			
	1.00			

(a) Determine the expected prices for a share of each stock.

(b) Should the investor select the stock with the highest expected value? Discuss.

THE SAMPLING DISTRIBUTION OF THE MEAN   6-3

We are now ready to make the transition from probability theory to statistical planning and analysis. For now, we will assume that many details regarding the population are known, and we will study the kinds of *possible* results that may be obtained for samples taken from the population. Such a deductive viewpoint is not usually encountered in a real sampling situation, because the sample results are known and the population details are unknown. But if we know how sample results are generated and what values are likely,

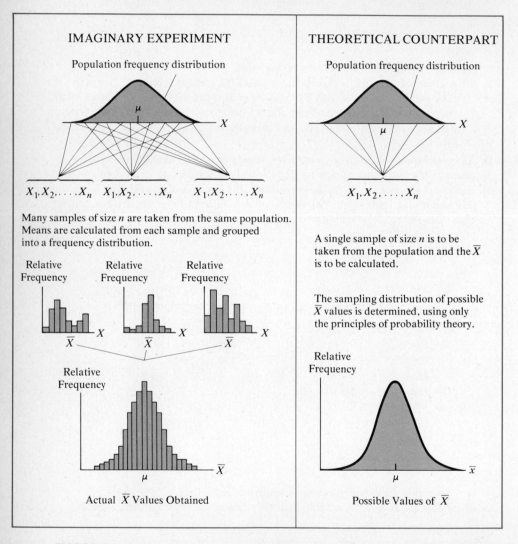

IMAGINARY EXPERIMENT

Population frequency distribution

Many samples of size n are taken from the same population. Means are calculated from each sample and grouped into a frequency distribution.

Actual $\overline{X}$ Values Obtained

THEORETICAL COUNTERPART

Population frequency distribution

A single sample of size n is to be taken from the population and the $\overline{X}$ is to be calculated.

The sampling distribution of possible $\overline{X}$ values is determined, using only the principles of probability theory.

Possible Values of $\overline{X}$

FIGURE 6-1

Interpretation of a theoretically derived sampling distribution. The procedure outlined at the left is never necessary but shows the concepts underlying the theory. The histogram at bottom left is an experimental representation of the sampling distribution at bottom right.

we can easily learn how to generalize about an unknown population from a known sample.

Estimating the population mean μ is an important inference. Usually, we would employ the corresponding sample mean $\bar{X}$ to make this estimate.

In the planning stage of a sampling study, before the data are collected, we can speak of $\bar{X}$ only in terms of probability. Its value is not determined and will depend on which particular elementary units happen to be randomly selected. Thus, before the sample results are obtained, $\bar{X}$ is a random variable. We emphasize that at one stage $\bar{X}$ is viewed as a random variable and later as a statistic whose value can be calculated from observed data. The probability distribution of $\bar{X}$ is of key importance. It is called a *sampling distribution*.

The basic concepts underlying the sampling distribution of $\bar{X}$ are illustrated in Figure 6-1. This distribution may be viewed as roughly analogous to the results that would be obtained if many different samples were taken and the sample mean $\bar{X}$ were calculated each time. In practice only one sample will ever be taken. Probability concepts are used to establish the sampling distribution of $\bar{X}$, represented in the lower right-hand corner of Figure 6-1.

We next illustrate how probability concepts can be extended to finding the sampling distribution of the mean. As a matter of convenience, a tiny population is considered.

Illustration: GMAT Scores

Table 6-6 shows the population of Graduate Management Admissions Test (GMAT) scores of four hypothetical Slippery Rock University seniors applying to Harvard. Suppose that the Harvard Business School will admit

TABLE 6-6
Hypothetical Population of GMAT Scores

Name	GMAT Score
Chen	600
Jones	500
O'Hara	700
Sandor	600

$$\mu = \frac{600 + 500 + 700 + 600}{4} = 600$$

$$\sigma = \sqrt{\frac{(600 - 600)^2 + (500 - 600)^2 + (700 - 600)^2 + (600 - 600)^2}{4}}$$

$$= \sqrt{5,000} = 70.71$$

TABLE 6-7

Possible Sample Results for Randomly Selecting
Two GMAT Scores Without Replacement

Applicants Selected	GMAT Scores	Sample Mean $\bar{x}$
Chen, Jones	600,500	550
Chen, O'Hara	600,700	650
Chen, Sandor	600,600	600
Jones, O'Hara	500,700	600
Jones, Sandor	500,600	550
O'Hara, Sandor	700,600	650

TABLE 6-8

Sampling Distribution of $\bar{X}$ for GMAT Scores
Selected Without Replacement

Possible Value $\bar{x}$	Probability $\Pr[\bar{X} = \bar{x}]$	Weighted Value $\bar{x}\Pr[\bar{X} = \bar{x}]$
550	1/3	550/3
600	1/3	600/3
650	1/3	650/3
	1	$\mu(\bar{X}) = 600$

exactly two of these applicants by drawing their names from a hat. The scores of the two chosen will be a random sample from the population of four.

Table 6-7 shows the six possible sample results and the corresponding values of the sample mean. We can easily determine the probabilities for each possible value. For example, the mean GMAT score for two potential samples is 600, so the probability of this result is $2/6 = 1/3$. The sampling distribution of $\bar{X}$ is shown in Table 6-8.

The procedure illustrated here serves primarily to establish the conceptual framework for understanding sampling distributions. Populations ordinarily involve too many units for us to catalogue possibilities in the above manner. Chapter 7 provides a more direct method for establishing the sampling distribution of $\bar{X}$.

Expected Value of the Sample Mean

The expected value of $\bar{X}$ is calculated in Table 6-8. Note that $\mu(\bar{X}) = 600$, which is equal to the population mean μ. In general, this is true for any random sampling situation, so we have the following relation for the

EXPECTED VALUE OF $\bar{X}$

$$\mu(\bar{X}) = \mu$$

This conclusion is quite plausible. In effect, this expression says that the long-run average value of sample means is the same as the mean of the population from which the sample observations are taken. The rationale for this is simple. If $\bar{X}$ is calculated over and over for different samples taken from the same population, the successive sample means tend to cluster about the population mean. If a sample of 100 men is taken from a population where the mean height is $\mu = 5'10''$, the sample mean $\bar{X}$ is expected to be $5'10''$. This does not imply that $\bar{X}$ cannot be $5'9\frac{1}{2}''$, but *on the average* $\bar{X}$ will equal $5'10''$.

Sampling With and Without Replacement

Our business-school admissions illustration is an example of *sampling without replacement:* Once an elementary unit (an applicant) is chosen for the sample, it has no chance of being chosen again. *As a practical matter, statistics almost always involves sampling without replacement.* However, to help us understand some essential concepts, we will also consider *sampling with replacement,* in which every elementary unit has an equal chance of being selected for each successive observation. In this case some population units may be represented more than once in the sample. Table 6-9 shows the possible sample results for randomly selecting two of the four GMAT scores when sampling with replacement. Here there are $4 \times 4 = 16$ equally likely outcomes.

TABLE 6-9
Possible Sample Results for Randomly Selecting Two GMAT Scores
with Replacement

Applicants Selected	GMAT Scores	Sample Mean $\bar{x}$	Applicants Selected	GMAT Scores	Sample Mean $\bar{x}$
Chen, Chen	600,600	600	O'Hara, Chen	700,600	650
Chen, Jones	600,500	550	O'Hara, Jones	700,500	600
Chen, O'Hara	600,700	650	O'Hara, O'Hara	700,700	700
Chen, Sandor	600,600	600	O'Hara, Sandor	700,600	650
Jones, Chen	500,600	550	Sandor, Chen	600,600	600
Jones, Jones	500,500	500	Sandor, Jones	600,500	550
Jones, O'Hara	500,700	600	Sandor, O'Hara	600,700	650
Jones, Sandor	500,600	550	Sandor, Sandor	600,600	600

TABLE 6-10

Sampling Distribution of $\bar{X}$ for GMAT Scores Selected with Replacement

$\bar{x}$	$\Pr[\bar{X} = \bar{x}]$
500	1/16
550	4/16
600	6/16
650	4/16
700	1/16
	1

Sample outcomes are statistically independent when sampling with replacement and statistically dependent when sampling without replacement. Whether or not sample outcomes are independent influences many statistical procedures that we will encounter later. For example, consider the population of the heights of all adult American males. If the height of the first man randomly selected for the sample is 7', independence does not permit this height to influence the probability distribution for the height of any other man selected. But if the selection is made without replacement, choosing a man 7' tall first will influence any probability for the second man's height.

No practical difficulty exists unless the population is small. For instance, if our population contains 1,000 men, exactly one of whom is over 7' tall, then independence is violated when sampling without replacement. For a very large population, the degree of dependence will be so slight that it can be ignored. For the United States as a whole, the removal of one seven-footer from the sample would not appreciably change the remaining proportion of persons who are 7' tall, but in a small population it could seriously affect the sampling outcome.

TABLE 6-11

Calculation of Mean and Standard Deviation of $\bar{X}$ for GMAT Scores Selected with Replacement

$\bar{x}$	$\Pr[\bar{X} = \bar{x}]$	$\bar{x}\,\Pr[\bar{X} = \bar{x}]$	$\bar{x} - \mu(\bar{X})$	$[\bar{x} - \mu(\bar{X})]^2$	$[\bar{x} - \mu(\bar{X})]^2\Pr[\bar{X} = \bar{x}]$
500	1/16	500/16	-100	10,000	10,000/16
550	4/16	2,200/16	-50	2,500	10,000/16
600	6/16	3,600/16	0	0	0
650	4/16	2,600/16	50	2,500	10,000/16
700	1/16	700/16	100	10,000	10,000/16
		$\mu(\bar{X}) = 600$			$\sigma^2(\bar{X}) = 2,500$

$$\sigma_{\bar{X}} = \sigma(\bar{X}) = \sqrt{2,500} = 50$$

Standard Error of the Sample Mean

The sampling distribution of $\bar{X}$ is summarized in Table 6-10. The expected value and the standard deviation of $\bar{X}$ are calculated in Table 6-11. Note that, as earlier, $\mu(\bar{X}) = 600$. The standard deviation of $\bar{X}$ is $\sigma(\bar{X}) = 50$. In many statistical applications, the standard deviation of a statistic is referred to as its *standard error*. Because we use it often, the standard error of $\bar{X}$ is represented by the special symbol $\sigma_{\bar{x}}$. Here, $\sigma_{\bar{x}} = 50$.

We have found that $\mu(\bar{X}) = \mu$. Similarly, $\sigma_{\bar{x}}$ relates to the population standard deviation. *When sampling with replacement or from large populations,* we can use the following expression to calculate the

STANDARD ERROR OF $\bar{X}$

$$\sigma_{\bar{X}} = \frac{\sigma}{\sqrt{n}}$$

Both the population standard deviation σ and sample size n influence the level of $\sigma_{\bar{x}}$. In Table 6-6 we found that the population of GMAT scores had a standard deviation of $\sigma = 70.71$. In this illustration, $n = 2$, so that

$$\sigma_{\bar{X}} = \frac{70.71}{\sqrt{2}} = \frac{70.71}{1.414} = 50$$

which is the same value of $\sigma_{\bar{x}}$ calculated in Table 6-11 directly from the sampling distribution.

When a sample is taken without replacement from a small population, $\sigma_{\bar{x}}$ does not equal $\sigma/\sqrt{n}$. In such cases, a canceling effect arises from the potential early selection of extremely large or small population values. In Chapter 7, we will learn how to correct for this.

EXERCISES

6-11 A sample of size 2 is to be randomly selected *without replacement* from five persons who have the following monthly incomes:

Identity	Income
Mr. A	$1,000
Mrs. D	1,200
Mr. J	1,200
Ms. P	1,000
Miss R	900

Find the sampling distribution of the sample mean.

6-12 Assume that the sample in Exercise 6-11 is to be chosen *with replacement*.
   (a) Find the sampling distribution of the sample mean.
   (b) Calculate the expected value, variance, and standard deviation of the sample mean.

6-13 The number of children in each of five families is given below:

Family	Number of Children
Chavez	2
Luke	5
Markowitz	3
Rogers	4
Williams	1

A sample of size 2 is to be randomly selected *without replacement*. Find the sampling distribution of the sample mean.

6-14 Repeat Exercise 6-12 using the data in Exercise 6-13.

6-4   BINOMIAL PROBABILITIES: THE SAMPLING DISTRIBUTION OF THE PROPORTION

Until now, we have been considering the sampling distribution of the mean, which involves quantitative data. For qualitative populations, the *sample proportion P* is of primary interest, and at this point we will consider the sampling distribution of *P*. We will begin by describing the *binomial distribution*—one of the most important concepts in statistics. This distribution is concerned with the *number of outcomes* in a particular category.

A Coin-Tossing Illustration

An evenly balanced coin is tossed fairly five times. The corresponding probability tree diagram appears in Figure 6-2, where the sample space is also listed. Our initial problem is to find the probability of obtaining exactly two heads. As each of the 32 outcomes is equally likely, the basic definition of probability allows us to find the answer by counting the number of elementary events involving two heads and dividing this result by the total number of equally likely elementary events. The sample space contains 32 elementary events, and Figure 6-2 shows that 10 of these are two-head outcomes. Thus,

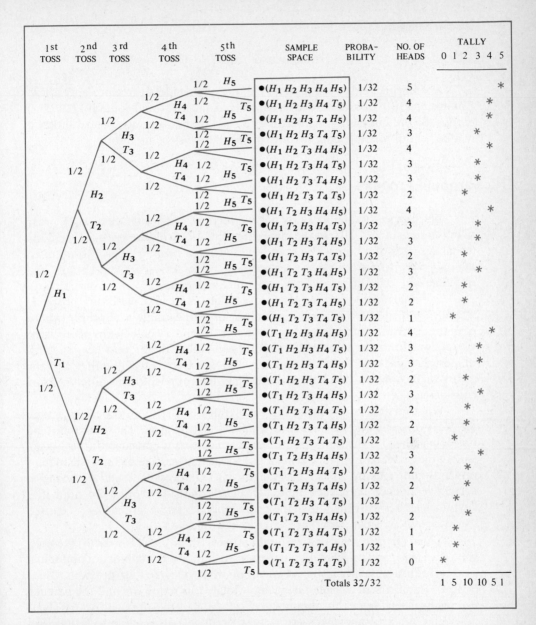

FIGURE 6-2
Probability tree diagram for five tosses of a fair coin.

we can determine that

$$Pr[\text{exactly two heads}] = 10/32$$

It is impractical to list all possible outcomes unless there are only a few. For instance, if 10 tosses were to be considered, the list would contain $1,024\,(2^{10})$ entries. Before discussing a procedure to simplify finding such probabilities, it will be helpful to relate coin tossing to a similar class of situations.

The Bernoulli Process

A sequence of coin tosses is one example of a *Bernoulli process*. A great many circumstances fall into the same category: All involve a series of situations (such as tosses of a coin), which are referred to as *trials*. For each trial, *there are only two possible complementary outcomes*, such as head or tail. Usually one outcome is referred to as a *success*; the other as a *failure*.

Examples include giving birth to a single child in a maternity hospital, where each birth is a trial resulting in a boy or girl; canning a vegetable, where each trial is a full can that is slightly overweight or underweight (cans of precisely correct weight are so improbable that we can ignore them); and keypunching numerical data, where each completed card is a trial that will either contain errors or be correct. In all these cases, only two opposite trial outcomes are considered.

What further distinguishes these situations as Bernoulli processes is that *the success probability remains constant* from trial to trial. The probability of obtaining a head is the same, regardless of which toss is considered; this is also true of the probability of delivering a girl for any successive birth in the maternity hospital, picking up an overweight can of vegetables, and receiving a correctly punched card each time. (The last condition would not hold if a keypuncher tires over time; then the probability would be larger that an earlier card would be correct than that a later card would be.)

A final characteristic of a Bernoulli process is that *successive trial outcomes must be independent events*. Like a fairly tossed coin, the probability of obtaining a success (head) must be independent of what occurred in previous trials (tosses). The births in a *single family* may violate this requirement if the parents use medical techniques to obtain a second child of the opposite sex of their first child. Or a keypuncher's errors may occur in batches due to fatigue, so that once an error is made it is more likely to be followed by another.

Sampling to determine the impact of advertising, voter preference, or response to drug treatment can all be classified as Bernoulli processes. To preserve the requirements of independence and constant probability of success, we must sample with replacement, thereby allowing each person the same chance of being selected each time and perhaps being chosen more than once.* In each case, the probability of a trial success would be the proportion of persons in the respective population who would provide the desired response.

The Number of Combinations

Now we will derive an algebraic expression for computing binomial probabilities. Looking at Figure 6-2 again, we see that 10 elementary events involve exactly two heads:

$$H_1H_2T_3T_4T_5 \qquad T_1H_2H_3T_4T_5 \qquad T_1T_2H_3H_4T_5$$
$$H_1T_2H_3T_4T_5 \qquad T_1H_2T_3H_4T_5 \qquad T_1T_2H_3T_4H_5$$
$$H_1T_2T_3H_4T_5 \qquad T_1H_2T_3T_4H_5 \qquad T_1T_2T_3H_4H_5$$
$$H_1T_2T_3T_4H_5$$

Each of these outcomes represents one path of branches in the probability tree. They differ only in terms of which particular two tosses are heads.

If we want to find out the number of two-head outcomes without constructing an entire probability tree we can determine how many different ways there are to pick the two tosses to be heads from the total of five. It will help if each toss is represented by a fork:

FIRST	SECOND	THIRD	FOURTH	FIFTH
TOSS	TOSS	TOSS	TOSS	TOSS

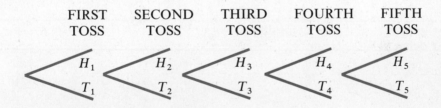

We want to save one branch from each fork. Two of the saved branches will be H's; three will be T's. Our problem is to determine how many ways there are to pick two H branches to save from the five.

If we pick the H branches one at a time, we have 5 possibilities for the first choice. No matter which H branch is chosen first, 4 H branch possibilities remain for the second choice. Multiplying the possibilities gives us

$$5 \times 4 = 20$$

* When sampling without replacement, a different probability distribution, the *hypergeometric distribution* (discussed in Chapter 17), should be used. In practice, when the population is large, the conditions of a Bernoulli process are very nearly met and the binomial distribution is acceptable.

This answer is twice as large as it should be because the order of selection is considered. But we don't care whether an H_3H_5 resulted from H_3 being the first or the second choice, so we divide by 2 to avoid accounting for order of selection:

$$\frac{5 \times 4}{2} = 10$$

This calculation can be used to find the number of *combinations* of items for a variety of situations. It will be helpful if we reexpress this fraction in the equivalent form

$$\frac{5 \times 4}{2} = \frac{5 \times 4}{2 \times 1} = \frac{5 \times 4 \times 3 \times 2 \times 1}{2 \times 1 \times 3 \times 2 \times 1} = 10$$

Multiplying both the numerator and the denominator of the middle fraction by 3, then by 2, and finally by 1 leaves the result unchanged. The final fraction contains factorial terms. A *factorial* is the product of successive integer values ending with 1. Such a product is denoted by placing an exclamation point after the highest number:

$$2! = 2 \times 1 \quad (= 2)$$
$$3! = 3 \times 2 \times 1 \quad (= 6)$$
$$5! = 5 \times 4 \times 3 \times 2 \times 1 \quad (= 120)$$

We define

$$1! = 1$$
$$0! = 1$$

In factorial notation, the number of two-head sequences in five coin tosses is

$$\frac{5 \times 4 \times 3 \times 2 \times 1}{2 \times 1 \times 3 \times 2 \times 1} = \frac{5!}{2!3!} = 10$$

This result suggests the general procedure for finding the

NUMBER OF COMBINATIONS
The number of combinations of r objects taken from n objects may be determined from

$$\frac{n!}{r!(n-r)!}$$

In our illustration, there are $n = 5$ tosses and we are considering exactly $r = 2$ heads occurring in those tosses. Thus, the number of two-head sequence combinations is

$$\frac{5!}{2!(5 - 2)!} = \frac{5!}{2!3!} = 10$$

The Binomial Formula

When the trial outcomes are the results of a Bernoulli process, the number of successes is a random variable having a binomial distribution. The following expression may then be used to find the probability values. It is referred to as the

BINOMIAL FORMULA

$$\Pr[R = r] = \frac{n!}{r!(n - r)!} \pi^r (1 - \pi)^{n - r}$$

where R = number of successes achieved

n = number of trials

π = trial success probability

$r = 0, 1, \ldots, n$

The binomial formula can be used to determine the probability we found earlier for obtaining $r = 2$ heads in $n = 5$ tosses of a fair coin. In this case, $\pi = \Pr[H] = \frac{1}{2}$ and $1 - \pi = \Pr[T] = \frac{1}{2}$, so that

$$\Pr[R = 2] = \frac{5!}{2!(5 - 2)!} \left(\frac{1}{2}\right)^2 \left(1 - \frac{1}{2}\right)^{5 - 2} = \frac{5!}{2!3!} \left(\frac{1}{2}\right)^2 \left(\frac{1}{2}\right)^3$$

$$= 10 \left(\frac{1}{2}\right)^5 = \frac{10}{32}$$

The product involving 1/2 represents the probability of obtaining any one of the 10 two-head sequences shown in the probability tree in Figure 6-2. Each of these positions is reached by traversing a particular path of 2 head and $5 - 2 = 3$ tail branches. The probability of doing this may be obtained by applying the multiplication law. Since a two-head result can occur in any one of 10 equally likely ways, the addition law of probability tells us to add 10 of the identical product terms together or more simply to multiply by 10. The entire binomial distribution for the number of heads is given in Table 6-12.

Labeling one attribute the "success" is a completely arbitrary designation, but we must be sure to use the appropriate value of π. An interesting feature of

TABLE 6-12

Binomial Distribution for the Number of Heads
Obtained in Five Coin Tosses

Possible Number of Heads r	$\Pr[R = r]$
0	$\dfrac{5!}{0!5!}\left(\dfrac{1}{2}\right)^{0}\left(\dfrac{1}{2}\right)^{5} = \dfrac{1}{32} = .03125$
1	$\dfrac{5!}{1!4!}\left(\dfrac{1}{2}\right)^{1}\left(\dfrac{1}{2}\right)^{4} = \dfrac{5}{32} = .15625$
2	$\dfrac{5!}{2!3!}\left(\dfrac{1}{2}\right)^{2}\left(\dfrac{1}{2}\right)^{3} = \dfrac{10}{32} = .31250$
3	$\dfrac{5!}{3!2!}\left(\dfrac{1}{2}\right)^{3}\left(\dfrac{1}{2}\right)^{2} = \dfrac{10}{32} = .31250$
4	$\dfrac{5!}{4!1!}\left(\dfrac{1}{2}\right)^{4}\left(\dfrac{1}{2}\right)^{1} = \dfrac{5}{32} = .15625$
5	$\dfrac{5!}{5!0!}\left(\dfrac{1}{2}\right)^{5}\left(\dfrac{1}{2}\right)^{0} = \dfrac{1}{32} = .03125$
	1.00000

the binomial formula is that it can also be used to obtain the probability that some number of failures will occur. For instance, the probability of obtaining exactly three tails in five tosses of the coin has the same value as the probability of obtaining two heads, because whenever there are two heads there must be three tails. In general, *when there are r successes, there must be n − r failures.*

Different Bernoulli processes will have different binomial probability values. Note that the probabilities for all possible values of R depend on the value of π. Different sizes for n will result in a larger or smaller number of possible values for R and will also affect each probability value. *For purposes of calculating probabilities, one Bernoulli process differs from another only by the values of π and the sizes of n.*

EXAMPLE: AIRCRAFT DELAY PROBABILITY

An airport administrator wishes to determine the number of aircraft departure delays that are attributable to inadequate control facilities. A random sample of 10 aircraft takeoffs is to be investigated. If the true proportion of such delays in all departures is .40, what is the probability that 4 of the sample departures will be delayed due to control inadequacies?

Letting a control-caused delay be a success, the trial success probability is each to the proportion of such outcomes, so that $\pi = .40$. The probability of $R = 4$ control-caused delays (successes), using the binomial formula with $\pi = .40$, $n = 10$, and $r = 4$, is

$$\Pr[R = 4] = \frac{10!}{4!(10 - 4)!} (.4)^4 (1 - .4)^{10-4}$$

$$= 210(.4)^4(.6)^6$$

$$= 210(.0256)(.046656)$$

$$= .2508$$

The Proportion of Successes

Ordinarily, the number of successes is a less useful random variable than the *proportion of successes*. The proportion of control-caused aircraft delays would be more meaningful to the administrator in our example than the number of delays. This proportion could be used to determine the total number of delays for all predicted levels of traffic.

The ratio of the number of successes to the number of trials, denoted by P, is the

PROPORTION OF SUCCESSES

$$P = \frac{R}{n}$$

The probabilities of various possible values of P can also be calculated from the binomial formula:

$$\Pr\left[P = \frac{r}{n}\right] = \frac{n!}{r!(n - r)!} \pi^r(1 - \pi)^{n-r}$$

The probability distribution for the proportion of control-caused aircraft delays is shown in Table 6-13. Each probability value is calculated from the binomial formula, with $n = 10$ and $\pi = .4$.

TABLE 6-13
Binomial Distribution
for the Proportion of Aircraft
Delays ($\pi = .4$)

r	$\dfrac{r}{n} = \dfrac{r}{10}$	$\Pr\left[P = \dfrac{r}{n}\right]$
0	.0	.0060
1	.1	.0404
2	.2	.1209
3	.3	.2150
4	.4	.2508
5	.5	.2007
6	.6	.1114
7	.7	.0425
8	.8	.0106
9	.9	.0016
10	1.0	.0001
		1.0000

The Binomial Distribution Family

To illustrate the concept of a family of binomial distributions, we construct Table 6-14, showing the binomial probability distributions for different trial success probabilities π. Each entry in the table is obtained from the binomial formula, and a sequence of $n = 5$ trials applies in each case. The same possible outcomes exist, but the probabilities for the values of P differ due to the different values of π. When $\pi = .9$, a higher proportion of successes is more likely than

TABLE 6-14
Binomial Probability Distributions with $n = 5$

$\dfrac{r}{5}$	$\Pr\left[P = \dfrac{r}{5}\right]$				
	$\pi = .1$	$\pi = .3$	$\pi = .5$	$\pi = .7$	$\pi = .9$
.0	.59049	.16807	.03125	.00243	.00001
.2	.32805	.36015	.15625	.02835	.00045
.4	.07290	.30870	.31250	.13230	.00810
.6	.00810	.13230	.31250	.30870	.07290
.8	.00045	.02835	.15625	.36015	.32805
1.0	.00001	.00243	.03125	.16807	.59049

when $\pi = .1$. Graphs of these binomial probability distributions, constructed for the values in Table 6-14, appear in Figure 6-3.

The graphs in Figure 6-4 illustrate another interesting feature of the binomial distribution. Here, π is a fixed value (.3) and n varies. Starting with $n = 5$, the number of trials is increased to 20 and then to 100. With n trials, there are $n + 1$ possible values of P (corresponding to 0, 1, 2, . . . , or n successes). For larger n values, we cannot show all of the possibilities as spikes on the graphs, because some of the probabilities are extremely tiny. (For example, when $\pi = .3$ and $n = 20$, the probability of all successes, or $\Pr[P = 1]$, is less than 40 trillionths.)

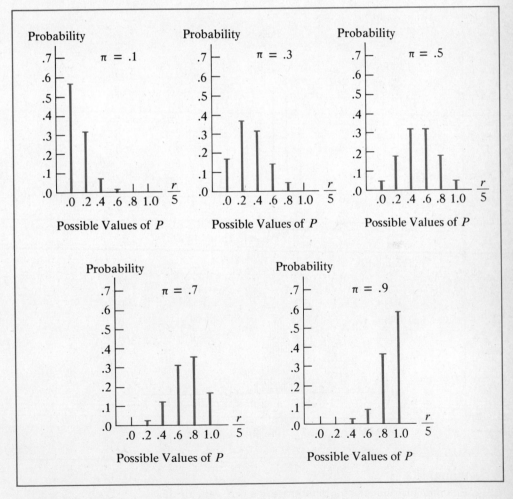

FIGURE 6-3
Binomial probability distributions for several values of π, with $n = 5$.

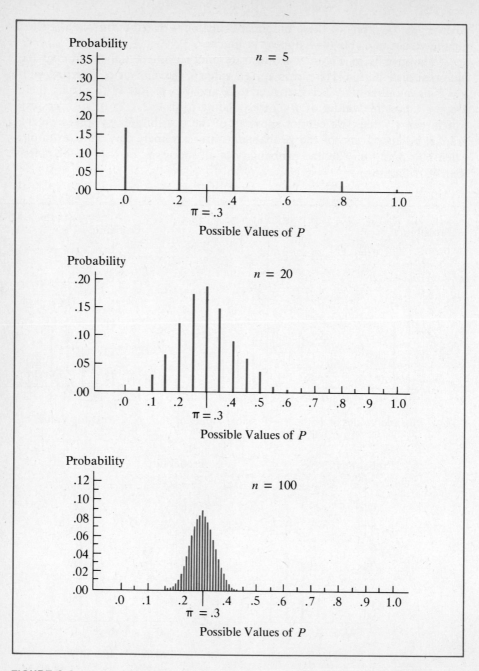

FIGURE 6-4
Binomial distributions for different levels of n, with $\pi = .3$.

Note that the number of spikes increases with n and that the spikes become more closely bunched as n increases. A significant feature is the tendency for the spikes to assume the "bell shape." In Chapter 7, we will learn to use this property to approximate the binomial distribution by the normal distribution.

The Cumulative Probability Distribution

Just as it can be convenient to deal with cumulative frequencies, which are readily determined from the population frequency distribution, we may wish to use *cumulative probabilities* for random variables. These can be simply obtained from the probability distribution for the number R of aircraft delays by creating column (3) for $Pr[R \leq r]$, as shown in Table 6-15. The values in this column are obtained by adding all the preceding entries for the values of $Pr[R = r]$. Thus

$$Pr[R \leq 0] = Pr[R = 0] = .0060$$

and

$$Pr[R \leq 1] = Pr[R = 0] + Pr[R = 1]$$

$$= .0060 + .0404$$

$$= .0464$$

and

$$Pr[R \leq 2] = Pr[R = 0] + Pr[R = 1] + Pr[R = 2]$$

$$= .0060 + .0404 + .1209$$

$$= .1673$$

TABLE 6-15

Cumulative Probability Distribution
for the Number of Aircraft Delays ($\pi = 0.4$)

(1) r	(2) $Pr[R = r]$	(3) $Pr[R \leq r]$
0	.0060	.0060
1	.0404	.0464
2	.1209	.1673
3	.2150	.3823
4	.2508	.6331
5	.2007	.8338
6	.1114	.9452
7	.0425	.9877
8	.0106	.9983
9	.0016	.9999
10	.0001	1.0000
	1.0000	

The values in columns (1) and (3) constitute the *cumulative probability distribution* of the random variable R. The binomial probability distribution of R and the cumulative probability distribution of R are graphed in Figure 6-5. The cumulative probability value corresponding to any particular proportion value is obtained from the *highest point* on the "stairway" directly above. For instance, the cumulative probability of values 5 or less is .8338 (not .6331, which belongs to the lower "step"). Note that the size of each step is the same as the

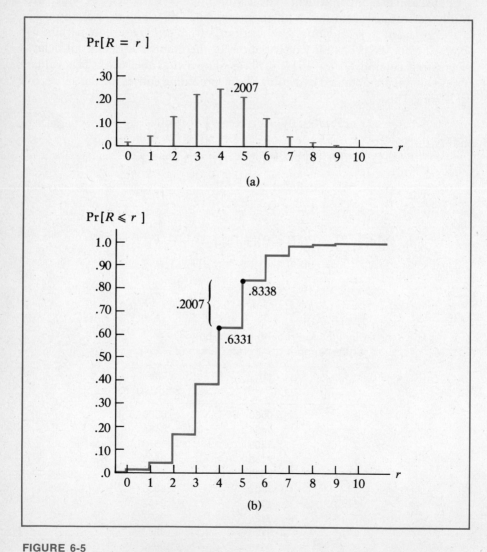

FIGURE 6-5

(a) Binomial probability distribution and (b) cumulative probability distribution for the number of successes from a Bernoulli process with $n = 10$ and $\pi = .4$.

height of the respective spike. Thus, the underlying probability distribution may be obtained from the cumulative probability distribution by finding these step sizes. For example, to find the probability that $R = 5$, we find the difference

$$\Pr[R = 5] = \Pr[R \leq 5] - \Pr[R \leq 4]$$
$$= .8338 - .6331 = .2007$$

Using Binomial Probability Tables

Because calculating binomial probabilities involves working with large factorial values and small numbers raised to large powers, it is convenient to have them already computed. Appendix Table C provides cumulative binomial probability values computed for various sizes of n (with separate tabulations for several π's).

We use this table to compute probabilities for the number of successes in the variety of situations described in the following list. Suppose we wish to find probabilities regarding the number of $n = 100$ patients who will respond favorably (success) to treatment with a new drug. We assume the drug will be successful 30% of the time, so that $\pi = .30$. To ease our discussion, a portion of Appendix Table C is reproduced here.

$n = 100$

π • • •	.20	.30	.40 • • •
r			
•	•	•	•
•	•	•	•
•	•	•	•
19	.4602	.0089	.0000
20	.5595	.0165	.0000
21	.6540	.0288	.0000
22	.7389	.0479	.0001
23	.8109	.0755	.0003
24	.8686	.1136	.0006
25	.9125	.1631	.0012
26 • • •	.9442	.2244	.0024 • • •
27	.9658	.2964	.0046
28	.9800	.3768	.0084
29	.9888	.4623	.0148
30	.9939	.5491	.0248
31	.9969	.6331	.0398
32	.9984	.7107	.0615
33	.9993	.7793	.0913
•	•	•	•
•	•	•	•
•	•	•	•

1. **Obtaining a result less than or equal to a particular value.** The probability that 20 or fewer patients will respond favorably is a cumulative probability value that may be read directly from the table when $r = 20$ successes:

$$\Pr[R \leq 20] = .0165$$

2. **Obtaining a result exactly equal to a single value.** Recall that cumulative probabilities represent the sum of individual probability values and are portrayed graphically as a stairway (see Figure 6-5). A single value probability may be obtained by determining the size of the step between two neighboring cumulative probabilities. For example, the probability that exactly 32 of the patients will respond favorably would be

$$\Pr[R = 32] = \Pr[R \leq 32] - \Pr[R \leq 31]$$
$$= .7107 - .6331$$
$$= .0776$$

3. **Obtaining a result strictly less than some value.** The probability that fewer than 30 successes are achieved is the same as the probability that exactly 29 or less successes are obtained, or

$$\Pr[R < 30] = \Pr[R \leq 29] = .4623$$

4. **Obtaining a result greater than or equal to some value.** To find the probability that at least 20 patients will respond favorably, we look up the cumulative probability that 19 or less will respond and subtract this value from 1:

$$\Pr[R \geq 20] = 1 - \Pr[R \leq 19]$$
$$= 1 - .0089 = .9911$$

Note that the situation is similar when the result must be *strictly greater than* some value. Then we find the cumulative probability of r itself and subtract this from 1. For example, the probability that more than 20 patients will respond to treatment is 1 minus its complementary probability that 20 or fewer patients will respond, or

$$\Pr[R > 20] = 1 - \Pr[R \leq 20]$$
$$= 1 - .0165 = .9835$$

5. **Obtaining a result that lies between two values.** Suppose we want to find the probability that the proportion of successes will lie somewhere between 25

and 35, inclusively. Thus, we want to determine

$$\Pr[25 \le R \le 35]$$

In this case, we obtain the difference between two cumulative probabilities:

$$\Pr[R \le 35] - \Pr[R \le 24] = .8839 - .1136 = .7703$$

The first term on the left represents all outcomes of 35 or fewer successes. But we do not want to include outcomes of 24 or fewer successes, so we subtract the second cumulative probability from the first, thereby accounting for only outcomes of between 25 and 35 successes.

6. **Finding probabilities when the trial success probability exceeds .50.** For brevity, our binomial table stops at $\pi = .50$. If we wish to find probabilities for the number of successes when the trial success probability is larger (say, .7), we can still use Appendix Table C by letting R represent the number of failures and π represent the trial failure probability.

Consider a keypuncher who correctly punches cards 99% of the time. Suppose we want to find the probability that at least 95 of $n = 100$ cards have been punched correctly. Here, a success represents a correct card and $\pi = .99$. At least 95 correct cards is the same as 5 or less incorrect cards (failures). Using $\pi = 1 - .99 = .01$ as the trial failure probability, from Appendix Table C we find

$$\Pr[R \le 5] = .9995$$

This makes use of the fact that for every success event there is a corresponding failure event with the same probability.

The Sampling Distribution of the Proportion

Often random samples are taken from qualitative populations. The sample proportion P of observations falling in a particular category may be used to estimate the population proportion π. When sampling with replacement or when dealing with a large population, *the successive observations are trials in a Bernoulli process.* The probability that a particular observation is a "success" is equal to π. For example, if one-half the population is men so that the overall proportion of men is $\pi = .50$, there is a .50 chance that any particular person chosen at random will be a man. The binomial distribution provides probabilities for various possible levels for the sample proportion P of men observed.

On the average, P will equal π, even though it may differ for a particular sample. The following property applies for the

EXPECTED VALUE OF P

$$\mu(P) = \pi$$

P may turn out to be many possible values. As we did with $\bar{X}$, we may summarize the variability in the sample proportion by its standard deviation or standard error. The following expression, applicable whenever sampling with replacement or from large populations, provides the

STANDARD ERROR OF P

$$\sigma_P = \sqrt{\frac{\pi(1 - \pi)}{n}}$$

This expression tells us that the variability in P depends on both the level of the population proportion π and the sample size n.

As an illustration, suppose that a large population of voters is 80% Democrat ($\pi = .80$). A sample of size $n = 100$ is taken. If P represents the sample proportion of Democrats,

$$\mu(P) = .80$$

$$\sigma_P = \sqrt{\frac{.80(1 - .80)}{100}} = \sqrt{\frac{.16}{100}} = .04$$

Knowing how to find the values of $\mu(P)$ and σ_P will be quite helpful later when we consider a variety of statistical procedures.

EXERCISES

6-15 Calculate the following factorial products:
    (a) 4!   (b) 6!   (c) 7!   (d) 8!

6-16 Calculate the following quantities:

    (a) $\dfrac{4!}{2!2!}$ (b) $\dfrac{6!}{3!3!}$ (c) $\dfrac{7!}{5!2!}$ (d) $\dfrac{6!}{2!4!}$

6-17 Can each of the following situations be classified as a Bernoulli process? If not, state why.
    (a) The outcomes of successive rolls of a die, considering only the events "odd" and "even."

(b) A crooked gambler has rigged a roulette wheel so that whenever the player loses, a mechanism is released that gives the player better odds; whenever a player wins, the chance of winning on the next spin is somewhat smaller than before. Consider the outcomes of successive spins.

(c) The measuring mechanism that determines how much dye to squirt into paint being mixed occasionally violates the required tolerances. The mechanism is highly reliable when it is new, but with use it continually wears, becoming less accurate. Consider the outcomes (within or not within tolerance) of successive mixings.

(d) A machine produces items that are sometimes too heavy or too wide to be used. The events of interest express the quality of each successive item in terms of both weight and width.

6-18 An evenly balanced coin is fairly tossed seven times.
  (a) Determine the probabilities for obtaining (1) exactly two heads; (2) exactly four heads; (3) no tails; (4) exactly three tails.
  (b) What do you notice about your anwers to (2) and (4)? Why is this so?

6-19 From a production process that yields 5% defectives, n parts are randomly chosen. What is the expected proportion of defectives?

6-20 A consumer testing agency has contacted a random sample of $n = 50$ persons. The actual proportion of all persons favoring a new package is $\pi = .30$. Assuming that the binomial distribution applies, use Appendix Table C to find the probabilities for the following outcomes relating to the number of favorable responses R:
  (a) $R \leq 11$ (c) $R < 18$ (e) $R > 25$
  (b) $R = 15$ (d) $R \geq 20$ (f) $17 \leq R \leq 23$

6-21 A fair coin is tossed 20 times in succession. Using Appendix Table C, determine the probability that the number of heads obtained is
  (a) less than or equal to 8.   (d) greater than or equal to 12.
  (b) equal to 10.   (e) greater than 13.
  (c) less than 15.   (f) between 8 and 14, inclusively.

6-22 Using the probability distribution of P provided in Table 6-14 with $\pi = .1$, calculate $\mu(P)$ as a weighted average. (Your answer should equal π, which is .1 here.)

6-23 Using the probability values in Table 6-14, construct the cumulative probability distribution of P when $n = 5$ and $\pi = .7$.

6-24 A form of malnutrition occurs in 10% of all persons. Determine the probabilities for the following malnutrition outcomes for five randomly chosen persons:
  (a) All have it.   (d) At least 2 have it.
  (b) None has it.   (e) Between 2 and 4, inclusively, have it.
  (c) At least 1 has it.

6-25 The chief engineer in a chemical plant has established a testing procedure using five sample vials drawn from the final stage of a chemical process at random times over a 4-hour period. If one or more vials contain impurities, all the setting tanks are cleaned. Find the probability that the tanks must be cleaned when the process is so clean that the probability of a dirty vial is
  (a) $\pi = .01$.   (b) $\pi = .05$.   (c) $\pi = .20$.   (d) $\pi = .50$.
  (*Hint:* Use the fact that Pr[at least one dirty vial] $= 1 - $ Pr[no dirty vials].)

6-26 A process produces defective parts at a rate of .05. If a random sample of 5 items is chosen, what is the probability that at least 4 of the sample will be defective?

6-27 A batch of 100 items in which 5% are defective is sampled *without replacement*. Five items are chosen. Does the binomial distribution apply here? Explain.

6-28 A lopsided coin provides a 60% chance of a head on each toss. If the coin is tossed 18 times, find the probability that the number of heads obtained is

(a) less than or equal to 8.   (d) greater than or equal to 12.
(b) equal to 9.   (e) greater than 13.
(c) less than 15.   (f) between 8 and 14, inclusively.

6-5 CONTINUOUS PROBABILITY DISTRIBUTIONS

We will begin our discussion of continuous probability distributions by considering an uncertain situation involving the selection of a sample from a population. The frequency distribution obtained from grouping the years of company service of each of the 22,000 employees of the Wheeling Wire Works is given in Figure 6-6. The height of each bar of this histogram represents the

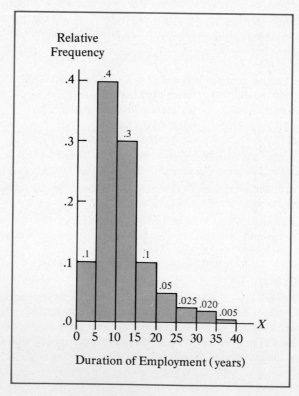

FIGURE 6-6
Frequency distribution for years of employment.

relative frequency for duration of employment in the interval covered by the bar. The random variable of interest, denoted by X, is the length of service of one person chosen randomly from the population. The time an employee has worked for the company can be measured on a continuous scale and therefore can be expressed to any desired fraction of the year. This makes X in our example a *continuous random variable*.

The probability that X lies within a particular class interval will be the same as the relative frequency of persons employed for that duration. Recall that the bars of the histogram for a relative frequency distribution have areas proportional to the relative frequency of the variate values in the corresponding interval. Since the relative frequencies must sum to 1, we can consider the total area under the histogram to be equal to 1. There, each class interval is one 5-year unit wide. Thus, the probability that X lies inside a particular class interval is equal to the area of the bar covering that interval. Therefore, the probability that our employee served between 10 and 15 years must be .3 (the area under the bar or rectangle of the histogram that covers these values, or width × height = $1 \times .3 = .3$).

But what if we want to consider an event that is more specific? For example, we may wish to find the probability that our selected person was employed between 10.95 and 11.05 years—a range of values lying totally within a single class interval. In this case, the frequency distribution does not provide sufficiently detailed information for us to find the probability directly.

Smoothed Curve Approximation

Our problem arises from the fact that the histogram artificially forces us to treat population variates in discrete lumps. Each value is arbitrarily placed into one of very few class intervals. To accurately consider intervals of any width, we may construct a smoothed, continuous approximation to the histogram, like the one in Figure 6-7, where the curve is superimposed on the histogram from Figure 6-6. This approximation may then be used to generate any probability value we desire.

If the smoothed curve is drawn properly, the area under portions of the curve will correspond closely to the area of the histogram bars for that part of the horizontal axis, and the total area under the curve can also be taken as unity. Thus, in Figure 6-7, the shaded area under the portion of the curve covering the points between A and B is nearly the same as the corresponding area under the bar covering the values from A to B. Here, the curve has been drawn so that the wedge-shaped area described by the points a, b, c is nearly the same as the area described by c, d, e. The sum of the areas for portions that are cut off the histogram by the curve should equal the sum of those areas that the curve adds by smoothing the corners.

The relative frequency of the population values lying between A and B in Figure 6-7 can be approximated by the area under the portion of the curve

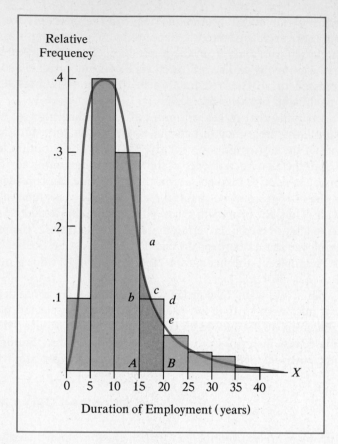

FIGURE 6-7
Frequency distribution from Figure 6-6 approximated by a continuous curve.

covering these values. We may therefore conclude that the probability that a continuous random variable assumes some value in an interval is represented by the area under the portion of the continuous curve covering that interval.

Probability Density Function

In Chapter 2, we learned that population frequency distribution graphs can be categorized according to their general shapes. We can construct smoothed curve approximations for any of these distributions. Essentially, we select a curve by choosing the appropriate shape.

Curves that have a particular shape can be defined by an equation or function. This allows us to group population frequency distributions according to the mathematical function that defines the curve corresponding most closely to the shape of the population histogram. If we know the function that defines the curve, we can mathematically compute the area over an interval to obtain the required probability.

A unique set of probabilities is obtained from a particular curve. For example, consider the two curves in Figure 6-8 for two random variables, X and Y, whose values range over the same scale. The areas covering the values between points A and B are different for the two curves, because the shapes of two curves are different. Thus

$$\Pr[A \leq X \leq B] \neq \Pr[A \leq Y \leq B]$$

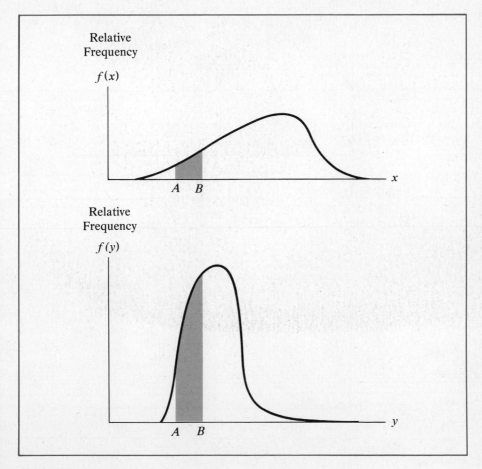

FIGURE 6-8
Two probability density function curves.

so that X and Y have different probability distributions. Note that the thickness or density of the two curves is distributed differently. For this reason, the mathematical expression describing such a curve is sometimes called a *probability density function*. In general, this function is denoted by $f(x)$, where x represents the possible values of the random variable X. (Since density is measured on the same scale, we will continue to label the vertical axis "relative frequency.") The probability density function therefore describes the probability distribution of a continuous random variable, so that a random variable may be categorized by the form of its function.

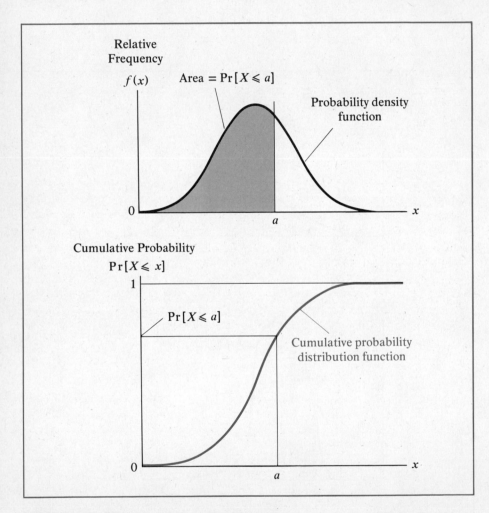

FIGURE 6-9

Graphical expressions of the probability distribution of a continuous random variable.

Cumulative Probability Distribution

The graphic expressions of the probability distribution of a continuous random variable are presented in Figure 6-9. Although the cumulative probability distributions of discrete and continuous random variables are defined identically, the cumulative probability distribution of a continuous random variable resembles a smoothed curve (Figure 6-9) instead of a stairway (Figure 6-5). Both distributions may be expressed algebraically, graphically, or as a table of values.

It follows that for a continuous random variable,

$$\Pr[X \le x] = \Pr[X < x]$$

because $\Pr[X = x] = 0$. (There is zero area under the portion of the density curve that covers a single *point*, since there is no width.)

The Expected Value and the Variance

The expected value of a continuous random variable, like that of a discrete random variable, can be viewed as the long-run average of many repetitions. This is also true of the variance. Both the expected value and the variance can be calculated from the probability density function, but the calculations are beyond the scope of this book. These values have already been determined for the common distributions that we will encounter.

Random variables that have the same basic density function and differ only in terms of the values of the parameters specifying that particular function are said to belong to the same *distribution family*. In Chapter 7, we will discuss one of these families—the normal distribution. Other continuous probability distributions will be introduced in later chapters.

REVIEW EXERCISES

6-29 A coin is tossed three times. Letting X represent the number of heads obtained, determine the probability distribution for X and calculate $\mu(X)$.

6-30 In canasta, points are assigned to cards in the following manner; red 3 = 100; joker = 50; ace or deuce = 20; 8 through king = 10; 4 through 7 and black 3 = 5. A canasta deck is composed of two ordinary decks of playing cards, each containing 52 cards and 2 jokers. Determine the probability distribution for the point value of the first card dealt from a shuffled canasta deck.

6-31 Calculate the mean, variance, and standard deviation of the number of heads X in Exercise 6-29, using your probability distribution from that exercise.

6-32 The number of persons arriving at a movie theater during any specified minute between 8 and 9 P.M. has the following probability distribution:

Persons	Probability
0	.4
1	.3
2	.2
3	.1
	1.0

Calculate the expected value and the variance of the number of persons arriving between 8:30 and 8:31 P.M.

6-33 Suppose that the proportion of adult U.S. citizens who approve of presidential policy is $\pi = .50$. A random sample of $n = 5$ persons is chosen. Using the binomial formula, determine the probability that

(a) exactly 5 approve.   (b) none approve.   (c) exactly 3 approve.

6-34 Find the standard deviation of the proportion of successes P found in n Bernoulli trials for the following situations:

(a) $n = 100, \pi = .8$ (b) $n = 25, \pi = .5$ (c) $n = 100, \pi = .10$

6-35 Blacks comprise 40% of the voters in a city. A jury of 12 persons has been impaneled for a trial. Assuming that every registered voter has an equal chance of being chosen, use Appendix Table C to find the following probabilities for the representation of blacks on the jury:

(a) at least 5 black jurors   (c) all black jurors
(b) no black jurors   (d) at most 10 black jurors

6-36 The ages of the straight-A students in a certain statistics class are

Mr. C	24
Miss H	19
Mr. J	21
Mr. M	20
Mrs. T	23

(a) A random sample of two persons is taken *without replacement*. Determine the sampling distribution for the mean age $\bar{X}$, and from this determine the values of $\mu(\bar{X})$ and $\sigma(\bar{X})$.

(b) Calculate the standard deviation of the age *population* of straight-A students. Using this value, determine the standard error of $\bar{X}$ if the random sample of two persons in (a) had been taken *with replacement*.

The Normal Distribution

The *normal distribution* may be the most important distribution used in statistical applications. The observed frequency distributions for many physical measurements and natural phenomena closely resemble the normal distribution. These include distributions for physical measurement, such as height and weight, as well as other human characteristics, such as IQ. The frequency distributions for these and many more populations closely resemble the *normal curve* in Figure 7-1. But there is a more fundamental reason why the normal distribution is so important in statistics. A theoretical property of the sample mean allows us to use the normal distribution to find probabilities for various sample results. Thus, the normal curve plays a basic role when inferences are made regarding the population mean and only the sample mean can be calculated directly.

CHARACTERISTICS OF THE NORMAL DISTRIBUTION   7-1

Several features of the normal curve in Figure 7-1 are interesting. Note that the curve is shaped like a bell with a single peak, making it *unimodal*, and that it is *symmetrical* about its center. The mean of a normally distributed population lies at the center of its frequency curve. Due to symmetry, the *median* and the

195

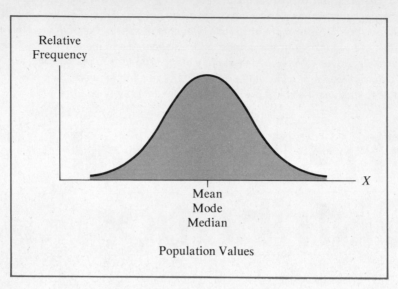

FIGURE 7-1
Frequency curve for the normal distribution.

mode of the distribution also occur at the curve's center, so that the mean, median, and mode of a normal curve all have the same value. Although it is impractical to show this on the graph, the tails of a normal curve extend indefinitely in both directions, never quite touching the horizontal axis.

We say that a population with a frequency distribution approximating the shape of the normal curve is *normally distributed*. If the random variable X is the value of an elementary unit chosen at random from a normally distributed population, then we say that X is normally distributed. The normal curve also represents the probability density function for X.* The normal curve depends on only two parameters: the mean μ and the standard deviation σ. Whatever the values of μ and σ are, the total area under the normal curve is always 1.

We know that the mean is a measure of central tendency or location. When normal curves for populations having different means are graphed together, they are located at different positions along the horizontal axis. This

---

* The function denoting the height of the curve is

$$f(x) = \frac{1}{\sqrt{2\pi\sigma^2}} e^{-[(x-\mu)^2/2\sigma^2]}$$

where π is the ratio of the circumference to the diameter of a circle (3.1416), and e is the base of natural logarithms (2.7183).

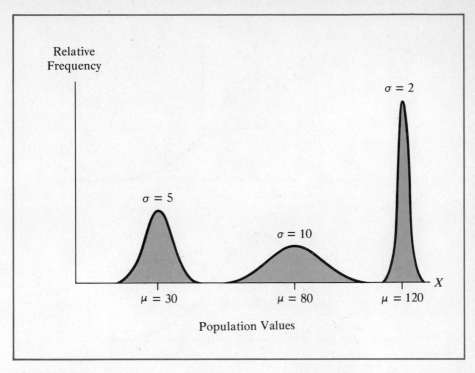

FIGURE 7-2
Three different normal distributions graphed on a common axis.

is illustrated in Figure 7-2 for three different populations with means of 30, 80, and 120, respectively.

Figure 7-2 also illustrates that the shape of a normal curve is determined by the population's standard deviation σ. Distributions with small standard deviations have narrow, peaked "bells," and those with large σ's have flatter curves with less pronounced peaks. The three populations in Figure 7-2 have standard deviations of 5, 10, and 2, respectively. A large class of populations belong to the normal family, and each member differs only by its mean and its standard deviation.

The probability that a normally distributed random variable will assume values within a particular interval is equal to the area of the portion of the curve covering that interval. Before we learn to use a table to find the areas under the normal curve, we should note a useful property of any normal distribution (see Figure 7-3): **The area under the normal curve covering an interval that is symmetrical about the mean is determined solely by the distance that separates the end points from the mean, measured in standard deviations.** For instance, the values of about 68% of the population lie within one standard

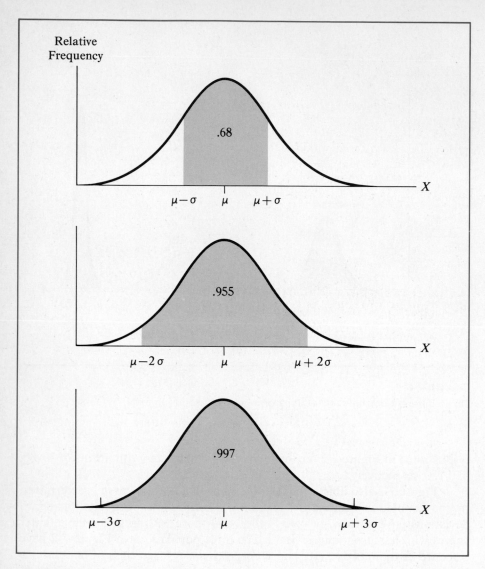

FIGURE 7-3

Relationship between the area under the normal curve and the distance from the mean, expressed in units of standard deviation.

deviation in either direction from the mean; that is, the area under the curve over the interval $\mu - \sigma$ through $\mu + \sigma$ is .68. This is true no matter what the values of μ and σ are. The values of about 95.5% of the population lie within two standard deviations of the mean, and approximately 99.7% fall within three standard deviations.

EXAMPLE: ARE MARINES TALLER THAN SOLDIERS?

The commander of an Army division has wagered with the commander of a Marine division that his Army troops are taller. To verify this, the Army commander's aide compiles the heights of the Army soldiers from their medical records and calculates that the mean height is 70 inches and the standard deviation is 2 inches. The aide also constructs a histogram, which exhibits an almost perfect bell shape, indicating that the heights of the Army soldiers may be described in terms of the normal distribution.

    The Marine commander arrives at similar conclusions. The mean height of his men is 69.5 inches, with a standard deviation of 2.5 inches. The two populations are compared below. Figure 7-4 shows the frequency curves for these two populations.

	Army Height	Marine Height
Mean (μ)	70 inches	69.5 inches
Standard deviation (σ)	2	2.5
68% from $\mu - \sigma$ to $\mu + \sigma$	68 to 72	67 to 72
95.5% from $\mu - 2\sigma$ to $\mu + 2\sigma$	66 to 74	64.5 to 74.5
99.7% from $\mu - 3\sigma$ to $\mu + 3\sigma$	64 to 76	62.5 to 77

    The Army commander claims to have won the bet, because the mean height of the Army soliders is .5 inch greater than the mean Marine height. The Marine commander objects, noting that his division contains a greater percentage of overly

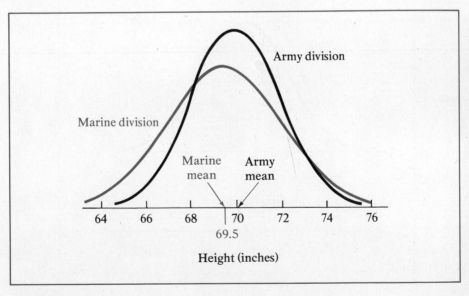

FIGURE 7-4
Normal curves for the heights of the Marine and Army divisions.

tall men. (Note that the upper tail of the Marines' frequency curve lies above the Army division's curve). The Army commander agrees but retorts that there is also a higher percentage of shorter Marines. (Observe the lower tails of the two frequency curves.)

This example illustrates how difficult it is to compare populations by using a measure of location, such as the mean. The fact that some Marines are taller than the tallest Army soldiers is due to the difference in population variabilities. The Marine division is a more diverse group with a larger standard deviation (2.5 inches compared to 2.0 inches for the Army division).

7-2   FINDING AREAS UNDER THE NORMAL CURVE

Before we can obtain probability values for normally distributed random variables, we must find the appropriate area lying under the normal curve by using Appendix Table D. To illustrate, we will determine the desired areas for the time a particular typesetter takes to compose 500 lines of standard type. We will assume that the population of times is normally distributed, with a mean of $\mu = 150$ minutes and a standard deviation of $\sigma = 30$ minutes. The time it takes to set any given 500 lines, such as the next 500 to be composed, represents a randomly chosen time from this population.

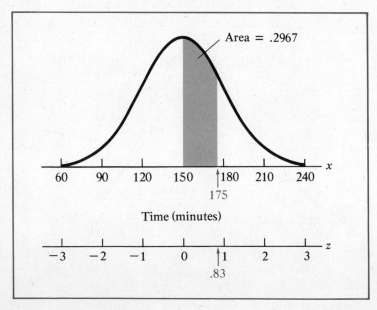

FIGURE 7-5
Determining the area under a normal curve.

The probability that it takes between 150 and 175 minutes to set 500 lines is represented by the shaded area under the normal curve in Figure 7-5. We know that the area beneath the normal curve between the mean and a certain point depends only on the number of standard deviations separating the two points. We see that 175 minutes is equivalent to a distance above the mean of .83 standard deviation. This figure is determined by observing that 175 minutes minus the mean of 150 minutes is equal to 25 minutes. Since the standard deviation is 30 minutes, 25 minutes is only a fraction, $25/30 = .83$, of the standard deviation.

Appendix Table D has been constructed for a special curve, called the *standard normal curve*, which provides the area between the mean and a point above it at a specified distance measured in standard deviations. Because this distance will vary, as a convenience it is represented by the letter z. Sometimes the value of z is referred to as a *normal deviate*. The distance z that separates a possible normal random variable value x from its mean, expressed in terms of standard deviations, is given by the following expression for the

NORMAL DEVIATE

$$z = \frac{x - \mu}{\sigma}$$

A negative value will be obtained for z when x is smaller than μ.

To ease our discussion, a portion of Table D is reproduced here:

Normal Deviate z	.00	.01	.02	.03	.04	.05	.06	.07
⋮				⋮				
.6	.2257	.2291	.2324	.2357	.2389	.2422	.2454	.2486
.7	.2580	.2612	.2642	.2673	.2704	.2734	.2764	.2794
.8	.2881	.2910	.2939	.2967	.2995	.3023	.3051	.3078
.9	.3159	.3186	.3212	.3238	.3264	.3289	.3315	.3340 ⋯
1.0	.3413	.3438	.3461	.3485	.3508	.3531	.3554	.3577
1.1	.3643	.3665	.3686	.3708	.3729	.3749	.3770	.3790
1.2	.3849	.3869	.3888	.3907	.3925	.3944	.3962	.3980
				⋮				

The first column of Table D lists values of z to the first decimal place. The second decimal place value is located at the head of one of the remaining ten columns. The area under the curve between the mean and z standard deviations is found at the intersection of the correct row and column. For example, when $z = .83$, we find the area of .2967 by reading the entry in the .8 row and the .03 column. The area under the normal curve for a completion time between 150 and 175 minutes is thus .2967, which represents the probability that it will take this long to set the next 500 lines of print.

Using the Normal Curve Table

Appendix Table D provides areas only between the mean and some point above it, but we can also use this table to find areas encountered in other common probability situations. Each of these areas is described in this section.

(a) **Area between the mean and some point lying below the mean.**

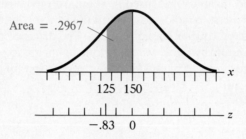

Area = .2967

125 150

−.83 0

To find the probability that the completion time lies between 125 and 150 minutes, first we must calculate the normal deviate:

$$z = \frac{x - \mu}{\sigma} = \frac{125 - 150}{30} = -.83$$

Here, z is negative because 125 is a point lying below the mean. Since the normal curve is symmetrical about the mean, this area must be the same as it would be for a positive value of z of the same magnitude (in this case .2967, as before). It is therefore unnecessary to tabulate areas for negative values of z. The area between the mean and a point lying below it will be equal to the area between the mean and a point lying the same distance above it.

(b) **Area to the left of a value above the mean.**

Total area = .5000 + .3790
= .8790

Area = .5000   Area = .3790

150   185

0   1.17

To find the probability that 500 lines can be set in 185 minutes or less, we must find the entire shaded area below 185. Here, we must consider the lower half of the normal curve separately. Since the entire area of the normal curve is 1, the area under the half to the left of 150 must be .5. The area between 150 and 185 is found from Table D, with $z = (185 - 150)/30 = 1.17$, to be .3790. The entire shaded area below 185 is the sum of the two areas, or .5000 + .3790 = .8790.

(c) **Area in upper tail.**

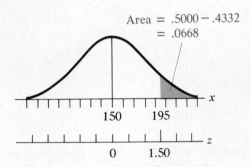

Area = .5000 − .4332
    = .0668

To find the probability that the number of minutes required exceeds 195, we must first determine the area between the mean and 195. The normal deviate is $z = (195 - 150)/30 = 1.50$; the area from Table D is .4332. Since the area under the upper half of the normal curve is .5, we find the area above 195 by subtracting the unwanted portion: .5000 − .4332 = .0668.

Because the total area under the normal curve is 1, we may use the value of the upper tail area to calculate the area to the left of 195. Subtracting the upper tail area of .0668 from 1 gives us 1.0000 − .0668 = .9332 for the area to the left of 195 (or the probability that the time will be less than 195 minutes). Similarly, when the area to the left of a point is known, the area to its right can be found by subtracting this value from 1. For example, in (b) we found that the area to the left of 185 is .8790; thus, the area in the upper tail to the right of 185 must be 1 − .8790 = .1210.

(d) **Area in lower tail.**

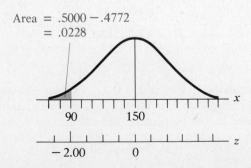

Area = .5000 − .4772
    = .0228

To find the probability that it will take 90 minutes or less to set the type, we follow two steps similar to those in (c). First, we find the area between 90 and 150. Using $z = (90 - 150)/30 = -2.00$, we obtain .4772 from Table D. Subtracting this value from .5 yields .5000 − .4772 = .0228.

(e) **Area to the right of a value below the mean.**

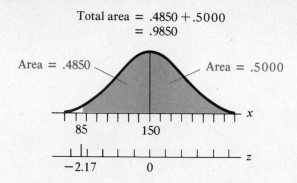

Total area = .4850 + .5000
= .9850

Area = .4850   Area = .5000

85   150

−2.17   0

To find the probability that the completion time will be equal to or greater than 85 minutes, the area between 85 and the mean is added to the area to the right of the mean, which is .5. Here, we calculate $z = (85 - 150)/30 = -2.17$. Adding the area from Table D (.4850) to .5, the combined area is .5000 + .4850 = .9850.

We can find the area in the lower tail below 85 by subtracting .9850 from 1, or 1 − .9850 = .0150. We can also find the area to the right of 90 by subtracting the lower tail area found in (d) from 1, or 1 − .0228 = .9772.

(f) **Area under portion overlapping the mean.**

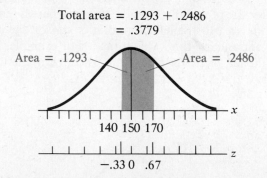

Total area = .1293 + .2486
= .3779

Area = .1293   Area = .2486

140 150 170

−.33 0 .67

To find the probability that it will take between 140 and 170 minutes to set the 500 lines, we simply add the portion of the shaded area below the mean to the portion above it. The respective normal deviate values are $z = (140 - 150)/30 = -.33$ and $z = (170 - 150)/30 = .67$. From Table D, the lower area is .1293 and the upper area is .2486, so that the combined area is .1293 + .2486 = .3779.

If we wish to find the probability that it will take between 120 and 180 minutes to set the type, the normal deviates are $z = (120 - 150)/30 = -1$ and $z = (180 - 150)/30 = 1$. From Table D, .3413 is the same area for both sides, so that the combined area is .3413 + .3413 = .6826. Because z expresses the number of standard deviation units from the mean, we see that .6826 is a more precise value for the area between $\mu \pm \sigma$ than the value used in the top graph of Figure 7-3.

(g) **Area between two values lying above or below the mean.**

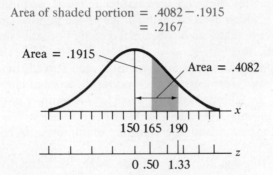

To find the probability that the composition time is between 165 and 190 minutes, we must first determine the areas between the mean and each of these values. The respective normal deviates are $z = (165 - 150)/30 = .50$ and $z = (190 - 150)/30 = 1.33$. From Table D, the area between the mean and 190 is .4082, and the area between the mean and 165 is .1915. Thus, the shaded area is found by subtracting the smaller area from the larger one, or .4082 - .1915 = .2167.

A similar procedure can be applied to an area lying below the mean. It is also possible to find the value for a complementary situation—that the time will be either below 165 minutes or greater than 190 minutes—by subtracting the shaded area from 1, or 1 - .2167 = .7833.

The normal curve represents values that lie on a continuous scale, such as height, weight, and time. There is zero probability that a specific value, such as 129.40 minutes, will occur (there is zero area under the normal curve covering a single point). Thus, in finding probabilities, it does not matter whether we use "strict" inequalities, such as the composition time is "less than" (<) 129.40 minutes, or an ordinary inequality, such as the time is "less than or equal to" (≤) 129.40 minutes. Using $z = (129.40 - 150)/30 = -.69$, the area is the same in either case: .5000 - .2549 = .2451.

Cumulative Probabilities and Percentiles

Cases (b) and (d) in our discussion of normal curve areas show us how all the values for a normal cumulative probability distribution can be determined. We have seen that the probability of a time of 185 minutes or less is .8790 and that the probability of 90 minutes or less is .0228. For values above the mean, when z is positive, we obtain the cumulative probabilities from Table D simply by adding .5 to the tabulated areas. For values below the mean, when z is negative, we subtract the area given in the table from .5 to obtain the cumulative probability.

It is frequently necessary to find the percentile values of a normally distributed population. Recall from Chapter 3 that a percentile is the population value below which a certain percentage of the population lies and that a percentile or fractile value can be obtained directly from the cumulative relative frequency distribution.

To find a population percentile, we must read Table D in *reverse*, since the specified percentage represents an area under the normal curve. For instance, in our example, the 90th percentile is a particular number of minutes, and the area under the normal curve to the left of this value will be .90. Thus, the area between the mean and the number of minutes to be found will be $.90 - .50 = .40$.

Searching through the body of the table, we select the area that lies closest to this figure—in this case, .3997. Since .3997 is in the $z = 1.2$ row and the .08 column, the corresponding normal deviate is $z = 1.28$. This means that the desired time is 1.28 standard deviations, or $30 \times 1.28 = 38.4$ minutes, above the mean. Adding this to the mean, we find that the 90th percentile is $150 + 38.4 = 188.4$ minutes.

In general, to determine a percentile, we begin by reading Table D in reverse to find z. We can then calculate the corresponding population value x from the following expression, which provides the

PERCENTILE FOR A NORMAL POPULATION

$$x = \mu + z\sigma$$

Note that below the 50th percentile, the lower tail area is used to find z, which will be negative in these cases.

EXAMPLE: DESIGNING WITHIN HUMAN LIMITATIONS

Engineers who are designing an aircraft cockpit want to arrange the controls so that 95% of all pilots can reach them while seated. This involves finding the maximum reach radius exceeded by 95% (but not by 5%) of all pilots. Thus, the engineers must find the reach that corresponds to the 5th percentile.

The maximum reach radii of airline pilots are assumed to be approximately normally distributed, with a mean of $\mu = 48$ inches and a standard deviation of $\sigma = 2$ inches. The engineers seek the point below which the area under the normal curve is equal to .05. This means that Table D must be searched to find the area closest to $.50 - .05 = .45$. Two areas, .4495 and .4505, are equally close. Therefore, the desired figure lies somewhere between the corresponding normal deviates 1.64 and 1.65. For simplicity, the engineers choose 1.64, preferring to err on the side of a larger rather than a smaller tail area. Since the 5th percentile is below the 50th, they are dealing with a lower tail area and their normal deviate is negative, so that $z = -1.64$. (The 5th percentile lies *below* the mean and is smaller, so that a distance of 1.64 standard deviations must be *subtracted* from the mean.) The 5th percentile is therefore

$$x = \mu + z\sigma = 48 - 1.64(2) = 44.72 \text{ inches}$$

The Standard Normal Random Variable

The area under any normal curve can be found by using the standard normal curve. This curve provides the probability distribution for the *standard normal random variable.*

In our typesetting illustration, we essentially transformed the original random variable X, the time to complete 500 lines, into the standard normal random variable whenever we used Table D to find areas under the normal curve. This transformation can be accomplished physically by shifting the center of the curve and then stretching or contracting it. To shift the original curve so that its center lies above the point $x = 0$, we subtract μ from each point on the x axis. Then the repositioned curve can be stretched or squeezed until the scale on the horizontal axis matches the scale for the standard normal distribution. If all values of the random variable are divided by its standard deviation, the transformed curve will have the same shape as the standard normal curve in Figure 7-6. The net effect will always be the same, no matter what the values of μ and σ are. The horizontal scale may be either expanded or contracted.

Fortunately, we do not need to physically transform the original random variable X into the standard normal random variable, because we can manipulate the possible values of x algebraically.

IQ tests illustrate this concept. These tests are designed so that the scores achieved by a cross section of persons are normally distributed for all practical purposes. One of the most popular, the Stanford-Binet test, has a mean of $\mu = 100$ points and a standard deviation of $\sigma = 16$. Letting X represent the IQ score achieved by a randomly chosen person, it is possible to determine various probabilities for the value achieved.

For example, we find the probability that this person's IQ is at most 140:

$\Pr[X \leqslant 140] =$
$.5000 + .4938 = .9938$

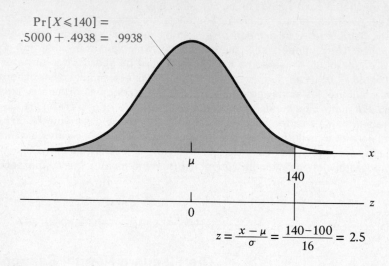

$$z = \frac{x - \mu}{\sigma} = \frac{140 - 100}{16} = 2.5$$

Here, we see that there is better than a 99% chance that this person's IQ falls below 140 on the Stanford-Binet scale.

Other IQ tests have different standard deviations than the Stanford-Binet, although the mean of most IQ tests is 100 points. If we ask the same question using another test with a standard deviation of $\sigma = 10$ points, we have

$$z = \frac{x - \mu}{\sigma} = \frac{140 - 100}{10} = 4.0$$
$$\Pr[X \leq 140] = .50000 + .49997 = .99997$$

Note that IQs above 140 are considerably rarer using this test.

Because various types of IQ tests are scaled differently, it is hard to compare IQ scores. For this reason, intelligence-test results are often transformed into *standard scores*, or normal deviate values. In our example, 140 points on the Stanford-Binet provides a standard score of 2.50. Thus, a person scoring 140 points on the second test would exhibit a much higher level of intelligence. (To obtain a standard score equal to 4.0 on the Stanford-Binet test, a person would have to be 4 standard deviations above the mean and have an IQ of $100 + 4(16) = 164$ on that scale.)

If σ is smaller than the 10 applicable to the second IQ test, the absolute value of z will be too large to obtain the area from Table D. Our table has to stop somewhere, but the tails of the normal curve extend indefinitely. In general, whenever z exceeds 4.00, the area between the mean and z is so close to .5 that for practical purposes .5 is used. In such cases, the upper and lower tail areas are so close to 0 that they are *negligible*. For example, the area below $z = 5$ is approximately 1, as is the area above $z = -5$; the areas above $z = 5$ and below $z = -5$ are each approximately 0.

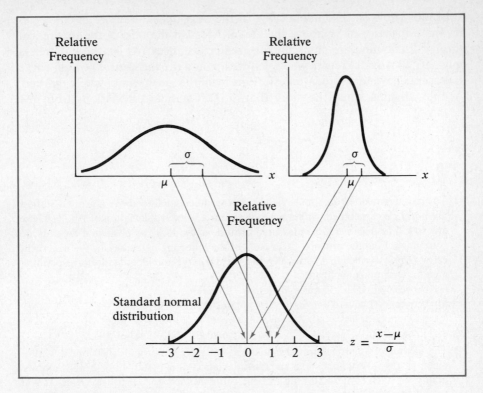

FIGURE 7-6
Illustration of the linear transformation of normal random variables into the standard normal distribution.

Concluding Remarks

The fact that the tails of the normal curve never touch the horizontal axis implies a probability (although tiny) that the random variable can exceed any value. Consider the distribution for heights of men, for instance. The normal curve literally assigns a probability to the event that a man will grow to be more than one mile tall. This seems impossible, but according to the normal curve the probability for this event is not zero, although it is quite remote. The normal curve also assigns a probability to the fact that a man will be negatively tall. Although these implications are absurd, they should not detract from the utility of the normal distribution. They are byproducts of the use of a convenient mathematical expression to describe a particular curve that fits some empirical frequency distributions quite accurately. According to

Appendix Table D, only about .13% of the area under the normal curve lies beyond a distance of 3σ above the mean. Areas in the tails of the theoretical normal distribution are miniscule for greater distances. (At 4σ, the tail area is only .003%.) Any discrepancy between reality and the theoretical normal curve that occurs beyond 4 standard deviations from the mean can be safely ignored. (A mile-tall man would be more than 40,000 standard deviations from the mean.)

EXERCISES

7-1 The measurement errors for the height of a weather satellite above a ground station are normally distributed, with a mean of 0 and a standard deviation of 1 mile. These errors will be negative if the measured altitude is too low and positive if the altitude is too high. Find the probability for each of the following outcomes for the next orbit error. (*Hint:* It will help if you sketch the normal curve and identify the corresponding area for each outcome first.)

(a) between 0 and $+1.55$ miles   (e) -1.25 miles or less
(b) between -2.45 and 0 miles   (f) greater than -1.25 miles
(c) $+.75$ miles or less   (g) between $+.10$ and $+.60$ miles
(d) greater than $+.75$ miles   (h) between $+1$ and $+2$ miles

7-2 The lifetime of a particular model of stereo cartridge is normally distributed, with a mean of $\mu = 1,000$ hours and a standard deviation of $\sigma = 100$ hours. Find the probability for each of the following lifetime outcomes for one cartridge. (*Hint:* It will help if you sketch the normal curve and identify the corresponding area for each outcome first.)

(a) between 1,000 and 1,150 hours   (e) 870 hours or less
(b) between 950 and 1,000 hours   (f) longer than 780 hours
(c) 930 hours or less   (g) between 700 and 1,200 hours
(d) longer than 1,250 hours   (h) between 750 and 850 hours

7-3 The time it takes a bank teller to cash a check is normally distributed, with a mean of $\mu = 30$ seconds and a standard deviation of $\sigma = 10$ seconds. Find the following percentiles: (a) 10th; (b) 25th; (c) 50th; (d) 75th; (e) 90th.

7-4 The heights of the male students at a particular university are normally distributed, with a mean of $\mu = 5'10''$ and a standard deviation of $\sigma = 2.5''$. Find the height below which the following percentages of men lie: (a) 1; (b) 5; (c) 30; (d) 60; (e) 95; (f) 99.

7-5 An architect designing a men's gymnasium wants to make the interior doors high enough so that 95% of the men using them will have at least a 1-foot clearance. Assuming that the heights will be normally distributed, with a mean of 70 inches and a standard deviation of 3 inches, how high must the architect make the doors?

7-6 A quality-control manager shuts down an automatic lathe for corrective maintenance whenever a sample of the parts it produces has an average diameter greater than 2.01 inches or smaller than 1.99 inches. The lathe is designed to produce parts with a mean diameter of 2.00 inches, and the sample averages have a standard deviation of .005 inches. Assume that the normal distribution applies.

(a) What is the probability that the quality-control manager will stop the process when the lathe is operating as designed, with $\mu = 2.00$ inches? (The manager does not know that the lathe is operating correctly.)

(b) If the lathe begins to produce parts that on the average are too wide, with $\mu = 2.02$ inches, what is the probability that the lathe will continue to operate?

(c) If an adjustment error causes the lathe to produce parts that on the average are too narrow, with $\mu = 1.99$ inches, what is the probability that the lathe will be stopped?

SAMPLING DISTRIBUTION OF THE SAMPLE MEAN FOR A NORMAL POPULATION   7-3

The normal curve describes the frequency distributions for a great many populations, notably measurements of persons and things. Some important questions can be answered by applying the normal curve. For instance, a new testing procedure may be analyzed by a counseling center wishing to replace its present test. A sample of persons can be given the new test to estimate the mean score to be achieved by all those who may ultimately take the test. Because scores obtained with the current test closely fit the normal distribution, the normal curve will probably apply to the new test scores as well.

It has been established mathematically that under certain circumstances when the population frequencies are described by the normal curve, the sampling distribution of $\overline{X}$ itself is also normally distributed.

SAMPLING DISTRIBUTION OF $\overline{X}$

If $\overline{X}$ is the mean of a random sample taken from a normally distributed population having a mean of μ and a standard deviation of σ, then the sampling distribution for $\overline{X}$ is also normal, its mean is also μ, and its standard deviation is $\sigma_{\overline{X}} = \sigma/\sqrt{n}$. This is true no matter what the size of the sample happens to be.*

The area under the normal curve between μ and a possible level of the sample mean is found by transforming $\overline{x}$ into the corresponding

NORMAL DEVIATE FOR THE SAMPLE MEAN

$$z = \frac{\overline{x} - \mu}{\sigma_{\overline{X}}}$$

---

* As we will see in Chapter 8, when σ is *also* of unknown value and n is small, it is necessary to use the Student t statistic instead of $\overline{X}$. This statistic has a different sampling distribution.

As an illustration, suppose that $n = 100$ scores are chosen as a random sample from a normally distributed population of IQ test scores having an unknown mean of μ and a standard deviation of $\sigma = 16$ points. The sample mean $\bar{X}$ is normally distributed with the same mean and has the standard deviation

$$\sigma_{\bar{X}} = \frac{16}{\sqrt{100}} = \frac{16}{10} = 1.6$$

Assuming that the population mean IQ is $\mu = 100$, consider the following cases:

1. The probability that $\bar{X}$ is less than or equal to 101.5:

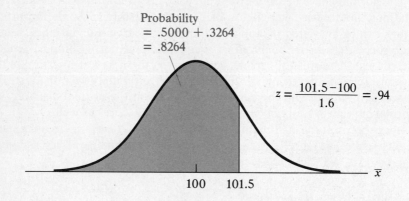

Probability
= .5000 + .3264
= .8264

$z = \dfrac{101.5 - 100}{1.6} = .94$

100   101.5

$\bar{x}$

2. The probability that $\bar{X}$ lies between 97 and 102:

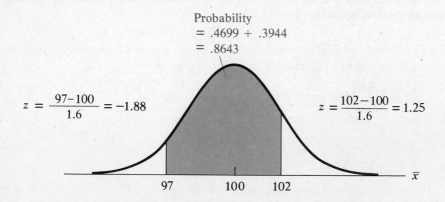

Probability
= .4699 + .3944
= .8643

$z = \dfrac{97 - 100}{1.6} = -1.88$

$z = \dfrac{102 - 100}{1.6} = 1.25$

97   100   102

$\bar{x}$

3. The probability that $\bar{X}$ is greater than 102.5:

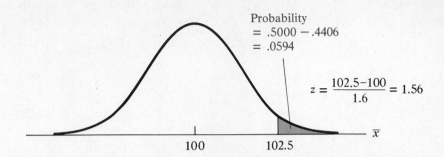

and for $n = 400$,

The Role Played by the Standard Error of $\bar{X}$

The Role Played by the Standard Error of $\bar{X}$

The standard deviation of $\bar{X}$ is sometimes referred to as the *standard error of* $\bar{X}$. Since $\sigma_{\bar{x}}$ is equal to the population standard deviation σ divided by the square root of the sample size n, $\sigma_{\bar{x}}$ will be *smaller* than σ whenever n exceeds 1. This reflects the fact that values of $\bar{X}$ are *more alike* than individual population values are. The normal curve for the population will therefore be *flatter* than the one we use to calculate probabilities for $\bar{X}$.

Since the standard error of the sample mean is inversely proportional to the square root of the sample size, σ_X becomes smaller as the sample size becomes larger. This is reflected by a tighter cluster in possible values of $\bar{X}$ about the central level μ for larger sample sizes. This is also reflected in the shape of the normal curve, which is more peaked when $\sigma_{\bar{x}}$ is small.

The normal curves for the possible values of the sample mean to be calculated from the IQ tests are shown in Figure 7-7 for $n = 100$ and $n = 400$. Although only one value of n will actually be used, we can compare the normal curves to determine which sample size is more accurate. The curve for the larger sample size is much denser near μ. This means that when $n = 400$, it is more likely that $\bar{X}$ will be close to μ. The shaded portions represent the probability that $\bar{X}$ will lie within one point of μ. For $n = 100$, the standard error of $\bar{X}$ is

$$\sigma_{\bar{X}} = \frac{\sigma}{\sqrt{n}} = \frac{10}{\sqrt{100}} = 1$$

and for a distance of $\bar{x} - \mu = 1$ point,

$$z = \frac{\bar{x} - \mu}{\sigma_{\bar{X}}} = \frac{1}{1} = 1.00$$

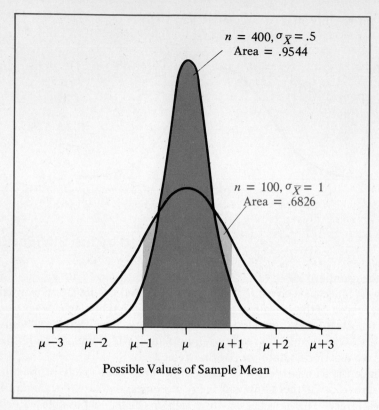

FIGURE 7-7
The effect of increased sample size on the sampling distribution of $\bar{X}$.

The probability that $\bar{X}$ will lie within one point of μ is therefore $2(.3413) = .6826$. For $n = 400$, a different standard error applies: $\sigma_{\bar{x}} = 10/\sqrt{400} = .5$. The corresponding normal deviate is

$$z = \frac{\bar{x} - \mu}{\sigma_{\bar{x}}} = \frac{1}{.5} = 2.00$$

The analogous probability is $2(.4772) = .9544$.

We see that the area covering ± 1 point from μ is larger for $n = 400$. This indicates that a larger sample size provides a more reliable estimate of μ for the same level of precision. Since larger sample size is reflected by a smaller value of $\sigma_{\bar{x}}$, we can conclude that the smaller the value of $\sigma_{\bar{x}}$, the more accurate the sample result will be.

The standard error of $\bar{X}$ is also affected by the population standard deviation σ. Thus, the size of $\sigma_{\bar{X}}$ depends on the degree of dispersion in the population values. This means that the standard error of $\bar{X}$ is a single gauge of sample accuracy that incorporates the effects of both the size of the sample and the variability of the population itself.

To illustrate the effect of population dispersion on sample accuracy, we will compare two samples taken from different populations.

EXAMPLE: ESTIMATING GASOLINE MILEAGES

The mean gasoline mileages of compact and intermediate cars made by the same manufacturer are to be estimated by a consumer testing service. The two car sizes are assumed to present different normally distributed populations, with standard deviations of 2 miles per gallon (mpg) for the compacts and 4 mpg for the intermediates. Thus, the mileages of the intermediate cars show more population dispersion.

Assuming that a separate random sample of $n = 100$ is taken from each population, the standard errors of the sample mean for each type of car are

$$\text{For compacts}: \quad \sigma_{\bar{X}} = \frac{2}{\sqrt{100}} = .2$$

$$\text{For intermediates}: \quad \sigma_{\bar{X}} = \frac{4}{\sqrt{100}} = .4$$

Normal curves are shown below for the respective sampling distributions of $\bar{X}$. The two curves are drawn on separate graphs to emphasize the fact that different

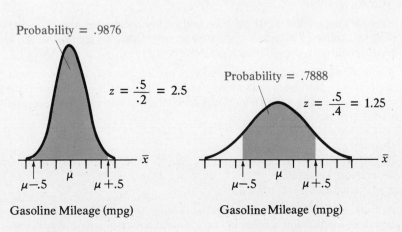

Probability = .9876

$$z = \frac{.5}{.2} = 2.5$$

Gasoline Mileage (mpg)

Compact Cars

Probability = .7888

$$z = \frac{.5}{.4} = 1.25$$

Gasoline Mileage (mpg)

Intermediate Cars

populations are involved, with the compacts having the smaller standard deviation. The shaded areas provide the probabilities that each sample mean will lie within $\pm.5$ mpg of the respective unknown population means.

A more reliable result is obtained for the compacts, since that normal curve for possible sample means provides a tighter clustering of $\bar{X}$ about μ. This is because the standard error of the sample mean is smaller for the compact cars.

We should be careful not to confuse $\sigma_{\bar{X}}$ with s. They are both standard deviations, but they represent entirely different phenomena—another reason why we usually call $\sigma_{\bar{X}}$ the standard error of $\bar{X}$. Remember that s is the standard deviation of a particular sample result and is actually calculated from the observed sample values at the final stage of sampling; s measures the variability among the observations that are actually made.

In contrast, $\sigma_{\bar{X}}$ measures the variability of the *possible* $\bar{X}$ values that *might be* obtained and is mainly used in the planning stage of sampling. Even though $\bar{X}$ will be assigned only one numerical value, depending on the sample results, we have demonstrated the need to know the pattern of variation among the possible values that $\bar{X}$ might assume. This variation is summarized by the standard error of the sample mean.

EXERCISES

7-7 A random sample of size n is selected from a normally distributed population. When $\sigma = 10$, find $\sigma_{\bar{X}}$ for the following values of n: (a) 4; (b) 9; (c) 25; (d) 100; (e) 400; (f) 2,500; (g) 10,000.

7-8 Random samples of size 100 are selected from five normally distributed populations with the following standard deviations σ: (a) .1; (b) 1.0; (c) 5.0; (d) 20; (e) 100. Calculate the value of $\sigma_{\bar{X}}$ for each sample.

7-9 The times taken to install bumpers on cars passing along a particular assembly line are normally distributed with mean time $\mu = 1.50$ minutes and standard deviation $\sigma = .25$ minute. A sample of $n = 25$ cars is obtained and the mean bumper installation time $\bar{X}$ is to be calculated.
(a) Calculate $\sigma_{\bar{X}}$.
Determine the probability that $\bar{X}$ will
(b) lie between 1.40 and 1.50 minutes.   (e) lie between 1.52 and 1.58 minutes.
(c) exceed 1.35 minutes.   (f) exceed 1.65 minutes.
(d) fall at or below 1.45 minutes.

7-10 The city public health department closes down a certain beach when the concentration of *E. coli* bacteria, present in raw sewage, becomes too high. Assume that on a particular day the contamination-level index is normally distributed, with a mean of $\mu = 160$ and a standard deviation of $\sigma = 20$ for each liter of water. A sample of $n = 25$ liters is taken, and the mean index value $\bar{X}$ is found.

(a) Calculate $\sigma_{\bar{X}}$.

Determine the probability that $\bar{X}$ will

(b) lie between 150 and 160.   (e) lie between 165 and 170.

(c) exceed 148.   (f) exceed 162.

(d) fall at or below 153.

7-11 The population of lifetimes of guinea pigs injected from birth with a carcinogenic substance is normally distributed, with a mean of $\mu = 800$ days and a standard deviation of $\sigma = 100$ days. A random sample of $n = 25$ animals is chosen.

(a) What is the probability that the sample mean will differ from the population mean by more than 20 days in either direction?

(b) What is the probability that $\bar{X}$ will fall below 740 days?

(c) Suppose that the dosage is increased and the population mean lifetime is undetermined. Assuming that the standard deviation remains $\sigma = 100$ days, what is the probability that $\bar{X} \leq 700$ if (1) $\mu = 750$; (2) $\mu = 700$; (3) $\mu = 650$; (4) $\mu = 600$?

SAMPLING DISTRIBUTION OF $\bar{X}$ WHEN THE POPULATION IS NOT NORMAL

7-4

Even when a sample is selected from a population whose frequency distribution is not normal, the sampling distribution of the mean will still be approximately normal for large sample sizes. Before discussing this result, called the *central limit theorem*, we will consider how the sample size affects the sampling distribution of $\bar{X}$.

Effect of Sample Size n on the Sampling Distribution of $\bar{X}$

Figure 7-8 illustrates the sampling distributions of $\bar{X}$ for samples of various sizes. Here, $\bar{X}$ represents the mean number of bedrooms in a sample of randomly selected homes in a certain city. The same population frequency distribution in Table 7-1 was used to generate all the cases shown. Note that as the sample size increases, the number of possible values of $\bar{X}$ increases and the sampling distributions become more symmetrical about μ. Although the number of possible $\bar{X}$ values increases as n increases, the total variability declines. This is indicated by the tendency of the values farther from μ to decrease in probability as the number of spikes of significant height nearer μ increases. This tendency is reflected in the standard error of $\bar{X}$, which becomes smaller as n becomes larger. It is even more interesting that the pattern presented

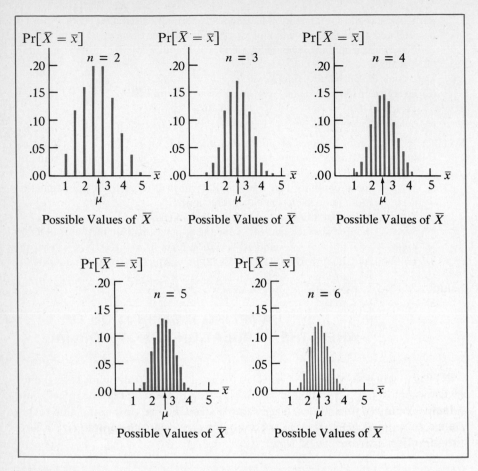

FIGURE 7-8
Comparative sampling distributions of $\bar{X}$ for different sample sizes.

by the spikes resembles the bell shape that we ordinarily associate with the normal distribution. This tendency becomes more pronounced as n increases, illustrating a crucial point of statistical theory, which is substantiated by the central limit theorem (to be discussed next). The tendency of the sampling distribution of $\bar{X}$ to assume the bell shape permits us to obtain the sampling distribution of $\bar{X}$ even when the population is not normally distributed.

The Central Limit Theorem

As the foregoing illustration suggests, it has been established mathematically that the normal distribution may be used as a basis for approximating the sampling distribution of $\bar{X}$. This fact alone makes the normal distribution

TABLE 7-1

Population Relative Frequency Distribution for
the Number of Bedrooms in Residential Units

Number of Bedrooms X	Relative Frequency f	fX
1	.2	.2
2	.3	.6
3	.2	.6
4	.2	.8
5	.1	.5
	1.0	$\mu = 2.7$

so fundamentally important in statistics.

We now state the

CENTRAL LIMIT THEOREM

As sample size _n_ becomes large, when each observation is independently selected from a population with a mean of μ and a standard deviation of σ, the sampling distribution of $\bar{X}$ tends to assume a normal distribution with a mean of μ and a standard deviation of $\sigma_{\bar{x}}$.

Note that the central limit theorem can be applied regardless of the shape of the population frequency distribution. It may be used whether the observation random variable is discrete or continuous.* Figure 7-9 shows the sampling distributions of $\bar{X}$ obtained mathematically for samples taken from populations having frequency distributions of various shapes. Note that the sampling distribution of $\bar{X}$ becomes more bell shaped as sample size increases.

The central limit theorem is useful because it permits us to make an inference about a population without knowing anything about its frequency distribution except what we can determine from a sample, although we must still presume a specific value for σ.

---

* One restriction is that the population must have a finite variance. This is a theoretical limitation of no practical significance to populations ordinarily encountered in statistical study.

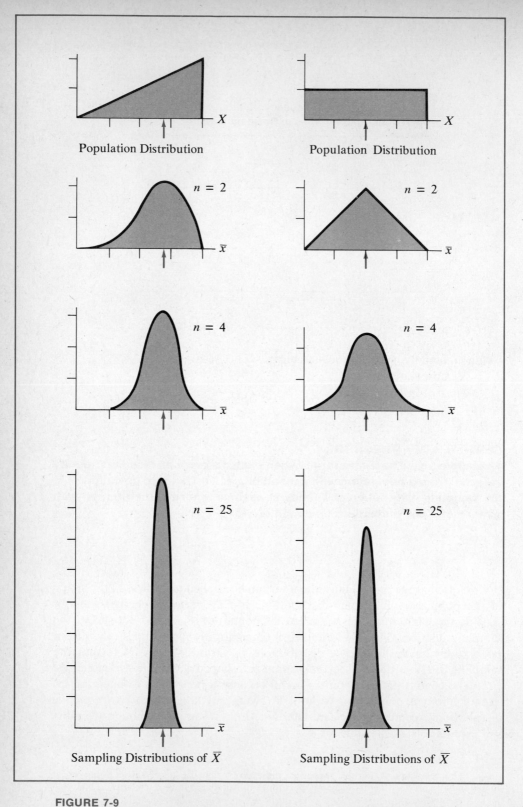

FIGURE 7-9

Illustration of the central limit theorem, showing the tendency toward normality in the sampling distribution of $\bar{X}$ as n increases for various populations.

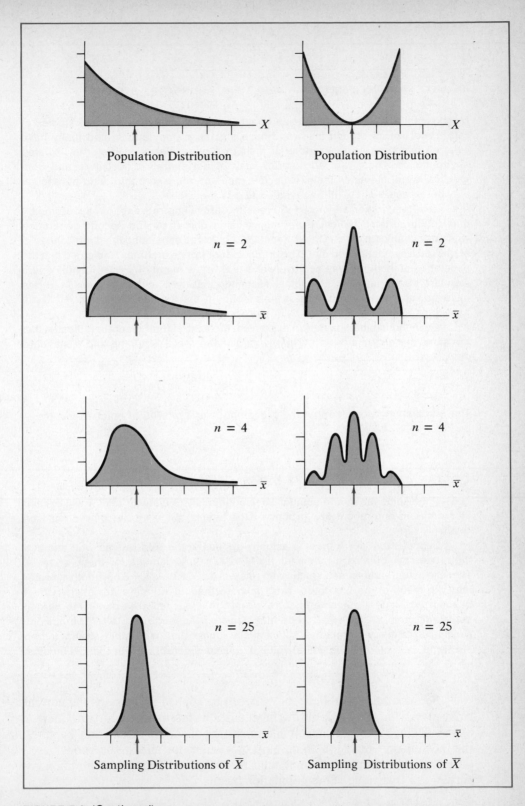

Population Distribution   Population Distribution

$n = 2$ $n = 2$

$n = 4$ $n = 4$

$n = 25$ $n = 25$

Sampling Distributions of $\overline{X}$ Sampling Distributions of $\overline{X}$

FIGURE 7-9 (Continued)

EXAMPLE: ARE WILD RATS HARDIER THAN LABORATORY STRAINS?

Much scientific research on human health and psychology is rooted in studies of rats. Laboratories around the world have stable rat populations, and many have cultured hundreds of generations of their own genetic strains in sterile, isolated environments. Thus, two laboratories may obtain entirely different results from identical experiments on their respective races of rats, and a third kind of reaction may be expected when "ill-behaved" wild rats are used.

One researcher attempted to break the prevailing rat-research syndrome by duplicating certain medical experiments using only first- and second-generation wild rats captured in cities. One of her studies involved estimating the mean longevity of rats fed a synthetic-food diet. The original laboratory experiment indicated that the population of lifetimes was positively skewed, with a mean of $\mu = 24$ months and a standard deviation of $\sigma = 6$ months. Assuming that the lifetimes of these wild rats have the same distribution, what is the probability that a sample of size $n = 100$ will have a mean lifetime of 25 months or more?

Even though the population is skewed, the central limit theorem indicates that the sampling distribution is normal, with a mean μ of 24 months and a standard deviation of

$$\sigma_{\bar{X}} = \frac{\sigma}{\sqrt{n}} = \frac{6}{\sqrt{100}} = .6$$

The researcher calculates that the probability that $\bar{X}$ will be at least 25 months is

$$z = \frac{\bar{X} - \mu}{\sigma_{\bar{X}}} = \frac{25 - 24}{.6} = 1.67$$

$$\Pr[\bar{X} \geq 25] = .5000 - .4525 = .0475$$

Thus, if the same mean and standard deviation apply to the wild rats, $\bar{X}$ will exceed 25 months in only about 5% of these experiments. Such an outcome should be relatively rare.

Suppose that $\sigma = 6$ months actually applies to the wild rats as well and that the researcher obtains a mean longevity of $\bar{X} = 30.53$ months for these animals. Then the actual sample mean of the wild rats will be $30.53 - 24 = 6.53$ months greater than the mean of the laboratory rats, or more than 10 standard deviations ($10\sigma_{\bar{X}}$) beyond $\mu = 24$. This outcome is so rare that the upper tail area under the normal curve will be much too small to be obtained from Appendix Table D. The obvious conclusion is that wild rats live substantially longer than laboratory strains fed on the same diet, and that the actual value of μ is considerably larger than 24 months.

Because the central limit theorem can be applied universally, the normal curve plays a fundamental role in a great portion of statistical theory. In Chapter 8, we will make use of this result in estimating population means. In Chapter 9, the theorem will enable us to develop procedures for making decisions based on the sample mean, which may tend to confirm or deny an assumption regarding the true value of the population mean.

EXERCISES

7-12 A sample of size $n = 49$ is taken from a population having a mean of $\mu = 100$ and a standard deviation of $\sigma = 14$. Find the probability that $\bar{X}$ will lie between 96 and 104.

7-13 An economist serving as a consultant to a large teamsters' local wishes to estimate the mean annual earnings μ of the membership. He will use the mean $\bar{X}$ of a sample of $n = 100$ drivers as an estimate. Assume that the standard deviation of the membership's annual earnings is $\sigma = \$1,500$.
  (a) Find the probability that the estimate will lie within $200 of the actual population mean.
  (b) Increasing n to 400, find the probability that the estimate will lie within $200 of the actual population mean. What is the percentage increase in probability over the probability you found in (a)?
  (c) Determine the probability that the estimate will lie within $100 of the true population mean when $n = 100$. What is the percentage reduction in probability over the probability you found in (a)?

7-14 A bus transportation agency wishes to estimate the mean mileage obtained by a new type of radial tire. Due to operating methods, a tire may be used on several buses during its useful life, so that a separate mileage log must be kept for each tire in the sample. Since this procedure is costly, only $n = 100$ tires are to be checked. In a previous study on another type of tire, a standard deviation of $\sigma = 2,000$ miles was determined. It is assumed that the same figure will apply to the new tires.
  (a) What is the probability that the sample mean will be within 500 miles of the population mean?
  (b) The maintenance superintendent states that the new tires may exhibit greater mileage variability than the tires in current use. Assuming that the standard deviation is increased to $\sigma = 4,000$ miles, find the probability that the sample mean will differ from the population mean by no more than 500 miles.
  (c) Compare your answers to (a) and (b). What can you conclude about the effect of increased population variability on the reliability of the sample mean as an estimator?
  (d) Suppose that a sample of $n = 200$ tires is used instead. If a standard deviation of $\sigma = 2,000$ miles is determined, find the probability that the sample mean will differ from the population mean by no more than 500 miles.
  (e) Compare your answers to (a) and (d). What can you conclude about the effect of increased sample size on the reliability of the sample mean as an estimator?

7-15 The operations manager of a port authority ordered an extensive study to determine the optimal number of toll booths to open during various times of the week. One unanticipated finding is that the mean time to collect a toll decreases as traffic becomes heavier. For example, on late Friday afternoons, collection times were found to have a mean of $\mu = 10$ seconds, with a standard deviation of $\sigma = 2$ seconds. On less busy Wednesday mornings, the mean was $\mu = 12$ seconds and the standard deviation was $\sigma = 3$ seconds. A consistency check is now being made to determine whether the season of the year affects efficiency. Random samples of $n = 25$ cars are taken on Wednesday mornings and Friday afternoons. Assume that these results are true population parameters.

(a) What is the probability that the Wednesday sample mean will differ by no more than 1 second from the assumed mean?

(b) What is the same probability for Friday?

(c) Why do the probabilities you found in (a) and (b) differ?

7-5   SAMPLING DISTRIBUTION OF P AND THE NORMAL APPROXIMATION

In Chapter 6, we established that the proportion of successes obtained from a Bernoulli process is a random variable having the binomial distribution. A random sample selected from a population of two complementary attributes may be construed to be a Bernoulli process when the population is very large or when sampling with replacement.

Advantages of Approximating the Binomial Distribution

Calculating probabilities from the binomial formula can be a tedious chore that may seem insurmountable in many cases. For example, consider calculating the probability that $R = 324$ convictions are obtained in the next $n = 2,032$ criminal cases brought to trial in a state, when the average rate of convictions is $\pi = .537$. This would involve evaluating

$$\frac{2,032!}{324!(2,032 - 324)!} (.537)^{324}(1 - .537)^{2,032 - 324}$$

Such computations would require working with both extremely large and small numbers, making the use of logarithms or other approximations almost mandatory.

Of course, using a high-speed digital computer might make the task manageable. Binomial probabilities can be calculated and tabulated, but then a new problem arises: For what values of π and n should tables be constructed? Obviously, not all π values can be accommodated, because they are infinite in number (all possible values between 0 and 1), and the probability table for a particular π would be quite long for even a moderately large n. Clearly, no table can be constructed that would contain all the binomial probabilities.

The Normal Distribution as an Approximation

The difficulties just discussed can be avoided if a satisfactory approximation to the binomial distribution can be used. Remember that a graph of the binomial distribution tends to become bell shaped as n increases (see Figure 6-4, page 180). This suggests that for large sample sizes, the binomial distribution

TABLE 7-2

Commonly Accepted Guidelines for Using the
Normal Approximation to the Binomial Distribution

Whenever π Equals			Use the Normal Approximation Only If n Is No Smaller Than
	.5		10
.40	or	.60	13
.30	or	.70	17
.20	or	.80	25
.10	or	.90	50
.05	or	.95	100
.01	or	.99	500
.005	or	.995	1,000
.001	or	.999	5,000

approaches the shape of the normal curve. It can even be established that the central limit theorem applies to the proportion of successes *P*. This is an advantage, because the areas under the normal curve are conveniently tabulated.

The commonly accepted guidelines for using the normal approximation are given in Table 7-2. These guidelines have been constructed according to the popular rule that the normal approximation to the binomial distribution is adequate whenever both of the following hold:

$$n\pi \geq 5$$

$$n(1 - \pi) \geq 5$$

Some statisticians insist that even larger sample sizes than those listed in Table 7-2 must be used before the approximation is acceptable. In some cases, a very large sample size should be used, because the skew of the binomial distribution is so pronounced for large or small π that the bell shape is assumed by the binomial distribution only for very large *n*.

In Chapter 6, we established expressions for the mean and the standard deviation of *P*. The expected value of *P* is π and the standard error of *P* is

$$\sigma_P = \sqrt{\frac{\pi(1 - \pi)}{n}}$$

Using a normal curve centered at π with a standard deviation of σ_P, we can determine for any possible level *p* of the sample proportion the corresponding

NORMAL DEVIATE FOR THE SAMPLE PROPORTION

$$z = \frac{p - \pi}{\sigma_P}$$

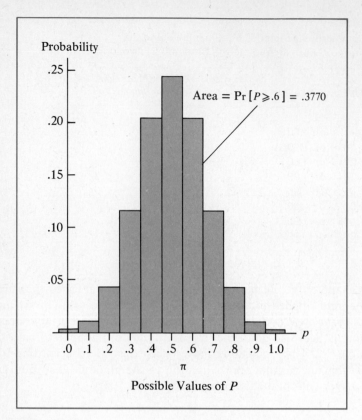

FIGURE 7-10
Binomial sampling distribution of P.

To illustrate the suitability of the normal approximation, we will consider the two distributions when $n = 10$ and $\pi = .5$. The actual sampling distribution is plotted in Figure 7-10. (The probability values were obtained from Appendix Table C.) To compare this discrete distribution to the continuous normal distribution, the probabilities are presented as bars rather than spikes. The height of each bar represents the probability that P assumes the value at the midpoint of the bar's base. Since the base of each bar may be considered to be of unit width (letting each unit represent a .10 increment on the scale of p), the area of a bar also represents the probability of the P value at the midpoint. The area .3770 of all the bars at or above the point $P = .6$ therefore represents the probability that P will be greater than or equal to .6 (the gray area in Figure 7-10).

We can obtain the normal approximation to the sampling distribution of P. Here, the mean and the standard error are

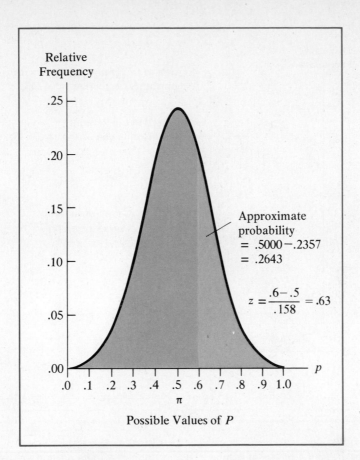

FIGURE 7-11

Normal approximation to the sampling distribution of *P*, with $n = 10$ and $\pi = .5$.

$$\mu(P) = \pi = .5$$

$$\sigma_P = \sqrt{\frac{.5(1 - .5)}{10}} = .158$$

The corresponding normal curve is graphed in Figure 7-11. Treating *P* as an approximately normal random variable, the probability that *P* lies at or above .6 is .2643 (the gray area under the approximating normal curve in Figure 7-11).

Note that the area under the normal curve above .6 only approximates the true probability indicated by the gray area under the bars in Figure 7-10. The true probability is .3770, but the normal curve provides a probability of .2643. The discrepancy between the binomial probabilities and the probabilities found by using the normal curve are negligible for larger sample sizes. It is easier to assess the nature of the approximation in Figure 7-12, where

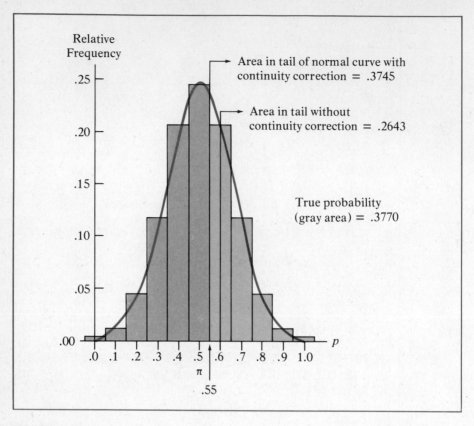

FIGURE 7-12
A comparison of the binomial sampling distribution of P to its normal approximation.

the graph for the binomial distribution of P is superimposed on the normal curve. Here, a continuity correction has been applied to obtain a better approximation.*

* In Figure 7-12, we could have obtained a value from the normal curve area above .55 rather than above .60:

$$z = \frac{.55 - .50}{.158} = .32$$

$$\Pr[P \geq .55] = .5000 - .1255 = .3745$$

The value .3745 is much closer to the true binomial probability of .3770. Using .55 instead of .60 applies a continuity correction of $.5/n = .5/10 = .05$. A continuity correction is needed because the *discrete* binomial distribution is being approximated by a *continuous* normal distribution. Thus, we may improve our normal curve approximation by subtracting $.5/n$ from the lower limit for P and adding $.5/n$ to the upper limit. The continuity correction will be ignored in this book to make discussions conceptually simpler. For most sample sizes where the normal approximation will be made, n is large enough so that $.5/n$ is relatively small, and the continuity correction only slightly improves the approximation.

EXAMPLE: EVALUATING VOTER PREFERENCES

A candidate for mayor wants to gauge potential voter reaction to an increase in recreational services by estimating the proportion of voters who now use city services. If we assume that 50% of the voters require city recreational services, what is the probability that 40% or fewer voters in a sample of $n = 100$ actually will use these city services?

Because the population is large, we assume that the sample proportion of users has a binomial sampling distribution. Here, $\pi = .50$ and we wish to evaluate the probability that $P \leq .40$. Referring to Table 7-2, we see that $n = 100$ is sufficiently large to use the normal approximation, so we calculate

$$\sigma_P = \sqrt{\frac{.5(1 - .5)}{100}} = .05$$

The normal approximation provides the following:

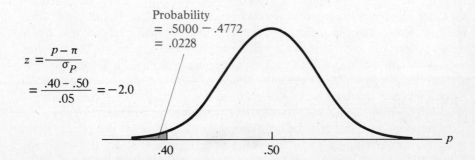

$$z = \frac{p - \pi}{\sigma_P}$$

$$= \frac{.40 - .50}{.05} = -2.0$$

Probability
= .5000 − .4772
= .0228

Thus, the probability is quite small that 40% or fewer of the sample will use city recreational services. Obtaining a sample result of $P = .40$ would be a rare outcome when the actual proportion of users is assumed to be $\pi = .50$. Such sample evidence would strongly indicate that the unknown value of π is smaller than .50.

Fortunately, the actual cumulative binomial probability values for $n = 100$ are given in Appendix Table C. For $\pi = .50$ and $n = 100$, we find that

$$\Pr[P \leq .40] = \Pr\left[\frac{R}{n} \leq \frac{40}{100}\right] = .0284$$

In this case, the normal approximation provides a value quite close to the actual binomial distribution.

If a sample of $n = 5$ were used, this probability calculation would not have been obtained using the normal approximation, because Table 7-2 indicates that n is too small for $\pi = .50$. This desired probability would have to have been obtained directly from the binomial distribution.

An Application of the Normal Approximation to Acceptance Sampling

We can illustrate the usefulness of the normal approximation to the binomial distribution with an example of *acceptance sampling*. A manufacturer who depends on outside suppliers for raw materials or components for production will accept only shipments of some set minimum quality. Due to the expense involved, the decision to accept or reject an incoming shipment is based on a sample of the items received. The procedure for accomplishing this is an acceptance sampling plan—a common approach in quality control.

Consider an automobile assembly plant that procures headlamps. The supplier has designed its production operation in such a way that, for various reasons, about 5% of its output is defective. The assembly plant quality-control manager inspects a sample of $n = 100$ from each shipment. If 7 or fewer (that is, 7% or fewer) items in the sample are defective, the manufacturer will accept the shipment; otherwise, the batch will be rejected and returned to the supplier.

The number 7 is referred to as the *acceptance number*. It forms the basis for a *decision rule*, which specifies the action to be taken regarding a particular batch. We can determine the implications for both the consumer and the producer by using such a rule to calculate various probabilities.

Suppose a large batch of headlamps contains $\pi = .10$ defectives. If the quality-control manager knew the value of π, he would reject the shipment. But because only a sample will be taken, he may accept the poor batch. What is the probability that the manager will accept the shipment?

The shipment will be accepted if the sample proportion of defectives P is $\leq .07$. Employing the normal approximation, we first find the standard error of P:

$$\sigma_P = \sqrt{\frac{\pi(1-\pi)}{n}} = \sqrt{\frac{.10(1-.10)}{100}} = .03$$

The probability for accepting the bad batch is then computed:

$$z = \frac{p - \pi}{\sigma_P} = \frac{.07 - .10}{.03} = -1.00$$

$$\Pr[P \leq .07] = .5000 - .3413 = .1587$$

Thus, there is a .1587 probability that this poor batch of headlamps will be accepted. Because accepting an inferior shipment is an incorrect decision that will hurt some product users, such a probability value is referred to as the *consumer's risk*.

The supplier can be hurt by another kind of erroneous decision. Suppose that an acceptable lot having $\pi = .05$ defectives is shipped. If, by chance, more than 7 defectives are sampled, the good shipment will be rejected. To find the probability that this will occur, we again use the normal approximation. The standard error of *P* is

$$\sigma_P = \sqrt{\frac{.05(1 - .05)}{100}} = .0218$$

The probability for rejecting the good batch is therefore

$$z = \frac{.07 - .05}{.0218} = .92$$

$$\Pr[P > .07] = .5000 - .3212 = .1788$$

A probability that an acceptable shipment will be erroneously rejected is referred to as the *producer's risk*.

The producer's risk and the consumer's risk are illustrations of the more general Type I and Type II errors of hypothesis testing that we will discuss in Chapter 9. There, we will describe how a decision rule may be constructed to balance both risks and to keep the probabilities for making erroneous decisions at desired levels.

EXERCISES

7-16 Use the normal approximation to find the probability that the proportion *P* of successes in $n = 100$ trials of a Bernoulli process with a trial success probability of $\pi = .5$ will lie between each of the following pairs of values:
(a) .40, .60   (b) .35, .65   (c) .50, .65   (d) .50, .90

7-17 Find the approximate probability that the proportion *P* of successes obtained from a Bernoulli process with a trial success probability of $\pi = .5$ will lie between .4 and .6 when the number *n* of trials is
(a) 64   (b) 100   (c) 400
(d) What relationship do you notice between the probability values you obtained and the trial size *n*? How do you explain this relationship?

7-18 A cannery accepts a shipment of tomatoes whenever *P* (the proportion that is ripe in a sample of $n = 100$) is $\geq .90$. Assume that the shipments are sufficiently large that the sampling process can be represented (with only minor error) as a Bernoulli process.

(a) What is the approximate probability for accepting a poor shipment in which the proportion of ripe tomatoes is $\pi = .8$?

(b) What is the approximate probability for rejecting a good shipment in which the proportion of ripe tomatoes is $\pi = .95$?

(c) How do you account for the fact that these probabilities indicate some shipments of poor quality will be accepted and some shipments of good quality will be rejected?

7-19 Random samples of voter preferences are obtained for candidates to the U.S. Senate. One sample of $n = 100$ is taken from a precinct's registered voters. There are only two candidates—a Republican and a Democrat. In both (a) and (b), each voter in the sample favors one of the candidates.

(a) Assuming that the proportion of Republicans in the precinct is $\pi = .40$ and that the voters will adhere strictly to party lines, find the approximate probability for each sample that the majority of the voters polled will favor the Republican candidate (that is, $P \geq .50$).

(b) Repeat (a), using samples of $n = 49$.

(c) Suppose some of the voters are "undecided" when they are polled. Can any sample that includes undecideds accurately reflect the votes that the precinct population will cast on election day? Explain.

REVIEW EXERCISES

7-20 The gestation period for a particular herd of cows can be represented by the normal distribution, with a mean of $\mu = 281$ days and a standard deviation of $\sigma = 5$ days. Assuming that the length of pregnancies for the next 100 births is a random sample from this population, find the probability that the mean gestation period will

(a) exceed 282 days.   (c) be less than 282.5 days.

(b) lie between 280 and 282 days.   (d) be less than 280.5 days.

7-21 Find the following percentiles for a normal distribution for women's heights with a mean of $\mu = 66$ inches and a standard deviation of $\sigma = 1.5$ inches: (a) 50th; (b) 15th; (c) 95th; (d) 75th; (e) 33rd.

7-22 The annual incomes of surgeons constitute a highly positively skewed population. Nevertheless, according to the central limit theorem, it is still possible to find the sampling distribution of the mean for a random sample of these incomes by using the normal distribution. Suppose the population has an unknown mean of μ and a standard deviation of $\sigma = \$10,000$. An estimate of μ is to be made using the value of the sample mean $\bar{X}$. This estimate must fall within $\pm \$1,000$ of the true mean.

(a) If $n = 100$ incomes, find the probability that the estimate will meet the desired accuracy.

(b) If a sample $n = 625$ incomes is used instead, find the probability that the estimate will meet the desired accuracy.

7-23 The weights of "1-ounce" gold ingots cast by a rare-metals refinery are normally distributed, with a mean of $\mu = 28$ grams and a standard deviation of $\sigma = 1$ gram. Consider a shipment of 25 ingots.

(a) What is the probability that the mean ingot weight will be less than 27.5 grams?

(b) Find the probability that the entire shipment will weigh more than 705 grams.

7-24 A parapsychologist is testing the extrasensory perception of a purported clairvoyant. Using a deck of cards, half of which are red and half of which are black, she selects the top card and asks her subject (who is in another room) to identify its color. The procedure is repeated 100 times. Each successive card is replaced in the deck, and the deck is shuffled before the next card is drawn. A subject with no powers of ESP should identify the correct color 50% of the time, so that $\pi = .5$. Under this assumption, what is the probability that a correct response will be obtained for 65 or more of the 100 cards? Use the normal approximation.

7-25 The IQ scores achieved by first-grade students in a certain state are normally distributed, with a mean of $\mu = 105$ and a standard deviation of $\sigma = 10$ points. A random sample of n scores is to be selected. Find the standard error of the sample mean and the probability that this mean IQ will exceed 106 points, when (a) $n = 100$; (b) $n = 25$; (c) $n = 900$.

7-26 A political-polling firm believes that the proportion of persons favoring a certain presidential policy is $\pi = .4$. A sample of $n = 100$ persons is selected at random from the entire electorate, and the proportion favoring the presidential policy can be approximated by the normal distribution.
(a) If the assumed parameter value holds, what is the probability that 50% or more of the persons polled will favor the policy?
(b) If the actual parameter value is only $\pi = .37$, what is the probability that 50% or more of the sample will favor the policy?

7-27 For each of the following sampling situations, indicate (by *yes* or *no*) whether the sampling distribution of the sample proportion can be approximated by the normal curve.
(a) $\pi = .50; n = 9$ (d) $\pi = .45; n = 20$
(b) $\pi = .10; n = 36$ (e) $\pi = .01; n = 1,000$
(c) $\pi = .83; n = 100$

7-28 If the population of members of the U.S. Army has a mean height of 70 inches and a standard deviation of 3 inches, determine the probability that the mean height for a random sample of 100 soldiers will be
(a) between 70 and 70.5 inches.   (d) less than 68 inches.
(b) less than 69.5 inches.   (e) between 69.4 and 70.8 inches.
(c) greater than 72 inches.

7-29 A students marks an examination consisting of 36 true-or-false questions by tossing a coin. For each question, he answers "true" for a head or "false" for a tail. Assuming that half the correct answers should be marked "true," what is the probability that the student will pass the examination by marking at least 75% of the answers correctly? Use the normal approximation.

7-30 A purchasing clerk measures the quantities in all bulk products purchased. On the average, 1,000-foot rolls of transparent tape should be close to the specified length. In evaluating a shipment, a random sample of 100 rolls of tape is selected, and the mean length is determined. If the mean is greater than or equal to 990 feet, the entire shipment is accepted; otherwise, the shipment is rejected and returned to the manufacturer.
(a) If a shipment has a mean roll length of only 980 feet, with a standard deviation of 50 feet, find the consumer's risk (probability of accepting these inferior tapes).

(b) If a shipment has mean roll length of 1,005 feet, with a standard deviation of 60 feet, find the producer's risk (probability of rejecting these good tapes).

7-6   OPTIONAL TOPIC: SAMPLING DISTRIBUTION OF $\bar{X}$ WHEN THE POPULATION IS SMALL

We have established that for independent sample observations, the standard error of $\bar{X}$ may be obtained from the expression $\sigma_{\bar{X}} = \sigma/\sqrt{n}$. Observations are always independent when sampling with replacement, but we usually sample *without replacement*. When populations are large compared to sample size, the probability distributions for successive observations barely change when earlier items are removed. In these cases, we can generally assume independence between observations.

But when the population is small compared to sample size, this fact must be reflected in computing $\sigma_{\bar{X}}$.

STANDARD ERROR OF $\bar{X}$ (SMALL POPULATIONS)

$$\sigma_{\bar{X}} = \frac{\sigma}{\sqrt{n}} \sqrt{\frac{N-n}{N-1}}$$

where σ = population standard deviation, N = population size, and n = sample size.

Finite Population Correction Factor

The term $\sqrt{(N-n)/(N-1)}$ is referred to as the *finite population correction factor*. When n is small in relation to N, the factor is very close to 1. Thus, the standard error obtained from a sample without replacement is close in value to one obtained from a sample with replacement. Note that the numerator in the expression will never be greater than the denominator, so that $\sigma_{\bar{X}}$ will be smaller when a sample is taken without replacement. *In practice, the finite population correction is usually ignored whenever n is less than 10% of N.*

For samples from small populations, the normal distribution approximately describes the sampling distribution of $\bar{X}$, even though the observations are not independent. As an illustration, consider a survey of incomes in two cities. City A has a population of $N = 10,000$; city B has a population of $N = 2,500$. Standard deviation of $\sigma = \$1,500$ is assumed to be common to both populations. A sample of $n = 1,000$ is taken from each. The following calculation provides the standard error of $\bar{X}$ for city A:

$$\sigma_{\bar{X}} = \frac{\$1,500}{\sqrt{1,000}} \sqrt{\frac{10,000 - 1,000}{10,000 - 1}} = \$45.00$$

Analogously, for city B,

$$\sigma_{\bar{X}} = \frac{\$1,500}{\sqrt{1,000}} \sqrt{\frac{2,500 - 1,000}{2,500 - 1}} = \$36.75$$

Note that the standard error of $\bar{X}$ for city B is smaller than that for city A. This is to be expected, because B is a smaller city. But the standard error of $\bar{X}$ for city A is only about 23% greater, although its population is four times as large.

The probability that $\bar{X}$ *exceeds* the mean by 100 or more can be calculated for each city. For A,

$$z = \frac{\bar{x} - \mu}{\sigma_{\bar{X}}} = \frac{100}{45.00} = 2.22$$

$$Pr[\bar{X} - \mu \geq 100] = .5000 - .4868 = .0132$$

and for B,

$$z = \frac{100}{36.75} = 2.72$$

$$Pr[\bar{X} - \mu \geq 100] = .5000 - .4967 = .0033$$

In either case, the probability for obtaining a sample mean that exceeds the respective mean by \$100 or more is relatively small. However, the same sample size provides about the same protection against obtaining extremely large sample values in *both* cities, regardless of the size of either.

When n is small in relation to N, the finite population correction factor approaches 1. Thus, a sample of $n = 1,000$ would yield almost the same finite population factor for a population of $N = 100,000$ that it would for one of $N = 1,000,000$ (.9949 versus .9995). Thus, if the standard deviations of the two populations are identical, there would be an imperceptible difference in the respective values of $\sigma_{\bar{X}}$, and the probabilities for possible sample results would be nearly identical. In this case, a sample of $n = 1,000$ would be almost as reliable in a large population as in a population one-tenth its size. This is why we concluded in Chapter 4 (page 89) that a sample of Alaska voters in a presidential election poll would have to be about the same size as a sample of California voters for the sample results to be equally reliable.

OPTIONAL EXERCISES

7-31 A random sample of 100 soldiers is chosen without replacement from a regiment of 900 men. If the mean regiment height is 71 inches, with a standard deviation of 2.5 inches, determine the probability that the sample mean height will be
(a) between 71 and 71.4 inches.   (c) greater than 71.6 inches.
(b) less than 70.5 inches.   (d) between 70.7 and 71.5 inches.

7-32 Suppose that a finite population has a standard deviation of $\sigma = 10$. Calculate $\sigma_{\bar{X}}$ for each of the following value pairs of population and sample size:
(a) $N = 1,000; n = 500$ (d) $N = 10,000; n = 1,000$
(b) $N = 1,000; n = 50$ (e) $N = 10,000; n = 500$
(c) $N = 10,000; n = 5,000$

7-33 A customs official wishes to estimate the mean weight of each of three batches of copper ingots obtained from different smelters. $N = 500$ ingots have arrived from Chile, $N = 1,000$ from Bolivia, and $N = 700$ from Arizona. Samples of $n = 100$ ingots from each batch are weighed. Each smelter is assumed to produce batches of ingots with a mean weight of $\mu = 100$ pounds and a standard deviation of $\sigma = 5$ pounds.
(a) Find the probability that the sample mean of each batch lies between 99 and 101 pounds.
(b) Using the same sample size, find the analogous probability values when the sample means are obtained from batches ten times as large.
(c) Do your answers differ significantly between smelters? Explain.

7-34 Suppose that the customs official in Exercise 7-33 establishes a rule to enable her to decide when an entire batch of ingots should be individually weighed. Because it is such a time-consuming procedure, she wants to weigh the ingots only when there is strong sample evidence that they are underweight.

Assume that an acceptable batch of $N = 1,000$ ingots has a mean weight of $\mu = 100$ pounds, but that the official does not know this value. She takes a sample of $n = 200$ and calculates $\bar{X}$. Suppose that the batch standard deviation is $\sigma = 5$ pounds. The official must consider three decision rules:

(1) Accept batch (do not weigh all 1,000 ingots) if $\bar{X} \geq 100$ pounds.
(2) Accept batch if $\bar{X} \geq 99.5$ pounds.
(3) Accept batch if $\bar{X} \geq 99.0$ pounds.

(a) Find the probability according to each of these decision rules that the customs official will accept the batch without weighing all the ingots.
(b) If she prefers the rule that maximizes the probability for not weighing all 1,000 ingots, which rule should she choose?

The more numerous the number of observations and the less they vary among themselves, the more their results approach the truth.

MARQUIS DE LAPLACE (1820)

8

Statistical Estimation

In this chapter, we will be concerned with the kinds of estimates we make from sample data and the procedures we use to make these estimates. Problems of estimation are crucial to practically every statistical application. This is especially true in business decision making, where often only sample results are available for establishing vital information. For example, samples are used to estimate a product's share of the market—a useful number for many important marketing decisions. Also, statistical estimates provide management with the average times required to complete various production operations, which allows for intelligent planning and better day-to-day control. Similarly, sample data yield median earnings of subscribers to various magazines, a demographic feature useful to merchandisers in choosing where to place advertisements. Estimation problems are perhaps most difficult in economic planning, where the impact of policies can affect the lives of literally every person in an entire nation.

ESTIMATORS AND ESTIMATES 8-1

Using samples to estimate a population parameter is one of the more common forms of statistical inference. A sample statistic that serves this purpose is called an *estimator*. An important segment of statistical theory is

237

concerned with finding statistics that are appropriate estimators. For instance, the sample mean is a particularly good estimator for the population mean. We will learn why this is so.

The Estimation Process

Parameter estimation requires a great deal of planning. First, we must choose an estimator. Then a major concern is to control the sampling error. The choice of a sampling method—judgment or convenience sampling, or some version of random sampling—will affect this. Our discussion will center on random samples, because we know they are free from sampling bias. We also know that small samples tend to be less reliable, so that sampling error can be controlled by using a sufficiently large sample. A practical consideration is to balance the level of reliability, which is influenced by sampling error, against the costs of obtaining the sample.

Due to chance, the value of the sample statistic we obtain may not be close to the population parameter. But unless we take a census of the population, we cannot determine the true parameter value. Therefore, statisticians must be able to assess the reliability of their results even though they can only theoretically determine what values are reasonable.

We are presently concerned only with estimates based on samples taken from already existing populations or from identifiable target populations. This excludes predictions based on other kinds of information—most notably, economic forecasts based on time-series data. For example, although sampling may be involved in forecasting the gross national product, the major role is played by other factors, such as judgment, current trends, contemplated changes in government policies, and the state of international affairs. Some procedures useful in making such predictions will be discussed in Chapter 10. Another category of excluded predictions concerns parameters of some future population. For example, the median annual family income for the United States in 1990 cannot be estimated from the usual sample, because the elementary units cannot be presently defined and the variate values themselves lie in the future.

Choosing an Estimator

Several alternative statistics can be used as estimators. For instance, we can use one of three sample statistics—the mean, median, or mode—to estimate the population mean. We may think that the sample mean is the most suitable estimator, but how can we substantiate our choice?

The most desirable feature of an estimator is that it has a value close to the unknown value of the population parameter. Under certain circumstances, the sample mean, median, and mode will each have a value that lies close to the population mean. The basic questions are therefore: Which statistic will be the most reliable estimator? And which will require the least expenditure of resources in terms of sample size? An important ancillary question is: Is there an easily applied theoretical procedure for determining the probability that the value obtained for the estimator statistic will lie sufficiently close to the parameter?

Criteria for Statistics Used as Estimators

Three criteria have been developed to compare statistics in terms of their worth in estimating a parameter. One determines whether there is a tendency, on the average, for the statistic to assume values close to the parameter in question. Another considers the reliability of the estimator. The third indicates whether this reliability improves as sample size increases.

Unbiased Estimators

An unbiased estimator is a statistic that has an expected value equal to the population parameter being estimated. A statistic that does not, on the average, tend to yield values equal to the parameter is said to be biased. The sample mean $\bar{X}$ is an unbiased estimator of the population mean μ, because we have established that

$$\mu(\bar{X}) = \mu$$

Now we can see why we define the sample variance s^2 as

$$s^2 = \frac{\sum(X - \bar{X})^2}{n-1}$$

Here, the divisor is $n - 1$ instead of n. However, it can be proved that s^2 defined using the $n - 1$ divisor is an unbiased estimator of σ^2, since

$$\mu(s^2) = \sigma^2$$

An intuitive reason why $n - 1$ is used as the divisor is because it provides a somewhat larger value for the sample variance than the n divisor. The estimator that results from using s^2 is larger, reflecting the fact that a sample is ordinarily less diverse than its population (the part is rarely more varied than the whole).

Efficient Estimators

One statistic is a more efficient estimator than another if its standard error is smaller for the same sample size. The sample median is an unbiased estimator of μ when the population is normally distributed. But its standard error is 1.2533 times as great as that of the sample mean. This implies that the sample mean is a more efficient estimator of μ than the sample median. Medians of samples taken from normally distributed populations tend to be more unlike each other than means of the same samples. Thus, the sample mean is always a more reliable estimator of μ.

Consistent Estimators

A statistic is a consistent estimator of a parameter if the probability that it will be close to the parameter's true value approaches unity with increasing sample size. The sample mean is a consistent estimator of the population mean. In general, *a statistic whose standard error becomes smaller as n becomes larger will be consistent*. In other words, a consistent estimator is more reliable when larger samples are used.

Consistency alone does not guarantee reliable sample results; reliability can be achieved only by increasing the sample size. But consistency is a necessary condition if a larger sample is to be more reliable. The net effect is that the use of a consistent estimator allows the statistician to buy greater reliability for the price of a larger sample.

The criteria do not always clearly indicate which statistic will be the most reliable estimator. Usually, the better of any two estimators that are both unbiased and consistent is the more efficient. But if the better statistic is cumbersome or theoretically difficult to apply, the second choice is generally used.

Commonly Used Estimators

We have seen that $\bar{X}$ is the most desirable estimator of μ, because it is unbiased, consistent, and more efficient than other estimators, including the sample median. $\bar{X}$ also has a readily obtainable *normal* sampling distribution when the sample is sufficiently large. As we will see in this chapter, knowing the sampling distribution enables us to choose a large enough sample to achieve an estimate with a desired level of reliability. It also allows us to qualify the estimate that we actually obtain after the sample results have been collected and tabulated.

EXAMPLE: ESTIMATING THE LENGTH OF A BOOK

A publisher wishes to determine the number of words in a book. An estimate of the mean number of words per page μ is obtained by counting the number of words on each of 30 randomly chosen pages. The sample mean number of words per page $\bar{X}$

is then calculated. Since there are 600 pages in the book, the estimated number of words is calculated by multiplying $\bar{X}$ by 600. The sample results are shown below.

Number of Words X	X^2	Number of Words X	X^2
383	146,689	175	30,625
325	105,625	278	77,284
411	168,921	351	123,201
416	173,056	423	178,929
395	156,025	327	106,929
372	138,384	381	145,161
293	85,849	317	100,489
361	130,321	362	131,044
216	46,656	338	114,244
431	185,761	411	168,921
406	164,836	371	137,641
394	155,236	393	154,449
402	161,604	388	150,544
376	141,376	295	87,025
268	71,824	421	177,241
		Totals 10,680	3,915,890

$$\bar{X} = \frac{\sum X}{n}$$

$$= \frac{10,680}{30}$$

$$= 356 \text{ words}$$

$$s = \sqrt{(\sum X^2 - n\bar{X}^2)/(n-1)}$$

$$= \sqrt{[3,915,890 - 30(356)^2]/29}$$

$$= \sqrt{3,924.48}$$

$$= 62.65 \text{ words}$$

The estimate of μ is 356 words. This results in an estimate of

$$356 \times 600 = 213,600$$

words in the entire book.

The sample variance s^2 is an unbiased and consistent estimator of the population variance σ^2. We can take the square root of s^2 to estimate the population standard deviation, since σ is the square root of σ^2. In the preceding example, the sample standard deviation is calculated as $s = 62.65$ words per page. This figure can then be used to estimate σ, the standard deviation of words per page for the entire book.

The population proportion π of elementary units that have a particular attribute can be estimated by the corresponding sample proportion P. We know that $\mu(P) = \pi$, so that P is an unbiased estimator of π. It is also consistent.

EXAMPLE: BUYING ON IMPULSE

A consumer information service measured the tendency of the average shopper to purchase whatever she or he picks up, whether it is needed or not. Shoppers in a random sample prepared detailed shopping lists. At the store, testers followed each sample shopper and recorded each item touched. The items on the shopper's original list were compared to those purchased. The number of items bought but not on the list was added to the number of items touched but not bought. This sum represented the total number of impulse items not included in the original list.

The proportion P of impulse items bought was then found for every sample shopper. One woman accrued a total of 132 impulse items; of these, she purchased 48. Her value of P was therefore determined to be

$$P = \frac{48}{132} = .364$$

which may be taken as an estimate of the actual proportion of impulse items she would buy under similar circumstances.

The average proportion of impulse purchases was estimated to be .55. The consumer service therefore advised its clients to keep their hands on their carts and rely on their initial grocery-buying decisions.

EXERCISES

8-1 For each of the following situations, indicate whether or not a sample may be selected from some population to make a current estimate. Explain your answers.
   (a) An estimate is to be made of the mean annual lifetime earnings of recent college graduates.
   (b) The taste preferences of the potential buyers of a new product not currently on the market are to be determined.
   (c) The responses of heart patients to a new drug, previously untried on humans, are to be measured.
   (d) The Federal Power Commission wishes to estimate the kilowatt-hour usage by customers at the end of the coming decade.

8-2 A survey is taken of hair-styling franchises to determine annual earnings. A random sample of $n = 10$ shops resulted in the following figures (in thousands of dollars, rounded): 15, 5, 2, 7, 25, 19, 11, 9, 13, 42.
   (a) Find a point estimate of the mean expenditures for all shops as a group.
   (b) Find a point estimate of the standard deviation of earnings for the population.
   (c) Estimate the proportion of earnings in excess of $10,000.
   (d) Estimate the proportion of earnings under $5,000.

8-3 Tossing a die twice in succession can be viewed as taking a random sample of $n = 2$ with replacement from a population of six observable values (corresponding to the number of dots showing on each side of the die).

(a) Calculate the range of this population.
(b) We define the *sample range* to be the difference between the highest and the lowest values obtained from the two tosses. The sampling distribution of the sample range is provided in the table below. Calculate the expected value of the sample range.
(c) Comparing your answers to (a) and (b), do you conclude that the sample range is an unbiased estimator of the population range? Explain.

Possible Sample Range	Elementary Events	Probability
0	(1, 1) (2, 2) (3, 3) (4, 4) (5, 5) (6, 6)	6/36
1	(1, 2) (2, 1) (2, 3) (3, 2) (3, 4) (4, 3) (4, 5) (5, 4) (5, 6) (6, 5)	10/36
2	(1, 3) (3, 1) (2, 4) (4, 2) (3, 5) (5, 3) (4, 6) (6, 4)	8/36
3	(1, 4) (4, 1) (2, 5) (5, 2) (3, 6) (6, 3)	6/36
4	(1, 5) (5, 1) (2, 6) (6, 2)	4/36
5	(1, 6) (6, 1)	2/36
		36/36

INTERVAL ESTIMATES OF THE MEAN USING A LARGE SAMPLE

8-2

An *interval estimate* in the form $a \leq \mu \leq b$ is preferred for reporting the results of sampling studies for several reasons:

1. It provides an estimated range of values for the unknown parameter.
2. It acknowledges that the parameter's actual value is uncertain.
3. It shows very simply the degree of precision achieved by the results obtained.

Thus, the interval estimate has a distinct advantage over a single value or *point estimate*.

Because it represents a range of possible values, an interval estimate implies the presence of uncertainty. It is ordinarily accompanied by a statement indicating the likelihood that the parameter being estimated actually does lie within the stated interval.

Whether a point or an interval estimate is to be made depends on the objective of the statistical study. Interval estimates would be unsuitable to a manufacturer as the basis for placing orders for raw materials, for the supplier cannot accept an order for "something between 2,000 and 3,000 units." Likewise, knowing only that the wattage of a stereo amplifier lies between 50 and 80 makes it difficult to compare that model to other brands. But the superiority of an interval estimate is evident in many kinds of planning or reporting. For example, knowing that next year's sales might lie between 10,000 and 11,000 units, a company president is able to draw up a realistic budget taking into account uncertainties about the sales level. He can visualize the best and worst cases, and thereby obtain a reasonable basis for evaluating the risks involved.

In this section, we will discuss the uses of the interval estimate and learn how it is constructed.

Confidence Interval Estimate of the Mean When σ Is Known

To illustrate how the interval estimate is obtained, we will begin by using $\bar{X}$ to estimate a population mean when the standard deviation of the population is known. To do this, we choose the interval end points in such a way that the interval is centered on the computed value of $\bar{X}$. Also, we attach to the interval a measure of likelihood (say, .90). The sampling distribution of $\bar{X}$ is normal, and we know there is a 90% chance that $\bar{X}$ will fall within ± 1.64 standard deviations from the mean:

$$\Pr[\mu - 1.64\sigma_{\bar{X}} \leq \bar{X} \leq \mu + 1.64\sigma_{\bar{X}}] = .90$$

We can transform this inequality by rearranging the terms inside the brackets, so that we obtain

$$\Pr[\bar{X} - 1.64\sigma_{\bar{X}} \leq \mu \leq \bar{X} + 1.64\sigma_{\bar{X}}] = .90$$

The event in this expression is in the form of an interval estimate. Substituting $\sigma/\sqrt{n}$ for $\sigma_{\bar{X}}$, we can use the following expression to calculate the interval estimate for the population mean:

$$\mu = \bar{X} \pm z\frac{\sigma}{\sqrt{n}} \quad \text{or} \quad \bar{X} - z\frac{\sigma}{\sqrt{n}} \leq \mu \leq \bar{X} + z\frac{\sigma}{\sqrt{n}}$$

This interval is centered at $\bar{X}$, and the end points are partly determined by the value selected for the normal deviate z.

The Meaning of the Interval Estimate

Remember that the interval just expressed may not actually contain μ. This is because the end points depend on the observed value of $\bar{X}$, which may differ greatly from the true μ due to sampling error. *Before* sample results are obtained, there is a probability that μ will lie within $\bar{X} \pm z\sigma_{\bar{X}}$. But our interval here is determined by a known value of $\bar{X}$ calculated *after* the sample is taken, so we cannot attach a probability to it. Once we calculate $\bar{X}$, the interval estimated from it is certain to either contain or not contain μ. But at this point, we cannot know which case applies. Figure 8-1 presents the conceptual implications of the interval estimate.

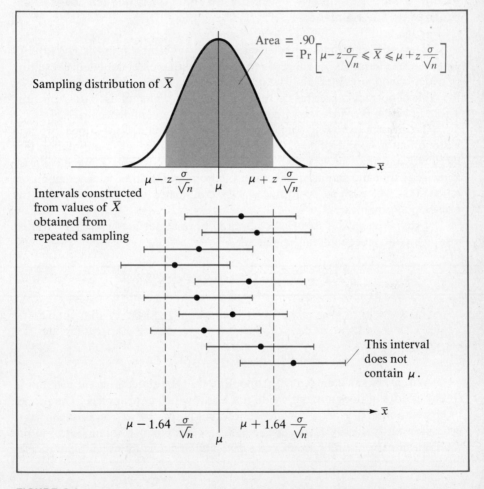

FIGURE 8-1

Illustration of the interval-estimate concept. Intervals have been obtained for ten values of $\bar{X}$ calculated for different samples taken from the same population. Note that nine of the intervals contain μ. If many repeated samples were to be taken from a population, the percentage of intervals obtained that contained μ would be approximately the same as the .90 probability that $\bar{X}$ lies within $\mu \pm 1.64\ \sigma/\sqrt{n}$.

Because we cannot attach a probability value to the truthfulness of our interval estimate, we employ a related term—*the confidence level*—to which we assign the following

DEFINITION
The confidence level is the percentage of interval estimates—obtained from many repeated samples, each of size *n*, from the same population— that will contain the actual value of the parameter being estimated. Although a single sample is ordinarily taken, the confidence level is expressed as a percentage.

The interval estimate is referred to as a *confidence interval*. The level of confidence determines the value of z to be used, with greater confidence resulting in wider confidence intervals. The required value of z may be found by reading the table of normal-curve areas in reverse. But it is simpler for us to use Appendix Table E, which provides the values of z for common confidence levels.

To illustrate how a confidence interval estimate is obtained, we will assume that a sample of $n = 100$ families is selected from the population of a large city and that the sample mean income is $\bar{X} = \$15{,}549.63$. We have *prior knowledge* that the standard deviation of the population of family incomes is $\sigma = \$5{,}000$. We wish to determine a 99% confidence interval estimate of the true population mean μ.

From Appendix Table E, the required normal deviate is $z = 2.57$, so that our 99% confidence interval estimate is

$$\mu = \bar{X} \pm z \frac{\sigma}{\sqrt{n}} = \$15{,}549.63 \pm 2.57 \frac{\$5{,}000}{\sqrt{100}}$$

$$= \$15{,}549.63 \pm \$1{,}285.00$$

and

$$\$14{,}264.63 \le \mu \le \$16{,}834.63$$

We *cannot* say there is a .99 probability that the stated interval contains μ. (It either does or does not, but we do not know which case applies.) The proper interpretation of this 99% confidence interval is this: **If we repeated the same procedure often, each time selecting a different sample of 100 families from the same population, then 99 out of every 100 similar intervals we obtained would, on the average, contain μ.**

Confidence Interval Estimate of the Mean When σ Is Unknown

The end points of a confidence interval estimate of μ depend on the population standard deviation σ, which usually is unknown and, like μ, must be estimated from the sample data. We may use the computed sample standard deviation s as the estimator of σ. For *large samples*, we can therefore use the following expression when σ is unknown to calculate the

CONFIDENCE INTERVAL ESTIMATE OF THE MEAN

$$\mu = \bar{X} \pm z\frac{s}{\sqrt{n}} \quad \text{or} \quad \bar{X} - z\frac{s}{\sqrt{n}} \le \mu \le \bar{X} + z\frac{s}{\sqrt{n}}$$

In effect, we ignore the sampling error of using s to estimate σ. Large value of s will tend to stretch the interval beyond the desired width, and small values of s will have the opposite effect. In practice, ignoring the variability in s is not serious, since about as many intervals obtained from repeated samples will contain μ on the average as would be the case if σ were known.

Illustration: Estimating Mean Bill Payment Times

A bank operating a credit card system wishes to find the time of the month for mailing customer bills that will provide the shortest average time to receipt of cash payment. Currently the bills are mailed at the end of the month, and the average time to receipt is about 10 days. Three samples of customers are randomly selected, one group to be billed on the 5th of the month, another on the 15th, and the final group on the 25th. In order to avoid any bias due to the transition in billing cycle, figures from the fourth month under the new dates constitute the samples. A total of $n = 500$ customers are selected for each sample group. Since the payment of bills is a continuous process, the populations may be treated as infinite. The following results are obtained:

5th of Month	15th of Month	25th of Month
$\bar{X} = 18.30$ days	$\bar{X} = 17.41$ days	$\bar{X} = 5.42$ days
$s = 6.2$	$s = 5.8$	$s = 2.3$

A confidence interval is to be constructed for each group. The desired level of confidence is 99%. Thus $z = 2.57$. For the 5th of the month, the confidence interval is calculated:

$$18.30 - \frac{2.57(6.2)}{\sqrt{500}} \le \mu \le 18.30 + \frac{2.57(6.2)}{\sqrt{500}}$$

or

$$\mu = 18.30 \pm .71$$

so that

$$17.59 \le \mu \le 19.01 \text{ days} \quad \text{(for 5th)}$$

Analogously, the following confidence intervals are obtained for the other groups:

$$\mu = 17.41 \pm .67 \quad \text{or} \quad 16.74 \leq \mu \leq 18.08 \text{ days} \quad \text{(for 15th)}$$
$$\mu = 5.42 \pm .26 \quad \text{or} \quad 5.16 \leq \mu \leq 5.68 \text{ days} \quad \text{(for 25th)}$$

Note that there is so little difference between the results for the 5th and 15th that the confidence intervals overlap. Of the three times of the month, the 25th appears to be the superior billing time. By switching its billing cycle so that bills are mailed on the 25th instead of at the end of the month, the bank *might* (remember, we do not *know* μ) be able to reduce the average collection time from its present 10 days to some value between 5.17 and 5.68 days, thereby allowing the firm to have longer use of the funds received.

Features Desired in a Confidence Interval

Two features are desirable in a confidence interval estimate. One is the level of confidence, which expresses the degree of *credibility* that may be attached to the results. The other is the *precision* of the estimate, which is gauged by the width of the interval. But credibility and precision are competing ends. For a fixed sample size, a reduction in the interval width, which causes greater precision, can be achieved only at the expense of reducing z, which is the same as using a lower confidence level. Conversely, greater confidence can be obtained only at the expense of precision. Thus, a confidence interval may be very precise but not at all credible, or vice versa. For example, if z were .5 in the bill payment illustration, then the confidence interval for mailing on the 25th would become

$$\mu = 5.42 \pm .5(2.3)/\sqrt{500} = 5.42 \pm .05$$

or

$$5.37 \leq \mu \leq 5.47 \text{ days}$$

which is more precise than before, but the confidence level would be only 38%. Likewise, increasing the confidence level to 99.8%, so that $z = 3.08$, would yield the less precise interval estimate

$$\mu = 5.42 \pm 3.08(2.3)/\sqrt{500} = 5.42 \pm .32$$

or

$$5.10 \leq \mu \leq 5.74 \text{ days}$$

The only way to increase both confidence and precision is to collect a larger sample to begin with. For example, suppose that $n = 1,000$ customers had been used for the billings on the 25th and the same sample results were obtained. For a 99.8% level, the confidence interval would be

$$\mu = 5.42 \pm 3.08(2.3)/\sqrt{1,000} = 5.42 \pm .22$$

or

$$5.20 \le \mu \le 5.64 \text{ days}$$

This is a *more precise* interval with a *greater confidence level* than the one originally obtained.

EXERCISES

8-4 For each of the following parameters, indicate whether you would use a point estimate or an interval estimate. Explain each choice.

(a) The mean age of product buyers, to be used in comparing alternative advertising plans.

(b) The proportion of an evening's television-viewing population watching a particular program, to be presented to advertisers as evidence of the program's drawing power.

(c) The mean pH value (a measure of acidity) of intermediary ingredients in a chemical process, so that a formula can be developed that indicates the amount of neutralizing agent to apply.

(d) The median income of doctors, to be published in a government report.

8-5 A corporation's board of directors is evaluating the performance of the corporation president in order to decide whether to retain her. One director proposes that each director rate the president—on a scale from 1 to 5—as to the quality of her performance in five areas: (1) sales growth, (2) personnel relations, (3) operating efficiencies achieved, (4) attainment of profit potential, and (5) successful new-product development. The ratings will be summarized in terms of frequency distributions, with the mean rating calculated for each category and submitted to the directors at the next meeting, when a vote on retention will take place. The board agrees to the procedure.

The chairman is concerned with the average ratings. He is worried about their precision and wonders what level of confidence should be attached to the results. A trustee who happens to be an expert in statistics diplomatically informs the chairman that these considerations are irrelevant. Explain the basis for the expert's stand.

8-6 Construct a 95% confidence interval for the population mean, given the following sample results:

(a) $n = 100$; $\bar{X} = 100.53$ minutes; $s = 25.3$ minutes.

(b) $n = 200$; $\bar{X} = 69.2$ inches; $s = 1.08$ inches.

(c) $n = 350$; $\bar{X} = \$12.00$; $s = \$7.00$.

8-7 A personnel director wishes to estimate the mean scores for a proposed aptitude test that may be used in screening applicants for clerical positions. Using a sample of $n = 100$ applicants, she found $\bar{X} = 75.6$ and $s = 14.7$. Assuming that the population is large, construct a 95% confidence interval estimate for the true mean.

8-8 A hotel manager desires to improve the level of room service by increasing personnel. To justify the increase, he must estimate the mean waiting time experienced by guests before being served. From a random sample of $n = 100$ orders, the sample mean waiting time was established at 30.3 minutes, with a standard deviation of 8.2 minutes. Construct a 99% confidence interval that the manager could use to estimate the actual mean waiting time.

8-9 A hospital administrator intends to improve the level of emergency-room service by increasing support personnel. To justify the increase, she must estimate the mean time that patients must wait before being attended by a physician. For a random sample of $n = 100$ previously recorded emergencies, the sample mean waiting time was 70.3 minutes, with a standard deviation of 28.2 minutes. Construct a 99% confidence interval that the administrator could use to estimate the actual mean waiting time.

8-10 A tire manufacturer has obtained the following sample results for the tread life of $n = 50$ radial tires tested: $\bar{X} = 52{,}346$ miles; $s = 2{,}911$ miles.
(a) Construct a 99% confidence interval estimate of the mean tread life for all such tires manufactured.
(b) What is the interpretation of the interval you found in (a)?

8-11 A government agency analyzing the rising cost of medical care over the past decade must estimate the mean fee for various operations. A random sample of 250 splenec-tomies yields the following results; $\bar{X} = \$374.00$; $s = \$56.25$.
(a) Construct a 99% confidence interval estimate of the current mean splenectomy fee.
(b) Ten years ago, a similar study based on 150 operations yielded the following results at a 95% confidence level: $\$130.00 \leq \mu \leq \$150.00$. Find the value of $\bar{X}$ used to construct this confidence interval.
(c) Using the values of $\bar{X}$ in (a) and (b) as point estimates of the mean fees in the two years, estimate the percentage increase in splenectomy fees over the ten-year period.

8-3 INTERVAL ESTIMATES OF μ USING SMALL SAMPLE SIZES

We have learned to construct confidence intervals when the standard deviation of the population is unknown by using s (computed from the sample) in place of σ. For the desired level of confidence, we used the corresponding normal deviate z. Actually, *the normal distribution is only approximately correct* and is suitable only for large sample sizes. Another distribution applies to small sample sizes, so we cannot use values of z to construct confidence intervals from them.

The Student t Distribution

W.S. Gosset first studied this problem in the early 1900s. Because his employer, a brewery, forbade him to publish, he chose the nom de plume "Student." Gosset derived a probability distribution, now referred to as the Student t distribution, for a random variable t. Whenever the population being sampled is normally distributed,

$$t = \frac{\bar{X} - \mu}{s/\sqrt{n}}$$

In practice, however, the t distribution may be used when samples are taken from any population that is not highly skewed.

Like the normal distribution, the Student t distribution has a relative frequency curve that is bell-shaped and symmetrical, as shown in Figure 8-2. The single parameter that determines the shape of its curve is called the *number of degrees of freedom*, which is $n - 1$ (because for a fixed value of $\bar{X}$, there are only $n - 1$ "free choices" for the values of the n observations used to calculate $\bar{X}$ and s).

To find the probability that t exceeds some value, we must consult the table of areas under the Student t distribution in Appendix Table F. There is a separate distribution—and therefore a separate set of values—for each number of degrees of freedom; these appear in separate rows in the table. It is traditional to use α (the lowercase Greek *alpha*) to represent the upper tail area. The values in the body of the table are the points t_α corresponding to the respective upper tail area α (column) and degrees of freedom (row). To find the point that corresponds to a given upper tail area t_α, so that

$$\Pr[t \geq t_\alpha] = \alpha$$

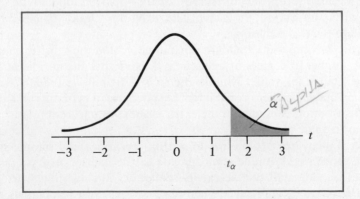

FIGURE 8-2
The Student t distribution.

we read down the column headed by the probability value α, stop at the row corresponding to the number of degrees of freedom, and read the desired t_α value. (Like the normal curve table, there is no need to list separate entries for the left-hand portion of the t curve.) A partial reproduction of Appendix Table F will be helpful.

Degrees of Freedom		Upper Tail Area α			
	$\cdots$	.05	.025	.01	$\cdots$
$\vdots$			$\vdots$		
19		1.729	2.093	2.539	
20		1.725	2.086	**2.528**	
21		1.721	2.080	2.518	
22	$\cdots$	1.717	2.074	2.508	$\cdots$
23		1.714	2.069	2.500	
24		**1.711**	2.064	2.492	
25		1.708	2.060	2.485	
26		1.706	2.056	2.479	
$\vdots$			$\vdots$		

For example, suppose that we wish to find $t_{.01}$ (the value for which the probability is .01 that t is greater than or equal to that value) and that the number of degrees of freedom is 20. From Appendix Table F, $t_{.01} = 2.528$. So

$$\Pr[t \geq 2.528] = .01$$

A common guideline is that the Student t distribution must only be used when the sample size is less than 30. For larger samples, the normal distribution is ordinarily used. The t distribution actually applies regardless of the magnitude of n. But for large values of n, the curves for the t distribution and the standard normal curve are almost the same. The arbitrary choice of $n = 30$ as the demarcation point is traditional.

Figure 8-3 shows the standard normal curve and the t distribution curve when the number of degrees of freedom is 4. Note that the normal curve falls below the t curve for values lying in the tails. This reflects the fact that the t distribution assigns higher probabilities to outcomes of extreme value than the normal distribution does, due to the extra element of uncertainty arising from the unknown value of σ.

The Student t distribution is important because small sample sizes are common. Often small sample sizes are due to the high cost of sampling, but sometimes only a small sample can be collected quickly enough to provide the required answers. This is true when observations must be obtained sequentially from a manufacturing process with a low output rate, such as commercial

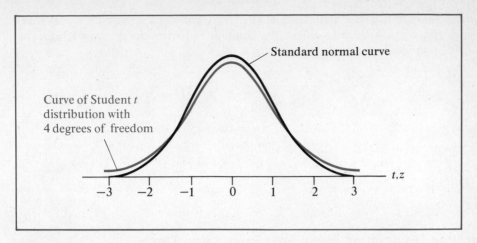

FIGURE 8-3
A comparison of the relative frequency curves of the standard normal distribution and the Student t distribution.

jet-aircraft production. Or perhaps sample observations must be made of rare phenomena, such as nuclear reactor accidents. Experiments in medicine, education, and psychology, in which people are observed, are often limited to small samples.

Constructing the Confidence Interval

We can now use t_α (just as we used z with large samples) to construct the

CONFIDENCE INTERVAL ESTIMATE OF THE MEAN

$$\mu = \bar{X} \pm t_\alpha \frac{s}{\sqrt{n}} \quad \text{or} \quad \bar{X} - t_\alpha \frac{s}{\sqrt{n}} \le \mu \le \bar{X} + t_\alpha \frac{s}{\sqrt{n}}$$

We can read value of t_α from Appendix Table F, using

$$\alpha = (1 - \text{confidence-level proportion})/2$$

(The middle area under the Student t curve included within $\pm t_\alpha$ corresponds to the confidence level. We must therefore divide by 2 to get α, because the areas in the *two* tails are excluded.)

To illustrate, we will find a 90% confidence interval for the sample results obtained from a survey, so that $\alpha = (1 - .90)/2 = .05$. The results obtained using $n = 25$ are $\bar{X} = 10.0$ and $s = 5.0$. The degrees of freedom are $n - 1 = 24$, and for an $\alpha = .05$ area under the Student t distribution, $t_{.05} = 1.711$. The 90% confidence interval estimate of μ is

$$\mu = \bar{X} \pm t_{.05} \frac{s}{\sqrt{n}} = 10 \pm 1.711 \frac{5.0}{\sqrt{25}}$$

$$= 10 \pm 1.7$$

so that

$$8.3 \leq \mu \leq 11.7$$

Illustration: Estimating Mean Lost Production Time

The production manager of an appliance manufacturing company wishes to develop data for a study to determine the quantities of various types of production equipment that should be purchased. An essential element of the study is maintenance costs. The production manager is concerned that too little attention has been paid to the cost of labor lost due to breakdowns. For instance, one assembly production line depends on parts fabricated by a stamping machine. Occassionally this machine breaks down, taking several hours to repair, and when the buffer stocks of stampings are exhausted, the line must be stopped and many of the workers dismissed for the day. The union contract requires that they receive a full day's wages regardless of the hours actually worked.

A larger buffer stock, the size of which currently fluctuates widely throughout the day, would alleviate this problem, but since so many different items may be made in the immediate future, a tremendous investment in work-in-progress inventory would be required to handle all contingencies. An alternative course of action would be to acquire another stamping machine for back-up. In the one year in which the plant has been in operation, the stamping machine has failed nine times. Table 8-1 shows the direct labor hours lost for each breakdown. An estimate of the mean number of labor hours lost per breakdown is to be obtained. This estimate may then be multiplied by the expected number of breakdowns per year to obtain an estimate of the hours that will be lost annually in the future.

The production manager desires a 95% confidence interval to estimate μ, the mean labor hours lost. The number of degrees of freedom is $9 - 1 = 8$. Here $\alpha = (1 - .95)/2 = .025$, so that we obtain $t_{.025} = 2.306$. The following confidence interval is obtained:

$$512.56 - 2.306(490.1)/\sqrt{9} \leq \mu \leq 512.56 + 2.306(490.1)/\sqrt{9}$$

TABLE 8-1

Labor Hours Lost Due to Stamping Machine Breakdowns

Labor Hours Lost X	X^2
205	42,025
1,123	1,261,129
528	278,784
359	128,881
1,421	2,019,241
723	522,729
57	3,249
172	29,584
25	625
Totals 4,613	4,286,247

$\bar{X} = \sum X/n = 4{,}613/9 = 512.56$ hours

$s = \sqrt{(\sum X^2 - n\bar{X}^2)/(n-1)} = \sqrt{(4{,}286{,}247 - 9(512.56)^2)/8}$

$\quad = 490.1$ hours

so that

$$\mu = 512.56 \pm 376.72$$

and

$$135.84 \le \mu \le 889.28 \text{ hours}$$

Notice that the confidence interval is quite wide, which is to be expected with such a large sample standard deviation and such a small sample.

EXERCISES

8-12 Determine the value t_α that corresponds to each upper tail area of the Student t distribution specified:

	(a)	(b)	(c)	(d)	(e)
α:	.05	.01	.025	.01	.005
Degrees of freedom:	10	13	21	120	30

8-13 Construct 95% confidence intervals for estimates of the means of populations yielding the following sample results:
(a) $n = 6$; $\bar{X} = \$8.00$; $s = \$2.50$.
(b) $n = 15$; $\bar{X} = 15.03$ minutes; $s = .52$ minutes.
(c) $n = 25$; $\bar{X} = 27.30$ pounds; $s = 2.56$ pounds.

8-14 Construct 99% confidence intervals for (a), (b), and (c) in Exercise 8-13.

8-15 A biological researcher wishes to estimate the longevity of wild rats raised under laboratory conditions. A random sample of $n = 25$ second-generation offspring was reared from birth and monitored until all of the rats were deceased. Lifetime statistics of $\bar{X} = 31.2$ months and $s = 5.4$ months were obtained. Construct a 95% confidence interval estimate of the population mean longevity.

8-16 A medical researcher has a contract with a state prison that provides volunteers to test new drugs. To test a drug to be marketed for the treatment of an exotic virus, each patient was given injections of the live virus until he or she was infected, and then the patient was treated with the drug. Due to obvious dangers, only 15 volunteers could be obtained. One parameter of interest is the mean recovery time. The following recovery times (in days) were obtained:

$$
\begin{array}{ccc}
7 & 12 & 17 \\
18 & 13 & 6 \\
9 & 5 & 11 \\
8 & 9 & 12 \\
32 & 8 & 18
\end{array}
$$

(a) To present the case to the Food and Drug Administration, a 95% confidence interval estimate of the mean recovery time must be obtained. Find the values of $\bar{X}$ and s, and then calculate the confidence interval that was used.

(b) Would it be appropriate to apply the result in (a) to the national population as a whole? Explain.

8-17 A psychologist tested a sample of $n = 17$ volunteers who had earlier watched Walt Disney's movie *Fantasia* and obtained the following results for the duration of rapid-eye-movement (REM) sleep: $\bar{X} = 90$ minutes; $s = 35.1$ minutes. Construct a 95% confidence interval estimate for the population mean REM duration.

8-18 A large casualty insurance company is revising its rate schedules. A staff actuary wishes to estimate the average size of claims resulting from fire damage in apartment complexes with between 10 and 20 units. The current year's claim settlements are to be used as a sample. Considering the 19 claim settlements for buildings in this category, the average claim size was $73,249, with a standard deviation of $37,246. Construct a 90% confidence interval estimate of the mean claim size.

8-19 To compare the IQ scores of professional persons, a sociologist administered the Stanford-Binet test to random samples of 16 doctors and 25 lawyers. The following results were obtained:

Doctors	Lawyers
$\bar{X} = 121.4$	$\bar{X} = 128.6$
$s = 12.0$	$s = 15.0$

(a) Construct a 95% confidence interval for the mean IQ of doctors.
(b) Construct a 95% confidence interval for the mean IQ of lawyers.
(c) Do the sample data indicate that lawyers have higher IQs than doctors? Explain.

8-20 An electronics firm currently trains its chassis assemblers on the job. It has been proposed that new employees be given one week's formal training. To test the quality of the training program, the output rate of ten trainees (Group A) was compared to that of ten new employees receiving on-the-job training (Group B). Each trainee from Group A was matched with one from Group B, based on aptitude test scores. After training, each person was given a specific number of chassis to wire, and the completion time was recorded. For each of the $n = 10$ pairs of assemblers, the time of the trainee from Group A was subtracted from that of the matching trainee from Group B. The following results were obtained: $\bar{X} = 3$ minutes; $s = 2$ minutes.

(a) Calculate a 95% confidence interval estimate of the mean time difference.

(b) If there is no difference between the two training methods and if the population standard deviation is $\sigma = 2$, use the normal distribution to find the probability that the sample mean will be 3 or more.

INTERVAL ESTIMATES 8-4
OF THE POPULATION PROPORTION

We can estimate the population proportion π by an interval, using the sample proportion P, in much the same way that we can estimate μ, using $\bar{X}$. To estimate π, we take advantage of the normal approximation to the binomial sampling distribution of P (which we learned in Chapter 7 is allowable only for certain sample sizes).

We will use P to estimate π by an interval in the form

$$P - z\sigma_P \leq \pi \leq P + z\sigma_P$$

Again, we do not know the value of σ_P, because it depends on π—the value we are estimating. So we must determine the value of σ_P from the sample results. We therefore replace π with P in

$$\sigma_P = \sqrt{\frac{\pi(1 - \pi)}{n}}$$

to obtain the point estimate

$$\sqrt{\frac{P(1 - P)}{n}}$$

We can then use this substitution to obtain the form of our

CONFIDENCE INTERVAL FOR THE POPULATION PROPORTION

$$\pi = P \pm z\sqrt{\frac{P(1 - P)}{n}} \quad \text{or} \quad P - z\sqrt{\frac{P(1 - P)}{n}} \leq \pi \leq P + z\sqrt{\frac{P(1 - P)}{n}}$$

EXAMPLE: PREFERENCE FOR CANNED CORN

To estimate the proportion of buyers who preferred the quality of its canned corn, a food processor randomly selected a panel of $n = 100$ persons. Each panelist was given three cans of brand X and three cans of brand Y, with the labels removed. The testers were asked to rate each can of corn on the basis of each of four factors: tenderness, sweetness, consistency, and color. The brand receiving the highest aggregate score by a tester was the preferred one. The test results were

Number preferring brand X:	59
Number preferring brand Y:	37
Number of ties:	4
Total	100

A 99% confidence interval is desired in estimating π, the proportion of the buying population preferring brand X. The normal deviate obtained from Appendix Table E is $z = 2.57$. The sample proportion of buyers preferring brand X is $P = 59/100 = .59$. The 99% confidence interval is

$$\pi = .59 \pm 2.57 \sqrt{\frac{.59(1 - .59)}{100}}$$

$$\pi = .59 \pm 0.13$$

so that

$$.46 \leq \pi \leq .72$$

EXERCISES

8-21 In a poll taken to estimate the President's current popularity, each person in a random sample of $n = 1,000$ voters was asked to agree with one of the following statements:

    (1) The President is doing a good job.
    (2) The President is doing a poor job.
    (3) I have no opinion.

A proportion of .59 chose statement (1). Assuming that the actual number of voters is quite large, construct a 95% confidence interval for the population proportion of voters who will choose statement (1).

8-22 Find the 99% confidence interval for π, the proportion of undersize boards passing out of a sawmill's planing machine, under the following assumptions:
    (a) The number of boards measured is $n = 500$; $P = .25$.
    (b) $n = 1,000$; $P = .01$.
    (c) $n = 300$; $P = .05$.

8-23 The highway-patrol director in a certain state has ordered a crackdown on drunk drivers. To see if his safety campaign is working, the director has ordered a sampling study to estimate the proportion of all fatal traffic accidents caused by drinking. In a random sample of $n = 100$ accidents, 42% were attributed to alcohol. Assuming that the accident population is large, construct a 95% confidence interval for the population proportion.

8-24 Find the 95% confidence interval for the proportion of defective items in large shipments of parts when
(a) $n = 1,000$; $P = .2$.   (b) $n = 2,000$; $P = .04$.   (c) $n = 500$; $P = .01$.

8-25 A printer has negotiated a contract with a telephone company to print directories. Because disgruntled customers whose names are misspelled or listed with the number of, say, a local mortuary will complain bitterly for a whole year, the telephone company is anxious to minimize directory errors. Its contract therefore states that the printer must forfeit $100 for every page that contains errors introduced in the printing process.

Counting the errors in a completed directory is an onerous chore. The majority are found by customer complaint, and most of these turn out not to be the printer's fault. Thus, a sample will be taken to determine what penalty fee to charge the printer. It is so difficult to find errors that many persons must search for them independently. A company accountant selects a sample of 50 pages from a 600-page directory and asks 20 people in succession to review each page for errors. An estimate is to be made of the proportion π of pages that contain errors.
(a) Construct a 99.8% confidence interval estimate of π if $P = .37$.
(b) The contract specifies that the estimate of π must be a point estimate. What are the losses to the printer if the estimate is too large by .02?
(c) Calculate the probability that the estimator P will be .37 or over if the actual error rate is only $\pi = .35$.

DETERMINING THE REQUIRED SAMPLE SIZE   8-5

Error and Reliability

Costs are usually incurred when an estimate is erroneous. In many statistical applications, it is quite difficult to place a monetary value on the results of an incorrect estimate. Ideally, a sample size should be chosen that achieves the most desirable balance between the chances of making errors, the costs of those errors, and the costs of sampling. Figure 8-4 illustrates the concepts involved in finding the optimal sample size, which minimizes the total cost of sampling. The costs of collecting the sample data increase with n. But larger samples are more reliable, so that the losses from chance sampling error decline. The total cost of sampling—the sum of collection costs and error costs—will achieve a minimum value for some optimal n. This is the sample size that should be used.

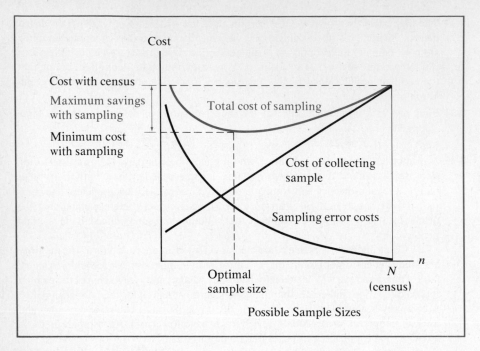

FIGURE 8-4
Relationships among sampling costs.

EXAMPLE: AIRLINES SHARING TICKET REVENUES

Lake Airlines is a small regional carrier feeding the larger airlines. Passengers purchasing a ticket for a cross-country trip are issued a single ticket for the entire journey at the point of embarkation. Lake is entitled to keep only its portion of the revenue and must forfeit shares to the other airlines involved. The reverse occurs for passengers originating with other airlines who use Lake for a single leg. In these cases, Lake must collect its share of the ticket revenue from the other airline. At the end of each month, the airlines balance their accounts based on a detailed enumeration of the interairline tickets collected. This has proved to be a very costly and time-consuming process.

A consulting statistician has suggested that the airlines balance their accounts by means of random samples of collected tickets, arguing that only a small portion of the tickets must then be analyzed. There would be a certain amount of risk, however, for the samples obtained could provide erroneous figures.

Using a sample has drawbacks. Because of chance sampling error, Lake's claims against other airlines will be understated in some months and overstated in others. Viewing the long-run average or expected revenue losses as the cost of sampling error, the statistician has computed (by a procedure too detailed to discuss here) the costs for various sample sizes:

Sample Size n	Cost of Collecting Sample	Sampling Error Costs	Total Cost of Sampling
100	$ 100	$40,222	$40,322
1,000	1,000	13,059	14,059
5,000	5,000	5,666	10,666
5,200	5,200	5,454	10,654
6,000	6,000	5,202	11,202
10,000	10,000	4,028	14,028

Assuming that each sample ticket processed costs Lake $1, the costs of sampling have also been calculated. The sample size yielding the minimum cost to Lake is about $n = 5,200$.

Due to the difficulties associated with finding the costs of sampling error, the procedure illustrated in Figure 8-4 is not usually used. Instead, we focus on a number that separates insignificant errors from decidedly undesirable ones. This is called the *tolerable error level*. Accepting some error is the price we pay for using a sample instead of a census.

We denote the tolerable error—the maximum amount the estimate should extend above or below the parameter—by e, which is a number in the same units as the parameter being estimated. As applied in traditional statistics, we define an estimate's reliability in terms of e.

DEFINITION

The reliability associated with using $\bar{X}$ to estimate μ is the probability that $\bar{X}$ differs from μ by no more than the tolerable error level e, or

$$\text{Reliability} = \Pr[\mu - e \leq \bar{X} \leq \mu + e]$$

Figure 8-5 illustrates the relationships among reliability, tolerable error, and the sampling distribution of $\bar{X}$ when reliability is .95. The shaded area provides the probability that $\bar{X}$ will lie within $\mu \pm e$.

For a fixed e, only one normal curve has a shape that provides the desired area. If the reliability is lowered to .80, another curve, shown in Figure 8-6, applies. Note that compared to the curve in Figure 8-5, this second curve is flatter and exhibits more variability, so that it must have a larger value of $\sigma_{\bar{X}}$.

Thus, for any specified tolerable error and corresponding reliability, there is a unique level for $\sigma_{\bar{X}}$. Since $\sigma_{\bar{X}} = \sigma/\sqrt{n}$ and σ is a fixed value, only the sample size n can be adjusted to achieve the required level for $\sigma_{\bar{X}}$. This means that a given e and reliability require a unique value of n.

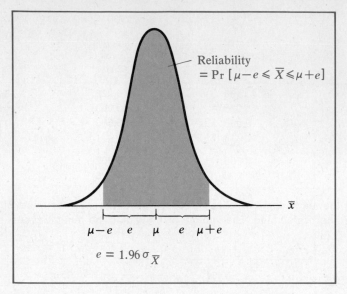

FIGURE 8-5
Relationships among reliability, tolerable error, and the sampling distribution of $\bar{X}$ when reliability (shaded area) = .95.

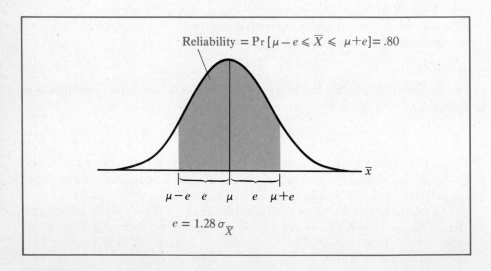

FIGURE 8-6
Relationships among reliability, tolerable error, and the sampling distribution of $\bar{X}$ when reliability (shaded area) = .80.

Steps for Finding the Required Sample Size to Estimate the Mean

To illustrate how to determine the appropriate sample size to use in estimating the mean, consider the problem faced by a government pension analyst who wishes to estimate the mean age at which federal employees retire.

STEP 1
Set a tolerable error level e. The same units apply to *e* as to the observed units. The analyst chooses *e* = .5 year as adequate precision for the estimate.

STEP 2
Set a target reliability probability and find z. Although the analyst would like to be certain of achieving the desired precision, such perfection cannot be achieved using a sample. He determines that a reliability of .95 will suffice. From Appendix Table E, this corresponds to a normal deviate of *z* = 1.96.

STEP 3
Establish a value for the population standard deviation σ. Ordinarily, this value must be a *guess.* As an aid, a value from a similar study may be used for *σ*. The analyst is aware that in a recent study of postal employee retirements, the sample standard deviation for the age of retirement was 3.5 years. Believing that a similar age dispersion applies to federal employees, he uses *σ* = 3.5 years.

STEP 4
Calculate the required sample size n:

REQUIRED SAMPLE SIZE FOR ESTIMATING THE MEAN

$$n = \frac{z^2 \sigma^2}{e^2}$$

Using the values

$$e = .5 \qquad z = 1.96 \qquad \sigma = 3.5$$

the pension analyst finds the required sample size:

$$n = \frac{(1.96)^2 (3.5)^2}{(.5)^2} = 188.2, \text{ or } 189$$

(When *n* is fractional the next larger whole number is chosen.)

Three Influences on Sample Size

1. *The required sample size is directly proportional to the population variance.* A larger sample size will be required for a population with high dispersion than for a population that exhibits little variability.

EXAMPLE: PROFESSORS AND CADILLACS

Estimating the mean distance traveled by vacationing college professors can be compared to estimating the mean number of miles Cadillacs can travel on a tank of gasoline. The mileage of the professors will be quite varied: Some will spelunk in Tasmania; others will travel no further than the local golf course. Similarly, there will be some variability in total gasoline mileage for Cadillacs, depending on the routes traveled, the amount of traffic, the driver, the brand of gasoline used, and other factors. The assumed standard deviation of $\sigma = 500$ miles for the population of distances traveled by professors is 50 times the value of $\sigma = 10$ miles that we will assume for the driving distance of Cadillacs without refueling.

If we wish to obtain estimates that differ from the true means by ± 5 miles with a reliability of .99, then in either case

$$e = 5 \text{ miles}$$

$$z = 2.57$$

Thus, the sample sizes will be

$$n = \frac{(2.57)^2(500)^2}{5^2} = 66{,}049 \text{ for professors}$$

$$n = \frac{(2.57)^2(10)^2}{5^2} = 26.42, \text{ or } 27 \text{ for Cadillacs}$$

We see that about 2,500 (the square of 50) times as many professors as Cadillacs will be needed in the sample to provide the same reliability for a tolerable error of 5 miles.

This does not imply that estimating the mean length of professorial odysseys would actually require such a large sample. It is unreasonable to assume that the tolerable error should be as small as the one that might be required by General Motors for its Cadillacs.

---

2. *The required sample size is inversely proportional to the square of the tolerable error.* Thus, if we reduce e by one-half, we need four times as large a sample if reliability remains constant. Similarly, if we increase e ten times, the sample size will be 100th as large. Therefore, if we let $e = 50$, ten times the

tolerable error used for the professors in our example, the sample size will be one-100th as large as before:

$$n = \frac{(2.57)^2(500)^2}{50^2} = 660.49, \text{ or } 661$$

3. *The required sample size is directly related to the square of the reliability-level normal deviate z.* Thus, doubling z will cause n to be four times as large. We cannot say that doubling z will ordinarily raise reliability by a factor of 2, however, because z does not have a linear relationship to normal-curve areas. For example, decreasing z from 2.57 to 1.28 (a reduction in z of about one-half, so that n is about one-fourth as large) lowers the reliability from .99 to .80 (not quite a 20% drop).

The choices of e and z are interrelated, because the targeted reliability usually depends on the selected tolerable error. Because the tolerable error serves merely as a convenient cutoff point between serious and insignificant errors, its choice may be inseparable from the selection of the reliability level.

Steps for Finding the Required Sample Size to Estimate the Proportion

Using the same type of analysis that we applied to the sample mean, we can find the number of observations required to estimate a population proportion from its sample counterpart. The roles of tolerable error and reliability, as they affect n, are completely analogous. Tolerable error e represents the maximum allowable deviation between π and its estimator P, and e is expressed as a *decimal fraction*. Reliability is the probability that P differs from π by no more than e.

We follow the same procedure for finding the required sample size as we did before. To illustrate, we will consider the problem at a gas company, where there is a continual discrepancy between the cubic feet of gas leaving the storage tanks and the actual consumption reported by company meter readers. Some of the discrepancy may be caused by inaccurate meters and minor leaks, but much of it is believed to be due to the inattention of the meter readers. The proportion of misread meters is to be estimated for each reader. A sample of homes from each meter reader's route is selected to be audited. A company supervisor, whose presence is unknown, will follow each reader on a portion of his or her route and reread the meters. Later, the figures will be compared.

STEP 1
Set a tolerable error level e. This will be a decimal fraction. A tolerable error of $e = .02$ is chosen.

STEP 2
Set a target reliability probability and find z. A reliability of .90 is considered adequate. The normal deviate is $z = 1.64$.

STEP 3
Establish a preliminary value for the population proportion π. Initially, it is assumed that 10% of the readings are in error. The preliminary value for the population proportion is therefore $\pi = .10$. (The true value of π remains unknown.)

STEP 4
Calculate the required sample size n:

REQUIRED SAMPLE SIZE FOR ESTIMATING PROPORTION

$$n = \frac{z^2 \pi (1 - \pi)}{e^2}$$

Using the values

$$e = .02 \qquad z = 1.64 \qquad \pi = .10$$

the required sample size is

$$n = \frac{(1.64)^2 (.10)(1 - .10)}{(.02)^2} = 605.16, \text{ or } 606$$

Note that the required n depends on the value of π, which we must assume before we select the sample from which we will estimate it. We cannot use the actual value of π in the calculation because it is still unknown.

Work-Sampling Applications

One very important concern of production management is work measurement. Because of the great expense involved in collecting the required data, statistical tools have been developed to make the task more efficient.

A generation ago, procedures for obtaining measurements involved stopwatch methods that required continuous monitoring. Sometimes there was one watcher for each worker, so that the cost of gathering the data was high. The development of a statistical technique called *work sampling* relieved management of the burden of the stopwatch and at the same time removed the worker from its purported tyranny.

One goal of work measurements is to establish work standards. These are ordinarily expressed as the time needed to complete a particular task, such as drilling a hole, installing a car bumper, or wiring a circuit board. Standards are established over a period of time by measuring a typical cross section of workers under normal production circumstances. The mean time to completion is estimated for each worker. A frequency distribution of these means is then compiled to set the standard.*

The classical procedure for finding the mean time to assemble a bumper on a car required that the worker be watched continuously and the installation times be duly recorded. If a sample of 20 completion times was required, each of about 10 minutes duration, about 200 minutes—or half a day—would be spent by the timer watching just this one worker. If ten workers in the automobile assembly plant install bumpers, and a sample is to be obtained for each, the timer must spend about 5 days measuring this one operation. But there are hundreds of operations involved in assembling cars, so that a great many timers would be required just to set standards and evaluate performance. Moreover, the very presence of a timer may bias the observations. It is widely acknowledged that performance is affected if the employee is aware of being measured.[†]

Work sampling is an alternative to continuous work measurement that can achieve a manyfold increase in the efficiency of the observer. The advantage of work sampling is that the same information obtainable from continuous observation may be achieved instead by many more shorter observations or "glimpses" taken randomly at scattered moments for an extended duration. Suppose that for a week the observer simply walks by the bumper installation station, making a note of what the worker is doing at that moment. It would be possible to do this many times, placing a check mark on a tally sheet for the particular task being performed at that instant. When the tallies have been accumulated at the end of the week, they would form the basis for estimating the proportion of time a worker spends at each task. The underlying principle of work sampling is that *the amount of time spent in a particular state of activity is proportional to the number of observations made of that state.*

Let us suppose that 100 observations have been made of an automobile bumper assembler; 86 times he was busy installing, and the rest of the time he was idle. Let us take this figure to calculate $P = 86/100 = .86$, which is a point estimate of the actual proportion of time π that the employee was busy installing bumpers. Assuming that all observations were made during the 35 hours in which the installer was officially on station and that he installed 222 bumpers, we may estimate the mean time taken for an installation.

---

* For a detailed discussion of setting work standards, see Elwood S. Buffa, *Modern Production Management* (New York: John Wiley & Sons, 1977).

† The famous Hawthorne studies highlight this point. Discussions are contained in many textbooks on management.

The total time actually spent installing is estimated to be

$$35(.86) = 30.1 \text{ hours}$$

By dividing the time duration by the number of cars completed, the mean installation time μ may be estimated as follows:

$$\frac{30.1}{222} = .136 \text{ hour per bumper}$$

$$= 8.2 \text{ minutes per bumper}$$

This is the same type of result that would have been obtained through many hours of continuous observation of the single worker. But the 100 work-sampling observations represent only minutes of the observer's time. On each pass through the plants he may collect tally marks for dozens of workers, each of whom performs a variety of tasks. Tallies would be obtained at all hours of the day and for every day of the week. No worker need feel uncomfortable about being watched, and the observer need carry no stopwatch.

The above results are only estimates. The same would be true of stopwatch timing. But work sampling is free of the many sources of potential bias of stopwatch studies. Thus the results may be both more reliable and cheaper.

To find the number of work-sampling observations required to estimate the mean time the worker takes to install bumpers, a very "rough" estimate would first have to be made of π. Suppose that we use $\pi = .80$. If a 5% tolerable error is desired with a 95% reliability, then $e = .05$ and $z = 1.96$. We obtain

$$n = \frac{(1.96)^2(.8)(1 - .8)}{(.05)^2} = 245.86, \text{ or } 246$$

We may compare this value to the number of stopwatch observations necessary to estimate μ, the mean bumper installation time. Suppose that we know this worker's standard deviation to be $\sigma = 2$ minutes per bumper. With a true value of μ near 8 minutes, a 5% tolerable error would correspond to $8(.05) = .4$ minute. Using $e = .4$ and $z = 1.96$, the required sample size would be

$$n = \frac{(1.96)^2(2)^2}{(.4)^2} = 96.04, \text{ or } 97$$

The number of continuous stopwatch observations would be 97 as compared to 246 for work sampling. But recall that the latter would take a few seconds each, whereas the former require several minutes each.

Work sampling has been used in many applications where traditional methods have proven impracticable. Perhaps the greatest advantage occurs when several tasks are performed, with small amounts of time spent on each. For example, consider timing a waitress as she performs dozens of short tasks in seemingly random sequence. A timer would have quite a job timing her as she pours coffee, clears a table, handles the cash register, makes a sundae, makes toast, wipes a table, and so forth. Yet a suitable estimate could be made by work-sampling methods of the amount of time spent per customer on each of these tasks.

Reliability Versus Confidence and Tolerable Error Versus Precision

It may prove helpful at this point to clarify the key concepts that have been introduced in this chapter. An obvious question is: What is the difference between reliability and confidence? Another might be: How do we distinguish between tolerable error and precision? In both cases, are we not really dealing with two terms that are synonymous?

In one sense, confidence and reliability mean the same thing. But the distinction between these concepts is that they apply in *different stages* of a sampling study. The notion of reliability is mainly a concern in the *planning stage*, where an appropriate sample size is found to satisfy the statistical analyst's goals. Greater reliability may only be achieved either by increasing the sample size, and hence the data-collection costs, or by reducing the tolerable error level. In selecting a reliability level, the statistician seeks to find a satisfactory balance between risks of error and cost. The concept of confidence applies in the last phase of a sampling study—the *data analysis and conclusions stage*. At this final point in an investigation, the statistician is concerned with communicating results, so that the overriding concern is with *credibility*. Although often of the same value, the reliability need not be equal to the reported confidence level. Indeed, there are instances when the sample size n is not under the statistician's control, so that reliability considerations do not even enter into the planning.

We have already encountered the notion of precision in constructing confidence intervals. As the limits of the interval estimate become narrower, we can say that the resulting estimate becomes more precise. Like confidence, precision applies in the final stage of a sampling study. At this point in time, all sample data have been collected and precision can be improved only by a compensating reduction in confidence, or vice versa. The tolerable error level, on the other hand, is merely a convenient point of demarcation that separates serious errors from those that may be acceptable. It is used only in the planning stage and, with reliability, serves primarily to guide the statistician in selecting an appropriate sample size. Only when the reliability coincidentally equals the confidence level, and the planning guess as to σ or π is on target, will tolerable error be of equal magnitude to reported precision.

EXERCISES

8-26 The mean time required by automobile assemblers to hang a car door is to be estimated. Assuming a standard deviation of 10 seconds, determine the required sample size under the following conditions:

(a) The desired reliability of being in error by no more than 1 second (in either direction) is .99.

(b) The desired reliability is .95 for a tolerable error of 1 second.

(c) A reliability of .99 is desired with a tolerable error of 2 seconds. How does the sample size you obtain compare with your answer to (a)?

(d) Answer (a), assuming that $\sigma = 20$ seconds. By how much does the sample size increase or decrease?

8-27 The purchasing officer for a government hospital that orders supplies from many vendors has the following policy: The quality of each shipment is to be estimated by means of a sample that will always include 10% of the items, regardless of the size of the shipment. Do you think this is a good policy? Discuss your answer.

8-28 The cost of obtaining the sample in Exercise 8-26 is $1.50 per observation.

(a) For a tolerable error of 1 second, find the reliability that can be achieved by an expenditure of $300.

(b) For a reliability of .95, find the tolerable error that must correspond to an expenditure of $300.

8-29 A quality-control manager has a flexible sampling policy. Because there are different quality requirements for different parts, there is a separate policy for each type of part. For example, a pressure seal used on some assemblies must be able to withstand a maximum load of 5,000 pounds per square inch (psi) before bursting. If the mean maximum load of a sample of seals taken from a shipment is less than 5,000 psi, then the entire shipment must be rejected.

The sampling policy for this item requires that a sample be large enough so that the probability that the sample mean differs from the population mean by no more than 10 psi (in either direction) is equal to .95. Historically, it has been established that the standard deviation for bursting pressures of this seal is 100 psi.

(a) Find the required sample size.

(b) If the true mean maximum bursting pressure of the seals in a shipment is actually $\mu = 5,010$ psi, use your sample size from (a) to determine the probability that the shipment will be rejected.

8-30 The engineer responsible for the design of a communications satellite wishes to determine how many extra batteries to include in the power supply so that the satellite's useful life will be likely to exceed minimum specifications. A random sample of a particular type of battery is selected for testing. The engineer wishes to find the sample size that will provide an estimate of the true proportion π of batteries whose performance will be satisfactory to an accuracy of $e = .05$ with a .95 reliability. We will assume that the population of batteries is quite large.

(a) If π is initially assumed to be .5, find the necessary sample size.

(b) After testing the number of batteries found in (a), only 250 prove satisfactory. What is the point estimate of π? If this were the true value of π, how many more batteries were tested unnecessarily?

8-31 A waiter has been observed at 200 random times while on his station. The following tabulation provides the frequency distribution for the tasks he was performing at each observation:

Cleaning tables	10
Taking orders	35
Walking	57
Placing orders	8
Preparing food	20
Picking up orders	15
Delivering orders	10
Calculating bills	10
Preparing or delivering drinks	35

If during the time period over which he was observed he worked a total of 25 hours, exclusive of idle time, and served a total of 1,000 customers, estimate the mean time (in minutes) spent on each of the above tasks per customer.

REVIEW EXERCISES

8-32 To estimate the mean number of square feet of scrap and wasted wood per log in the manufacture of one-half-inch exterior plywood, a random sample of 100 logs of various sizes from several mills was chosen. Each log's volume was carefully measured, and the theoretical number of square feet of plywood was computed. This value was compared to the actual quantity achieved at the end of production, and the wood loss was calculated. When the study data were assembled, the sample mean and the standard deviation were computed as $\bar{X} = 1,246$ square feet and $s = 114.6$ square feet, respectively. Construct a 95% confidence interval estimate for the mean wood loss.

8-33 The following salaries were determined for a sample of ten university presidents:

$35,000	$67,500	$51,500	$53,000	$38,000
$42,000	$29,500	$31,500	$46,000	$37,500

(a) Find the value of an efficient, unbiased, and consistent estimator of the mean of the salaries of all university presidents.
(b) Find the value of an unbiased and consistent estimator of the variance of the salaries of all university presidents.
(c) Find the value of an unbiased estimator of the proportion of all university presidents' salaries above $40,000.

8-34 The respective proportions of viewers watching each of the earliest prime-time television programs on one network on four successive nights are to be estimated within a tolerance of $\pm.01$ so that there is a 95% chance this precision will be achieved (reliability = .95). The sample proportion P is to be used. Determine the required

sample sizes when the assumed population proportion for
(a) Tuesday is .3.   (c) Thursday is .5.
(b) Wednesday is .6.   (d) Friday is .1.

8-35 A psychologist has designed a new aptitude test for life insurance companies to use
to screen applicants for beginning actuarial positions. To estimate the mean score
achieved by all future applicants who will eventually take the test, the psychologist
has administered it to nine persons. The sample mean and the standard deviation
have been calculated as $\bar{X} = 83.7$ and $s = 12.9$ points, respectively. Determine the confidence interval estimates of the mean for each of the following levels of confidence:
(a) 99%   (b) 95%   (c) 90%   (d) 99.5%

8-36 The following IQ scores have been obtained for a random sample of persons taken
from a large population:

70	110	110
110	100	80
120	110	90

Construct a 95% confidence interval estimate of the population mean IQ.

8-37 A random sample of $n = 100$ widths for two-by-fours was selected. The results show
that $\bar{X} = 3.5$ inches and $s = .1$ inches. Construct a 95% confidence interval estimate
of the boards in the entire shipment.

8-38 A pharmaceutical house wants to estimate the mean number of milligrams (mg) of
drug its machinery inserts into each capsule in four different sizes of pills. In each
case, a standard deviation of 1 mg is assumed and a reliability of 95% is desired.
However, the amount of error tolerated varies with pill size. Find the required
sample size for
(a) tiny pills, .05 mg error.   (c) medium pills, .2 mg error.
(b) small pills, .1 mg error.   (d) large pills, .5 mg error.

8-39 The proportion of blue-eyed children is to be estimated, with a tolerable error of .1
and a reliability of .99. A very conservative guess is that 20% of all children born
have blue eyes.
(a) Find the necessary sample size.
(b) Suppose that 13 blue-eyed children were found. Using the sample size in (a),
construct a 95% confidence interval estimate of the population proportion of
blue-eyed children.

8-40 The example on page 74 shows how work sampling can be used to allocate the
expense of keypunching. Suppose that two departments in a company have data
keypunched. Three alternatives are possible for determining the charges assigned to
the respective departments: (1) Detailed record keeping (census) at a cost of $200
per day. (2) Stopwatch sampling to estimate the mean daily keypunching time per
operator for Department A; then estimating the total daily time for Department A
by multiplying this sample mean by the number of operators. (Department B's usage
can then be estimated from employee time-clock records.) (3) Work sampling to
estimate the proportion of all keypunching time for Department A, which when
applied to total daily keypunching time yields an estimate equivalent to that in (2).
(a) Suppose that the daily operator keypunching time for Department A, determined
by stopwatch sampling, has a standard deviation of $\sigma = .5$ hour. A .95 reliability

probability is desired, with a tolerable error of $e = .08$ hour. Find the required sample size n.

(b) Suppose that the proportion of time spent on Department A, determined by a work-sampling study, is presumed to be $\pi = .50$ (a guess). A .95 reliability probability is desired, with a tolerable error of $e = .02$. Determine the required number of observations n.

(c) Suppose that each stopwatch observation costs $1.00, but that each work-sampling observation costs only $.05. Determine the daily costs for each method. Which one is least expensive?

They are entirely fortuitous you say? Come! Come! Do you really mean that?
. . . When the four dice produce the venus-throw, you may talk of accident:
But suppose you made a hundred casts and the venus-throw appeared a
hundred times; could you call that accidental?

QUINTUS, in Cicero's *De Divinatione*, Book I

Hypothesis Testing

I n Chapter 8, we learned how to use sample information to estimate a population parameter. In this chapter, we will examine another basic type of inference: using a sample to choose between two complementary courses of action. For example, a medical researcher must often decide whether a new treatment is an improvement over an existing one. Or a shipment of supplies must be accepted or rejected, depending on overall quality. Frequently, such choices must be made only on the basis of sample data.

The statistical methods we use to make these types of decisions are based on two complementary assumptions regarding the true nature of a population. A new treatment will reduce a patient's recovery time, or it will not. A shipment of supplies will contain 10% or fewer defective items, or a greater percentage. These assumptions are traditionally referred to as *hypotheses*. The observation of the sample is viewed as a test. Thus, the procedures for making such decisions fall into an area of statistical inference called *hypothesis testing*.

BASIC CONCEPTS OF HYPOTHESIS TESTING   9-1

We will develop the many concepts that link hypotheses together via an extensive example based on whether or not the government should approve the manufacture of a new drug.

Illustration: Testing a Dietary Supplement for Approval

A dietary supplement is being tested to determine if it can dissolve cholesterol deposits in arteries. A major cause of coronary ailments is hardening of the arteries caused by the accumulation of cholesterol. The Food and Drug Administration will not allow the product to be marketed unless there is strong evidence that it is effective. A sample of 100 middle-aged men is selected for the test, and each man is given a standard daily dosage of the supplement. At the end of the experiment, the change in each man's cholesterol level will be recorded.

On the basis of the sampling experiment, a government administrator must decide between two sources of action:

1. Approve the supplement.
2. Disapprove the supplement (and request further research).

Some risks are involved due to uncertainty regarding the supplement's true effectiveness (the sample may not be representative of the population as a whole).

The Null and Alternative Hypotheses

The possible effectiveness states of the supplement are referred to as *hypotheses*. Because the drug administrator is considering only two possibilities—ineffective versus effective—the hypotheses are

1. The supplement is ineffective.
2. The supplement is effective.

The true effectiveness of the supplement is defined by the population of the percentage reductions in cholesterol levels experienced by all middle-aged men who took the supplement. We will summarize the supplement's effectiveness in terms of the population mean:

$$\mu = \text{population mean percentage reduction in cholesterol level}$$

The government will classify the drug as effective only if it produces more than a 20% improvement ($\mu > 20\%$). The percentage reductions constitute a *target population* that does not exist until (and will not exist unless) the supplement is widely available. However, it is still possible to draw a sample from this "imaginary" population.

The first hypothesis is customarily referred to as the *null hypothesis*. Here, "null" represents no change from the natural variation in cholesterol level. In terms of the population mean μ, we will use the following expression for the

NULL HYPOTHESIS

$$H_0: \mu \leq 20\% \text{ (supplement ineffective)}$$

The second hypothesis is called the *alternative hypothesis*. We express the

ALTERNATIVE HYPOTHESIS

$$H_A: \mu > 20\% \text{ (supplement effective)}$$

The hypotheses will always be formulated so that H_0 and H_A are opposites: When one is true, the other is false.

To achieve another special notation in hypothesis testing, we add the subscript 0 to the symbol for the population parameter. In the present example, this gives us

$$\mu_0 = 20\%$$

which is the pivotal value of the population mean. In general, the null hypothesis holds for all levels of μ that are equal to μ_0 or that fall on one side of this value, and the alternative hypothesis applies otherwise. Unfortunately, we do not know which side actually applies, because the true value of μ remains unknown.

Making the Decision

The decision as to whether or not to approve the dietary supplement will coincide with the test results from the sample patients. This decision will be based on the level of

$$\bar{X} = \text{Sample mean percentage reduction in cholesterol levels}$$

We refer to $\bar{X}$ as the *test statistic*.

The chosen action must be consistent with the sample evidence. The two choices may be expressed in terms of the hypotheses:

1. *Disapprove the supplement.* Conclude that the null hypothesis is true (the supplement is ineffective). This is expressed as *accept the null hypothesis.*
2. *Approve the supplement.* Conclude that the null hypothesis is false (the supplement is effective). This is expressed as *reject the null hypothesis.*

The government administrator will approve the supplement if it provides a large mean reduction $\bar{X}$ in the cholesterol levels of the sample patients, because this result would seem to deny H_0. Similarly, she will disapprove the supplement if $\bar{X}$ is small and the supplement has not been very effective during the sample test—evidence that tends to confirm H_0.

In effect, the administrator will decide what action to take when she sets the point of demarcation for $\bar{X}$. For now, we will suppose the value 22 is chosen. This value is commonly referred to as the *critical value* or the *acceptance number*. The test statistic is then used to formulate a decision rule. A *decision rule* allows us to translate sample evidence into a plan for choosing a course of action.

DECISION RULE

Accept H_0 (disapprove the supplement) if $\bar{X} \leq 22$

Reject H_0 (approve the supplement)   if $\bar{X} > 22$

The decision rule identifies *acceptance and rejection regions* for the possible value of $\bar{X}$, which can be illustrated as follows:

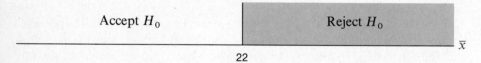

Having established the decision rule, the ultimate action to be taken by the government agency will be determined by the observed sample results. The calculated value of $\bar{X}$ will fall within either the acceptance or the rejection region, and H_0 will be accepted or rejected accordingly. When H_0 is rejected, the results are said to be *statistically significant*. Here, we use the word *significant* in a special sense. All hypothesis-testing results are important, but the test is statistically significant only when the null hypothesis is rejected.

The structure of the decision is shown in Table 9-1. Two courses of action are considered: *Disapprove* or *approve* the supplement. There are also two population states: The supplement is either *ineffective* or *effective*. For each combination of an action and a supplement quality, the outcome represents a correct decision or an error. Two correct decisions are (1) to accept a true null hypothesis (disapprove an ineffective supplement) and (2) to reject a false null hypothesis (approve an effective supplement). The decision rule may also result in two errors. The *Type I error is to reject H_0 when it is true* (approve an ineffective supplement). The *Type II error is to accept H_0 when it is false* (disapprove an effective supplement).

To evaluate the decision rule, we find the probability for committing each error. It is conventional to denote the probabilities for these errors by α and

TABLE 9-1

The Decision Structure for Dietary Supplement

Possible States of the Population	Courses of Action	
	ACCEPT NULL HYPOTHESIS *resulting $\bar{X} \leq 22$* (disapprove)	**REJECT NULL HYPOTHESIS** *resulting $\bar{X} > 22$* (approve)
NULL HYPOTHESIS TRUE $\mu \leq 20\ (\mu_0)$ (ineffective supplement)	**CORRECT DECISION** Probability $= 1 - \alpha$ (disapprove ineffective supplement)	**TYPE I ERROR** Probability $= \alpha$ Significance level $= \alpha$ (approve ineffective supplement)
NULL HYPOTHESIS FALSE $\mu > 20\ (\mu_0)$ (effective supplement)	**TYPE II ERROR** Probability $= \beta$ (disapprove effective supplement)	**CORRECT DECISION** Probability $= 1 - \beta$ (approve effective supplement)

β (the lower case Greek *beta*), where

$$\alpha = \Pr[\text{Type I error}] = \Pr[\text{reject } H_0 | H_0 \text{ is true}]$$

$$\beta = \Pr[\text{Type II error}] = \Pr[\text{accept } H_0 | H_0 \text{ is false}]$$

The value α is also called the *significance level* of the test.

Finding the Error Probabilities

How good is the choice of 22 as the critical value? When establishing the decision rule, the decision maker's attitude toward the two types of errors should be reflected in the levels of the α and β probabilities.

To find α, we consider the case when the null hypothesis is true: The population mean $\mu \leq 20$. The null hypothesis holds for any value of $\mu = 20$ or less, such as 0, 9.7, or 17.5. We focus on the extreme possibility, when the pivotal value of μ_0 applies, and set the population mean at $\mu = 20$.

The normal curve at the top of page 280 shows the sampling distribution of $\bar{X}$ when the null hypothesis is true and $\mu = \mu_0$ exactly. Although we do not know its actual value, for purposes of illustration we assume that the population standard deviation for reductions in cholesterol levels is $\sigma = 10$. The probability for committing the Type I error (the shaded area to the right of 22 under the top curve) is

$$\alpha = \Pr[\bar{X} > 22 | H_0 \text{ is true } (\mu = \mu_0 = 20)] = .5 - .4772 = .0228$$

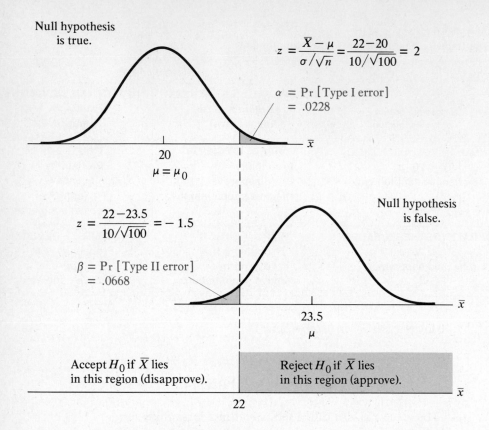

Null hypothesis is true.

$$z = \frac{\overline{X} - \mu}{\sigma/\sqrt{n}} = \frac{22-20}{10/\sqrt{100}} = 2$$

$$\alpha = \Pr\left[\text{Type I error}\right]$$
$$= .0228$$

20
$\mu = \mu_0$

$$z = \frac{22-23.5}{10/\sqrt{100}} = -1.5$$

Null hypothesis is false.

$$\beta = \Pr\left[\text{Type II error}\right]$$
$$= .0668$$

23.5
μ

Accept H_0 if $\overline{X}$ lies in this region (disapprove).

Reject H_0 if $\overline{X}$ lies in this region (approve).

22

When the null hypothesis is false, a different sampling distribution of $\overline{X}$ applies, as shown in the bottom curve. The dietary supplement will be effective for any mean percentage reduction in cholesterol levels greater than 20. In our illustration, $\mu = 23.5$ (and the population standard deviation is assumed to be the same as before). The corresponding probability for the Type II error (the shaded area to the left of 22 under the bottom curve) is

$$\beta = \Pr[X \le 22 \,|\, H_0 \text{ is false } (\mu = 23.5)] = .5 - .4332 = .0668$$

A similar β probability can be obtained for any other level of μ greater than 20. For instance, when $\mu = 24.5$ is used, we obtain $\beta = .0062$.

Interpreting the Sample Results

Once we have formulated a suitable decision rule, all we need to do is wait for the sample results. Suppose that the sample mean reduction in cholesterol levels is calculated to be $\bar{X} = 25.7\%$:

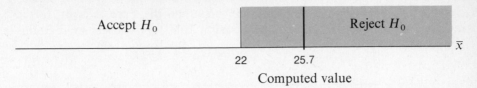

This result is greater than the critical value and falls in the rejection region. This evidence strongly refutes the null hypothesis of an ineffective supplement, *and the decision maker must reject H_0 and approve the supplement.* Stated another way, the sample results are statistically significant.

Choosing the Decision Rule

In hypothesis-testing situations, the decision maker's only real choice is selecting the decision rule. In testing the dietary supplement, the administrator could use a different critical value, such as 23 or 24. Different α and β probabilities would then apply. An ideal decision rule provides a balance between α and β.

In general, the Type I error (rejecting the null hypothesis when it is true) is more serious. In our example, the Type I error is to approve an ineffective supplement, because great care is taken in the United States to avoid marketing any drug or food additive that could have serious undesirable side effects. *Decision rules are usually selected to guarantee a target level for α.*

Avoiding the Type I error by keeping α small tends to increase the chances of committing a Type II error (accepting a false null hypothesis). Disapproving an effective supplement could result in many deaths that could have been prevented by the use of an anticholesterol agent. Both the α and β error probabilities can be kept tolerably small by using a large sample size.

Formulating Hypotheses

The null and alternative hypotheses are opposites: When one is true, the other is false. This allows us to consider just two actions and two outcomes for each hypothesis, depending on whether H_0 is true or false. One area of potential confusion is deciding which hypothesis to label H_0. In earlier statistical applications, the null hypothesis corresponded to the assumption that no change occurs, which accounts for the adjective *null*. For example, in testing a drug, the usual hypothesis would be that it yields little or no improvement over nature. The H_0 used in the dietary supplement illustration corresponds to this designation. Another frequently used guideline is to designate the hypothesis that the decision maker wishes to disprove as H_0. *A common practice is to designate as H_0 the hypothesis for which rejection when true is the more serious error.* Thus, the Type I error would be worse than the Type II error. In the dietary supplement illustration, the decision rule was chosen so that this happened to be the case.

EXAMPLE: TESTING GRADE POINT AVERAGES

A college president wants to show that special admissions students have lower grade point averages (GPAs) than regular students, whose mean GPA is 2.75. Letting μ represent the mean GPA of the special admissions, the following hypotheses apply:

$$H_0: \mu \geq 2.75 \quad \text{(GPAs are as high)}$$

$$H_A: \mu < 2.75 \quad \text{(GPAs are lower)}$$

Here, the null hypothesis places $\mu \geq$ (greater than or equal to) 2.75, because this is what the decision maker wishes to disprove.

EXAMPLE: TESTING CANNERY MACHINERY

A cannery inspector must determine whether or not cans are being filled with a mean $\mu = 16$ ounces of tomato paste. Machinery must be adjusted if the cans are being over-filled or underfilled. The inspector uses the following hypotheses:

$$H_0: \mu = 16 \text{ ounces} \quad \text{(machinery is satisfactory)}$$

$$H_A: \mu \neq 16 \text{ ounces} \quad \text{(machinery must be adjusted)}$$

Here, the null hypothesis is $\mu = 16$ ounces, so that the machinery is assumed to be satisfactory. The Type I error of rejecting H_0 when it is true (unnecessarily adjusting the machinery when it is satisfactory) is the more serious error. The alternative hypothesis states that $\mu \neq$ (is not equal to) 16 ounces. This can also be stated as $\mu > 16$ ounces (cans are being overfilled) or $\mu < 16$ ounces (cans are being under-filled). This is an example of a *two-sided alternative*.

EXERCISES

9-1 A new slow-dissolving aspirin tablet is being evaluated. The null hypothesis is that the tablet dissolves too fast, with a mean dissolving time of μ less than or equal to 10 seconds. The new aspirin will replace the old one if the sample mean $\bar{X} > 12$ seconds; otherwise, the old tablet will continue to be marketed.

Express the null and alternative hypotheses as inequalities involving μ. Then construct an appropriate hypothesis-testing decision table similar to Table 9-1. Be sure to identify the Type I and Type II errors. (Do *not* try to find the values of α and β.)

9-2 For each of the following situations, indicate the Type I and Type II errors and the correct decisions.

(a) H_0: New system is no better than the old one.
  (1) Adopt new system when new one is better.
  (2) Retain old system when new one is better.
  (3) Retain old system when new one is not better.
  (4) Adopt new system when new one is not better.

(b) H_0: New product is satisfactory.
  (1) Introduce new product when unsatisfactory.
  (2) Do not introduce new product when unsatisfactory.
  (3) Do not introduce new product when satisfactory.
  (4) Introduce new product when satisfactory.

(c) H_0: Batch of transistors is of good quality.
  (1) Reject good-quality batch.
  (2) Accept good-quality batch.
  (3) Reject poor-quality batch.
  (4) Accept poor-quality batch.

9-3 Indicate whether the following statements are true or false. If false, explain why.

(a) Committing the Type II error is the same as accepting the alternative hypothesis when it is false.

(b) Either a Type I or a Type II error must occur.

(c) The significance level is the probability of rejecting the null hypothesis when it is true.

9-4 For each of the following hypothesis-testing situations, state the Type I and Type II errors in nonstatistical terms (for example, approving a poor drug or disapproving a good drug).

(a) The null hypothesis is that a new manufacturing process is no improvement over the existing one. If sample evidence indicates that it is better, the new process will be adopted.

(b) A union is negotiating with management for a 4-day workweek with 9-hour working days at a slightly higher wage. The union argues that efficiency and morale will be improved, thereby increasing member output. Choosing as its null hypothesis that productivity will not increase, management will sign the contract if there is adequate evidence that the union claims are true. Otherwise, some hard negotiating sessions will follow.

(c) The objective of a statistical study is to determine whether a stronger consumer protection law is favored by more voters than the present one. Taking as its null hypothesis that the new law does not meet greater voter approval, a consumer action group will lobby for the new law if there is significant evidence to the contrary.

9-5 A university introduces a new degree program.

(a) What null and alternative hypotheses are being tested if the Type I error is to incorrectly conclude that the program will succeed?

(b) What hypotheses are being tested if the Type II error is to incorrectly conclude that the program will succeed?

9-6 The following hypotheses are to be tested, with $\mu_0 = 150$:

$$H_0: \mu \leq \mu_0$$

$$H_A: \mu > \mu_0$$

Suppose that the population standard deviation is $\sigma = 30$ and a sample of $n = 100$ is used. Find the Type I error probability α when $\mu = 150$ and the Type II error probability β when $\mu = 160$, assuming that the following decision rule applies:

Accept H_0 if $\bar{X} \leq 154$

Reject H_0 if $\bar{X} > 154$

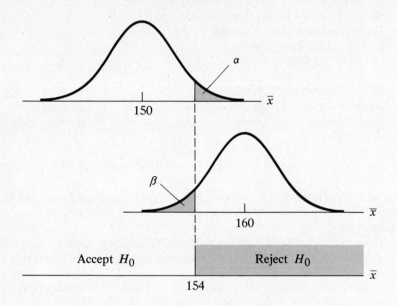

9-7 Repeat Exercise 9-6 when 156 is the critical value. Is the new α larger or smaller? Is the new β larger or smaller?

9-8 Repeat Exercise 9-6 using a larger sample of $n = 200$. Is the new α larger or smaller? Is the new β larger or smaller?

9-2   LARGE-SAMPLE TESTS USING THE MEAN

Now that we have mastered the basic concepts of hypothesis testing, we are ready to apply the procedure. The sampling study begins with the formulation of the hypotheses. After the decision maker establishes a Type I error probability α (significance level) and an appropriate sample size n, everything that follows is largely mechanical. The sample data are collected, and a critical value corresponding to the desired value of α is found. The computed value of the test statistic is then compared to the critical value, and the decision rule indicates the appropriate course of action.

Further Concepts: Acceptance Sampling

Quality control frequently requires decisions about the disposition of batches or lots of items procured from a supplier or produced internally. A manufacturer wishes to protect the finished product by keeping materials supplied from outside at an acceptable level of quality, buying only lots considered to be "good." Because a sample is used to decide whether a lot is good or bad, hypothesis-testing procedures may be used to determine a choice of action. These also apply when a manufacturer must choose between marketing his own production lot through regular channels or as inferior quality "seconds."

Typically there are two actions involved. Accepting the lot is the same as accepting a hypothesis that the lot is good, whereas rejecting the lot is equivalent to rejecting the hypothesis that the lot is good. A good lot may be one having a high mean μ, as would be the case for lifetimes experienced in a shipment of batteries, or one having a low mean, the case for the drying times of a batch of interior house paint.

In Chapter 7 we discussed the two kinds of errors: rejecting a good shipment and accepting a bad shipment. We referred to the probability for the first error as the producer's risk (because a producer desires most to avoid this error), whereas the chance of the second error (which the user wants most to avoid) is the consumer's risk. In acceptance sampling we ordinarily designate as H_0 the hypothesis that the lot is good. This makes the Type I error probability α correspond to the producer's risk, whereas the Type II probability β is the consumer's risk.

To illustrate how a decision rule is obtained in acceptance sampling, we consider a quality-control manager's problem of deciding whether or not to fill an office-furniture maker's order with a current batch of springs produced in his plant. Springs meeting user specifications must withstand a mean stretching force of 50 pounds or more before permanent distortion is induced. The following hypotheses apply:

$$H_0 : \mu \geq \mu_0 \ (= 50 \text{ pounds}) \quad \text{(springs are good)}$$

$$H_A : \mu < \mu_0 \qquad\qquad\qquad \text{(springs are bad)}$$

Here μ represents the unknown population mean stretching force. The manager decides that a sample of $n = 100$ springs will be subjected to a special fatigue test. The test statistic will be the sample mean stretching force $\overline{X}$ when the sample springs first become permanently distorted. Because a low value of the sample mean stretching force would tend to refute H_0 that the population is good, a value for $\bar{X}$ *lower* than the critical value should lead to rejection of the null hypothesis.

To allow a 5% chance for the Type I error of incorrectly rejecting the null hypothesis, the manager has chosen his critical value so that the shaded area in the lower tail of the sampling distribution below is $\alpha = .05$.

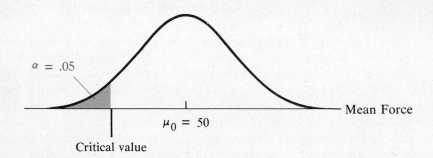

The null hypothesis allows μ to be any force ≥ 50. But for purposes of setting α, the smallest satisfactory stretching force of exactly 50 pounds serves as the assumed value for μ. (In hypothesis testing we set μ_0 by ignoring the "<" or ">" portions, so that the value of α always reflects the extreme case.)

Thus, the normal curve with a mean of $\mu_0 = 50$ represents the sampling distribution of $\bar{X}$. The standard deviation of this curve is $\sigma_{\bar{X}} = \sigma/\sqrt{n}$. (Even though the population standard deviation σ has an unknown value, we can still use the normal curve because the sample size is large.) The quality-control manager estimates the value of σ from the sample data by using the sample standard deviation $s = 5.1$. Thus $\sigma_{\bar{X}}$ is estimated by

$$\frac{s}{\sqrt{n}} = \frac{5.1}{\sqrt{100}} = .51$$

Finding the Acceptance and Rejection Regions

In employing the normal curve, any value of $\bar{X}$ corresponds to a particular normal deviate by the relationship

$$z = \frac{\bar{X} - \mu_0}{\sigma_{\bar{X}}}$$

We denote the *critical normal deviate* by the symbol z_α. For simplicity, z_α will always be a *positive quantity*. From Appendix Table E, when $\alpha = .05$, we find the value of z_α (which we can also denote by $z_{.05}$) to be 1.64. Because H_0 will be rejected when the sample mean falls in the lower tail, this indicates that the critical value of $\bar{X}$ should lie 1.64 standard deviations *below* the mean. Using .51 as the value for $\sigma_{\bar{X}}$, we find the critical value

$$\bar{X} = \mu_0 - z_\alpha \sigma_{\bar{X}} = 50 - 1.64(.51) = 49.16$$

The resulting decision rule has the following acceptance and rejection regions:

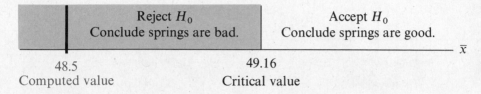

| Reject H_0 | Accept H_0 |
| Conclude springs are bad. | Conclude springs are good. |

48.5
Computed value

49.16
Critical value

$\bar{x}$

After compiling the test data, the quality-control manager computes $\bar{X} = 48.5$ for the mean stretching force at which the sample springs were permanently distorted. Because this value falls in the rejection region, the manager must *reject* the null hypothesis and conclude that the entire lot of springs is bad. Expressed another way, the test results are statistically significant at a level of $\alpha = .05$.

The above decision rule was not actually determined until *after* the sample data were collected. Sometimes a decision rule can be established before the data are collected. In such cases, it is possible to see what Type II error probabilities β result and to make modifications that provide a better balance between α and β. But unless σ is known in advance (or can be accurately guessed), the decision rule can be found only after the sample statistics have been computed.

Testing with z Instead of $\bar{X}$

A variety of test statistics other than $\bar{X}$ may be used in hypothesis testing, and many of these have sampling distributions represented by the normal curve. For simplicity, the normal deviate itself is often used as the test statistic. Then the quality-control manager would use the following decision rule instead:

| Reject H_0 | Accept H_0 |
| Conclude springs are bad. | Conclude springs are good. |

-2.94
z
Computed value

-1.64
$-z_\alpha$

z

where $z_\alpha = 1.64$ is the critical normal deviate we obtained earlier.

The computed value of z can be obtained from the sample results:

$$z = \frac{\bar{X} - \mu_0}{\sigma_{\bar{X}}} = \frac{48.5 - 50}{.51} = -2.94$$

This value falls in the rejection region, so we must make the same conclusion we reached earlier.

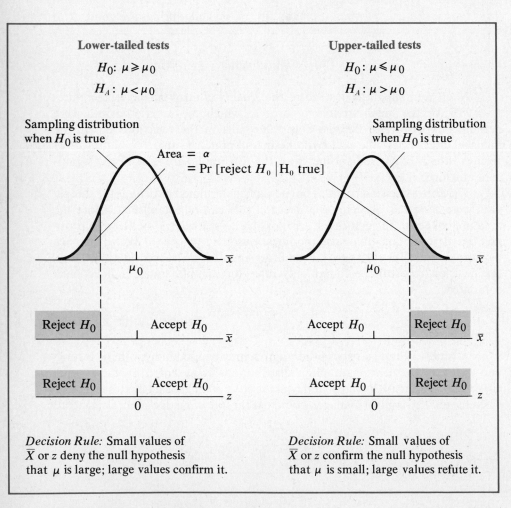

FIGURE 9-1

Upper-tailed and lower-tailed testing situations.

Upper-Tailed and Lower-Tailed Tests

It is convenient to categorize hypothesis-testing situations in terms of the portion of the test statistic's sampling distribution that corresponds to α. In the preceding illustration, α represents the lower tail area under the normal curve for $\bar{X}$, and this situation is therefore referred to as a *lower-tailed test*. In our earlier example, the dietary supplement hypotheses were designated so that an *upper-tailed test* resulted. It will be convenient to establish the following

DEFINITION
The direction of the inequalities in the hypotheses determines the type of test.

$$\text{Upper-tailed test}: H_0:\mu \le \mu_0 \qquad H_A:\mu > \mu_0$$

$$\text{Lower-tailed test}: H_0:\mu \ge \mu_0 \qquad H_A:\mu < \mu_0$$

Upper- and lower-tailed testing situations are summarized in Figure 9-1.

Illustration: The Trend in Human Growth

Over the past 100 years, a pronounced trend in the United States has been for successive generations to grow taller than their parents. This tendency has been attributed to improved health care and nutrition. One researcher believes that this trend has recently reversed, due to degraded environments, diminished exercise, and diets dominated by packaged convenience foods that lack sound nutritional value. To test her theory, the researcher has collected a random sample of heights for $n = 100$ young men. As her null hypothesis, she assumes that the past trend is unchanged, so that the mean height of men is at least as great now as it was 20 years ago, or $\mu_0 = 70$ inches.

We will use this example to explain the steps in the hypothesis-testing procedure.

Summary of Hypothesis-Testing Steps

STEP 1
Formulate the null and alternative hypotheses.

$$H_0:\mu \ge 70 \text{ inches} \qquad \text{(growth trend continues)}$$

$$H_A:\mu < 70 \text{ inches} \qquad \text{(growth trend is reversed)}$$

The null hypothesis contains ≥ because a continuing growth trend includes all possible levels for the mean 70 inches or more.

STEP 2
Select the test statistic. Here, the normal deviate z will be used.

STEP 3
Obtain the significance level α and identify the acceptance and rejection regions. The researcher wants to allow only a 5% chance of wrongly concluding that the growth trend is reversed, so that $\alpha = .05$. From Appendix Table E, we read $z_\alpha = z_{.05} = 1.64$. This test is lower-tailed, and the two decision regions are given below.

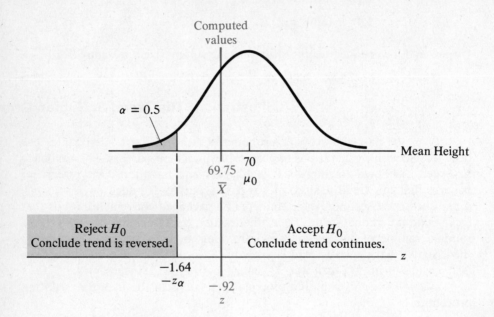

STEP 4
Collect the sample and compute the value of the test statistic. The researcher computes the following sample results:

$$\bar{X} = 69.75 \text{ inches}$$

$$s = 2.71 \text{ inches}$$

The computed value for the test statistic is then calculated.

NORMAL DEVIATE TEST STATISTIC

$$z = \frac{\bar{X} - \mu_0}{s/\sqrt{n}}$$

$$= \frac{69.75 - 70}{2.71/\sqrt{100}} = -.92$$

STEP 5

Make the decision. Because $z = -.92$ falls in the acceptance region, the null hypothesis must be *accepted.* The results are not significant at the 5% level, and the researcher must discard her theory and conclude that the trend in human growth continues.

Type II Error Considerations

The human growth illustration raises an important issue in hypothesis testing. In accepting the null hypothesis that the present trend in successively taller generations is continuing, the researcher has used a rule that protects her with a probability of .05 from committing the Type I error of wrongly concluding that the growth trend is over when it is not. *But she may still have made an incorrect decision by committing the Type II error of accepting the null hypothesis when it is not true.* In other words, the growth trend may have ended and the true mean height for the present generation may actually be some smaller value, such as $\mu = 69.5$ inches. If this figure happens to be the true mean, the probability for incorrectly accepting H_0 is substantial. The traditional procedure outlined in this section always guarantees that the desired level will be achieved for α, but the danger of ignoring the Type II error probability β is ever-present. As a practical matter, we obtain a smaller β by using a larger sample size. We will discuss this point in greater detail later.

EXERCISES

9-9 In testing the hypotheses

$$H_0 : \mu \leq \mu_0 \qquad \text{and} \qquad H_A : \mu > \mu_0$$

with $\mu_0 = 100$, the desired Type I error probability is $\alpha = .10$. The values $\bar{X} = 102.6$ and $s = 25$ have been computed from a sample of $n = 100$.
(a) Is this a lower- or an upper-tailed test? Sketch this situation; include the appropriate normal curve.

(b) Find the critical normal deviate z_α and indicate the acceptance and rejection regions on your sketch.

(c) Calculate the value of the test statistic z and indicate in which region it falls. Should the null hypothesis be accepted or rejected?

9-10 In testing the hypotheses

$$H_0 : \mu \geq \mu_0 \qquad \text{and} \qquad H_A : \mu < \mu_0$$

with $\mu_0 = 30$, the desired Type I error probability is $\alpha = .05$. The values $\bar{X} = 25$ and $s = 10$ have been computed for a sample of $n = 100$. Referring to Exercise 9-9, answer (a), (b), and (c).

9-11 For each of the following testing situations, state whether the test is upper- or lower-tailed and find the value of the critical normal deviate z_α. Then, for each sample result you obtain, calculate the normal deviate value z and indicate whether the null hypothesis should be accepted or rejected. Finally, for each outcome, determine the approximate probability for obtaining a value of the test statistic as rare as or rarer than the one you obtained by finding the tail area under the normal curve corresponding to the computed z.

(a) $H_0 : \mu \leq 105$; $H_A : \mu > 105$; $\alpha = .05$; $n = 100$; $\bar{X} = 108$; $s = 15$.
(b) $H_0 : \mu \geq 5$; $H_A : \mu < 5$; $\alpha = .01$; $n = 100$; $\bar{X} = 4.8$; $s = 2$.
(c) $H_0 : \mu \geq .8$; $H_A : \mu < .8$; $\alpha = .10$; $n = 36$; $\bar{X} = .6$; $s = .6$.
(d) $H_0 : \mu \leq 10$; $H_A : \mu > 10$; $\alpha = .001$; $n = 900$; $\bar{X} = 10.5$; $s = 6$.

9-12 A psychologist believes that age influences IQ. A random sample of 100 middle-aged persons whose IQs had been tested at age 16 were tested again. Subtracting their earlier scores from their new scores resulted in a mean difference of $\bar{X} = 5$ points and a standard deviation of $s = 8$ points. Using $\alpha = .01$ as the significance level, the psychologist wishes to test the null hypotheses $H_0 : \mu \leq 0$ that no improvement in IQ occurs with age.

(a) Is this an upper- or a lower-tailed test? Find the critical normal deviate and sketch the acceptance and rejection regions.

(b) Calculate the normal deviate for the test results. Does this value indicate that H_0 must be accepted or rejected? Does IQ improve with age?

9-13 The superintendent of a printing plant has selected a random sample of 100 rolls of paper from a large shipment. The average length of the rolls is $\bar{X} = 531$ feet, with a standard deviation of $s = 52$ feet. The null hypothesis that $\mu \geq 525$ is to be tested against the alternative hypothesis that $\mu < 525$. Should the null hypothesis be accepted or rejected?

9-14 Fire department officials have modified their aptitude test to eliminate sex bias. The mean score of all persons who will take the test is uncertain. The values $\bar{X} = 83$ and $s = 15$ have been computed for a random sample of $n = 100$ persons who have taken the test. If μ is higher than 80 (the civil service eligibility level), the difficulty of the test is to be increased. The department will base its decision on the sample results, allowing only a 1% chance of unnecessarily raising the test's difficulty when the true population mean score does not exceed the civil service eligibility level.

Complete all of the hypothesis-testing steps. What should the fire department do about its test?

9-15 A new power cell will be used in future communications satellites if sample evidence refutes the null hypothesis that it exhibits a mean-time-between-failure no greater than the present cell, which is rated at $\mu = 550$ hours. A sample test of 100 new cells provides a sample mean of $\bar{X} = 565$ hours, with a standard deviation of $s = 200$ hours. Assuming a significance level of 5%, perform all the hypothesis-testing steps. Should the new power cell be used?

TWO-SIDED HYPOTHESIS TESTS USING THE MEAN   9-3

In Section 9-2, we learned how to conduct the *one-sided hypothesis test*, which may be either upper-tailed or lower-tailed. In the *two-sided hypothesis test*, both extremes in the sample mean—the large and the small values of $\bar{X}$—lead to a rejection of the null hypothesis.

Quality control is an important branch of applied statistics in which two-sided tests are frequently used. We will illustrate two-sided procedures for testing the mean with a detailed illustration of such an application.

The Testing Procedure: A Quality-Control Illustration

A food company advertises that there are 105 chunks of beef in every 15-ounce can of its chili. The company has spent a great deal of money to create the image that its canned foods are high-quality products containing generous portions of expensive ingredients. To verify that its standards are being met in chili production, periodic checks of product quality are made by measuring the quantity of beef in several random samples of 15-ounce cans of chili.

Management cannot be sure that 105 chunks of beef will be added to every can, unless each can is carefully filled individually. This is not practical, because the chili is cooked in large kettles and the ingredients are added proportionately at that time. Each kettle contains several hundred pounds of chili, and the actual number of chunks of beef distributed to each can will vary. Management can control only the *mean number* of chunks per can. How close the mean actually is to the desired goal depends on how precisely the ingredients are mixed during preparation.

Management's goal is to see that the mean number of beef chunks is very close to the advertised 105. If current production has a lower mean, then some remedial action is to be taken. However, because beef is by far the most expensive chili ingredient, a mean higher than 105 will also require remedial action. Management wishes to avoid both high and low extreme values in the mean beef quantity. A sample of $n = 100$ cans is to be tested.

STEP 1
Formulate the null and alternative hypotheses.

$$H_0 : \mu = 105 \qquad \text{(no correction is required)}$$
$$H_A : \mu \neq 105 \qquad \text{(correction is required)}$$

Here, $\mu_0 = 105$ beef chunks. The two-sided alternative hypothesis states that the quantity of beef differs from μ_0, so that either $\mu > 105$ (too much beef) or $\mu < 105$ (too little beef).

STEP 2
Select the test statistic. Here, the normal deviate z will be used.

STEP 3
Obtain the significance level α and identify the acceptance and rejection regions. Here, there are two ways for the Type I error to occur: to unnecessarily correct for overfilling or for underfilling.

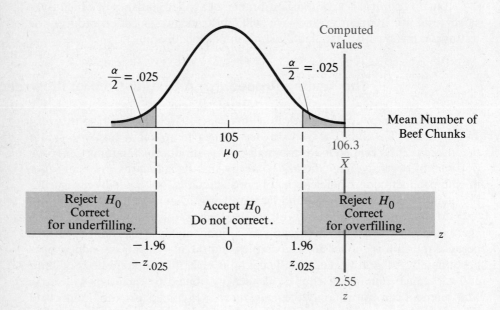

Management desires a significance level of $\alpha = .05$. This is split: Half of this value is assigned to the left tail and half to the right tail. The critical normal deviate is denoted by $z_{\alpha/2}$. For $\alpha/2 = .025$, Appendix Table E provides the value $z_{.025} = 1.96$.

STEP 4
Collect the sample and compute the value of the test statistic. The sample results provide

$$\bar{X} = 106.3 \text{ beef chunks}$$

$$s = 5.1$$

and the computed value of the normal deviate is

$$z = \frac{\bar{X} - \mu_0}{s/\sqrt{n}} = \frac{106.3 - 105}{5.1/\sqrt{100}} = 2.55$$

STEP 5
Make the decision. Since $z = 2.55$ falls in the upper rejection region, the null hypothesis must be *rejected.* The test results are significant at the 5% level, and management must take corrective action to remedy overfilling.

Explanation of Hypothesis Testing in Terms of Confidence Intervals

We can begin the two-sided hypothesis test by constructing a confidence interval. *If μ_0 is a value lying inside the confidence interval constructed from the sample results, H_0 must be accepted.* Only 5% of these intervals can ever be so distant that they do not bracket μ_0 when it is the true mean. In other words, the Type I error of rejecting H_0 when it is true can be committed only 5% of the time that this procedure is employed.

In our quality-control illustration, $\bar{X} = 106.3$ beef chunks, $s = 5.1$, and $n = 100$. Using these data to construct a 95% confidence interval gives us

$$\mu = \bar{X} \pm z_{\alpha/2} \frac{s}{\sqrt{n}}$$

$$= 106.3 \pm 1.96 \frac{5.1}{\sqrt{100}}$$

$$= 106.3 \pm 1.0$$

or

$$105.3 \leq \mu \leq 107.3 \text{ beef chunks}$$

Because this interval does not contain the value $\mu_0 = 105$, H_0 must be *rejected.* This is the same conclusion we reached earlier.

EXERCISES

9-16 In testing the hypotheses

$$H_0: \mu = \mu_0 \ (=0) \qquad \text{and} \qquad H_A: \mu \neq \mu_0$$

the values $\bar{X} = 1.2$ and $s = 9.5$ have been computed from a sample of $n = 100$. Should the null hypothesis be accepted or rejected at the $\alpha = .05$ significance level?

9-17 For each of the following two-sided testing situations, construct a 95% confidence interval for μ and use it to indicate whether H_0 must be accepted or rejected.
(a) $\mu_0 = 10$; $\bar{X} = 10.1$; $s = .3$; $n = 100$.
(b) $\mu_0 = .72$; $\bar{X} = .705$; $s = .13$; $n = 169$.
(c) $\mu_0 = 0$; $\bar{X} = -.7$; $s = 2.5$; $n = 625$.

9-18 A researcher believes that marijuana has a latent effect on dream duration, but he does not know whether dreaming is increased or decreased. The night's sleep of 36 volunteers is monitored 24 hours after each has experienced a marijuana high. Previous research indicates that the population mean dream duration is 1.5 hours under ordinary conditions. Taking that figure as his null hypothesis, the researcher tests H_0 at the .01 significance level against the two-sided alternative. With $\bar{X} = 1.8$ hours and $s = .6$ hours, what should the researcher conclude?

9-19 The output of a chemical process is checked periodically to determine the level of impurities in the final product. If it contains too many impurities, the product will be reclassified; if it contains too few, an expensive catalyst is being consumed in undesirably large quantities. The process is designed to tolerate impurities of $\mu = .05$ gram per liter.

The level of impurities in a sample of $n = 100$ 1-liter vials is measured at random times during each 4-hour period. The process is stopped if there are too many impurities (the tanks are purged) or too few impurities (the control valves are re-adjusted). The chief engineer will accept a Type I error probability of $\alpha = .01$ for unnecessarily stopping the process when the mean level of impurities is at the target level.
(a) Perform hypothesis-testing steps 1–3.
(b) What action should be taken in the following cases?
(1) At 8 A.M., $\bar{X} = .042$ and $s = .023$.
(2) At 12:30 P.M., $\bar{X} = .052$ and $s = .024$.
(3) At 5 P.M., $\bar{X} = 5.6$ and $s = .025$.

9-4   TESTS OF THE MEAN USING SMALL SAMPLES

Often a statistical decision must be made using data from a small sample. When people are tested, sample data are frequently incomplete and sample sizes of less than 30 are quite common. The population standard deviation σ is usually not known and must be estimated by its sample counterpart s. In Chapter 8, we saw that the normal curve does not perform satisfactorily as the sampling distribution of $\bar{X}$ under these circumstances. In its place, therefore, we employ the Student t distribution.

The Student t statistic is found in the same manner as the normal deviate z. The following expression is used to compute the

STUDENT t TEST STATISTIC

$$t = \frac{\bar{X} - \mu_0}{s/\sqrt{n}}$$

We will use a medical experiment with artificial blood to illustrate testing the mean using small samples.

Illustration: Artificial Blood

A blood substitute would be highly advantageous for patients undergoing major surgery, which is severely complicated by the need for whole blood. The substitute might also be used to flush out and cleanse diseased body organs like the liver and kidneys.

A team of medical researchers wishes to determine what harmful effects will result from completely substituting a chemical solution for the blood in a rat's circulatory system.

The scientists insert a plastic tube into a white rat's jugular vein and slowly drain away all its blood. Simultaneously, they pump a milky-white artificial blood into the rat's circulatory system. The substitute is a man-made chemical concoction containing fluorocarbons, or compounds of fluorine and carbon that are more commonly used in aerosol propellants, refrigerants, and fire-extinguishing agents.

Shortly after the blood exchange has taken place, the rat begins to move about its cage, washing itself or perhaps taking a drink of water. It not only survives, but it thrives, illustrating for the first time anywhere that animals can live without any element of natural blood in their bodies.*

To decide whether or not the same chemical agent should be used in future experiments, a sample of $n = 25$ "middle-aged" rats has been selected. Each rat's blood has been totally replaced by the same fluorocarbon solution. Over a period of several days, this artificial blood has been gradually supplanted internally by naturally regenerated blood. A hypothesis-testing procedure will determine whether or not the new blood is harmful, thereby enabling the researchers to decide whether to test the fluorocarbon solution further or to seek a new blood substitute.

STEP 1

Formulate the null and alternative hypotheses. The null hypothesis is that the artificial blood is not harmful, so that the mean remaining life span of injected rats is as long as that of untreated rats of the same age. Thus

$$H_0: \mu \geq \mu_0 \, (= 4 \text{ months}) \quad \text{(artificial blood is not harmful)}$$

$$H_A: \mu < \mu_0 \qquad\qquad\qquad \text{(artificial blood is harmful)}$$

The one-sided alternative has been chosen, because the researchers intend to change chemical solutions only if the fluorocarbon solution proves harmful in the long run.

---

* Jeffery A. Perlman, "Artificial Blood, Long a Goal of Scientists, Sustains Lab Animals," *The Wall Street Journal* (January 29, 1974), p. 1. Although the experiments themselves are real, the decision situation and the data presented in this example are strictly hypothetical.

STEP 2

Select the test statistic. The Student t statistic is used, because n is small and the population standard deviation s must be estimated from the sample results.

STEP 3

Obtain the significance level α and identify the acceptance and rejection regions. The Type I error, rejecting H_0 when it is true, will unnecessarily retard further research while new chemical substitutes are evaluated; thus, a fairly low significance level of $\alpha = .01$ has been established. Appendix Table F provides the critical value t_α. For $n - 1 = 25 - 1 = 24$ degrees of freedom, we read $t_{.01} = 2.492$. Because a small sample mean—and therefore a negative level for t—will tend to refute H_0, the test is lower-tailed. The acceptance and rejection regions are shown below.

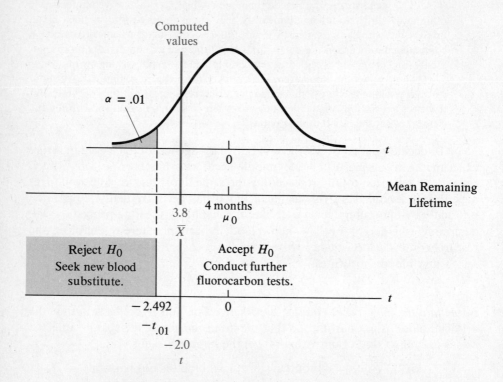

STEP 4

Collect the sample and compute the value of the test statistic. The researcher obtains the following sample results:

$$\bar{X} = 3.8 \text{ months}$$

$$s = .5$$

The computed value of the test statistic is

$$t = \frac{\bar{X} - \mu_0}{s/\sqrt{n}} = \frac{3.8 - 4}{.5/\sqrt{25}} = -2.0$$

STEP 5
Make the decision. The computed value of t falls in the acceptance region. The null hypothesis must be *accepted*, and the test results are not significant at the .01 level. The evidence supports the assumption that fluorocarbon-based artificial blood is not harmful, and the solution will be used in further testing.

An Assumption of the *t* Test

The Student t distribution is based on the assumption that the underlying population frequency distribution is represented by the normal curve. As a practical matter, this requirement poses no difficulties, as long as the population frequency distribution is unimodal and fairly symmetrical. Thus, the t test is basically insensitive to this requirement. Statisticians would say that the t test is, to some degree, *robust* with respect to the non-normality of the population.

To avoid difficulties that may arise when the normality assumption does not hold, certain nonparametric tests may be employed. We will discuss these tests in Chapter 18.

EXERCISES

9-20 In testing the hypotheses

$$H_0 : \mu \le 160 \quad \text{and} \quad H_A : \mu > 160$$

a significance level of $\alpha = .05$ is desired. The test is upper-tailed, as shown below.

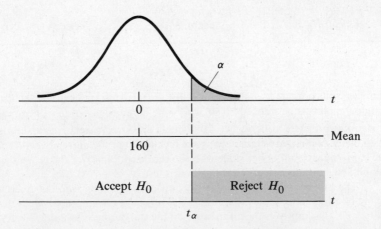

The values $\bar{X} = 170$ and $s = 30$ have been computed for a sample of $n = 25$. Complete the hypothesis-testing steps to determine whether the null hypothesis should be accepted or rejected.

9-21 A structural engineer is testing the strength of a newly designed steel beam required in cantilever construction. As his null hypothesis, he assumes that the mean strength will be at most as great as the 100,000 pounds per square inch (psi) for traditional beams. A test sample of $n = 9$ new beams has provided $\bar{X} = 105,000$ psi and $s = 10,000$ psi. At the $\alpha = .05$ significance level, should the engineer accept or reject his null hypothesis?

9-22 The average length of a random sample of $n = 20$ rolls of wire is $\bar{X} = 2,031$ meters (m), with a standard deviation of $s = 47$ m. The null hypothesis that $\mu \geq 2,000$ m is to be tested. Assuming that $\alpha = .01$, should the null hypothesis be accepted or rejected?

9-23 The output of a chemical process is monitored by taking a sample of $n = 25$ vials to determine the level of impurities. The null hypothesis is that the mean level is exactly .05 gram per liter. If the mean level of impurities in the sample is too high, the process will be stopped and the tanks will be purged; if the sample mean is too low, the process will also be stopped and the valves will be readjusted. Otherwise, the process will continue. A two-sided test applies, and the testing situation is illustrated below.

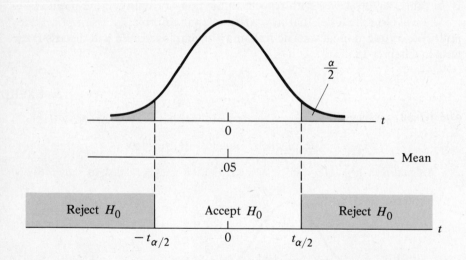

Sample results provide $\bar{X} = .064$ gram with $s = .017$. At a significance level of $\alpha = .01$, should the process be stopped? If so, what type of remedial action will be required?

9-24 A statistics professor believes her 25 students are a representative random sample of all the beginning statistics students she has taught at the university. At the end of the term, she gives all of her students a standardized test. Over the years, they have established a mean score of 75 points on this test.

The professor wishes to assess the impact of spending twice as much time as before on probability theory and taking the time away from other "core" statistical material. If this improves the mean score on the standardized test, she will adopt the new policy in the future. As her null hypothesis, she assumes that there will be no score improvement, so that $H_0: \mu \leq 75$ points. The professor assigns an $\alpha = .01$ probability of making the Type I error of adopting the new policy (rejecting H_0) when it does not actually improve test scores.

The sample results provide $\bar{X} = 86$ points, with $s = 10$ points. Should the new policy be adopted?

9-25 A cannery inspector insists that the mean weight of ingredients in 16-ounce cans be precisely accurate. If the mean weight falls above or below this figure, some remedial action is taken. The inspector has established a significance level of $\alpha = .05$ for a sampling test, so that the Type I error (taking unnecessary action) is avoided 5% of the time. Each hour, a sample of $n = 25$ cans is taken, each can is opened, and the contents are weighed.

(a) Perform hypothesis-testing steps 1–3.

Should the inspector take remedial action if

(b) $\bar{X} = 15.9$ ounces and $s = .4$ ounce?

(c) $\bar{X} = 16.2$ ounces and $s = .3$ ounce?

HYPOTHESIS TESTS USING THE PROPORTION 9-5

The procedures described earlier for hypothesis tests using the mean can be adapted to hypothesis tests using the proportion. These tests are used when a decision must be made regarding a qualitative population. In such cases, the unknown parameter of interest is the population proportion π. For example, the effectiveness of a new flu vaccine can be determined by the proportion π of treated persons who do not catch the disease after they are exposed to it. Similarly, the proportion π of users of an existing vaccine who prefer a new vaccine can determine whether the new vaccine will be marketed. If π is too small, it may be unprofitable to introduce the new vaccine.

Illustration: Testing a Theory on Right-Handedness

Have you ever wondered why most people are right-handed? Right-handers outnumber lefties by 19 to 1, regardless of culture, race, or nationality. One researcher has proposed the following theory to partially justify this phenomenon:

1. Handedness would be strictly a "coin-flipping" proposition, except for environmental influences that bias the development of the right hand.
2. Mothers' hearts tend to be on their left sides.
3. Babies feel secure when they sense heartbeat rhythms and are happier when placed close to their mothers' hearts. Unconsciously accommodating this

behavior, babies are held predominantly on the mothers' left side *supported by her left arm.*
4. This leaves only the mother's right arm free. Because the mother's right hand is used for most hand-to-hand contact with the baby, her infant begins life using the right hand more than the left.
5. In effect, heart placement helps to determine handedness. If the human heart were on the right side instead of the left, more babies would be left-handed, according to this theory.

To test her theory, the researcher collected a random sample of $n = 100$ new mothers. To eliminate the bias toward right-handedness, each mother was asked to wear a tiny amplifier to make her heartbeat louder on the right side when holding her baby—in effect repositioning the apparent heart location. To avoid biasing their behavior, the mothers were not told the purpose of the device or the experiment. At the end of the experiment, the babies were tested to determine their dominant hand. If the researcher's theory is correct, the results of this experiment should favor the motor development of the babies' left hands.

STEP 1
Formulate the null and alternative hypotheses. As her null hypothesis, the researcher assumes that the proportion π of babies who become left-handed when reared under these experimental conditions will be no greater than the proportion of left-handed babies in the general population:

$$H_0: \pi \leq \pi_0 \ (.05) \quad \text{(theory is untrue)}$$

The alternative hypothesis is that experimental conditions will yield a higher proportion of left-handed babies:

$$H_A: \pi > \pi_0 \quad \text{(theory is true)}$$

STEP 2
Select the test statistic. The test results will provide a value for the sample proportion P of left-handed babies. For large sample sizes, the sampling distribution of P may be approximated by the normal curve. The normal deviate z can therefore be used as the test statistic.

STEP 3
Obtain the significance level α and identify the acceptance and rejection regions. Because the researcher does not want to lead scientists down a "blind alley," she is extremely cautious about potential errors and chooses a very small

significance level of $\alpha = .001$. This offers her considerable protection against proposing her theory when it really does not apply. Appendix Table E provides a critical normal deviate of $z_\alpha = z_{.001} = 3.08$. Because large values of the sample proportion P (and therefore of z) will refute the null hypothesis, the test is upper-tailed. The test is summarized in the figure below:

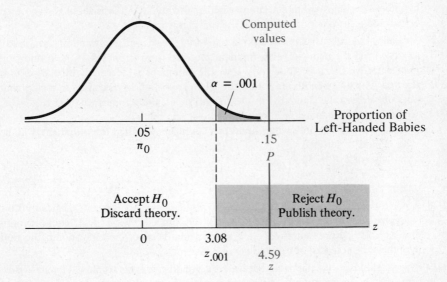

STEP 4
Collect the sample and compute the value of the test statistic. At the end of the experiment, the researcher finds that 15 babies are left-handed, so that $P = 15/100 = .15$. The computed value of the test statistic is the

NORMAL DEVIATE FOR TESTING THE PROPORTION

$$z = \frac{P - \pi_0}{\sqrt{\pi_0(1 - \pi_0)/n}}$$

$$= \frac{.15 - .05}{\sqrt{.05(1 - .05)/100}} = 4.59$$

STEP 5
Make the decision. Because the computed value of z exceeds z_α and falls within the rejection region, the researcher must *reject* H_0. She will publish her findings and promulgate her theory.

EXERCISES

9-26 In testing the hypotheses

$$H_0 : \pi \leq \pi_0 \ (=.5) \quad \text{and} \quad H_A : \pi > \pi_0$$

for a sample of $n = 100$, the desired Type I error probability is $\alpha = .10$. Complete the hypothesis-testing steps to determine whether the null hypothesis should be accepted or rejected if the computed value of the sample proportion is $P = .55$.

9-27 A quality-control inspector at a government supply depot tests random samples of $n = 100$ from the items in each shipment received from suppliers. If the proportion P of defective items is sufficiently low, the shipment is accepted; otherwise, it is rejected. The null hypothesis is that the true proportion of defectives in a shipment does not exceed the standard: $\pi \leq \pi_0$. In all cases, the significance level is $\alpha = .05$.
(a) Does an upper-tailed or a lower-tailed test apply?
(b) For each of the following situations, indicate whether the shipment will be accepted or rejected.
   (1) $\pi_0 = .05$; $P = .08$.
   (2) $\pi_0 = .10$; $P = .08$.
   (3) $\pi_0 = .01$; $P = .04$.

9-28 A congressman wishes to test the null hypothesis that at least 50% of the voters recognize his name. Only 10 people from a random sample of $n = 25$ could identify him as a U.S. Representative. At a significance level of $\alpha = .05$, should the null hypothesis be accepted or rejected?

9-29 A professional group claims that at least 40% of all engineers employed by aerospace firms switch jobs within three years of being hired. The alternative hypothesis is that the rate of job changing is below 40%. At a significance level of .01, should the claim be accepted or rejected if the sample results show that 25 out of $n = 100$ engineers changed jobs?

9-6 SELECTING THE TEST

Some Important Questions

We have covered the basic concepts of hypothesis testing that apply to means and proportions. It is now time to consider some fundamental questions.

1. *What other test statistics can we use?* Many test statistics other than $\bar{X}$ and P can be applied to hypothesis-testing situations. Usually, the nature of the test determines the type of test statistic to be employed. In later chapters, we will learn to use many new statistics to compare the parameters of two or more populations. In lieu of testing with $\bar{X}$, a number of so-called *non-parametric tests* (to be described in Chapter 18) may be used.

2. *How can we determine which test statistic will work best?* Often, we must choose from several alternative procedures. Perhaps the most important factor to consider in making this choice is the relative efficiency of each test.
3. *How can we evaluate a particular decision rule in terms of the protection it provides against both kinds of incorrect decisions, the Type I and Type II errors?* Both types of decision errors should be avoided. But we have seen that when the choice is based on sample data, some probability for each error (α and β) must be tolerated. Ordinarily, only α—the probability for committing the more serious Type I error—is explicitly considered in selecting a decision rule. But the statistician may revise the initial decision rule if it proves to be inadequate in protecting against the Type II error.
4. *What assumptions underlie the proposed testing procedure, and how do they affect its applicability?* As we have seen, the *t* test assumes that the population itself is normally distributed. Just how critical is such an assumption? At one extreme, an assumption can limit a test so severely that it cannot be applied to many decision-making situations. On the other hand, an assumption may prove to be relatively unimportant. A test that basically accomplishes what it is intended to do even when a requirement is not exactly met is said to be *robust* with respect to a violation of that assumption.

The Power Curve

Usually there are several possible values of β, the Type II error probability. Remember that β is the probability for accepting H_0 when H_0 is false (and H_A is true). Using the alternative hypothesis $H_A:\mu > 20$ in our dietary supplement illustration, we can compute a different β for any particular level of μ greater than 20. On page 280, we computed $\beta = .0668$ when $\mu = 23.5$. Some other values are

$$\beta = .8413 \text{ when } \mu = 21$$

$$\beta = .5000 \text{ when } \mu = 22$$

$$\beta = .0228 \text{ when } \mu = 24$$

These probabilities are represented as points on the graph in Figure 9-2. The curve connecting the points is called a *power curve*. The height of the curve provides the probability $(1 - \beta)$ of rejecting the null hypothesis at the indicated possible level for μ.

The power curve can be useful in evaluating a hypothesis test. The usual hypothesis-testing procedure begins with a prescribed significance level α (the Type I error probability) and a set sample size n. These parameters fix the levels for the β probabilities.

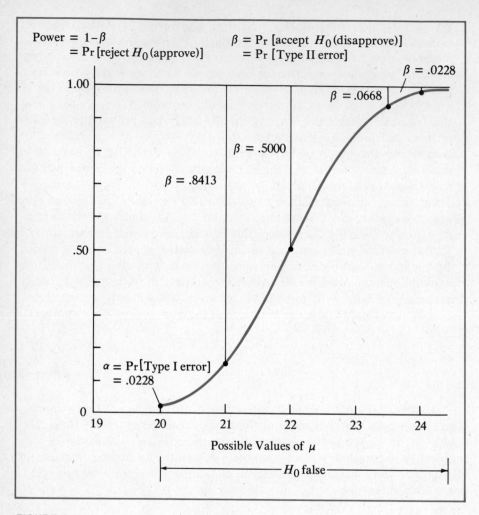

FIGURE 9-2
Power curve for dietary supplement decision.

The decision maker may study the power curve and conclude that the β's are higher than the desired levels. In our dietary supplement illustration, two remedies are available:

1. Raise the significance level α, so that there is a higher Type I error probability for approving an ineffective supplement. This shifts the critical value of the test statistic, lowering all the β's and providing a lower Type II error probability for disapproving an effective supplement for each μ.

2. Increase the sample size n. This will decrease both the α and β probabilities for all levels of μ.

Efficiency and Power

The choice of a test statistic is an important question in advanced applications of hypothesis-testing theory. In many decision situations, a number of statistical tests can be performed. For instance, rather than find the actual levels of cholesterol reduction in patients treated in the dietary supplement illustration, the statistician could simply determine the proportion P of patients who achieved a major reduction in their cholesterol levels. Instead, the more detailed measure $\bar{X}$—the mean percentage reduction in cholesterol levels—was chosen. Generally, $\bar{X}$ provides more information than P, so that $\bar{X}$ is a more efficient test statistic than P. Thus, for the same sample size and Type I error probability α, the decision rule obtained for $\bar{X}$ should provide smaller Type II error probabilities β (and higher probabilities for rejecting H_0 when it is false). In this case, the $\bar{X}$ test is said to be more *powerful* than the P test.

LIMITATIONS OF HYPOTHESIS-TESTING PROCEDURES   9-7

In this chapter we presented the *classical* statistical decision analysis. Its applicability to the rather large class of decision problems faced by the modern manager is limited in several ways. Paramount among these is the dependence on samples for obtaining supportive or contradictory evidence. Many significant decisions must be made without first taking a sample, because a sample is too costly or simply unobtainable. For example, in deciding whether to drill for oil, the wildcatter does not collect direct sample evidence by drilling holes at random as the basis for rejecting or accepting the null hypothesis that there is oil. The only way such evidence may be obtained is to sink the shaft, and this can be done only after the decision has already been made to drill. What evidence is obtained must be of an indirect nature, such as results of geological surveys, seismic tests, and nearby drilling experience. Another limitation of hypothesis testing is its restricted applicability to decisions involving two alternatives when so many decisions involve choosing among a multitude of alternatives.

Even when classical hypothesis testing is applicable, there are improvements that can be made. One is to expand the process so that more information can be utilized. For instance, the fact that certain values for the population decision parameter may be more likely than others can be used by the decision maker to reduce the probability for an incorrect decision. Methods for doing this will be described in Chapter 24. A further limitation of the classical approach is that it does not in any way help to resolve the problem faced by a decision maker in assessing the risks associated with the incorrect decisions; this subject is covered further in Chapter 21. The cost of collecting the sample is not explicitly considered in the classical treatment. Chapter 24 shows how we can relate the costs of selecting the sample to the costs of the ensuing risks, so that an optimal course of action can be achieved.

REVIEW EXERCISES

9-30 For each of the following situations, find the Type I error probability α that corresponds to the stated decision rule. In all cases, use a sample of $n = 100$ and assume that the population is large.

(a) *Accept H_0 if $\bar{X} \geq 12$*   *Reject H_0 if $\bar{X} < 12$*   when $\mu_0 = 14$ and $\sigma = 18$

(b) *Accept H_0 if $\bar{X} \leq 100$*   *Reject H_0 if $\bar{X} > 100$*   when $\mu_0 = 97$ and $\sigma = 20$

9-31 For each of the following hypothesis-testing situations, $n = 100$ and $\alpha = .05$. In each case, state whether H_0 should be accepted or rejected:

(a) $H_0: \mu \geq 25$; $H_A: \mu < 25$; $\bar{X} = 23.0$; $s = 4.7$.

(b) $H_0: \mu \leq 14.7$; $H_A: \mu > 14.7$; $\bar{X} = 15.3$; $s = 1.5$.

(c) $H_0: \mu = 168$; $H_A: \mu \neq 168$; $\bar{X} = 169$; $s = .92$.

9-32 A presidential candidate plans to campaign in only those primaries where she is preferred by at least 20% of the voters in her party. A random sample of 100 voter preferences is to be obtained from each state. In each case, the null hypothesis will be that the state meets the 20% criterion. If an $\alpha = .05$ Type I error probability is desired, indicate whether the null hypothesis should be accepted or rejected in each of the following states. Also indicate whether or not the candidate will enter the campaign in each of these states.

	State	Number in Sample Preferring Candidate
(a)	New Hampshire	12
(b)	Florida	23
(c)	Wisconsin	15
(d)	Massachusetts	10

9-33 To assess the time required to cure a disease using an experimental drug, the drug is administered to a sample of $n = 25$ persons. The medical team is basing its decision to continue or terminate its research on whether or not the drug is deemed effective. An $\alpha = .01$ Type I error probability is desired for not continuing the research when the drug actually yields a mean advantage in cure time of at least two days over the present treatment time. If the sample results for the cure-time advantage are $\bar{X} = 1.5$ and $s = .5$ day, should the null hypothesis be accepted or rejected? Should the drug experiment be continued or terminated?

9-34 The plant manager of a shoe manufacturer wishes to determine whether or not the sizing department's work is satisfactory. Due to ordinary variability, a shoe does not fall into a precise size category like $8\frac{1}{2}$EEE. Finished shoes must be individually measured and then classified to the nearest appropriate size. To determine whether or not to take corrective action, the manager asks that a sample of 100 randomly selected shoes be meticulously resized. The proportion of shoes that were originally correctly sized is to be determined. As his null hypothesis, the manager assumes that the true proportion of all shoes incorrectly sized will be $\leq.05$, the desirable level. He must construct a decision rule that will reject this hypothesis when it is true with a probability of .05.

The sample results show that $P = .062$. Should the null hypothesis be accepted or rejected? Should corrective action be taken?

9-35 A psychologist wishes to give her standard screening examination to some recent job applicants to decide whether or not today's job seekers achieve higher scores than applicants did ten years ago. If they do, she will use a new examination in the future. The test is to be administered to a random sample of $n = 25$ persons. In analyzing the results, the psychologist wishes to allow only a 1% chance of incorrectly changing procedures when the actual mean screening examination score is ≤ 86, the historical mean figure ten years ago. If the sample group achieves a mean score of 88, with a standard deviation of 10 points, should the present screening examination be retained or changed?

10

Regression and Correlation

In many important business decisions it is necessary to predict the values of unknown variables. A personnel manager is concerned with predicting the success of a job applicant, which may be expressed in terms of the applicant's productivity. A production manager for a chemical company may wish to predict the levels of impurities in a final product. Many economists make predictions of gross national product (GNP). In each case, knowledge of one factor may be used to better predict another factor. The personnel manager may use screening examination scores as the basis for predicting future on-the-job performance. The production manager may use process temperatures or concentrations of ingredient chemicals to forecast impurities. The economist may use current interest rates, unemployment levels, and government spending to make GNP prognostications.

Regression analysis tells us how one variable is related to another by providing an equation that allows us to use the known value of one or more variables to estimate the unknown value of the remaining variable. For instance, a medical researcher can use regression analysis to estimate a laboratory animal's longevity (a variable of unknown value) resulting from the caloric content of its daily diet (a variable of known value). This information can lead to further research on how diet affects the human life span. Similarly, an economist may use regression analysis to show how one variable, such as percentage unemployment, can be used to predict the percentage inflation rate. The resulting mathematical relationship provides a graphical display called the *Phillips curve.* More than one variable can be used to estimate an unknown variable. When several variables are used to make a prediction, the technique is called *multiple regression.* (We will discuss this topic in Chapter 11.)

The term *regression analysis* is derived from studies made by Sir Francis Galton at the turn of this century. Galton compared the heights of persons to the heights of their parents. His major conclusion was the offspring of unusually tall persons tend to be shorter than their parents, whereas children of unusually short parents tend to be taller. In a sense, the successive generations of offspring from tall persons "regress" downward toward the mean height of the population, and the reverse is true of the offspring from short families.* Since Galton used one variable (the height of the parent) to predict another (the height of the child), the original term *regression* was eventually applied to more general analyses in which one variable was used to predict another. Aside from the context of prediction, the term *regression* as used in this chapter has little relationship to Galton's original notion of regressing toward the mean.

Correlation analysis tells us the degree to which two variables are related. It is useful in expressing how efficiently one variable has estimated the value of another variable. Correlation analysis can also identify the factors of a multiple-characteristic population that are highly related, either directly or by a common connection to another variable.

10-1 REGRESSION ANALYSIS

The primary goal of regression analysis is to obtain predictions of one variable using the known values of another. These predictions are made by employing an equation such as $Y = a + bX$, which provides the estimate of an unknown variable Y when the value of another variable X is known. Such an expression is referred to as a *regression equation.* Knowing the regression equation, we can readily predict Y from a given X. Unlike the results from ordinary mathematical equations (such as $A = b \times h$ for the area of a rectangle, or interest $= i \times P \times t$, where i is the rate of interest, P is the principal, and t is the time), we cannot be certain about the value of Y that we obtain from the regression equation. This is due to inherent statistical variability. Predictions made from the regression equation are subject to error and are only *estimates* of the true values.

---

* But the distribution of heights for the total population continues to exhibit the same variability from generation to generation. This is because the more prevalent parents of near-average height produce more tall offspring than the relatively rare tall parents do.

TABLE 10-1

Sample Observations of Rail Distances
and Transportation Times for Ten Shipments by a Parts Supplier

Customer	Rail Distance to Destination (miles) X	Transportation Time (days) Y
1. Muller Auto Supply	210	5
2. Taylor Ford	290	7
3. Auto Supply House	350	6
4. Parts 'n' Spares	480	11
5. Jones & Sons	490	8
6. A. Hausman	730	11
7. Des Moines Parts	780	12
8. Pete's Parts	850	8
9. Smith Dodge	920	15
10. Gulf Distributors	1,010	12

Regression analysis begins with a set of data involving pairs of observed values, one number for each variable. Table 10-1 shows the observations of transportation times and distances for a sample of ten rail shipments made by an automobile parts supplier. These data will be used to arrive at predictions of transit times for future shipments. From these the regression equation will be determined. Because using a sample gives rise to sampling error, the regression equation obtained may not be truly representative of the actual relationship between the variables. In order to reduce the chances of large sampling error, a sample size considerably greater than ten ought to be used. We have taken such a small number here merely for ease in showing calculations.

The Scatter Diagram

A first step in regression analysis is to plot the value pairs as points on a graph, as Figure 10-1. The horizontal axis corresponds to values of the variable distance, denoted by the letter X. The vertical scale represents values of the variable time, for which we use the designation Y. A point is found for each shipment. For example, the shipment to Jones & Sons, located at a distance of $X = 490$ miles from the plant, took $Y = 8$ days to arrive. This is represented by the point ($X = 490$, $Y = 8$) on the graph. The points obtained in Figure 10-1 are spread in an irregular pattern. For this reason, such a plot is referred to as a *scatter diagram*.

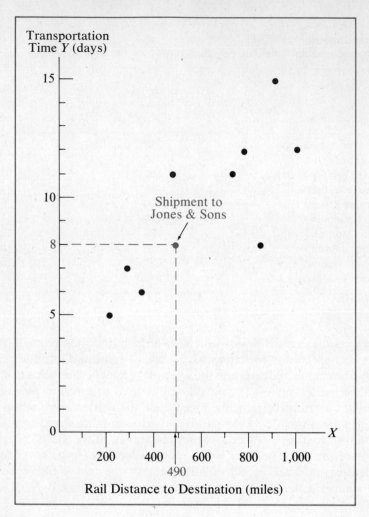

FIGURE 10-1
Scatter diagram for shipments by a parts supplier (values from Table 10-1).

The parts supplier wishes to use the known rail distance to the customer as the basis for predicting an order's unknown transportation time. It is customary to refer to the variable whose value is known as the *independent variable*, the possible values of which are represented on the X axis of the scatter diagram. The variable whose value is being predicted is called the *dependent variable*, the possible magnitudes of which are represented on the Y axis. Thus, rail distance X is the independent variable, because it is determined by the shipment's destination only, whereas the transportation time Y is the dependent variable, because it is in part predictable from rail distance.

These designations follow from simple algebra, where the X axis represents the independent variable and the Y axis provides values for the dependent variable by means of a function or an equation. Thus, Y is a function of X. The dependence of Y on X does not necessarily mean that Y is *caused* by X. The type of relationship found by regression analysis is a statistical one. (We can find a *statistical* relationship expressing family household expenditures as a dependent variable that is a function of an independent variable, family disposable income. But having money to spend does not mean that appliances will be purchased. A purchase is a voluntary decision that is merely allowed to occur because money has been made available; a purchase need not be made just because there is sufficient income.)

The Data and the Regression Equation

The second step in regression analysis is to find a suitable function to use for the regression equation, so that it will provide the predicted value of Y for a given value of X. The clue to finding an appropriate regression equation can be found in the general pattern presented by the points in the scatter diagram. A quick examination of the parts supplier data in our example indicates that a straight line, like the one shown in Figure 10-2, might be a meaningful summary of the information provided by the sample. This line seems to "fit" the rough scatter pattern of the data points.

A linear relationship between the variables X and Y is the simplest to visualize. The general equation for a straight line is $Y = a + bX$. The constant a is the value of Y obtained when $X = 0$, so that $Y = a + b(0) = a$. Because this is the value of Y at which the line intersects the Y axis, a is usually referred to as the Y *intercept*. The constant b, or the *slope* of the line, represents the change in Y due to a one-unit increase in the value of X. Figure 10-3 shows the line for the equation $Y = 3 + 2X$. Here, the Y intercept is $a = 3$ and the slope is $b = 2$. Y increases by two units for every one-unit change in X.

To review how we determine the value of Y for X, suppose we wish to find the Y that corresponds to $X = 5$. Substituting 5 for X in the expression gives us $Y = 3 + 2(5) = 13$. The same value may be read directly from Figure 10-3 by following the vertical black line from $X = 5$ to the line relating Y to X. The vertical distance represents the value of Y. Any point on the line, for example $(X = 5, Y = 13)$, can be described by the horizontal distance (5) and the vertical distance (13) from the origin.

The line used to describe the average relationship between the variables X and Y, or the *estimated regression line*, is generally obtained from sample data. The estimated regression line provides an estimate of the mean level of the dependent variable Y when the value of X is specified. We use the symbol $\hat{Y}_X$

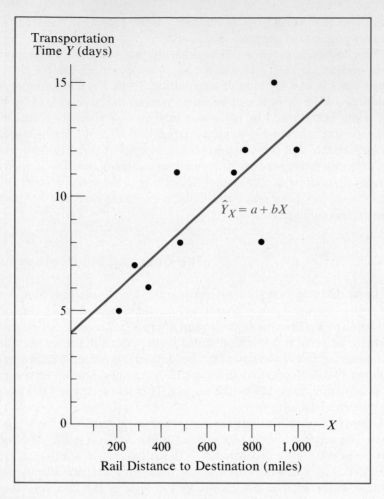

FIGURE 10-2
Fitting a regression line to the parts supplier data.

("Y-hat sub X") to represent the values obtained from the linear

ESTIMATED REGRESSION EQUATION

$$\hat{Y}_X = a + bX$$

The symbol $\hat{Y}_X$ distinguishes estimates of the dependent variable from the observed data points, which, for simplicity, we denote by the symbol Y. For a specific X, the resulting $\hat{Y}_X$ is a *predicted* value of the dependent variable. As we will see, the values a and b in this expression are found from sample data and are referred to as *estimated regression coefficients*.

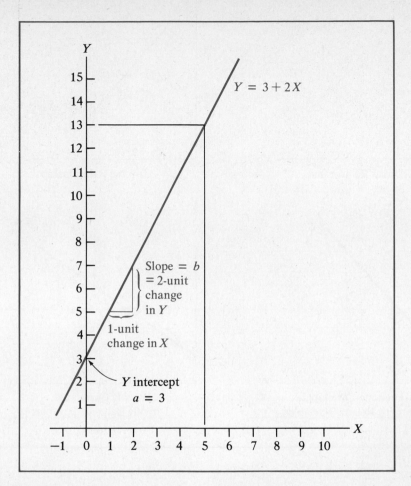

FIGURE 10-3
Slope and Y intercept for a straight line.

Some Characteristics of the Regression Line

Some important general properties of the regression line and its fit to the data are illustrated in Figure 10-4. First, we will consider the manner in which Y is related to X. There are two basic kinds of regression lines. If the values of the dependent variable Y increase for larger values of the independent variable X, then Y is *directly related* to X, as shown in Figure 10-4(a). Here, the slope of the line is positive, so that $b > 0$; this is because Y will increase as X becomes larger. Figure 10-2 shows that transportation time and rail distance are directly related variables. Other examples of directly related variables are

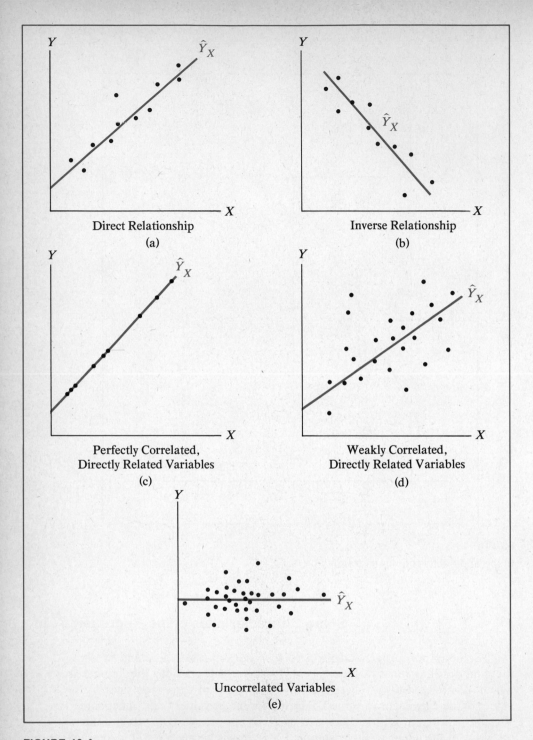

FIGURE 10-4
Properties of the regression line and possible relationships between variables.

age and salary (during employment years), weight and daily caloric intake, and the number of passengers and the quantity of luggage on commercial aircraft flights. In Figure 10-4(b), the slope of the regression line is negative and $b < 0$. Here, Y becomes smaller for larger values of X, so that the variables X and Y are *inversely related*. Examples of inverse relationships include remaining tire tread and miles driven, crop damage by insects and the quantity of insecticide applied, and the typical economic demand curve where demand decreases as price increases.

The degree to which two variables are related is indicated by the amount of scatter in the data points about the regression line. In Figure 10-4(c), the data points all lie on the regression line. This unusual "perfect fit" indicates that X and Y are *perfectly correlated*. In Figure 10-4(d), the data points are widely scattered about the regression line, indicating that X and Y are *weakly correlated*. Contrast this scatter diagram with the one in Figure 10-4(a), where the points cluster more closely about the line. Predictions of Y tend to be more accurate when there is less scatter, because then the sample results are less varied in relation to the regression line. As we saw in Chapter 8, small sample variability indicates a smaller standard error in the estimate, so that the sampling error is less pronounced and more reliable estimates are obtained. When there is less scatter, the degree of correlation is higher. This suggests that correlation analysis can be used to qualify the accuracy of estimates made from the regression line. When X and Y are *uncorrelated*, as in Figure 10-4(e), the regression line is horizontal, indicating that values of Y are independent of the values of X. This means that X is a worthless predictor of Y.

Fitting the Data by a Straight Line

How do we determine which particular regression line to use? One simple procedure would be to use our judgment in positioning a straightedge until it appears to summarize the linear pattern of the scatter diagram, and then draw the line. We can find the equation of this regression line by reading its Y intercept and slope directly from the graph. Although this procedure may be adequate for some applications, estimates of Y obtained in this way are often crude. A major drawback to using freehand methods in fitting a line is that two persons will usually draw different lines to represent the same data. Freehand fitting can therefore create unnecessary controversy about the conclusions of the data analysis. Another serious objection is that statistical methodology cannot be used to qualify the estimation errors (by confidence limits, for instance).

These difficulties can be overcome by using a statistical method to fit the line to the data. The most common technique is the *method of least squares*, which we will discuss in Section 10-2. From many different viewpoints, this procedure provides the best possible fit to a set of data and thereby the best possible predictions.

Regression Curves

It must be emphasized that a straight line is not always an appropriate function relating Y to X. The scatter diagrams in Figure 10-5 illustrate cases for which various types of nonlinear functions fit the data more closely. Notice that all the functions are curves. Such relationships between X and Y are called *curvilinear* relationships.

Figure 10-5(a) shows the relationship between crop yield Y and quantity X of fertilizer applied. The curve fits data obtained from test plots to which various amounts of fertilizer have been applied. As the amount of fertilizer increases to a certain point, it proves beneficial in increasing the harvest. But beyond this point, the benefits become negative, because additional fertilizer burns plant roots and causes the crop yield to decline. The greatest increases in Y occur for small values of X, with Y increasing at a decreasing rate. The peak value of Y may be referred to as the "point of negative returns," after which Y decreases at an increasing rate. This curve expressing Y as a function of X has the shape of an inverted U. A regression equation in the form $Y = a + bX - X^2$ could be used here.

This is contrasted to the U-shaped curve in Figure 10-5(b), where the incremental or marginal cost Y of production is plotted against volume X. This regression equation assumes the form $Y = a + bX + cX^2$. Here, data points have been obtained for the various levels of production activity in a plant. At low levels of plant activity, all factors of production are employed less efficiently. But as volume increases, the cost of additional units declines as factors are employed more efficiently. Increases in efficiency become less pronounced, until a point is reached after which extra production can be handled—but less efficiently than before—and marginal costs begin to rise.

In Figure 10-5(c), the size Y of a population of *Drosophila* (fruit flies) is shown as a function of the number of generations X since the colony was established. The data points represent observations made after successive hatchings. There are no natural checks on the population growth, because the flies are reared in an artificial environment. The curve obtained is an exponential or geometric growth curve, so that Y is related to X by an expression like $Y = a^X$ or $Y = X^c$. Plots of the world's human population growth during the past several centuries assume the same basic shape. With exponential growth, Y increases at an increasing rate for larger values of X.

This is contrasted to the negative exponential curve in Figure 10-5(d), where the percentage appreciation Y in the value of each share of a mutual fund is plotted against the total money invested X by the fund. The points represent several successful mutual funds at various stages of growth. The appropriate regression equation would be in the form $Y = ae^{-bX}$. Negative exponential curves correspond to values of Y that decrease at a decreasing rate as X becomes larger. A rationale for such a result is that a small mutual fund can be highly

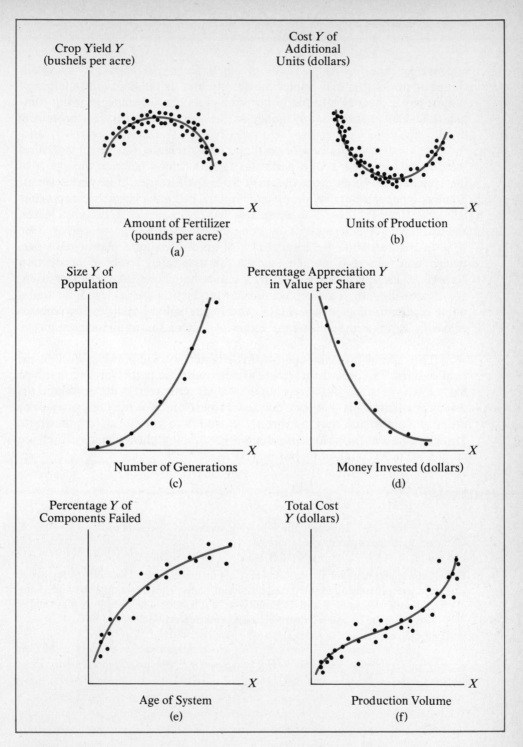

FIGURE 10-5

Examples of curvilinear relationships found for scatter diagrams.

selective in choosing its portfolio, giving it the opportunity to buy small-company stocks that may appreciate greatly. But as the fund grows larger, it cannot buy as heavily into the rather limited number of small, growing companies and must invest more money in the stocks of larger, less promising firms.

Figure 10-5(e) shows a logarithmic curve relating the percentage Y of the original components of a particular type that have failed to the age X of the system. The points were obtained from the histories of several systems. Some components survive almost indefinitely, but most original components fail early. Here, values of Y increase at a decreasing rate as X becomes larger. We can express Y in terms of X by using the equation $Y = a + b\log(1 + X)$.

In Figure 10-5(f), the total cost Y of production for a plant over a particular time period is plotted against the associated levels of production activity X. These points are fitted by a curve that shows total cost increasing at a decreasing rate as greater volumes of production are achieved. Beyond a point of diminishing returns for X, the larger volume reduces incremental efficiency, and the curve shows the values of Y increasing at an increasing rate.

Throughout the remainder of this chapter, we will consider only linear relationships. The procedure we use to determine the particular line that best fits the data is called *linear regression analysis*. One reason for our emphasis on linear equations is that they are easier to explain. The methods of analysis are also simpler and may be directly extended to curvilinear relationships. Straight lines are useful for describing a great many phenomena, so they are among the most common regression relationships.

EXERCISES

10-1 On graph paper, construct the following regression lines and indicate whether X and Y are directly or inversely related:

(a) $\hat{Y}_X = -5 + 2X$ (b) $\hat{Y}_X = 5 + 3X$ (c) $\hat{Y}_X = 20 - 2X$

10-2 Plot a scatter diagram for the following data on graph paper. Then use a straightedge to draw a line through the data points that appears to summarize the underlying relationship between X and Y. From your graph, determine the values for a and b. Then write the equation corresponding to your regression line.

Number of Pages X	Hours of Typing Time Y	Number of Pages X	Hours of Typing Time Y
10	20	100	50
20	10	40	40
20	30	60	40
30	30	110	30
50	20	120	50

10-3 Plot a scatter diagram for each of the following sets of data. Sketch the shape of the regression curve that seems to be the best fit for each relationship.

(a)		(b)		(c)	
Number of Workers	Output per Labor-Hour	Number of Units	Total Cost	Minutes Between Rest Periods	Pounds Lifted per minute
X	Y	X	Y	X	Y
5	10	600	$11,000	5.5	350
8	4	50	3,100	9.6	230
1	3	470	10,200	2.4	540
1	2	910	15,700	4.4	390
1	7	160	6,300	.5	910
8	8	950	19,500	7.9	220
7	10	690	13,900	2.0	680
10	2	90	1,800	3.3	590
3	5	310	8,800	13.1	90
3	8	1,000	25,700	4.2	520

10-4 Consider the two variables, family savings and income. For each of the following studies, indicate which of these two variables would be independent and which would be dependent.

(a) A bank wishes to predict the increase in its time deposits due to a 10% increase in the salaries of state employees, whose incomes are known.

(b) A mutual fund wishes to use its customers' purchases of stock (a form of savings) to predict their incomes. This information will be used by its salespersons to identify leads for increased business.

(c) An economist wishes to forecast increases in savings due to inflationary wage settlements.

THE METHOD OF LEAST SQUARES   10-2

Least-squares regression is the technique of fitting a regression equation to the observed data. Due to its many desirable properties, the least-squares criterion is the most commonly used tool in regression analysis. Although much of this chapter is restricted to linear relationships between two variables, the procedures that we will describe here can also be extended to a variety of situations where a curvilinear fit is desired. We will begin by applying the method of least squares to our previous parts supplier data, which are again plotted in Figure 10-6. The least-squares criterion requires that a line be chosen to fit our data so that the *sum of the squares of the vertical deviations separating the points from the line will be a minimum.* The deviations are represented by the lengths of vertical line segments that connect the points to the estimated regression line in the scatter diagram.

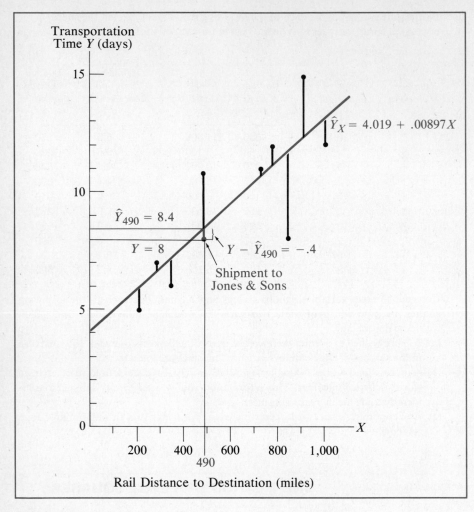

FIGURE 10-6
Fitting a regression line to the parts supplier data using the method of least squares.

Rationale for Least Squares

To explain how this procedure may be interpreted, we investigate the shipment to Jones & Sons at a distance of $X = 490$ miles from the supplier's plant. Our data (page 313) show that $Y = 8$ days were required for the shipment to arrive. This transportation time is represented on the graph by the vertical distance to the corresponding data point along the thin line from the X axis at $X = 490$. The predicted or estimated transportation time for the *next* shipment to Jones & Sons equals the vertical distance all the way up to the regression line, a total of $\hat{Y}_X = \hat{Y}_{490} = 8.4$ days. The difference between the observed transportation time, $Y = 8$ days, and the predicted value for Y is the deviation $Y - \hat{Y}_X = Y - \hat{Y}_{490} = 8 - 8.4 = -.4$ days. This is represented by the colored vertical line segment connecting the point to the regression line. Because the observed value of Y lies below the predicted value, a negative deviation is obtained; if the observed Y lay above the line, the deviation would be positive. The vertical deviation represents the amount of *error associated with using the regression line to predict* a future shipment's transportation time. We want to find values for a and b that will minimize the sum of the squares of these vertical deviations (or prediction errors).

One reason for minimizing the sum of the squared vertical deviations is that some of the deviations are negative and others are positive. For any set of data, a great many lines can be drawn for which the sum of the unsquared deviations is zero, but most of these lines would fit the data poorly.

Finding the Regression Equation

The sum to be minimized is

$$\sum (Y - \hat{Y}_X)^2 = \sum (Y - a - bX)^2$$

which involves two unknowns, a and b. Mathematically, it may be shown that the required values must simultaneously satisfy the following expressions, referred to as the *normal equations** :

$$\sum Y = na + b \sum X$$
$$\sum XY = a \sum X + b \sum X^2$$

Solving these equations algebraically gives us the expression for b :

$$b = \frac{n \sum XY - \sum X \sum Y}{n \sum X^2 - (\sum X)^2}$$

We can obtain the equation for a by substituting the value of b into the first normal equation. Then

$$a = \frac{1}{n} \left(\sum Y - b \sum X \right)$$

The expressions for a and b can be further simplified by using the mean values $\bar{X} = (\sum X)/n$ and $\bar{Y} = (\sum Y)/n$ to compute the

---

* The word *normal* as used here has nothing to do with the normal curve. Rather, *normal equations* receive their name from a mathematical property of linear algebra.

ESTIMATED REGRESSION COEFFICIENTS

$$b = \frac{\sum XY - n\bar{X}\bar{Y}}{\sum X^2 - n\bar{X}^2}$$

$$a = \bar{Y} - b\bar{X}$$

Note that b must be calculated before a.

The advantage of calculating a and b from these expressions is that every step involves computations with values of moderate size. Although this may increase the danger of rounding errors, such errors are usually negligible.

Illustration of the Method

We are now ready to find the regression equation for the parts supplier's regression line obtained from the $n = 10$ observations. In order to evaluate the expressions for a and b, we must perform a set of intermediate calculations (shown in Table 10-2). To find b, we must calculate $\bar{X}$, $\bar{Y}$, $\sum XY$, and $\sum X^2$. Columns of values for X, Y, XY, and X^2 are used for this purpose. An extra column for the squares of the dependent variable observations, Y^2, is computed for use in later discussions of the regression line. Using the intermediate values obtained, we can first find the value for b:

$$b = \frac{\sum XY - n\bar{X}\bar{Y}}{\sum X^2 - n\bar{X}^2} = \frac{64,490 - 10(611.0)(9.5)}{4,451,500 - 10(611.0)^2} = \frac{6,445}{718,290} = .00897$$

Substituting $b = .00897$, we obtain

$$a = \bar{Y} - b\bar{X} = 9.500 - .00897(611.0)$$

$$= 9.500 - 5.481 = 4.019$$

Thus we have determined the following equation for the estimated regression line graphed in Figure 10-6:

$$\hat{Y}_X = 4.019 + .00897X$$

We may now use the above regression equation to predict the transportation time $\hat{Y}_X$ for a shipment of known rail distance X from the parts supplier's plant. For instance, when $X = 490$, we have

$$\hat{Y}_{490} = 4.019 + .00897(490) \cdot$$

$$= 8.4$$

TABLE 10-2

Intermediate Calculations for Obtaining the Parts Supplier's Estimated Regression Line

(1)	(2)	(3)	(4)	(5)	(6)
	Rail	Transportation			
Shipment	Distance	Time			
Destination	X	Y	XY	X^2	Y^2
1. Muller Auto Supply	210	5	1,050	44,100	25
2. Taylor Ford	290	7	2,030	84,100	49
3. Auto Supply House	350	6	2,100	122,500	36
4. Parts 'n' Spares	480	11	5,280	230,400	121
5. Jones & Sons	490	8	3,920	240,100	64
6. A. Hausman	730	11	8,030	532,900	121
7. Des Moines Parts	780	12	9,360	608,400	144
8. Pete's Parts	850	8	6,800	722,500	64
9. Smith Dodge	920	15	13,800	846,400	225
10. Gulf Distributors	1,010	12	12,120	1,020,100	144
Totals	6,100	95	64,490	4,451,500	993
	$= \sum X$	$= \sum Y$	$= \sum XY$	$= \sum X^2$	$= \sum Y^2$

$$\bar{X} = \frac{\sum X}{n} = \frac{6,110}{10} = 611.0 \qquad \bar{Y} = \frac{\sum Y}{n} = \frac{95}{10} = 9.5$$

Thus the prediction for the transportation to a customer 490 miles away is $\hat{Y}_{490} = 8.4$ days. This is the same value previously read from the graph of the regression line in Figure 10-6.

The least-squares regression line has two important features. First, it goes through the point $(\bar{X}, \bar{Y})$ that corresponds to the mean of the X and Y observations. Second, the sum of the deviations of the Y's from the regression line is zero, or

$$\sum (Y - \hat{Y}_X) = 0$$

Thus, the positive and negative deviations from the regression line cancel one another, so that the least-squares line goes through the center of the data scatter. This can be a useful check to determine if any miscalculations have been made in finding a and b.

Meaning and Use of the Regression Line

Once the regression equation has been obtained, predictions or estimates of the dependent variable may be made. For purposes of planning, the parts supplier now has a basis for determining his order-filling priorities so that there will be a reasonable chance that shipments will be received by customers at the required times. His estimated regression equation is $\hat{Y}_X = 4.019 + .00897X$. The value $a = 4.019$ is an estimate of the Y intercept. An interpretation of this value is that about 4 days of "overhead" are built into all rail shipments. This roughly corresponds to the time a shipment spends in delivery to and from the railhead, being loaded and unloaded, and waiting between various stages of the shipping process. The slope $b = .00897$ reflects the impact of distance alone on the total transportation time. For each additional mile, an estimated .00897 day is added to the total time. Stated another way, each additional 100 miles of distance adds roughly .9 day to the transportation time. Note that b is positive, indicating that transportation time varies directly with distance—the greater the distance, the longer, on the average, it will take for a shipment to be delivered.

Knowing that Jones & Sons is 490 miles distant, the $\hat{Y}_{490} = 8.4$ days may be used as a point estimate of the transportation time required for the shipment. But the proper interpretation of 8.4 days is that *on the average* all future shipments to Jones & Sons will require about this much time in transit. It is an average because the conditions existing for successive shipments to this customer will vary due to a host of factors, such as freight-train schedules, total freight to be handled, loading and unloading conditions, and routing of the box car. The times will vary from shipment to shipment. In general, the dependent variable Y will vary for a given X. Furthermore, the same regression equation will be used for other customers at a 490-mile distance; shipments to these others may travel on other roads having different characteristics. This leads us to the next important consideration of regression analysis.

Measuring Variability of Results

The fundamental indication of variability provided by the sample data is the measure of the spread or scatter about the estimated regression line. As we have noted, estimates made from the regression line will be more precise when the data are less scattered. Thus, we can investigate the degree of scatter to determine an expression for the error involved in making estimates through regression. We wish to employ a measure that fits naturally into the scheme of least-squares regression. Recall that we obtained the Y intercept a and the slope b of the regression line by minimizing $\sum (Y - \hat{Y}_X)^2$, the *sum* of the squared deviations about the regression line. As we saw in selecting the fundamental measure of a population's variability, the variance is the average of the squared deviations from the mean. This suggests that we can select as our measure of variablity the *mean* of the squared deviations about the regression line:

$$\frac{\sum (Y - \hat{Y}_X)^2}{n}$$

The square root of the mean squared deviations is referred to as the *standard error of the estimate* about the regression line. As this suggests, we will use it to estimate the true variability in Y. For convenience, we will modify the preceding expression before taking its square root, using as our standard error of the estimate

$$s_{Y \cdot X} = \sqrt{\frac{\sum (Y - \hat{Y}_X)^2}{n - 2}}$$

The subscript $Y \cdot X$ indicates that the deviations are about the regression line, which provides values of Y for given levels of X. We divide the sum of the squared deviations by $n - 2$, which will make $s_{Y \cdot X}^2$ an unbiased estimator of the true variance of the Y values about the regression line. We subtract 2 from n to indicate that 2 degrees of freedom are lost, since the values of a and b contained in the expression for $\hat{Y}_X$ have been calculated from the same data.

The standard error of the estimate resembles the *sample standard deviation* calculated for the individual Y's, which we designate as s_Y:

$$s_Y = \sqrt{\frac{\sum (Y - \bar{Y})^2}{n - 1}} = \sqrt{\frac{\sum Y^2 - n\bar{Y}^2}{n - 1}}$$

Here, X does not appear in the subscript because s_Y makes no reference to the values of X. The sample standard deviation is the square root of the mean of the squared deviations about the center $\bar{Y}$ of the sample: $(Y - \bar{Y})^2$. Thus, s_Y represents the *total variability* in Y. Ordinarily, the deviations about $\bar{Y}$ are larger than their counterparts about the estimated regression line, so that s_Y will be larger than $s_{Y \cdot X}$. Figure 10-7 illustrates this concept. Note that s_Y and $s_{Y \cdot X}$ summarize the dispersions of separate sample frequency distributions.

In calculating $s_{Y \cdot X}$, the following mathematically equivalent expression is usually used for the

STANDARD ERROR OF THE ESTIMATE

$$s_{Y \cdot X} = \sqrt{\frac{\sum Y^2 - a \sum Y - b \sum XY}{n - 2}}$$

We will use this expression to calculate $s_{Y \cdot X}$ in our parts supplier illustration. Using the values obtained previously for a and b and taking the intermediate calculations from Table 10-2, we have

$$s_{Y \cdot X} = \sqrt{\frac{993 - 4.019(95) - .00897(64,490)}{10 - 2}}$$

$$= 2.02 \text{ days}$$

When knowledge of X is ignored, and no regression line is used, the total variability in Y is summarized by the sample standard deviation s_Y,

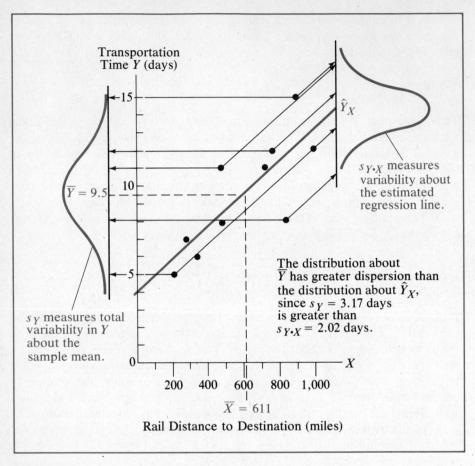

FIGURE 10-7
Illustration of the difference between total variability and variability about the estimated regression line.

which we calculate for the same data:

$$s_Y = \sqrt{\frac{993 - 10(9.5)^2}{10 - 1}} = 3.17 \text{ days}$$

This value of s_Y is larger than the value found for $s_{Y \cdot X}$, reflecting the fact that the variability about the regression line is smaller than the total variation in Y. This is illustrated in Figure 10-7, where the underlying frequency curve for deviations about the estimated regression line is more compact than the one that might be constructed for the Y's without a knowledge of X. Thus, a prediction interval calculated using s_Y would be wider. A general conclusion is that predictions tend to be more reliable and accurate when X is used as a predictor than when it is not.

EXERCISES

10-5 A statistician for the Civil Aeronautics Board wishes to construct an equation relating destination distance to freight charge for a standard-sized crate. She obtains the following results from a random sample of 10 freight invoices:

Distance (hundreds of miles) X	Charge (to nearest dollar) Y
14	68
23	105
9	40
17	79
10	81
22	95
5	31
12	72
6	45
16	93

(a) Plot a scatter diagram for these data.
(b) Using the method of least squares, determine the equation for the estimated regression line.
(c) Check your calculations by computing $\sum(Y - \hat{Y}_X)$. (Allowing for rounding errors, this should equal zero. If it does not, find your error.) Then plot the regression line on the scatter diagram.

10-6 For each of the following sets of data, determine the estimated regression equation $\hat{Y}_X = a + bX$.
(a) $\bar{X} = 10$; $\bar{Y} = 20$; $\sum XY = 3{,}000$; $\sum X^2 = 2{,}000$; $n = 10$.
(b) $\bar{X} = 10$; $\bar{Y} = 20$; $\sum XY = 1{,}000$; $\sum X^2 = 2{,}000$; $n = 10$.
(c) $\bar{X} = 50$; $\bar{Y} = 10$; $\sum XY = 30{,}000$; $\sum X^2 = 135{,}000$; $n = 50$.

10-7 A food processor uses least-squares regression to predict the total cost for production runs of Crunchola. The following data apply.

Production Quantity (tons) X	Total Cost (thousands of dollars) Y
400	200
150	85
220	115
500	200
300	140
100	65
150	70
150	65
240	125
350	190

(a) Using the method of least squares, determine the equation for the estimated regression line.

(b) Using your answer to (a), predict the total cost of production runs involving the following tonnages: (1) 200; (2) 350; (3) 400.

10-8 A personnel manager wishes to evaluate various employee aptitude tests to determine if one will predict productivity. A sample of $n = 100$ electronics assembly workers has been selected. Each worker is administered one of the tests currently being evaluated, and his or her aptitude score X is determined. The production foreman has previously evaluated the performance of each worker by means of a productivity index Y. The following intermediate calculations have been obtained:

$$\sum X = 5,000 \qquad \sum Y = 600 \qquad \sum XY = 50,000$$
$$\sum X^2 = 350,000 \qquad \sum Y^2 = 8,600$$

Suppose the test is adopted. Determine the regression equation $\hat{Y}_X = a + bX$ and plot it on graph paper.

10-9 A government economist wishes to establish the relationship between annual family income X and savings Y. A sample of $n = 100$ low-income families has been randomly chosen from various income levels between \$5,000 and \$20,000. After a thorough investigation of these families, the following intermediate calculations are obtained (X and Y are measured in thousands of dollars):

$$\sum X = \$1,239 \qquad \sum Y = \$79$$
$$\sum XY = 1,613 \qquad \sum X^2 = 17,322 \qquad \sum Y^2 = 293$$

(a) Determine the equation for the estimated regression line.
(b) State the meanings of the slope b and the Y intercept a.
(c) Calculate $s_{Y \cdot X}$ and s_Y. Does a comparison between these two values indicate that the regression line may be a useful tool for predicting family savings? Why or why not?

10-10 A stereo-cartridge manufacturer has conducted a regression analysis to estimate the average cartridge lifetime (in hours) at various record tracking forces X (in grams). The regression equation $\hat{Y}_X = 1,300 - 200X$ has been obtained for a sample of $n = 100$ cartridges that were played at different tracking forces until they were worn out. The standard error of the estimate for cartridge lifetimes about this line is $s_{Y \cdot X} = 100$ hours.

(a) Plot the regression line on graph paper.
(b) State the meaning of the slope of the regression line.
(c) Calculate $\hat{Y}_X$ when $X = 1$, $X = 2$, and $X = 3$ grams.

10-3 ASSUMPTIONS AND PROPERTIES OF LINEAR REGRESSION ANALYSIS

In Section 10-2, we were introduced to the mechanical process of fitting a regression line to the data. In this section, we will examine the assumptions and properties of the theoretical model for regression analysis.

Assumptions of Linear Regression Analysis

Suppose that in our parts supplier illustration we consider each possible transportation time Y for *all* shipments, past and future, to customers at a specified distance X from the plant. For this fixed X, the values of Y represent a population, and they will fluctuate and cluster about a central value. Similarly, for any other rail distance X, there will be a corresponding population of Y values. Since the means of these populations depend on the respective values for X, we may represent them symbolically by $\mu_{Y \cdot X}$, where, as before, the subscript $Y \cdot X$ signifies that the values of Y are for a given value of X.

Figure 10-8 illustrates how several populations for Y fit into the context of linear regression. This graph is three dimensional, having an extra axis perpendicular to the XY plane. This vertical axis represents the relative frequency for Y at a specified level X. The curves are drawn with their centers at a distance $\mu_{Y \cdot X}$ from the X axis. Thus, we may refer to $\mu_{Y \cdot X}$ as the *conditional mean* of Y given X. There will be a different frequency curve for each X. Here, we show these curves for all shipments to destinations of $X = 300$, $X = 500$, and $X = 700$ miles. The respective conditional means are denoted by $\mu_{Y \cdot 300}$, $\mu_{Y \cdot 500}$, and $\mu_{Y \cdot 700}$.

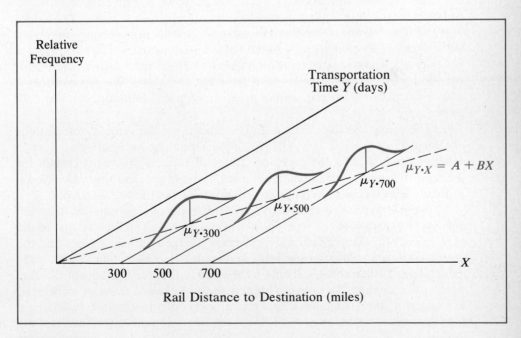

FIGURE 10-8
Populations for Y at various given values of X.

In linear regression analysis, we make four theoretical assumptions about the populations for Y:

1. All populations have the same standard deviation, denoted by $\sigma_{Y \cdot X}$, no matter what the value of X is.
2. The means $\mu_{Y \cdot X}$ all lie on the same straight line, given by the equation

$$\mu_{Y \cdot X} = A + BX$$

which is the expression for the *true regression line*.
3. Successive sample observations are independent.
4. The value of X is known in advance.

The additional assumption that the population for Y is normally distributed is sometimes made. But many regression analyses do not require such a strong and sometimes unrealistic condition.

Estimating the True Regression Equation

We have learned how to use the method of least squares to derive the estimated regression equation $\hat{Y}_X = a + bX$. Now we will investigate how this equation is related to the true regression equation $\mu_{Y \cdot X} = A + BX$. The values of A and B, which we call the *true regression coefficients*, are generally unknown. We will use $\hat{Y}_X$ to estimate $\mu_{Y \cdot X}$ in the same way that we used $\bar{X}$ to estimate μ in Chapter 8. The estimated regression equation differs from the true regression equation only in the values of the Y intercept and slope. We can consider a and b, calculated from the sample data, to be point estimates of A and B, respectively.

The values calculated for a and b depend on the sample observations obtained. The equation $\hat{Y}_X = 4.019 + .00897X$ relating transportation time to distance resulted from the particular times of the ten shipments chosen for the sample. A regression equation obtained from ten different shipments would probably have differed—perhaps considerably—from the one we found.

When a and b are calculated by the method of least squares, they are *unbiased estimators* of the true coefficients A and B. This means that if the experiment of collecting samples is repeated a large number of times and the regression line is found by the least-squares method each time, the average value of the Y intercepts a will tend to be close to the true Y intercept A. Likewise, on the average, the values of b will tend to be close to B. As we mentioned in Chapter 8, unbiasedness is a desirable property of an estimator. In addition, the least-squares criterion provides the estimators of smallest variance, making the method of least squares the *most efficient* of all unbiased estimators for linear regression coefficients. Therefore, a and b minimize chance sampling error, so that estimates made from the regression line $\hat{Y}_X = a + bX$ are the most reliable ones that can be attained for a fixed sample size. The values of a and b obtained by the least-squares method are also *consistent* estimators of A and B.* Recall from Chapter 8 that a consistent estimator becomes progressively closer to the target parameter with increasing sample size. This can be attributed to the sampling distributions of a and b, whose variances decrease with n.

PREDICTIONS AND STATISTICAL INFERENCES **10-4**
USING THE REGRESSION LINE

We use the estimated regression line to make a variety of inferences, which can be grouped into two broad categories: (1) predictions of the dependent variable, and (2) inferences regarding the regression coefficients A and B. Because predictions are made more often, we will discuss them first.

Making Predictions with the Regression Equation

The major goal of regression analysis is to predict Y from the regression line at given levels of X. This may be done in two ways: by predicting the value of the conditional mean $\mu_{Y \cdot X}$, or by predicting an individual Y value, rather than a mean.

As an example of the first kind of prediction, the parts supplier might want to predict the mean transportation time that will be achieved by all shipments over a distance of 500 miles. In this case, $X = 500$, and the best point estimate for $\mu_{Y \cdot X}$ will be the fitted Y value from the regression line:

$$\hat{Y}_{500} = a + b\,(500)$$

$$= 4.019 + .00897(500)$$

$$= 8.50 \text{ days}$$

This same value may be used to estimate the transportation time for a particular shipment over the same distance. To distinguish a mean value from an *individual value*, both of which can only be estimated from the sample, we use the special symbol Y_I.

---

* When the Y's are normally distributed, the least-squares estimators fall into a broad class referred to as *maximum likelihood estimators* (MLE). In addition to being consistent, the MLE are most efficient and are normally distributed.

Either kind of estimate will involve sampling error, which can be acknowledged and expressed in terms of confidence intervals. Because of the special nature of regression analysis, the numbers obtained are usually referred to as *prediction intervals*.

Prediction Intervals for the Conditional Mean

We determine prediction intervals in a similar manner to the way we derived confidence intervals in Chapter 8. There, we used

$$\mu = \bar{X} \pm z\frac{s}{\sqrt{n}} \qquad \text{and} \qquad \mu = \bar{X} \pm t_\alpha\frac{s}{\sqrt{n}}$$

$$\text{for large samples} \qquad\qquad \text{for small samples}$$

where $s/\sqrt{n}$ is the estimator of $\sigma_{\bar{X}}$.

In regression analysis, Y (not X) is the variable being estimated. Intervals of analogous form are required to estimate a conditional mean of Y:

$$\mu_{Y \cdot X} = \hat{Y}_X \pm (z \text{ or } t_\alpha) \text{ estimated } \sigma_{\hat{Y}_X}$$

The standard error of $\hat{Y}_X$, denoted by $\sigma_{\hat{Y}_X}$, represents the amount of variability in possible $\hat{Y}_X$ values at the particular level for X that a prediction is desired. In the context of transportation time, a somewhat different line $\hat{Y}_X = a + bX$ (such as $\hat{Y}_X = 3.5 + .011X$ or $\hat{Y}_X = 4.2 + .009X$) might have fitted the least-squares method if some other random sample of ten shipments has been selected. Thus, for some other sample $\hat{Y}_{500}$ might have computed to a different value than 8.50 days. For every level of X, the potential set of $\hat{Y}_X$ values would have a distribution with a standard deviation of $\sigma_{\hat{Y}_X}$.

There are two components of the variability in $\hat{Y}_X$:

$$\sigma_{\hat{Y}_X}^2 = \frac{\text{Variability in the}}{\text{mean of } Y\text{'s}} + \frac{\text{Variability caused by the}}{\text{distance of } X \text{ from } \bar{X}}$$

The first source is analogous to the variability in the sample mean, which, as we saw in Chapter 8, depends on the potential standard deviation and the sample size. The second source of variation is associated with the distance that X lies from $\bar{X}$. Figure 10-9 shows why this is so. Here, several estimated regression lines have been plotted, each representing different samples of shipments taken from the same population. (Although each sample involves the same X's and thus the same level for $\bar{X}$, each involves slightly different values for a, b, and $\bar{Y}$.) Note that these lines tend to diverge and that their separations become greater as the distance between X and $\bar{X}$ increases. Therefore, the values of $\hat{Y}_X$ become more varied the farther they are from $\bar{X}$.

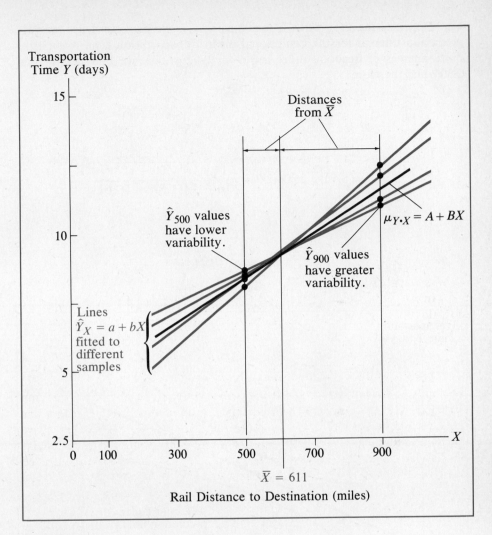

FIGURE 10-9
Illustration of how variability in $\hat{Y}_X$ increases for larger distances separating X from $\bar{X}$.

With small samples (generally, when $n < 30$), we use the following expression to calculate the

PREDICTION INTERVAL FOR THE CONDITIONAL MEAN USING SMALL SAMPLES

$$\mu_{Y \cdot X} = \hat{Y}_X \pm t_\alpha s_{Y \cdot X} \sqrt{\frac{1}{n} + \frac{(X - \bar{X})^2}{\sum X^2 - n\bar{X}^2}}$$

where t_α is found from Appendix Table F for $n - 2$ degrees of freedom.

Representing $\mu_{Y \cdot X}$ when $X = 500$ as $\mu_{Y \cdot 500}$, we may construct the 95% prediction interval for the conditional mean transportation time using $10 - 2 = 8$ degrees of freedom. Since $\alpha = (1 - .95)/2 = .025$, we find that $t_{.025} = 2.306$, and therefore

$$\mu_{Y \cdot 500} = \hat{Y}_{500} \pm t_{.025} s_{Y \cdot X} \sqrt{\frac{1}{n} + \frac{(5000 - \bar{X})^2}{\sum X^2 - n\bar{X}^2}}$$

$$= 8.50 \pm 2.306(2.02) \sqrt{\frac{1}{10} + \frac{(500 - 611)^2}{4,451,500 - 10(611)^2}}$$

$$= 8.50 \pm 1.59$$

or

$$6.91 \leq \mu_{Y \cdot 500} \leq 10.09 \text{ days}$$

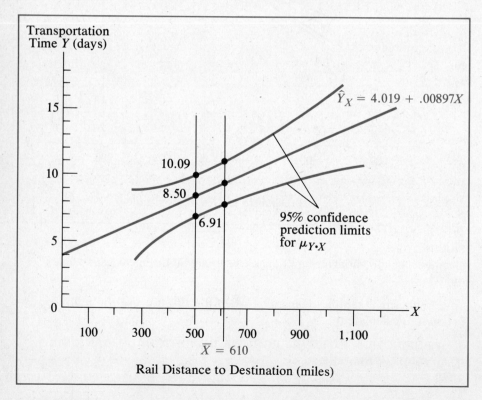

FIGURE 10-10

Confidence limits for predictions of mean transportation time.

We would therefore conclude that the transportation time for destinations 500 miles from the plant is *on the average* somewhere between 6.91 and 10.09 days. Our confidence that this statement is correct rests on the procedure used, which provides similar intervals containing the true mean about 95% of the time.

We may calculate 95% prediction intervals for the other values of X, thus obtaining prediction limits for $\mu_{Y \cdot X}$ over the entire range of X. In Figure 10-10, this has been done for the parts supplier data. Note that the width of the confidence band depends on the distances of the X values from the mean.

When sample size is large ($n \geq 30$), the normal curve applies and the normal deviate z replaces t_α. However, further simplification is usually required for large samples. We use an abbreviated but approximate expression for the

PREDICTION INTERVAL FOR THE CONDITIONAL MEAN USING LARGE SAMPLES

$$\mu_{Y \cdot X} = \hat{Y}_X \pm z \frac{s_{Y \cdot X}}{\sqrt{n}}$$

This expression resembles the one we use to construct a confidence interval for the population mean. Here, $\hat{Y}_X$ is analogous to the sample mean, and $s_{Y \cdot X}$ replaces s.

Prediction Intervals for an Individual Value of Y Given X

Predicting an individual value of Y given X is similar to predicting the mean. If our parts supplier wished to predict the transportation time of the next shipment over a distance of $X = 500$ miles, then the same point estimate, $\hat{Y}_{500} = 8.50$ days would be made from the regression equation. The following expression provides the

PREDICTION INTERVAL FOR AN INDIVIDUAL Y USING SMALL SAMPLES

$$Y_I = \hat{Y}_X \pm t_\alpha s_{Y \cdot X} \sqrt{\frac{1}{n} + \frac{(X - \bar{X})^2}{\sum X^2 - n\bar{X}^2} + 1}$$

This expression is the same as the one for $\mu_{Y \cdot X}$, except for the addition of "$+1$." This reflects the fact that when $\hat{Y}_X$ is used to estimate Y_I, a third source of variability is present—the dispersion of individual Y values about the regression line. (Even if the true regression line were known, the Y's at a particular level for X would have a variance of $\sigma_{Y \cdot X}^2$).

We may now construct a 95% prediction interval for Y_I, using the parts supplier data when $X = 500$:

$$Y_I = \hat{Y}_{500} \pm t_{.025} s_{Y \cdot X} \sqrt{\frac{1}{n} + \frac{(500 - \bar{X})^2}{\sum X^2 - n\bar{X}^2} + 1}$$

$$= 8.50 \pm 2.306(2.02) \sqrt{\frac{1}{10} + \frac{(500 - 611)^2}{4{,}451{,}500 - 10(611)^2} + 1}$$

$$= 8.50 \pm 4.92$$

or

$$3.58 \le Y_I \le 13.42 \text{ days}$$

Note that this interval is considerably wider than the interval obtained previously for $\mu_{Y \cdot 500}$. This is to be expected, because Y_I is the estimate for the transportation time of *a particular shipment*, not a mean, and the greater width is attributable to the added variability that would be present even if the true regression line were available in making the prediction.

For *large samples*, the normal curve can be assumed, so that an appropriate normal deviate value z replaces t_α. Ordinarily, we use the following abbreviated equation to compute the

PREDICTION INTERVAL FOR AN INDIVIDUAL Y USING LARGE SAMPLES

$$Y_I = \hat{Y}_X \pm z s_{Y \cdot X}$$

Dangers of Extrapolation in Regression Analysis

The set of observations used to establish a regression equation covers a limited range of values of the independent variable X. Caution must be exercised in making predictions of the dependent variable Y whenever X falls outside this range. Such predictions are called *extrapolations*.

In the parts supplier illustration, the regression line was computed for ten shipments whose distances ranged from 210 to 1,010 miles. Suppose we wished to use our results to predict the transportation time for a shipment to a customer separated from the plant by 2,000 miles, a distance considerably greater than the largest used in calculating the regression line. How reliable would such an extrapolation be? This depends on whether our assumption of a linear relationship between time and distance is valid for shipments over great distances. There is simply no way to know this without including some longer shipments in the sample. As noted earlier in this chapter, the selection of a straight line was motivated by the general appearance of the data's scatter. Perhaps with a few additional points for distances between 1,000 and 2,000 miles, a curvilinear relationship might provide a better fit. There may be a good logical basis for assuming that a line will prevail—but then we might expect its slope to be flatter or steeper because of the influence from additional data points. On the other hand, there are reasonably sound arguments that longer distances ought to involve less time-consuming recomposition of trains (adding and removing box cars), so that the additional time per extra mile (that is, the slope of the regression curve) decreases for longer trips. This might be the case for 2,000-mile shipments to the Pacific Coast from a midwestern plant. Railroads in the western United States generally cover greater distances than eastern ones do, and because of the lower industrial concentrations in the West, they may require less switching of box cars at intermediate stops. There would also be greater distances covered between stops. Thus we might reasonably expect that a higher proportion of the total transportation time would actually be spent rolling instead of sitting, and that the rolling would be at higher speeds. Because longer shipments would invariably go to the Pacific Coast, we might therefore justify a curvilinear relationship between time and distance.

Regression analysis is limited only to the range of actual observations. These observations—and not qualitative reasoning—are used to quantify the relationship between X and Y. Qualitative reasoning is useful initially in selecting the form of the regression equation (linear versus curvilinear) and later in interpreting the results, but it cannot be used in place of actual observation. We are not ruling out extrapolation here but are merely indicating its potential pitfalls. If there are no data available beyond the range of required predictions, then extrapolation may be the only suitable alternative. Keep in mind that *extraplation assigns values using a relationship that has been measured for circumstances differing from those for the prediction.*

Circumstances may differ for reasons other than extrapolation. Should the underlying populations change over time—as would be the case for changes in railroad operations or technology—then the regression line would not represent the shipments in subsequent years. Often, a regression analysis is short lived in its applicability because underlying relationships can change over time.

Inferences Regarding the Slope of the Regression Line

Second in importance to prediction intervals are inferences regarding slope B of the true regression line. This is especially true in statistical applications where the underlying relationship between X and Y is more important than predicting Y at a particular level for X. For example, much economic theory relies on regression analysis to substantiate hypothetical models requiring supply and demand curves. The *coefficients* of these demand and supply equations—not the predicted quantities for a given price—are our main interest. Similarly, a metallurgist might use regression analysis to develop a mathematical relationship between alloy concentrations and strength properties. He might be more concerned with predicting how much extra shearing force is needed to break a metal for each unit increase in alloy material than with predicting a particular force (that is, he might wish to estimate B rather than $\mu_{Y \cdot X}$).

A prediction interval can be constructed for B, and hypotheses regarding the true level for B can be tested. A optional discussion of these procedures appears in Section 10-8.

EXERCISES

10-11 The estimated regression line $\hat{Y}_X = 50 + .1X$ provides the yield (bushels) per acre of corn when X pounds of nitrate fertilizer is applied. This result was obtained from a sample of $n = 100$ acres for which $s_{Y \cdot X} = 5$ bushels. Calculate 95% prediction intervals for $\mu_{Y \cdot X}$ and Y_I when $X = 150$ pounds.

10-12 The relationship between the total weight Y (in pounds) of luggage stored in an aircraft's baggage compartment and the number of passengers X on the flight manifest is $\hat{Y}_X = 250 + 27X$. Airport superintendents will use this to determine how much additional freight can be stored safely on a flight, after considering the fuel load and the weight of the passengers themselves. The data were obtained from a sample of $n = 25$ flights. Other results were $s_{Y \cdot X} = 100$ pounds, $s_Y = 300$ pounds, $\sum X^2 = 64,000$, and $\bar{X} = 50$. Construct 95% prediction intervals for $\mu_{Y \cdot X}$ and Y_I when (a) $X = 50$; (b) $X = 75$; (c) $X = 100$.

10-13 A city planner has determined the number of square feet in each home X and the size of each family Y for a sample of 400 families. The planner intends to use the estimated regression line $\hat{Y}_X = .5 + .002X$ to predict the family size for given square-footage levels. The values $s_Y = .30$ and $s_{Y \cdot X} = .10$ have been obtained. For a home with 1,000 square feet, construct 95% prediction intervals for (a) the mean family size and (b) the number of persons in a particular family.

10-14 The credit manager of a department store has determined that the regression equation for a customer credit-rating index X and the proportion of customers Y who eventually incur bad debts is $\hat{Y}_X = .09 - .002X$. The index values range from 0 to 40. A sample of $n = 25$ is taken, and other calculations show that $\sum X^2 = 23,000$, $s_{Y \cdot X} = .02$, and $\bar{X} = 30$.
(a) Determine the 90% prediction interval for the mean proportion of customers with ratings of $X = 20$ who will incur bad debts.
(b) The proper interpretation of Y_I in this case is the *probability* that a particular individual will incur a bad debt. Determine the 90% prediction interval for Y_I when $X = 20$.

CORRELATION ANALYSIS 10-5

Regression analysis provides an equation for estimating the value of one variable from the value of another variable. Correlation analysis is used to measure the *degree* to which two variables are related—to show how closely two variables can move together.

Correlation analysis is a useful auxiliary tool in regression analysis, because it can indicate how well the regression line explains the variation in the values of the dependent variable. Correlation is used instead of regression when the only question is how strongly two variables are related. One application is isolating statistically related characteristics of a population to explain their differences. For example, a pharmacologist may wish to identify the chemicals that can be formulated into a drug to alleviate various symptoms of a particular disease, such as anemia, pain, and poor appetite. A high positive correlation between dosage X of a specific chemical and appetite Y (measured by the quantity of food consumed) may make the chemical a good candidate for inclusion in the final drug.

The central focus of correlation analysis is to find a suitable index that indicates how strongly X and Y are related. Initially, it is convenient to treat correlation as an adjunct to regression analysis, so this index will be explained here in relation to the regression line. Later in this section, a parallel explanation will be given that does not require prior knowledge of the estimated regression line.

Measuring the Degree of Association Between *X* and *Y*

In Section 10-1, we saw that the degree to which X and Y are related can be explained in terms of the magnitude of data scatter about the regression line. One extreme case occurs when the scatter is so great that the regression line has a slope of zero and is parallel to the X axis, as shown in Figure 10-11(a). Here, levels of Y have no relationship to the value of X. We say that the degree of correlation is *zero*, since knowledge of X cannot add to the accuracy of predictions of Y. Figure 10-11(b) illustrates the opposite extreme—a perfect fit between the X and Y observations is achieved because all of the data points happen to lie on the same line. Since there is no scatter about the regression line, the data indicate that Y will change by some predetermined amount for each increment in X, showing the strongest possible relationship between X and Y. We can say that the degree of correlation is *perfect*, so that knowing X allows us to make perfect predictions of Y.

We will now develop two indexes to summarize the strength of association between X and Y. The more important—the *coefficient of determination*—expresses the relative reduction in the variation in Y that can be attributed to a knowledge of X and its relationship to Y via the regression line. From this, another useful index—the *correlation coefficient*—can be obtained.

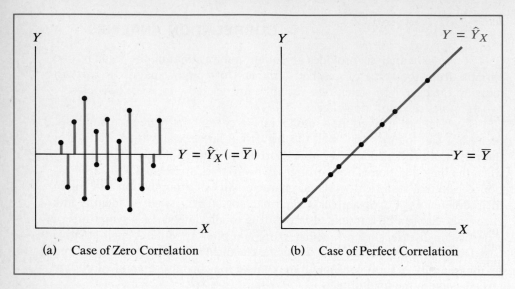

(a)   Case of Zero Correlation   (b)   Case of Perfect Correlation

FIGURE 10-11
Two extreme cases illustrating degrees of correlation.

The Coefficient of Determination

In Section 10-2, we summarized the data scatter about the estimated regression line by using the standard error of the estimate $s_{Y \cdot X}$. Its square, $s_{X \cdot Y}^2$, is the *mean* of the squared vertical deviations of the data points around the regression line. It is now more convenient to summarize the scatter with the *sum* of the squared deviations about $\hat{Y}_X$, or

$$\sum (Y - \hat{Y}_X)^2$$

This can be compared to the scatter of the sample observations about their mean, represented by

$$\sum (Y - \bar{Y})^2$$

which is the sum of the squared vertical deviations around the horizontal line $Y = \bar{Y}$. We can easily recognize the terms in the above sums of squares as the numerators of $s_{Y \cdot X}^2$ and s_Y^2, respectively.

We will use the sums of squares to construct the indexes for measuring strength of association. When we compared $s_{Y \cdot X}$ to s_Y previously, we saw that their relative sizes indicate the predictive power of knowing X. Using the corresponding sums of squares, we can construct the *sample coefficient of determination* to express how strongly X is associated with Y:

$$r^2 = 1 - \frac{\sum(Y - \hat{Y}_X)^2}{\sum(Y - \bar{Y})^2}$$

When X and Y have zero correlation, as they do in Figure 10-11(a), the regression line has a slope of zero and $\hat{Y}_X = \bar{Y}$. In this case, the deviations from $\hat{Y}_X$ are the same as the deviations from $\bar{Y}$. This makes the numerator the same as the denominator in the expression for r^2, so that the fraction must be equal to 1; thus, $r^2 = 1 - 1 = 0$. If X and Y are perfectly correlated, as they are in Figure 10-11(b), then $\sum(Y - \hat{Y}_X)^2 = 0$, so that $r^2 = 1 - 0 = 1$. The value of r^2 therefore always lies between 0 and 1.

In actual practice, calculating r^2 from this expression can be cumbersome. Instead, we can use the estimated regression coefficients and the intermediate values we obtained in finding these coefficients to calculate r^2 from the mathematically equivalent expression for the

SAMPLE COEFFICIENT OF DETERMINATION

$$r^2 = \frac{a\sum Y + b\sum XY - n\bar{Y}^2}{\sum Y^2 - n\bar{Y}^2}$$

This equation uses the values we previously obtained from regression analysis.

In the parts supplier illustration, we may calculate r^2 from the above expression, using $a = 4.019$, $b = .00897$, and the intermediate calculations from Table 10-2:

$$r^2 = \frac{4.019(95) + .00897(64,490) - 10(9.5)^2}{993 - 10(9.5)^2}$$

$$= \frac{57.78}{90.50} = .64$$

The sample coefficient of determination can be interpreted in terms of variation in levels of the dependent variable. The total variation in Y has two components:

Total variation = Explained variation + Unexplained variation

Much of the difference between the observed values of Y can be *explained* in terms of the regression line. For example, transportation times Y tends to be higher when rail distance X is great, as indicated by the regression line. But individual observations do not fit perfectly on the line, and the resulting scatter in times cannot be explained solely by distances. In effect, the coefficient of determination expresses the proportion of total variation that can be explained by regression:

$$r^2 = \frac{\text{Explained variation}}{\text{Total variation}}$$

For our parts supplier illustration, we calculated $r^2 = .64$. This signifies that 64% of the total variation or scatter of transportation time Y around their mean can be explained by the relationship between this variable and the corresponding rail distance X, as estimated by the regression line for X and Y.

Because the explained variation can never exceed the total variation, their ratio is at most 1, and, again, the greatest possible value of r^2 is 1. Similarly, when the explained variation is zero, so that knowledge of the regression line cannot reduce prediction errors, the value of r^2 must be zero. This corresponds to our previous example where the regression line was horizontal and had both a slope and a correlation of zero.

Because the sample coefficient of determination r^2 will vary in value, depending on the sample results, it is only an estimate of the underlying *population coefficient of determination*, which is represented by the symbol ρ^2 (the lowercase Greek *rho*, squared). Inferences may be made regarding ρ^2 (see the Selected References section for a detailed discussion).

The Correlation Coefficient

Although the rationale for using the coefficient of determination to express the degree of relationship between X and Y is well justified, statisticians sometimes employ another value calculated from the sample data. This value is called the *sample correlation coefficient*, which is the square root of the coefficient of determination, calculated from

$$r = \sqrt{r^2}$$

Analogously the *population correlation coefficient* ρ is defined

$$\rho = \sqrt{\rho^2}$$

We can consider r to be a point estimator of ρ.

Because the square root of any number can be positive or negative, the correlation coefficient may be a more useful expression of the strength of association. The symbol r, positive or negative, can be used to signify the *direction* of the relationship between X and Y. Thus, when Y varies *directly* with X, r is *positive*; when Y varies *inversely* with X, r is *negative*.

For the parts supplier data, the sample correlation coefficient is

$$r = \sqrt{r^2} = \sqrt{.64} = +.80$$

We choose a positive sign for r, because the data indicate that transportation times should be greater with increasing distance, so that X and Y are directly related.

Figure 10-12 shows the values of the correlation coefficient r calculated for various sets of data. The values of $r = \sqrt{r^2}$ can range from -1 to 1. Negative values of r are obtained for the data sets in Figures 10-12(b) and (d), where Y varies inversely with X. The slope of the estimated regression lines fitted to each set of these data will be negative, because Y decreases with increasing X. In Figures 10-12(a) and (c), the correlation coefficients are positive, because each set of data indicates a direct relationship between X and Y. Thus, the slope of the estimated regression lines will be positive. It can be seen that *the signs for the correlation coefficient and the slope of the regression line must agree*.

Just like the coefficient of determination, the value of the correlation coefficient approaches zero as the degree of scatter becomes greater. Diagrams (a) and (c) in Figure 10-12 illustrate this point. In (a), the data are perfectly correlated, so that $r = 1$; in (c), some scatter indicates a smaller degree of correlation, so that $r = .95$. Likewise, for the perfectly correlated data in diagram (b), $r = -1$, whereas the less correlated, more scattered data in (d) yield $r = -.57$. As with r^2, $r = 0$ signifies a zero correlation.

Diagrams (e) and (f) represent two instances where $r = 0$. In (e), there is no apparent relationship between X and Y. But in (f)—although $r = 0$, indicating no statistical relationship between X and Y—the data clearly exhibit a well-pronounced curvilinear relationship. Thus, it is incorrect to conclude that there is no relationship between X and Y in this case. Therefore, our *correlation coefficient must be restricted to instances where the underlying relationship between X and Y is believed to be linear*. Different procedures are required to calculate the strength of association for data that have a curvilinear relationship.

As we have already seen, the sample coefficient of determination r^2 shows the proportion of the total variation in Y that is explained by the regression line. As an adjunct to regression analysis, r^2 is a more useful measure of association. Because r is a decimal fraction, it will always be larger (in absolute value) than r^2. Thus, $r = .70$ may give us the false impression that there is a substantial explanation of the variation in Y through regression. The reduction in total variation actually amounts to only $r^2 = (.70)^2 = .49$, or less than $1/2$. Statisticians usually prefer to use the coefficient of determination r^2 to explain the strength of association between X and Y when employing regression analysis.

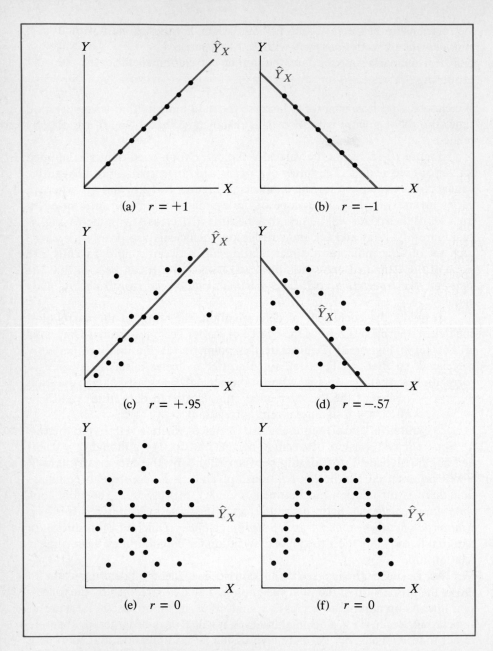

FIGURE 10-12
Scatter diagrams for sample results having various degrees of correlation.

When regression analysis is not used, however, the correlation coefficient r may provide more meaningful results. In such cases, r is calculated from a different expression (to be explained next) that automatically provides the proper sign indicating whether a direct or an inverse relationship exists.

Correlation Explained Without Regression

The value of the correlation coefficient may be computed without performing regression analysis first. For many statistical applications, only the correlation coefficient is required. For example, in determining if a chemical's concentration X improves patient recovery time Y, a medical researcher may only be interested in determining whether or not raising the drug's dosage reduces symptoms; that is, she would be concerned only with the *degree* to which X and Y are related, which she could determine from the correlation coefficient. The researcher may not be at all interested in *predicting* the number of days that it takes a patient given a 50-cc dosage to recover, which is what she would learn from a regression line.

We can use the following expression to directly calculate the Pearson

SAMPLE CORRELATION COEFFICIENT

$$r = \frac{\sum XY - n\bar{X}\bar{Y}}{\sqrt{(\sum X^2 - n\bar{X}^2)(\sum Y^2 - n\bar{Y}^2)}}$$

For example, we may use the intermediate calculations found earlier for rail distance X and transportation time Y:

$$n = 10 \qquad \bar{X} = 611 \qquad \bar{Y} = 9.5$$
$$\sum XY = 64{,}490 \qquad \sum X^2 = 4{,}451{,}500 \qquad \sum Y^2 = 993$$

We substitute these values into the expression for the sample correlation coefficient:

$$r = \frac{64{,}490 - 10(611)(9.5)}{\sqrt{[(4{,}451{,}500 - 10(611)^2][993 - 10(9.5)^2]}} = .80$$

This is the same value that we found earlier by taking the positive square root of r^2.

Correlation and Causality

The correlation coefficient measures only the strength of *association* between two variables. This is a statistical relationship, and a large positive or negative value of r does not indicate that a high value of one variable necessarily *causes* the other variable to be large. Examples of nonsense or *spurious* correlations abound.

EXAMPLE: SPERM WHALES AND STOCK PRICES

Consider the correlation between the number of sperm whales Y caught between 1964 and 1968 and *Standard and Poor's Price Index* for 500 stocks X during the same time period. In the table below, $r = -.89$ is calculated, indicating a high negative correlation between X and Y. How do we interpret this result? Clearly, there is no apparent logical connection between whaling and the New York Stock Exchange prices for common stocks. The values of X and Y have simply moved in opposite directions by approximately the same relative amounts for the five years considered. Statistically, large values of X have occurred with small values of Y, and vice versa. Obviously, r does not measure *causality* here, because we cannot say that increasing stock prices have caused a decline in the number of sperm whales caught, or the reverse.

Year	Number of Sperm Whales Caught (thousands) X	Standard and Poor's Common-Stock Price Index Y	X^2	Y^2	XY
1964	29	81	841	6,561	2,349
1965	25	88	625	7,744	2,200
1966	27	85	729	7,225	2,295
1967	26	92	676	8,464	2,392
1968	24	99	576	9,801	2,376
Totals	131	445	3,447	39,795	11,612
	$= \sum X$	$= \sum Y$	$= \sum X^2$	$= \sum X^2$	$= \sum XY$

$$\bar{X} = 262 \qquad \bar{Y} = 89$$

$$r = \frac{\sum XY - n\bar{X}\bar{Y}}{\sqrt{(\sum X^2 - n\bar{X}^2)(\sum Y^2 - n\bar{Y}^2)}} = \frac{11,612 - 5(26.2)(89)}{\sqrt{[3,447 - 5(26.2)^2][39,795 - 5(89)^2]}} = -.89$$

DATA SOURCE: *Yearbook of Fishery Statistics,* Food and Agricultural Organization, United Nations, and *Economic Report of the President,* February 1970.

EXERCISES

10-15 Determine the value of the sample correlation coefficient for each of the following situations:

(a) $\hat{Y}_X = 10 + 2X; r^2 = .974$ (c) $\hat{Y}_X = 1.1 + 5.5X; r^2 = .950$

(b) $\hat{Y}_X = .3 - .1X; r^2 = .640$ (d) $\hat{Y}_X = 20 - 2X; r^2 = .810$

10-16 For each of the following situations, indicate whether a correlation analysis, a regression analysis, or both would be appropriate. In each case, give the reasons for your choice.

(a) To choose advertising media, an agency account executive is investigating the relationship between a woman's age and her annual expenditures on a client firm's cosmetics.

(b) A trucker wishes to establish a decision rule that will enable him to determine when to inspect or replace his tires, based on the number of miles driven.

(c) A government agency wishes to identify which field offices of various sizes (based on numbers of employees) do not conform to the prevailing pattern of working days lost due to illness.

(d) A research firm conducts attitude surveys in two stages. The first stage identifies coincident factors, such as age and income. The second stage is more detailed and involves a separate study to predict the values of one variable using the known values of other variables associated with it in the initial stage.

10-17 The following data represent disposable personal income and personal consumption expenditures (in billions of dollars) for the United States during the five-year period from 1964 through 1968. (Source: *Economic Report of the President*, February 1970.)

Year	Disposable Personal Income X	Personal Consumption Expenditures Y
1964	438	401
1965	473	433
1966	512	466
1967	547	492
1968	590	537

(a) Use the method of least squares to find the estimated linear regression equation that provides consumption expenditure predictions for specified levels of disposable income.

(b) Is the Y intercept negative or positive? Why do you think this is so?

(c) State in words the meaning of the slope of the regression line.

(d) Use your intermediate calculations from (a) and the values found for the estimated regression coefficients to calculate the coefficient of determination.

10-18 In helping to evaluate the effectiveness of a retraining program, a personnel manager is studying the following results obtained for a sample of $n = 10$ employees. The data provide productivity indexes before and after retraining.

Student	Prior Score X	Current Score Y
Grace Brown	90	83
Patrick Gray	75	72
Lisa White	80	84
Homer Black	65	76
John Green	85	77
Linda Jones	90	82
Carl Smith	95	95
Freddy Tyler	75	68
Lisa Adams	70	78
Karen Johnson	60	55

(a) Use the method of least squares to determine the equation for the estimated regression line.

(b) Calculate the sample coefficient of determination.

(c) What percentage of the total variation in Y is explained by the estimated regression line?

10-19 Referring to Exercise 10-17, answer the following questions:

(a) What percentage of the total variation in Y is explained by the regression line?

(b) What is the value of the correlation coefficient?

(c) Calculate the correlation coefficient directly, using the regression results.

10-20 A statistics instructor wants to know if there is a correlation between his students' homework point totals and their average examination scores. A random sample of five of his students has produced the following results:

Homework Point Total X	Average Examination Score Y
140	90
80	80
90	60
150	80
110	70

Calculate the correlation coefficient for these data.

10-6 COMMON PITFALLS AND LIMITATIONS OF REGRESSION AND CORRELATION ANALYSIS

There are several limitations and pitfalls associated with linear regression and correlation analysis that should be given special mention. These may be categorized as being due to (1) violations of the theoretical assumptions, (2) improper use of the regression equation and correlation coefficient, and (3) mis-interpretation of these coefficients. Regression and correlation are powerful tools, but ludicrous conclusions may be drawn if they are not used properly.

Relevancy of Past Data

Care must be taken in using past data to determine a future relationship. Regression analysis is often used in forecasting with time series, which we will discuss in Chapter 12. There the independent variable is time, with the dependent variable Y being forecast. The source of data is history, so that often a time span of many years is encompassed by the observed Y values. A very serious danger is that the underlying Y populations may change character over time. The regression equation assumes that variance in Y is constant. Thus if there is reason to believe that the values of Y are becoming more or less volatile over time, a critical assumption of regression analysis is violated. In addition to difficulties with the variance, the observations over the years are usually statistically *dependent*, violating another assumption of regression theory. The net effect of such violations is that probabilistic interpretations of inferences are invalid and cannot be used.

Because relationships are apt to change over time, special care should be taken when past data are used to predict a future value of the dependent variable. The hazards are most pronounced in forecasting with time series, but similar difficulties may arise whenever the data are collected over a long time period. This would be the case in attempting to express the cost of a production run by means of a regression line relating it to quantity produced. A very useful result of such an analysis is identification of fixed and variable cost components, so that the Y intercept is the estimated fixed cost per production run, while the slope of the regression line provides an estimate of variable costs per unit. Since production runs may be infrequent, obtaining sufficient data points for predictions precise enough to be useful may require historical experience from several years. But during this period, a great many changes—in labor and material costs, in efficiency (due to technological improvements), or in facilities—can cause unknown changes in the relationship between cost and production volume.

The regression assumptions presume that the same conditions prevail at all levels of production, which is probably not the case. For example, low production volumes may predominate in early time periods, with higher volumes in later years. The slope of the regression line obtained may be unduly steep, exaggerating the true cost–volume relationship if prices have been increasing over the duration. If there is inflation, labor rates and per-unit material prices will rise with time, so that all basic historical accounting data ought to be recalculated at present wage and price levels prior to regression analysis, if that is at all possible.

Cause-and-Effect Relationships

Earlier in this chapter we noted that regression and correlation analysis only provides a statistical relationship between two variables. It cannot tell us whether the values of X cause the values of Y. There may be a cause-and-effect relationship, but not necessarily. Possibly there is not even a logical reason for a relationship. (Although for any practical problem there is ordinarily some reasonable connection.) We have illustrated this by the spurious correlation found for the sperm-whale catch and stock prices. Often the relationship between two variables may be explained by their interactions with a common factor. Thus we might find that a straight line provides a very close fit to the scatter diagram that relates the number of major oil slicks polluting our coasts to the number of electrical power failures. Clearly, one set of events is not caused by the other, but both may be explained by a common linkage: the demand for energy.

10-7   CURVILINEAR REGRESSION ANALYSIS

As we have seen in the earlier sections of this chapter, the assumption of a linear relationship between two variables is the simplest, but it does not always provide the best explanation. We have shown several examples where the relationship between X and Y is curvilinear.

The method of least squares may be applied to curvilinear relationships. The simplest of these finds a regression equation of the form

$$\hat{Y}_X = a + bX + cX^2$$

which is the equation for a parabola. It differs from a linear equation by the term containing X^2. Sometimes a curve may be used that involves higher powers of X, such as X^3 or X^4. An additional regression coefficient is required for each higher power of X. The regression equations so obtained are referred to mathematically as polynomials. It is unusual in applications of regression analysis to go beyond a polynomial involving X^3.

As with linear regression, the values for a, b, and c may be determined from a simultaneous solution of normal equations:

$$\sum Y = na + b\sum X + c\sum X^2$$
$$\sum XY = a\sum X + b\sum X^2 + c\sum X^3$$
$$\sum X^2 Y = a\sum X^2 + b\sum X^3 + c\sum X^4$$

For curves involving X^3 terms, one more normal equation is required. In general, if the highest power of X is X^n, then $n + 1$ unknown regression coefficients must be found by simultaneously solving $n + 1$ normal equations.

REVIEW EXERCISES

10-21 For each of the following results, calculate (1) the sample coefficient of determination and (2) the sample correlation coefficient.
(a) $\hat{Y}_X = 10 + .5X; \sum Y = 200; \sum XY = 5,000; \sum Y^2 = 5,250; n = 10.$
(b) $\hat{Y}_X = 15 - .5X; \sum Y = 500; \sum XY = 8,000; \sum Y^2 = 4,500; n = 100.$

10-22 A security analyst wants to be convinced that the efficiency of capital utilization, expressed by the annual turnover of inventory, actually does have an effect on a manufacturer's earnings. A sample of five firms is chosen, and the following results are obtained:

Company	Inventory Turnover X	Earnings as a Percentage of Sales Y
A	3	10
B	4	8
C	5	12
D	6	15
E	7	13

Calculate the correlation coefficient for the above data.

10-23 A statistician has obtained the following data pertaining to the mean litter size of mice Y from mothers of various ages X:

Age (months) X	Mean Litter Size Y	Age (months) X	Mean Litter Size Y
12	10	2	14
7	9	9	12
4	13	5	10
9	8	8	12
7	13	4	11

(a) Find the estimated regression equation that can be used to predict mean litter size from a mother mouse's age.
(b) Use your results from (a) to find (1) the sample coefficient of determination and (2) the correlation coefficient.
(c) What percentage of the variation in mean litter size is explained by the regression line?

10-24 A linear least-squares relationship is used to determine how the grade point average Y of a particular student in a certain university relates to his or her average weekly study time X (in hours per week). GPA is measured on a scale where 4.0 is "straight 'A'." A random sample of 100 students was selected, and the following intermediate calculations were obtained:

$$\sum X = 3,000 \qquad \sum Y = 260 \qquad \sum XY = 8,050$$
$$\sum X^2 = 92,500 \qquad \sum Y^2 = 775$$

(a) Find the estimated regression equation.
(b) State in words the meaning of the slope of your estimated regression line.

10-25 An economist has established that personal income X may be used to predict personal savings Y by the relationship $\hat{Y}_X = 24.0 + .06X$ (billions of dollars). For each of the following levels of personal income, calculate the predicted value of personal savings:
(a) $X = \$300$ billion (b) $X = \$500$ billion (c) $X = \$700$ billion

10-26 A California rancher has kept records over the past $n = 10$ years of the amount of rainfall X (in inches) in his county and of the number of alfalfa bales Y he has had to buy to supplement grazing grass for his herd until he has been able to sell his excess cattle. The following estimated regression line has been obtained: $\hat{Y}_X = 20,000 - 500X$. The rancher has calculated $s_Y = 1,000$ bales, $s_{Y \cdot X} = 500$ bales, and $\sum X^2 = 2,500$. The mean rainfall for the time period considered has been $\bar{X} = 15$ inches. To arrange bank financing, the rancher wishes to predict how much alfalfa he must buy for the remainder of the current year. Since the dry season has arrived, he knows how much rain has fallen. Construct 95% prediction intervals for the required number of bales if this year's rainfall is (a) 15 inches; (b) 20 inches; (c) 10 inches.

10-27 An agronomist believes that over a limited range of fertilizer-application levels X (in gallons per acre) she can obtain a prediction of crop yield Y (in bushels per acre) using linear least-squares regression. For a sample of $n = 25$ plots, she has established that $\hat{Y}_X = 50 + .05X$, with $s_Y = 8$ bushels per acre, $s_{Y \cdot X} = 2$ bushels per acre, $\bar{X} = 210$ gallons, and $\sum X^2 = 1,105,000$.
(a) Construct a 95% prediction interval for the conditional mean bushels per acre $\mu_{Y \cdot X}$ for an application of 200 gallons of fertilizer per acre.
(b) Construct a 95% prediction interval for the yield of a particular 1-acre plot where 200 gallons of fertilizer will be applied.

10-28 Suppose that the estimated regression line providing total ingredient cost for chemical batches of size X (in thousands of liters) is $\hat{Y}_X = \$30,000 + \$5,000X$. This result was obtained from a sample of $n = 100$ production runs, for which $s_{Y \cdot X} = \$400$.
(a) Construct the 95% prediction interval for the conditional mean batch cost $\mu_{Y \cdot X}$ when $X = 10$ thousand liters.
(b) Construct the 99% prediction interval for the total cost of the next 10-thousand-liter batch.

OPTIONAL TOPIC: 10-8

INFERENCES REGARDING REGRESSION COEFFICIENTS

In some applications of regression analysis, the major concern is not predictions but instead the nature of the regression equation itself. The slope and intercept of the true regression equation

$$\mu_{Y \cdot X} = A + BX$$

can be estimated or tested, so that inferences regarding A or B could be made. Such investigations are usually limited to the slope B.

Confidence Interval Estimate of B

An unbiased estimate of slope B of the true regression line may be obtained from its sample counterpart b. We construct the

CONFIDENCE INTERVAL ESTIMATE FOR B

$$B = b \pm t_\alpha \frac{s_{Y \cdot X}}{\sqrt{\sum X^2 - n\bar{X}^2}}$$

where t_α is found from Appendix Table F for $n - 2$ degrees of freedom.*

We may use our parts supplier illustration to apply this procedure. Suppose that a 95% confidence interval is desired for the true regression coefficient B. Using $\alpha = (1 - .95)/2 = .025$ and $10 - 2 = 8$ degree of freedom, we have, from Appendix Table F, $t_\alpha = t_{.025} = 2.306$. Using the intermediate calculations from Table 10-2 and the values previously determined for b and $s_{Y \cdot X}$,

---

* Here we use the principle that

$$B = b \pm t_\alpha(\text{estimated } \sigma_b)$$

It can be established that

$$\sigma_b = \frac{\sigma_{Y \cdot X}}{\sqrt{\sum X^2 - n\bar{X}^2}}$$

Using $s_{Y \cdot X}$ in place of $\sigma_{Y \cdot X}$, we obtain an estimate of the standard error for b.

we have as our 95% confidence interval:

$$B = .00897 \pm 2.306 \frac{2.02}{\sqrt{4,451,500 - 10(611)^2}} = .00897 \pm .00550$$

$$.00347 \leq B \leq .01447$$

This means that we are 95% confident that the true value of B, the mean number of days required for transporting a shipment each additional mile, lies between .00347 and .01447. Repeating this procedure with 100 different samples, then, about 95% of the time we will construct an interval containing the true value of B.

This interval estimate of B is not very precise and would probably be of little use to the parts supplier. As with the confidence interval for the population mean discussed in Chapter 8, precision can be increased by using a large sample—considerably larger than the one used in this illustration.

In practice, when n is large, the Student t distribution closely fits the normal distribution. In these cases, the normal deviate z may replace t_α in calculating the confidence interval for B.

Testing Hypotheses About B

We may extend tests of hypotheses to inferences about B. Ordinarily, the fact of greatest importance in testing for the value of B is whether it equals zero. Figure 10-13 illustrates a regression line having zero slope. Note that no matter what the value of X, $\mu_{Y \cdot X}$ remains at A, parallel to the X axis. Thus, if $B = 0$, then since the population distributions of Y have the same mean and variance, we may usually conclude that the Y distributions are identical for all values of X. This means that there is no statistical relationship between X and Y. Thus, if $B = 0$, regression analysis will be of no value in making predictions of Y.

In making the two-sided test, our hypotheses are

$$H_0 : B = 0$$

$$H_A : B \neq 0$$

If we choose .05 (1 minus the previous confidence level) as our significance level, then we only need to find out if our confidence interval contains the point $B = 0$. If it does not, we may reject the null hypothesis. Because the lower limit of the 95% confidence interval calculated previously is .00347, a number greater than $B = 0$, we may reject the null hypothesis at a .05 significance level, concluding that rail distance does affect transportation time.

It may be appropriate instead to employ a one-sided upper-tailed test with $H_A : B > 0$. This would be better if Y varies directly with X, as seems natural in

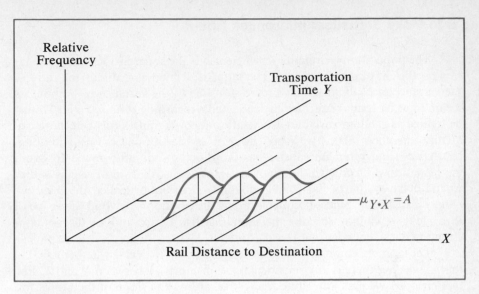

FIGURE 10-13
Illustration of a true regression line having zero slope, so that X and Y are uncorrelated.

the case of transportation time and distance. The first step is to calculate the t

STUDENT t STATISTIC FOR THE SLOPE

$$t = \frac{b}{\dfrac{s_{Y \cdot X}}{\sqrt{\sum X^2 - n\bar{X}^2}}}$$

The value obtained is then compared to the critical value t_α, that corresponds to the prescribed significance level α. If t is smaller than t_α, the null hypothesis is accepted. The reverse is true for a lower-tailed test, where the alternative is that $B < 0$ (the case when Y bears an inverse relationship to X).

We may illustrate the one-sided test with the parts supplier's results. For example, if we use $\alpha = .005$ and $10 - 2 = 8$ degrees of freedom, then from Appendix Table F we obtain the critical value $t_{.005} = 3.355$. The calculated test statistic is

$$t = \frac{.00897}{\dfrac{2.02}{\sqrt{4,451,500 - 10(611)^2}}} = 3.769$$

Since $t = 3.769$ exceeds 3.355, the null hypothesis is *rejected* at a significance level of $\alpha = .005$, which indicates that the slope of the true regression line is greater than zero.

Using *B* to Make Statistical Inferences About ρ

The population correlation coefficient ρ is the square root of the population coefficient of determination. The latter is defined as 1 minus the ratio of the variance of Y about the true regression line to the variance of Y about its mean. In order to make probability statements regarding ρ, both X and Y must be treated as random variables. Correlation theory requires this before we can qualify inferences about ρ. When using r as the estimator, one additional restrictive—and often unrealistic—assumption is ordinarily made to avoid mathematical difficulties in finding the sampling distribution of r. This is that X and Y have a particular joint probability distribution called the *bivariate normal distribution*. Since discussion of this distribution is beyond the scope of this book, we cannot describe the sampling distribution of r or discuss construction of the confidence interval estimates of ρ.

The far more common inference desired is a test to decide whether $\rho = 0$—that is, whether there is any statistical relationship at all between X and Y. For this purpose, we may substitute the test regarding the slope of the regression line. Recall that a slope of zero for the regression line indicates a zero correlation. Thus, a test of whether $B = 0$ may be used to reject or to accept the null hypothesis $\rho = 0$. *A test rejecting the null hypothesis that $B = 0$ will also reject the assumption that $\rho = 0$.*

OPTIONAL EXERCISES

10-29 Referring to the information provided in Exercise 10-12:
- (a) Construct a 99% confidence interval estimate of the slope B of the true regression line.
- (b) In testing H_0 that $B = 0$ against the two-sided alternative that $B \neq 0$, should the null hypothesis be accepted or rejected at the .01 significance level?
- (c) In testing H_0 that $B \leq 0$ against the one-sided alternative that $B > 0$, should the null hypothesis be accepted or rejected at the .01 significance level?

10-30 Referring to the information provided in Exercise 10-14:
- (a) Construct a 95% confidence interval estimate of slope B of the true regression line.
- (b) In testing H_0 that $B = 0$ against the two-sided alternative that $B \neq 0$, should the null hypothesis be accepted or rejected at the .05 significance level?
- (c) In testing H_0 that $B \geq 0$ against the one-sided alternative that $B < 0$, should the null hypothesis be accepted or rejected at the .05 significance level?

11

Multiple Regression and Correlation

In Chapter 10, we learned to construct a regression equation to make predictions of a dependent variable when the value of a single independent variable is known. The techniques we used there are referred to as *simple regression and correlation analysis*. We have already noted that such predictions may be too imprecise for practical application, because a substantial amount of the variation in Y cannot be explained by X. In this chapter, these techniques will be expanded to include *multiple regression and correlation analysis*, which involves *several* independent predictor variables. The total variation in Y can then be explained by two or more variables, which permits us to make a more precise prediction in many situations than is possible in simple regression analysis.

The essential advantage in using two or more independent variables is that it allows greater use of available information. For example, a regression line expressing a new store's sales in terms of the population of the city it serves should yield a poorer sales forecast than an equation that also considers median income, number of nearby competitors, and the local unemployment rate. A plant manager ought to predict more precisely the cost for processing a new order if he considers, in addition to the size of that order, the total volume of orders, his current manpower level, or the production capacity of available equipment. A marketing manager ought to gauge more finely the sales response to a magazine advertisement by considering, in addition to its circulation, the demographical features of its readers, such as median age, median income, or proportion of urban readers.

11-1 LINEAR MULTIPLE REGRESSION INVOLVING THREE VARIABLES

In linear multiple regression analysis, we extend simple linear regression analysis to consider two or more independent variables. In the case of two independent variables, denoted by X_1 and X_2, we use the

ESTIMATED MULTIPLE REGRESSION EQUATION

$$\hat{Y} = a + b_1 X_1 + b_2 X_2$$

As before, $\hat{Y}$ denotes values of Y calculated from the estimated regression equation. Here, we must consider two independent variables and one dependent variable, or a total of three variables. The sample data will consist of three values for each sample unit observed, so that a scatter diagram of these observations will be three dimensional.

Regression in Three Dimensions

To explain how sample data can be portrayed in three dimensions, we will use the analogy of the walls and the floor of a room. Letting a corner of the room represent the situation when all three variables have a value of zero, the data points may be represented by marbles suspended in space at various distances from the floor and the two walls. A marble's height above the floor can be the value of Y for that observation. Its distance along the right wall can represent the observed value of X_1, and its distance along the left wall can express the value of X_2. Figure 11-1 is a pictorial representation of one three-dimensional scatter for a hypothetical set of data.

The estimated multiple regression equation corresponds to a *plane*,* which must be slanted in the way that provides the best fit to the sample data. This results in a three-dimensional surface called the *regression plane*. Choosing

---

*Although Y is related to X_1 and X_2 by a plane instead of a line, we still say that the relationship is linear. The three-dimensional extension of a two-dimensional line is a plane. Although a line can also exist in three dimensions, it is defined as the intersection of two planes. Thus, a three-dimensional line is like a point in two dimensions, which can be defined by the intersection of two lines.

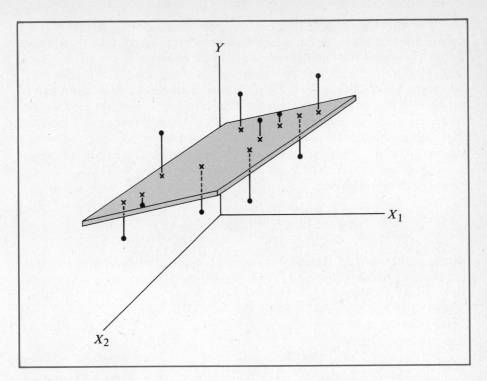

FIGURE 11-1
Scatter diagram and regression plane for multiple regression, using three variables.

this plane is analogous to determining how to position a pane of glass through the suspended marbles so that its incline approximates that of the pattern of scatter. Of course, everyone should obtain the same regression plane, so we will adapt the method of least squares to three dimensions.

The constants a, b_1, and b_2 in the equation of the regression plane $\hat{Y} = a + b_1 X_1 + b_2 X_2$ are called the *estimated regression coefficients*. As with $\hat{Y}_X = a + bX$, a is the value of Y where the regression plane intersects the Y axis, so that we still refer to a as the Y intercept. However, the interpretations of b_1 and b_2 are somewhat different in multiple regression. The constant b_1 expresses the net change in Y for a one-unit increase in X_1, holding X_2 at a constant value. We can view b_1 as the slope of the edges obtained by slicing the regression plane with cuts made parallel to the X_1 axis at a distance X_2 from the origin. Likewise, b_2 is the net change in X_2, holding X_1 constant. Because b_1 or b_2 by itself only indicates part of the total movement in Y in response to increases in its respective independent variable, b_1 and b_2 are referred to as *estimated partial regression coefficients*.

For example, suppose that college grade point average (GPA) Y (on a scale of 4) is related to high-school GPA, denoted by X_1, and Scholastic Aptitude Test (SAT) score X_2 (in hundreds of points) by the equation $\hat{Y} = -1.8 + .8X_1 + .3X_2$. Here $a = -1.8$, so that the regression plane cuts the Y axis at $Y = -1.8$. The regression coefficient $b_1 = .8$ signifies that a student's college GPA can be predicted to be .8 point higher for each additional point in her high-school GPA, regardless of how well she has done on the SAT. The constant $b_2 = .3$ indicates that each additional 100 points on the SAT is estimated to add .3 point to a student's college GPA, no matter what her high-school GPA. Thus, a "straight A" high-school student (GPA = 4.0) scoring 750 on the SAT would have a predicted college GPA of

$$\hat{Y} = -1.8 + .8(4.0) + .3(7.5) = 3.65$$

which is an "A minus" average. Because the sample of students used to obtain the above regression equation will exhibit individual differences in background, motivation, study load, high-school grading levels, and so forth, we cannot assume that this particular student will necessarily achieve a 3.65 GPA. As noted in Chapter 10, we must account for variability in Y, both inherent and due to sampling error.

11-2   MULTIPLE REGRESSION USING THE METHOD OF LEAST SQUARES

Mathematically, we use the method of least squares to position the regression plane so that the sum of the squared vertical deviations from its surface to the data points is minimized. The dots in Figure 11-1 represent the observed data points, and the vertical deviations of these points from the plane are shown as line segments. The crosses indicate the corresponding points on the regression plane that have identical values of X_1 and X_2. The height of a cross above the X_1X_2 plane (or the floor in our analog) represents the value of $\hat{Y}$ computed from $\hat{Y} = a + b_1X_1 + b_2X_2$. The deviation between the observed and the computed heights, $Y - \hat{Y}$, will be positive for dots lying above the regression plane and negative for dots lying below, so that the method of least squares minimizes

$$\sum(Y - \hat{Y})^2$$

As in simple regression, we determine the coefficients of the estimated regression plane by solving a set of three equations in three unknowns. As before, these are refered to as the

NORMAL EQUATIONS

$$\sum Y = na + b_1 \sum X_1 + b_2 \sum X_2$$
$$\sum X_1 Y = a \sum X_1 + b_1 \sum X_1^2 + b_2 \sum X_1 X_2$$
$$\sum X_2 Y = a \sum X_2 + b_1 \sum X_1 X_2 + b_2 \sum X_2^2$$

These equations may be obtained from the equation for a plane $Y = a + b_1 X_1 + b_2 X_2$. The first normal equation is found by summing each term of the plane equation, using the fact that $\sum a = na$. The second normal equation is obtained by multiplying every term in the plane equation by X_1 and summing the results. The third equation is found by multiplying every term by X_2 and summing.*

We will not introduce separate expressions for the regression coefficients here, because the resulting equations would be too complicated and cumbersome. Instead, we find a, b_1, and b_2 by calculating the required the required sums from the data for the various combinations of Y, X_1, and X_2 and substituting these sums into the normal equations, which are solved simultaneously.

A Supermarket Illustration

In order to illustrate how to find the estimated multiple regression equation, we consider the problem of predicting profit Y (in thousands of dollars) for supermarkets in a large metropolitan area. As independent variables we use the total sales (in tens of thousands of dollars) X_1 of foods and X_2 of nonfoods. For simplicity of calculation we use only the ten hypothetical observations shown in Table 11-1.

One reason for splitting total sales into food and nonfood categories is that stores will differ significantly from each other in their offerings of nonfood items. Virtually all will handle tobacco and cleaning agents. But some will have liquor departments, offer lines of convenience hardware and small appliances, or handle clothing articles. Separate sales figures for the two categories are easily obtainable, because the food items (in this particular locale) are not subject to sales tax, whereas the nonfood items are. The nonfood items typically have higher markups and move more slowly off the shelf. Thus, treating food and nonfood sales separately ought to provide a better prediction of a store's profit than total sales.

---

* This procedure is a mnemonic aid. Calculus was used to actually derive these equations.

TABLE 11-1

Profit and Sales of Food and Nonfood Items
for Ten Hypothetical Supermarkets

Supermarket Number	Profit (thousands of dollars) Y	Sales of Foods (tens of thousands of dollars) X_1	Sales of Nonfoods (tens of thousands of dollars) X_2
1	20	305	35
2	15	130	98
3	17	189	83
4	9	175	76
5	16	101	93
6	27	269	77
7	35	421	44
8	7	195	57
9	22	282	31
10	23	203	92

The intermediate calculations necessary to find the regression equation are shown in Table 11-2. The first eight columns contain the individual variable values and the squared and product terms. Column (9) contains the values for Y^2, which are not needed to obtain the regression coefficients but will be used later. Substituting the appropriate column totals, we obtain the following

TABLE 11-2

Intermediate Calculations for Obtaining Regression Coefficients

(1) Y	(2) X_1	(3) X_2	(4) X_1Y	(5) X_2Y	(6) X_1X_2	(7) X_1^2	(8) X_2^2	(9) Y^2
20	305	35	6,100	700	10,675	93,025	1,225	400
15	130	98	1,950	1,470	12,740	16,900	9,604	225
17	189	83	3,213	1,411	15,687	35,721	6,889	289
9	175	76	1,575	684	13,300	30,625	5,776	81
16	101	93	1,616	1,488	9,393	10,201	8,649	256
27	269	77	7,263	2,079	20,713	72,361	5,929	729
35	421	44	14,735	1,540	18,524	177,241	1,936	1,225
7	195	57	1,365	399	11,115	38,025	3,249	49
22	282	31	6,204	682	8,742	79,524	961	484
23	203	92	4,669	2,116	18,676	41,209	8,464	529
191 $= \sum Y$	2,270 $= \sum X_1$	686 $= \sum X_2$	48,690 $= \sum X_1Y$	12,569 $= \sum X_2Y$	139,565 $= \sum X_1X_2$	594,832 $= \sum X_1^2$	52,682 $= \sum X_2^2$	4,267 $= \sum Y^2$

normal equations:

$$191 = 10a + 2{,}270b_1 + 686b_2$$
$$48{,}690 = 2{,}270a + 594{,}832b_1 + 139{,}565b_2$$
$$12{,}569 = 686a + 139{,}565b_1 + 52{,}682b_2$$

Solving these equations simultaneously for the unknowns a, b_1, and b_2, we obtain the solutions:

$$a = -23.074$$
$$b_1 = .1148$$
$$b_2 = .2349$$

The simultaneous solution of three normal equations by hand can be quite a chore. This task, as we will see later in this chapter, can be enormously simplified by using a computer.

The above values for a, b_1, and b_2 provide the estimated multiple regression equation

$$\hat{Y} = -23.074 + .1148X_1 + .2349X_2$$

This equation may then be used to forecast the profit of a particular store. Suppose that a new store is built that will have estimated sales of $2,500,000 for foods and $750,000 for nonfoods, so that $X_1 = 250$ and $X_2 = 75$. The estimated store profit for these sales levels would be

$$\hat{Y} = -23.074 + .1148(250) + .2349(75)$$
$$= 23.244 \text{ (thousand dollars), or } \$23{,}244$$

We may interpret the value $b_1 = .1148$ as follows: For each $10,000 increase in food sales there will be an estimated increase of .1148 thousand dollars, or $114.8, in profits, holding the sales of nonfood items at any fixed level. Likewise, each additional $10,000 from the sale of nonfood items results in an estimated profit increase of $b_2 = .2349$ thousand dollars, or $234.90, for any fixed amount of food sales. Thus the marginal contribution of nonfoods to profits is, dollar for dollar, slightly more than twice that of the foods. The result $a = -23.074$ signifies that $23,074 of "fixed" costs, on the average, must be absorbed before a store can show a profit. These correspond to those expenses, such as rent, that continue whether or not the store is open. The $23,074 figure may vary considerably from the actual fixed costs of any particular store, because of sampling error and due to different store characteristics. It also depends on (1) the linear model being appropriate and (2) the validity of extrapolation down to zero sales for both types of items.

Advantages of Multiple Regression

Two interesting questions may be posed regarding the use of multiple rather than simple regression analysis. First, is the simultaneous analysis of two independent variables through multiple regression any improvement over that obtained through two separate simple regressions? Second, how can we show that the accuracy of predictions is improved by the use of multiple regression analysis? We will begin to answer these questions by continuing with our supermarket illustration.

Three two-dimensional diagrams of the supermarket data are shown in Figure 11-2. In Figure 11-2(a), profit is plotted against sales of food items. Food sales X_1 by itself might be a fairly reliable predictor of profit Y, since there is a high value of .76 for the sample correlation coefficient. Denoting this particular sample correlation coefficient by the double-subscripted symbol r_{Y1}, where $Y1$ indicates that the strength of association between the variables Y and X_1 is being measured, we have $r_{Y1} = .76$. A similar figure may be obtained for any pair of the three variables. Thus, analysis relating profit Y to the sales X_2 of nonfood items provides another sample correlation coefficient, $r_{Y2} = -.29$. A cursory examination of this second relationship, shown in Figure 11-2(b), seems to indicate that nonfood sales have little effect on profits. Does this indicate that nonfood sales would be a poor predictor of a store's income? Two-variable analysis seems to contradict our previous finding that profit will increase by an estimated \$234.90 for each additional \$10,000 in nonfood sales. How might we explain this?

A comparison of food sales X_1 to nonfood sales X_2, as represented by the scatter diagram in Figure 11-2(c), shows a pronounced negative correlation between these two variables. We distinguish the sample correlation coefficient in this case from the others by using the subscript 1 2; here, $r_{1\,2} = -.77$. Supermarkets having higher than average food sales tend to have lower than average nonfood sales, whereas the reverse is true for stores with below-average food sales. By specializing heavily in nonfood items, the low–food-volume operators have managed to survive and achieve profits in face of competition from stores more successful at selling food. The two stores having annual food sales below \$1.5 million have extraordinarily large profits, which are not very well explained by food sales alone. Their data points lie considerably above the regression line in Figure 11-2(a). The impact of nonfood sales is camouflaged by the interaction of food and nonfood sales.

This illustrates a general inadequacy in using simple regression to separately determine how a dependent variable relates to several independent variables. The same is true of correlation. Separate simple correlations of Y versus X_1 and Y versus X_2 seem to indicate that food-item sales would be a far better predictor of profit than nonfood sales. A two-variable analysis indicates that X_2 has a negligible correlation with Y of $r_{Y2} = -.29$. But it would be a

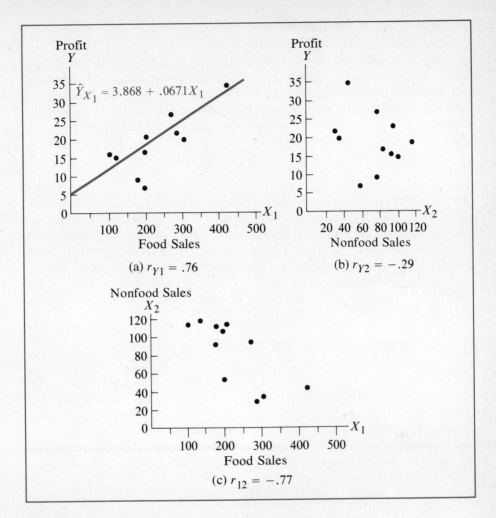

FIGURE 11-2

Scatter diagrams for variable pairs using supermarket data.

blunder to discard X_2 because it shows such a small correlation with Y. This is because X_2 is highly negatively correlated with X_1, $r_{12} = -.77$, so that its influence on Y can be explained only by considering how Y relates to both X_1 and X_2 through multiple regression. The interactions between the predictor variables themselves must first be determined before a variable can be discarded. We will later present techniques of multiple regression and correlation analysis used to provide this information.

The Standard Error of the Estimate

We have seen that the errors in making predictions from the regression line are smaller when the scatter of the data is less pronounced. The degrees of such variation in Y can be expressed by the standard error of the estimate for values of Y. The same is true in multiple regression. The standard error of the estimate for values of Y about the regression plane is defined by

$$S_{Y \cdot 12} = \sqrt{\frac{\sum (Y - \hat{Y})^2}{n - 3}}$$

We use a different subscript notation, $Y \cdot 12$, to show that two independent variables X_1 and X_2 are being used to predict Y. As in simple regression, we find the standard error of the estimate by taking the square root of the mean squared deviations of observed Y values from the estimated regression plane. The divisor $n - 3$ is chosen because 3 degrees of freedom are lost in estimating the regression coefficients. This makes $S_{Y \cdot 12}^2$ an unbiased estimator of the variance of Y about the true regression plane. In practice, it is simpler to use the following expression to calculate the

STANDARD ERROR OF THE ESTIMATE

$$S_{Y \cdot 12} = \sqrt{\frac{\sum Y^2 - a \sum Y - b_1 \sum X_1 Y - b_2 \sum X_2 Y}{n - 3}}$$

because it does not require $\hat{Y}$ to be calculated for every data point.

Using the data in Table 11-2 and the regression constants found previously, we calculate the standard error of the estimate for the supermarket data:

$$S_{Y \cdot 12} = \sqrt{\frac{4{,}267 - (-23.074)(191) - .1148(48{,}690) - .2349(12{,}569)}{10 - 3}}$$

$$= 4.343 \text{ (thousand dollars)}$$

This value may be compared to the corresponding standard error found in a simple regression for profit Y and food sales X_1. Using X_1 as the only independent variable, and then applying the procedures of Chapter 10 to the supermarket data, we find the regression line

$$\hat{Y}_{X_1} = 3.891 + .0670 X_1$$

The scatter of data about this regression *line* is summarized by the standard error of the estimate

$$s_{Y \cdot X_1} = 5.719 \text{ (thousand dollars)}$$

Comparing this with the standard error of the estimate about the regression plane, $S_{Y \cdot 1\ 2} = 4.343$, we see that multiple regression provides a better interpretation of the data. As we noted earlier, the standard error expresses the amount of variation in Y that is left unexplained by regression analysis. Since $S_{Y \cdot 1\ 2}$ is smaller than $s_{Y \cdot X_1}$, our regression plane (incorporating both food and nonfood sales as *two* independent variables) explains more of the variation in Y than the regression line does (where food sales serves as the *single* independent variable and there is more unexplained variation). This indicates that inclusion of nonfood sales data in the analysis will indeed provide better predictions of store profit. Before we see how such predictions can be made from multiple regression results, we must clarify some theoretical points.

Assumptions of Multiple Regression

The assumptions of linear multiple regression are similar to those of simple regression. The least-squares method provides a plane that is an estimate of

$$\mu_{Y \cdot 1\ 2} = A + B_1 X_1 + B_2 X_2$$

where $\mu_{Y \cdot 1\ 2}$ denotes the conditional mean of Y, given X_1 and X_2. The values of the true regression coefficients A, B_1, and B_2 remain unknown and are estimated by a, b_1, and b_2, respectively. For each combination of X_1 and X_2 there is a corresponding point on the true regression plane. The height of this point is $\mu_{Y \cdot 1\ 2}$, the mean of the corresponding population for Y. As in simple regression, each Y population is assumed to have the same variance $\sigma_{Y \cdot 1\ 2}^2$.

It is possible to extend the theoretical developments of simple regression to multiple regression and construct prediction intervals for $\mu_{Y \cdot 1\ 2}$ and individual values of Y.

Prediction Intervals in Multiple Regression

The procedures used to construct prediction intervals to qualify estimates made from the regression plane are similar to the ones used in simple regression. The following equation provides the

PREDICTION INTERVAL ESTIMATE
OF THE CONDITIONAL MEAN

$$\mu_{Y \cdot 1\ 2} = \hat{Y} \pm t_\alpha \frac{S_{Y \cdot 1\ 2}}{\sqrt{n}}$$

where $\hat{Y}$ is the value of Y computed from the regression plane at the given levels for X_1 and X_2.

We may illustrate this with the supermarket data. Suppose that $\mu_{Y \cdot 1\,2}$ is to be estimated by a 95% prediction interval, so that $\alpha = .025$. This is done for all stores having food sales of \$2,000,000 and nonfood sales of \$500,000. Thus (in \$10,000 units), $X_1 = 200$ and $X_2 = 50$, and the corresponding height of the regression plane is

$$\hat{Y} = -23.074 + .1148(200) + .2349(50) = 11.631 \text{ (thousand dollars)}$$

From Appendix Table F, the critical value for $n - 3 = 7$ degrees of freedom is $t_\alpha = t_{.025} = 2.365$. Substituting these values and $S_{Y \cdot 1\,2} = 4.343$ into the above expression, we obtain the prediction interval:

$$\mu_{Y \cdot 1\,2} = 11.631 \pm \frac{2.365(4.343)}{\sqrt{10}}$$

$$= 11.631 \pm 3.248$$

$$8.383 \le \mu_{Y \cdot 1\,2} \le 14.879 \text{ (thousand dollars)}$$

Thus we are 95% confident that the true mean profit for all such stores lies between \$8,383 and \$14,879. This interval is probably too wide for practical planning purposes, largely due to the small sample size used.

As in any sampling situation, a larger sample size would provide an even more precise estimate of mean supermarket profit. *When the sample size is large* $(n \ge 30)$, *the normal distribution can be used*, so that t_α is replaced by the normal deviate z for an upper tail area of α under the normal curve.

An individual value of Y is denoted here by Y_I. The following expression provides the

PREDICTION INTERVAL ESTIMATE
OF AN INDIVIDUAL VALUE OF Y

$$Y_I = \hat{Y} \pm t_\alpha S_{Y \cdot 1\,2} \sqrt{\frac{n+1}{n}}$$

We illustrate this for the college GPA example (page 364) by finding the 95% prediction interval for the college GPA, Y, of a student whose high-school GPA is $X_1 = 2.75$ and whose SAT score is 480, so that $X_2 = 4.80$. The college GPA may be computed from the regression equation:

$$\hat{Y} = -1.8 + .8X_1 + .3X_2$$

$$= -1.8 + .8(2.75) + .3(4.80)$$

$$= 1.84$$

We will use $n = 20$ and $S_{Y \cdot 1\,2} = .2$. From Appendix Table F, using $20 - 3 = 17$ degrees of freedom, the critical value is $t_\alpha = t_{.025} = 2.110$. The student's college GPA is predicted with 95% confidence by

$$Y_I = 1.84 \pm 2.110(.2) \sqrt{\frac{20 + 1}{20}}$$

$$= 1.84 \pm .43$$

$$1.41 \leq Y_I \leq 2.27$$

When n is large, the above expression is modified by replacing t_α by the appropriate normal deviate z.

Inferences Regarding Regression Coefficients

Although we can use b_1 and b_2 to estimate the true partial regression coefficients B_1 and B_2 from sample data, a sampling error may occur. Because two or more coefficients are involved in a multiple regression, special procedures must be used to make inferences about them. We will consider these procedures in Chapter 16 when we discuss the analysis of variance. As an optional topic to that chapter, we will apply hypothesis-testing concepts to multiple regression coefficients.

EXERCISES

11-1 An economist wishes to predict the incomes of restaurants that are more than two years old from a regression equations, using total floor space and number of employees. The following data have been obtained for a sample of $n = 5$ restaurants:

Income (thousands of dollars) Y	Floor Space (thousands of square feet) X_1	Number of Employees X_2
20	10	15
15	5	8
10	10	12
5	3	7
10	2	10

(a) Calculate the estimated regression equation for these data.

(b) State in words the meaning of the partial regression coefficients b_1 and b_2.

11-2 A college admissions director uses high-school GPA X_1 and IQ score X_2 to predict college GPA Y via the regression equation

$$\hat{Y} = .5 + .8X_1 + .003X_2$$

Calculate the predicted college GPA for each of the following students:

	(a)	(b)	(c)	(d)
High-school GPA	2.9	3.0	2.7	3.5
IQ score	123	118	105	136

11-3 A food processor uses least-squares regression to predict the total cost Y for production runs of its products. For Crunchola the independent variables used for prediction are production quantity X_1 and wheat price index X_2. The following data apply:

Total Cost (thousands of dollars) Y	Production Quantity (tons) X_1	Wheat Price Index X_2
200	400	120
85	150	134
115	220	115
200	500	90
140	300	85
65	100	140
70	150	95
65	150	80
125	240	96
190	350	125

(a) Calculate the estimated regression equation for the given data.

(b) By how much do these results indicate that total cost will increase if one additional ton of the final product is made? If the wheat price index goes up one point?

11-4 The editor of a statistics journal wishes to use the intermediate sample-data calculations provided below to determine a regression equation that predicts the total typing hours Y for article drafts. As independent variables, she uses the number of words in the draft X_1 (expressed in tens of thousands) and an index X_2 for level of difficulty on a scale from 1 (least difficult) to 5 (most difficult).

$$n = 25 \qquad \sum Y = 200 \qquad \sum X_1 = 100 \qquad \sum X_2 = 75$$

$$\sum X_1 Y = 1,000 \qquad \sum X_2 Y = 800 \qquad \sum X_1^2 = 600$$

$$\sum X_2^2 = 325 \qquad \sum Y^2 = 3,800 \qquad \sum X_1 X_2 = 200$$

(a) Determine the equation for the regression plane $\hat{Y} = a + b_1 X_1 + b_2 X_2$.

(b) Explain the meaning of the values you obtain for b_1 and b_2.

11-5 A record manufacturer uses special machines to press recording grooves onto blank disks from a die. Each die can be used in about 1,000 pressings. Due to time constraints, it is sometimes necessary to operate several pressing machines simultaneously. Since an expensive die disk must be used on each machine and many production runs are completed before the lifetime of each disk has been expended, this raises production costs. For $n = 100$ production runs, the total manufacturing cost Y (in thousands of dollars) has been determined for the number of pressings made X_1 (in thousands) and the number of die disks required X_2. The following results have been calculated:

$$\hat{Y} = 1.082 + 1.2X_1 + .553X_2$$

$$\bar{Y} = 32.0 \qquad \bar{X}_1 = 23.0 \qquad \bar{X}_2 = 6.0$$

$$\sum Y^2 = 104,000 \qquad \sum X_1 Y = 75,000 \qquad \sum X_2 Y = 15,000$$

(a) Calculate the estimated cost $\hat{Y}$ of each of the following production runs:

	Number of Pressings (thousands)	Number of Die Disks
(1)	15	5
(2)	20	3
(3)	15	4
(4)	100	10

(b) Calculate the standard error of the estimate of values of Y about the regression plane.
(c) For each production run in (a), construct the 95% prediction interval for the *mean* production costs.
(d) Repeat (c) for *individual* production runs.

11-6 Referring to Exercise 11-4, answer the following:
(a) Determine the equation for the estimated regression line $\hat{Y}_{X_1} = a + bX_1$, using a *single* independent variable X_1.
(b) Calculate $s_{Y \cdot X_1}$, using X_1 instead of X in the expression for $s_{Y \cdot X}$ on page 329. Then compute $S_{Y \cdot 1 2}$, using the expression on page 370. Do you think more accurate predictions result if the level-of-difficulty variable X_2 is included than result if X_1 is used as the sole predictor? Why?
(c) A particular article containing 50,000 words is rated at a difficulty level of 3. Determine the prediction interval for the hours required to type the article at a a 95% confidence level. Then compute the same interval for the mean typing time of all articles of the same length and difficulty.

REGRESSION WITH MANY VARIABLES: COMPUTER APPLICATIONS **11-3**

Multiple regression may involve more than two independent variables. For example, in predicting college GPA, we could also include the number of extracurricular high-school activities, since this information is usually considered in making college admissions decisions. Including this third independent variable, denoted by X_3, would provide the regression equation

$$\hat{Y} = a + b_1 X_1 + b_2 X_2 + b_3 X_3$$

To determine the regression coefficients, additional product sums involving X_3 must be computed and four normal equations must be solved simultaneously. *Ordinarily, the necessary calculations are made on a digital computer rather than by hand*, so a detailed discussion of the computations will not be provided here.

In multiple regression using X_1, X_2, and X_3, we compute the standard error of the estimate about the regression *hyperplane*. (Including Y, there are now four variables, and the regression surface can no longer be graphed in only three dimensions. Mathematicians call a four-dimensional plane a "hyperplane.") The standard error, denoted by $S_{Y \cdot 1\,2\,3}$, is calculated in the same way that we calculate $S_{Y \cdot 1\,2}$, but with the extra term $-b_3 \sum X_3 Y$ in the numerator. The denominator will be $n - 4 = 10 - 4 = 6$. The number of degrees of freedom will also be $n - 4$, and we calculate the prediction intervals as we did before, except using this reduced value.

When more predictor variables are added, we can expand the regression analysis by including X_4, X_5, and so on. When there are m total variables (one dependent and $m - 1$ independent), the number of degrees of freedom will be $n - m$, which is used as the denominator in computing the standard error and in finding t_α when constructing prediction intervals.

A computer application will now be discussed to illustrate multiple regression using three independent variables.

Using the Computer in Multiple Regression Analysis

By hand, the computations required to perform a regression analysis are at best tedious, even with the assistance of a hand calculator. Although hand calculation may be feasible when the number of observations and the number of independent variables are small, tedium can be alleviated by using a digital computer. The computer not only saves time and energy but also prevents the inevitable cascade of errors that occurs whenever one error, such as mispunching a calculator button, is committed. Also a computer generally provides greater levels of accuracy due to its superior capability of handling a large number of significant figures.

It is not necessary for users to prepare their own computer programs to perform regression and correlation analysis. Library programs have been written for this purpose and are widely available. These "canned" programs vary considerably in format and type of output data, but they usually include determinations of the regression coefficients (a, b_1, b_2, and so on), standard errors, and correlation coefficients. These programs can be elaborate enough to include lower-dimensional regression equations for several variable combinations, values of the t statistic (used in making inferences regarding regression coefficients), and coefficients of partial determination (to be discussed in Section 11-5). The most complex programs provide inference-making computations that are too sophisticated to discuss here.

TABLE 11-3

Sample Observations of Supermarket Profits, Using Food Sales,
Nonfood Sales, and Size of Store as Independent Variables

Supermarket Number	Food Sales (tens of thousands of dollars) X_1	Nonfood Sales (tens of thousands of dollars) X_2	Store Size (thousands of square feet) X_3	Profit (thousands of dollars) Y
1	305	35	35	20
2	130	98	22	15
3	189	83	27	17
4	175	76	16	9
5	101	93	28	16
6	269	77	46	27
7	421	44	56	35
8	195	57	12	7
9	282	31	40	22
10	203	92	32	23

Now we will describe how to use a particular program: the Multiple Linear Regression Routine of the International Timesharing Corporation. Although this is a time-sharing program, similar programs are available that operate in a batch-processing mode. In those cases, the input data and operating parameters are punched onto cards rather than fed directly into a teletype terminal.

To explain the use of the computer, we will expand our supermarket illustration to include a third independent variable—the size of the store in thousands of square feet. Our previous sample data are augmented by including the ten observations for X_3 provided in Table 11-3.

Based on the data in Table 11-3, Figure 11-3 shows the log of terminal communications, how the data are entered, and the output of the regression analysis. The computer results indicate that profit predictions may be made from the estimated regression equation

$$\hat{Y} = -10.170 + .027X_1 + .097X_2 + .525X_3$$

We may interpret the values for the partial regression coefficients. The value $b_1 = .027$ indicates that a store's profit will increase by an estimated .027 thousand dollars $= \$27$ for each additional $\$10,000$ in food sales, holding other variables fixed. This is smaller than the value for b_1 found in the previous three-variable regression, but b_1 has a different meaning here. Instead of a single additional variable, two variables, both nonfood sales and store size, are accounted for and are being held constant. Likewise, $b_2 = .097$ indicates that holding store size and food sales constant, store profits will increase by an

```
I T S   STATISTICAL PACKAGE

WANT TO SEE STAT INDEX? NO
WANT ADVANCED USER STATUS(YES OR NO)?  NO
TYPE THE CODE OF THE ROUTINE YOU WISH TO USE  MU
TYPE NO. OF VARIABLES AND OBSERVATIONS  4.,10
DATA ORDERED BY OBSERVATION OR BY VARIABLE(OBS OR VAR)?  OBS
ENTER DATA FILE NAME OR 'TERMINAL' -- TERMINAL
INPUT DATA IN FORM XX.X,XX.X,XX.X,...XX.X (RET)
305,35,35,20
130,98,22,15
189,83,27,17
175,76,16,9
101,93,28,16
269,77,46,27
421,44,56,35
195,57,12,7
282,31,40,22
203,92,32,23
LISTING(YES OR NO)?  YES
VARIABLE
OBS.       1 X₁        2 X₂        3 X₃       4 Y
   1    305.0000    35.0000    35.0000    20.0000
   2    130.0000    98.0000    22.0000    15.0000
   3    189.0000    83.0000    27.0000    17.0000
   4    175.0000    76.0000    16.0000     9.0000
   5    101.0000    93.0000    28.0000    16.0000
   6    269.0000    77.0000    46.0000    27.0000
   7    421.0000    44.0000    56.0000    35.0000
   8    195.0000    57.0000    12.0000     7.0000
   9    282.0000    31.0000    40.0000    22.0000
  10    203.0000    92.0000    32.0000    23.0000
DO YOU WISH TRANSFORMATIONS OF DATA?  NO
DO YOU WISH TO SAVE THIS DATA ON A DISK FILE?  NO
ENTER NUMBER OF INDEPENDENT VARIABLES IN THE FORM XX ⌐RET⌐    3
ENTER COLUMN NUMBERS OF THE INDEPENDENT VARIABLES IN
THE FORM XX,XX,... ETC. ⌐RET⌐  1,2,3
ENTER COLUMN NUMBER OF DEPENDENT VARIABLE IN THE FORM XX  4
```

VARIABLE NO.	MEAN	STANDARD DEVIATION	CORRELATION X VS Y	REGRESSION COEFFICIENT	STD. ERROR OF REG.COEF.	COMPUTED T VALUE
1 $\bar{X}_1$	227.000	94.011 s_{X_1}	.760 r_{Y1}	.027 b_1	.012	2.246
2 $\bar{X}_2$	68.600	24.994 s_{X_2}	-.286 r_{Y2}	.097 b_2	.030	3.219
3 $\bar{X}_3$	31.400	12.492 s_{X_3}	.979 r_{Y3}	.525 b_3	.059	3.369
DEPENDENT						
4 $\bar{Y}$	19.100	8.293 s_Y				

```
INTERCEPT                                       -10.170243 a
MULTIPLE CORRELATION COEFFICIENT      R_{Y·123}    .992399
STANDARD ERROR OF ESTIMATE                       1.249868 S_{Y·123}

WANT TABLE OF RESIDUALS(YES OR NO)?   YES
DO YOU WANT TO OUTPUT RESIDUALS TO A FILE(YES OR NO)?   NO
```

TABLE OF RESIDUALS

OBSERVATION	Y OBSERVED	$\hat{Y}$ Y ESTIMATED	$\hat{Y}-Y$ RESIDUAL	RES. PCNT. OF EST.
1	20.000	19.837	-.163	-.822
2	15.000	14.399	-.601	-4.176
3	17.000	17.162	.1621	.941
4	9.000	10.332	1.332	12.894
5	16.000	16.277	.277	1.704
6	27.000	28.711	1.711	5.960
7	35.000	34.865	-.135	-.387
8	7.000	6.930	-.070	-1.006
9	22.000	21.450	-.550	-2.563
10	23.000	21.037	-1.963	-9.332

```
WANT ANOTHER MULTIPLE LINEAR REGRESSION(YES OR NO)?  NO
```

FIGURE 11-3

Computer run for multiple regression, using sample supermarket data. (Variables in color are not on the computer printout but are added here to clarify the data.)

estimated \$97 for each \$10,000 in nonfood sales. The independent variable X_3, store size, has a partial regression coefficient of value $b_3 = .525$, indicating that a supermarket can increase its profits by an estimated \$525 for each thousand square feet of additional floor space.

The substantial changes in the values a, b_1, and b_2 from the earlier multiple regression indicate that including store size X_3 as one of the independent variables should lead to better profit predictions. This improvement can be seen by comparing the standard error of the estimate of $S_{Y \cdot 1\,2\,3} = 1.250$ (shown in the computer printout) to the earlier one of 4.343. Including the additional variable X_3 for store size considerably reduces the standard error of the estimate for Y. This indicates that profit can be predicted with greater precision when store size is included as well as food and nonfood sales.

Stepwise Multiple Regression

Modern data centers have computer programs in their libraries that can be used to handle a variety of statistical problems. One group of such programs is used to perform *stepwise multiple regression*, which includes additional predictor variables one at a time, in successive stages, each raising the dimensions of the analysis by one. A great many independent variables can be handled on the computer using this procedure. One such program selects the most promising independent variable—the one that provides the greatest reduction in the unexplained variation in Y—at each stage. In doing this, the computer performs simple regression separately for each independent variable, printing the results for the best one. The next step of the program perform separate multiple regressions, each combining one of the remaining independent variables with those selected in the previous stages. Again, the one that reduces unexplained variation the most is chosen to be included in all future stages. The process continues in successively higher dimensions either until every variable has been included in a multiple regression involving them all, or until no further reduction in the unexplained variation is possible. Such a program efficiently saves all previous calculations necessary for the higher-dimensional analysis.

The output from such a program can be rich—providing at each step the various multiple and partial coefficients of determination. In addition, it may provide simple correlations for all variable pairs.

Many assumptions concerning the linearity of the data are automatically made by such computer programs. As with any other application, in multiple regression analysis the computer does not eliminate the need for good judgment. It is best used in intermediate stages of the analysis. Computer printouts must be thoroughly evaluated to determine if there is a meaningful explanation of why some variables are excluded and others are included. A variable may be rejected because it does not reduce the unexplained variance, but this may actually be due to a strong curvilinear relation between the variable and Y.

The use of a digital computer may be extended to consider curvilinear multiple regression. This involves even longer computations than the linear case considered in this chapter. The most complex of multivariate techniques are well suited to the digital computer.

EXERCISES

11-7 The following family data are given:

(a) Determine the coefficients for the estimated regression line $\hat{Y}_{X_1} = a + bX_1$, and calculate $s_{Y \cdot X_1}$.

Family	Total Spending Y	Income X_1	Size X_2	Additional Savings X_3
A	8,000	10,000	3	1,000
B	7,000	10,000	2	1,500
C	7,000	9,000	2	700
D	12,000	16,000	4	1,800
E	6,000	7,000	6	200
F	7,000	9,000	4	500
G	8,000	8,000	6	0
H	7,000	8,000	5	100
I	10,000	12,000	6	200
J	8,000	11,000	2	1,000

(b) Determine the coefficients for the estimated regression plane $\hat{Y} = a + b_1X_1 + b_2X_2$, and calculate $S_{Y \cdot 1 2}$.

(c) Does a comparison of the values $S_{Y \cdot 1 2}$ and $s_{Y \cdot X_1}$ suggest that including family size has been worthwhile in predicting total spending? Explain.

11-8 *Computer exercise:* Use the data from Exercise 11-7.

(a) Determine the coefficients for the estimated regression hyperplane

$$\hat{Y} = a + b_1X_1 + b_2X_2 + b_3X_3$$

(b) Determine the value of $S_{Y \cdot 1 2 3}$.

(c) Compare the values of $S_{Y \cdot 1 2 3}$ and $S_{Y \cdot 1 2}$. Does your comparison suggest that including the additional savings variable has proved worthwhile in predicting total spending? Explain your answers.

11-9 A government analyst used the following data, obtained with one popular car model, to predict gasoline mileage based on driving speed, altitude, and passenger weight.

(a) Determine the coefficients for the estimated regression line $\hat{Y}_{X_1} = a + bX_1$, and calculate $s_{Y \cdot X_1}$.

(b) Determine the coefficients for the estimated regression plane $\hat{Y} = a + b_1X_1 + b_2X_2$, and calculate $S_{Y \cdot 1 2}$.

(c) Does a comparison of the values $S_{Y \cdot 1 2}$ and $s_{Y \cdot X_1}$ suggest that including altitude has been worthwhile in predicting gasoline mileage? Explain.

Trip	Miles per Gallon Y	Average Speed X_1	Altitude (thousands of feet) X_2	Passenger Weight (hundreds of pounds) X_3
1	28	50	2	2
2	26	55	0	3
3	28	44	1	2
4	18	46	5	4
5	31	57	0	2
6	25	53	2	3
7	24	54	4	3
8	21	60	4	2
9	23	59	1	5
10	26	52	1	4

11-10 *Computer exercise:* Use the data from Exercise 11-9.
   (a) Determine the coefficients for the estimated regression hyperplane

$$\hat{Y} = a + b_1 X_1 + b_2 X_2 + b_3 X_3$$

   (b) Determine the value for $S_{Y\cdot 1\,2\,3}$.
   (c) Compare the values of $S_{Y\cdot 1\,2\,3}$ and $S_{Y\cdot 1\,2}$. Does your comparison suggest that including the additional weight variable has proved worthwhile in predicting mileage? Explain your answer.

DUMMY VARIABLE TECHNIQUES   11-4

It is sometimes necessary to determine how a dependent variable is related to quantitative independent random variables when there are nonhomogeneous factors influencing their interaction. Often such factors are qualitative in nature. For example, in printing a book, the cost of typesetting can be expressed by an equation involving the size of the book and the number of figures and tables. But this cost is also affected by the kind of book. A technical book using many special symbols and characters (like this one) requires a considerable amount of hand work and is, page for page, more expensive to set than a literary anthology. "Kind of book" can be viewed as a qualitative variable that influences composition cost. Although regression analysis requires all variables to be quantitative, qualitative variables may be incorporated into the framework of multiple regression through the introduction of dummy variables.

There are two major kinds of statistical studies where dummy variables may be used. Those made with sample data collected at a single point in time are sometimes referred to as *cross-sectional studies*. Studies made using observations collected over an extended time period use *time-series data*. In business and economics, time-series data are very important because observations of a dependent variable, such as consumption of electricity or gross national product, can only be obtained periodically. Often, time itself becomes the independent variable. Unavoidable historical heterogeneous influences, such as war or recession, may be treated as qualitative variables, and their effects must be identified somehow. Another important use for qualitative variables in time series is to identify seasonal influences and relate them to the dependent variable.

Using a Dummy Variable

We illustrate the dummy variable technique in Figure 11-4 for data relating composition costs to total number of words for a hypothetical sample of textbooks. The books have been printed during one year by a firm under contract to various publishers. The data points for technical books are shown as crosses; the dots represent nontechnical texts. As we have noted, the technical books tend to be more costly, so that the crosses cluster higher than the dots. Two parallel lines have been constructed for the scatter diagram. The top one, fitting the data for the technical books, represents an upward shift in composition costs due to extensive hand typesetting.

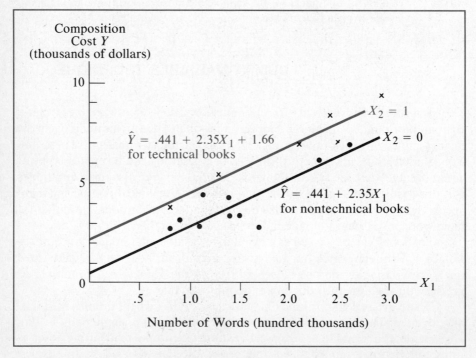

FIGURE 11-4
Results from multiple regression with dummy variable, showing projected lines.

These lines have been obtained by means of a linear least-squares *multiple* regression with two independent variables, word length X_1 and type of book X_2. X_2 is a *dummy variable* that takes on only two values, 1 or 0, depending on whether or not a book is technical. We may envision X_2 as a "switching" variable that is "on" when an observation is made of a technical book and "off" for an ordinary text. The estimated multiple regression equation is of the form

$$\hat{Y} = a + b_1X_1 + b_2X_2$$

where $X_2 = 0$ for nontechnical books and $X_2 = 1$ for technical books. This is the equation for a plane in three-dimensional space. However, we have restricted X_2 in such a way that only two parallel slices through the plane at $X_2 = 0$ and $X_2 = 1$ are possible. These are projected onto the X_1Y plane as the two estimated regression lines in Figure 11-4, the equations for which are

$$\hat{Y} = a + b_1X_1 + b_2 \quad \text{for technical books}$$
$$\hat{Y} = a + b_1X_1 \qquad\quad \text{for nontechnical books}$$

The value of a is the Y intercept for nontechnical books. The value for b_1 is the incremental or variable cost per additional word for either type of book. The amount b_2 represents the additional estimated cost—here assumed to be fixed in nature—associated with setting the type for a technical book. Thus the Y intercept for technical books is a total fixed cost estimated to be $a + b_2$.*

Table 11-4 shows the data for this example. Solving the corresponding normal equations, the following regression constants are obtained:

$$a = .441 \qquad b_1 = 2.35 \qquad b_2 = 1.66$$

and the estimated regression equation is

$$\hat{Y} = .441 + 2.35X_1 + 1.66X_2$$

For the two values of X_2, the multiple regression equation provides

$$\hat{Y} = .441 + 2.35X_1 \quad \text{for nontechnical books}$$

and

$$\hat{Y} = .441 + 2.35X_1 + 1.66$$

or

$$\hat{Y} = 2.101 + 2.35X_1 \quad \text{for technical books}$$

* A more general model would allow for different variable costs for technical books, which might be more realistic and would allow lines of different slope for the two types of texts.

TABLE 11-4

Intermediate Calculations for Multiple Regression with Dummy Variable
in Textbook Illustration

(Cost Y in thousands of dollars; length X_1 in hundreds of thousands of words)

Cost Y	Length X_1	X_2	X_1Y	X_2Y	X_1X_2	X_1^2	X_2^2	Y^2
4.2	1.4	0	5.88	0	0	1.96	0	17.64
2.8	1.1	0	3.08	0	0	1.21	0	7.84
3.3	1.4	0	4.62	0	0	1.96	0	10.89
3.9	.8	1(T)	3.12	3.9	.8	.64	1	15.21
6.7	2.6	0	17.42	0	0	6.76	0	44.89
7.0	2.5	1(T)	17.50	7.0	2.5	6.25	1	49.00
9.3	2.9	1(T)	26.97	9.3	2.9	8.41	1	86.49
2.7	.8	0	2.15	0	0	.64	0	7.29
6.9	2.1	1(T)	14.49	6.9	2.1	4.41	1	47.61
2.8	1.2	0	3.36	0	0	1.44	0	7.84
5.4	1.3	1(T)	7.02	5.4	1.3	1.69	1	29.16
6.1	2.3	0	14.03	0	0	5.29	0	37.21
8.3	2.4	1(T)	19.92	8.3	2.4	5.76	1	68.89
3.3	1.5	0	4.95	0	0	2.25	0	10.89
3.1	.9	0	2.79	0	0	.81	0	9.61
75.8	25.2	6	147.30	40.8	12.0	49.48	6	450.46
$= \sum Y$	$= \sum X_1$	$= \sum X_2$	$= \sum X_1Y$	$= \sum X_2Y$	$= \sum X_1X_2$	$= \sum X_1^2$	$= \sum X_2^2$	$= \sum Y^2$

$$\bar{Y} = 75.8 \qquad \bar{X}_1 = 1.680 \qquad \bar{X}_2 = 6$$

We may interpret these lines as follows. The value $a = .441$ represents an estimated setup or fixed cost of \$441 applicable, on the average, to every book set in type. The partial regression coefficient value $b_1 = 2.35$, which is also the slope of each of the lines for the two kinds of books, is the estimated average variable cost of \$2,350 for each additional 100,000 words in length, or 2.35 cents per additional word. This applies to both technical and nontechnical books. The other partial regression coefficient, $b_2 = 1.66$, applies only to technical texts and indicates the estimated additional average fixed cost of handsetting special characters, \$1,660 per book, so that these books involve an average fixed cost estimated to be \$441 + \$1,660 = \$2,101 each.

The importance of treating "kind of book" as a separate variable is that this allows us to properly identify the relationship between composition cost and book length. Using the dummy variable X_2, we can treat the two kinds of books

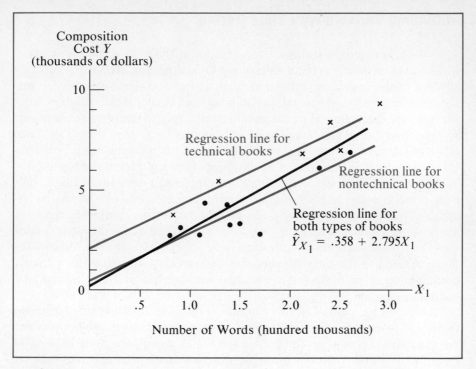

FIGURE 11-5

Graphical comparison of linear least-squares regression and multiple regression with dummy variable.

separately and yet still glean the common information that both types yield in relating book length to cost. We certainly should not ignore the kind of book in determining the relationship between these primary variables. Doing so can lead to very erroneous conclusions, as is shown in Figure 11-5, where a simple linear regression line relates Y to X_1. This line does not fit the data nearly as well as the two lines obtained from multiple regression, and the amount of scatter about this single line is substantially greater than that achieved for the parallel lines obtained from multiple regression analysis using a dummy variable.

But why can't we do just as well by making two separate linear regression analyses, splitting the observations into separate samples for technical and non-technical books? If the sample sizes are large, this would be more desirable than multiple regression with dummy variables, because then we could have lines of different slopes reflecting the possibility that technical books may also have higher (or lower) variable costs. However, in this illustration the number of technical books observed was small, so that there may be considerable error in doing a separate regression analysis for each type of book.

Using a Dummy Variable with Time Series

We have noted situations where regression analysis using dummy variables with time-series data may be advantageous. One of these involves isolation of the effect of nonhomogeneous influences, such as war and peace or recession and economic expansion, on the relationship between two or more variables. Such variables are characterized by the uncontrollability of the sampling experiment, so that observations can be collected only as they occur over time. The statistician or economist cannot make a large number of "peacetime-only" observations in order to determine a relationship between personal consumption expenditures and disposable personal income, for the data are available only with peacetime and war years intermingled. Separate analyses would be less reliable, because of the scarcity of observations, than a multiple regression analysis treating war as a dummy variable. The procedure for the latter is completely analogous to the one illustrated for the printing of textbooks. A regression plane is fitted to the three-dimensional data array, with values for personal consumption expenditure and disposable personal income variables, and with the dummy variable taking the value 0 for peacetime and 1 for war years.

Figure 11-6 shows the two projected regression lines obtained from the regression plane using disposable personal income and war to predict personal consumption expenditures for the United States, using data from 1935–1949. Because of World War II's dramatic disruptions of personal lives and the economy, there was a significantly different pattern of consumer spending during the 1942–1945 period. This was due to a host of factors, such as rationing and the appropriation of production facilities for military materiel.

The equation for the regression plane is

$$C = 1.49 + .915Y_D - 22.66W$$

where C represents personal consumption expenditures, Y_D is the traditional economists' designation for disposable income (taken here to be an independent variable), and W is a dummy variable ($W = 0$ for peacetime and $W = 1$ for wartime). When $W = 0$, the regression plane becomes the line

$$C = 1.49 + .915Y_D \quad \text{(in peacetime)}$$

which may be interpreted as an estimate of the *aggregate consumption function*. The slope of this line, .915, is the personal consumption expenditure for each dollar increase in disposable income. For the economy as a whole, 91.5 cents from each extra dollar of personal income is estimated to be spent on consumption. In the terminology of economics, .915 is the *marginal propensity to consume*. When $W = 1$, then with $b_2 = -22.66$ the regression line (the consumption function) shifts downward by an amount of $22.66 billion, reflecting a general

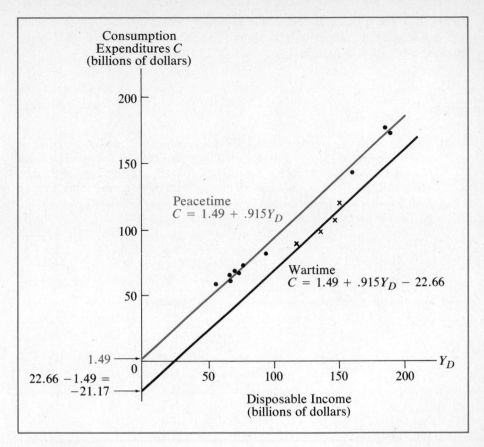

FIGURE 11-6
Illustration of multiple regression with dummy variable: consumption function during wartime and peacetime.

drop in personal expenditures due to the shortage of many goods formerly purchased by individuals. Notice that this shifted regression line has the same slope, reflecting the fact that the marginal propensity to consume is the same as before.*

If the time-series data were expanded to include the 1950s and 1960s, the effects of two other wars, in Korea and Vietnam, would be much less pronounced. These smaller wars were not very disruptive to consumer spending habits, since there were no major shortages or rationing. For this reason, a

---

* As we have noted, this may not be realistic, but it is a necessary consequence of the linear multiple regression model.

better fit to the time series could be obtained by treating these later war years like peacetime (so that perhaps the dummy variable, wartime *rationing*, would then be a better designation).

EXERCISES

11-11 Six women and four men have taken a test that measures their manual dexterity and patience in handling tiny objects. Each has then gone through a week of intensive training as electronics assemblers, followed by a month at actual assembly, during which their productivity was measured by a relative index having values ranging from 0 to 10 (with 10 for the most productive worker). The results obtained are provided in the table below:

Subject	Productivity Index Y	Test Score X_1	Sex
A	5.2	5.8	F
B	6.0	8.5	M
C	6.5	8.2	F
D	2.0	3.5	F
E	2.7	6.5	M
F	10.0	9.5	F
G	6.4	9.8	M
H	6.6	9.2	M
I	3.5	4.0	F
J	4.0	5.5	F

(a) Plot the data on a scatter diagram, using dots for women and crosses for men.

(b) Using a dummy variable having value $X_2 = 1$ for women and $X_2 = 0$ for men, determine the coefficients for the equation $\hat{Y} = a + b_1 X_1 + b_2 X_2$ of the estimated regression plane. Draw the lines corresponding to $X_2 = 0$ and $X_2 = 1$ on your scatter diagram.

(c) State in words the meaning of the partial regression coefficients.

(d) Determine the estimated regression line obtained when the sex of the subjects is ignored, and plot it on your scatter diagram.

11-12 The data in the following table provide personal savings and personal income (in billions of dollars) for the time period 1935–1949.

(a) Plot the data as a scatter diagram showing peacetime years with dots and wartime years with crosses.

(b) Using as values for the dummy variable $X_2 = 0$ for peacetime and $X_2 = 1$ for wartime, determine the estimated regression equation $\hat{Y} = a + b_1 X_1 + b_2 X_2$.

(c) Plot the two lines obtained from this equation corresponding to wartime and peacetime on your scatter diagram.

Year	Personal Savings Y	Personal Income X_1	War Year
1935	2	60	
1936	4	69	
1937	4	74	
1938	1	68	
1939	3	73	
1940	4	78	
1941	11	96	
1942	28	123	yes
1943	33	151	yes
1944	37	165	yes
1945	30	171	yes
1946	15	179	
1947	7	191	
1948	13	210	
1949	9	207	

SOURCE: *Economic Report of the President,* February 1970.

11-13 Referring to Exercise 11-11, and your answers to that exercise, answer the following:
  (a) Compute the average test scores separately for the men and the women. Then compute the average productivity index for men and for women.
  (b) Based on these averages, what conclusion can you draw with respect to the average test-score results for the sample men as compared to those for the women? Can the same conclusion be drawn for the productivity index?
  (c) How do you reconcile your conclusion in part (b) with the interpretation of the partial regression coefficient b_2?

11-14 A statistician used the following data to illustrate the cows–acres paradox:

Farm	Number of Cows Y	Number of Acres X_1	Primary Product
1	280	650	beef
2	250	400	beef
3	350	200	milk
4	370	350	milk
5	200	480	beef
6	420	340	milk
7	300	600	beef
8	320	130	milk
9	320	280	milk
10	250	450	beef

   (a) Plot the above data on a scatter diagram, using dots for data from farms whose main product is beef and crosses for those producing primarily milk.

   (b) Find the estimated regression equation expressing number of cows Y in terms of the single independent variable, number of acres X_1. Then plot this line on your scatter diagram. Your line should indicate that Y (number of cows) decreases as X_1 (number of acres) increases. How might you explain this?

   (c) Let X_2 represent a dummy variable with level $X_2 = 0$ for beef as the primary product and $X_2 = 1$ for milk. Find the estimated regression equation $\hat{Y} = a + b_1 X_1 + b_2 X_2$. Then plot the two corresponding lines on your graph for the respective farm types. (In each case, a positive slope indicates that the predicted number of cows should be greater for larger acreages.)

11-5   MULTIPLE CORRELATION

Sample Coefficient of Multiple Determination

The concept of correlation can be extended to multiple variables. As we did in our earlier discussion of simple linear relationships, we will begin by describing the *sample coefficient of multiple determination* as an index of association. When a multiple regression involves only two independent variables, X_1 and X_2, this is denoted symbolically by $R_{Y \cdot 1 2}^2$, which represents the ratio of the variation in Y that is explained by the regression plane to the total variation, or

$$R_{Y \cdot 1 2}^2 = \frac{\text{Explained variation}}{\text{Total variation}} = 1 - \frac{\sum(Y - \hat{Y})^2}{\sum(Y - \bar{Y})^2}$$

The *sample multiple correlation coefficient* is defined as the square root of the coefficient of multiple determination, or $R_{Y \cdot 1 2} = \sqrt{R_{Y \cdot 1 2}^2}$. The sign of $R_{Y \cdot 1 2}$ is always considered positive.

A shorter, equivalent expression is used to calculate the

SAMPLE COEFFICIENT OF MULTIPLE DETERMINATION

$$R_{Y \cdot 1 2}^2 = 1 - \frac{S_{Y \cdot 1 2}^2}{s_Y^2}\left(\frac{n - m}{n - 1}\right)$$

As before, m represents the number of variables used. Here, $m = 3$.

To illustrate this computation, we will again use our earlier results from the multiple regression for store profit Y when food sales X_1 and nonfood sales

X_2 are the only independent variables. The standard deviation in sample profit is

$$s_Y = \sqrt{\frac{\sum Y^2 - n\bar{Y}^2}{n-1}} = \sqrt{\frac{4{,}267 - 10(19.1)^2}{9}} = 8.293$$

From before we have $s_{Y \cdot 1\,2} = 4.343$. Using these values we calculate

$$R^2_{Y \cdot 1\,2} = 1 - \frac{(4.343)^2}{(8.293)^2}\left(\frac{10-3}{10-1}\right) = .787$$

The interpretation of $R^2_{Y \cdot 1\,2} = .787$ is that an estimated 78.7% of the total variation in Y may be explained by knowledge of the regression plane and the values for X_1 and X_2. Only $100 - 78.7 = 21.3\%$ of the variation in Y is estimated due to other causes, such as chance, plus factors not explicitly considered—for example, store size, location, and advertising.

When the supermarket data were expanded to include an additional independent variable for store size, X_3, the estimated regression equation was recalculated to be $\hat{Y} = -10.170 + .027X_1 + .097X_2 + .525X_3$. As noted earlier, the standard error of the estimate for Y about this higher-dimensional regression hyperplane was significantly smaller than when store size was ignored. This indicates that, in this case, more of the total variation in Y may be explained by raising the number of regression variables. To see how far the total variation in Y that is explained by regression has risen, we obtain the coefficient of multiple determination when X_3 is included in the analysis. The computer printout in Figure 11-3 provides the value $R_{Y \cdot 1\,2\,3} = .9924$. Squaring this, we have

$$R^2_{Y \cdot 1\,2\,3} = \frac{\text{Explained variation}}{\text{Total variation}} = (.9924)^2 = .985$$

Thus the total variation in Y that is explained by regression increases from 78.7% to 98.5% when store size X_3 is included as an additional independent variable. This further illustrates that expanding regression analysis to include more variables may increase the reliability of predictions.

We now consider a set of indexes helpful in determining whether or not it is worthwhile to include additional variables in the regression analysis.

Partial Correlation

Let's begin by comparing the results of a simple regression analysis relating store profit Y to food sales X_1 alone. As we have seen, the equation for the estimated regression line is

$$\hat{Y}_{X_1} = 3.981 + .0670X_1$$

and the simple coefficient of determination is $r_{Y1}^2 = (.760)^2 = .578$. Suppose we now incorporate nonfood sales X_2 as a second independent variable. We have seen that better predictions for Y will result from doing this. To quantify the improvement achieved with a higher-dimensional regression analysis, we can calculate the proportional reduction in previously explained variation. This provides a further index of association, $r_{Y2 \cdot 1}^2$, called the *coefficient of partial determination*, that measures the correlation between Y and X_2 when the other independent variable X_1 is still considered but held constant. The subscript $Y2 \cdot 1$ indicates this.

COEFFICIENT OF PARTIAL DETERMINATION

$$r_{Y2 \cdot 1}^2 = \frac{\text{Reduction in unexplained variation}}{\text{Previously unexplained variation}} = \frac{R_{Y \cdot 12}^2 - r_{Y1}^2}{1 - r_{Y1}^2}$$

Substituting our earlier values, we obtain

$$r_{Y2 \cdot 1}^2 = \frac{.787 - .578}{1 - .578} = .495$$

The difference, $R_{Y \cdot 12}^2 - r_{Y1}^2$, represents the reduction in unexplained variation (which is also the increase in explained variation). The value $r_{Y2 \cdot 1}^2 = .495$ tells us that 49.5% of the variation in store profit Y that was left unexplained by a simple regression with food sales X_1 alone can be explained by the regression obtained from including nonfood sales X_2.

In multiple regression analysis, the coefficient of partial determination or its square root, the *partial correlation coefficient*, provides an index of the correlation between two variables *after* the effects of all the other variables have been considered. Thus, the value .495 expresses the net association between store profit Y and nonfood sales X_2 when food sales X_1 are held constant but are still accounted for.

The value $r_{Y2 \cdot 1}^2 = .495$ may be compared to the *simple* coefficient of determination between Y and X_2, $r_{Y2}^2 = (-.286)^2 = .082$. The former provides a truer impact of including X_2 in the multiple regression analysis than would be indicated by a simple correlation analysis using only X_2. When evaluating the potential merits of a predictor variable, the coefficients of partial determination (or the partial correlation coefficients) are sometimes employed in determining which independent variables to include (a superior method to using the simple counterpart coefficient). We can illustrate how the coefficient of partial determination may be used to do this.

In discussing a third independent variable X_3 (size of store), a multiple regression was run on the computer. The coefficient of partial determination may be found from

$$r^2_{Y3 \cdot 1\,2} = \frac{\text{Reduction in unexplained variation}}{\text{Previously unexplained variation}} = \frac{R^2_{Y \cdot 1\,2\,3} - R^2_{Y \cdot 1\,2}}{1 - R^2_{Y \cdot 1\,2}}$$

Substituting the coefficients of multiple determination obtained from the previous two multiple regressions using the supermarket data, we have

$$r^2_{Y3 \cdot 1\,2} = \frac{.985 - .787}{1 - .787} = .930$$

Thus, $r^2_{Y3 \cdot 1\,2} = .930$ tells us that there is a 93.0% reduction in previously unexplained variation by including X_3. This additional variable has been quite effective in increasing the sharpness of the multiple regression analysis.

Table 11-5 shows how the amount of variation in Y that is explained by regression has increased through successive inclusions of additional variables. Note how the variation in Y is reduced from $s_Y = 8.293$, which was obtained when no regression was performed. Variation is expressed by the successive standard errors of the estimates of Y about the respective regression surfaces (line, plane, and hyperplane) shown in column (3). Paralleling the declining standard errors are increases in the proportion of variation explained by

TABLE 11-5
Summary of How Successively Higher-Dimensional Regression Equations Increase the Explained Variation in Y When Additional Variables Are Included in the Supermarket Illustration

(1) Independent Variables Included	(2) Additional Variable	(3) Standard Error of Estimate (thousands of dollars)	(4) Proportion of Variation Explained	(5) Proportional Reduction in Previously Unexplained Variation
none	—	$s_Y = 8.293$	0.000	—
X_1	X_1	$S_{Y \cdot X_1} = 5.719$	$r^2_{Y1} = .578$*	$\dfrac{.578 - 0}{1 - 0} = .578 = r^2_{Y\,1}$
X_1, X_2	X_2	$S_{Y \cdot 1\,2} = 4.343$	$R^2_{Y \cdot 1\,2} = .787$	$\dfrac{.787 - .578}{1 - .578} = .495 = r^2_{Y2 \cdot 1}$
X_1, X_2, X_3	X_3	$S_{Y \cdot 1\,2\,3} = 1.250$	$R^2_{Y \cdot 1\,2\,3} = .985$	$\dfrac{.985 - .787}{1 - .787} = .930 = r^2_{Y\,3 \cdot 1\,2}$

* This value is the simple coefficient of determination calculated for the regression line $\hat{Y}_{X_1} = 3.891 + .0670X_1$.

regression, shown in column (4). Column (5) shows the proportional reduction in previously unexplained variation in Y resulting from the additional variable.

The total explained variations in Y will not always be reduced by including additional variables. It is not always obvious that an additional factor will sharpen rather than cloud the predictive powers of regression analysis. Good common sense must be exercised in selecting variables for which there is a meaningful nonstatistical explanation of their influence on the dependent variable. Sales is one factor used to calculate profit, so it has a logical connection. Likewise, a larger store requires a greater investment in facilities and may be operated more efficiently than a smaller one, so store size can also be reasoned to have a direct relation to profits.

EXERCISES

11-15 The following intermediate calculations, made by an economist for a sample of $n = 100$, relate annual family spending Y to income X_1, size X_2, and annual savings X_3. All monetary figures are in thousands of dollars.

$$\bar{Y} = 10 \qquad \bar{X}_1 = 12 \qquad \bar{X}_2 = 5 \qquad \bar{X}_3 = 1$$

$$\sum Y^2 = 11,400 \qquad \sum X_1 Y = 13,000 \qquad \sum X_2 Y = 6,000 \qquad \sum X_3 Y = 500$$

$$\text{Plane } A: \quad \hat{Y} = 1.7 + .4X_1 + .7X_2 \qquad\qquad (\text{excluding } X_3)$$

$$\text{Plane } B: \quad \hat{Y} = 4.3 + .3X_1 + .6X_2 - .9X_3 \quad (\text{including } X_3)$$

(a) Calculate $S_{Y.12}$ and $R^2_{Y.12}$. What percentage of the variation in Y can be explained by regression plane A?

(b) If $S_{Y.123} = .722$ and $R^2_{Y.123} = .964$, what percentage of the variation in Y can be explained by regression hyperplane B?

(c) Has the inclusion of X_3 in the regression analysis reduced the standard error of the estimate of Y?

(d) Use the regression equations for planes A and B to determine the proportions of unexplained variation in Y. What proportional change in previously unexplained variation is achieved by adding X_3 to the analysis? Does this increase or decrease the unexplained variation?

11-16 From the data in Exercise 11-13, the following results have been obtained for predicting Crunchola production cost Y from quantity X_1 and wheat price index X_2.

$$r^2_{Y1} = .9245 \qquad r^2_{Y2} = .0009 \qquad R^2_{Y.12} = .9596$$

(a) Compute the coefficient of partial determination $r^2_{Y2.1}$ for production cost and wheat price index, holding quantity constant.

(b) Compute the coefficient of partial determination $r^2_{Y1.2}$ for production cost and quantity, holding wheat price index constant.

11-17 From the data in Exercise 11-19, the following regression results have been obtained for predicting gasoline mileage Y from average speed X_1, altitude X_2, and passenger weight X_3.

$$s_Y = 3.742$$

$$s_{Y \cdot X_1} = 3.968 \qquad r_{Y1}^2 = .0003$$

$$s_{Y \cdot X_2} = 2.426 \qquad r_{Y2}^2 = .6262$$

$$s_{Y \cdot 1\,2} = 2.523 \qquad R_{Y \cdot 1\,2}^2 = .6466$$

$$s_{Y \cdot 1\,2\,3} = 1.794 \qquad R_{Y \cdot 1\,2\,3}^2 = .8468$$

Calculate the following coefficients of partial determination:
(a) $r_{Y2 \cdot 1}^2$ (b) $r_{Y1 \cdot 2}^2$ (c) $r_{Y3 \cdot 1\,2}^2$
(d) What is the percentage reduction in unexplained Y variation achieved by adding X_3 as the third independent variable? Does this suggest that including passenger weight should improve mileage predictions? Explain.

11-18 Suppose that the economist in Exercise 11-15 has determined *for another sample* that the simple coefficient of determination is .50 between Y and X_1 and that it is .60 between Y and X_2. She has also found that the coefficient of multiple determination is .75, considering only X_1 and X_2.
(a) Calculate the coefficients of partial determination $r_{Y1 \cdot 2}^2$ and $r_{Y2 \cdot 1}^2$.
(b) State in words the meaning of the values that you obtained in (a).
(c) Compare the values in (a) to their respective simple coefficients of determination. Why may the corresponding values differ?

11-19 A wholesale distributor of business forms has developed three predictors of sales Y (in thousands of dollars) achieved by its representatives in the field. For each salesperson, X_1 represents the score on a sales aptitude test, X_2 the score on a motivation test, and X_3 the total years of sales experience. The following intermediate calculations have been obtained for sample data representing $n = 30$ salespersons.

$$\bar{Y} = 50 \qquad \bar{X}_1 = 10 \qquad \bar{X}_2 = 5 \qquad \bar{X}_3 = 5$$

$$\sum Y^2 = 80,600 \qquad \sum X_1 Y = 16,000 \qquad \sum X_2 Y = 8,500 \qquad \sum X_3 Y = 8,100$$

$$\text{Plane } A: \quad \hat{Y} = 10 + 3X_1 + 2X_2 \qquad (\text{excluding } X_3)$$

$$\text{Plane } B: \quad \hat{Y} = 5 + 2X_1 + 1X_2 + 4X_3 \quad (\text{including } X_3)$$

(a) Calculate $s_{Y \cdot 1\,2}$ and $R_{Y \cdot 1\,2}^2$. What percentage of the variation in Y may be explained by the regression plane A?
(b) $R_{Y \cdot 1\,2\,3}^2 = .964$. What percentage of the variation in Y may be explained by the regression hyperplane B?
(c) Calculate the coefficient of partial determination for adding X_3 to the analysis. What is the percentage reduction in previously unexplained variation from adding X_3 to the analysis?
(d) Do you believe that years of sales experience ought to be included in predictions of sales performance? Why?

REVIEW EXERCISES

11-20 A college admissions director wishes to predict the college GPA of applicants using high-school GPA and verbal SAT score as independent variables. The following data apply:

Student	College GPA Y	High-School GPA X_1	Verbal SAT Score X_2
1	3.8	4.0	750
2	2.7	3.7	380
3	2.3	2.2	580
4	3.2	3.8	510
5	3.5	3.8	620
6	2.4	2.8	440
7	2.6	3.0	540
8	3.0	3.4	650
9	2.7	3.3	480
10	2.8	3.0	550

(a) Determine the estimated regression equation for the given data.

(b) Using your anwer to (a), compute the predicted college GPA for the following applicants:

	High-School GPA	SAT
(1)	3.4	550
(2)	2.8	400
(3)	3.2	650
(4)	3.7	450

11-21 From the data in Exercise 11-20, the estimated regression equation for predicting college GPA Y using only high-school GPA X_1 is

$$\hat{Y}_{X_1} = .4957 + .7286X_1$$

The values for the standard error of the estimate and coefficient of determination are

$$s_{Y \cdot X_1} = .265 \qquad r_{Y1}^2 = .723$$

with the sample standard deviation in Y having value $s_Y = .478$. The standard error of the estimate when verbal SAT score X_2 is included as an independent variable is

$$s_{Y \cdot 1 2} = .111$$

(a) Compute the sample coefficient of multiple determination when both X_1 and X_2 are used to predict Y.

(b) Compute the coefficient of partial determination ($r^2_{Y2 \cdot 1}$) for adding SAT score X_2 to the regression analysis. What is the percentage reduction in the unexplained variation in college GPA that can be attributed to including SAT score as the second independent variable? Does this suggest that SAT results might be useful in predicting college success? Explain.

11-22 *Computer exercise.* An additional independent variable, number of extracurricular activities X_3, is added to the college GPA prediction problem in Exercise 11-20. The following data apply:

Student	Number of Activities
1	7
2	6
3	5
4	6
5	3
6	2
7	8
8	2
9	7
10	4

Determine the coefficients for the estimated regression hyperplane

$$\hat{Y} = a + b_1 X_1 + b_2 X_2 + b_3 X_3$$

11-23 *(Continuation of Exercise 11-22)*:

(a) Determine the value of $S_{Y \cdot 1 2 3}$.

(b) Compared to the value $S_{Y \cdot 1 2} = .111$, does your answer suggest that including the number of high-school activities improves the accuracy of prediction of college GPA? Explain.

(c) Compute the coefficient of partial determination ($r^2_{Y3 \cdot 1 2}$) for including X_3 in the regression analysis. What is the percentage reduction in previously unexplained variation in college GPA that can be attributed to including number of activities as the third independent variable?

12

Time-Series Analysis and Forecasting

E very successful business and organization must plan for the future. Quite a dazzling array of statistical tools are available to facilitate this task. In this chapter we focus on statistical methodology that can transform past experience into forecasts of future events. Thus, a government economist can forecast annual personal income five years into the future by projecting from the trend indicated by the levels of personal income in previous years, and use this information to predict future tax revenue. An electric company can decide that the demand for power will grow at a rate similar to that which has prevailed during the previous decade in order to project its generating capacity requirements five, ten, or twenty years into the future. A department store buyer can use past experience to decide when to make purchases and in what quantities. In each case values of the variable being predicted are available for several past periods of time. Such data are called *time series*. Statistical procedures that use such values are called *time-series analysis*.

THE TIME SERIES AND ITS COMPONENTS 12-1

Time series are best described in terms of a graph like the one shown in Figure 12-1, in which the gross national product (GNP) of the United States is plotted against time for the period 1929–1975. The graph shows that the GNP

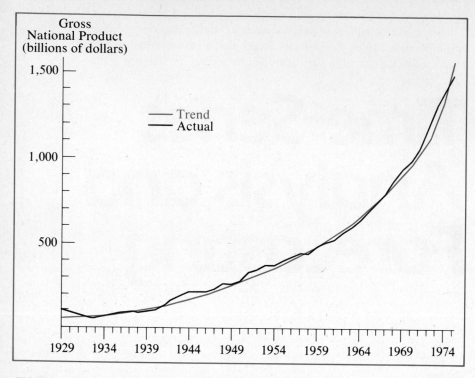

FIGURE 12-1

U.S. gross national product. (SOURCE: *Economic Report of the President,* 1970, 1976.)

has grown over the years, but that this growth has been erratic, being faster in some years than in others. Wide swings are evident, the GNP declining during the Great Depression of the 1930s and then rising rapidly with the advent of World War II. One goal of time-series analysis is to identify the swings and fluctuations of a time series and sort them into various categories. This is done through arithmetic manipulation of the numerical values obtained.

Several models may be used to characterize time series. The classical model used by economists provides the clearest explanation of the following four time-series components of variation and their relation to each other:

1. Secular trend (T_t) **3.** Seasonal fluctuation (S_t)
2. Cyclical movement (C_t) **4.** Irregular variations (I_t)

These components can be related to the forecast variable by mathematical equations. The forecast variable is denoted by the symbol Y_t, where the subscript t refers to a period of time. Examples of Y_t are annual sales, passenger miles flown by domestic airlines, and acre-feet of water supplied to a city.

Secular trend is defined as the long-range general movement in Y_t over an extended period of time. In this chapter, we will develop methods for isolating trend from other variational components in a time series. *Cyclical movement* in time-series data is characterized by wide swings—usually a year or more in duration—upward or downward from the secular trend. Cyclical movements are temporary in nature and are typified by alternating periods of economic expansion and contraction or recession. *Seasonal fluctuation* is a generally recurring upward and downward pattern of movement in Y_t, usually on an annual basis. A classic example of seasonal fluctuation is the household consumption of fossil fuels, such as oil, coal, or natural gas. *Irregular variations* are characterized by events that are completely unpredictable. These variations, sometimes referred to as *random factors*, can be the most perplexing.

THE CLASSICAL TIME-SERIES MODEL 12-2

The classical time-series model originally used by economists combines the four components of time-series variation in the equation

$$Y_t = T_t \times C_t \times S_t \times I_t$$

This equation states that factors associated with each of these components can be multiplied together to provide the value of the dependent variable.

This model can be explained by means of a hypothetical time series—the sales Y_t of stereo speakers by the Speak E-Z Company. To construct this time series, we will begin with the trend component T_t, shown in Figure 12-2 as a straight line relating sales to time period. Each successive quarter raises the

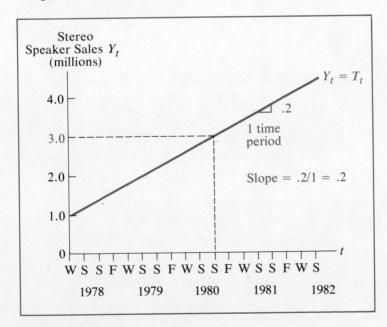

FIGURE 12-2
Trend line T_t for Speak E-Z sales.

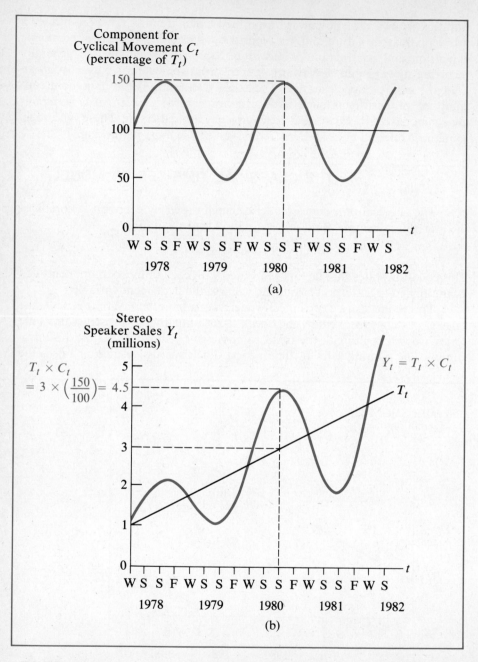

FIGURE 12-3

(a) Cyclical movement component of Speak E-Z sales, and (b) time series showing only trend and cyclical components.

level of sales by .2 million. Thus, initial sales of 1 million units have grown by the summer of 1980 (10 periods later) to $1 + .2(10) = 3$ million units.

Now let's consider the influence on the series of cyclical movement that produces sales temporarily above the trend in good years or below it in lean years. For convenience, we express the cyclical effect as a percentage of trend, as shown in Figure 12-3(a). The average value of C_t throughout an entire cycle is considered to be 100%. In the summer of 1980, the value of C_t is 150%, indicating that sales are 50% above the trend during that quarter. The time series for speaker sales with this added component is graphed in Figure 12-3(b), where the curve for C_t has been superimposed on the trend line.

We can follow the same procedure in dealing with seasonal fluctuations in stereo speaker sales, which can be viewed as short-term swings about the longer-term sales level indicated by the combined trend and cyclical components. Sales will lie above this level during the busy season and below it during slack times. Thus, we can consider S_t to be a proportion of the long-term sales level established by T_t and C_t, also expressed as a percentage.

The values of S_t, referred to as *seasonal indexes*, are shown in Figure 12-4. For the summer of 1980, the seasonal index is $S_t = 60\%$, indicating that sales

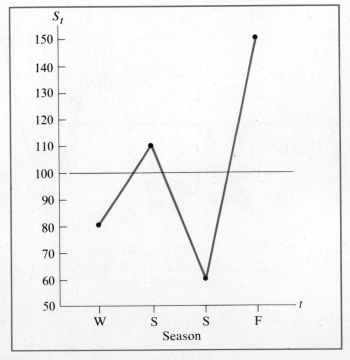

FIGURE 12-4

Seasonal fluctuations of Speak E-Z sales. S_t = seasonal index of fluctuation about the long-term sales level, expressed as a percentage of $T_t \times C_t$.

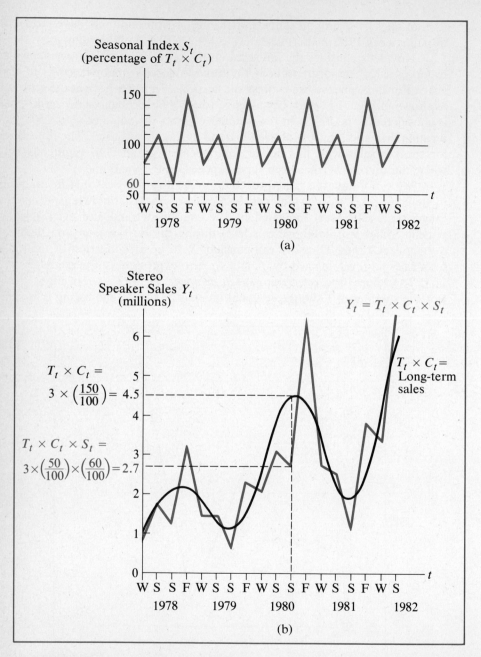

FIGURE 12-5

(a) Seasonal index component for Speak E-Z sales, and (b) time series showing only trend, cyclical, and seasonal components.

were only 60% of "normal" during this period. The time series obtained by superimposing the seasonal fluctuations on longer-term sales is shown in Figure 12-5.

We must still consider the irregular component of variation in speaker sales. As in the case of C_t and S_t, this variation is represented by a percentage I_t. Viewing all irregular movement as short term in nature, we can consider I_t to be the last factor—the one that raises or lowers sales from the level established by the regular pattern of systematic factors that we have already considered. The values used to construct the graph for I_t (shown in Figure 12-6) have been obtained randomly, so that the long-run average value of I_t is 100%.

For the summer of 1980, I_t is 97% of "normal." These irregular variations can be superimposed on the curve that we obtained previously for T_t, C_t, and S_t to provide a complete hypothetical time series for Speak E-Z's sales. The variational influence of each component is summarized in Figure 12-7, where the final time series is obtained.

This illustration has shown us how a hypothetical time series can be synthesized from the assumed characteristics of the four components. In actual applications, however, we may not know anything about T_t, C_t, S_t, or I_t. Usually, we begin with the raw time-series data and reverse the procedure, shifting the data to sort out and identify the components. We will discuss some examples of this technique in this chapter.

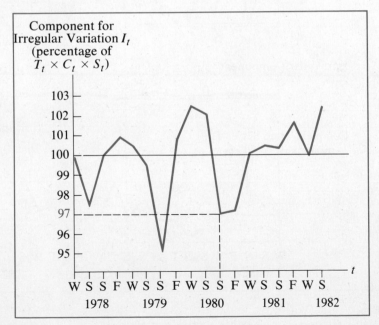

FIGURE 12-6

Component of irregular fluctuation in Speak E-Z sales.

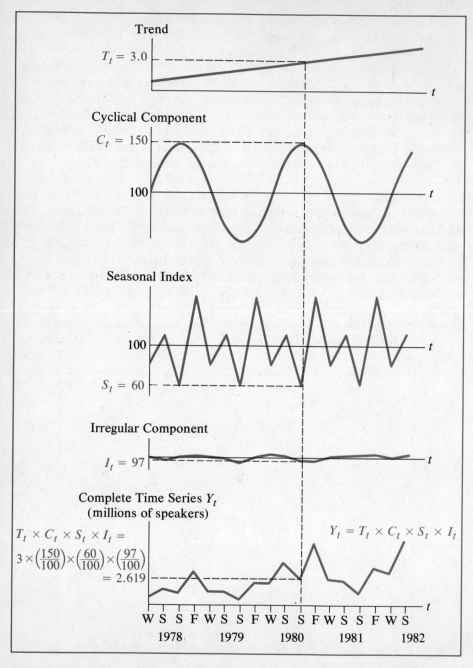

FIGURE 12-7

Construction of complete time series for Speak E-Z's sales, using individual components.

The classical time-series model has limitations. The multiplicative relationship between trend, cyclical, seasonal, and irregular components has been criticized as being oversimplified and unrealistic. A large number of possible relationships other than multiplicative can exist between T_t, C_t, S_t, and I_t, and some of these may be more appropriate. Other models use more factors; for example, Y_t is sometimes assumed to also depend on Y_{t-1}, the value from the preceding time period. As we will see, several different techniques are available for isolating the components themselves, so that for a specific time series the classical model yields no unique solution. In spite of these difficulties, the model is easy to understand and to explain and can provide a basis for useful time-series analysis.

More sophisticated time-series models have been developed. The field of *econometrics*, which employs specialized statistical tools to explain and predict economic activity, embodies regression techniques for analyzing time-series data. An especially powerful tool uses variables from several different time series to predict values of other variables of interest. To do this, variables called *leading indicators*, which have historically moved upward or downward ahead of the time series being analyzed, are identified and combined mathematically. The general level of economic activity in a future time period may be predicted by a relationship between such leading indicators as business inventories, housing starts, and new durable goods orders.

Other models explain time-series movement in terms of its past behavior only, without isolating seasonal and cyclical components. Later in the chapter we describe one technique for doing this, *exponential smoothing*, which involves the averaging of past data, giving greater weight to current values.

EXERCISES

12-1 Higbee Mills has projected that sales for 19X2 will grow by $100,000 per quarter. Assuming that sales during the fall quarter of 19X1 are $1,000,000, determine the trend sales levels for the winter, spring, summer, and fall quarters of 19X2.

12-2 Bixbee Mills had sales of $1,000,000 in the fall of 19X1. The trend indicates that sales are growing at a compound rate of 9% per quarter. Determine the trend levels for the winter, spring, summer, and fall quarters of 19X2.

12-3 The Variety Galore Store wishes to forecast its sales for the next calendar year. The following components have been determined by its accountant:

Quarter t	Trend T_t	Cyclical Component C_t	Seasonal Index S_t
Winter	$100,000	90	80
Spring	110,000	110	70
Summer	121,000	100	100
Fall	133,100	90	150

Determine the forecasts of sales Y_t for each quarter.

12-4 The sales data for Humpty Dumpty Toys, Inc., have been analyzed, and the trend, cyclical, seasonal, and irregular components have been determined for the preceding four quarters' operations. Find the missing values in the table below, assuming that the time-series components are multiplicative.

Quarter t	Trend T_t	Cyclical Component (percentage) C_t	Seasonal Index (percentage) S_t	Irregular Component (percentage) I_t	Sales Y_t
Winter	$1,000,000	107	50	101	
Spring	1,100,000	105	70		$ 820,000
Summer	1,200,000	105		98	987,840
Fall	1,300,000		200	97	2,622,880

12-3 ANALYSIS OF SECULAR TREND

The trend component of a time series is perhaps the most valuable one in making forecasts using time-series data. Trend analysis focuses on finding an appropriate trend line that summarizes the movement of the series over an extended period of time. As we have seen, a trend line often takes the form of a curve. In this section, several approaches for determining trend are investigated. All of these involve treating the historical data points like a scatter diagram, which we first encountered in conjunction with regression analysis in Chapter 10. The objective of trend analysis is similar to that of regression: to find a trend curve that summarizes the historical scatter of the dependent variable Y_t over time.

The regression techniques discussed in Chapter 10 are severely limited when applied to time-series data. Most of the assumptions, such as independence, normality, and constant variance, are usually not relevant to time-series data. Thus it is impossible to attach a measure of statistical confidence to a prediction made from a trend line. This should be evident because a forecast must deal with the future, using the past only as a guide. There are no assurances that future influences on the time series will operate like those in the past. Furthermore, a prediction using a trend curve is a projection beyond the range of observations (there are no future data points), so that forecasting with the times series is subject to the major difficulty of extrapolation usually avoided in regression analysis.

Describing Trend

In the hypothetical time series for stereo speaker demand that was discussed in Section 12-2, secular trend was represented by a straight line. A straight-line trend assumes that Y_t changes at a constant rate. An increasing series will thus have a positively sloping trend line, whereas a declining one will be represented by a line with a negative slope.

Most time series do not involve long-term behavior that changes constantly over time. In business or economic situations, a variable will usually increase or decline at a rate that itself changes from period to period. Some basic shapes for nonlinear trend curves frequently encountered are provided in Figure 12-8. For example, consider the movement in GNP levels over a prolonged period. The GNP for the United States has increased by more in the recent past than it did following World War II, so that the trend is represented best by a curve with increasing and positive slope as in diagram (a). In absolute

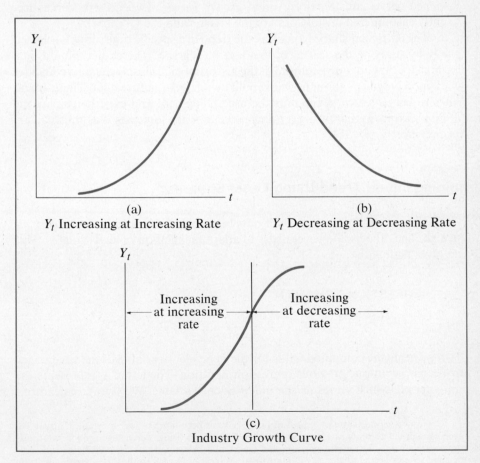

FIGURE 12-8
Shapes of commonly encountered trend curves.

terms, the American GNP has been increasing in recent times at an increasing rate.*

Diagram (b) shows the trend for a time series that decreases at a decreasing rate. The level of activity for a declining industry may sometimes be represented by such a curve. The decline is initially dramatic but becomes more gradual with time. An example of such a trend is the number of railroad passengers carried in the United States each year during the 1950s and 1960s.

The long-range growth of a firm or an industry may sometimes be explained in terms of a trend having the shape shown in diagram (c). Here, output increases at an increasing rate when innovative products are brought onto the market to satisfy emerging needs. As the industry matures, rapid growth is replaced by a period of gradual increases; product sales still increase, but at a decreasing rate. Ultimately, sales peak and a period of stagnation begins. Such S-shaped curves are therefore called *growth curves*. Two growth curves frequently encountered in statistics are the logistic and the Gompertz.

The particular shape to portray the trend of a specific time series is selected partly by studying the scatter of the data on a graph. The choice should be a basic shape that not only seems to fit the historical data closely but also coincides with good judgment about how it should be related to future data. There are no rules to tell us which shape must be used. Judgment and experience play the dominant role, so that fitting a trend curve to make forecasts is as much an art as a science.

Determining Linear Trend Using Least Squares

Time series covering a small number of years can usually be fitted by a straight line. The usual procedure is to adapt the method of least squares. This involves finding the

REGRESSION EQUATION

$$\hat{Y}_X = a + bX$$

$\hat{Y}_X$ represents the computed value for the dependent variable of the time series, in keeping with our previous regression notation. The letter X represents the time period, which serves as the independent variable. We use X instead of t

---

* The *percentage* rate of *growth* in real GNP has been more or less constant at about 3%, even though the absolute rate of increase in GNP has been rising. Percentage growth rate is expressed relative to the current position and works like compound interest paid by a bank. The original savings-account balance will increase at an increasing rate, so that the shape in diagram (a) applies over time, even though the percentage rate of interest or growth remains constant.

because the task of computing the regression coefficients may be considerably simplified by expressing time relative to some base period. For example, suppose that the time series comprises observations for the years 1978, 1979, 1980, and 1981. Rather than using the four-digit year numbers for X, we can instead let $X = 0$ represent 1978, $X = 1$ stand for 1979, which is one time period later than 1978, and so forth. Thus, the successive values used for X in these four years would be 0, 1, 2, and 3. The values of $\hat{Y}_X$ are used to obtain the corresponding trend values T_t.

The regression coefficients a and b are obtained by the method of least squares by solving the normal equations

$$\sum Y = na + b\sum X$$
$$\sum XY = a\sum X + b\sum X^2$$

The summations are carried out for successive periods of time. The letters X and Y represent the time periods observed and their associated value for the dependent variable. It may be simpler to obtain a and b from the following equations for the

ESTIMATED REGRESSION COEFFICIENTS

$$b = \frac{\sum XY - n\bar{X}\bar{Y}}{\sum X^2 - n\bar{X}^2}$$

$$a = \bar{Y} - b\bar{X}$$

We illustrate this procedure for the total level of civilian employment in the United States from 1966 through 1975. The time-series data are provided in the first two columns of Table 12-1. The results are $b = 1.47$ and $a = 72.95$, so that the trend equation is $\hat{Y}_X = 72.95 + 1.47X$, with $X = 0$ at 1966.

This equation indicates that the trend value for 1966 is an employment level of 72.95 million and that Y_t increases by 1.47 million per year. Because we have transformed the calendar years, it is important to indicate the base year: $X = 0$ at 1966. The trend line and time series obtained are plotted in Figure 12-9.

The trend line may be used to project the level of employment for 1986. We must use $X = 20$, because this year is $X = 1986 - 1966 = 20$ periods beyond the base year. From the trend equation, we project employment for 1986 as

$$\hat{Y}_X = 72.95 + 1.47(20)$$

$$= 102.35 \text{ (million)}$$

TABLE 12-1

Computations for Fitting Least-Squares Line to Employment Data

Year	Year in Transformed Units X	Total Civilian Employment (millions) Y	XY	X^2
1966	0	72.9	0	0
1967	1	74.4	74.4	1
1968	2	75.9	151.8	4
1969	3	77.9	233.7	9
1970	4	78.6	314.4	16
1971	5	79.1	395.5	25
1972	6	81.7	490.2	36
1973	7	84.4	590.8	49
1974	8	85.9	687.2	64
1975	9	84.8	763.2	81
	45	795.6	3,701.2	285
	$= \sum X$	$= \sum Y$	$= \sum XY$	$= \sum X^2$

$$n = 10 \qquad \bar{X} = 4.5 \qquad \bar{Y} = 79.56$$

$$b = \frac{\sum XY - n\bar{X}\bar{Y}}{\sum X^2 - n\bar{X}^2} = \frac{3,701.2 - 10(4.5)(79.56)}{285 - 10(4.5)^2} = 1.47$$

$$a = \bar{Y} - b\bar{X} = 79.56 - 1.47(4.5) = 72.95$$

$$\hat{Y}_X = 72.95 + 1.47X \qquad (X = 0 \text{ at } 1966)$$

SOURCE OF DATA: *Economic Report of the President, 1976.*

The projected trend line is shown in Figure 12-9 as the portion of the colored line extending beyond the time periods actually observed. It is emphasized that the estimate of 102.35 million is an *extrapolation*; its validity depends on the assumption that the ensuing ten years will exhibit growth similar to that existing in the past. The assumption of linearity is perhaps a poor one to use here, for the labor force grows in the same manner as the population, which for the United States has been nonlinear. Had more past years been used, a curve with slope increasing over time would have provided a closer fit to the actual time-series data. Even then, good judgment would be an essential element. Ordinarily many more than ten years are required to determine a basic pattern for trend.

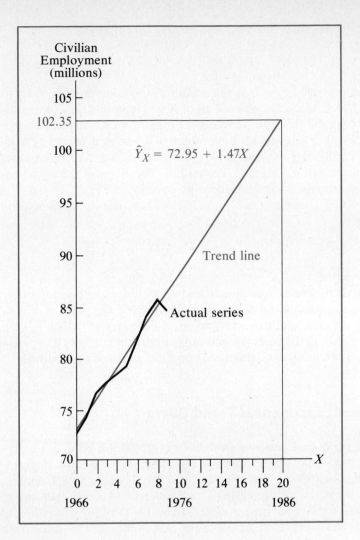

FIGURE 12-9
Actual data and trend line for civilian employment in the United States.

Modifying Trend for Periods Shorter Than One Year

Trend values are usually obtained from annual data to avoid distortions brought about by seasonal and irregular fluctuations. When time-series analysis is extended to consider the seasonal and cyclical components, it may be necessary to modify a trend equation to obtain monthly or quarterly values.

Consider, for example, the trend line for civilian employment:

$$\hat{Y}_X = 72.95 + 1.47X$$

with X in years and $X = 0$ at the *middle* of 1966. Suppose we wish to obtain monthly trend values. The new values of X will represent the number of months from the base and will be centered at the middle of each month. Thus, we want to have $X = 0$ at January 15, 1966, which is 5.5 months prior to the middle of 1966 (July 1, 1966). Each successive month will increase $\hat{Y}_X$ by one-twelfth the annual increment: $b/12 = 1.47/12 = .1225$. The slope of the new line will be .1225. The intercept a will be reduced by an amount $5.5(.1225) = .67$ (rounded), so that the new Y intercept is $72.95 - .67 = 72.28$. The equation of the modified trend line is

$$\hat{Y}_X = 72.28 + .1225X$$

with X in months and $X = 0$ at January 15, 1966.

The same general approach would apply to finding quarterly trend equations. The slope would be one-fourth as large, and the point $X = 0$ would be shifted to the middle of the winter quarter, February 15, which is 1.5 quarters from the middle of the year. When trend is itself fitted directly to deseasonalized quarterly or monthly data, the trend equation need not be modified.

Nonlinear Trend: Exponential Trend Curve

For many time series, a straight line provides a poor fit to the data. A straight line assumes that Y_t increases (or decreases) by a constant amount each year. This assumption is not valid for most time series, where Y_t may change at either an increasing or decreasing rate. Figure 12-10 shows the time series for domestic airline fare-paying passengers from 1946 to 1969. Here the number of passengers has increased at an increasing rate over time. This is contrasted with the number of railroad passengers carried, also shown, for which the trend is decreasing but at a seemingly decreasing rate.

Part of the railroad-passenger drop may be explained by a preference for airplane travel and part by wider use of the automobile. Few passenger trains run over long distances in the United States, and these have been declining so very fast that the government now runs railroad-passenger service. The leveling-off at around 300 million passengers may be attributed to the growing significance of commuters, who now comprise a high proportion of rail passengers.

Either of these time series may be fitted by a J-shaped or exponential curve. For the airline-passenger data, the equation would be of the form

$$\hat{Y}_X = ab^X$$

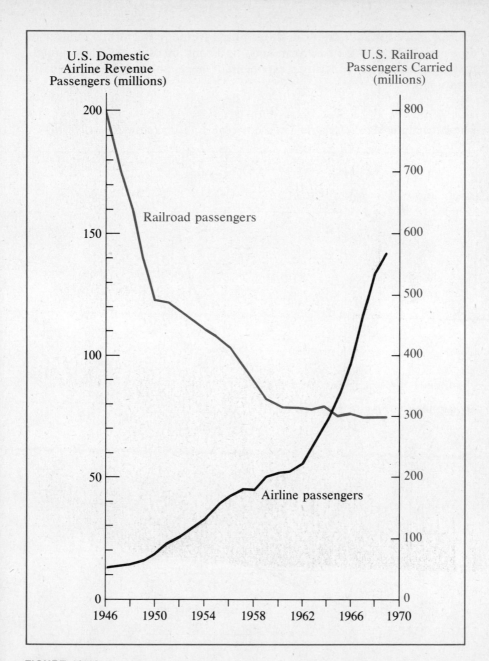

FIGURE 12-10

Airline and railroad passenger data. (SOURCE: *Moody's Transportation Manual,* 1970.)

where b is a constant (always positive) raised to the power of the number of time periods beyond the base year, and a is a constant multiple. The railroad-passenger series has a negative exponential shape, so that the appropriate equation is

$$\hat{Y}_X = ab^{-X}$$

where the minus sign before the X is used because $\hat{Y}_X$ decreases for increasing X.

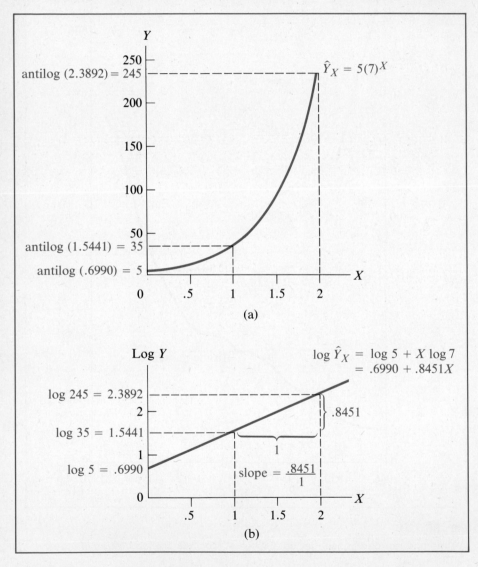

FIGURE 12-11

Graphs of exponential curve and corresponding logarithmic trend line.

The chief advantage of using an exponential curve for fitting a trend is evident when taking the logarithm of both sides of $\hat{Y}_X = ab^X$. When an exponential trend curve is converted in this manner, we have the equation for the

LOGARITHMIC TREND LINE

$$\log \hat{Y}_X = \log a + X \log b$$

This equation has the form of a straight line using log $\hat{Y}_X$ for the values on the vertical scale, with intercept of log a and slope of log b.

Figure 12-11(a) shows the graph of the exponential trend curve $\hat{Y}_X = 5(7)^X$, so that $a = 5$ and $b = 7$. We have

$$\log \hat{Y}_X = \log 5 + X \log 7$$
$$= .6990 + .8451X$$

The logarithm values are obtained from Appendix Table G. The logarithmic trend line in diagram (b) is drawn with log $\hat{Y}_X$ as the dependent variable, so that the vertical axis is expressed in log Y units. Both the exponential trend curve and the corresponding logarithmic trend line provide the same information, and values of $\hat{Y}_X$ may be obtained from the value for log $\hat{Y}_X$, and vice versa.

This suggests the possibility of fitting historical time-series data to an exponential trend curve, using the method of least squares to first find the logarithmic trend line. This may be accomplished by taking the logarithm of the observed time-series values Y, determining the regression coefficients log a and log b. For this purpose we modify the earlier equations for a and b, obtaining the

LOGARITHMIC REGRESSION COEFFICIENTS

$$\log b = \frac{\sum X \log Y - \bar{X} \sum \log Y}{\sum X^2 - n\bar{X}^2}$$

$$\log a = \frac{\sum \log Y}{n} - \bar{X} \log b$$

These expressions replace a, b, and Y values by log a, log b, and log Y. To avoid possible confusion of symbols, we use the fact that $n\bar{Y} = \sum Y$ and replace $n\bar{Y}$ with $\sum \log Y$.

The procedure is illustrated by the calculations in Table 12-2 using the airline-passenger data. The logarithmic regression coefficients are

$$\log b = \frac{497.7083 - 11.5(38.7612)}{4,324 - 24(11.5)^2} = .0452$$

$$\log a = (38.7612)/24 - 11.5(.0452) = 1.0953$$

TABLE 12-2
Computations for Fitting Exponential Trend Curve to Data for Number of U.S. Domestic Airline Passengers

Year	Years Beyond Base Period X	Number of Passengers (millions) Y	log Y	X log Y	X^2
1946	0	12.2	1.0864	0	0
1947	1	12.9	1.1106	1.1106	1
1948	2	13.2	1.1206	2.2412	4
1949	3	15.1	1.1790	3.5370	9
1950	4	17.4	1.2405	4.9620	16
1951	5	22.7	1.3560	6.7800	25
1952	6	25.0	1.3979	8.3874	36
1953	7	28.7	1.4579	10.2053	49
1954	8	32.5	1.5119	12.0952	64
1955	9	38.0	1.5798	14.2182	81
1956	10	41.7	1.6201	16.2010	100
1957	11	44.8	1.6513	18.1643	121
1958	12	44.4	1.6474	19.7688	144
1959	13	50.5	1.7033	22.1429	169
1960	14	51.6	1.7126	23.9764	196
1961	15	52.0	1.7160	25.7400	225
1962	16	55.3	1.7427	27.8832	256
1963	17	63.2	1.8007	30.6119	289
1964	18	72.1	1.8579	33.4422	324
1965	19	83.6	1.9222	36.5218	361
1966	20	97.8	1.9903	39.8060	400
1967	21	118.7	2.0745	43.5645	441
1968	22	134.4	2.1284	46.8248	484
1969	23	142.3	2.1532	49.5236	529
	276		38.7612	497.7083	4,324
	$= \sum X = 276$		$= \sum \log Y$	$= \sum X \log Y$	$= \sum X^2$

$$n = 24 \qquad \bar{X} = 11.5$$

SOURCE OF DATA: *Moody's Transportation Manual*, 1970.

The logarithmic trend line has the equation

$$\log \hat{Y}_X = 1.0953 + .0452X$$

The values for a and b may be found by taking the antilogs of the above coefficients:

$$a = \text{antilog} \ (1.0953) = 12.45$$

$$b = \text{antilog} \ (.0452) = 1.11$$

The equation for the exponential trend curve is then

$$\hat{Y}_X = 12.45(1.11)^X \qquad (X = 0 \text{ at } 1946)$$

In computing trend values, the equation form $\hat{Y}_X = ab^X$ would not ordinarily be used. Instead, the log $\hat{Y}_X$ values would be calculated from the fitted logarithmic trend line. The $\hat{Y}_X$ values can be easily obtained from the antilogs of these. For example, to find the $\hat{Y}_X$ for 1969, we have $X = 1969 - 1946 = 23$. Thus,

$$\log \hat{Y}_{23} = 1.0953 + .0452(23) = 2.1349$$

$$\hat{Y}_{23} = \text{antilog } (2.1349) = 136.4 \text{ (million passengers)}$$

This value is fairly close to the actual figure of 142.3 million passengers carried in 1969.

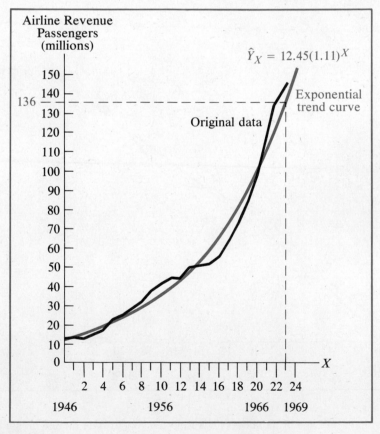

FIGURE 12-12
Exponential trend and original data for airline passengers.

The trend curve and the original time series are shown in Figure 12-12. This graph has an *arithmetic* vertical scale. The time-series behavior in the early years is hardly noticeable, whereas the points in later years rise by progressively larger amounts. So as not to exaggerate the importance of the later data, exponential trend curves are sometimes graphed with a compressed vertical scale called the *semilogarithmic*. The ruling on such a scale is made so that equal vertical distances correspond to values of Y that have increased by the same percentage. This is accomplished by spacing the scale marks at distances that correspond to the logarithm of their heights.

Figure 12-13 shows the airline-passenger time series and trend plotted with a semilogarithmic vertical scale. The exponential trend curve plots as a straight line, indicating that the number of passengers grew at a constant percentage

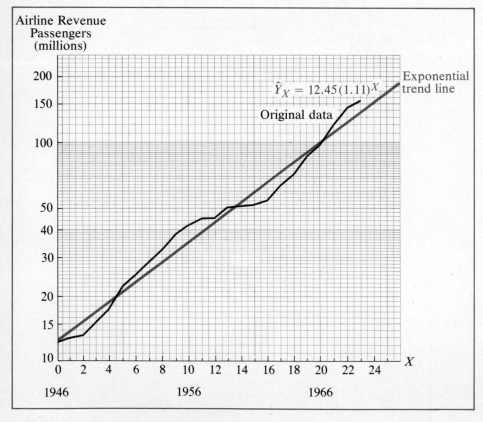

FIGURE 12-13

Exponential trend and original data for airline-passenger data plotted on semi-logarithmic graph paper.

rate during the period shown. The value $b = 1.11$ indicates that this rate was .11, or 11%, per year.*

12-5 A certain large retailing corporation has experienced the following annual sales:

Year	Sales	Year	Sales
1971	$18 million	1977	$ 82 million
1972	28	1978	89
1973	26	1979	108
1974	43	1980	121
1975	55	1981	155
1976	54		

(a) Plot the above time-series data on ordinary graph paper.
(b) Would a straight line provide a suitable summary of the trend in sales? Explain.

12-6 The electricity usage (in million of kilowatt-hours) in a region served by a certain utility company from 1972 through 1981 is provided below:

Year	Consumption	Year	Consumption
1972	205	1977	241
1973	206	1978	267
1974	223	1979	268
1975	234	1980	277
1976	231	1981	290

(a) Plot the above time-series data on a graph.
(b) Using the method of least squares, determine the equation for the estimated regression line $\hat{Y}_X = a + bX$ with X in years and $X = 0$ at 1972. Draw this line on your graph.
(c) Using your answer to part (b), estimate the electricity consumption for 1990.

12-7 Using the data provided in Exercise 12-5, answer the following:
(a) Representing sales by Y, use least-squares regression of $\log Y$ on X, where X is in years and $X = 0$ at 1971, to find the logarithmic trend line, $\log \hat{Y}_X = \log a + X \log b$.
(b) Determine the exponential trend equation $\hat{Y}_X = ab^X$.
(c) What is the annual percentage growth in sales indicated by the trend equation found in part (b)?
(d) Using the trend equation from part (a), what is the projected sales level for 1986?

---

* Recall that the expression for determining the amount A to which a principal amount P will grow at an interest rate i per year is $A = P(1 + i)^n$, where n is the number of time periods. The trend starts in 1946 at 12.45, and $\hat{Y}_X = 12.45(1.11)^X = 12.45(1 + .11)^X$. The .11 represents the percentage rate of growth in passengers and is analogous to i, whereas 12.45 is similar to P, and X, like n, is the number of years beyond 1946.

12-8 For each of the following trend equations, make the required modifications.

(a) $\hat{Y}_X = 1,000,000 + 120,000X$, with X in years and $X = 0$ at July 1, 1970. This expresses the number of employees in a utility industry. Find (1) $\hat{Y}_X$ for monthly employment, with X in months and base January 15, 1970, and (2) $\hat{Y}_X$ for quarterly employment, with X in quarters and base Winter 1970 (February 15, 1970).

(b) $\hat{Y}_X = 10,000 + 600X$, with X in years and $X = 0$ at July 1, 1971. This expresses the number of welfare recipients in a particular county. Find (1) $\hat{Y}_X$ for the number of welfare recipients, with x in months and base January 15, 1969, and (2) $\hat{Y}_X$ for the number of welfare recipients, with X in quarters and base February 15, 1969.

12-4 FORECASTING USING SEASONAL INDEXES

In this section, we will examine a procedure for isolating seasonal fluctuations in time-series data. Identifying seasonal patterns is a necessary first step in short-range planning. The management of a firm whose business drops in May is not alarmed if it is only the beginning of an annual seasonal trough. Likewise, government economists recognize that the Consumer Price Index will rise or fall in certain months solely due to the influence of seasonal factors, such as changing varieties of produce on the market. To monitor the performance of a business or an economy, it is useful to "deseasonalize" time-series data to determine whether a current drop or rise is greater than normal. A technique for doing this will be described at the end of this section.

Ratio-to-Moving-Average Method

The ratio-to-moving-average method is widely used to isolate seasonal fluctuations. This approach is the reverse of the sequence we followed in constructing the Speak E-Z time series earlier. Beginning with the actual time series, the trend and cyclical elements are isolated together in what is referred to as a "smoothed" time series. The isolation of the long-term elements is accomplished by means of *four-quarter moving averages*.

In the context of the classical time-series model, the ratio-to-moving-average method is summarized by the expression

$$\frac{Y_t}{\text{Moving average}} = \frac{T_t \times C_t \times S_t \times I_t}{T_t \times C_t} = S_t \times I_t$$

The moving average provides both trend and cycle, so that $T_t \times C_t$ is obtained for each time period. Dividing Y_t by the moving average is therefore equivalent to canceling the $T_t \times C_t$ terms from the multiplicative model, so that only the seasonal and irregular components, expressed by $S_t \times I_t$, remain.

To illustrate this procedure, data for average weekly freight-car loadings are provided in Table 12-3. The original data are listed in column (2), and the moving totals for four successive quarters appear in column (3). For 1968, the quarterly figures are 508.0, 565.7, 549.7, and 543.7 thousands of carloads. The total of these four quarters is 2,167.1, which represents all of the carloading figures for 1968. This is not an annual total, however, because the original data are weekly averages. The entries in column (3) are positioned in the table to fall *between* the quarters. These numbers must be adjusted to arrive at a true value for each quarter.

We will consider the first two entries, 2,167.1 and 2,166.4, from column (3). Their sum, 4.333.5, represents two overlapping years of carloadings and appears in column (4) in the row for the summer quarter of 1968. Because three quarters have been counted twice, the value 4,333.5 represents a total of eight quarterly figures. The four-quarter moving average for Summer 1968 is obtained by dividing the sum 4,333.5 by 8. This yields 541.7, which is entered in column (5), where four-quarter moving averages have been similarly obtained for the remaining quarters.

The four-quarter moving averages from column (5) of Table 12-3 are graphed in Figure 12-14 along with the original time-series data. Note that two quarters at the beginning and end of the series are "lost." If the series curve provided by the moving averages (the colored line in the graph) represents only the trend and cyclical elements, the fluctuations in the original data about this curve illustrate the seasonal and irregular components. The combination $S_t \times I_t$ can be obtained by dividing the original data by the corresponding four-quarter moving average. For Summer 1968, the actual average weekly carloading value is 549.7 thousand. Dividing this number by the corresponding moving average of 541.7 and multiplying by 100 gives us the *percentage of moving average*:

$$\frac{549.7}{541.7} \times 100 = 101.48$$

The percentages of moving averages for the remaining quarters, provided in column (6) of Table 12-3, are plotted in Figure 12-15. Note the repetitive nature of the oscillations, which are more or less regular from one time period to the next. The overall pattern is not precisely the same for all years, however, due to the irregular variation I_t. The remaining step in the ratio-to-moving-average method is to isolate a seasonal index completely by removing the irregular component.

The classical model assumes that short-term random influences will either increase or decrease the value of Y_t from its expected level for a particular quarter. If the summer quarters for several different years are considered, the irregular fluctuations will have a positive effect in some years and a negative effect in others. For the duration of the time series, we can assume that the average effect of random factors will be zero.

TABLE 12-3

Ratio-To-Moving-Average and Deseasonalization Calculations for Average Weekly Carloadings

(1) Quarter	(2) Average Weekly Carloadings (thousands)	(3) Four-Quarter Moving Total	(4) Sum of Two Successive Four-Quarter Totals	(5) Four-Quarter Moving Average [(4) ÷ 8]	(6) Original Data as a Percentage of Moving Average [(2) ÷ (5)] × 100	(7) Seasonal Index	(8) Deseasonalized Data [(2) ÷ (7)] × 100
1968							
Winter	508.0					96.11	528.7
Spring	565.7	2,167.1				103.66	545.7
Summer	549.7	2,166.4	4,333.5	541.7	101.48	100.25	548.3
Fall	543.7	2,162.4	4,328.8	541.1	100.48	99.98	543.8
1969							
Winter	507.3	2,158.0	4,320.4	540.1	93.93	96.11	527.8
Spring	561.7	2,171.3	4,329.3	541.2	103.79	103.66	541.9
Summer	545.3	2,162.7	4,334.0	541.8	100.65	100.25	543.9
Fall	557.0	2,151.3	4,314.0	539.3	103.28	99.98	557.1
1970							
Winter	498.7	2,129.7	4,281.0	535.1	93.20	96.11	518.9
Spring	550.3	2,086.0	4,215.7	527.0	104.42	103.66	530.9
Summer	523.7	2,073.0	4,159.0	519.9	100.73	100.25	522.4
Fall	513.3	2,048.7	4,121.7	515.2	99.63	99.98	513.4

1971							
Winter	485.7	2,002.0	4,050.7	506.3	95.93	96.11	505.4
Spring	526.0	1,946.0	3,948.0	493.5	106.59	103.66	507.4
Summer	477.0	1,928.3	3,874.3	484.3	98.49	100.25	475.8
Fall	457.3	1,916.0	3,844.3	480.5	95.17	99.98	457.4
1972							
Winter	468.0	1,939.3	3,855.3	481.9	97.12	96.11	486.9
Spring	513.7	1,993.3	3,932.6	491.6	104.50	103.66	495.6
Summer	500.3	2,028.0	4,021.3	502.7	99.52	100.25	499.1
Fall	511.3	2,047.6	4,075.6	509.5	100.35	99.98	511.4
1973							
Winter	502.7	2,077.3	4,124.9	515.6	97.50	96.11	523.0
Spring	533.3	2,100.3	4,177.6	522.2	102.13	103.66	514.5
Summer	530.0	2,109.3	4,209.6	526.2	100.72	100.25	528.7
Fall	534.3	2,108.3	4,217.6	527.2	101.35	99.98	534.4
1974							
Winter	511.7	2,085.3	4,193.6	524.2	97.62	96.11	532.4
Spring	532.3	2,030.7	4,116.0	514.5	103.46	103.66	513.5
Summer	507.0					100.25	505.7
Fall	479.7					99.98	479.8

SOURCE OF DATA : *Moody's Transportation Manual*, 1975.

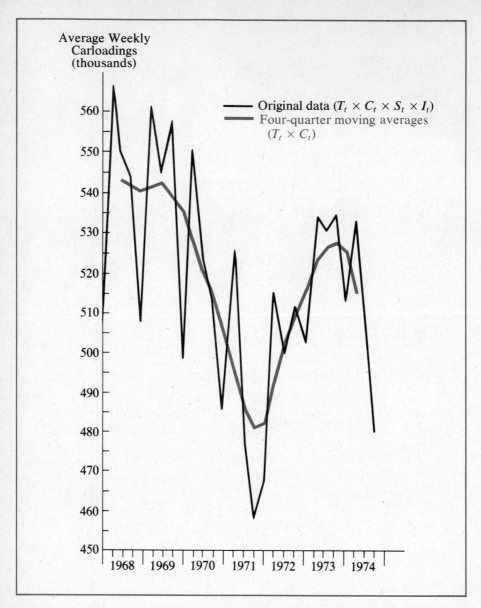

FIGURE 12-14

Four-quarter moving averages and original data for average weekly carloadings.

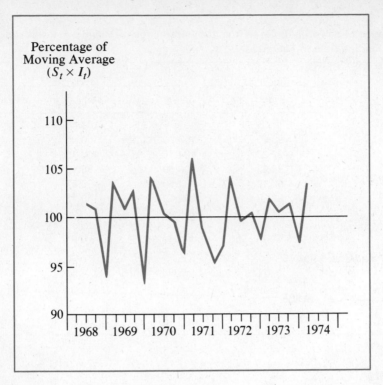

FIGURE 12-15

Original average weekly carloading data as percentages of four-quarter moving averages.

We can isolate S_t by averaging the $S_t \times I_t$ values for the same season. This is illustrated in Table 12-4, where the percentage-of-moving-average values from column (6) of Table 12-3 are grouped into the four seasonal categories. Note that the values for winter range from a low of 93.20 for 1970 to a high of 97.62 for 1974. The median will be used to represent the true seasonal factor. For winter, this value is 96.53, which is found by averaging the two middle-sized values:

$$\frac{95.93 + 97.12}{2} = 96.53$$

(For an odd number of years, the median is simply the middle-sized value itself.)

The successive medians for the remaining groups are 104.11, 100.69, and 100.42. From these medians, a seasonal index can be obtained for each quarter. The sum of the medians is 401.75. A further adjustment (multiplying each median by 400/401.75) is necessary for the seasonal indexes to sum to 400. For instance, the seasonal index for the winter quarter is 96.53(400/401.75) = 96.11.

TABLE 12-4

Calculation of Seasonal Indexes from the Percentage-of-Moving-Average Values for Carloading Data in Table 12-3

| Year | Quarter | | | |
	Winter	Spring	Summer	Fall
1968			101.48	100.48
1969	93.93	103.79	100.65	103.28
1970	93.20	104.42	100.73	99.63
1971	95.93	106.59	98.49	95.17
1972	97.12	104.50	99.52	100.35
1973	97.50	102.13	100.72	101.35
1974	97.62	103.46		
Median	96.53	104.11	100.69	100.42

Sum of medians = 401.75

$$\text{Seasonal index} = \frac{400}{401.75} \times \text{median}$$

	96.11	103.66	100.25	99.98

Monthly Data

Because a quarter is too long a time period to provide an accurate seasonal index for many businesses, fluctuations are often predicted on the basis of a 12-month seasonal cycle instead. The general procedures illustrated for quarterly data can also be applied to monthly figures. First, 12-month moving totals are obtained and centered by finding the sum of two successive totals. The 12-month moving average is determined by dividing the centered totals by 24 (the number of monthly figures included). The percentages of moving averages are then found by dividing the original monthly data by the respective moving averages. Seasonal indexes are obtained by grouping all the percentages for the same month in each year together to obtain 12 groups. The median of each monthly group is then calculated, and the sum of the medians is determined. Finally, these medians are adjusted so that their sum is 1,200 (12 months per year × 100%).

Deseasonalized Data

The ratio-to-moving-average method isolates the seasonal indexes. The seasonal indexes can be used to deseasonalize the original time-series data, which is often an asset in analyzing long-term movement in a time series.

A deseasonalized series is sometimes used in place of smoothed data to identify cyclical activity and can be used to determine whether a turning point has been reached. For example, inflation may be expressed in terms of the Consumer Price Index (CPI), which is computed by the U.S. Bureau of Labor Statistics. The CPI measures the percentage change in current prices beyond a base period. Government economists base current monetary and fiscal policies on the level of the CPI after the seasonal influence has been removed. Thus, a slower increase in the deseasonalized CPI may indicate that the peak of an inflationary period has been reached. Monthly deseasonalized CPI values may also be converted on an annual basis. Thus, an increase of .5% over the previous month's value would be multiplied by 12 to provide an annual rate of inflation of 6%.

Seasonal fluctuations are removed by dividing the original time-series data points by the corresponding seasonal indexes. In terms of the classical model, this is expressed symbolically as

$$\text{Deseasonalized value} = \frac{Y_t}{\text{Seasonal index}} = \frac{T_t \times C_t \times S_t \times I_t}{S_t} = T_t \times C_t \times I_t$$

Dividing Y_t by the seasonal index removes the S_t component from the series, leaving only the trend, cyclical, and irregular components, $T_t \times C_t \times I_t$. The seasonal indexes—the values corresponding to the respective quarter in each year—are entered in column (7) of Table 12-3. The original data values in column (2) are then divided by the seasonal index in column (7) and multiplied by 100. The resulting values, shown in column (8), are deseasonalized. For instance, the actual average number of weekly carloadings for Winter 1968 is 508.0 thousand, and the seasonal index is 96.11. Thus, for this time period

$$\text{Deseasonalized value} = \frac{508.0}{96.11} \times 100 = 528.7 \text{ (thousand)}$$

Making the Forecast

Seasonal indexes are useful in making short-term forecasts. First, a trend over the annual period is determined, and then seasonal adjustments are made for each period within the year. For example, suppose that the managers of a department store who wish to forecast monthly sales for the next calendar year determine the following trend equation:

$$\hat{Y}_X = 1,025,000 + 50,000X$$

where X is in months and $X = 0$ at January 15 of next year. Calculations of the monthly forecasts of department store sales are provided in Table 12-5.

TABLE 12-5
Calculations of the Monthly Forecasts of Department Store Sales

(1) Month	(2) X	(3) Seasonal Index	(4) Monthly Sales Trend Level $\hat{}_x$	(5) Monthly Sales Forecast [(3) × (4)] ÷ 100
January	0	56 7	$1,025,000	581,175
February	1	64.5	1,075,000	693,375
March	2	62.1	1,125,000	698,625
April	3	99.9	1,175,000	1,173,825
May	4	83.6	1,225,000	1,024,100
June	5	67.4	1,275,000	859,350
July	6	58.2	1,325,000	771,150
August	7	100.1	1,375,000	1,376,375
September	8	110.6	1,425,000	1,576,050
October	9	137.7	1,475,000	2,031,075
November	10	167.3	1,525,000	2,551,325
December	11	191.9	1,575,000	3,022,425
		1,200.0		

The 12 monthly seasonal indexes appear in column (3), and the monthly sales trend levels are calculated in column (4) from the managers' trend equation. For January, the trend value of $1,025,000 is multiplied by 56.7% to obtain the forecast sales of $581,175. The sales forecasts for all 12 months are listed in column (5).

EXERCISES

12-9 The following percentage-of-moving-average values have been obtained for a dairy's ice cream sales. Determine the seasonal index for each quarter.

	Quarter			
Year	Winter	Spring	Summer	Fall
1975			156	111
1976	49	92	137	109
1977	53	93	148	108
1978	52	91	162	104
1979	51	89	153	110
1980	51	90	151	112
1981	48	88		

12-10 A certain men's clothing chain has experienced the following quarterly sales data (in millions of dollars) for the years 1977–1981.

	1977	1978	1979	1980	1981
Winter	3.9	7.8	12.9	13.9	13.5
Spring	6.1	10.6	15.2	14.4	18.2
Summer	4.3	6.9	10.3	10.2	14.2
Fall	10.8	13.5	18.7	17.3	20.7

(a) Plot the above time-series data on a graph.

(b) Determine the four-quarter moving averages.

(c) Calculate the percentage-of-moving-average values and use them to determine the seasonal indexes.

(d) Deseasonalize the original sales data and plot them on your graph.

12-11 In arranging for short-term credit with its bank, the Make-Wave Corporation must project its cash needs on a monthly basis. To help in doing this, seasonal indexes must be developed from the following historical data on cash requirements (in hundreds of thousands of dollars).

(a) Using 12-month moving averages, determine the seasonal index number for each month by means of the ratio-to-moving-average method.

(b) Make-Wave's cash needs for the coming year are forecast to be on the average equal to $1,000,000 per month. Using the seasonal indexes calculated in (a), estimate the cash requirements for each month.

	1977	1978	1979	1980	1981
January	2.7	2.9	3.5	2.5	4.2
February	5.4	6.4	7.3	8.1	9.6
March	9.3	10.1	11.3	7.9	12.4
April	2.4	4.1	3.8	5.2	6.2
May	6.1	7.8	8.1	9.2	8.7
June	7.3	7.4	6.9	8.1	8.3
July	6.5	5.5	6.5	7.6	6.6
August	9.7	9.6	8.9	9.3	9.8
September	13.4	13.5	14.3	15.8	16.3
October	10.6	10.7	11.5	12.6	13.4
November	5.1	4.8	6.5	7.2	6.9
December	3.4	2.8	3.9	4.1	6.1

12-12 Suppose that the following seasonal indexes have been obtained by Make-Wave Corporation:

Month	Index	Month	Index
January	40	July	90
February	100	August	120
March	140	September	180
April	60	October	140
May	110	November	80
June	90	December	50
		Total	1,200

(a) Use these indexes to deseasonalize the data provided in Exercise 12-11.

(b) Plot the original data for 1980 and 1981 on a graph. Then plot the deseasonalized data for both years on the same graph.

(c) Is the rise in 1981 cash requirements from August to September abnormally large? Explain.

12-5 IDENTIFYING CYCLES AND IRREGULAR FLUCTUATION

Cyclical movement is the most troublesome systematic component of time-series data to analyze. Unlike seasonal fluctuation, which is repetitious and fairly regular from year to year, longer-term oscillations, or *cycles*, tend to be erratic. It is virtually certain, for example, that a department store will achieve greater sales in the fall quarter than in any other period within a particular year. With cyclical variation, however, the peak may occur at any point within a calendar year, and the cycle's duration may vary from one to several years. Cyclical activity usually varies in both intensity and distance from peak to successive peak, making the task of forecasting a cycle formidable—if not impossible.

Identifying past cycles in time-series data is difficult in itself. One problem is that it is normally impossible to achieve a true separation of trend and cyclical movement using the statistical tools available. Determining a trend curve is largely a matter of judgment. For instance, other methods in addition to the least-squares method may be used to fit the data to a regression line. The number of time periods to be included and the particular procedure to be used are matters of individual choice. Different trend forecasts will result from the various techniques used. Procedures for identifying cycles are described in the bibliography in the back of the book.

Because irregular fluctuations reflect no systematic influence, they are not ordinarily computed and are of little practical use in traditional forecasting methods.

12-6 EXPONENTIAL SMOOTHING

Exponential smoothing is a popular forecasting procedure that offers two basic advantages: It simplifies forecasting calculations, and its data-storage requirements are small. Exponential smoothing produces self-correcting forecasts with built-in adjustments that regulate forecast values by increasing or decreasing them in the opposite direction of earlier errors, much like a thermostat.

Single-Parameter Exponential Smoothing

The basic exponential-smoothing procedure provides the next period's forecast directly from the current period's actual and forecast values. This is summarized by the expression

$$F_{t+1} = \alpha Y_t + (1 - \alpha)F_t$$

where t is the current time period, F_{t+1} and F_t are the forecast values for the next period and the current period, respectively, and Y_t is the current actual value. α (the lowercase Greek *alpha*) is the *smoothing constant*—a chosen value lying between 0 and 1. Since only one smoothing constant is used, we refer to this procedure as *single-parameter exponential smoothing.*

To illustrate, we will suppose that actual sales of Blitz Beer in period 10 (October) were $Y_{10} = 5,240$ barrels, and that $F_{10} = 5,061.6$ had been forecast earlier for this period. Using a smoothing constant of $\alpha = .20$, the forecast for period 11 (November) sales can be calculated as

$$F_{11} = .20(5,240) + (1 - .20)(5,061.6) = 5,097.3 \text{ barrels}$$

Elementary exponential smoothing is extremely simple, because only one number—last period's forecast—must be saved. But, in essence, the entire time series is embodied in that forecast. If we express F_t in terms of the preceding actual Y_{t-1} and forecast F_{t-1} values, then the equivalent expression for next period's forecast is

$$F_{t+1} = \alpha Y_t + \alpha(1 - \alpha)Y_{t-1} + (1 - \alpha)^2 F_{t-1}$$

Continuing this for several earlier periods shows us that all preceding Y's are reflected in the current forecast. The name for this procedure is derived from the successive weights α, $\alpha(1 - \alpha)$, $\alpha(1 - \alpha)^2$, $\alpha(1 - \alpha)^3, \ldots$, which *decrease exponentially*. Thus, the more current the actual value of the time series is, the greater its weight is. Progressively less forecasting weight is assigned to older Y's, and the oldest Y's are eventually wiped out. The forecasting procedure can be modified at any time by changing the value of α.

Table 12-6 provides the actual and forecast Blitz Beer sales for 20 periods when $\alpha = .20$. There, the actual sales figure for period 1 has been used for the initial forecast for period 2. (Eventually, the same F's will be achieved in later time periods, regardless of the initial value.) The errors in this procedure are determined by subtracting the forecasts from their respective actual values. α should be set at a level that minimizes these errors. Often several trial periods are required to "tune" the smoothing constant to past data. Large α levels assign more weight to current values, whereas small α levels emphasize past data. By trial and error, an optimal level can be found for α that minimizes *variability* in forecasting errors.

TABLE 12-6

Forecast of Blitz Beer Sales
by Single-Parameter Exponential Smoothing ($\alpha = .20$)

Period t	Actual Sales Y_t	Forecast Sales F_t	Error $Y_t - F_t$
1	4,890	—	—
2	4,910	4,890.0	20.0
3	4,970	4,894.0	76.0
4	5,010	4,909.2	101.8
5	5,060	4,929.4	130.6
6	5,100	4,955.5	144.5
7	5,050	4,984.4	65.6
8	5,170	4,997.5	172.5
9	5,180	5,032.0	148.0
10	5,240	5,061.6	178.4
11	5,220	5,097.3	122.7
12	5,280	5,121.8	158.2
13	5,330	5,153.5	176.5
14	5,380	5,188.8	191.2
15	5,440	5,227.0	213.0
16	5,460	5,269.6	190.4
17	5,520	5,307.7	212.3
18	5,490	5,350.2	139.8
19	5,550	5,378.1	171.9
20	5,600	5,412.5	187.5

Two-Parameter Exponential Smoothing

Note that the forecast sales for Blitz Beer are smaller than (lag behind) the actual sales. Whenever there is a pronounced upward trend in actual data (here, increasing sales), forecasts resulting from single-parameter exponential smoothing will be consistently low.

Two-parameter exponential smoothing eliminates such a lag by explicitly accounting for trend by using a second smoothing constant for the trend itself. A total of three equations are employed:

$$V_t = \alpha Y_t + (1 - \alpha)(V_{t-1} + b_{t-1}) \quad \text{(smooth the data)}$$

$$b_t = \gamma(V_t - V_{t-1}) + (1 - \gamma)b_{t-1} \quad \text{(smooth the trend)}$$

$$F_{t+1} = V_t + b_t \quad \text{(forecast)}$$

Here, V_t represents the smoothed value for period t. The difference between the current and the prior smoothed values provides the current trend: $V_t - V_{t-1}$. The second equation contains the *trend-smoothing constant* γ (the lowercase

Greek letter gamma), which is used to obtain smoothed-trend values, represented by b_t. The third equation provides the forecast.

Table 12-7 lists the forecasts of Blitz Beer sales when $\alpha = .20$ and $\gamma = .30$. (The initial smoothed-data value of $V_2 = 4,890$ is the actual sales for period 1. The first smoothed-trend value of $b_2 = 20$ is the difference in actual sales for periods 1 and 2.) As an illustration, to forecast sales for period 8, we first obtain the smoothed-data value for period 7:

$$V_7 = .20Y_7 + (1 - .20)(V_6 + b_6)$$
$$= .20(5,050) + .80(5,045 + 35.7)$$
$$= 5,075 \text{ barrels}$$

Then we compute the smoothed-trend value for period 7:

$$b_7 = .30(V_7 - V_6) + (1 - .30)b_6$$
$$= .30(5,075 - 5,045) + .70(35.7)$$
$$= 34.0$$

TABLE 12-7
Forecast of Blitz Beer Sales by Two-Parameter Exponential Smoothing
($\alpha = .20$ and $\gamma = .30$)

Period t	Actual Sales Y_t	Smoothed Data V_t	Smoothed Trend b_t	Forecast Sales F_t	Error $Y_t - F_t$
1	4,890	—	—	—	—
2	4,910	4,890	20.0	—	—
3	4,970	4,922	23.6	4,910.0	60.0
4	5,010	4,958	27.3	4,945.6	64.4
5	5,060	5,000	31.7	4,985.3	74.7
6	5,100	5,045	35.7	5,031.7	68.3
7	5,050	5,075	34.0	5,080.7	−30.7
8	5,170	5,121	37.6	5,109.0	61.0
9	5,180	5,163	38.9	5,158.6	21.4
10	5,240	5,210	41.3	5,201.9	38.1
11	5,220	5,245	39.4	5,251.3	−31.3
12	5,280	5,283	39.0	5,284.4	−4.4
13	5,330	5,324	39.6	5,322.0	8.0
14	5,380	5,367	40.6	5,363.6	16.4
15	5,440	5,414	42.5	5,407.6	32.4
16	5,460	5,457	42.7	5,456.5	3.5
17	5,520	5,504	44.0	5,499.7	20.3
18	5,490	5,536	40.4	5,548.0	−58.0
19	5,550	5,571	38.8	5,576.4	−26.4
20	5,600	5,608	38.3	5,609.8	−9.8

which indicates that sales were increasing at a rate of 34.0 barrels per period at that time. The forecast for period 8 is the sum of the preceding period's smoothed data and trend values:

$$F_8 = V_7 + b_7 = 5{,}075 + 34.0 = 5{,}109.0 \text{ barrels}$$

The forecasts that result from this procedure are close to the actual sales values. The current trend itself is readjusted for each period to coincide with the latest growth in the raw data.

Further Exponential-Smoothing Procedures

A wide variety of exponential-smoothing procedures can be used. One adjusts the smoothing constant α itself from period to period. Others consider nonlinear relationships between values. A somewhat more complicated procedure than the one just described involves three parameters and provides for *seasonal smoothing* in addition to trend smoothing.

EXERCISES

12-13 Use single-parameter exponential smoothing with $\alpha = .40$ to forecast sales levels for the actual data given in Table 12-6.

12-14 Repeat Exercise 12-13 with $\alpha = .50$.

12-15 Use two-parameter exponential smoothing with $\alpha = .30$ and $\gamma = .20$ to forecast sales levels for the actual data given in Table 12-7.

REVIEW EXERCISES

12-16 Consider the following sales data (in thousands) for bottles of Malabug, a mosquito repellent made by Albers, Crumbly, and Itch.

	1977	1978	1979	1980	1981
Winter	21	35	39	78	54
Spring	42	54	82	146	114
Summer	60	91	117	136	160
Fall	12	14	38	30	29
Total	135	194	276	390	357

(a) Determine the trend-line regression equation using annual sales.

(b) What is the trend level of sales for 1981?

12-17 Use the Malabug sales data in Exercise 12-16.

(a) Find the seasonal index values for each quarter.

(b) Plot the original time-series data and the four-quarter moving averages on a graph.

12-18 Suppose that the trend line $\hat{Y}_X = 35 + 3X$ ($X = 0$ at February 15, 1977), applies for sales of Malabug at individual quarters, and that the seasonal indexes are

Winter	60
Spring	120
Summer	180
Fall	40

Calculate the quarterly trend values and deseasonalize the data given in Exercise 12-16.

12-19 Find the equation for the semilogarithmic trend line for the Malabug quarterly sales data given in Exercise 12-16. What sales level is forecast for Winter 1984?

12-20 A complete time-series analysis of the unit sales of Bri-Dent toothpaste has resulted in the following conclusions:

(1) Quarterly trend (in millions) may be determined from

$$\hat{Y}_X = 10 + .5X \qquad (X = 0 \text{ at February 15, 1977})$$

(2) The cyclical component percentages for 1981 are

Winter	105.7
Spring	103.2
Summer	101.4
Fall	98.6

(3) The seasonal indexes are

Winter	105
Spring	110
Summer	85
Fall	100

(4) The irregular time-series component percentage values for 1981 are

Winter	98
Spring	101
Summer	100
Fall	102

Find the original sales data for the quarters of 1981.

The whole art of statistical inference lies in the reconciliation of random mathematics with biased samples. Every new problem has some fresh kind of bias and might contain some new pitfall.

W. S. GOSSETT (1933)

Index Numbers

Inflation is a decrease in the purchasing power of money, evident to consumers as a pervasive increase in the prices of goods and services over a period of time. Many factors may contribute to inflation, but whatever the root causes at any particular time, the first step in prescribing possible remedies is to measure the amount of inflation. This is most commonly accomplished by measuring changes in price levels.

How do we measure changes in price? We only have to shop for groceries to notice that it is natural for prices to fluctuate. Some items, such as watermelon, are seasonal and sell at "outrageous" prices before they reach their peak supply and then sell for a small fraction of their previous cost. Other items have no seasonal explanation for their price fluctuations. To obtain a meaningful measure of inflation, the prices of all goods and services must somehow be represented.

PRICE INDEXES 13-1

A numerical value that summarizes price levels is called a *price index*. For such an index, the seasonal rise in the price of a commodity like strawberries should not have any significant influence. But increases in strawberry prices

TABLE 13-1

Percentage Price Relatives Index for
Frying Chicken

Year	Price per Pound	Price Relative to 1971	Index
1971	$0.94	1.00	100
1972	0.90	.96	96
1973	0.82	.87	87
1974	0.78	.83	83
1975	0.74	.79	79
1976	0.76	.81	81
1977	0.80	.85	85
1978	0.84	.89	89
1979	1.02	1.09	109
1980	1.14	1.21	121
1981	1.26	1.34	134

from one May to the next should. Likewise, only goods that are important to a great many people ought to be considered. Van Gogh paintings have increased steadily in price over the past 60 years, but such a price rise affects a minuscule percentage of persons and may be ignored.

Measuring price changes in the United States is an official task of the Department of Labor. Its Bureau of Labor Statistics periodically publishes the well-known Consumer Price Index (CPI).

It was the question of price changes that led to the development of the first price index more than 200 years ago by G.R. Carli, an Italian who compared the change in the prices of grain, wine, and oil from the years 1500 to 1750.* Carli's original index was a crude one compared to today's CPI, which is quite elaborate and complicated.

Price indexes have other uses besides being barometers of inflation. Economic models deal with *real* wages and *real* income, so that wages and income are adjusted to account for changing price levels by means of index numbers. Actual economic growth is ordinarily expressed in terms of change in real national product, which is found by using a price index to adjust annual gross national product (GNP) values computed at current price levels to values comparable to those of a prior period.

The simplest price index measures the change over time in the price of a single item. Such an index is illustrated by Table 13-1, which gives the retail prices per pound of frying chicken sold at a particular store from 1971 through

* Bruce D. Mudgett, *Index Numbers* (New York: John Wiley & Sons, 1951), p. 6.

1981. The *base period*, the point in time to which all later prices are compared, is 1971. This is expressed by the equation 1971 = 100, which means that the index expresses all prices relative to that of 1971. Index numbers are ordinarily expressed in percentages. In 1971 the price of chicken was $0.94 per pound. The *price relative* for 1972, when chicken cost $0.90 per pound, is the ratio of the two prices, .90/.94 = .96, which indicates that in 1972 the price of chicken was 96% as high as in 1971. The price index for 1972 is thus 96. In a similar manner, the 1973 price of $0.82 is divided by the 1971 price, .82/.94 = .87, so that for 1973 the index is 87, or 87% of the price that prevailed in 1971; this is a 100 − 87 = 13% price reduction. Index values so obtained are called *percentage price relatives*.

Expressed symbolically in terms of the base period price p_0 and the price p_n in a given period, we have the

PERCENTAGE PRICE RELATIVES INDEX

$$I = \frac{p_n}{p_0} \times 100$$

It is sometimes convenient to use the calendar years as subscripts, so that for 1979 the price index would be expressed as

$$I = \frac{p_{79}}{p_{71}} \times 100 = \frac{1.02}{.94} \times 100 = 109$$

Price relatives for a single item are limited in usefulness. Price changes in a single commodity such as chicken do not indicate the general movement in prices. Of greater importance in economic planning and in comparing conditions from year to year is a composite price index that covers many different items. The rise in the price of chicken will not significantly affect the lifestyles of most persons. But a rise in food prices may, especially for those whose incomes are fixed. Thus an index that considers several food items can help us measure the change in the standard of living of a large number of people.

AGGREGATE PRICE INDEXES 13-2

In Table 13-2 the prices for selected meats, poultry, and fish sold at the same retail outlet are provided for the years 1979, 1980, and 1981. A price index for several items may be determined in a manner similar to that used previously for frying chicken. An *aggregate* of the prices for each year is calculated by summing together the prices for the four items. Such an aggregate may be

TABLE 13-2

Prices per Pound for Selected Food Items

Year	Rib Roast	Pork Chops	Canned Tuna	Frying Chicken
1979	$2.40	$1.80	$2.24	$1.02
1980	2.70	1.84	1.98	1.14
1981	2.94	2.16	2.16	1.26

either unweighted or weighted. Extending the previous notation, we determine the

UNWEIGHTED AGGREGATE PRICE INDEX

$$I = \frac{\sum p_n}{\sum p_0} \times 100$$

Using 1979 = 100, we obtain the price aggregate:

$$\sum p_{79} = \$2.40 + 1.80 + 2.24 + 1.02 = \$7.46$$

For 1980, the price aggregate is

$$\sum p_{80} = \$2.70 + 1.84 + 1.98 + 1.14 = \$7.66$$

so that the unweighted aggregate price index is

$$I = \frac{\sum p_{80}}{\sum p_{79}} \times 100 = \frac{\$7.66}{\$7.46} \times 100 = 103 \text{ (rounded)}$$

Thus the aggregate price of these foods has increased by 3% from 1979 to 1980.

Simple unweighted price aggregates are subject to certain difficulties. One problem is that the price index obtained is affected by a factor other than price, the *units* on which the prices are based. In computing the above index numbers, each food item was expressed in the same units of 1 pound. Except for canned tuna, the foods listed are sold by the pound. If instead we had used $6\frac{1}{2}$ ounces, the weight of a standard can of tuna, as the unit of measurement, we would have obtained (rounded to the nearest cent) $0.92 for 1979, $0.80 for 1980, and $0.88 for 1981. Thus the price index for 1980 would have been 106 instead of 103.

One way to eliminate potential distortion caused by choice of units is to weight items in such a way that any units may be used. As a bonus, a weight may be given to each item in relation to its importance, thereby yielding a more meaningful measure of price change. For most households, the prices of rib roast and other beef products are much more significant than the price of canned tuna, since beef is more of a staple item and thus should weigh more heavily in the index.

The weights traditionally used in price indexes are quantities, denoted by the letter q. The product $p \times q$ represents total value and is unaffected by the units chosen, because p is price per unit and q is the number of units. A meaningful price index is one that uses quantities in proportion to the importance or usage of the item. Thus, in establishing an index intended to measure a family's purchasing power, the quantities of each item purchased by a typical family would be used. When the $p \times q$ terms are summed for several items, the resulting value is referred to as a *weighted price aggregate*, which is denoted by $\sum pq$.

Various indexes may be computed from weighted price aggregates. The differences between those presented in this book are due to the choice of quantities used. We first consider the

LASPEYRES INDEX

$$I = \frac{\sum p_n q_0}{\sum p_0 q_0} \times 100$$

The numerator $\sum p_n q_0$ represents the value of all the items purchased in time period n. The Laspeyres index uses quantity weights q_0 from the base period, so that only the prices are allowed to change. The denominator $\sum p_0 q_0$ provides the value of the same quantities of these commodities when they are purchased in the base period.

Table 13-3 shows the weekly quantities (in pounds) of selected food items assumed to be purchased by a typical family in 1979, the base period. The

TABLE 13-3

Price and Quantity Data for Food Items Used in Calculating
Laspeyres Index, 1979 = 100

Item	Price per Pound 1979 p_{79}	Price per Pound 1980 p_{80}	Quantity (pounds) 1979 q_{79}	Value of Items 1979 $p_{79}q_{79}$	Value of Items 1980 $p_{80}q_{79}$
Rib roast	$2.40	$2.70	5	$12.00	$13.50
Pork chops	1.80	1.84	2	3.60	3.68
Canned tuna	2.24	1.98	1	2.24	1.98
Frying chicken	1.02	1.14	4	4.08	4.56
				$21.92	$23.72

Laspeyres index for 1980 is calculated using the quantities for 1979 and prices for 1979 and 1980:

$$I = \frac{\sum p_{80} q_{79}}{\sum p_{79} q_{79}} \times 100 = \frac{\$23.72}{\$21.92} \times 100 = 108$$

Thus the prices for the selected items in 1980 are 8% higher than in the 1979 base period, which has index value 100.

Another commonly used weighted aggregate price index is the

PAASCHE INDEX

$$I = \frac{\sum p_n q_n}{\sum p_0 q_n} \times 100$$

The Paasche index is similar to the Laspeyres, but the quantities q_n from the *current* period are used in calculating the price aggregate. If the quantities of the various items consumed by a typical family change over time, the two indexes will provide different percentage changes.

For example, in 1980 the market for canned tuna was seriously disturbed by concern over the porpoises killed in tuna nets. This possibly caused a drop in the level of consumption of tuna from the previous year. For the sake of illustration, suppose that the typical family in 1980 consumed only $\frac{1}{2}$ pound of tuna per week, raising its consumption of chicken to $4\frac{1}{2}$ pounds, while the quantities of rib roast and pork chops remained constant at 5 and 2 pounds. The weighted price aggregates are, for 1979,

$$\sum p_{79} q_{80} = \$2.40(5) + \$1.80(2) + \$2.24(\tfrac{1}{2}) + \$1.02(4\tfrac{1}{2}) = \$21.31$$

and for 1980,

$$\sum p_{80} q_{80} = \$2.70(5) + \$1.84(2) + \$1.98(\tfrac{1}{2}) + \$1.14(4\tfrac{1}{2}) = \$23.30$$

The Paasche index is

$$I = \frac{\sum p_{80} q_{80}}{\sum p_{79} q_{80}} \times 100 = \frac{\$23.30}{\$21.31} \times 100 = 109$$

According to this index, the prices in 1980 were 9%, not 8%, higher than in 1979.

Ordinarily the discrepancies between the Laspeyres and Paasche indexes can be expected to grow as the time differential between the base period and current period becomes greater. There are two basic causes for widening disagreement. The primary one is the changing tastes of consumers. Newer items used in the Paasche index will, over time, tend to supplant the older ones in importance. We need only list some of the meat products heavily purchased 50 years ago to illustrate this point. Kidneys, tripe, heart, brains, and tongue are today hard to find in the ordinary supermarket, and are now largely used in processed meats, such as frankfurters. The Laspeyres index would use the earliest quantities for all items, giving high weight to some items hardly consumed at all and little weight to newer but more popular ones.

A second explanation for the discrepancy between the Paasche and Laspeyres indexes is that in a period of inflation there may be a downward shift in quantities consumed of higher-priced items. Persons of fixed income will buy less beef and eat more chicken as prices rise. If wages rise as fast as or faster than prices, the reverse may be true.

The Paasche index may thus be more realistic than the Laspeyres, because changes in taste and earnings—as reflected in revised quantity weights—are considered as well as price. However, for reasons indicated below, the Paasche index is more difficult to work with than the Laspeyres, so that a generalized version of the latter is usually preferred. To overcome the difficulties of shifting consumer emphasis on various items, the base period may be revised from time to time to bring an index up to date. The Consumer Price Index, for example, is revised to a new base period about every ten years.

Generally, the use of base-period quantities to weight prices, as with the Laspeyres index, is preferred to the use of current quantities, as with the Paasche index. One reason for this is that it is expensive to collect data to find quantity weights, so that to do so every year would require a greater expenditure of resources than to use a single set of quantities for all years. Another difficulty in using current weights is that the base period's weighted aggregate must be recomputed for each new period considered. As we will see, this is an important consideration in shifting the base period of an index. When different quantity weights have been used for each period, the change of base can be accomplished only by recomputing the index for all periods. The data required to make these calculations may not be available, and the various shortcut procedures depend on the use of the same weights in all time periods. A final advantage of using the same quantity weights in all years is that the percentage change in prices between two periods, neither of which need be a base, can then be determined from the index numbers obtained for each of the years.

It is convenient to consider a price index similar to the Laspeyres that uses arbitrarily chosen quantity weights q_a. This is the

FIXED-WEIGHT AGGREGATE PRICE INDEX

$$I = \frac{\sum p_n q_a}{\sum p_0 q_a} \times 100$$

This equation provides the Laspeyres price index when $q_a = q_0$. The major advantages of this more general index are that, like the Laspeyres, it is free from period-to-period changes in the quantity weights and that, unlike the Laspeyres, its weights are not tied to a single base period.

EXERCISES

13-1 A random sample of 100 households has been selected in order to establish a price index for housing utilities. The following average annual figures have been obtained.

	Prices (dollars per unit)				Quantities			
	1978	1979	1980	1981	1978	1979	1980	1981
Electricity	1.97	2.05	2.09	2.10	62	64	68	70
Gas	7.90	8.25	8.60	8.80	8.7	9.0	9.5	10.1
Water	.29	.30	.31	.32	296	297	298	300
Telephone	2.40	2.45	2.50	2.50	55	56	58	60

The units are as follows: thousands of kilowatt-hours for electricity, hundreds of therms for gas, hundreds of cubic feet for water, and hundreds of message-unit equivalents for telephone.

Calculate unweighted aggregate price indexes for 1979–1981, using 1978 = 100.

13-2 Using the 1978 quantities as weights, calculate the *Laspeyres price indexes for* 1979–1981, using the data in Exercise 13-1, with 1978 = 100. What is the percentage increase in 1981 prices over those of 1978?

13-3 Using the quantities for the current year as weights, calculate the Paasche indexes for 1979–1981, using the data in Exercise 13-1, with 1978 = 100. What is the percentage increase in 1981 prices over those of 1978?

13-4 Use the following quantity weights q_a: 65, 10, 300, 50 for electricity, gas, water, and telephone. Answer the following, using the data from Exercise 13-1.
(a) For a 1980 base, calculate the fixed-weight aggregate price index for 1981. What is the percentage increase in 1981 prices over those of 1980?
(b) For a 1981 base, calculate the fixed-weight aggregate price index for 1980. What is the percentage increase in 1981 prices using these index numbers? Is your answer the same as in (a)? Explain.

13-5 A consumer finance-rate index will be made using the following hypothetical interest-rate data. As quantities, total loans made that year are used.

	Rates				Quantities (billions)			
	1978	1979	1980	1981	1978	1979	1980	1981
Home mortgages	9%	10%	13%	15%	100	150	100	70
Car loans	12	12	15	15	50	55	60	60
Credit cards	15	16	17	18	30	35	40	45
Signature loans	18	19	20	22	5	6	8	10

Calculate the unweighted consumer finance-rate (price) indexes for 1979–1981, using 1978 = 100.

13-6 Using the 1978 quantities as weights, calculate the Laspeyres price indexes for 1979–1981, using the data in Exercise 13-5, with 1978 = 100. What is the percentage increase in 1981 rates over those of 1978?

13-7 Using the quantities for the current year as weights, calculate the Paasche indexes for 1979–1981, using the data in Exercise 13-5, with 1978 = 100. What is the percentage increase in 1981 rates over those of 1978?

13-8 Use the following quantity weights q_a: 100, 60, 40, 7 for home mortgages, car loans, credit cards, and signature loans. Answer the following, using the data in Exercise 13-5.

(a) For a 1980 base, calculate the fixed-weight aggregate price index for 1981. What is the percentage increase in 1981 rates over those of 1980?

(b) For a 1981 base, calculate the fixed-weight aggregate price index for 1980. What is the percentage increase in 1981 rates using these index numbers. Is your answer the same as in (a)? Explain.

PRICE RELATIVES INDEXES   13-3

The simplest index encountered at the outset of this chapter was the percentage price relative for a single commodity. This is provided by the ratio of the current price p_n to the base price p_0 times 100:

$$\text{Percentage price relative} = \frac{p_n}{p_0} \times 100$$

The indexes presented so far for two or more quantities have involved price aggregates. Price indexes may also be constructed by taking averages of price relatives. As with price aggregates, these may be either weighted or unweighted. The simplest index computed from p_n/p_0 ratios for N items is the

UNWEIGHTED PRICE RELATIVES INDEX

$$I = \frac{\sum \left(\frac{p_n}{p_0} \times 100 \right)}{N}$$

Table 13-4 illustrates how this index is computed for the four food items previously considered. For 1980 the unweighted price relatives index is 103.7. Note that this is smaller than two of the aggregate price indexes previously obtained for the same data: 108 for the Laspeyres and 109 for the Paasche.

Aside from its simplicity, the unweighted price relatives index has one positive feature. It is independent of the units in which prices are expressed. The

TABLE 13-4

Calculation of Unweighted Price
Relatives Index for Selected Items

| | Price per Pound | | Percentage |
| | 1979 | 1980 | Price Relative |
Item	p_{79}	p_{80}	$(p_{80}/p_{79}) \times 100$
Rib roast	$2.40	$2.70	112.5
Pork chops	1.80	1.84	102.2
Canned tuna	2.24	1.98	88.4
Frying chicken	1.02	1.14	111.8
			414.9

$$I = \frac{\Sigma\left(\dfrac{p_{80}}{p_{79}} \times 100\right)}{4} = \frac{414.9}{4} = 103.7$$

same value would be obtained whether tuna was priced in units of $6\frac{1}{2}$ ounces or 1 pound. But, like the unweighted price aggregate index, all items are treated as equally important. As we have seen, this can make such an index less meaningful as a measure of how households are affected by general price increases.

The importance of the individual items is handled in the aggregate price indexes by using quantity weights. For price relatives, the analogous measure for the importance of each commodity is its total value. For a consumer index, this would correspond to the total amount paid for each item. The following expresses the

WEIGHTED PRICE RELATIVES INDEX

$$I = \frac{\Sigma\left(\dfrac{p_n}{p_0} \times 100\right)v}{\Sigma v}$$

where v = value of the item, which is here equal to price × quantity. Various combinations of prices and quantities may be used. Only three are mentioned here.

When $v = p_0 q_0$, then the index

$$I = \frac{\Sigma\left(\dfrac{p_n}{p_0} \times 100\right)p_0 q_0}{\Sigma p_0 q_0}$$

TABLE 13-5

Calculation of Weighted Price Relatives Index for Selected Items

(1)	(2)		(3)	(4)	(5)	(6)
	Price per Pound		Percentage Price Relative	Quantity (pounds) 1979		Weighted Price Relative
	1979	1980	$\dfrac{p_{80}}{p_{79}} \times 100$		Value	
Item	p_{79}	p_{80}		q_{79}	$v = p_{79}q_{79}$	(3) × (5)
Rib roast	$2.40	$2.70	112.5	5	$12.00	$1,350.00
Pork chops	1.80	1.84	102.2	2	3.60	367.92
Canned tuna	2.24	1.98	88.4	1	2.24	198.02
Frying chicken	1.02	1.14	111.8	4	4.08	456.14
					$21.92	$2,372.08

$$I = \frac{\sum \left(\dfrac{p_{80}}{p_{79}} \times 100 \right) p_{79}q_{79}}{\sum p_{79}q_{79}} = \frac{\$2,372.08}{\$21.92} = 108$$

is obtained. This is algebraically equivalent to the Laspeyres index, because the p_0 terms in the numerator cancel, so that the numerator is 100 times the price aggregate for period n when base-period weights q_0 are used.

The computation for the weighted price relatives index when $v = p_0q_0$ is provided in Table 13-5. The index number obtained, 108, is exactly the same as the value obtained on page 444 for the Laspeyres index.

When $v = p_0q_n$, then the following index is obtained:

$$I = \frac{\sum \left(\dfrac{p_n}{p_0} \times 100 \right) p_0q_n}{\sum p_0q_n}$$

This index is equivalent to the Paasche index.

When $v = p_0q_a$, then the index is

$$I = \frac{\sum \left(\dfrac{p_n}{p_0} \times 100 \right) p_0q_a}{\sum p_0q_a}$$

which is identical in value to the fixed-weight aggregate price index.

Other values of v could be used, but they would not be as meaningful. We see that the three weighted price relatives indexes are equivalent to the weighted price aggregate indexes discussed, so that numerical values calculated from the two sets will be equal. Whether to use price relatives instead of aggregates is a matter of the availability of data, and in some cases price relatives are simplest to acquire. The Bureau of Labor Statistics uses price relatives in computing the Consumer Price Index because of the tremendous substitution of items involved.

EXERCISES

13-9 From the price data of Exercise 13-1 (page 446), calculate the percentage price relatives for each utility item for the years 1979–1981, using 1978 = 100. Then, assuming value weights of $150, $90, $100, and $150 for electricity, gas, water, and telephone, determine the weighted price relatives indexes.

13-10 From the rate data of Exercise 13-5, calculate the percentage price relatives for each loan type for the years 1979–1981 using 1978 = 100. Then assuming value weights equal to typical household loan levels of $50,000, $5,000, $1,000, and $500 for home mortgages, car loans, credit cards, and signature loans, determine the weighted price relatives indexes.

13-11 Critics of consumer price indexes complain that too much weight is placed on home mortgages. Although rates on new loans have risen dramatically in past years, long-term homeowners are unaffected by rate increases. This distortion might be eliminated by using a smaller weight. One suggested value is $5,000 (one-tenth the typical mortgage value, reflecting the assumption that only one in ten households buys a new home in any year). Recompute the indexes in Exercise 13-10, substituting this value for home mortgages.

13-4   THE CONSUMER PRICE INDEX

The Consumer Price Index measures changes in prices that affect the cost of living for a large fraction of the U.S. population.* This is achieved by means of a "market basket" of goods and services that consists of about 400 items. These comprise most major expenses incurred by a typical city wage earner, including food, clothing, medical treatment, entertainment, rent, and transportation. Table 13-6 shows the major goods and services categories comprising the CPI and the number of items in each. Also shown is the relative importance, as a percentage of the total market basket, given to the items in each major grouping.

The CPI has been published continuously since 1913, although major modifications have been made in the method of computation, data collection, contents of the market basket, and base period. Since World War II there have

---

* A nontechnical discussion of the CPI is contained in U.S. Department of Labor, *The Consumer Price Index: Concepts and Content Over the Years*, 1978.

TABLE 13-6

Groups of Goods and Services Priced for the Consumer Price Index and Their December 1977 Relative Importance

Major Group	Relative Importance	Major Group	Relative Importance
All items	100.000	Apparel and upkeep	5.836
		Apparel commodities	5.200
Food and beverages	20.480	Men's and boy's	
Food	19.297	apparel	1.644
Food at home	13.493	Women's and girls'	
Cereals and bakery		apparel	2.081
products	1.692	Infants' and toddlers'	
Meals, poultry, fish,		apparel	.144
and eggs	4.399	Footwear	.757
Dairy products	1.821	Other apparel	
Fruits and		commodities	.575
vegetables	1.837	Apparel services	.636
Sugar and sweets	.466		
Fats and oils	.390	Transportation	20.233
Nonalcoholic		Private transportation	19.249
beverages	1.728	New cars	4.275
Other prepared		Used cars	3.855
food	1.161	Gasoline	4.786
Food away from		Maintenance and repair	1.664
home	5.804	Other private	
Alcoholic beverages	1.183	transportation	4.668
		Commodities	.815
Housing	40.683	Services	3.854
Shelter	26.373	Public transportation	.985
Rent, residential	5.322		
Other rental costs	.488	Medical care	4.492
Homeownership	20.563	Medical care commodities	.780
Home purchase	8.753	Medical care services	3.712
Financing, taxes,		Professional services	1.916
and insurance	8.507	Other medical care	
Maintenance and		services	1.796
repairs	3.303		
Services	2.322	Entertainment	3.910
Commodities	.981	Entertainment	
Fuel and other utilities	6.398	commodities	2.497
Fuels	4.268	Entertainment services	1.413
Fuel oil, coal and			
bottled gas	.892	Other goods and services	4.367
Gas (piped) and		Tobacco products	1.454
electricity	3.375	Personal care	1.813
Other utilities and		Toilet goods and	
public services	2.130	personal care	
Household furnishings		appliances	.871
and operation	7.912	Personal care services	.942
Housefurnishings	4.735	Personal and educational	
Housekeeping		expenses	1.100
supplies	1.616	School books and	
Housekeeping		supplies	.166
services	1.560	Personal and	
		educational services	.934

SOURCE : U.S. Bureau of Labor Statistics.

been three base periods: 1947–1949, 1957–1959, and the single year, 1967, or about one change every ten years. The CPI closely resembles the Laspeyres index, which, as we have noted, uses fixed quantities for weights. Because of the shifting importance of items, it is advantageous to revise the weights periodically. The weight revisions for the CPI do not necessarily occur at the same time as the shift of base, which is one reason why the CPI is not a Laspeyres index.

The CPI is based on sample data, so sampling error is present. Furthermore, only a few key items are included, because it would be prohibitively expensive to collect data on all products and services provided in the United States. Even for the limited number of items, it would be impossible to monitor every transaction in order to obtain exact prices. Instead, average prices are used as estimates. These are obtained for all of the larger metropolitan areas and for a select group of smaller ones. Separate price indexes are computed for each city and then averaged by means of population weights, so that price levels in the larger cities have greater importance.

The data-collecting procedures are far too elaborate for detailed discussion in this book.* A few key features are important. The price data are obtained from a few randomly selected stores in each area. The store types are chosen to be representative of the merchandise classes from a recent Census of Retail Trade. The selection is made by *quota sampling*, where a specified number or quota of stores from each category is sought.

A very significant problem encountered in obtaining pricing data is that of product quality changes. When a product's quality declines while its price is fixed, it amounts to the same thing as a price increase. The nickel candy bar, now gone, is a good example. For over 50 years the price of many brands of candy bars had been held at 5 cents, even though inflation had been recurrent. The amount of candy per bar was gradually reduced in order to cover the rising costs of manufacture. The Bureau of Labor Statistics has published detailed quality specifications on the items included in the price index in order to measure the real changes in price.

The CPI is a hybrid of two indexes discussed previously. In the early periods following a switch of base periods, it is the same as the Laspeyres index. But because the value (or quantity) weights used in recent times have been switched several periods after the base period was revised, the computations for a fixed-weight index must be used. The reason for this complicated scheme is that CPI base periods have been changed whenever the current index level was judged too high for further meaningful comparisons. The value data necessary for revising weights are obtained from special studies that may occur in a different year than the base period. For example, at the press time for this

---

* For a comprehensive discussion, see Doris P. Rothwell, *The Consumer Price Index Pricing and Calculation Procedures* (Preliminary), (Washington, D.C.: U.S. Bureau of Labor Statistics, March 17, 1964).

TABLE 13-7

Consumer Price Index Values (1967 = 100)

1940	42.0	1950	72.1	1960	88.7	1970	116.3
1941	44.1	1951	77.8	1961	89.6	1971	121.3
1942	48.8	1952	79.5	1962	90.6	1972	125.3
1943	51.8	1953	80.1	1963	91.7	1973	133.1
1944	52.7	1954	80.5	1964	92.9	1974	147.7
1945	53.9	1955	80.2	1965	94.5	1975	161.2
1946	58.5	1956	81.4	1966	97.2	1976	170.5
1947	66.9	1957	84.3	1967	100.0	1977	181.5
1948	72.1	1958	86.6	1968	104.2	1978	195.4
1949	71.4	1959	87.3	1969	109.8	1979	217.4
						1980	247.7

book the current base period is 1967 = 100. The latest change in weights was made in December 1977 in accordance with the 1972–1974 Consumer Expenditure Survey.

Table 13-7 shows the consumer price index values over 1940–1980 when 1967 = 100.

DEFLATING TIME SERIES USING INDEX NUMBERS 13-5

A major use of a price index such as the CPI is to measure the "real" values of economic time-series data expressed in monetary amounts. For example, the U.S. gross national product is calculated in current dollars, so that price changes over time are reflected in the original data. GNP is the basic measure of economic growth, so that to determine how much the physical goods and services have grown over time, increases in value due to price should not be included. It is possible for the quantity of goods and services to actually fall during a recession, whereas inflation causes the GNP to rise due solely to price increases. Real economic growth may be determined by using a price index to *deflate* GNP values.

Table 13-8 provides the values for GNP and its price deflator index during 1963–1980. Note that the index values have been continually increasing. The GNP has also been rising during this period of inflation. To isolate the changes in real GNP, the effect of rising prices may be found by transforming the actual values obtained for current dollars into their equivalent in 1972 base dollars. This is achieved by dividing the original GNP data by the respective deflator index numbers. Thus in 1963 the GNP was 594.7 billion dollars, and the index

TABLE 13-8

Actual U.S. Gross National Product, Price Deflator Index, and Real GNP

Year	(1) Price Deflator Index (1972 = 100)	(2) U.S. GNP in Current Dollars (billions)	(3) Real GNP (billions) (2) ÷ (1) × 100
1963	71.59	$ 594.7	$ 830.7
1964	72.71	635.8	874.4
1965	74.32	688.1	925.9
1966	76.76	753.0	981.0
1967	79.02	796.3	1,007.7
1968	82.57	868.5	1,051.8
1969	86.72	935.5	1,078.8
1970	91.36	982.4	1,075.3
1971	96.02	1,063.4	1,107.5
1972	100.00	1,171.1	1,171.1
1973	105.80	1,306.6	1,235.0
1974	116.02	1,412.9	1,217.8
1975	127.15	1,528.7	1,202.3
1976	133.71	1,702.1	1,273.0
1977	141.70	1,899.5	1,340.5
1978	152.05	2,127.5	1,399.2
1979	165.50	2,368.5	1,431.1
1980	177.40	2,627.4	1,480.7

SOURCE : *Economic Report of the President*, 1980.

was 71.59. Since prices were lower in 1963 than for the average base period, *real* GNP should have been higher and is found by dividing 594.7 by 71.59%:

$$\text{Real 1963 GNP} = \frac{594.7}{71.59} \times 100 = 830.7 \text{ billion dollars}$$

The original data show that in 1974, a recession year, the GNP grew from 1,306.6 to 1,412.9 billion dollars. But prices grew faster. The high levels of unemployment in 1974 actually caused a drop in real GNP, from 1,235.0 billion 1972 dollars to 1,217.8 billion 1972 dollars, the latter being found by dividing the 1974 current-dollar GNP by the deflator index value of 116.02.

Economic models are often expressed in terms of real or price-adjusted values. Thus, we may use price indexes to find real wages, real income, or real production. To determine whether a typical worker's salary has risen enough to provide an increased standard of living, a price index can be used to compare take-home wages to prices. Thus if wages grow by 5% when prices grow by

6%, there has been a change in real wages (and hence, in living standard) of $105/106 - 1 = -.01$, or a 1% decline. If a union has enough clout, its workers' standard of living may rise even during a period of inflation. Thus a 10% wage increase when there is a 6% CPI rise results in $110/106 - 1 = .038$, or a 3.8% increase in real wages.

EXERCISES

13-12 The average gross weekly earnings by persons employed in manufacturing, as reported in the 1976 *Economic Report of the President*, are provided below:

Year	Earnings	Year	Earnings
1963	$ 99.63	1970	$133.73
1964	102.97	1971	142.44
1965	107.53	1972	154.69
1966	112.34	1973	166.06
1967	114.90	1974	176.40
1968	122.51	1975	189.51
1969	129.51		

Using the CPI values in column (1) of Table 13-7, deflate the above time series to obtain "real" gross carnings with a base $1967 = 100$.

13-13 The following data pertain to a large manufacturer's equipment:

Year	Book Value	Replacement Price Index
1977	$ 954,000	120
1978	1,236,000	135
1979	1,544,000	148
1980	2,020,000	165
1981	2,400,000	187

Deflate the book-value data to determine the "real" equipment worth with a base $1976 = 100$.

QUANTITY INDEXES   13-6

Up to this point we have dealt only with price index numbers. The other major type of index number is the quantity index, which measures changes in physical volume over time. One of the most important is the Index of Production, published monthly by the U.S. Federal Reserve Board. This index provides a summary of the total amount of goods produced in a given period relative to production during a base period. The index is composed of products in three broad categories or market groupings—consumer goods, equipment, and materials—with separate indexes published for major items. Index numbers are also reported in terms of major industry grouping—durable, nondurable, mining, and utilities. This index is an important barometer of economic conditions and is used by government agencies and businesses in short-range planning.

Quantity indexes are constructed in the same manner as price indexes. Because it is quantity that is being measured, the q values are changed each period. Prices are used as weights, so that the more expensive products are weighted more heavily. As with price indexes, the weights ordinarily remain constant from one reporting period to the next, so that the p values are usually not changed. Thus, algebraic expressions for computing quantity indexes can be obtained from those developed in the preceding sections for price changes by reversing the roles of p and q. For instance, the Laspeyres procedure may be adapted to measure change in quantities, providing the equation

$$I = \frac{\sum q_n p_0}{\sum q_0 p_0} \times 100$$

The numerator and denominator are weighted quantity aggregates. As with price indexes, these are both total values of items. Since prices are held constant at the base-period level, an increase in quantities from one period to the next will increase total value (there are then more items), and the quantity index will therefore rise.

EXERCISES

13-14 The following data apply to a county's orchard production.

	Quantity (bushels)				Price per bushel			
	1978	1979	1980	1981	1978	1979	1980	1981
Almonds	10,000	11,000	11,500	12,000	$1.00	$1.50	$1.25	$1.70
Cherries	60,000	70,000	75,000	80,000	2.00	1.50	2.25	2.50
Apricots	20,000	15,000	17,000	18,000	.90	1.50	1.30	1.60
Walnuts	55,000	48,000	45,000	57,000	1.00	1.50	2.00	2.25

Using the 1978 quantities as weights, calculate the Laspeyres quantity indexes for 1979–1981, with 1978 = 100. What is the percentage change in 1981 production over that of 1978?

13-15 Using the prices for the current year as weights, calculate the Paasche quantity indexes for 1979–1981, using the data in Exercise 13-14, with 1978 = 100. What is the percentage change in 1981 production over that of 1978?

SHIFTING AND SPLICING **13-7**
INDEX-NUMBER TIME SERIES

We will conclude with a discussion of two related technical points concerning time series of index numbers: (1) how the base period of the index may be *shifted*, and (2) how two indexes covering different periods of time may be combined into a single series by *splicing*.

Shifting the Base

An index such as the Consumer Price Index is considered by most persons to be a measure of the dollar's purchasing power. When discussing inflation, journalists often translate the CPI with statements like "in January 1971 a dollar was worth only 73 cents in 1959 purchasing power." For persons who experienced the rapid inflation of 1970–1971, 1959 seemed like such a long time ago that comparisons with prices during this base period became hard to relate to. The U.S. Bureau of Labor Statistics in 1971 shifted the base period of the CPI forward to 1967, thereby making inflationary comparisons more meaningful and easier. At the time of writing, further inflation is causing similar problems, and a newer base period is expected.

What are the implications of a shifted base, and how is such a shift accomplished? Table 13-9 shows a price index for sporting goods with a base period of

TABLE 13-9
Original and Shifted Price Indexes
for Sporting Goods

Year	(1) *Original Price Index* (1970 = 100)	(2) *Price Index with Shifted Base* (1980 = 100)
1970	100	71.4
1971	104	74.3
1972	106	75.7
1973	107	76.4
1974	110	78.6
1975	112	80.0
1976	115	82.1
1977	117	83.6
1978	125	89.3
1979	131	93.6
1980	140	100.0
1981	147	105.0

1970. The index numbers in column (1) were calculated using the Laspeyres method. The base period of this index is shifted to 1980 in column (2). The numbers must be transformed so that the index number for the new base year is equal to 100. This is accomplished by dividing each value in the original series by 140, the original index number for 1980, and then multiplying by 100. The index numbers for years preceding 1980 are thus projected backward from the new base. The shifted index for 1970 is $(100/140) \times 100 = 71.4$, and for 1975 it is $(112/140) \times 100 = 80.0$.

The values in column (2) may be viewed as the purchasing power of 1980 dollars. Thus a 1980 dollar buys the same quantity of sporting goods that could have been purchased in 1975 for 80 cents. The original and shifted series provide the same year-to-year percentage price increases. For instance, in 1971, prices increased by 4% over 1970, as the index number in column (1) is 104 for 1971 and 100 for 1970. In column (2) the 1970 index is 71.4 and the 1971 index is 74.3; thus, using the new base, 1971 prices are still $74.3/71.4 = 1.04$, or 104% of those for 1970.

Symbolically, ignoring the 100 factor, the shift in index number for 1980 is achieved by dividing $\sum p_{75}q_{70}/\sum p_{70}q_{70}$ by $\sum p_{80}q_{70}/\sum p_{70}q_{70}$, so that the sums involving $p_{70}q_{70}$ cancel, and the resulting index is

$$\frac{\sum p_{75}q_{70}}{\sum p_{80}q_{70}} \times 100 = 80.0$$

The quantity weights q_{70} in the above expression are still those used in 1970. The new 1980 base index shows prices relative to 1970 and is therefore no longer a Laspeyres index. As noted in Section 13-2, changing tastes and new products may make the weights given to the items in 1970 obsolete. Shifting the base period does not by itself bring an index up to date. One way to avoid this difficulty would be to revise to 1980 weights. In this case, the above shifting procedure would no longer apply, and the weighted aggregates would have to be recalculated for every year. In practice, revision of quantity weights would not be carried backward. Real problems, for example, would be created by using 1980 weights for sporting equipment not yet invented in 1970. Instead, the index for 1980 and future years would be computed using 1980 weights. The original series and the new series would then be spliced.

Splicing Index-Number Time Series

Splicing of two sets of price index numbers covering different periods of time is usually required when there is a major change in quantity weights. It may also be necessary due to a new method of calculation or the inclusion of new commodities in the index. Such a change is usually accompanied by a change in base period as well. In Table 13-10, old and revised price index numbers with 1970 = 100 are combined into a new series with 1975 = 100.

TABLE 13-10
Splicing Two Index-Number Time Series

Year	(1) Old Price Index (1970 = 100)	(2) Revised Price Index (1975 = 100)	(3) Spliced Price Index (1975 = 100)
1970	100.0		87.6
1971	102.3		89.6
1972	105.3		92.2
1973	107.6		94.2
1974	111.9		98.0
1975	114.2	100.0	100.0
1976		102.5	102.5
1977		106.4	106.4
1978		108.3	108.3
1979		111.7	111.7
1980		117.8	117.8
1981		121.6	121.6

The procedure for projecting the earlier index number backward from 1975 is identical to shifting the base from 1970 to 1975. For 1975 there is a value for the old and new series. The two series must overlap before they can be spliced to provide a continuous new series. The value of 114.2 for 1975 has been calculated with the weights used in the old series; the value of 100.0 for 1975 is used to obtain the weights for the new series. For all years after 1975, the new weights are used in calculating the index numbers.

EXERCISES

13-16 The CPI values for 1967–80, with 1967 = 100, are given below.

Year	Price Index	Year	Price Index
1967	100.0	1974	147.7
1968	104.2	1975	161.2
1969	109.8	1976	170.5
1970	116.3	1977	181.5
1971	121.3	1978	195.4
1972	125.3	1979	217.4
1973	133.1	1980	247.7

Suppose that a new base period of 1977 is chosen. Determine the new 1967–80 values of the shifted CPI series.

13-17 Two index-number time series are given below for the U.S. GNP Price Deflator Index.

Year	Old Price Index (1958 = 100)	Revised Price Index (1972 = 100)
1958	100.0	
1959	101.6	
1960	103.3	
1961	104.6	
1962	105.8	
1963	107.2	
1964	108.8	
1965	110.9	
1966	113.9	
1967	117.6	
1968	122.3	
1969	128.1	
1970	135.2	
1971	141.6	
1972	146.1	100.0
1973		105.8
1974		116.0
1975		127.2
1976		133.7
1977		141.7
1978		152.1
1979		165.5
1980		177.4

Determine the values of the spliced index, with 1972 = 100.

14

Inferences Using Two Samples

An important area of statistical inference involves the use of data from two samples, each representing a different population. A comparison of the two sample means may lead to the conclusion that one population tends to have larger values than the other. Two-sample inferences are important in a great many experiments where the impact of a change is assessed. For example, a medical researcher may compare the effectiveness of a new treatment to that of an existing procedure. Thus, an experimental chemotherapy agent for leukemia and an existing drug could be compared in terms of their respective remission rates. Two separate samples—one for each drug—would ordinarily be required, because this controls treatment conditions by limiting the explanation of potential findings to differences between the drugs rather than to other factors that could influence recovery.

Generalizations from two samples may take the form of an estimate or a test. Estimates can be made of the differences in the respective population means, proportions, or variances. Two-sample hypothesis-testing procedures indicate either that one population has greater values than the other or that the differences between the two populations are not significant. Two-sample procedures follow the same dichotomies we identified in Chapter 9: one-sided versus two-sided tests; the proportion versus the mean; and small versus large samples. But the existence of two populations can involve a further choice—whether to use independent or matched-pairs samples. Matching the sample observations usually results in a considerable reduction in the number of observations that might otherwise be required.

14-1 CONFIDENCE INTERVALS FOR THE DIFFERENCE BETWEEN MEANS USING LARGE SAMPLES

The simplest two-sample inference involves estimating the difference between two population parameters. We begin by constructing a confidence interval estimate of the difference between the two population means.

For convenience, we will designate the populations as A and B.

POPULATION MEANS

$$\mu_A = \text{Mean of population } A$$
$$\mu_B = \text{Mean of population } B$$

The population that is presumed to have the larger mean is usually denoted by A, although this designation is completely arbitrary. The difference between population means $\mu_A - \mu_B$ can be estimated from the sample results. There are two ways to do this: The observations can either be made independently or by matching pairs.

Independent Samples

It is a simpler procedure to select the sample observations independently. Sample means from the respective populations are denoted by $\bar{X}_A$ and $\bar{X}_B$. As an unbiased estimator of $\mu_A - \mu_B$, we use the

DIFFERENCE BETWEEN THE SAMPLE MEANS

$$d = \bar{X}_A - \bar{X}_B$$

The samples can vary in size, and the numbers of observations made for the respective samples are denoted by n_A and n_B.

The sampling distributions of $\bar{X}_A$ and $\bar{X}_B$ can be approximated by the normal curve for sufficiently large samples. Because the samples are chosen independently, *d will also be approximately normally distributed, with a mean of $\mu_A - \mu_B$ and a standard deviation of*

$$\sigma_d = \sqrt{\frac{\sigma_A^2}{n_A} + \frac{\sigma_B^2}{n_B}}$$

where σ_A^2 and σ_B^2 are the respective population variances.* Ordinarily, when μ_A and μ_B are unknown, σ_A^2 and σ_B^2 will also be unknown. The respective sample variances s_A^2 and s_B^2 can be used to estimate these parameters. For *large samples* (in general, $n \geq 30$), we thereby obtain

$$s_d = \sqrt{\frac{s_A^2}{n_A} + \frac{s_B^2}{n_B}}$$

as an estimate of σ_d.

We are now ready to construct the confidence interval estimate of the difference between population means. We select a normal deviate z corresponding to the desired confidence level, so that we estimate the difference between population means by the interval

$$\mu_A - \mu_B = d \pm z s_d$$

However, it is more convenient to use the following expression for the

CONFIDENCE INTERVAL FOR THE DIFFERENCE BETWEEN MEANS USING INDEPENDENT SAMPLES

$$\mu_A - \mu_B = \bar{X}_A - \bar{X}_B \pm z \sqrt{\frac{s_A^2}{n_A} + \frac{s_B^2}{n_B}}$$

Illustration: Comparing the Effectiveness of Two Textbooks

A statistics professor wishes to compare a new textbook with the textbook she currently uses. She teaches two sections of the same course. The professor continues to use the old book in section B and begins to use the new book in section A. Both sections contain 30 students, and the respective classes are conducted almost identically. Because the quality of the new book is being evaluated, we will refer to section A as the experimental group and to section B as the control group. The students have been randomly assigned to section A or B, and each may be viewed as a sample of all students taking statistics in the professor's classes.

The effectiveness of each textbook can be measured in terms of the combined examination scores achieved by the students on common tests given throughout the term. Population A represents the potential scores of all

---

* Here, we make use of the fact that $\bar{X}_A$ and $\bar{X}_B$ are independent random variables, so that the variance of their sum or difference equals the sum of the individual variances:

$$\sigma_d^2 = \sigma^2(\bar{X}_A - \bar{X}_B) = \sigma^2(\bar{X}_A) + \sigma^2(\bar{X}_B) = \frac{\sigma_A^2}{n_A} + \frac{\sigma_B^2}{n_B}$$

students at the university who might use the new textbook in a course taught by this particular professor; population B represents the potential scores these same students would receive using the old textbook.

The sample results are provided in Table 14-1. We will use these data to estimate the difference between mean population scores, $\mu_A - \mu_B$. For this study, $n_A = n_B = 30$, although it is not necessary for both samples to be the

TABLE 14-1

Sample Results for Students in Two Statistics Classes Using New and Old Textbooks

Students Using New Book (A)			Students Using Old Book (B)		
Student's Initials	Combined Exam Score X_A	X_A^2	Student's Initials	Combined Exam Score X_B	X_B
C. A.	95	9,025	C. A.	85	7,225
L. A.	87	7,569	J. A.	74	5,476
A. B.	84	7,056	O. A.	59	3,481
I. B.	79	6,241	P. A.	71	5,041
P. B.	78	6,084	W. A.	48	2,304
B. C.	93	8,649	D. B.	78	6,084
M. C.	72	5,184	D. C.	94	8,836
D. D.	92	8,464	M. C.	75	5,625
E. E.	89	7,921	B. E.	92	8,464
S. F.	68	4,624	F. E.	85	7,225
T. G.	73	5,329	J. F.	76	5,776
V. G.	53	2,809	S. G.	69	4,761
G. I.	81	6,561	G. H.	82	6,724
F. L.	90	8,100	F. J.	76	5,776
M. L.	77	5,929	J. K.	85	7,225
N. L.	86	7,396	S. L.	70	4,900
G. M.	83	6,889	K. M.	69	4,761
H. M.	91	8,281	M. M.	71	5,041
T. M.	68	4,624	D. N.	80	6,400
N. N.	81	6,561	A. O.	91	8,281
Q. P.	75	5,625	L. P.	77	5,929
P. R.	76	5,776	A. S.	90	8,100
P. S.	64	4,096	B. S.	88	7,744
S. S.	63	3,969	T. S.	60	3,600
W. S.	61	3,721	W. S.	47	2,209
M. T.	83	6,889	L. T.	74	5,476
S. T.	72	5,184	N. T.	63	3,969
W. T.	69	4,761	T. T.	63	3,969
T. W.	59	3,481	L. V.	71	5,041
W. W.	53	2,809	P. W.	60	3,600
	2,295	179,607		2,223	169,043

same size. The following statistics are obtained:

$$\bar{X}_A = \frac{2{,}295}{30} = 76.5 \qquad\qquad \bar{X}_B = \frac{2{,}223}{30} = 74.1$$

$$s_A^2 = \frac{179{,}607 - 30(76.5)^2}{30 - 1} = 139.29 \qquad s_B^2 = \frac{169{,}043 - 30(74.1)^2}{30 - 1} = 148.92$$

The professor wants to construct a 95% confidence interval of $\mu_A - \mu_B$. The required normal deviate is $z = 1.96$, giving us

$$\mu_A - \mu_B = \bar{X}_A - \bar{X}_B \pm z\sqrt{\frac{s_A^2}{n_A} + \frac{s_B^2}{n_B}} = 76.5 - 74.1 \pm 1.96\sqrt{\frac{139.29}{30} + \frac{148.92}{30}}$$

$$= 2.4 \pm 6.08$$

so that

$$-3.68 \le \mu_A - \mu_B \le 8.48$$

Thus, the mean difference between scores is estimated to lie between -3.68 and 8.48. The limit -3.68 indicates the *disadvantage* in mean scores using book A might be 3.68 points (so that book B might be better), and 8.48 tells us that the *advantage* in mean scores using book A might be as great as 8.48 points.

This interval indicates that if this experiment were repeated several times, the resulting interval estimates would contain the true $\mu_A - \mu_B$ difference about 95% of the time. Note that this estimate is not very precise and does not clearly indicate much difference in effectiveness between the two books. Now we will investigate another procedure that we could also use to estimate the difference between population means for this example.

Matched-Pairs Samples

To obtain a *matched-pairs sample*, each observation about population A is made in such a way that the selected elementary unit is matched with a "twin" from population B. For example, in evaluating a new drug, each patient in the control group would have an experiment partner of the same sex, with a similar occupation, family environment, and medical history, as well as age, weight, height, build, and other physiological characteristics. In educational testing, partners would be matched by aptitude and intelligence, family background, socioeconomic level, and achievement. In every case, matching would be based on factors that might influence the characteristic being measured.

The individual differences between each pair can be used to estimate population differences. Thus, matched-pairs sampling attempts to explain the differences between individual pairs in terms of differences caused by the factor under examination, while minimizing variability from extraneous influences. The net effect is to reduce the impact of sampling error and thereby obtain greater sampling efficiency.

To illustrate how we obtain matched pairs, we will expand the preceding sampling study. Suppose that the statistics professor decides to match the

TABLE 14-2

Matching Students in Two Statistics Classes in Terms of Mathematics Grades and Quantitative SAT Scores

	Students Using New Book (A)			Students Using Old Book (B)		
Pair	Student's Initials	Mathematics Grade	Quantitative SAT Score	Student's Initials	Mathematics Grade	Quantitative SAT Score
1	A.B.	A	706	A.O.	A	713
2	B.C.	A	702	A.S.	A	698
3	C.A.	A	674	B.E.	A	685
4	D.D.	A	660	B.S.	A	659
5	E.E.	A	622	C.A.	A	610
6	F.L.	B	685	D.C.	A	507
7	G.M.	B	683	D.N.	B	671
8	G.I.	B	623	D.B.	B	654
9	H.M.	B	583	F.J.	B	612
10	I.B.	B	582	F.E.	B	596
11	L.A.	B	581	G.H.	B	575
12	N.L.	B	525	J.K.	B	533
13	M.T.	B	489	J.F.	B	477
14	M.L.	B	425	J.A.	B	454
15	M.C.	C	523	K.M.	B	426
16	N.N.	C	512	L.P.	C	544
17	Q.P.	C	510	L.T.	C	523
18	P.R.	C	468	L.V.	C	481
19	P.B.	C	455	M.C.	C	454
20	P.S.	C	421	M.M.	C	409
21	S.F.	C	411	N.T.	C	408
22	S.S.	C	394	O.A.	C	402
23	S.T.	C	359	P.A.	C	383
24	T.M.	C	342	P.W.	C	336
25	T.G.	C	326	S.L.	C	326
26	T.W.	C	308	S.G.	C	315
27	W.S.	C	295	T.S.	D	435
28	W.T.	D	421	T.T.	D	386
29	V.G.	D	351	W.A.	D	321
30	W.W.	D	288	W.S.	D	317

students in the two classes. Each matched pair should be closely alike in statistical aptitude and achievement in related areas. Because the students *should be matched before the actual sample observations are made*—a safeguard that minimizes bias in selecting pairs—the professor cannot base her matching criteria on the two classes' performance on statistics tests given during the year.

However, the professor can use the quantitative SAT scores currently on file as a matching criterion for statistical aptitude. As a measure of achievement, she might use the students' grades in the prerequisite mathematics course. Suppose the professor uses these indicators to match her students. Table 14-2 shows one matching assignment, using mathematics grades as the main ranking device and quantitative SAT scores to distinguish between students who have identical mathematics grades. In terms of these criteria, two "A" students having respective SAT scores of 706 and 713 constitute the first pair; similarly, two "C" students scoring 455 and 454 are the nineteenth pair. The observations can be denoted by the symbols

X_{A_i} = Sample value for population A partner in the ith matched pair

X_{B_i} = Sample value for population B partner in the ith matched pair

For each sample, we can find the

MATCHED-PAIR DIFFERENCE

$$d_i = X_{A_i} - X_{B_i}$$

Note that d_i should not be confused with the unsubscripted d that we used earlier to express the difference between the means of two independent samples rather than individual pairs.

In the usual manner, we find the

MEAN OF THE MATCHED-PAIRS DIFFERENCES

$$\bar{d} = \frac{\sum d_i}{n}$$

where n is the *number of pairs* observed in the sampling study. Analogously, we determine the

STANDARD DEVIATION OF THE MATCHED-PAIRS DIFFERENCES

$$s_{d\text{-paired}} = \sqrt{\frac{\sum d_i^2 - n\bar{d}^2}{n - 1}}$$

Here, $\bar{d}$ represents the mean of n separate matched-pairs differences and has an expected value of $\mu_A - \mu_B$. Earlier, we noted the same fact about d in our examination of the differences between the means of entire independent samples. We can use both d and $\bar{d}$ to draw inferences about the difference between the means of populations A and B. However, each matched-pair difference d_i can be treated like a single observation, so that $\bar{d}$ has the same properties described in earlier chapters for $\bar{X}$. Thus, the central limit theorem indicates that for a sufficiently large sample ($n \geq 30$), $\bar{d}$ *is approximately normally distributed, with a mean of* $\mu_A - \mu_B$ *and a standard deviation estimated from*

$$\frac{s_{d\text{-paired}}}{\sqrt{n}}$$

This enables us to select a normal deviate z that corresponds to a desired confidence level. Thus, we can estimate $\mu_A - \mu_B$ by the

CONFIDENCE INTERVAL FOR THE DIFFERENCE BETWEEN MEANS USING MATCHED PAIRS

$$\mu_A - \mu_B = \bar{d} \pm z \frac{s_{d\text{-paired}}}{\sqrt{n}}$$

Returning to the statistics professor's problem of estimating the difference between population mean scores using the new and old books, the values of the matched-pairs differences are given in Table 14-3. For $n = 30$ pairs, the professor obtained the following results:

$$\bar{d} = 2.4 \qquad s_{d\text{-paired}} = 5.01$$

Note that 2.4 is the same point estimate for $\mu_A - \mu_B$ that we found earlier using the same data as independent samples. We can now construct a 95% confidence interval estimate of $\mu_A - \mu_B$, using the normal deviate value $z = 1.96$:

$$\mu_A - \mu_B = \bar{d} \pm z \frac{s_{d\text{-paired}}}{\sqrt{n}} = 2.4 \pm 1.96 \left(\frac{5.01}{\sqrt{30}} \right)$$

$$= 2.4 \pm 1.79$$

so that

$$.61 \leq \mu_A - \mu_B \leq 4.19$$

This confidence interval is more precise than the one we found previously using the student test data as independent samples. We will now compare these two procedures.

TABLE 14-3
Calculation of Matched-Pairs Differences for Student Scores
Using New and Old Books

	Student Pair		Combined Exam Scores			
i	Group A Student	Group B Student	Group A X_{A_i}	Group B X_{B_i}	Difference d_i	d_i^2
1	A.B.	A.O.	84	91	−7	49
2	B.C.	A.S.	93	90	3	9
3	C.A.	B.E.	95	92	3	9
4	D.D.	B.S.	92	88	4	16
5	E.E.	C.A.	89	85	4	16
6	F.L.	D.C.	90	94	−4	16
7	G.M.	D.N.	83	80	3	9
8	G.I.	D.B.	81	78	3	9
9	H.M.	F.J.	91	76	15	225
10	I.B.	F.E.	79	85	−6	36
11	L.A.	G.H.	87	82	5	25
12	N.L.	J.K.	86	85	1	1
13	M.T.	J.F.	83	76	7	49
14	M.L.	J.A.	77	74	3	9
15	M.C.	K.M.	72	69	3	9
16	N.N.	L.P.	81	77	4	16
17	Q.P.	L.T.	75	74	1	1
18	P.R.	L.V.	76	71	5	25
19	P.B.	M.C.	78	75	3	9
20	P.S.	M.M.	64	71	−7	49
21	S.F.	N.T.	68	63	5	25
22	S.S.	O.A.	63	59	4	16
23	S.T.	P.A.	72	71	1	1
24	T.M.	P.W.	68	60	8	64
25	T.G.	S.L.	73	70	3	9
26	T.W.	S.G.	59	69	−10	100
27	W.S.	T.S.	61	60	1	1
28	W.T.	T.T.	69	63	6	36
29	V.G.	W.A.	53	48	5	25
30	W.W.	W.S.	53	47	6	36
			2,295	2,223	72	900

$$\bar{d} = \frac{\sum d_i}{n} = \frac{72}{30} = 2.4$$

$$s_{d\text{-paired}} = \sqrt{\frac{\sum d_i^2 - n\bar{d}^2}{n-1}} = \sqrt{\frac{900 - 30(2.4)^2}{29}}$$

$$= \sqrt{25.0759} = 5.01$$

Matched Pairs Compared to Independent Samples

Is independent sampling or matched pairs the better procedure? We can begin to answer this question by comparing the sample sizes required to provide identical levels of estimation reliability for the same tolerable error.

In general, the comparative advantage of matched-pairs sampling depends on how closely the matching scheme correlates population A values with population B values. A good matching criterion should eliminate most extraneous sources of variation, so that this correlation would ordinarily be quite high.

It is easier to compare sample sizes if an equal number of observations are taken from A and B, so that $n_A = n_B$. We may also presume that population variabilities are similar, so that σ_A and σ_B are approximately equal. Under these conditions, it is possible to prove mathematically that *matched-pairs sampling requires a smaller sample size than independent sampling and that this reduction in required sample size is proportional to the population correlation coefficient*. Thus, if A and B have a correlation coefficient of .9, a 90% savings in sample size can be achieved by using matched pairs; that is, 10 times as many observations are required for independent samples to yield as precise and reliable results as matched pairs. If the correlation coefficient is .99, independent sampling would require 100 times as many observations as matched-pairs sampling.

Returning to our original question, we may ask why independent samples are used at all. In part, independent sampling is prevalent because it is often difficult to match pairs and because it is time-consuming to make each observation. Sometimes it is impossible to match elementary units advantageously. Successful matching generally requires that a large amount of data be collected for every unit. Not only must these data be studied to establish relevant matching criteria—perhaps using multiple regression and correlation techniques—but they must also be carefully applied when the sample units are paired. Once this has been accomplished, further sampling costs are comparable, observation by observation, to independent-sampling costs. For a study involving continued monitorship over a long period of time, matched-pairs samples have a distinct cost advantage. Thus, they are preferable in medical studies, where the fixed cost of establishing the sample pairs is small in comparison to total expenses.

EXERCISES

14-1 The business school dean at State U. wants to compare the GMAT scores of his graduates to those of nonbusiness students. Two independent random samples produced the following data:

Business	Nonbusiness
$n_A = 100$	$n_B = 150$
$\bar{X}_A = 620.1$	$\bar{X}_B = 576.4$
$s_A^2 = 1,874$	$s_B^2 = 2,057$

Construct a 95% confidence interval for the difference in population means.

14-2 A meteorologist wishes to find the difference between the mean annual tornado intensities (verified annual number of occurrences per 10,000 square miles) for two midwestern states. Assuming that past tornado events over a 10-year period in the two states represent independent random samples of long-run weather patterns, she has plotted all reported tornadoes on a map and obtained the following intensity data:

State A	State B
$n_A = 96$	$n_B = 114$
$\bar{X}_A = .53$	$\bar{X}_B = .34$
$s_A^2 = 1.2$	$s_B^2 = 1.5$

Construct a 99% confidence interval for the difference between the mean annual tornado intensities.

14-3 A Federal Aviation Agency statistician must determine how two airlines differ in their ability to meet flight schedules. From routes common to both airlines, he has randomly selected 100 actual arrival times for airline A and 100 for airline B. Each observation has been matched not only by route but by date and time of day. From these data, the number of minutes late in every matched pair was determined for each airline (negative values were used for early arrivals). The sample mean and the standard deviation of the difference (the late time for airline A minus the late time for airline B) were determined to be

$$\bar{d} = 2 \text{ minutes} \qquad s_{d\text{-paired}} = 3 \text{ minutes}$$

Construct a 95% confidence interval estimate for the difference between the mean amounts of late time for the two airlines.

14-4 A city manager wants to determine the amount of gasoline savings that would result if the tires were converted on all city cars. New tires were ordered for 200 cars; half of these cars were equipped with steel radial tires and the rest with the standard cord variety. Cars in the two groups were matched by make, model, age, use (police, building inspection, or whatever), and general condition. A month-long test revealed that the mean increase in gasoline mileage when radial tires were used was 2 miles per gallon. The standard deviation in gasoline savings for the paired cars was also 2 miles per gallon. Construct a 95% confidence interval estimate of the savings advantage in mean gasoline mileage that would result if radial tires were adopted for all city cars.

14-2

HYPOTHESIS TESTS FOR COMPARING TWO MEANS USING LARGE SAMPLES

In addition to estimating the difference between two population means, hypothesis-testing procedures can be used to determine whether or not population A values are significantly larger or smaller than population B values. These methods permit choices to be made. Thus, an experimental compensation program may be adopted because the statistical test concludes that workers paid under it attain a higher mean productivity than those paid under the existing program. A new drug may be rejected by a hospital staff because its mean relief level does not significantly exceed that of a drug currently in use. The owner of a taxicab fleet may find that sample results indicate a new tire supplier should be awarded a contract because the new tires wear longer than the present brand. In each case, sample data are required for both the control and the experimental groups.

In two-sample tests, the null hypothesis may be stated in any one of the following forms:

$$H_0: \mu_A \geq \mu_B \quad \text{or} \quad H_0: \mu_A - \mu_B \geq 0 \qquad \text{(lower-tailed test)}$$

$$H_0: \mu_A \leq \mu_B \quad \text{or} \quad H_0: \mu_A - \mu_B \leq 0 \qquad \text{(upper-tailed test)}$$

$$H_0: \mu_A = \mu_B \quad \text{or} \quad H_0: \mu_A - \mu_B = 0 \qquad \text{(two-sided test)}$$

The equivalent expressions on the right are found by subtracting μ_B on both sides of the original inequalities. As in estimating the difference between two means, two-sample hypothesis tests can be applied to independent samples as well as to matched pairs.

Independent Samples: Testing Exhaust Emissions

When samples are selected independently, an appropriate test statistic is $d = \bar{X}_A - \bar{X}_B$. We know that for sufficiently large samples, d is approximately normally distributed, with a mean of $\mu_A - \mu_B$ and a standard deviation of σ_d. Under any of the preceding null hypotheses, $\mu_A - \mu_B$ has an extreme value of zero. Therefore, we work with the

NORMAL DEVIATE USING INDEPENDENT SAMPLES

$$z = \frac{d}{s_d} = \frac{\bar{X}_A - \bar{X}_B}{\sqrt{\dfrac{s_A^2}{n_A} + \dfrac{s_B^2}{n_B}}}$$

We will illustrate this procedure by conducting a comparison of automobile exhaust emissions.

In evaluating how effectively a new automobile engine controls exhaust emissions, the U.S. Environmental Protection Agency might conduct a sampling study like the one we will describe here. The experimental group (A) consists of $n_A = 50$ prototype rotary engines of a type proposed to replace conventional engines. The control group (B) contains $n_B = 100$ piston engines equipped with catalytic converters. (Note that different sample sizes were used for the two groups.)

STEP 1

Formulate the null hypothesis. The EPA wishes to use the null hypothesis that the mean sulfur-dioxide emissions level in parts per million (ppm) for all rotary engines is at least as great as that for the current piston engines:

$$H_0 : \mu_A \geq \mu_B$$

The opposite alternative hypothesis applies.

STEP 2

Select the test statistic. The normal deviate z will be used.

STEP 3

Obtain the significance level α and identify the acceptance and rejection regions. Because rejecting the null hypothesis will permit a manufacturer to install the rotary engine in all its cars, the EPA has established an $\alpha = .01$ significance level for the test; that is, there is a 1% chance of making the incorrect decision

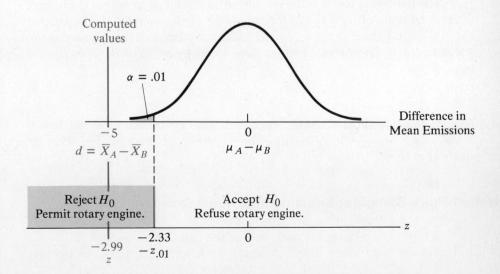

and concluding that rotary engines are cleaner than piston engines when the opposite is true. From Appendix Table E, we find $z_{.01} = 2.33$. Because a large negative difference between sample means will refute the null hypothesis, the test is lower-tailed. The two decision regions are shown on page 473.

STEP 4

Collect the sample and compute the value of the test statistic. The sample results for sulfur-dioxide emissions are

Rotary Engine	Piston Engine
$\bar{X}_A = 25$ ppm	$\bar{X}_B = 30$ ppm
$s_A^2 = 90$	$s_B^2 = 100$

The computed valued of the test statistic is

$$z = \frac{\bar{X}_A - \bar{X}_B}{\sqrt{\frac{s_A^2}{n_A} + \frac{s_B^2}{n_B}}} = \frac{25 - 30}{\sqrt{\frac{90}{50} + \frac{100}{100}}} = -2.99$$

STEP 5

Make the decision. Because $z = -2.99$ falls in the rejection region, the null hypothesis must be *rejected.* The EPA must permit the rotary engine to be installed.

This example involves a lower-tailed test. An upper-tailed test with $H_0: \mu_A \leq \mu_B$ has a rejection region with large positive values of z. A two-sided test (when $H_0: \mu_A = \mu_B$) has a double rejection region.

The two-sided test is often conducted in conjunction with a confidence interval estimate of $\mu_A - \mu_B$. In the beginning of this chapter, we found that the 95% confidence interval for the difference between population mean statistics scores for students using a new book *A* and students using an old book *B* was

$$-3.68 \leq \mu_A - \mu_B \leq 8.48$$

Because this interval contains the value 0, the null hypothesis for those data— $H_0: \mu_A = \mu_B$ or $\mu_A - \mu_B = 0$—must be *accepted* at the $\alpha = 1 - .95 = .05$ signficance level.

Matched-Pairs Samples: Smoking and Heart Disease

When matched pairs are used, we find the pair differences $d_i = X_{A_i} - X_{B_i}$ and calculate their mean $\bar{d}$ and standard deviation $s_{d\text{-paired}}$. For sufficiently large samples, we presume that the normal curve applies, so that we can work with the

NORMAL DEVIATE USING MATCHED PAIRS

$$z = \frac{\bar{d}}{s_{d\text{-paired}}/\sqrt{n}}$$

This procedure is illustrated in the following example about health and cigarette smoking.

One of the most controversial applications of statistics has been in cigarette-smoking studies. A researcher interested in the tendency of middle-aged American men to develop angina pectoris conducted an elaborate matched-pairs sampling investigation to determine if the onset of this disease is accelerated by heavy smoking. Random samples of male smokers and nonsmokers were chosen from medical records. To speed her investigation, the researcher decided not to follow each patient until the disease was detected, but to study only those men already suffering from angina.

Each smoker was matched to a nonsmoker, first by age and then according to lifestyle, occupation, medical history, and general physical characteristics. The actual matching was determined from carefully prepared dossiers from which all references to the status of the circulatory system and the heart were deleted by an independent physician. (It is very important that any data directly relevant to the variable being observed not be allowed to influence the matching process, thereby eliminating an obvious source of bias.)

STEP 1
Formulate the null hypothesis. As her null hypothesis, the researcher assumes that the mean age at which angina was first detected was the same for smokers *A* and nonsmokers *B*:

$$H_0: \mu_A = \mu_B$$

STEP 2
Select the test statistic. The normal deviate z will be used.

STEP 3
Obtain the significance level α and identify the acceptance and rejection regions. An $\alpha = .05$ significance level is desired. The test is two-sided, because either extreme positive or negative differences in the ages at which angina was detected will refute the null hypothesis. From Appendix Table E, the critical normal deviate is $z_{\alpha/2} = z_{.025} = 1.96$. The decision regions are shown on the next page.

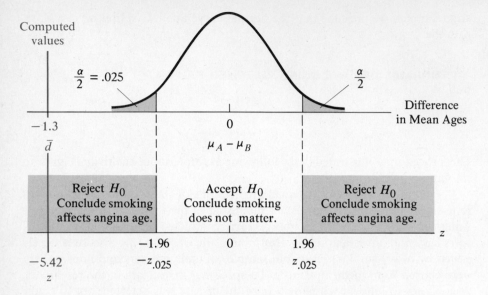

STEP 4

Collect the sample and compute the value of the test statistic. A total of $n = 100$ pairs were obtained by matching, and the sample ages X_{A_i} for smokers and X_{B_i} for nonsmokers when angina was first detected were determined. The individual matched-pair differences d_i were then computed, and the following statistics were obtained:

$$\bar{d} = -1.3 \text{ years} \qquad s_{d\text{-paired}} = 2.4 \text{ years}$$

The computed value of the test statistic is

$$z = \frac{\bar{d}}{s_{d\text{-paired}}/\sqrt{n}} = \frac{-1.3}{2.4/\sqrt{100}} = -5.42$$

STEP 5

Make the decision. Because $z = -5.42$ falls in the lower rejection region, the null hypothesis must be *rejected*. The researcher must conclude that smoking hastens the occurrence of angina.

The value $\bar{d} = -1.3$ serves as a point estimate for the difference $\mu_A - \mu_B$, indicating that for the sample results obtained, angina occurs on the average 1.3 years sooner among smokers than among nonsmokers. We can obtain a 95% confidence interval estimate of the difference between mean population ages directly from these results:

$$\mu_A - \mu_B = \bar{d} \pm z \frac{s_{d\text{-paired}}}{\sqrt{n}} = -1.3 \pm 1.96 \frac{2.4}{\sqrt{100}}$$

$$= -1.3 \pm .47$$

so that

$$-1.77 \le \mu_A - \mu_B \le -.83 \text{ years}$$

We could have used this confidence interval as the basis for testing the original null hypothesis. Because this interval lies totally below the value $\mu_A - \mu_B = 0$, again we can see that the null hypothesis must be rejected at the $\alpha = .05$ significance level.

The Efficiency of Matched-Pairs Testing

The earlier arguments for and against matched-pairs sampling apply to hypothesis testing as well. A matched-pairs test is much more efficient or powerful than a test of independent samples. In matched-pairs sampling, a much smaller sample size can generally be used to provide similar protection against committing the Type I and Type II errors. However, matched-pairs sampling is more costly and should be used only when suitable matching criteria and supporting data are available.

EXERCISES

14-5 A physician must compare two treatments for veneral disease. Because new bacteria strains have evolved that are resistant to antibiotics, a formerly obsolete chemical cure has been revived. The doctor wishes to test the null hypothesis that patient recovery time using penicillin to treat gonorrhea is less than or equal to that of a sulfa-drug treatment. If this hypothesis is rejected, the doctor will use sulfa drugs on future patients; if the sample results are not significant, she will continue to administer penicillin. A random sample of $n_A = 100$ patients was treated with penicillin, and a sample of $n_B = 50$ patients received the sulfa treatment. The following results were obtained:

Penicillin	Sulfa
$\bar{X}_A = 10$ days	$\bar{X}_B = 9$ days
$s_A^2 = 10$	$s_B^2 = 3$

If an $\alpha = .01$ significance level is desired, which treatment should the doctor use in the future?

14-6 To combat aphid infestation, a farmer must decide whether he should spray his fruit trees with pesticide or inundate them with ladybugs. He sprays 250 trees and uses ladybugs on another 250 trees. The farmer's null hypothesis is that spraying (*A*) produces a mean yield at least as great as the yield obtained using natural predators (*B*). His sample results for spraying provide a mean of 10 bushels of good fruit per tree, with a standard deviation of 3 bushels. Using ladybugs, the mean is 10.5 bushels of good fruit per tree, with a standard deviation of 1 bushel. Should the farmer accept or reject the null hypothesis at the .05 significance level?

14-7   A soft-drink bottler wants to determine whether or not noncaloric drinks help people with weight problems. Although these drinks do reduce caloric intake, they do not reduce overall appetite levels as sugared drinks do. The company dietician matched 36 pairs of obese persons in terms of weight, sex, diet, lifestyle, eating habits, and prior weight problems. The experimental group (A) was asked to use only diet soft drinks and artificial sweeteners. Both groups were to continue with their regular diets, but the control group (B) was asked to avoid noncaloric beverages of any kind. At the end of six months, the weight change was determined for each person. As her null hypothesis, the dietician assumes that the mean weight reduction will be the same for both groups. The mean of the matched-pairs differences was a 1-pound reduction in favor of noncaloric drinks, and the standard deviation was 2 pounds. If an $\alpha = .01$ significance level is desired for this two-sided test, what conclusion should the dietician reach?

14-8   Referring to the confidence interval you found in Exercise 14-2 for the difference between mean annual tornado intensities in two states, should the meteorologist accept or reject the null hypothesis that the mean intensity is the same in both states at the $\alpha = .01$ significance level?

14-9   Referring to the confidence interval you found in Exercise 14-3 for the difference between the mean amounts of late time for two airlines, should the FAA accept or reject the null hypothesis that the mean number of minutes late is the same for both airlines at the $\alpha = .05$ significance level?

14-10   Referring to the confidence interval you found in Exercise 14-4 for the savings in mean gasoline mileage using radial tires, should the city manager accept or reject the null hypothesis that mean gasoline consumption is the same for radial and standard tires at the $\alpha = .05$ significance level?

14-3   INFERENCES FOR TWO MEANS USING SMALL SAMPLES

Until now, we have considered inferences about two populations made from large samples. In these cases, the normal curve provides the sampling distributions of d and $\bar{d}$. But when small sample sizes are used, the normal approximation is inadequate in testing situations where the population variance is unknown, so the Student t distribution must be used instead.

Estimating the Difference Between Means: Independent Samples

When independent samples are used, confidence intervals can be constructed for the difference between population means by applying almost the same procedure presented earlier. Again, $d = \bar{X}_A - \bar{X}_B$ is used to estimate $\mu_A - \mu_B$, and t_α replaces the normal deviate z. In addition, the standard error of d must be calculated somewhat differently.

When small samples are used (as a practical rule, whenever *both* n_A and n_B are less than 30), *both populations must have the same variance* to apply the Student t distribution, so we must assume that $\sigma_A^2 = \sigma_B^2 = \sigma^2$. We can therefore express the standard error of d in a slightly different form than we did before:

$$\sigma_d = \sqrt{\frac{\sigma^2}{n_A} + \frac{\sigma^2}{n_B}} = \sigma \sqrt{\frac{1}{n_A} + \frac{1}{n_B}}$$

To fulfill this requirement, s_A^2 and s_B^2 must serve as unbiased estimators of σ^2. We can obtain an even better estimate of σ^2 by pooling the sample results. Thus, σ_d can be estimated by taking the square root of a weighted average of the sample variances, giving the following expression for the

STANDARD ERROR OF THE DIFFERENCE IN SAMPLE MEANS

$$s_{d\text{-small}} = \sqrt{\frac{(n_A - 1)s_A^2 + (n_B - 1)s_B^2}{n_A + n_B \quad 2}} \sqrt{\frac{1}{n_A} + \frac{1}{n_B}}$$

(Note that the denominator of the first term has been reduced by 2 to yield an unbiased estimator.)

The following expression is used to compute the

**CONFIDENCE INTERVAL FOR THE DIFFERENCE
BETWEEN MEANS USING INDEPENDENT SAMPLES**

$$\mu_A - \mu_B = \bar{X}_A - \bar{X}_B \pm t_\alpha s_{d\text{-small}}$$

We find t_α from Appendix Table F, using the α that corresponds to the confidence level. The number of degrees of freedom is $n_A + n_B - 2$. In effect, the combined sample size $n_A + n_B$ must be reduced by 2, because one degree of freedom is lost in calculating each of the two sample variances.

EXAMPLE: COMPARING TELESCOPES

An astronomer wishes to compare the resolution (magnification accuracy) of two large telescopes. One telescope, located in the Southern Hemisphere, is refractive and is equipped with a thick transparent lens that focuses light from distant galaxies. Purported to be of equal power, a reflective telescope in the Northern Hemisphere focuses starlight by means of polished mirrors. Although different galaxies may be viewed from each hemisphere, the density of galaxies in space is considered to be the same in both cases.

The astronomer has taken $n_A = 16$ pictures with the refractive telescope at randomly chosen coordinates in the South; a random sample of $n_B = 26$ frames has been exposed on the reflective telescope in the North. Camera and film type were the same at both sites, and all exposures were made for exactly one hour on clear nights. The detectable galaxies in each frame were counted, and the following sample means and variances of the respective counts were obtained:

$$\bar{X}_A = 29.5 \qquad \bar{X}_B = 21.4$$

$$s_A^2 = 96.3 \qquad s_B^2 = 112.4$$

The standard error of the differences between sample means, $d = \bar{X}_A - \bar{X}_B$, is

$$s_{d\text{-small}} = \sqrt{\frac{(16-1)96.3 + (26-1)112.4}{16+26-2}} \sqrt{\frac{1}{16} + \frac{1}{26}} = 3.277$$

Using these data, a 95% confidence interval was constructed for the difference between population mean galaxy counts. Using $\alpha = (1 - .95)/2 = .025$, for $16 + 26 - 2 = 40$ degrees of freedom, $t_{.025} = 2.021$. Thus

$$\mu_A - \mu_B = 29.5 - 21.4 \pm 2.021(3.277)$$

$$= 8.1 \pm 6.62$$

so that

$$1.48 \le \mu_A - \mu_B \le 14.72$$

The advantage in resolution of the refractive telescope (A) over the reflective telescope (B) is therefore estimated to fall somewhere between 1.48 and 14.72 galaxies.

Estimating the Difference Between Means:
Matched-Pairs Samples

In matched-pairs sampling, we use $\bar{d}$ to estimate $\mu_A - \mu_B$, the mean of the individual matched-pair differences. The only difference in using small samples (generally, when the number of pairs $n < 30$) is the substitution of t_α for the normal deviate z. The standard error of $\bar{d}$ remains the same, and it is still estimated from the standard deviation of individual paired differences $s_{d\text{-paired}}$.

The following expression is used to compute the

CONFIDENCE INTERVAL FOR THE DIFFERENCE
BETWEEN MEANS USING MATCHED PAIRS

$$\mu_A - \mu_B = \bar{d} \pm t_\alpha \frac{s_{d\text{-paired}}}{\sqrt{n}}$$

We choose a value t_α that corresponds to the confidence level, using $n - 1$ degrees of freedom.

EXAMPLE: EVALUATING A DATA-PROCESSING SYSTEM

A systems analyst for the Marlborough County Data Center has proposed adopting the dynamic-core allocation system used in neighboring Kent County. Both counties have identical computers and otherwise identical software (program systems). A special test is to be administered to estimate the savings in mean "throughput" time required to process regular job batches under the proposed system. Designating Marlborough's present system A and the new system B, a random sample of $n = 25$ batches has been run twice, once in each county. Thus, the two processing times for one batch constitute the observations of a matched pair. For the ith batch, d_i represents the throughput time on the Marlborough computer run minus the throughput time on the Kent run. The following results were obtained:

$$\bar{d} = 25 \text{ minutes} \qquad s_{d\text{-paired}} = 20 \text{ minutes}$$

A 95% confidence interval estimate of the savings in mean batch throughput time from using dynamic-core allocation can then be constructed. With $\alpha = (1 - .95)/2 = .025$, for $25 - 1 = 24$ degrees of freedom $t_{.025} = 2.064$. The estimated mean time advantage from using dynamic-core allocation is therefore

$$\mu_A - \mu_B = 25 \pm 2.064 \frac{20}{\sqrt{25}}$$

$$= 25 \pm 8.3$$

so that

$$16.7 \leq \mu_A - \mu_B \leq 33.3 \text{ minutes per batch}$$

This interval indicates that Marlborough's present computer software can be improved by using dynamic-core allocation, which will reduce mean batch processing time between 16.7 and 33.3 minutes.

Comparing Means Using Independent Samples: Effect of Coffee on Sleep

When small independent samples are used to test hypotheses about μ_A and μ_B, we calculate the

t STATISTIC FOR INDEPENDENT SAMPLES

$$t = \frac{\bar{X}_A - \bar{X}_B}{\sqrt{\dfrac{(n_A - 1)s_A^2 + (n_B - 1)s_B^2}{n_A + n_B - 2}} \sqrt{\dfrac{1}{n_A} + \dfrac{1}{n_B}}}$$

We can find the critical value corresponding to the desired significance level α from Appendix Table F for $n_A + n_B - 2$ degrees of freedom. The following example illustrates this procedure.

Drinking coffee affects many people's sleep. One investigator compared the effects of coffee on habitual drinkers with its effects on occasional drinkers. Independent random samples of $n_A = 10$ light and $n_B = 15$ heavy coffee drinkers were obtained. One night's sleep was monitored for each subject, and before retiring, every person was given a cup of coffee.

STEP 1
Formulate the null hypothesis. As his null hypothesis, the researcher assumes that the light coffee drinkers will go to sleep at least as fast as the heavy coffee drinkers. In terms of mean time to onset of sleep

$$H_0: \mu_A \leq \mu_B$$

STEP 2
Select the test statistic. The Student t statistic is used because both n_A and n_B are less than 30.

STEP 3
Obtain the significance level α and identify the acceptance and rejection regions. The researcher chooses an $\alpha = .05$ significance level. For $10 + 15 - 2 = 23$ degrees of freedom, Appendix Table F provides $t_{.05} = 1.714$. Because a large positive value of $d = \bar{X}_A - \bar{X}_B$ will refute the null hypothesis, the test is upper-tailed. The acceptance and rejection regions are shown below.

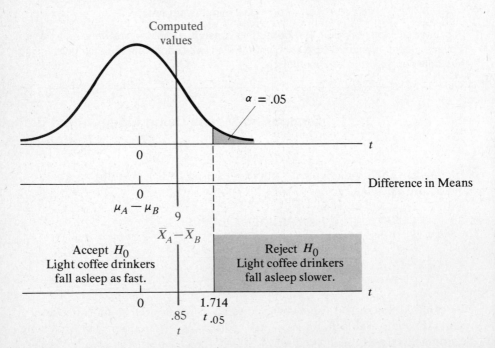

STEP 4

Collect the sample and compute the value of the test statistic. The following test data were obtained:

Light Coffee Drinkers	Heavy Coffee Drinkers
$\bar{X}_A = 45$ minutes	$\bar{X}_B = 36$ minutes
$s_A^2 = 706$	$s_B^2 = 654$

The value obtained for the test statistic is

$$t = \frac{45 - 36}{\sqrt{\dfrac{9(706) + 14(654)}{23}} \sqrt{\dfrac{1}{10} + \dfrac{1}{15}}} = .85$$

STEP 5

Make the decision. The computed value $t = .85$ falls within the acceptance region. The null hypothesis that light coffee drinkers fall asleep as quickly as heavy drinkers must be *accepted.*

Comparing Means Using Matched-Pairs Samples

When small samples are used, matched pairs involve the

t STATISTIC USING MATCHED PAIRS

$$t = \frac{\bar{d}}{s_{d\text{-paired}}/\sqrt{n}}$$

We can find the critical value t_α of the desired significance level α from Appendix Table F, using $n - 1$ degrees of freedom. For a two-sided test, we would use $t_{\alpha/2}$ as the critical value.

To illustrate this procedure, we will conduct a one-tailed test using the data from the data-processing system evaluation on page 481.

STEP 1

Formulate the null hypothesis. The null hypothesis being tested is that Kent County's new dynamic-core allocation system (B) is as slow as or slower (not faster) than Marlborough's present system (A):

$$H_0 : \mu_A \leq \mu_B$$

STEP 2
Select the test statistic. Because $n < 30$, the Student t statistic is used.

STEP 3
Obtain the significance level α and identify the acceptance and rejection regions.
A significance level of $\alpha = .05$ is desired. For $n - 1 = 24$ degrees of freedom,
the critical value of the test statistic is $t_{.05} = 1.711$. Because a large positive
value of $\overline{d}$ (and therefore t) will refute the null hypothesis, the test is upper-tailed.
The acceptance and rejection regions are shown below.

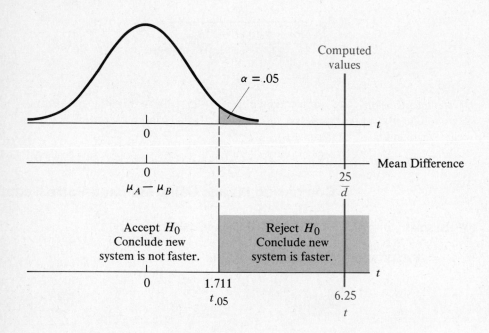

STEP 4
Collect the sample and compute the value of the test statistic. The test statistic
calculated from the data on page 481 is

$$t = \frac{25}{20/\sqrt{25}} = 6.25$$

STEP 5
Make the decision. Because $t = 6.25$ falls in the rejection region, the null
hypothesis must be rejected. The analyst must conclude that Kent County's
new data-processing system is faster than Marlborough's present system.

EXERCISES

14-11 Construct a 95% confidence interval estimate of the difference between the mean gasoline mileages of two car models if the following results have been obtained for independent samples:

Car A	Car B
$n_A = 10$	$n_B = 10$
$\bar{X}_A = 24$ mpg	$\bar{X}_B = 22.5$ mpg
$s_A^2 = 2$	$s_B^2 = 2.5$

14-12 Suppose that the drivers of the two cars in Exercise 14-11 have been matched according to their driving skills. The mean and the standard deviation of the difference between gasoline mileages found by subtracting car B's mpg from car A's are

$$\bar{d} = 1.5 \text{ mpg} \qquad s_{d\text{-paired}} = 1.2 \text{ mpg}$$

Construct a 95% confidence interval for the difference between mean gasoline mileages.

14-13 A city manager wants to estimate the mean lifetime of a new type of tire compared to an existing tire. In a test, 10 taxis from the same company were equipped with two of each type of tire. Each car's total mileage was recorded until both tires of the same type had to be replaced. For each car, the replacement mileage for the present tire (B) was subtracted from the corresponding figure for the new tire (A). The mean difference between mileages was 2,000 miles, with a standard deviation of 2,000 miles.
(a) Construct a 95% confidence interval estimate of the difference between mean tire lifetimes.
(b) Should the null hypothesis that the mean tire lifetimes are identical be accepted or rejected at the 5% significance level?

14-14 Suppose that a test similar to the one in Exercise 14-13 was performed on another group of cars. Here, 15 cars were fully equipped with the new tire (A), and 10 cars were fully equipped with the old tire (B). No attempt was made to match the cars. All four tires were replaced when driving each car became hazardous. The following mileage results were obtained:

$$\bar{X}_A = 35,000 \text{ miles} \qquad \bar{X}_B = 33,000 \text{ miles}$$
$$s_A^2 = 6,500,000 \qquad s_B^2 = 7,000,000$$

(a) Construct a 95% confidence interval estimate of the difference between mean tire lifetimes.
(b) Should the null hypothesis that the mean tire lifetimes are identical be accepted or rejected at the 5% significance level?

14-15 One aspect of horticulture is to find and nurture genetic strains of plants with favorable characteristics. In producing trees for timber, fast-growing seeds are desirable. Two parent strains of a certain species are being compared in test plots.

The following data have been obtained for the third-year growth of seedlings:

Strain A	Strain B
$n_A = 14$	$n_B = 28$
$\bar{X}_A = 2.4$ feet	$\bar{X}_B = 2.6$ feet
$s_A^2 = .25$	$s_B^2 = .35$

Should the null hypothesis that seedling growth for strain A is at least as great as seedling growth for strain B be accepted or rejected at the 5% significance level?

14-16 A real-estate property manager wishes to compare net monthly incomes under two different leasing agreements. Plan A has tenants sign for a lower rent but requires them to pay for all small repairs themselves. Under plan B, the tenant pays more rent but is not responsible for repairs. As the null hypothesis the manager assumes that plan B yields at least as great a mean income as plan A. The following data have been obtained for independent random samples of 15 leases each:

Plan A	Plan B
$\bar{X}_A = \$132.45$	$\bar{X}_B = \$128.06$
$s_A^2 = 123$	$s_B^2 = 95$

Can the manager reject the null hypothesis using a 5% significance level?

14-17 Suppose that the astronomer in the example on page 479 wishes to compare the near-range resolution of the Southern Hemisphere refractive telescope (A) with the resolution of the Northern Hemisphere reflective telescope (B). Because Jupiter can be seen from both sites simultaneously, shots of this planet were made from both telescopes at the same time on 15 different nights throughout the year. The photographs were then analyzed to determine the number of Jovian moons that could be detected in each. The mean difference between moon counts was found to be .3 moon, with a standard deviation of .7 moon. Must the null hypothesis that the reflective telescope provides at least as high a moon count as the refractive telescope be accepted or rejected at the 5% significance level?

14-18 A company's telephone salespeople are all currently salaried. To improve incentives, management will convert the compensation either to a commission-only basis or to a salary-plus-commission plan. The null hypothesis is that the mean sales per solicitor will be the same under either plan. To test the schemes, random samples of 25 salespeople from each of two different offices were matched into pairs according to past sales records. After a trial six months, the sales of the salary-plus-commission partner were subtracted from the corresponding figure for the commission-only person. A mean difference of $100 was obtained with a standard deviation of $25. At the 5% significance level, which compensation plan, if any, is the better one?

14-4   HYPOTHESIS TESTS FOR COMPARING PROPORTIONS

Qualitative populations can be compared in terms of their respective proportions. We can extend the hypothesis-testing concepts that we applied in

Section 14-3, using sample proportions as the basis for comparison. This type of testing may be useful, for example, to a politician who must determine whether or not the popularity of new legislation differs between two groups of constituents. Similarly, a television-network director, debating whether or not to reschedule programs, might wish to compare the proportions of potential audiences that prefer two of the network's comedies over a third, all shown on different nights.

As before, the two population proportions are represented by π_A and π_B. The respective sample proportions are designated as P_A and P_B. The difference between population proportions, $\pi_A - \pi_B$, may be estimated by the

DIFFERENCE IN SAMPLE PROPORTIONS

$$d = P_A - P_B$$

One of the following null hypotheses is to be tested:

$$H_0 : \pi_A \geq \pi_B \quad \text{or} \quad H_0 : \pi_A - \pi_B \geq 0 \qquad \text{(lower-tailed test)}$$
$$H_0 : \pi_A \leq \pi_B \quad \text{or} \quad H_0 : \pi_A - \pi_B \leq 0 \qquad \text{(upper-tailed test)}$$
$$H_0 : \pi_A = \pi_B \quad \text{or} \quad H_0 : \pi_A - \pi_B = 0 \qquad \text{(two-sided test)}$$

Ordinarily, the sampling distributions of P_A and P_B can be approximated by the normal curve when sample sizes are large enough. Because the samples are selected independently, it follows that d can be assumed to be normally distributed, with a mean of $\pi_A - \pi_B$ and a standard deviation of*

$$\sigma_d = \sqrt{\frac{\pi_A(1 - \pi_A)}{n_A} + \frac{\pi_B(1 - \pi_B)}{n_B}}$$

Under any of the preceding null hypotheses, we can assume that $\pi_A = \pi_B$, so that $\pi_A - \pi_B = 0$, and we can treat the sample results as if they applied to the same population. As an estimator of either π_A or π_B, the sample results can be pooled to compute the

COMBINED SAMPLE PROPORTION

$$P_C = \frac{n_A P_A + n_B P_B}{n_A + n_B}$$

---

* As with the difference between independent sample means, it follows that the variance of $d = P_A - P_B$ is the sum of the respective variances of the individual sample proportions.

The standard deviation of d is then estimated by

$$s_d = \sqrt{P_C(1 - P_C)\left(\frac{1}{n_A} + \frac{1}{n_B}\right)}$$

The following expression is used to compute the

NORMAL DEVIATE FOR COMPARING TWO PROPORTIONS

$$z = \frac{P_A - P_B}{\sqrt{P_C(1 - P_C)\left(\frac{1}{n_A} + \frac{1}{n_B}\right)}}$$

Illustration: Television Programming

To illustrate the application of this procedure, we will examine a decision involving television programming. A television program director has decided to replace his network's Monday-night comedy show "Grundy's Mondays" with one of two candidate comedy series: "The Sky Is Falling" (A) or "Mr. McGregor's Garden" (B). Both of the new programs should have a wider appeal than the current comedy, but the director wants to pick the new comedy best suited to the present Monday-night audience. To determine this, special pilots of "Sky" and "McGregor" were shown on nonsuccessive weeks in the "Grundy" time slot. Two independent random samples of regular "Grundy" viewers who also watched both comedy pilots were then selected, and the sample proportion favoring each of the new shows was determined.

STEP 1
Formulate the null hypothesis. As his null hypothesis, the program director assumes that the respective population proportions of viewers preferring each of the program pilots over "Grundy's Mondays" are the same. Therefore,

$$H_0 : \pi_A = \pi_B$$

STEP 2
Select the test statistic. The normal deviate z serves as the test statistic.

STEP 3
Obtain the significance level α and identify the acceptance and rejection regions. Because either extreme positive or extreme negative differences in sample proportions tend to refute the null hypothesis, the test is two-sided. Even though

"Grundy" is not well liked by its current audience, the program director will use the limited sample results to choose a replacement only if one new show is significantly preferred over the other. Otherwise, he will investigate further before choosing a new show. The director therefore selects a significance level of $\alpha = .10$ for this test, which provides only a 10% chance of choosing the new show that was not more preferred by "Grundy" addicts. The acceptance and rejection regions are shown below.

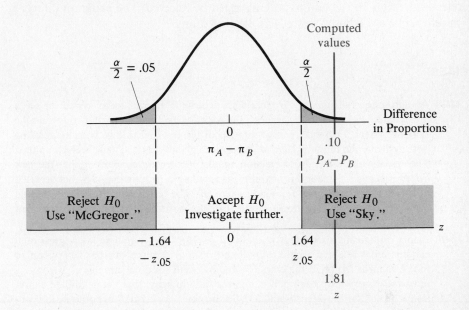

STEP 4
Collect the sample and compute the value of the test statistic. With $n_A = 150$ and $n_B = 100$, the respective sample proportions favoring the new programs were

$$P_A = .80 \quad \text{and} \quad P_B = .70$$

These results provide the difference between proportions

$$d = P_A - P_B = .80 - .70 = .10$$

The combined sample proportion favoring a change in programs is

$$P_C = \frac{150(.80) + 100(.70)}{150 + 100} = .76$$

The computed value of the normal deviate is

$$z = \frac{.80 - .70}{\sqrt{.76(1-.76)\left(\frac{1}{150}+\frac{1}{100}\right)}} = 1.81$$

STEP 5
Make the decision. The computed normal deviate $z = 1.81$ falls in the upper rejection region, so the null hypothesis must be rejected. The program director should replace "Grundy" with "The Sky Is Falling."

EXERCISES

14-19 An aerospace statistician is evaluating a prototype aircraft radar system to see if it can detect a plane passing through the edges of the radar horizon better than the existing system. Two independent samples of 100 test flights have been made using both systems, and the respective proportions of planes detected have been computed. The statistician desires 5% protection against incorrectly concluding that the new system (A) has a detection probability (a population proportion of flights detected at the horizon) higher than the old system (B). The test results yield $P_A = .80$ and $P_B = .70$. Based on the sample results, can the statistician conclude that the new system is better?

14-20 Before organizing his strategy in the final weeks of the campaign, a congressional candidate has retained a political polling firm to determine whether he is stronger in the suburbs (A) or in the cities (B). Independent random samples of 100 voters in each category have been polled, and the data indicate that the candidate is preferred by 48% in the suburbs and 53% in the cities. As his null hypothesis, the candidate assumes that he is equally strong in both areas. At the $\alpha = .05$ significance level, what should he conclude from the sample results?

14-5   FURTHER REMARKS

In this chapter, we have considered two-sample inferences regarding population means and proportions. Either independent or matched-pairs samples can be employed to test means, and each procedure has its advantages and its disadvantages. Generally, matched-pairs sampling makes more efficient use of sample information, but it is more elaborate and costly.

Again, we have encountered the basic dichotomy between large and small samples. With large samples, we make use of the fact that when the population standard deviations are known, the central limit theorem indicates that the respective estimators and test statistics are approximately normally distributed.

Ordinarily, however, σ_A^2 and σ_B^2 are unknown and must be estimated by s_A^2 and s_B^2. In all such cases, the Student t distribution applies instead of the normal distribution. For large samples, the normal curve is nearly identical to the Student t distribution, so we use it anyway.

A requirement of the Student t distribution is that the parent population be normally distributed. As we saw in Chapter 9, this usually presents no problem unless the population is highly skewed. In independent, two-sample tests with small n's, an additional requirement is that both population variances must be equal. If there is reason to believe that σ_A^2 differs greatly from σ_B^2, the t statistic should not be used on small, independent samples.

Other statistical procedures can be employed to make inferences regarding σ_A^2 and σ_B^2. Up to this point, we have largely avoided the issue of making inferences about variances. We will discuss these procedures in Chapter 15, after the necessary sampling distribution has been described in its more traditional applications.

REVIEW EXERCISES

14-21 Consider the following sample data for the weights and ages of athletic and non-athletic men, all approximately 6 feet tall and of medium build:

Athletic Men (A)		Nonathletic Men (B)	
Weight	Age	Weight	Age
151	22	152	21
148	24	153	26
156	31	149	33
155	37	162	35
157	41	165	39
161	42	168	44
158	49	157	47
168	51	178	49
149	55	161	57
174	61	186	59

(a) Calculate the sample means and variances of the weights of the two groups.
(b) Treating the weights as independent random samples, should the null hypothesis that the population mean weight of athletic men is at least as large as that of nonathletic men be accepted or rejected at the $\alpha = .10$ significance level?

14-22 A statistician who is also a baseball fan is interested in how major league batting averages differ between home and away games. She has selected 25 batters at random and has kept detailed records of their performances. As her null hypothesis, the statistician assumes that the mean batting average of hitters away from home is at least as large as the mean average they attain playing on their home field. After

compiling her data and separately computing the batting averages, the statistician finds that the mean difference between the players' batting averages is .028 in favor of the home games. The standard deviation in the differences is computed to be .056.
(a) At the $\alpha = .05$ significance level, what should the statistician conclude?
(b) Construct a 95% confidence interval for the advantage in mean batting average attained by playing at home.

14-23 A company tested two cereals to determine the taste preferences of potential buyers. Two different panels of persons were selected: The individuals on one panel tasted brand A; the others tasted brand B. Then each person was asked if he or she would buy the product tasted. The following results were obtained:

	Cereal A	Cereal B
Number who would buy	75	80
Number who would not buy or who were undecided	50	60
	125	140

Consider the null hypothesis that there is no difference in the proportion of potential buyers who would prefer to buy either product. At the 5% significance level, should H_0 be accepted or rejected?

14-24 Referring to Exercise 14-21, assume that the data represent observation pairs matched by age.
(a) Compute the weight differences between each pair, subtracting the nonathlete's weight from the athlete's. Then find the mean and the standard deviation of these matched-pairs differences.
(b) Treating the weights as matched-pairs samples, should the null hypothesis that the population mean weight of athletic men is at least as great as that of non-athletic men be accepted or rejected at the $\alpha = .10$ significance level?

14-25 A car rental company wishes to determine which of two types of gasoline provides more miles per gallon. Two independent random samples of 15 cars were selected, and each car was driven 2,000 miles by employees on normal business. A different type of gasoline was used in each sample. The following results were obtained:

Type A	Type B
$\bar{X}_A = 25.0$ mpg	$\bar{X}_B = 24.8$ mpg
$s_A^2 = 1.44$	$s_B^2 = .81$

(a) Construct a 95% confidence interval estimate of the difference between mean gasoline mileages.
(b) Should the null hypothesis that there is no difference between the gasolines be accepted or rejected at the $\alpha = .05$ significance level?

14-26 An agronomist wishes to compare the corn yield per acre on which a nitrate-based fertilizer was used with the corn yield per acre on which a sulfate-based fertilizer was used. A sample of 100 acres fertilized with nitrates (A) yields an average of 56.2 bushels per acre, with a variance of 156.25. A sample of 150 acres fertilized with sulfates (B) yields an average of 52.6 bushels per acre, with a variance of 190.44.

For each of the null hypotheses (a), (b), and (c), indicate whether H_0 must be accepted or rejected at a significance level of $\alpha = .05$.
(a) There is no difference in crop yields between the two fertilizers.
(b) The nitrates provide at least the yield of the sulfates.
(c) The sulfates provide at least the yield of the nitrates.

14-27 Suppose that the company in Exercise 14-25 matches the cars using gasoline A with those using gasoline B, that the cars in each pair are nearly identical in many respects (make, age, engine size, accessories, and so on), and that both groups of cars are driven over the same route by the same professional driver. The results indicate that the mean difference between the gasoline mileages of the cars using gasoline A and the cars using gasoline B is .2 mile per gallon, with a standard deviation of .1.
(a) Construct a 95% confidence interval estimate of the difference between mean gasoline mileages.
(b) Should the null hypothesis that there is no difference between gasoline mileages be accepted or rejected at the $\alpha = .05$ significance level?

14-28 Suppose that the plots used to test the nitrate- and sulfate-based fertilizers in Exercise 14-26 were selected so that the two types of fertilizers were placed on neighboring 1-acre plots and that the yield of each sulfate-fertilized acre (B) was subtracted from the yield of its neighboring nitrate-fertilized acre (A). From a total of 100 pairs, the mean matched-pairs difference is 1.3 bushels, with a standard deviation of 3.2 bushels. Should the null hypothesis that there is no difference in yields due to choice of fertilizer be accepted or rejected at the $\alpha = .05$ significance level?

The frequent observation that things of a sort are disposed together in any place would lead us to conclude, upon discovering there any object of a particular sort, that there are laid up with it many others of the same sort.

THE REVEREND THOMAS BAYES (1763)

15

Chi-Square Applications

U ntil now, we have described procedures using the normal and Student t distributions, but many important statistical applications cannot be represented by these distributions. In this chapter, we will expand our basic repertoire considerably by investigating the *chi-square distribution*.

We will begin by applying the chi-square distribution to sample data *to test whether or not two variables are independent*. We have seen in applications of probability that statistical independence between sample observations permits us to streamline procedures markedly. Perhaps it is most important to know if variables are independent when alternative methods and treatments are being evaluated.

When population variables are qualitative characteristics (such as marital status, political affiliation, sex, state of health, type of treatment, or kind of response), the presence or absence of independence between variables can be used to draw important conclusions. A doctor who knows that preventive measures (vaccinated, unvaccinated) and resistance (diseased, not diseased) are dependent can conclude whether a vaccine is effective. A marketing researcher who learns that highest education level (elementary, high school, college) and brand preference (candidates A, B, C) are dependent might use this information to help choose advertising media. If severity of automobile accidents (property damage, injury, fatalities) is found to be dependent on where the accidents

495

occur (city streets, rural roads, highways), a public safety director may decide to revise traffic law-enforcement procedures.

In this chapter, we will also examine a second area of statistical inference— *comparing several population proportions*. This application is closely related to testing for independence. We have already learned to compare two populations in Chapter 14, but when three or more populations are tested, we follow a different procedure that makes use of the chi-square distribution.

Although these two applications of the chi-square distribution are the most common in business decision making, others exist. Inferences regarding the population variance are discussed as an optional topic later in this chapter. Chapter 17 considers a chi-square procedure called the goodness-of-fit test.

15-1   TESTING FOR INDEPENDENCE AND THE CHI-SQUARE DISTRIBUTION

The first new application of statistical inference that we will investigate is the testing procedure used to determine if two qualitative population variables are independent. In doing this, we will describe a special test statistic with the chi-square sampling distribution.

Independence Between Qualitative Variables

To describe what we mean by independence between two qualitative variables or population characteristics, we will briefly consider a basic concept of probability. Remember that two events A and B are statistically independent if the occurrence of either event does not affect the probability for the other. For example, drawing an *ace* from a fully shuffled deck of 52 ordinary playing cards is statistically independent of drawing a *club*:

$$\Pr[ace|club] = 1/13 = 4/52 = \Pr[ace]$$

This is true because the proportion of aces in the club suit (1/13) is the same as the proportion of aces in the entire deck (4/52).

We can extend this concept of statistical independence to populations and samples. Consider a population whose elementary units may be classified in terms of two qualitative variables A and B. If variable A represents, say, a person's sex, it will have two possible attributes—male and female; if variable B represents political party, its attributes will be Democrat, Republican, and so on. If we assume that men occur in the same proportion throughout the entire population as they do among Democrats, we can conclude that for the attributes of a randomly selected person,

$$\Pr[\text{male}|\text{Democrat}] = \Pr[\text{male}]$$

so that the *events* "male" and "Democrat" are statistically independent. If this is true, the multiplication law of probability for independent events tells us that

$$\Pr[\text{male } and \text{ Democrat}] = \Pr[\text{male}] \times \Pr[\text{Democrat}]$$

If a similar fact holds for all attribute combinations, the population variables will exhibit a very important property:

DEFINITION
Two qualitative population variables *A* and *B* are independent if the proportion of the total population having any particular attribute of *A* is the same as it is in the part of the population having a particular attribute of *B*, no matter which attributes are considered. This implies that the frequency of units having any particular attribute pair may be found by multiplying the respective frequencies for the individual attributes and dividing by the total number of observations.

EXAMPLE: SYMPHONY ORCHESTRA

There are 100 members in a symphony orchestra: 70 men and 30 women. Each member is categorized by the type of instrument he or she plays: woodwind, brass, string, or percussion. The following table shows the frequencies of the members in terms of two characteristics: sex and instrument. Here, the two characteristics are independent. The ratio of male woodwind players to the total number of woodwind players is 21/30 = .7. The same ratio is obtained for male brass, string, or percussion players: 14/20, 21/30, and 14/20, respectively. All equal .7, which is the proportion of men in the orchestra. Similarly, the ratio of female woodwind players to total woodwind players is .3, which is the same ratio obtained for any other instrument played by females and is also the proportion of females in the orchestra.

	Instrument Category				
Sex	(1) Woodwind	(2) Brass	(3) String	(4) Percussion	Total
(1) Male	21	14	21	14	70
(2) Female	9	6	9	6	30
Total	30	20	30	20	100

Because sex and instrument are independent, the frequency of male string players is equal to the product of the frequencies of the respective attributes divided by the total number of players in the orchestra, or

$$\frac{70 \times 30}{100} = 21$$

The frequency of female string players is found by multiplying the number of females by the number of string players and dividing by 100:

$$\frac{30 \times 30}{100} = 9$$

When the population frequencies are unknown, the presence or absence of independence between two variables is also unknown. A sample can be used to test for independence, but to do this we must extend the principles of hypothesis testing that we used earlier. It will be helpful to develop an example as an illustration.

Superior Oil Company markets its products through two distinctly different kinds of service station outlets. One is a company-owned chain, selling gasoline and other products under the Superior label. The other is a chain of franchised dealerships—independently owned and managed—advertised with Superior but also identified as Sentinel stations. The reason for the double identity is twofold.

1. By creating two types of stations, the company believes that it appeals to two segments of the market: those preferring to buy from a big operator, and those who like to buy from a local dealer (thereby receiving more individual attention and service).
2. The company-owned stations are major operations, located in high-volume traffic areas where the capital investment must be huge. A company-owned station in these locations is believed to be most effective and profitable. Franchise locations are mainly neighborhoods and small towns.

Another advantage is that when traveling on major highways, a customer who trades at a Sentinel station will stop at conveniently located company-owned stations. Likewise, an urban customer of Superior's company-owned stations will be more likely to trade at a Sentinel station when traveling on side roads or in rural areas.

Superior's new chairman is concerned about the value of operating in this fashion. He wonders whether the extra promotional costs are really justified and even if there are actually two separate market segments. He has requested

TABLE 15-1

Contingency Table for Actual Sample Results from a Superior Oil Customer Survey

| | Customer Preference | | |
Reason	(1) Sentinel Station	(2) Company Station	Total
(1) Location	32	8	40
(2) Quality of Service	12	2	14
(3) Cleanliness	13	3	16
(4) Personal Attention	56	35	91
(5) Mechanical Service	11	13	24
(6) Staff Appearance	6	9	15
Total	130	70	200

that his chief statistician conduct a survey to determine if there is any significant difference between the Sentinel customers and those trading at company-owned stations.

The statistician has mailed a questionnaire to a random sample of 200 credit card customers. The replies received have been partially tabulated in Table 15-1. Each subject has been categorized as preferring Sentinel or company stations. The primary reason given for each person's preference is placed in one of six categories: location, quality of service, cleanliness, personal attention, mechanical service, and staff appearance.

Contingency Tables and Expected Frequencies

In Table 15-1, the customer preference variable is represented by a column for each station type and the second variable by a row for each primary reason given for preferring a particular type of station. The value within each cell is the tally for those subjects classified as having the corresponding attributes. These cell numbers are referred to as *actual frequencies*. Such an arrangement of data is called a *contingency table*, because it accounts for all combinations of the factors being investigated—in other words, for all contingencies.

Using the statistician's data, we wish to find out whether or not the customer's primary stated reason for choosing where to trade actually matters in the final selection of the particular type of station. The null hypothesis is

TABLE 15-2
Contingency Table for Expected Sample Results under the Null Hypothesis of Independence for Superior Oil Variables

Reason	Customer Preference		Total
	(1) Sentinel Station	(2) Company Station	
(1) Location	40 × 130/200 = 26.00	40 × 70/200 = 14.00	40
(2) Quality of Service	14 × 130/200 = 9.10	14 × 70/200 = 4.90	14
(3) Cleanliness	16 × 130/200 = 10.40	16 × 70/200 = 5.60	16
(4) Personal Attention	91 × 130/200 = 59.15	91 × 70/200 = 31.85	91
(5) Mechanical Service	24 × 130/200 = 15.60	24 × 70/200 = 8.40	24
(6) Staff Appearance	15 × 130/200 = 9.75	15 × 70/200 = 5.25	15
Total	130	70	200

that the type of station preferred is not related to the stated reason. Stated another way, *the null hypothesis is that the two variables are independent.*

The first step in accepting or rejecting H_0 is to determine the kind of results that might be expected if the variables were truly independent. As a basis for comparison, we use the sample results that would be obtained on the average if the null hypothesis of independence was true. These hypothetical data are referred to as the *expected sample results.*

Table 15-2 is the contingency table for the expected results of this study. The total numbers of observations for each reason and type of station are the same. Because they are calculated in a manner consistent with the null hypothesis of independence, the cell entries are referred to as *expected frequencies.* These frequencies are denoted by the symbol f_e and appear in color in the table. The following expression summarizes the calculation of the

EXPECTED FREQUENCIES

$$f_e = \frac{\text{Row total} \times \text{Column total}}{n}$$

For example, the entry in row (1) and column (1) is

$$f_e = \frac{40 \times 130}{200} = 26.00$$

It is not necessary for the expected frequencies to be whole numbers. (Usually extending the frequency to two decimal places provides adequate accuracy.)

Now that the actual and the expected frequencies have been obtained, they must be compared before the null hypothesis of independence can be accepted or rejected. For this, a new statistic, the chi square, should be used.

The Chi-Square Statistic

A decision rule for determining independence must be established that provides a desirable balance between the probabilities for committing both the Type I error of rejecting independence when it actually exists and the Type II error of accepting independence when it does not exist. In practice, when the sample size is fixed in advance—as it is in our Superior Oil illustration—we can control only the Type I error probability α.

The test statistic we use must (1) measure the amount of deviation between the actual and the expected results and (2) have a sampling distribution that enables us to determine the Type I error probability α. Such a statistic is based on the individual differences between the actual and the expected frequencies in each cell. It is convenient to place both types of frequencies in a single combined contingency table, as shown in Table 15-3.

The following expression is used to calculate the test statistic, which we call the

CHI-SQUARE STATISTIC

$$\chi^2 = \sum \frac{(f_a - f_e)^2}{f_e}$$

where the symbol χ is the lowercase Greek *chi* (pronounced "kie") and χ^2 is read "chi square." The summation is taken over all cells in the contingency table. Table 15-4 shows the χ^2 calculations for the Superior Oil study results. There, we find that $\chi^2 = 16.929$.

The possible values of χ^2 range upwards from zero. If the deviation between f_a and f_e is large for a particular cell, the squared deviation $(f_a - f_e)^2$ is also large. In calculating χ^2, the squared deviations are divided by f_e before they are summed to ensure that any differences are not exaggerated simply because a large number of observations have been obtained. Large $(f_a - f_e)^2/f_e$ ratios occur when the actual and the expected results differ considerably, making

TABLE 15-3

Combined Contingency Table for Actual and Expected Frequencies

Reason	Customer Preference		Total
	(1) Sentinel Station	(2) Company Station	
(1) Location	f_a 32 / f_e 26.00	8 / 14.00	40
(2) Quality of Service	12 / 9.10	2 / 4.90	14
(3) Cleanliness	13 / 10.40	3 / 5.60	16
(4) Personal Attention	56 / 59.15	35 / 31.85	91
(5) Mechanical Service	11 / 15.60	13 / 8.40	24
(6) Staff Appearance	6 / 9.75	9 / 5.25	15
Total	130	70	200

TABLE 15-4

Chi-Square Calculations for Superior Oil Results

(Row, Column)	Actual Frequency f_a	Expected Frequency f_e	$f_a - f_e$	$(f_a - f_e)^2$	$\dfrac{(f_a - f_e)^2}{f_e}$
(1, 1)	32	26.00	6.00	36.0000	1.385
(1, 2)	8	14.00	−6.00	36.0000	2.571
(2, 1)	12	9.10	−2.90	8.4100	.924
(2, 2)	2	4.90	2.90	8.4100	1.716
(3, 1)	13	10.40	2.60	6.7600	.650
(3, 2)	3	5.60	−2.60	6.7600	1.207
(4, 1)	56	59.15	−3.15	9.9225	.168
(4, 2)	35	31.85	3.15	9.9225	.312
(5, 1)	11	15.60	−4.60	21.1600	1.356
(5, 2)	13	8.40	4.60	21.1600	2.519
(6, 1)	6	9.75	−3.75	14.0625	1.442
(6, 2)	9	5.25	3.75	14.0625	2.679
	200	200.00	0.00		$\chi^2 = 16.929$

the size of χ^2 large. Therefore, the more the sample results deviate from what would be expected if independence were true, the larger the value of χ^2 will be, and vice versa.

Before we can draw any conclusions regarding customer preference and reason variables, we must determine whether or not the value of χ^2 that we obtained from the sample is inconsistent with the null hypothesis that these variables are independent. As in earlier hypothesis-testing procedures, the sampling distribution of χ^2 will be used to obtain the critical value of this test statistic.

The Chi-Square Distribution

The chi-square distribution is a theoretical probability distribution that, under the proper conditions, may be used as the sampling distribution of χ^2. It is described by a single parameter—the number of degrees of freedom—that has much the same meaning as it did in our discussion of the Student t distribution. (The procedure used to determine the number of degrees of freedom will be given later in this section.)

Figure 15-1 shows curves for the chi-square distributions when the degrees of freedom are 2, 4, 10, and 20. Note that these curves are positively skewed

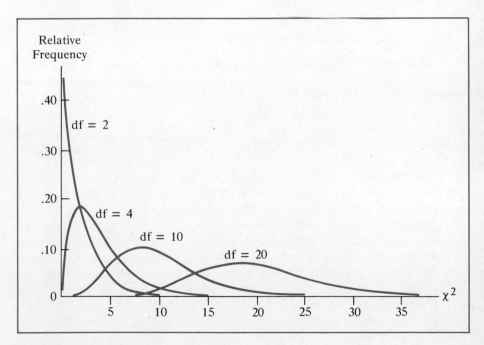

FIGURE 15-1

Various curves for the chi-square distribution.

but that the degree of skew declines as the number of degrees of freedom increases, until eventually the chi-square distribution approaches the normal distribution.

Appendix Table H provides upper tail areas for the chi-square distribution. We use the symbol χ_α^2 to represent the chi-square value for which the upper tail area is α:

$$\alpha = \Pr[\chi^2 \geq \chi_\alpha^2]$$

Like the Student t distribution, there is a separate chi-square distribution for each number of degrees of freedom, and a row corresponding to each of these numbers appears in the table of upper tail areas. The tail areas are given at the head of each column, and the entries in the body of the table are the corresponding values of χ_α^2. A portion of this table is reproduced here.

Degrees of Freedom		Upper Tail Area			
	$\cdots$	.90	$\cdots$	.10	.05	$\cdots$
1		.0158		2.706	3.841
2		.211		4.605	5.991
3		.584		6.251	7.815
4		1.064		7.779	9.488
5		1.610		9.236	11.070
6		2.204		10.645	12.592
$\vdots$		$\vdots$		$\vdots$	$\vdots$

The curve for the chi-square distribution with 5 degrees of freedom is shown in Figure 15-2. To find the upper 5% value of χ^2, denoted by $\chi_{.05}^2$, we read the entry in the row for 5 degrees of freedom and the column for area .05, obtaining $\chi_{.05}^2 = 11.070$. Thus, we have

$$.05 = \Pr[\chi^2 \geq 11.070]$$

We can also use Appendix Table H to find areas above points lying in the lower portion. For example, for an upper tail area of .90, we obtain the value $\chi_{.90}^2 = 1.610$ at 5 degrees of freedom (see Figure 15-2).

Remember that we used row and column totals to estimate the expected frequencies in Table 15-2 for the Superior Oil study. We found the cell entries in that contingency table by multiplying the respective marginal totals for each row and column and then dividing by the sample size. Because the six row totals must sum to n, the values of any five rows automatically fix the value of the sixth row. We can say that five of the row totals are "free" and the sixth is "fixed." Likewise, one of two column totals is free and the second is fixed. Although there are $6 \times 2 = 12$ cells, we are free to specify only $5 \times 1 = 5$ of these cells. Our number of degrees of freedom is therefore 5. In general, we

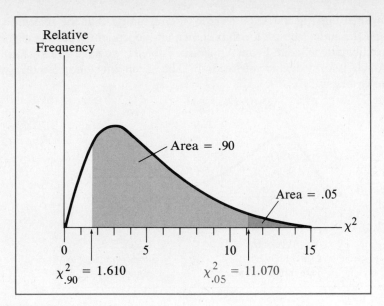

FIGURE 15-2
Chi-square curve for 5 degrees of freedom.

establish the

RULE FOR THE NUMBER OF DEGREES OF FREEDOM

(Number of rows − 1) × (Number of columns − 1)

The Hypothesis-Testing Steps

STEP 1
Formulate the null hypothesis. We are testing the null hypothesis that a customer's preferred station type and the reason given for choosing it are independent.

H_0: The two variables are independent.

STEP 2
Select the test statistic. The chi-square statistic will be used.

STEP 3
Obtain the significance level α and identify the acceptance and rejection regions.
Suppose that our researcher wants to protect himself at the $\alpha = .05$ significance

level against making the Type I error of concluding that the selected station type and the stated reason for that choice are dependent when these variables are actually independent. From Appendix Table H, we have already found that $\chi^2_{.05} = 11.070$ for 5 degrees of freedom. The acceptance and rejection regions are shown below.

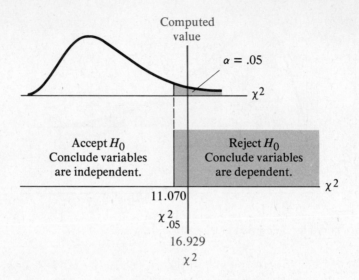

The interpretation of $\chi^2_{.05} = 11.070$ is that due to chance sampling error, this number will be exceeded on the average by only 5% of all χ^2 values calculated from repeated samples, each taken from a population where the null hypothesis is true. Thus, if the two variables are independent, about 5% of the sample results will disagree so much that independence will be rejected. Each of these samples will have a value, such as 13.5 or 14.2, that exceeds 11.070.

The preceding decision rule illustrates a convenient feature of this procedure: *Whenever a chi-square test is used to compare frequencies, it is upper-tailed.* Stated another way, large χ^2 values tend to refute H_0, whereas small χ^2 values confirm the null hypothesis.

STEP 4
Collect the sample and compute the value of the test statistic. This step includes finding the values of f_a and f_e for the combined contingency table and then computing χ^2. For the Superior Oil study results, we have already found $\chi^2 = 16.929$.

STEP 5
Make the decision. The decision maker's computed value of $\chi^2 = 16.929$ falls in the rejection region. The null hypothesis must be rejected, and the statistician

must conclude that the type of station preferred depends on the customer's stated primary reason for choosing where he or she trades.

Special Considerations

If a smaller significance level of, say, $\alpha = .001$ were desired, the researcher would have obtained $\chi^2_{.001} = 20.517$ from Appendix Table H for 5 degrees of freedom, and because this value is larger than the 16.929 computed from sample data, the null hypothesis of independent variables would have been accepted. In such cases, there may be a considerable chance of committing the type II error of accepting H_0 when the variables are not independent. (In our illustration, this would mean concluding that reason for choice use does not influence type preference when it does.) These errors can occur in many ways, and finding β probabilities is too difficult a task to undertake here. The chance of committing a Type II error can be kept within reasonable boundaries only by making the sample size large.

The true sampling distribution of χ^2 that is calculated for testing independence is only approximated by the chi-square distribution. This approximation is similar to substituting the normal distribution for the binomial distribution in certain instances. The approximation is usually adequate if the sample size is sufficiently large. In practice, the sample will be large enough when the expected frequencies for each cell are 5 or more. If some expected cell frequencies are smaller than 5, this requirement can be met by combining two rows or columns before calculating χ^2. A corresponding reduction in the number of degrees of freedom must then be made to account for the lower number of cells.

EXERCISES

15-1 Find the value of the chi-square statistic that holds for each upper tail area α and each number of degrees of freedom (df) stated below:

	(a)	(b)	(c)	(d)	(e)
α	.05	.01	.99	.10	.80
Degrees of freedom	10	20	18	6	29

15-2 A marketing researcher is testing to determine if a person's sex is independent of a preference for fruits. For each of the following situations, where the sample results have been summarized, find the number of degrees of freedom and the critical value. Then indicate whether the null hypothesis of independence should be accepted or rejected.
(a) Bananas, apples, and pears are considered. $\chi^2 = 7.85$ and $\alpha = .05$.

(b) Pineapples, guavas, papayas, and passionfruit are considered. $\chi^2 = 15.23$ and $\alpha = .01$.

(c) Boysenberries, blueberries, huckleberries, blackberries, and strawberries are considered. $\chi^2 = 7.801$ and $\alpha = .10$.

(d) Plums, apricots, peaches, cherries, persimmons, and nectarines are considered. $\chi^2 = 10.99$ and $\alpha = .05$.

15-3 Various random samples have been selected and the chi-square statistics have been computed for each of the following independence-testing situations. At the stated significance level, determine the number of degrees of freedom and the critical value in each situation. Then indicate whether or not the null hypothesis of independence should be accepted for each of the actual chi-square statistics obtained.

(a) Sex (male, female) versus marital status (single, married), with $\alpha = .05$ and $\chi^2 = 3.62$.

(b) College major (liberal arts, science, social science, professional) versus type of employment (manufacturing, service, government), with $\alpha = .01$ and $\chi^2 = 23.885$.

(c) Number of siblings (0, 1, 2, 3, or more) versus desired family size (0, 1, 2, 3, or more), with $\alpha = .10$ and $\chi^2 = 14.753$.

(d) Political affiliation (Republican, Democrat, other) versus sexual attitude (repressive, permissive), with $\alpha = .01$ and $\chi^2 = 8.063$.

15-4 A contingency table containing actual sample frequencies is given below:

Preference	Marital Status (1) Single	(2) Married	Total
(1) Brand A	20	10	30
(2) Brand B	20	50	70
Total	40	60	100

(a) Perform hypothesis-testing Step 3.
(1) Find the number of degrees of freedom.
(2) Find the critical value of the test statistic and identify the acceptance and rejection regions. Assume a significance level of $\alpha = .01$.

(b) Perform hypothesis-testing Step 4.
(1) Find the expected frequencies under the null hypothesis of independence between marital status and brand preference. Enter these in a combined contingency table.
(2) Calculate the chi-square statistic.

(c) Should the null hypothesis that martial status and brand preference are independent be accepted or rejected?

15-5 An advertiser wishes to determine whether there are any significant differences in regular series program preferences between male and female television viewers. A random sample of persons has been interviewed, and each person has been asked to indicate which one of five program types he or she prefers. The results are provided below:

Preferred Program Type	Viewer's Sex		Total
	(1) Male	(2) Female	
(1) Western	32	18	50
(2) Situation Comedy	17	13	30
(3) Drama	27	33	60
(4) Comedy	13	7	20
(5) Variety	24	16	40
Total	113	87	200

The null hypothesis is that a viewer's sex and preference are independent.
(a) Determine the expected results.
(b) Calculate the chi-square statistic.
(c) How many degrees of freedom are associated with this test statistic?
(d) If the advertiser wishes to protect herself against committing the Type I error of incorrectly concluding sex is dependent on program preference at a significance level of $\alpha = .01$, determine the critical value for the test statistic. Should the hypothesis of independence be accepted?
(e) Would your conclusion in (d) change if an $\alpha = .05$ significance level were chosen instead? Explain.

15-6 Repeat Exercise 15-5 if the following data were obtained instead:

Preferred Program Type	Viewer's Sex		Total
	(1) Male	(2) Female	
(1) Western	37	13	50
(2) Situation Comedy	21	23	44
(3) Drama	26	32	58
(4) Comedy	19	15	34
(5) Variety	22	18	40
Total	125	101	226

15-2   TESTING FOR THE EQUALITY OF SEVERAL PROPORTIONS

An important statistical question involves a comparison of several population parameters. In Chapter 14, we described the inferences regarding two populations and the tests used to compare *two* population means or proportions. There, we determined which of two television programs was preferred by using the proportions of two independent samples taken from different populations. If we wish to compare more than two programs, we can still make pair comparisons as we did in Chapter 14. But sometimes it is better to test the population proportions simultaneously to see if they really differ. In comparing three or more population proportions, we can take a different approach and apply the procedures that we used earlier in testing for independence.

Illustration: Remembering Advertisements

We will illustrate the procedure for testing several proportions with a situation commonly encountered in the advertising field. An advertising agency wishes to determine whether there are any differences in terms of reader recall among three kinds of magazine advertisements. One ad is humorous, the second is quite technical, and the third is a pictorial comparison with competing brands. A national magazine with three regional editions is chosen for the test, and a different quarter-page advertisement is placed in each of its editions. A random sample of persons is chosen from a list of subscribers in each region, and one month after the ads have appeared, each person in each sample is visited. All participants are shown five ads of similar format, four of which are fakes, and are asked if they remember any of the five ads. Those selecting the correct ad are included in the tally of rememberers, and those unable to select the correct ad are classified as nonrememberers. A brief quiz is administered after the identification test to determine whether or not the magazine was read (to avoid prestige bias by persons who falsely claim to have read the ad), and nonreaders are eliminated from the sample.

STEP 1

Formulate the null hypothesis. The null hypothesis is that there are no differences in the mnemonic properties of the three kinds of advertisements. This may be expressed in terms of the proportions of the readers of the three magazine editions who remembered the advertisements. Thus, letting π_1 represent the proportion remembering the first ad, π_2 the second, and π_3 the third, we can express the null hypothesis as

$$H_0 : \pi_1 = \pi_2 = \pi_3$$

The results are provided in Table 15-5. Recall that in testing for independence, the expected cell frequencies are determined by assuming equal ratios

TABLE 15-5

Results of Mnemonic Advertising Samples

	Type of Advertisement		
	(1) Humorous	(2) Technical	(3) Comparative
Number of rememberers	25	10	7
Number of nonrememberers	73	93	108
Number of readers	98	103	115
Proportion of rememberers	$P_1 = .255$	$P_2 = .097$	$P_3 = .061$

or proportions. In testing for independence, we are really testing for equality of proportions.

STEP 2

Select the test statistic. The chi-square statistic will be used.

STEP 3

Obtain the significance level α and identify the acceptance and rejection regions. Each of the three types of advertisements will be represented by a column in the contingency table. The rememberers and nonrememberers will be represented by rows. The number of degrees of freedom is

(Number of rows $- 1) \times$ (Number of columns $- 1) = (2 - 1) \times (3 - 1) = 2$

The chosen level of significance is $\alpha = .01$, and Appendix Table H provides the critical value of $X^2_{.01} = 9.210$. The acceptance and rejection regions are shown below.

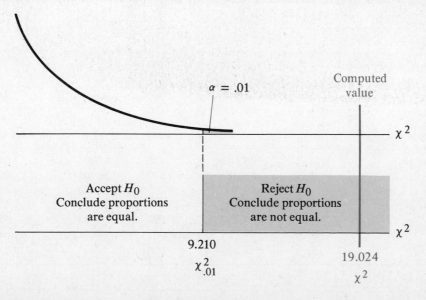

TABLE 15-6

Combined Contingency Table for Actual and Expected Frequencies for Mnemonic Advertisement Study

| | Type of Advertisement | | | |
	(1) Humorous	(2) Technical	(3) Comparative	Total
(1) Rememberers	25 / 13.025	10 / 13.690	7 / 15.285	42
(2) Nonrememberers	73 / 84.975	93 / 89.310	108 / 99.715	274
Total	98	103	115	316

STEP 4

Collect the sample and compute the value of the test statistic. Table 15-6 is the combined contingency table for the actual and the expected results. The chi-square statistic is calculated in the same manner as it is in testing for independence. These calculations are provided in Table 15-7.

STEP 5

Make the decision. Because the computed value of $\chi^2 = 19.024$ falls in the rejection region, *the null hypothesis of equal proportions must be rejected* at the .01 significance level. The agency must conclude that the three ads are not equally easy to remember. (The results are also significant at the .001 level, because $\chi^2_{.001} = 13.815$.)

TABLE 15-7

Chi-Square Calculations for Mnemonic Advertisement Study

(Row, Column)	Actual Frequency f_a	Expected Frequency f_e	$f_a - f_e$	$(f_a - f_e)^2$	$\dfrac{(f_a - f_e)^2}{f_e}$
(1, 1)	25	13.025	11.975	143.4006	11.010
(1, 2)	10	13.690	−3.690	13.6161	.995
(1, 3)	7	15.285	−8.285	68.6412	4.491
(2, 1)	73	84.975	−11.975	143.4006	1.688
(2, 2)	93	89.310	3.690	13.6161	.152
(2, 3)	108	99.715	8.285	68.6412	.688
	316	316.000	0.000		$\chi^2 = 19.024$

EXERCISES

15-7 In planning her campaign strategy to determine how to approach a particular issue, a congressional candidate wants to know if there are any differences in the proportion of voters who favor the issue among her rural, suburban, and urban constituents. She has collected sample opinions and has obtained the following results:

	Rural	Suburban	Urban
In favor	65	63	52
Not in favor	35	37	48

(a) Construct a combined contingency table for the actual and the expected frequencies.
(b) Calculate the chi-square statistic.
(c) Find the number of degrees of freedom and the critical value for this test at the $\alpha = .01$ significance level.
(d) Should the null hypothesis that the proportion in favor of the issue is equal among constituent types be accepted or rejected?

15-8 A pastry chef who wishes to determine whether the proportion of unsatisfactory bear claws is affected by oven temperature has baked batches at 350°, 400°, and 425° and has obtained the following results:

	Temperature		
	350°	400°	425°
Number satisfactory	132	128	111
Number unsatisfactory	14	17	35

The chef wishes to test the null hypothesis of identical proportions at an $\alpha = .05$ significance level. What conclusion should he reach?

15-9 The quality-control manager of an electronics assembly plant wants to know if the day of the week influences the number of erroneous cable harness assemblies. He suspects that on Mondays and Fridays the error rate is significantly higher. If he finds that the day of the week does make a difference, he will conduct a more detailed study to determine on which days the error rate is higher and will recommend that cables be assembled only on special days. The manager has collected a random sample of assemblies produced on different days of the week and obtained the following results:

	Day of the Week				
	Monday	Tuesday	Wednesday	Thursday	Friday
Number of erroneous assemblies	32	12	15	18	27
Number of correct assemblies	95	87	91	79	73

The manager wishes to test the null hypothesis that the proportions of erroneous assemblies are identical at the $\alpha = .05$ significance level. What conclusion should he reach?

15-10 The quality-control manager in an appliance factory wishes to determine if the proportion of defective toasters is affected by the speed of the assembly line. She has taken three samples of 100 toasters from the line when it was running at three different speeds. The proportions of defectives obtained are as follows: .05 for 100 parts per hour; .07 for 150 parts per hour; and .10 for 200 parts per hour.

  Will the manager conclude that the proportions differ if she wishes to protect herself at the $\alpha = .01$ level of significance against making the wrong decision? What is the lowest significance level at which the results will lead her to conclude that the proportion of defectives varies with line speed?

15-3   OTHER CHI-SQUARE APPLICATIONS

As mentioned earlier, testing for independence and comparing several population proportions are the most common uses of the chi-square distribution in business decision making. But this versatile distribution has wider applications.

Optional Section 15-4 tells us how the chi-square distribution can be used to make inferences about the population variance. Until now, we have not estimated the population variance σ^2 by means of *confidence intervals* or tested *hypotheses about* σ^2. Although important, inferences concerning σ^2 are less common than inferences regarding μ and π, where $\bar{X}$ and P are used as estimators and test statistics. We have waited until this point to examine this area of inference because the chi-square distribution plays a role in the sampling distribution of s^2—the test statistic used in making inferences about σ^2.

Chapter 17 considers a fourth chi-square application—the *goodness-of-fit test*. This procedure is useful in determining whether a variety of technical assumptions hold with respect to the characteristics of the population sampled. The goodness-of-fit test also allows us to decide whether a particular probability distribution actually applies.

REVIEW EXERCISES

15-11 A government population-planning official in an Asian country must determine if traditional contraceptive procedures will work with the rural poor. Using a sample of 250 women, each practicing one of four methods of contraception for two years, he has obtained the following results:

	Contraceptive Method				
	(1) *IUD*	(2) *Pill*	(3) *Mechanical*	(4) *Other*	Total
(1) *Pregnant*	20	10	30	30	90
(2) *Not Pregnant*	40	50	50	20	160
Total	60	60	80	50	250

Should the population planner conclude that any of the four contraceptive procedures will work at the $\alpha = .01$ significance level?

15-12 A marketing student wishes to test the null hypothesis that people's sexual attitudes are independent of the kinds of cars they own. A random sample of 100 persons was selected, and each was interviewed and placed into a dominant sexual-attitude category. The following results were obtained:

Dominant Sexual Attitude	Previously Purchased Car			
	(1) *Foreign*	(2) *Small*	(3) *Large*	Total
(1) *Repressed*	3	6	11	20
(2) *Ho-Hum*	13	25	12	50
(3) *Swinger*	14	9	7	30
Total	30	40	30	100

At the $\alpha = .05$ significance level, what should the student conclude?

15-13 A sales manager is determining a marketing strategy. To do this, she must determine whether her product's appeal is generally broad or whether it varies from region to region. A sampling study has provided the following regional data:

	East	Central	South	West
Number preferring	21	29	23	16
Number not preferring	39	41	27	24

Should the manager conclude that her product is equally preferred in all regions? (Use $\alpha = .01$.)

15-14 Suppose that the sales manager in Exercise 15-13 wishes to test the null hypothesis that her product is equally strong in the East and the West. At the $\alpha = .05$ significance level, what should she conclude?

15-15 A market researcher for a soap manufacturer wishes to determine whether the amount of time that people spend watching daytime television influences their choice of laundry bleach. From a random sample of 260 persons, he has determined their viewing habits and the type of bleach they use most often. The results are shown below:

Television Viewing Time	Bleach Types		
	Liquid Chlorine	Dry Chlorine	Oxygen
Light	53	5	38
Moderate	29	10	35
Heavy	45	8	37

Should the researcher conclude that television viewing time and type of bleach used are independent? Test at the $\alpha = .05$ level of significance.

15-16 Suppose that the market researcher in Exercise 15-15 has decided that the amount of television viewing time does affect the type of bleach used. He now wishes to determine whether a preference for "soap operas" or game shows affects the choice of bleach. Another sample is collected, and the following results are obtained:

Program Preference	Bleach Types		
	Liquid Chlorine	Dry Chlorine	Oxygen
Soap opera	58	5	55
Game show	48	13	47
Other	16	12	27

Are preference of daytime programming and type of bleach used independent? Test at the $\alpha = .05$ level of significance.

15-4 OPTIONAL TOPIC: INFERENCES REGARDING THE POPULATION VARIANCE

We have seen that knowledge of population variability is an important element of statistical analysis. For instance, an educator may choose a particular teaching method because it provides low dispersion in student achievement levels, thereby making the teacher's job less demanding than a high-variability procedure would. A bank policy might favor a single waiting line that feeds into several teller windows rather than separate lines, because it has been estimated that the single line will minimize variability in patron waiting times— even though the mean time spent in line is the same in either case. In a car pool, tires with a low variability in wear may be preferred to more durable tires with greater variability in useful lifetime, simply because it is cheaper to replace an entire set of tires periodically for each car.

Like the population mean, σ^2 *is ordinarily unknown and its value must be estimated using sample data.* Until now, we have been using the sample variance

$$s^2 = \frac{\sum X^2 - n\bar{X}^2}{n - 1}$$

to make point estimates of σ^2. These estimates have been made as an adjunct to inferences regarding μ, where confidence intervals and hypothesis-testing decision rules have been constructed using s^2 in place of σ^2.

Inferences about σ^2 are made in a similar way to inferences about μ, in that s^2 can be used as an unbiased and consistent estimator of the population variance, playing the analogous role with σ^2 that $\bar{X}$ does with μ. Thus, s^2 serves as the basis either for constructing confidence interval estimates or for testing hypotheses regarding σ^2.

Probabilities for the s^2 Variable

In the *planning stage* of a sampling study, s^2 must be treated as a random variable, because its value is not yet determined and is subject to chance variation. In discussing how to apply probability analysis to s^2, it will be helpful to briefly review our earlier treatment of $\bar{X}$.

Recall that when σ^2 is unknown, the Student t distribution is used to make inferences about the population mean. The chi-square distribution assigns a role to s^2 that is similar to the role the Student t distribution assigns to $\bar{X}$. *Before the sample data are collected, s^2 must be viewed as a random variable* (just like $\bar{X}$). Multiplying s^2 by $n - 1$ and dividing by σ^2 converts it into the

CHI-SQUARE RANDOM VARIABLE

$$\chi^2 = \frac{(n - 1)s^2}{\sigma^2}$$

which exhibits the chi-square distribution.* The number of degrees of freedom is $n - 1$.

To illustrate this, suppose that a sample of $n = 25$ heights has been randomly selected from a population of men with a standard deviation of $\sigma = 3$ inches. To find the probability that s^2 falls within a range of numbers, two limits must be chosen for the chi-square variable. This may be done so that χ^2 is just as likely to fall below its *lower limit*, denoted by χ_L^2, as it is to fall above

---

* As is true of the Student t distribution, to apply the chi-square distribution, the *population* must be normally distributed.

its *upper limit* χ_U^2. For example, if these limits are to be chosen for a 0.90 probability, then

$$.90 = \Pr\left[\chi_L^2 \leq \frac{(n-1)s^2}{\sigma^2} \leq \chi_U^2\right]$$

In this case, χ^2 values are found from Appendix Table H for $n - 1 = 25 - 1 = 24$ degrees of freedom. The upper tail areas are .95 for the lower limit and .05 for the upper limit, and the area under the portion of the chi-square curve between these limits is .90:

$$\chi_L^2 = \chi_{.95}^2 = 13.848$$

$$\chi_U^2 = \chi_{.05}^2 = 36.415$$

Two separate table values are required to find the limits χ_L^2 and χ_U^2, because the chi-square distribution is *positively skewed*. This is in contrast to our earlier applications involving the normal and the Student t curves, both of which are symmetrical.

With $\sigma = 3$ inches ($\sigma^2 = 9$),

$$.90 = \Pr\left[13.848 \leq \frac{24s^2}{9} \leq 36.415\right]$$

so that

$$.90 = \Pr\left[\frac{9(13.848)}{24} \leq s^2 \leq \frac{9(36.415)}{24}\right] = \Pr[5.19 \leq s^2 \leq 13.66]$$

Thus, there is a 90% chance that a sample of $n = 25$ observations from a population with $\sigma = 3$ inches will provide a sample variance of between 5.19 and 13.66. (In other words, there is a 90% probability that s will lie between $\sqrt{5.19} = 2.3$ and $\sqrt{13.66} = 3.7$ inches.)

Confidence Interval Estimate of σ^2

The foregoing probability analysis presumes that the value of the population variance is known. However, this is not usually the case in a sampling situation. *After the sample data are collected*, s^2 can be calculated and can serve as the basis for making inferences about σ^2. Treating σ^2 as the unknown and s^2 as the known quantities, we can transform the probability interval for s^2 into the form for making an interval estimate of σ^2. The following expression is used to construct the

CONFIDENCE INTERVAL ESTIMATE OF σ^2

$$\frac{(n-1)s^2}{\chi_U^2} \leq \sigma^2 \leq \frac{(n-1)s^2}{\chi_L^2}$$

Here, χ_L^2 and χ_U^2 must be chosen from Appendix Table H, using $n - 1$ degrees of freedom, to correspond to the confidence level desired.

EXAMPLE: VARIABILITY IN PRISONER IQ

A sociologist wishes to estimate the variability in IQ on the Stanford-Binet scale for the inmates at a certain state prison. For a sample of $n = 30$ prisoners, he has obtained a standard deviation of $s = 10.2$ points, which is considerably below the corresponding figure of 16 points for all test takers.

The sample variance is $s^2 = (10.2)^2 = 104.04$. Using this result, the sociologist can construct a 90% confidence interval estimate of the variance for the IQs of the prison population. For $30 - 1 = 29$ degrees of freedom, Appendix Table H provides the following lower and upper limits for the chi-square variable:

$$\chi_L^2 = \chi_{.95}^2 = 17.708$$

$$\chi_U^2 = \chi_{.05}^2 = 42.557$$

The following interval estimate is obtained for σ^2:

$$\frac{(30 - 1)104.04}{42.557} \leq \sigma^2 \leq \frac{(30 - 1)104.04}{17.708}$$

or

$$70.8969 \leq \sigma^2 \leq 170.3840$$

To estimate the population standard deviation of these results, the square root of these terms is taken, and the following 90% confidence interval for σ is obtained:

$$\sqrt{70.8969} \leq \sigma \leq \sqrt{170.3840}$$

or

$$8.42 \leq \sigma \leq 13.05$$

Because this interval lies well below 16, the sociologist may safely conclude that the prisoners have a lower variability in IQ than the population as a whole. Later in this section, we will learn to use hypothesis-testing concepts to make such comparisons.

In explaining why inmates exhibit less variability in IQ, the sociologist concludes: You can't be dull-witted and be a serious criminal; besides, retarded persons who commit crimes are incarcerated in other kinds of institutions. On the other hand, the smartest people, by and large, avoid a life of crime. In effect, the tails of the normal curve for IQ are underrepresented in prisons, which explains the lower variability in measured intelligence.

Testing Hypotheses Regarding σ^2

Hypothesis tests concerning σ^2 can be conducted in a manner similar to the tests described in earlier chapters. The null hypothesis regarding σ^2 may be either one- or two-sided, and decision rules can be constructed in terms of the chi-square variable, which serves as the test statistic. For a hypothesized value of the population variance, denoted by σ_0^2, we use the appropriate

TEST STATISTIC FOR THE VARIANCE

$$\chi^2 = \frac{(n-1)s^2}{\sigma_0^2}$$

The significance level α establishes the critical values of this statistic.

For one-sided tests, the decision rule is based on χ_α^2, which must correspond to the upper tail of the chi-square curve, or on $\chi_{1-\alpha}^2$, which must correspond to the lower tail of that curve. This is true because the critical value of χ^2 for a lower-tailed test must be the point *below* which the area is α. The cases may be summarized as follows:

Lower-Tailed Test	Upper-Tailed Test
$H_0 : \sigma^2 \geq \sigma_0^2$	$H_0 : \sigma^2 \leq \sigma_0^2$
Critical value $= \chi_{1-\alpha}^2$	Critical value $= \chi_\alpha^2$

That predictability and variability are related can be illustrated by the attitudes of people wait in lines. A lengthy waiting time is more acceptable if the variability is smaller, even though the average wait may be the same. When the variability is smaller, the inconvenience of waiting becomes more predictable. This accounts for the fact that many businesses and government offices dealing with the public have instituted a "single-line" policy to replace the chaotic procedure of maintaining independent lines at various service areas (like supermarket checkstands or windows at a post office). Although the mean waiting time is not greatly affected by the single-line policy, the variability in waiting time is.

Illustration: Variability in Waiting Time

A particular postmaster has determined that the current procedure of separate lines yields a standard deviation in waiting times on December mornings of $\sigma_0 = 10$ minutes per customer. She wishes to implement the single-line policy on a trial basis to see if a reduction in waiting-time variability is achieved. A sample of 30 customers were monitored, and their waiting times were determined.

STEP 1

Formulate the null hypothesis. As her null hypothesis, the postmaster assumes that the variability in waiting times will be at least as great under the experimental procedure, or

$$H_0 : \sigma^2 \geq \sigma_0^2 \quad (10^2 = 100)$$

STEP 2

Select the test statistic. The chi-square statistic will be used.

STEP 3

Obtain the significance level α and identify the acceptance and rejection regions. This test is lower-tailed, because the null hypothesis is refuted by a small value of s^2 (and therefore χ^2). A significance level of $\alpha = .01$ is desired. For $30 - 1 = 29$ degrees of freedom, the critical value of the chi-square statistic is

$$\chi_{1-\alpha}^2 = \chi_{.99}^2 = 14.256$$

The postmaster's acceptance and rejection regions are shown below.

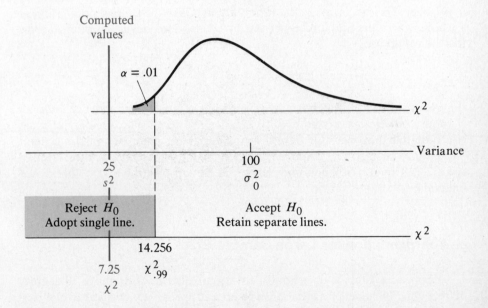

STEP 4

Collect the sample and compute the value of the test statistic. A sample standard deviation of $s = 5$ minutes is obtained. The computed value of the test statistic is

$$\chi^2 = \frac{(30 - 1)(5)^2}{(10)^2} = 7.25$$

STEP 5

Make the decision. Because the computed value of $\chi^2 = 7.25$ falls in the rejection region, the null hypothesis must be *rejected*. The postmaster should adopt the new single-line system.

When the null hypothesis takes the form $H_0:\sigma^2 = \sigma_0^2$, a two-sided test applies. Large or small values of the sample variance tend to refute H_0. Thus, either a large or a small calculation for χ^2 will result in a rejection of H_0. As with the two-sided tests that we discussed in earlier chapters, a convenient procedure for making a decision is to construct a confidence interval that corresponds to the significance level. If σ_0^2 falls outside this interval, H_0 must be rejected; if the interval contains σ_0^2, H_0 must be accepted.

To illustrate this procedure, we will again consider the variance for prisoner IQs in the example on page 519. There, we obtained the 90% confidence interval

$$70.8969 \leq \sigma^2 \leq 170.3840$$

Now suppose the sociologist wishes to test the null hypothesis that the variability in prisoner IQs is the same as the variability in IQs in the general population as a whole, where the Stanford-Binet test has a standard deviation of 16. To do this, he would use

$$H_0:\sigma^2 = \sigma_0^2 \quad (16^2 = 256)$$

Since $\sigma_0^2 = 256$ lies above the upper limit of the 90% confidence interval, H_0 must be *rejected* at the $\alpha = .10$ significance level. The sociologist must conclude that prisoner IQs have a lower variability than IQs in the general population. In fact, H_0 could be rejected at the lower significance level of $\alpha = .02$. (As an exercise, you may verify this by constructing a 98% confidence interval.)

Normal Approximation for Large Sample Sizes

Appendix Table H does not list chi-square distributions above 30 degrees of freedom. *For values of n larger than 30, we can use the normal curve to approximate the chi-square distribution.* It can be shown that χ^2 has a mean equal to the degrees of freedom and a variance equal to twice that amount. Thus, we can compute the applicable normal deviate from

$$z = \frac{\chi^2 - (n - 1)}{\sqrt{2(n - 1)}}$$

This expression is useful for testing hypotheses about σ^2. To construct a confidence interval, however, it is more convenient to express the critical values of χ^2 in terms of the tabled value of z by using the equivalent expressions

$$X_L^2 = \chi_{1-\alpha}^2 = n - 1 - z\sqrt{2(n-1)}$$

and

$$\chi_U^2 = \chi_\alpha^2 = n - 1 + z\sqrt{2(n-1)}$$

EXAMPLE: EFFECTIVENESS OF A TRANQUILIZER

A medical researcher who wishes to determine how effectively a particular tranquilizer induces sleep administers the drug to a sample of $n = 100$ patients. Based on the sample standard deviation in times required to onset of sleep, $s = 7.3$ minutes, the researcher can construct a 95% confidence interval for the variability in time to sleep. Using $z = 1.96$ (so that the area under the standard normal curve between $\pm z$ is .95 and the area in each tail is $\alpha = .025$), the limits for the chi-square variable are

$$\chi_L^2 = \chi_{.975}^2 = 99 - 1.96\sqrt{2(99)}$$

$$= 99 - 27.58 = 71.42$$

$$\chi_U^2 = \chi_{.025}^2 = 99 + 27.58 = 126.58$$

Using these values, with $s = 7.3$ and $n = 100$, the researcher can construct the 95% confidence interval:

$$\frac{(n-1)s^2}{\chi_U^2} \leq \sigma^2 \leq \frac{(n-1)s^2}{\chi_L^2}$$

$$\frac{99(7.3)^2}{126.58} \leq \sigma^2 \leq \frac{99(7.3)^2}{71.42}$$

$$41.68 \leq \sigma^2 \leq 73.87$$

Taking the square root of each term reveals a 95% confidence that the population standard deviation will fall between 6.46 and 8.59 minutes.

Suppose our researcher wishes to test the null hypothesis that the variability in the new sleeping drug is the same as the variability in the old drug, for which previous records show the standard deviation is 8 minutes. Then

$$H_0: \sigma^2 = \sigma_0^2 \quad (8^2 = 64)$$

Because this value falls inside the 95% confidence interval for σ^2, the researcher must *accept* H_0 at the $\alpha = .05$ significance level and conclude that the two drugs are identical in variability of time necessary to induce sleep.

OPTIONAL EXERCISES

15-17 For each of the following situations, construct a 90% confidence interval estimate of the population variance:
 (a) $s^2 = 20.3$; $n = 25$.
 (b) $s^2 = 101.6$; $n = 12$.
 (c) $s^2 = .53$; $n = 15$.
 (d) $s^2 = 7.78$; $n = 20$.

15-18 For each of the following hypothesis-testing situations, (1) indicate whether the test is lower- or upper-tailed; (2) find the critical value of the chi-square statistic at the indicated significance level; and (3) calculate χ^2 and state whether H_0 must be accepted or rejected.
 (a) $H_0:\sigma^2 \geq 16$; $n = 25$; $\alpha = .05$; $s^2 = 19$.
 (b) $H_0:\sigma^2 \leq 100$; $n = 10$; $\alpha = .01$; $s^2 = 105$.
 (c) $H_0:\sigma^2 \geq .64$; $n = 19$; $\alpha = .10$; $s^2 = .59$.
 (d) $H_0:\sigma^2 \leq 6.1$; $n = 17$; $\alpha = .05$; $s^2 = 10.1$.

15-19 A sociologist has administered the Stanford-Binet test to 30 randomly chosen persons in a particular city to draw conclusions regarding the IQ scores of professionals. The sample standard deviation is $s = 12.7$.
 (a) Assuming that the population standard deviation for the IQs of nonprofessionals is 14, should the null hypothesis that professionals exhibit at least as great a variability in IQ be accepted or rejected at the $\alpha = .05$ significance level?
 (b) If the population standard deviation for attorneys is 10.3, should the null hypothesis that professionals as a whole do not exhibit larger variability in IQ be accepted or rejected at the $\alpha = .05$ significance level?
 (c) Construct a 90% confidence interval estimate of the variance in professional IQ scores. Does your answer indicate that the variability in IQ for this group is different from the variability in IQ for the general population, for which the standard deviation is 16? (Use $\alpha = .10$.)

15-20 Suppose that the researcher in the example on page 523 administers another drug to $n = 200$ patients and that the standard deviation for time to sleep is $s = 8.2$ minutes.
 (a) Construct a 95% confidence interval for the population variance of the time to sleep.
 (b) Suppose that the researcher wishes to test the null hypothesis that this drug will exhibit the same variability as the current drug, for which the standard deviation is 8 minutes. What conclusion should he reach at the $\alpha = .05$ significance level?

15-21 A consumer agency has tested a random sample of $n = 100$ sets of radial tires. The sample standard deviation for tire-set lifetimes is $s = 2,000$ miles.
 (a) Construct a 99% confidence interval estimate of the population variance of radial tire lifetimes.
 (b) Should the agency conclude that radial tires have a different lifetime variability than nonradial tires, which have a population standard deviation of 2,500 miles? (Use $\alpha = .01$.)

In order to recognize the best of the treatments in use in the healing of a malady, it is sufficient to test each of them on an equal number of patients, making all the conditions exactly similar.

MARQUIS DE LAPLACE (1820)

16

Analysis of Variance

I n this chapter, we will examine procedures for analyzing several quantitative populations. Our central focus will be a statistical application called *analysis of variance*, which is actually a method for comparing the means of more than two populations.

This kind of analysis is useful in many areas of decision making and research. A manufacturer can use it in determining the most effective packaging for a product. It can indicate to a production manager whether various ingredient combinations really make a difference in the quality of the final product. In drug research, it can help to determine whether a patient's recovery is influenced by different treatments. In agriculture, it may help to determine if crop yields differ according to the type of fertilizer or pesticide used.

The basic questions that we will consider in this chapter have already been posed in Chapter 14, where we compared two population means using samples from control and experimental groups. When more than two samples are involved, however, a radically different approach is required.

Analysis of variance may appear to be misnamed, because it deals with means, but we will see that *this procedure achieves its goal by comparing sample variances*. To do this, a new probability distribution—the *F distribution*—is employed.

16-1   ANALYSIS OF VARIANCE AND THE F DISTRIBUTION

Analysis of variance uses sample data to compare several *treatments* to determine if they achieve different results. Here, we use the word *treatment* in a broad sense that includes not only medical therapy, but other factors that a researcher might investigate. Thus, an agronomist may consider several concentrations of fertilizer to be "treatments" for crops. Similarly, each reading program in elementary education can be a different "treatment" for a developing child. Sample data can be obtained by applying the respective treatments to different samples. These sample data can then be analyzed to determine if the treatments differ.

Testing for Equality of Means

As an illustration, consider the data in Table 16-1 obtained for three different fertilizer treatments being evaluated by a lawn-sod farmer. Here, the total dosages are the same, but the quantities applied and the fertilizing frequencies are varied. The sample observations represent the number of square yards of marketable sod obtained from random plots of 100 square yards, all seeded with the same variety of grass mix. In each case, a random sample of 5 plots was subjected to one of the three treatments.

TABLE 16-1
Sample Sod Yields (in square yards) from Three Fertilizer Treatments

| Sample Observation Number | Fertilizer Treatment | | |
	(1) Quarter Dosage, Once Weekly	(2) Half Dosage, Every Two Weeks	(3) Full Dosage, Every Four Weeks
1	77	83	80
2	79	91	82
3	87	94	86
4	85	88	85
5	78	85	80
Totals	406	441	413
Means	$\bar{X}_1 = 81.2$	$\bar{X}_2 = 88.2$	$\bar{X}_3 = 82.6$

$$\bar{\bar{X}} = \frac{406 + 441 + 413}{5(3)} = 84$$

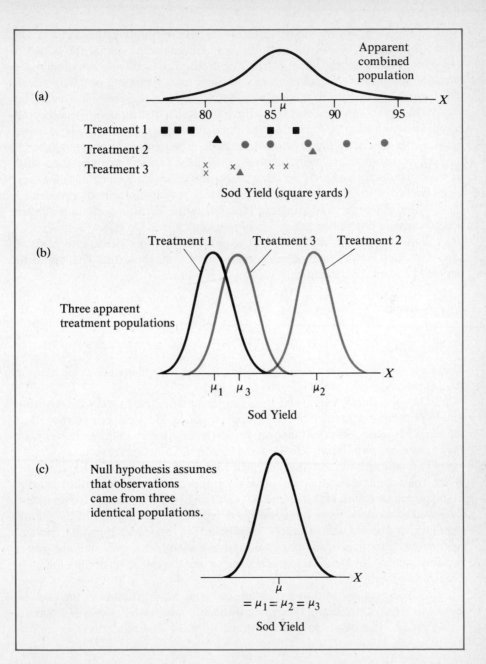

FIGURE 16-1
Concepts underlying analysis of variance.

The analysis of variance procedure used to determine whether fertilization affects sod yield involves two variables. We can note some similarities between this situation and regression analysis. The method of fertilizing is a qualitative variable, sometimes referred to as a *factor*, and each treatment is a *factor level*. The factor is analogous to the independent variable in regression analysis. The sod yield is the *response variable*, which must be quantitative. Because the response achieved may depend on the particular treatment used, this variable plays a role similar to the dependent variable in regression analysis. Although analysis of variance primarily involves *qualitative factors*, it can be used with quantitative factors as well; in such cases, the factors are fixed at a few key levels and constitute *quantitative categories* rather than continuous variates.

The sod farmer must compare the alternative fertilizer techniques to see if they provide different mean yields of marketable sod. In effect, he wishes to test the null hypothesis that the population mean yield per 100 square yards of seeded surface is the same for each treatment. Using subscripts 1, 2, and 3 to denote the respective treatments, we can express the

NULL HYPOTHESIS

$$H_0:\mu_1 = \mu_2 = \mu_3$$

The corresponding alternative hypothesis is that the means are not equal, or that at least one pair of the μ's differs.

The procedures described in this chapter actually test a somewhat stronger null hypothesis—that the treatment populations are identical, or that they have the same frequency distribution form. In particular, this assumption means that each treatment population has the same common value for its variance.

The concepts underlying this procedure are illustrated in Figure 16-1. When the sample data are combined, they appear to be observations from a single population with high dispersion, as shown in (a). But when each treatment is viewed separately, these same sod yields appear to belong to three separate populations with smaller variances, as indicated in (b). Under the null hypothesis, however, the treatment populations have identical means and the same variance, so that an identical frequency curve like the one in (c) applies for each fertilization method.

As in earlier hypothesis-testing procedures, we must convert the sample data into a test statistic and determine if the value achieved refutes H_0. Before we do this, it is necessary to establish some notation and concepts.

Summarizing the Data

Each sample plot yield in Table 16-1 can be represented by a symbol X_{ij}, where i refers to the row or observation number and j refers to the column. In each column, the values are the sample observations made from the corresponding treatment populations. For example, $X_{3\,2} = 94$ square yards—the sod yield for the third test plot in the second sample, where treatment (2) of half the fertilizer dosage every two weeks was used. In the usual manner, we can calculate the

SAMPLE MEAN FOR THE *j*th TREATMENT COLUMN

$$\bar{X}_j = \frac{\sum\limits_i X_{ij}}{r}$$

where all the observations in the *j*th column are summed and divided by r, the number of observations per treatment or the number of rows. In our example, $j = 1, 2$, or 3, depending on which treatment is being considered. The sample mean of the first treatment is found by summing the values in column (1) and dividing by $r = 5$:

$$\bar{X}_1 = \frac{77 + 79 + 87 + 85 + 78}{5} = 81.2 \text{ square yards}$$

The other sample means are calculated similarly as $\bar{X}_2 = 88.2$ and $\bar{X}_3 = 82.6$ square yards.

To facilitate testing the null hypothesis, the data from the three samples are pooled in calculating the

GRAND MEAN

$$\bar{\bar{X}} = \frac{\sum\limits_j \sum\limits_i X_{ij}}{rc}$$

Here, the grand mean is denoted by $\bar{\bar{X}}$ ("X double bar"). The double summation indicates that we obtain the column totals first and then sum these figures for all treatments. The resultant total is divided by the combined sample size. We denote the number of treatments or columns by the letter c. There are $c = 3$ treatments in this illustration, so the combined sample size is $rc = 5(3) = 15$. The grand mean in our example is computed in Table 16-1 as $\bar{\bar{X}} = 84$ square yards.

Using Variability to Identify Differences

In any testing problem, the test statistic should (1) highlight the differences between the observed sample results and the expected results under the null hypothesis, and (2) have a convenient sampling distribution to measure the effect of chance sampling error. We have seen that it is easy to estimate the amount of sampling error from the variability in the sample results. We can also use the variability in the results to express differences. For instance, if several values are unalike, their dispersion—expressed by a range, variance, or standard deviation—will be greater than it will be if the values are close to one another. When many values are involved, their collective differences can be summarized by one of these measures of variability.

Because we are dealing with several populations, it is convenient to use the sample data to measure three sources of variability: (1) *treatments* variation, which shows how the sample results differ under the various treatments; (2) *error*, which collectively summarizes how the observations vary within their respective samples; and (3) *total* variation of the sample observations without regard to their populations.

To summarize the variability in sample results, we use the following expression for the

TREATMENTS SUM OF SQUARES

$$SST = r \sum (\bar{X}_j - \bar{\bar{X}})^2$$

To calculate SST, we sum the squared deviations of the sample treatment means from the grand mean and then multiply by the number of observations r made under each treatment. Using the earlier results from our sod-fertilization experiment, we find

$$SST = 5[(81.2 - 84)^2 + (88.2 - 84)^2 + (82.6 - 84)^2]$$
$$= 5(7.84 + 17.64 + 1.96) = 137.2$$

We multiply by $r = 5$, the number of observations per treatment (or the number of rows), so that with $c = 3$ treatments or columns, all of the $rc = 15$ observations are represented.

The treatments sum of squares expresses the variation between columns. This is often called the *explained variation*, because SST is obtained from differences in the sample means. Thus, SST summarizes the differences in sample results that may be due to inherent differences in the treatment populations rather than to chance alone.

We summarize the variability within samples by summing the squared deviations of the individual observations about their respective sample means. This variability is referred to as the

ERROR SUM OF SQUARES

$$SSE = \sum \sum (X_{ij} - \bar{X}_j)^2$$

TABLE 16-2
Calculation of the Error Sum of Squares for the Sod-Fertilization Experiment

i	$(X_{i1} - \bar{X}_1)^2$	$(X_{i2} - \bar{X}_2)^2$	$(X_{i3} - \bar{X}_3)^2$
1	$(77 - 81.2)^2 = 17.64$	$(83 - 88.2)^2 = 27.04$	$(80 - 82.6)^2 = 6.76$
2	$(79 - 81.2)^2 = 4.84$	$(91 - 88.2)^2 = 7.84$	$(82 - 82.6)^2 = .36$
3	$(87 - 81.2)^2 = 33.64$	$(94 - 88.2)^2 = 33.64$	$(86 - 82.6)^2 = 11.56$
4	$(85 - 81.2)^2 = 14.44$	$(88 - 88.2)^2 = .04$	$(85 - 82.6)^2 = 5.76$
5	$(78 - 81.2)^2 = 10.24$	$(85 - 88.2)^2 = 10.24$	$(80 - 82.6)^2 = 6.76$
	80.80	78.80	31.20

$$SSE = \sum\sum(X_{ij} - \bar{X}_j)^2 = 80.80 + 78.80 + 31.20 = 190.80$$

Table 16-2 shows the calculation of the error sum of squares for the results of the sod-fertilization experiment. There, we obtain $SSE = 190.80$. In performing this calculation, it is convenient to maintain the same column arrangement we used initially: We total the squared deviations for each treatment first, then sum them to obtain SSE. Like SST, SSE accounts for each individual sample observation made.

The error sum of squares expresses the variation within columns. This is sometimes called *unexplained variation*, because the error sum of squares measures differences between sample values that are due to chance (or residual) variation, for which no identifiable cause can be found. This is in contrast to SST, which explains the variation between samples in terms of differences in treatment populations.

If we initially ignore the groupings of the sample observations, we can determine the sum of squares for a single combined sample. The result is called the

TOTAL SUM OF SQUARES

$$\text{Total } SS = \sum\sum(X_{ij} - \bar{\bar{X}})^2$$

We can calculate the total sum of squares for our 15 observations:

$$\text{Total } SS = (77 - 84)^2 + (79 - 84)^2 + \cdots + (80 - 84)^2 = 328$$

Note that if we add the treatments and error sums of squares, we obtain the same result:

$$SST + SSE = 137.20 + 190.80 = 328$$

In general, it can be mathematically shown that

$$\text{Total } SS = SST + SSE$$

for any set of sample results.

When sample sizes are large, sum of squares calculations can become quite burdensome and may be simplified considerably by using shortcut expressions for *SST* and *SSE*. Because the more straightforward procedures presented here work equally well for small samples, shortcut expressions are not included in this book. Instead, *it is recommended that large problems be run on a computer*. Canned programs that perform analysis of variance for a variety of testing situations are widely available from most large computer centers and time-sharing facilities.

The Basis for Comparison

We have just seen that the two components of total variation are explained (*SST*) and unexplained (*SSE*) variations. We must determine whether the explained variation is significant enough to warrant rejecting the null hypothesis that the treatment populations have identical means.

As our test statistic, we must use a summary measure to express how much the sample results deviate from what is expected when the null hypothesis is true. This may be achieved by comparing the explained and the unexplained variations. Whether or not H_0 is true, it is natural in any random-sampling experiment to expect some error or unexplained variation within each sample. But according to the H_0 that the population means are identical for each fertilizer treatment, the amount of unexplained variation between sample treatments should be small; that is, the respective sample means should be about the same. If, in fact, the population means are equal, then the components of the explained and the unexplained variations should be of comparable size.

We can use the ratio between the variation explained by treatments and the error or unexplained variation as a basis of comparison. To find this ratio, we cannot immediately divide the sums of squares. Recall that *SSE* is the sum of $r \times c$ squared differences (for the 15 observations in our example), but that only c squares (representing the 3 samples in our example) are used to calculate *SST*. Each sum of squares must be converted to an average before *SSE* and *SST* are comparable. This gives us sample variances, which we will call *mean squares* to avoid confusion with population variances. Mean squares may be viewed as estimators of the population variances, which, under the assumption that the samples are taken from identical populations, are equal to a common value of σ^2. These estimators are unbiased when the proper divisors are chosen. We express the

TREATMENTS MEAN SQUARE

$$MST = \frac{SST}{c-1}$$

Similarly, we define the

ERROR MEAN SQUARE

$$MSE = \frac{SSE}{(r-1)c}$$

Returning to our sod-fertilization illustration, we have

$$MST = \frac{137.2}{3-1} = 68.6 \quad \text{and} \quad MSE = \frac{190.8}{(5-1)3} = 15.9$$

Note that the treatments mean square is more than four times as large as the error mean square. Under the null hypothesis of identical population means, these mean squares should be nearly the same. Such a large difference seems unlikely, but it may be "explained" by differences between the populations. We have yet to determine exactly how unlikely this large a discrepancy would be if the populations were in fact identical.

The ANOVA Table

It is helpful to summarize the computations for analysis of variance in the format shown in Table 16-3. For simplicity, this is referred to as an ANOVA table. (ANOVA is a contraction of "Analysis Of Variance".) This table is important in organizing the ANOVA computations, because it provides the degrees of freedom, the sum of squares, and the mean square for each source of variation. The degrees of freedom are the divisors that we use to calculate the mean squares. The test statistic, which will be discussed next, is the value in the column labeled *F*.

TABLE 16-3
ANOVA Table for Sod-Fertilization Experiment

Variation	Degrees of Freedom	Sum of Squares	Mean Square	F
Explained by Treatments (between columns)	$c - 1 = 2$	$SST = 137.2$	$MST = 137.2/2$ $= 68.6$	MST/MSE $= 68.6/15.9$ $= 4.31$
Error or Unexplained (within columns)	$(r - 1)c = 12$	$SSE = 190.8$	$MSE = 190.8/12$ $= 15.9$	
Total	$rc - 1 = 14$	$SS = 328$		

The *F* Statistic

We now have two sample variances, MST and MSE. To conform to our earlier discussions, we will refer to the treatments mean square as the variance explained by treatments and to the error mean square as the unexplained variance. In calculating the ratio of these variances, we obtain our

TEST STATISTIC FOR ANALYSIS OF VARIANCE

$$F = \frac{\text{Variance explained by treatments}}{\text{Unexplained variance}} = \frac{MST}{MSE}$$

For our example, we calculate

$$F = \frac{68.6}{15.9} = 4.31$$

Under the null hypothesis, we would expect values of F to be close to 1, because MST and MSE are both unbiased estimators of the common population variance σ^2. Because they have the same expected value, similar values of MST and MSE should be obtained from the sample, and the ratio of these values should be near 1. To formulate a decision rule, we must establish the sampling distribution for this test statistic. From this, we can then determine a critical value that will tell us whether the calculated value of F is large enough to reject the null hypothesis. The probability distribution we use to do this is called the F distribution. Before describing this distribution, we must clarify what we mean by degrees of freedom.

We may view the divisor $c - 1$ in the calculation for MST as the number of degrees of freedom associated with using MST to estimate σ^2. Here, finding the sum of squares involves calculating $\bar{\bar{X}}$, which can be expressed in terms of the $\bar{X}_j$'s. For a fixed $\bar{\bar{X}}$, all except one of the $c \bar{X}_j$'s are free to vary.

Analogously, the $(r - 1)c$ divisor in the calculation for MSE is the number of degrees of freedom associated with using MSE to estimate σ^2. This is true because in finding the sum of squares, each term involves an $\bar{X}_j$ calculated from the X_{ij} values. For a given value of $\bar{X}_j$, only $r - 1$ of the X_{ij}'s are free to assume any value. For c treatments, the number of free variables is therefore only $(r - 1)c$.

Thus, a pair of degrees of freedom is associated with the F statistic. This pair sums to the total number of observations minus 1. In our sod-fertilization example, $r = 5$ and $c = 3$, so that the following degrees of freedom apply:

For the numerator: $3 - 1 = 2$

For the denominator: $(5 - 1)3 = 12$

and $2 + 12 = 14$, a value of 1 less than the total number of observations.

The *F* Distribution

Under the proper conditions, we can employ the *F* distribution to obtain probabilities for possible values of *F*. Like the *t* and chi-square distributions, the *F* distribution is characterized by degrees of freedom. Because the *F* statistic is defined as a ratio, however, the *F* distribution has two kinds of degrees of freedom—one associated with the numerator and one associated with the denominator.

Figure 16-2 illustrates curves for the *F* distribution when the degrees of freedom for the numerator and denominator, respectively, are 6 and 6, 20 and 6, and 30 and 30. Note that the *F* distribution curve is positively skewed, with

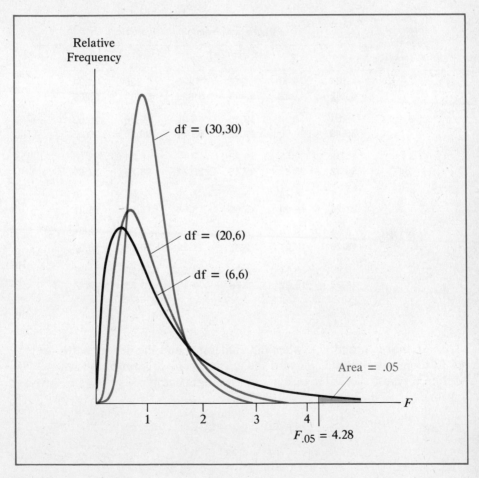

FIGURE 16-2
Various *F* distribution curves.

possible values ranging from zero to infinity. There is a different distribution and curve for each pair of degrees of freedom.

The F distribution is continuous, so that probabilities for the values of F are provided by the areas under the curves. The critical values of upper tail areas under the F distribution are given in Appendix Table I. The table is constructed in the same manner as the tables for the t and chi-square distributions. Due to space limitations, only two upper tail areas are considered.

In Appendix Table I, the rows correspond to the number of degrees of freedom in the denominator and the columns correspond to the number of degrees of freedom in the numerator. The entries in the body of the table are critical values, designated as F_α, where $\alpha = .01$ or $.05$. A portion of this table is reproduced here.

Degress of Freedom in Denominator	Degress of Freedom in Numerator						
	1	2	3	4	5	6	$\cdots$
1	161	200	216	225	230	234	
	4,052	**4,999**	**5,403**	**5,625**	**5,764**	**5,859**	
2	18.51	19.00	19.16	19.25	19.30	19.33	
	98.49	**99.00**	**99.17**	**99.25**	**99.30**	**99.33**	
3	10.13	9.55	9.28	9.12	9.01	8.94	
	34.12	**30.82**	**29.46**	**28.71**	**28.24**	**27.91**	$\cdots$
4	7.71	6.94	6.59	6.39	6.26	6.16	
	21.20	**18.00**	**16.69**	**15.98**	**15.52**	**15.21**	
5	6.61	5.79	5.41	5.19	5.05	4.95	
	16.26	**13.27**	**12.06**	**11.39**	**10.97**	**10.67**	
6	5.99	5.14	4.76	4.53	4.39	4.28	
	13.74	**10.92**	**9.78**	**9.15**	**8.75**	**8.47**	
$\vdots$			$\vdots$				

To find $F_{.01}$ and $F_{.05}$ when the numerator and the denominator degrees of freedom are 6 and 6, we read the entries from column 6 and row 6. The boldface type is the value of $F_{.01}$, and the lightface type is the value of $F_{.05}$, so that $F_{.01} = 8.47$ and $F_{.05} = 4.28$. Thus

$$\Pr[F \geq 8.47] = .01$$

$$\Pr[F \geq 4.28] = .05$$

The second probability is represented by the shaded area under the curve for this distribution, beginning at $F_{.05} = 4.28$. Because the tails of the F curve are long and narrow, the graph in Figure 16-2 does not show the area above $F_{.01}$. These values signify that in repeated experiments, the value of F will exceed 4.28 on an average of 5% of the time and will exceed 8.47 about 1% of the time.

Testing the Null Hypothesis

STEP 1
Formulate the null hypothesis. We are testing the null hypothesis that the mean yields of all fertilizer treatments are identical:

$$H_0:\mu_1 = \mu_2 = \mu_3$$

STEP 2
Select the test statistic. The F statistic will be used.

STEP 3
Obtain the significance level α and identify the acceptance and rejection regions. A Type I error probability of $\alpha = .05$ is desired for incorrectly concluding that the population means of the various fertilizer treatments are not identical. Earlier, we found the degrees of freedom to be 2 for the numerator and 12 for the denominator, and Appendix Table I provides $F_{.05} = 3.88$. The acceptance and rejection regions are shown below. Note that large values of F tend to refute the H_0 of equal means.

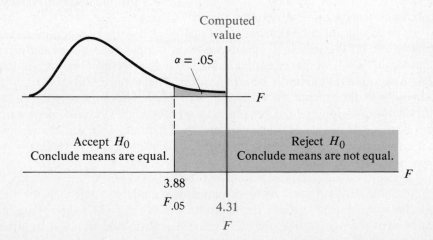

STEP 4
Collect the sample and compute the value of the test statistic. This step includes all the preliminary calculations for the construction of the ANOVA table. Earlier, we found the computed value of $F = 4.31$.

STEP 5

Make the decision. Because the computed value of $F = 4.31$ falls in the rejection region, the null hypothesis must be *rejected*. We must conclude that the fertilization methods have different mean yields.

If the decision maker required greater protection against committing the Type I error and used the smaller significance level of $\alpha = .01$, we would reach the opposite conclusion. The critical value would be $F_{.01} = 6.93$, which is larger than the computed F, and 4.31 would fall in the acceptance region.

Additional Comments

What should the sod farmer do after rejecting the null hypothesis of identical population means? Obviously, he should pick the fertilizer treatment that will maximize mean sod yield. But his results are only sample data, which are subject to sampling error. He could arrive at a decision by comparing individual treatments. (A procedure for doing this will be described in Section 16-2.)

The analysis of variance procedure that we have described thus far has considered only one type of treatment—sod fertilization. This one factor explained enough of the variation in sod yield to justify our conclusion that the fertilizer treatments produce different effects. But there may be other explanations for the differences in sample yields—for example, varying soil conditions or watering schedules. We will discuss procedures for analyzing two factors, in Section 16-3, and later we will see how as many as three factors may be incorporated into the analysis. As more factors are included, the testing procedures become more discriminating and efficient.

The theoretical conditions under which the F distribution applies to our problem are (1) that the populations for each sample must be *normally distributed* with identical means and variances (standard deviations), and (2) that all sample observations must be *independent*. The requirement of normality is shared with the Student t distribution. As we noted in our discussion of the t distribution, as long as the populations are not highly skewed, departures from normality are not considered serious. In Chapter 18, we will examine an alternative procedure that does not require the assumption of normality—the Kruskal-Wallis test.

Because we have already established the sample sizes, we are free only to set the value of the Type I error probability α for falsely rejecting the null hypothesis of identical fertilizer treatment means. We saw in Chapter 9 that no matter what the sample size, a critical value can always be obtained to guard against erroneously rejecting the null hypothesis. But when H_0 is accepted after comparing F to the critical value, there may be a very large chance of committing the Type II error and accepting a false hypothesis. To protect against erroneous acceptance, much larger samples should generally be used than those in our sod-fertilization experiment.

EXERCISES

16-1 For each pair of degrees of freedom (df) and each significance level provided below, indicate the critical value F_α for the *F* statistic above which the stated tail area holds.

	(a)	(b)	(c)	(d)
α	.01	.05	.05	.01
Numerator df	10	12	5	10
Denominator df	10	8	26	5

16-2 A chemical engineer is investigating different pressure settings to determine whether pressure affects particular synthetics in terms of quantity produced. Three sample batches, each of size 5, are run at low, medium, and high pressures once for every chemical. For each synthetic, H_0 is that the mean output is identical under all pressure settings. In the following cases, find the appropriate critical value and then indicate whether H_0 should be accepted or rejected:
(a) For synthetic *X*, $F = 5.32$ and $\alpha = .01$.
(b) For synthetic *Y*, $F = 4.97$ and $\alpha = .05$.
(c) For synthetic *Z*, $F = 6.43$ and $\alpha = .05$.

16-3 A statistics instructor teaches three small experimental sections of the same course. He uses a different book in each section. The instructor wishes to test the null hypothesis that the mean scores achieved on a standard examination by all students using a particular textbook at his university will be identical. He has administered this test to his students, and the nine scores obtained from each class constitute a random sample from the respective population of university-wide scores. Apply the *F* test and find the respective degrees of freedom. Then, in each of the following situations, determine whether the null hypothesis should be accepted or rejected at the stated significance level.
(a) $F = 4.86$ and $\alpha = .01$.   (c) $F = 5.91$ and $\alpha = .01$.
(b) $F = 4.59$ and $\alpha = .05$.   (d) $F = 3.19$ and $\alpha = .05$.

16-4 Quicker Oats is contemplating changing the shape of its box from the quaint cylinder presently in use. Different random samples were selected from five stores of similar size in the same region, and one of three candidate boxes was substituted for the cylinder for several days. The total number of boxes sold are shown below:

	Box Shape		
Sample	(1)	(2)	(3)
Store	Pyramid	Rectangle	Cube
1	110	57	92
2	85	65	81
3	69	73	66
4	97	49	71
5	78	77	70

(a) Perform hypothesis-testing Step 3:
(1) Determine the degrees of freedom for the numerator and the denominator.

(2) Find the critical value of the test statistic and identify the acceptance and rejection regions when the significance level is $\alpha = .05$.
(b) Perform hypothesis-testing Step 4:
(1) Calculate the individual sample means and the grand mean.
(2) Determine the treatments and the error sums of squares. Use these values to find the total sum of squares.
(3) Construct the appropriate ANOVA table and compute F.
(c) Should the null hypothesis that the mean sales are identical regardless of box shape be accepted or rejected?

16-5 In assessing the impact of the level of impurities in a particular ingredient on the solubility of aspirin tablets, a statistician wishes to test the null hypothesis that the mean dissolving time is the same regardless of the impurity level. In test batches the following dissolving times, measured in seconds, were obtained:

	Level of Impurities		
Observation	(1) 1%	(2) 5%	(3) 10%
1	2.01	1.95	2.30
2	1.82	2.21	2.29
3	1.74	2.14	2.17
4	1.90	1.93	2.06
5	2.03	2.07	2.58

What should the statistician conclude at the $\alpha = .05$ significance level?

16-6 A detergent manufacturer advertises that its product will remove all stains except oil-base paint in any kind of water. A consumer information service reporting on detergent quality is testing this claim. Batches of washings were run in five randomly chosen homes having a particular type of water—hard, moderate, or soft. Each batch contained an assortment of rags and cloth scraps stained with food products, grease, and dirt over a 100-square-inch area. After washing, the number of square inches that were still stained was determined, and the following results were obtained:

	Level of Water Hardness		
Observation	(1) Hard	(2) Moderate	(3) Soft
1	5	4	4
2	3	7	0
3	2	8	1
4	10	3	3
5	6	2	2

(a) At the $\alpha = .01$ level of significance, will the consumer service conclude that the type of water affects the effectiveness of the detergent?
(b) Are there any factors other than water type that might "explain" some of the sources of variation?

ESTIMATING TREATMENT MEANS AND DIFFERENCES 16-2

Once the null hypothesis of equal treatment population means has been rejected, what do we do next? Continuing with the sod farmer's experiment, we have concluded that the sod yields differ according to the fertilizer treatment used. But what should the farmer do? He might simply select the treatment that produces the greatest mean yield, but then chance sampling error will still cloud the results. In this section, we will examine additional procedures that extend the analysis of variance results and enable us to formulate a course of action.

There are two basic aspects to this extended analysis. First, we will individually estimate the means of the treatment populations. Then, we will compare treatments by estimating the differences between pairs of population means. In both cases, the confidence interval estimation techniques used earlier in this book may be applied.

Confidence Intervals for Treatment Population Means

Remember that a confidence interval for a single sample can be constructed for the population mean by using the expression

$$\mu = \bar{X} \pm t_\alpha \frac{s}{\sqrt{n}}$$

where α corresponds to the chosen confidence level, t_α is the critical value for which the upper tail area under the Student t curve is α, s is the sample standard deviation, and n is the sample size. When several samples have been taken from different populations, only slight modifications are necessary to obtain a similar expression for each of the respective population means $\mu_1, \mu_2, \dots$.

Remember that the treatment populations were presumed to have identical variance σ^2. Thus, we can obtain a better estimator for σ^2 by pooling the sample results, using the unexplained variance MSE in place of a single sample variance s^2. We can use the sample treatment mean $\bar{X}_j$ to estimate μ_j, the population mean for the jth treatment. We replace n with the number of observations r made under that treatment. Finally, the number of degrees of freedom used to find t_α is $(r-1)c$ (the same figure we used to compute MSE). Thus, we obtain the following expression for computing the

CONFIDENCE INTERVAL ESTIMATE OF μ_j

$$\mu_j = \bar{X}_j \pm t_\alpha \sqrt{\frac{MSE}{r}}$$

Returning to the results of the fertilizer experiment, we can construct a 95% confidence interval estimate of the mean sod yield for treatment (1) in Table 16-1. For $(r - 1)c = (5 - 1)3 = 12$ degrees of freedom, Appendix Table F provides $t_{.025} = 2.179$. With $\bar{X}_1 = 81.2$ and $MSE = 15.9$, we obtain the following estimated mean yield of marketable sod per 100 square yards:

$$\mu_1 = \bar{X}_1 \pm t_{.025} \sqrt{\frac{MSE}{5}}$$

$$= 81.2 \pm 2.179 \sqrt{\frac{15.9}{5}}$$

$$= 81.2 \pm 3.9$$

or

$$77.3 \leq \mu_1 \leq 85.1 \text{ square yards}$$

Similarly, the confidence intervals for the other treatments are

$$84.3 \leq \mu_2 \leq 92.1 \text{ square yards}$$
$$78.7 \leq \mu_3 \leq 86.5 \text{ square yards}$$

Comparing Treatment Means Using Differences

In Chapter 14, we examined various procedures for using two sample means to estimate the difference between population means. In the case of independent samples, we used a confidence interval of the form

$$\mu_A - \mu_B = \bar{X}_A - \bar{X}_B \pm t_\alpha s_{d\text{-small}}$$

where $s_{d\text{-small}}$ is an estimate of the standard error of the difference $d = \bar{X}_A - \bar{X}_B$ that is found by pooling the sample standard deviations. Although we are using number subscripts here, this procedure can be extended to any treatment pair. We can replace $s_{d\text{-small}}$ with $\sqrt{2MSE/r}$, again reflecting the fact that the common population variance is estimated by MSE. The 2 indicates that the variability is additive for two samples. Thus, we have the following

CONFIDENCE INTERVAL FOR THE DIFFERENCE $\mu_2 - \mu_1$

$$\mu_2 - \mu_1 = \bar{X}_2 - \bar{X}_1 \pm t_\alpha \sqrt{\frac{2MSE}{r}}$$

where t_α is chosen to correspond to the confidence level, and the degrees of freedom are $(r - 1)c$, as before. Applying this expression to the results of our sod-fertilization experiment, the 95% confidence interval for the difference between the mean sod yields of treatments (2) and (1) is

$$\mu_2 - \mu_1 = 88.2 - 81.2 \pm 2.179 \sqrt{\frac{2(15.9)}{5}}$$

$$= 7 \pm 5.5$$

or

$$1.5 \le \mu_2 - \mu_1 \le 12.5 \text{ square yards}$$

Thus, we estimate that the advantage in mean sod yield of using treatment (2) over treatment (1) is somewhere between 1.5 and 12.5 square yards.

   We can use this confidence interval to test the null hypothesis that the treatment means are equal. The fact that the interval does not overlap zero indicates that $H_0: \mu_1 = \mu_2$ can be *rejected*, and we can only conclude that μ_1 and μ_2 differ at the $1 - .95 = .05$ significance level. The sod farmer can therefore choose treatment (2) over treatment (1) with only a .05 probability that this action will be incorrect, with the treatments actually yielding identical results.

   We can construct similar 95% confidence intervals for the remaining treatment pairs:

$$\mu_3 - \mu_1 = 1.4 \pm 5.5 \quad \text{or} \quad -4.1 \le \mu_3 - \mu_1 \le 6.9 \text{ square yards}$$

$$\mu_3 - \mu_2 = -5.6 \pm 5.5 \quad \text{or} \quad -11.1 \le \mu_3 - \mu_2 \le -.1 \text{ square yard}$$

The first confidence interval contains zero, indicating that $H_0: \mu_3 = \mu_1$ must be accepted at the 5% significance level. However, the null hypothesis that $\mu_3 = \mu_2$ must be rejected.

Multiple Comparisons

   In applying the preceding methods, we simply extended our earlier hypothesis-testing concepts to encompass the analysis of variance results. But the interpretation of these extended methods may be somewhat misleading. The confidence and significance levels that we just obtained apply only to the *single* estimate or test and *not to the entire series* of estimates or tests. It would be incorrect to tie the preceding three confidence intervals together in a single statement, such as: "The greatest mean sod yield is obtained from treatment (2); treatments (1) and (3) are similar to each other but are both inferior to (2)." This is because *as a set*, these three confidence intervals correspond to less than a 95% confidence level. Stated another way, there is a greater than 5% chance that there will be at least one erroneous rejection of the individual null hypotheses that each pair of treatments has equal means. The separate inferences are interdependent.

This problem can be alleviated by constructing somewhat wider confidence intervals—a procedure referred to as *multiple comparisons*. Due to their relative complexity, a discussion of these methods is beyond the scope of this book.*

EXERCISES

16-7 The following ANOVA table and mean calculations apply to the results of a study to determine if a person's attitude toward his or her present job influences the number of years he or she has remained in that position. The treatments were three job attitudes: (1) dislike, (2) indifference, and (3) enjoyment. A sample of four persons was used for each treatment.

Variation	Degrees of Freedom	Sum of Squares	Mean Square	F
Treatments	2	104	52	8.7
Error	9	54	6	
Total	11	158		

$\bar{X}_1 = 2$ years $\bar{X}_2 = 4$ years $\bar{X}_3 = 9$ years $\bar{\bar{X}} = 5$ years

(a) Construct 95% confidence intervals for the treatment population means.
(b) Construct 95% confidence intervals for the differences $\mu_2 - \mu_1$, $\mu_3 - \mu_2$, and $\mu_3 - \mu_1$.
(c) Referring to your answers to (b), are the population means of attitudes (3) and (1) significantly different at the 5% level?

16-8 Referring to Exercise 16-4:
(a) Construct a 95% confidence interval estimate for the mean sales of the pyramid-shaped box.
(b) Construct a 95% confidence interval for the difference between the mean sales of the rectangular box and the mean sales of the pyramid box. Do the two styles have different mean sales at the $\alpha = .05$ significance level?

16-9 Referring to Exercise 16-5, determine whether the 1% and 10% impurity levels provide significantly different mean dissolving times at the $\alpha = .05$ significance level.

16-3   TWO-FACTOR ANALYSIS OF VARIANCE

In our analysis of the sod-yield data in Table 16-1, the test results were barely significant to warrant rejecting the H_0 of identical treatment means at the 5% level. At $\alpha = .01$, however, H_0 would have to be accepted, because there

---

* A complete description of three multiple comparison procedures developed by Tukey, Scheffé, and Bonferroni, respectively, are described in John Neter and William Wasserman, *Applied Linear Statistical Models* (Homewood, Ill.: Richard D. Irwin, 1974), Chapter 14.

would be too much unexplained variation in sod yields to justify rejecting H_0 at this lower level. Is it possible to lower the level of unexplained variation by explaining a portion of it in terms of another factor? Perhaps some differences in yield could be explained by varying soil qualities, slopes of plots, or watering methods—all of which could be influential factors if the plots were in homogeneous parcels of land.

Suppose that every sample observation consisted of three neighboring plots in each of five separate parcels owned by the farmer and that each treatment was randomly assigned to one plot in every parcel. We could then treat the parcel as a second factor in the analysis. Although the farmer wants to determine how sod yields are affected by fertilizer treatments (not by the parcels themselves, which are permanent features of the farm), considering the parcels as a second factor may explain some of the differences in yields. In this section, we will describe a procedure called *two-factor analysis of variance*, which is employed for this purpose.

There are two basic forms of two-factor analysis of variance. We will illustrate the first form by expanding our analysis of the sod-fertilization experiment, making inferences about the original factor only. Later, we will investigate situations in which inferences are made regarding both factors.

The Randomized Block Design

Each parcel of land is referred to as a *block*. In Table 16-4 the original sod-yield data are arranged by block and treatment. This arrangement is called a *randomized block design*, because treatments have been randomly assigned to units within each block. Although the blocks in our illustration are represented as contiguous, presumably rectangular areas (like "city blocks"), the term actually refers to a second factor in the analysis that is used primarily to reduce the unexplained variation by ensuring that the sample units in each block are homogeneous. This same design can also be used to evaluate several reading programs in terms of achievement. But if the students in a variety of schools are to be tested, the experimental results will be more discriminating if the schools are blocked according to size and region (either of which can influence achievement) and if each block is equally represented in the various reading programs.

The sod farmer is still faced with his earlier question: Are mean sod yields affected by fertilizer treatments? His null hypothesis is the same as before:

$$H_0 : \mu_1 = \mu_2 = \mu_3$$

Extending the procedure outlined in Section 16-1, which we now refer to as *one-factor analysis of variance*, provides us with another source of variation. In addition to treatments and error variation, *blocks variation*—attributable to differences between blocks—is a third component of total variation. Each of

TABLE 16-4

Two-Factor Randomized Block Design for Sod-Fertilization Experiment Using Parcels as Blocks and with Sod Yield (in square yards) as the Response Variable

Blocks	Fertilizer Treatment			Mean Yield (square yards)
	(1) Quarter Dosage, Once Weekly	(2) Half Dosage, Every Two Weeks	(3) Full Dosage, Every Four Weeks	
(1) Parcel A	77	83	80	$\bar{X}_{1.} = 80$
(2) Parcel B	79	91	82	$\bar{X}_{2.} = 84$
(3) Parcel C	87	94	86	$\bar{X}_{3.} = 89$
(4) Parcel D	85	88	85	$\bar{X}_{4.} = 86$
(5) Parcel E	78	85	80	$\bar{X}_{5.} = 81$
Mean Yield (square yards)	$\bar{X}_{.1} = 81.2$	$\bar{X}_{.2} = 88.2$	$\bar{X}_{.3} = 82.6$	$\bar{\bar{X}} = 84$

these three component variations can be measured by computing a sum of squares and a mean square. But before we can measure blocks variation, we must expand our earlier notation.

Sample Mean Calculations

As before, each observation is denoted by X_{ij}, but here i refers to the ith block and j refers to the treatment. In addition to the column means for each treatment, there are now row means for each block. The row means are designated by $\bar{X}_{i.}$'s, and the column means are represented by $\bar{X}_{.j}$'s. The subscript dots signify that more than one factor is being considered. The following calculation provides the

SAMPLE MEAN OF THE ith BLOCK

$$\bar{X}_{i.} = \frac{\sum_j X_{ij}}{c}$$

where all the observations in the ith row are summed and divided by the number of entries in that row, which is the number of columns c. In our example, $i = 1, 2, 3, 4,$ or 5, depending on which block we are considering. For block (4), representing plots in parcel D, the sample mean is found by summing the values in row (4) and then dividing by $c = 3$:

$$\bar{X}_{4\cdot} = \frac{85 + 88 + 85}{3} = \frac{258}{3} = 86$$

The sample means of the remaining rows are provided in the right margin of Table 16-4.

Similarly, we calculate the

SAMPLE MEAN OF THE *j*th TREATMENT

$$\bar{X}_{\cdot j} = \frac{\sum_i X_{ij}}{r}$$

These values are also provided in Table 16-4, in the bottom margin.

As before, the mean of all observations in the combined sample (which can be computed by averaging either the row or the column means) serves as the

GRAND MEAN

$$\bar{\bar{X}} = \frac{\sum \bar{X}_{i\cdot}}{r} = \frac{\sum \bar{X}_{\cdot j}}{c}$$

The Two-Factor ANOVA Table

The two-factor ANOVA table for our sod-fertilization experiment appears in Table 16-5. It is similar to Table 16-3, but contains an additional row for the blocks variation. The number of degrees of freedom for the blocks row is $r - 1$, or the number of rows minus 1. The number of degrees of freedom for the error row is $(r - 1)(c - 1)$, accounting for the fact that one degree of freedom each is lost from the rows and the columns.

Here, the treatments sum of squares SST expresses variation between column means; it is found, as before, by summing the squared deviations of the column means from the grand mean and then multiplying by the number of rows. Analogously, the variation between row means is expressed by summing their squared deviations from the grand mean and then multiplying by the number of columns to obtain the

BLOCKS SUM OF SQUARES

$$SSB = c \sum (\bar{X}_i. - \bar{\bar{X}})^2$$

TABLE 16-5

Two-Factor ANOVA Table of Sod-Fertilization Experiment
for Randomized Block Design

Variation	Degrees of Freedom	Sum of Squares	Mean Square	F
Explained by Treatments (between columns)	$c - 1 = 2$	$SST = 137.2$	$MST = 137.2/2$ $= 68.6$	MST/MSE $= 68.6/3.6$ $= 19.06$
Explained by Blocks (between rows)	$r - 1 = 4$	$SSB = 162$	$MSB = 162/4$ $= 40.5$	*
Error or Unexplained (residual)	$(r - 1)(c - 1) = 8$	$SSE = 28.8$	$MSE = 28.8/8$ $= 3.6$	
Total	$rc - 1 = 14$	$SS = 328$		

* In the randomized block design, *F* is not ordinarily calculated for blocks.

In our example

$$SSB = 3[(80 - 84)^2 + (84 - 84)^2 + (89 - 84)^2 + (86 - 84)^2 + (81 - 84)^2]$$
$$= 3(16 + 0 + 25 + 4 + 9) = 162$$

A New Source of Explained Variation

Together, SST and SSB account for the explained variation in sod yields. SST explains this variation in terms of differences between fertilizer treatments; SSB explains it in terms of differences between parcels of land. Left unexplained by either fertilization or parcels is the residual variation in sod yield, which is expressed by the error sum of squares SSE. It can be shown mathematically that the total sum of squares is the sum of the three components:

$$\text{Total } SS = SST + SSB + SSE$$

By calculating the total sum of squares and then subtracting SSB and SST, we can determine the

ERROR SUM OF SQUARES

$$SSE = \text{Total } SS - SST - SSB$$

Earlier, we found the total sum of squares for sod yields to be 328. Thus, for that experiment

$$SSE = 328 - 137.2 - 162 = 28.8$$

Note that SSE is smaller than it was in our earlier one-factor analysis of variance, because much of the formerly unexplained variation in sod yields can now be attributed to differences between parcels or blocks.

The mean squares for each component of variation are computed by dividing every respective sum of squares by the applicable number of degrees of freedom. We can calculate the

BLOCKS MEAN SQUARE

$$MSB = \frac{SSB}{r - 1}$$

In our example

$$MSB = \frac{162}{5 - 1} = 40.5$$

The value of the treatments mean square, $MST = 68.6$, remains unchanged from our earlier one-factor analysis of variance. Dividing SSE by a smaller number of degrees of freedom than before, we calculate the

ERROR MEAN SQUARE WITH TWO FACTORS

$$MSE = \frac{SSE}{(r - 1)(c - 1)}$$

For the sod-fertilization experiment,

$$MSE = \frac{28.8}{(5 - 1)(3 - 1)} = 3.6$$

Remember that each mean square is really a sample variance and serves as an estimator of the common variance for the treatment populations.

Testing the Null Hypothesis for One Factor

STEP 1
Formulate the null hypothesis. We are testing the null hypothesis that the mean yields of all fertilizer treatments are equal.

$$H_0 : \mu_1 = \mu_2 = \mu_3$$

STEP 2

Select the test statistic. The test statistic is calculated from

$$F = \frac{\text{Variance explained by treatments}}{\text{Unexplained variance}} = \frac{MST}{MSE}$$

STEP 3

Obtain the significance level α and identify the acceptance and rejection regions. A 1% (α = .01) significance level is assumed. The degrees of freedom are 2 for the numerator and 8 for the denominator. Appendix Table I provides the critical value of $F_{.01} = 8.65$. The acceptance and rejection regions are shown below.

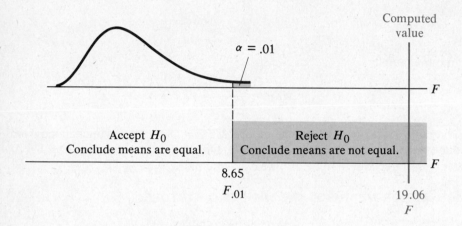

STEP 4

Collect the sample and compute the value of the test statistic. We obtained the ANOVA table values earlier. The computed value of the test statistic is

$$F = \frac{MST}{MSE} = \frac{68.6}{3.6} = 19.06$$

Although it is possible to use *MSB* to calculate a value of *F* for blocks, this variable is of secondary interest and is used primarily to permit a finer discrimination to be made between treatments. Next we will investigate two-factor experiments in which inferences are made regarding both factors.

STEP 5

Make the decision. The computed value of $F = 19.06$ falls in the rejection region, indicating that the sod farmer must *reject* the null hypothesis and conclude that the fertilizer treatment means are unequal.

Our example illustrates how two-factor analysis of variance can be more efficient than one-factor analysis. We obtained a much larger value of F for treatments than before, allowing us to reject H_0 (equal fertilizer population means) at a smaller significance level than we could in the one-factor analysis. This is because including parcels as a second factor considerably reduced the previously unexplained variation in sod yield.

Testing the Null Hypotheses for Both Factors: Evaluating Swimming-Pool Chemicals

Another kind of two-factor analysis of variance is performed when inferences are made about both factors. To illustrate this, consider the data

TABLE 16-6
Two-Factor Randomized Design and Sample Results for Relating Monthly Drop in Swimming-Pool Alkalinity (in ppm) to Acidity Levels and Chlorine Concentrations

Factor B (Chlorine Concentration)	Factor A (Acidity Level)				Sample Mean
	(1) pH 7.2	(2) pH 7.4	(3) pH 7.6	(4) pH 7.8	
(1) Low	23	18	9	7	$\bar{X}_1. = 14.25$
(2) Medium	10	12	8	4	$\bar{X}_2. = 8.50$
(3) High	9	9	7	4	$\bar{X}_3. = 7.25$
Sample Mean	$\bar{X}._1 = 14$	$\bar{X}._2 = 13$	$\bar{X}._3 = 8$	$\bar{X}._4 = 5$	$\bar{\bar{X}} = 10$

Total $SS = \sum\sum (X_{ij} - \bar{\bar{X}})^2$:

$$
\begin{array}{llll}
(23 - 10)^2 = 169 & (18 - 10)^2 = 64 & (9 - 10)^2 = 1 & (7 - 10)^2 = 9 \\
(10 - 10)^2 = 0 & (12 - 10)^2 = 4 & (8 - 10)^2 = 4 & (4 - 10)^2 = 36 \\
(9 - 10)^2 = 1 & (9 - 10)^2 = 1 & (7 - 10)^2 = 9 & (4 - 10)^2 = 36 \\
\hline
170 & 69 & 14 & 81
\end{array}
$$

Total $SS = 170 + 69 + 14 + 81 = 334$

$SSA = r \sum (\bar{X}._j - \bar{\bar{X}})^2$ (between-columns variation)

$\qquad = 3[(14 - 10)^2 + (13 - 10)^2 + (8 - 10)^2 + (5 - 10)^2]$

$\qquad = 3(16 + 9 + 4 + 25) = 162$

$SSB = c \sum (\bar{X}_i. - \bar{\bar{X}})^2$ (between-rows variation)

$\qquad = 4[(14.25 - 10)^2 + (8.50 - 10)^2 + (7.25 - 10)^2]$

$\qquad = 4(18.0625 + 2.25 + 7.5625) = 111.5$

$SSE = $ Total $SS - SSA - SSB = 334 - 162 - 111.5 = 60.5$

in Table 16-6, where factor A (acidity level) and factor B (chlorine concentration) are used to explain the monthly drop in alkalinity (in part per million, or ppm) for a random sample of swimming pools. Pool acidity and chlorine are set at various levels to satisfy sanitation requirements, and for each combination of these factor levels, the drop in alkalinity must be periodically offset by adding soda ash.

Here, we wish to make inferences regarding the underlying populations for the alkalinity-drop response variable at various levels for both factors. In this example, the sample units (swimming pools) have been randomly assigned with equal probability to each factor combination. This type of two-factor study is arranged in a *completely randomized design.* It contrasts with the randomized block design described earlier, in which the units are assigned to blocks on the basis of their particular location, characteristic, or some other fixed factor. In a completely randomized design involving two factors, there is a separate treatment for each combination of levels for the two factors, as represented by the cells in Table 16-6.

STEP 1
Formulate the null hypotheses. A completely randomized design allows us to *simultaneously test two null hypotheses:*

1. The response (alkalinity-drop) means of factor A populations (acidity levels) are identical.
2. The response (alkalinity-drop) means of factor B populations (chlorine concentrations) are identical.

A separate test may be performed on each hypothesis using the same data. Again, this shows that a two-factor analysis can be more efficient than two separate one-factor analyses. (As we will see in Section 16-4, the two-factor analysis sometimes answers questions that cannot even be considered in one-factor analyses.)

STEP 2
Select the test statistic. Two F statistics—one for each test—will be used.

STEP 3
Obtain the significance level α and identify the acceptance and rejection regions. A significance level of $\alpha = .05$ is chosen for each test. Because there are two tests, two separate acceptance and rejection regions apply as shown below. The two $F_{.05}$ values are obtained after the appropriate numbers of degrees of freedom are calculated (to be described later).

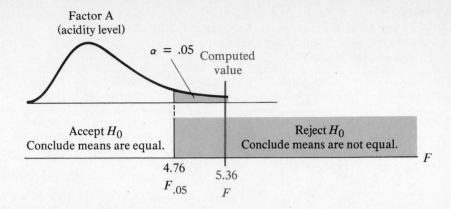

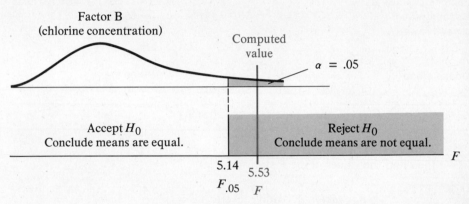

STEP 4

Collect the sample and compute the value of the test statistic. Because there is a separate treatment for each cell, it is necessary to modify our previous notation somewhat and use *SSA* and *SSB*, respectively, to denote the factor *A* and the factor *B* sums of squares. *SSA* and *SSB* calculated in Table 16-6 in the same manner that their counterparts are computed in the randomized block design: *SSA* is based on the deviations of *column means*, and *SSB* is based on the deviations of *row means*. The total and the error sums of squares are also calculated in Table 16-6 in the same way as before.

The ANOVA table for this experiment appears in Table 16-7. The respective mean squares, *MSA*, *MSB*, and *MSE*, are found by dividing the corresponding sum of squares by the applicable numbers of degrees of freedom, which are obtained from the ANOVA table:

	Numerator	Denominator
Factor A	3	6
Factor B	2	6

TABLE 16-7
Two-Factor ANOVA Table for Completely Randomized Design
of Experiment for Effects of Acidity and Chlorine Levels
on Drop in Swimming-Pool Alkalinity

Variation	Degrees of Freedom	Sum of Squares	Mean Square	F
Explained by Factor A: Acidity Level (between columns)	$c-1=3$	$SSA=162$	$MSA=162/3$ $=54$	MSA/MSE $=54/10.08$ $=5.36$
Explained by Factor B: Chlorine Concentration (between rows)	$r-1=2$	$SSB=111.5$	$MSB=111.5/2$ $=55.75$	MSB/MSE $=55.75/10.08$ $=5.53$
Error or Unexplained (residual)	$(r-1)(c-1)=6$	$SSE=60.5$	$MSE=60.5/6$ $=10.08$	
Total	$rc-1=11$	$SS=334$		

We then find the values of the F statistics:

$$\text{For factor } A \text{ (acidity level):} \quad F = \frac{MSA}{MSE} = 5.36$$

and

$$\text{For factor } B \text{ (chlorine concentration):} \quad F = \frac{MSB}{MSE} = 5.53$$

STEP 5
Make the decision. The computed value of $F = 5.36$ for factor A falls in the rejection region, and the null hypothesis of identical acidity-level population means must be *rejected*. The computed value of $F = 5.53$ for factor B also falls in the rejection region, and the null hypothesis of identical chlorine-concentration population means must be *rejected* as well.

Interactions

In the preceding example, we have made the important assumption that the factor effects are *additive*, so that the population treatment mean is raised or lowered from some "background" level by a different constant amount for each level of every treatment. But this may not be true. There may be an *inter-action* between the two factors that cannot be explained by either factor alone. For example, the mean alkalinity drop in swimming pools may be more severe if the acidity level is high when the chlorine concentration is decreased than would be indicated if the separate effects were combined. Because it does not account for such interactions, the testing procedure just outlined may be less discriminating.

EXERCISES

16-10 A randomized block design is being employed to test the null hypothesis that mean responses are identical under five treatments. Using four levels for the blocking factor, the following data are obtained:

$$SST = 84$$

$$SSB = 132$$

$$\text{Total } SS = 288$$

(a) Construct the ANOVA table.
(b) Should the null hypothesis of identical population means be accepted or rejected at the $\alpha = .01$ significance level?

16-11 Suppose the sample data in Exercise 16-4 represent three observations from stores in each of five regions. Treat "region" as the blocking variable, and refer to your answers to that exercise.

(a) Determine the sample means of each row, and then compute the blocks sum of squares.
(b) Construct the ANOVA table.
(c) Compare the value of SSE to the one you obtained in Exercise 16-4. What can you conclude about the effect on unexplained variation of adding the blocking variable?
(d) At the $\alpha = .01$ significance level, should the null hypothesis of identical mean sales for each box shape be accepted or rejected?

16-12 An economist wishes to assess the effects of factor A (education) with five levels and factor B (occupation) with four levels on a person's annual earnings. The following data have been obtained in each category for 20 randomly chosen persons:

$$SSA = 800,000$$

$$SSB = 900,000$$

$$\text{Total } SS = 2,000,000$$

(a) Construct a one-factor ANOVA table, using education as the only treatment. At the $\alpha = .05$ significance level, can you conclude that the treatment means differ?

(b) Construct a one-factor ANOVA table, using occupation as the only treatment. At the $\alpha = .05$ significance level, can you conclude that the treatment means differ?

(c) Construct a two-factor ANOVA table, using both education and occupation as treatments. What can you conclude regarding the respective null hypotheses for identical mean incomes for education levels and for occupations at the $\alpha = .01$ significance level?

(d) Do you notice any discrepancy between the one-factor and the two-factor results? Explain.

16-13 A computer-programming instructor conducted an experiment to determine the effects on student achievement of the computer languages taught and the types of computers used. Over a period of four terms, the instructor taught 16 classes and gave each class a standard achievement test. The mean scores of the respective classes were as follows:

	Language			
Type of Computer	(1) BASIC with No FORTRAN	(2) BASIC with Some FORTRAN	(3) FORTRAN with No BASIC	(4) FORTRAN with Some BASIC
(1) Batch/No Time Share	64	74	68	69
(2) Batch/Some Time Share	86	83	85	84
(3) Time Share/ No Batch	88	90	84	87
(4) Time Share/ Some Batch	84	92	69	89

(a) Computer the row, column, and grand means.

(b) Construct the two-factor ANOVA table.

(c) At the $\alpha = .01$ significance level, can the instructor conclude that *language* makes a difference in achievement?

(d) At the $\alpha = .01$ significance level, can the instructor conclude that *computer type* makes a difference in achievement?

16-4 LATIN SQUARES AND OTHER ANOVA DESIGNS

In the preceding sections of this chapter, we have considered basic analysis of variance procedures involving one and two factors. But analysis of variance is a very broad topic, and we have barely scratched the surface. In this section, we will briefly examine some other aspects of analysis of variance, and we will conclude by discussing a useful procedure for analyzing three factors.

Important ANOVA Considerations

Until now, our examples have covered all combinations of factors. In our sod-fertilization experiment, we investigated three levels for the fertilizer factor and five levels for the parcel blocking factor. Each level for both factors was represented in the sample data. Such experiments are called *complete factorial designs*. But sometimes only a fraction of the combinations can be studied, because a complete study would be prohibitively expensive or infeasible (in our example, the soil in some parcels might not absorb heavy applications of fertilizer). Such experiments are said to have a *fractional factorial design*. At times, it may be useful to omit certain combinations of factors but still to incorporate all factorial levels in the experiment. This is called an *incomplete factorial design*. A useful experiment in this category is conducted with a *Latin-square design*, the procedure for analyzing three factors that we will describe later in this section.

Also in our illustrations until now, we have set levels in advance for the factors under investigation. For instance, the three fertilizer treatments were established before the data were collected. That study involved a *fixed-effects experiment*, in contrast to a *random-effects experiment* in which the factor levels are not established in advance but are subject to chance, as when the levels constitute a sample from a larger population. For example, suppose a medical society is studying doctors' incomes, based on the factors "specialty" and "region where educated." Data might be compiled from one medical school in each region, so that the school selected is just a sample; the levels of the education factor would also be samples. A detailed discussion of random-effects experiments is beyond the scope of this book.

In Section 16-2, we saw that analysis of variance can be extended to answer other important questions. There, we learned to compare pairs of treatments. Often, interactions between two or more factor levels have a considerably important influence when combined that the factors do not exhibit individually. Techniques for analyzing interactions are so complex that we have limited the experiments in this book to applications where interactions can be ignored.

Statisticians use the word *replication* to indicate that an experiment is repeated. The number of repetitions is the number of replications. When a single sample is used, this number is the sample size. In hypothesis-testing situations, the chance of committing the Type II error of accepting a false null hypothesis becomes smaller as the level of n is raised. When two or more samples are used, we compare population parameters or estimate their differences. The advantages of using large sample sizes are accrued in either case, but large sample sizes are naturally more expensive.

In our analysis of variance illustrations, replication occurred only in the one-factor application when each fertilizer treatment was applied to five plots.

Our two-factor experiments involved a single observation for each combination of factors, and the designs for these experiments did not involve replication. More complicated designs (not described in this book) incorporate replication when multiple observations are made for each cell. Such a replication scheme is required to investigate interaction effects.

We classified our two-factor analysis of the sod-fertilization treatments as a randomized block design. The sample units within each parcel were randomly assigned to treatments, thereby nullifying slight differences between plots within parcels. Thus, no plot could systematically influence the results—as might be true, say, if the plot receiving the best irrigation was always heavily fertilized. Often an important element in sampling studies, the assignment of sample units to treatments is referred to as *randomization*. For example, in evaluating the effectiveness of training programs, it is important to account for the differences between instructors. This can be accomplished by randomly assigning a particular training program to each instructor. Such a randomization is less discriminating than treating "instructors" as a separate blocking variable, but the latter procedure would force each instructor to train once under each treatment program, which would be both time-consuming and burdensome.

The Latin-Square Design: Comparing Training Methods

Until this point, we have discussed only one- and two-factor designs. We will now consider a *three-factor analysis* involving one treatment factor of interest and two blocking variables.

A psychologist conducting an experiment to evaluate the effects of three different training methods on achievement-test performance may wish to use a design that incorporates other factors that might explain performance variability, such as a trainee's aptitude and age. To do this, the psychologist might block the sample subjects in terms of three levels of aptitude and three levels of age. A total of 9 blocks would be necessary—one for each aptitude- and age-level combination. In a complete design, three treatments must be considered for each case, so that a minimum of 27 subjects would be required. In the case of four treatments and two blocking variables, each also having four levels, a minimum of $4^3 = 64$ subjects would be necessary for a complete design. Because a large number of sample units may be required for a moderate number of levels, such experiments can be very expensive.

One way to reduce the number of sample units is to use an incomplete design that does not represent all combinations of treatments and blocks. An efficient procedure for this is the *Latin-square design*, which limits the number of treatments to the number of levels used for the blocks. In the training-method evaluation, only 9 instead of 27 subjects are required for a Latin-square design. The data obtained in this experiment for a sample of 9 persons appear in

TABLE 16-8

Layout and Sample Data for Latin-Square Design
Using Achievement-Test Scores to Evaluate Training Programs

Row Blocking Factor (Aptitude)	Column Blocking Factor (Age)			Sample Mean for Row
	(1) Young	(2) Middle	(3) Old	
(1) Low	A 82	B 87	C 80	$\bar{X}_{1.} = 83$
(2) Medium	B 92	C 82	A 81	$\bar{X}_{2.} = 85$
(3) High	C 90	A 83	B 88	$\bar{X}_{3.} = 87$
Sample Mean for Column	$\bar{X}_{.1} = 88$	$\bar{X}_{.2} = 84$	$\bar{X}_{.3} = 83$	$\bar{\bar{X}} = 85$
Treatment: Training Program Sample Mean	A $\bar{X}_A = 82$	B $\bar{X}_B = 89$	C $\bar{X}_C = 84$	

Table 16-8. The letters in the cells identify the particular program applied. These same letters form the

LATIN SQUARE

$$
\begin{array}{ccc}
A & B & C \\
B & C & A \\
C & A & B
\end{array}
$$

This arrangement of Latin letters is designed so that each letter appears exactly once in each column and in each row. Other Latin squares for four and five letters are

$$
\begin{array}{cccc}
A & B & C & D \\
B & A & D & C \\
C & D & A & B \\
D & C & B & A
\end{array}
\qquad
\begin{array}{ccccc}
A & B & C & D & E \\
B & A & E & C & D \\
C & D & A & E & B \\
D & E & B & A & C \\
E & C & D & B & A
\end{array}
$$

Because a letter represents a particular treatment, and because a different level of the blocking variables corresponds to each column and row, *the Latin-square design permits each treatment to be applied exactly once under each level of both blocking variables.*

The sample means of the rows and columns in Table 16-8 have been calculated in the usual manner. The treatment sample means are found by summing the cell responses with the same letter and then dividing by the number of cells having that letter. The values of the respective means, denoted by $\bar{X}_A$, $\bar{X}_B$, and $\bar{X}_C$, are given in Table 16-8. For training program A, the sample mean is calculated as

$$\bar{X}_A = \frac{82 + 81 + 83}{3} = 82$$

In the usual way, we calculate the

SUMS OF SQUARES FOR THE ROW AND COLUMN BLOCKING FACTORS

$$SSROW = r\sum(\bar{X}_{i\cdot} - \bar{\bar{X}})^2$$
$$= 3[(83 - 85)^2 + (85 - 85)^2 + (87 - 85)^2]$$
$$= 3(4 + 0 + 4) = 24$$

$$SSCOL = c\sum(\bar{X}_{\cdot j} - \bar{\bar{X}})^2$$
$$= 3[(88 - 85)^2 + (84 - 85)^2 + (83 - 85)^2]$$
$$= 3(9 + 1 + 4) = 42$$

In Latin-square designs, the numbers of rows and columns are the same, so that $c = r(= 3$, here). The number of treatments must also be equal to the number of rows, so that we have the

TREATMENTS SUM OF SQUARES

$$SST = r[(\bar{X}_A - \bar{\bar{X}})^2 + (\bar{X}_B - \bar{\bar{X}})^2 + (\bar{X}_C - \bar{\bar{X}})^2]$$
$$= 3[(82 - 85)^2 + (89 - 85)^2 + (84 - 85)^2]$$
$$= 3(9 + 16 + 1) = 78$$

The total sum of squares is found, as before, by squaring the deviations of the column responses from the grand mean and then summing these totals:

(1)	(2)	(3)
$(82 - 85)^2 = 9$	$(87 - 85)^2 = 4$	$(80 - 85)^2 = 25$
$(92 - 85)^2 = 49$	$(82 - 85)^2 = 9$	$(81 - 85)^2 = 16$
$(90 - 85)^2 = 25$	$(83 - 85)^2 = 4$	$(88 - 85)^2 = 9$
$\overline{83}$	$\overline{17}$	$\overline{50}$

$$\text{Total } SS = 83 + 17 + 50 = 150$$

The following difference provides the

ERROR SUM OF SQUARES

$$SSE = \text{Total } SS - SST - SSROW - SSCOL$$
$$= 150 - 78 - 24 - 42 = 6$$

The ANOVA table appears in Table 16-9. The degrees of freedom are $r - 1 = 2$ for all factors and $(r - 1)(r - 2) = 2$ for the error. The mean squares are calculated in the usual manner.

The test statistic is

$$F = \frac{MST}{MSE} = 13.0$$

From Appendix Table I, with the numerator and the denominator each having 2 degrees of freedom, $F_{.01} = 99.00$ and $F_{.05} = 19.00$. Because both critical values are larger than the calculated value of F, the null hypothesis of identical treatment population means must be *accepted*, and we must conclude that the training programs do not differ.

The primary advantage of the Latin-square design is that it reduces the number of sample units required to conduct a three-factor analysis of variance.

TABLE 16-9
Latin-Square Design ANOVA Table

Variation	Degrees of Freedom	Sum of Squares	Mean Square	F
Explained by Treatments (between programs)	$r - 1 = 2$	$SST = 78$	$MST = 78/2$ $= 39$	MST/MSE $= 39/3.0$ $= 13.0$
Explained by Column Blocks (between ages)	$r - 1 = 2$	$SSCOL = 42$	$MSCOL = 42/2$ $= 21.0$	*
Explained by Row Blocks (between aptitudes)	$r - 1 = 2$	$SSROW = 24$	$MSROW = 24/2$ $= 12$	*
Error or Unexplained (residual)	$(r - 1)(r - 2) = 2$	$SSE = 6$	$MSE = 6/2$ $= 3.0$	
Total	$r^2 - 1 = 8$	$SS = 150$		

* *F* is not usually calculated for blocking variables.

But our illustration also points out an important disadvantage of this design: The Latin-square design has a small number of degrees of freedom, making it difficult to reject the null hypothesis of identical population means unless SST is quite large in relation to SSE. However, this type of test becomes more discriminating if several observations are made for each cell or if the number of treatments is increased. Another drawback to the Latin-square design is that it is limited to testing situations where exactly the same number of levels are used for both the blocking variables and the treatments.

EXERCISES

16-14 Indicate whether or not each of the following arrangements forms a Latin square:

(a)	(b)	(c)
A B	A B C D	A B C
B A	B C D A	B C A
	C D A B	C A C

(d)	(e)
E A D C B	D C B A
A C B D E	C D A B
D E C B A	B A C D
B D A E C	A B D C
C B E A D	

16-15 The following sums of squares were obtained using a 4-by-4 Latin-square design:

$$SST = 97$$

$$SSCOL = 53$$

$$SSROW = 48$$

$$\text{Total } SS = 236$$

Construct the ANOVA table. What can you conclude about the treatment means at the $\alpha = .05$ significant level?

16-16 A Latin-square design can be used to make inferences regarding three separate factors. Suppose that the following sums of squares were obtained for a 5-by-5 design:

$$SST = 157 \text{ (factor } X)$$

$$SSCOL = 77 \text{ (factor } Y)$$

$$SSROW = 143 \text{ (factor } Z)$$

$$\text{Total } SS = 452$$

(a) Construct the ANOVA table, and calculate the appropriate F statistics for each factor.

(b) For which factors are the population means significantly different at the $\alpha = .01$ level?

16-17 The following sample data represent scores achieved on a classical-music appreciation test. The "treatment" of interest is the age of the subject:

$$A = \text{Preteen}$$

$$B = \text{Teenager to 25}$$

$$C = \text{Over 25}$$

Two blocking variables, "family background" and "intelligence," are used:

Family Background	Intelligence		
	(1) Low	(2) Medium	(3) High
(1) Blue Collar	A 78	B 88	C 68
(2) White Collar	B 90	C 82	A 74
(3) Professional	C 87	A 79	B 92

(a) Construct the ANOVA table for the Latin-square design.

(b) At the $\alpha = .05$ significance level, can you conclude that mean music appreciation scores differ for the various age levels?

REVIEW EXERCISES

16-18 A researcher conducting a study on human behavior offered a contest to the students in four psychology classes. Points were given for the number of nonsense syllables memorized during a class period. Rewards varied between classes. Using a random sample of 5 students from each class, the following numbers of correct syllables were obtained:

Student	Number of Points in Reward			
	(1) $\frac{1}{2}$	(2) 1	(3) $1\frac{1}{2}$	(4) 2
1	18	32	46	52
2	34	48	58	73
3	27	25	37	46
4	31	26	48	63
5	20	39	61	56

(a) Determine the sample means, calculate the sums of squares, and construct the ANOVA table.

(b) Should the researcher conclude that motivation to perform meaningless tasks is affected by the level of reward at the $\alpha = .01$ significance level?

(c) Construct 95% confidence interval estimates for $\mu_4 - \mu_1$. Do rewards of 1/2 and 2 points provide significantly different motivations at the 5% level?

16-19 The following results were obtained from a two-factor experiment in which six levels were used for factor A and five levels were used for factor B:

$$SSA = 78$$

$$SSB = 65$$

$$\text{Total } SS = 183$$

(a) Construct the ANOVA table.

(b) At the $\alpha = .01$ significance level, what can you conclude regarding the null hypotheses of equal population means for the levels of the respective factors?

16-20 An efficiency expert wishes to determine how workers' performance ratings are affected by their skill levels. Three levels are considered:

$$A = \text{Unskilled}$$

$$B = \text{Semiskilled}$$

$$C = \text{Skilled}$$

The following data were obtained from a sample of three workers from each level, with "productivity" and "attitude" serving as blocking variables:

	Attitude		
Productivity	(1) Poor	(2) Fair	(3) Good
(1) Low	A 50	B 79	C 105
(2) Medium	B 74	C 106	A 63
(3) High	C 98	A 61	B 93

(a) Calculate the sample means and the sums of squares.

(b) Construct the ANOVA table for the Latin-square design.

(c) At the $\alpha = .01$ significance level, what can you conclude about the effect of skill level on the mean performance ratings?

16-21 A school psychometrist arranged a special study to evaluate the effects of testing environment on the performance of high-school students taking the Scholastic Aptitude Test (SAT). Group A students took the test in their homerooms in the presence of their regular teachers. Group B students also took the test in their homerooms, but they were proctored by strangers. Group C students were examined in distant cities with their regular teachers as proctors. Group D students took the test in a distant city under the supervision of strangers. The students in the experiment were carefully blocked in terms of motivation and academic performance. The following data were obtained:

	Motivation			
	(1) Poor	(2) Fair	(3) Good	(4) Excellent
(1) Poor	A 342	B 313	C 325	D 324
(2) Fair	B 368	A 406	D 349	C 401
(3) Good	C 377	D 348	A 512	B 493
(4) Excellent	D 455	C 482	B 575	A 634

(a) Calculate the sample means and the sums of squares.
(b) Construct the ANOVA table.
(c) Do the data suggest that test surroundings affect SAT performance at the $\alpha = .01$ significance level?

16-22 Referring to Exercise 16-21, answer the following:
(a) Construct a one-factor ANOVA table, using "test surroundings" as the treatment and ignoring blocking factors. Can you reject the null hypothesis of identical population means at the $\alpha = .05$ significance level?
(b) With "academic performance" as the only blocking variable, construct a two-factor ANOVA table for a randomized block design, again using "test surroundings" as the treatment. Can you reject the null hypothesis of identical population means at the $\alpha = .05$ significance level?
(c) What does a comparison of your one-, two-, and three-factor ANOVA tables suggest regarding the discrimination of testing when "academic performance" and "motivation" are used as the blocking variables?

OPTIONAL TOPIC: ANALYSIS OF VARIANCE FOR MULTIPLE REGRESSION RESULTS   16-5

Regression analysis and analysis of variance are closely related topics. Regression analysis is concerned with finding an estimated regression equation that can be used to predict the mean values of a dependent variable for given levels of independent variables. With three independent variables, the estimated regression equation is expressed as

$$\hat{Y} = a + b_1 X_1 + b_2 X_2 + b_3 X_3$$

Analysis of variance presumes that a particular relationship—usually a linear one—exists between the mean response and given levels of factor variables, and it is concerned with determining whether the relationship differs for various levels of these variables. This is done by using an F ratio to compare explained and unexplained variation.

Regression analysis also involves a comparison of the variation in the dependent variable Y that is explained by the regression plane to the total variation. In regression analysis, this ratio provides the coefficient of multiple determination, which measures the strength of association between the variables. By considering the total variation in regression as the sum of two components:

Total variation = Variation explained by regression + Unexplained variation

we can extend analysis of variance procedures to testing various hypotheses about the underlying regression coefficients B_1, B_2, and B_3 of the true regression line.

In keeping with our present terminology, the variation explained by regression can be presented by the *regression sum of squares*, which we denote by SSR. The ratio of explained to total variation provides the coefficient of multiple determination:

$$R^2_{Y \cdot 123} = \frac{\text{Explained variation}}{\text{Total variation}} = \frac{SSR}{\text{Total } SS}$$

The *error sum of squares*, which here summarizes variation left unexplained, is the difference between total and explained variation:

$$SSE = \text{Total } SS - SSR$$

Unexplained variation = Total variation − Explained variation

The respective sums of squares can be computed using the following equations:

$$SSR = a \sum Y + b_1 \sum X_1 Y + b_2 \sum X_2 Y + b_3 \sum X_3 Y - n\bar{Y}^2$$

$$SSE = \sum Y^2 - a \sum Y - b_1 \sum X_1 Y - b_2 \sum X_2 Y - b_3 \sum X_3 Y$$

$$\text{Total } SS = \sum Y^2 - n\bar{Y}^2$$

These values may be used to test the null hypothesis that the multiple regression coefficients for X_1, X_2, and X_3 are all 0:

$$H_0 : B_1 = B_2 = B_3 = 0$$

This test is accomplished by computing the mean squares. The *regression mean square* is

$$MSR = \frac{SSR}{m - 1}$$

TABLE 16-10
ANOVA Table for Multiple Regression Using Supermarket Data from Chapter 11

Variation	Degrees of Freedom	Sum of Squares	Mean Square	F
Explained by Regression	$m - 1 = 3$	$SSR = 609.53$	$MSR = 609.5/3$ $= 203.18$	MSR/MSE $= \dfrac{203.18}{1.56} = 130.2$
Unexplained	$n - m = 6$	$SSE = 9.4$	$MSE = 9.4/6$ $= 1.56$	
Total	$n - 1 = 9$	$SS = 618.9$		

where m is the total number of variables in the regression analysis. The number of degrees of freedom for MSR is $m - 1$. The *error mean square* is

$$MSE = \frac{SSE}{n - m}$$

where $n - m$ is the number of degrees of freedom and represents the difference between the number of observations and the number of variables in the multiple regression. The test statistic is

$$F = \frac{\text{Variance explained by regression}}{\text{Unexplained variance}} = \frac{MSR}{MSE}$$

Ordinarily, a computer is used to perform multiple regression analysis, and these computations are made then. For example, consider the supermarket example of Chapter 11, where we obtained the following estimated regression equation using the computer printout in Figure 11-3 (page 378):

$$\hat{Y} = -10.170 + .027X_1 + .097X_2 + .525X_3$$

This equation involves $m = 4$ variables and was determined for a sample of size $n = 10$. The ANOVA table for these results appears in Table 16-10. There, $F = 130.2$.

From Appendix Table I with 3 numerator and 6 denominator degrees of freedom, $F_{.01} = 9.78$. Because the calculated value of F is considerably larger than this critical value, the null hypothesis that all multiple regression coefficients are equal to zero must be *rejected*.

*All events, even those which . . . do not seem to follow the great laws of
nature, are a result of it just as necessarily as the revolutions of the sun.*

MARQUIS DE LAPLACE (1820)

17

Further Probability Distributions and the Goodness-of-Fit Test

T here are several probability distributions of special importance in business decision making. Five of these have already been considered. This chapter describes four more, two discrete and two continuous.

Both of the discrete probability distributions to be described are similar to the binomial distribution but have different applications. The hypergeometric distribution is appropriate when sampling is done without replacement from a small qualitative population. The Poisson distribution provides probabilities for the number of events that will occur over a specified time interval or in some given space.

The two continuous distributions to be discussed are the uniform and the exponential. The uniform distribution applies to quantitative populations having frequency distributions shaped like a rectangle. It assumes that all possible values are equally likely, a common circumstance for many physical or random phenomena. The exponential distribution, which has the shape of a reversed *J*, is closely related to the Poisson and has similar applications.

These four distributions are tremendously important to sampling applications, and they also play important roles in some quantitative methods used in analyzing decisions. The Poisson, exponential, and uniform distributions are crucial in many models of waiting lines or queues. The uniform distribution also is fundamental to Monte Carlo simulation techniques.

Many quantitative techniques assume that a particular probability distribution applies. Such assumptions may not coincide with reality. This chapter concludes with a hypothesis-testing procedure, the goodness-of-fit test, which can verify whether a sample comes from a population having a particular frequency distribution or whether a sample result is consistent with certain theoretical assumptions.

17-1 THE HYPERGEOMETRIC DISTRIBUTION*

As we have seen, the binomial distribution may be used to find the probabilities for the number of successes from a Bernoulli process. Thus, the binomial distribution can only be applied when sampling is done *with replacement* from small populations. Ordinarily, however, a sample is chosen from a population *without replacement*. For instance, when items are selected from a shipment of parts in order to determine the quality of the shipment, the items inspected are usually set aside and thus cannot be chosen again. As we noted in Chapter 4, the sampling process itself may destroy the item, so that it is impossible to sample with replacement. The hypergeometric distribution allows us to overcome these difficulties and find the probabilities for the number of successes in samples taken without replacement from small populations.

Finding Probabilities When Sampling Without Replacement

To illustrate the basic principles underlying the hypergeometric distribution, we will use a simple illustration involving the random selection of marbles from a box. Our initial view of sampling without replacement will be one where sample items are removed one at a time. Then we will develop the hypergeometric distribution by viewing the sample items as if they were withdrawn as a group. Identical probabilities will be obtained under either method.

A box contains 6 black and 4 white marbles. Four marbles are selected randomly, one at a time, from the box. Once selected, a marble is set aside. We want to find the probability distribution for the number of black marbles obtained.

Figure 17-1 shows the probability tree diagram. At each stage the color composition of the marbles remaining in the box is provided inside the oval

* This section uses optional material from Section 5-7.

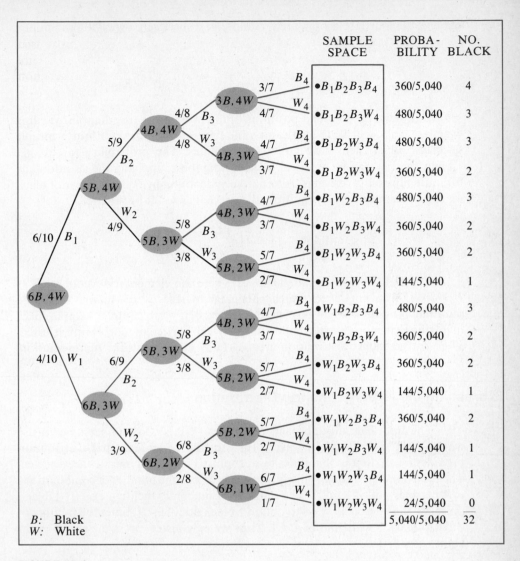

	SAMPLE SPACE	PROBA-BILITY	NO. BLACK
	•$B_1B_2B_3B_4$	360/5,040	4
	•$B_1B_2B_3W_4$	480/5,040	3
	•$B_1B_2W_3B_4$	480/5,040	3
	•$B_1B_2W_3W_4$	360/5,040	2
	•$B_1W_2B_3B_4$	480/5,040	3
	•$B_1W_2B_3W_4$	360/5,040	2
	•$B_1W_2W_3B_4$	360/5,040	2
	•$B_1W_2W_3W_4$	144/5,040	1
	•$W_1B_2B_3B_4$	480/5,040	3
	•$W_1B_2B_3W_4$	360/5,040	2
	•$W_1B_2W_3B_4$	360/5,040	2
	•$W_1B_2W_3W_4$	144/5,040	1
	•$W_1W_2B_3B_4$	360/5,040	2
	•$W_1W_2B_3W_4$	144/5,040	1
	•$W_1W_2W_3B_4$	144/5,040	1
	•$W_1W_2W_3W_4$	24/5,040	0
		5,040/5,040	32

B: Black
W: White

FIGURE 17-1

Probability tree diagram for sampling without replacement from a box containing 6 black and 4 white marbles.

next to the branching point. The branches on the path leading to the end position $B_1W_2W_3W_4$ are shown as black lines. Initially there are 6 black and 4 white marbles. The probability that the first selection is black is therefore 6/10. If the first marble is black, 5 black and 4 white marbles remain for the second selection. The conditional probability that the second marble is white given that the first is black is thus 4/9. Continuing along this path we see that the conditional probabilities for white as the third selection, followed by white again as the fourth selection, are 3/8 and 2/7, respectively. Applying the multiplication law, we obtain

$$\Pr[B_1 W_2 W_3 W_4] = \left(\frac{6}{10}\right)\left(\frac{4}{9}\right)\left(\frac{3}{8}\right)\left(\frac{2}{7}\right) = \frac{144}{5,040}$$

Note that there are four end positions of the probability tree diagram yielding exactly 1 black marble (which can be the first, second, third, or fourth marble selected); all of them have equal probabilities. The probability for exactly 1 black marble may be obtained by summing these, applying the addition law for mutually exclusive events. Alternatively, denoting by R the number of black marbles obtained, we may multiply the preceding result by 4 to get

$$\Pr[R = 1] = 4\left(\frac{144}{5,040}\right) = \frac{576}{5,040}$$

Figure 17-1 yields a sample space where the elementary events, such as $B_1 W_2 W_3 W_4$, do not reflect which particular marbles were chosen. The same sampling procedure might be viewed in terms of a more complex sample space involving elementary events that do reflect which particular marbles are selected. The hypergeometric distribution involves such a combinatorial representation.

Expressing the Hypergeometric Distribution

The tree diagram approach in Figure 17-1 is too cumbersome to use when larger samples are involved. Instead, the hypergeometric distribution provides the desired probabilities by accounting for which marbles are selected. The total number of equally likely outcomes equals the number of combinations of 4 marbles out of $10 : C_4^{10}$. The number of outcomes yielding exactly 1 black marble is the number of ways to obtain 1 black out of 6, C_1^6, times the number of ways to select 3 white out of 4, C_3^4. Thus we may also express the above probability as

$$\Pr[R = 1] = \frac{C_1^6 C_3^4}{C_4^{10}}$$

We can verify that this ratio is identical in value to our previous result by expressing it in factorials:

$$\Pr[R = 1] = \frac{\left(\dfrac{6!}{1!5!}\right)\left(\dfrac{4!}{3!1!}\right)}{\dfrac{10!}{4!6!}}$$

which, by rearranging and canceling terms, equals

$$\frac{6 \times 4 \times 4! \times 6!}{10!} = \frac{6 \times 4 \times 4!}{10 \times 9 \times 8 \times 7}$$

$$= 4\left(\frac{6}{10}\right)\left(\frac{4}{9}\right)\left(\frac{3}{8}\right)\left(\frac{2}{7}\right) = \frac{576}{5{,}040}$$

Thus, to find the probability for the number of successes, R, in a sample of size n taken *without replacement* from a population of size N, we use the

HYPERGEOMETRIC PROBABILITY DISTRIBUTION

$$\Pr[R = r] = \frac{C_r^S C_{n-r}^{N-S}}{C_n^N}$$

where $r = 0, 1, \ldots, n$ or S (whichever is smaller) and S = number of successes in the population.

In our example, $n = 4$, and $N = 10$. Since we consider drawing a black marble to be a successful event, $S = 6$. Using $r = 4$, so that $n - r = 4 - 4 = 0$, we find

$$\Pr[R = 4] = \frac{C_4^6 C_0^4}{C_4^{10}} = \frac{360}{5{,}040}$$

which is the same as $\Pr[B_1 B_2 B_3 B_4]$ in Figure 17-1.

Illustrations with Poker Hands

The hypergeometric distribution may be used to determine the probability of being dealt five cards that are all hearts from a deck of fully shuffled ordinary playing cards. Poker players call such a hand a flush. Since there are 13 hearts in the deck, $S = 13$. We have $r = 5$, $n = 5$, and $N = 52$, so that

$$\Pr[\text{heart flush}] = \frac{C_5^{13} C_0^{39}}{C_5^{52}}$$

$$= \frac{\dfrac{13!}{5!8!} \times \dfrac{39!}{0!39!}}{\dfrac{52!}{5!47!}}$$

$$= \left(\frac{13}{52}\right)\left(\frac{12}{51}\right)\left(\frac{11}{50}\right)\left(\frac{10}{49}\right)\left(\frac{9}{48}\right) = \frac{33}{66{,}640} \approx \frac{1}{2{,}019}$$

This probability applies for a flush in any specified suit, since each has 13 cards, so that the same constants may be used in the hypergeometric probability expression. To get the probability for a flush regardless of suit, which is of interest for poker odds, we may apply the addition law for mutually exclusive events to the 4 satisfying events (a flush in each suit—hearts, diamonds, clubs, or spades). Thus,

$$\Pr[\text{flush}] = 4\left(\frac{33}{66{,}640}\right) = \frac{132}{66{,}640}$$

$$= .00198 = \frac{1}{505}$$

The odds against receiving a flush are therefore 504 to 1.*

In a similar manner, we may find the probability for getting four of a kind in a five-card hand. Consider getting four aces. Here we have $r = 4$, $S = 4$, $n = 5$, and $N = 52$, so that

$$\Pr[\text{4 aces}] = \frac{C_4^4 C_1^{48}}{C_5^{52}}$$

$$= \frac{\dfrac{4!}{4!0!} \times \dfrac{48!}{1!47!}}{\dfrac{52!}{5!47!}}$$

$$= 5\left(\frac{4}{52}\right)\left(\frac{3}{51}\right)\left(\frac{2}{50}\right)\left(\frac{1}{49}\right)\left(\frac{48}{48}\right) = \frac{1}{54{,}145}$$

(One interpretation of the factor 5 in the third line of the above expression is that there are 5 different sequence positions possible for getting the non-ace.) Since there are 13 denominations in a deck of cards, we may multiply the above probability by 13 to get the probability for four of a kind where denomination is not specified. (Multiplying by 13 is really applying the addition law for 13 mutually exclusive, equally probable events.) We thus obtain

$$\Pr[\text{4 of a kind}] = \frac{13}{54{,}145} = .00024 = \frac{1}{4{,}165}$$

so that the odds against getting four of a kind are 4,164 to 1. This explains why such a hand is superior in poker to a flush (but not to a straight flush): A flush is more than 8 times as likely.

---

* When the straight and royal flushes are separately categorized, the above odds increase to 508 to 1. Our calculations did not eliminate flushes that are also straights.

Hypergeometric Distribution for the Proportion of Successes

As we noted in discussing the binomial distribution, we are usually more interested in the proportion of successes in the sample than in the number of successes. As before, π and P are the proportion of successes in the population and sample, respectively. Thus, since

$$P = \frac{R}{n} \quad \text{and} \quad \pi = \frac{S}{N}$$

we modify our earlier expression to provide the

HYPERGEOMETRIC PROBABILITY DISTRIBUTION FOR THE PROPORTION

$$\Pr\left[P = \frac{r}{n} \right] = \frac{C_r^{\pi N} C_{n-r}^{(1-\pi)N}}{C_n^N}$$

with $r = 0, 1, 2, \ldots, n$ or πN (whichever is smaller).

It may be established mathematically that the expected value and variance of P are

$$\mu(P) = \pi$$

$$\sigma^2(P) = \frac{\pi(1 - \pi)}{n} \left(\frac{N - n}{N - 1} \right)$$

Note that $\mu(P)$ is the same as for the binomial distribution, whereas $\sigma^2(P)$ differs from its binomial counterpart by the $(N - n)/(N - 1)$ term. In Chapter 7, we saw that this factor plays an important role in analyzing samples taken from finite populations. Note that as N gets very large, the term approaches 1. This suggests that for large N, the hypergeometric distribution probabilities will not differ much from those calculated using the binomial formula.

EXAMPLE: PROBABILITIES FOR FIREWORKS

Diablo Pyrotechnics, Ltd., is investigating a new compound to use in its Fantastic Spiroid rockets. It is supposed to produce a larger and brighter explosive pattern. Unfortunately, the new compound is highly unstable, so that some unknown proportion of the rockets will produce a loud pop instead of the intended dazzling display. The company sales manager wishes to know what the proportion of duds will be so that she can set prices to account for the uncertain results the customer must expect. She has decided to use a sample of rockets from the upcoming test shots in estimating this proportion.

Diablo has produced 100 Fantastic Spiroids for a variety of tests. A sample of 5 is chosen at random. Supposing that 20% of the rockets produced will be duds, we may determine the probability that the proportion of duds in the sample will be between 0 and 40%, using $n = 5$, $\pi = .20$, and $N = 100$:

$$\Pr[0 \le P \le .40] = \Pr[P = 0] + \Pr[P = .2] + \Pr[P = .4]$$

$$= \Pr\left[P = \frac{0}{5}\right] + \Pr\left[P = \frac{1}{5}\right] + \Pr\left[P = \frac{2}{5}\right]$$

$$= \frac{C_0^{20} C_5^{80}}{C_5^{100}} + \frac{C_1^{20} C_4^{80}}{C_5^{100}} + \frac{C_2^{20} C_3^{80}}{C_5^{100}}$$

$$= \left(\frac{80}{100}\right)\left(\frac{79}{99}\right)\left(\frac{78}{98}\right)\left(\frac{77}{97}\right)\left(\frac{76}{96}\right) + 5\left(\frac{20}{100}\right)\left(\frac{80}{99}\right)\left(\frac{79}{98}\right)\left(\frac{78}{97}\right)\left(\frac{77}{96}\right)$$

$$+ \frac{5 \times 4}{2}\left(\frac{20}{100}\right)\left(\frac{19}{99}\right)\left(\frac{80}{98}\right)\left(\frac{79}{97}\right)\left(\frac{78}{96}\right)$$

$$= .3193 + .4201 + .2073 = .9467$$

We may compare this result to that obtainable from the binomial distribution (which would be *improper* to use here). Consider the second term in the above probability sum; this is the probability for getting exactly one dud in a sample of size 5. Using the binomial formula, we obtain

$$\frac{5!}{1!4!}(.2)^1(.8)^4 = .4096$$

which is close to the .4201 calculated above. Note that the binomial here understates the true probability; for some other values of r, it will overstate it.

EXERCISES

17-1 A personnel manager randomly selects $n = 3$ different names from a file of $N = 10$ job applicants, $S = 5$ of whom are experienced. Construct a probability tree diagram with a separate stage for each person chosen, with forks containing branches for the events "experienced" and "inexperienced." Enter the appropriate probability values on each branch. Then use the multiplication law to determine the probability for each end position. From your tree, determine the following probabilities:
(a) No selected persons are experienced.
(b) Only one selected person is experienced.
(c) Exactly two selected persons are experienced.
(d) All persons selected are experienced.

17-2 A personnel manager selects $n = 5$ names from a file containing $N = 10$ job applicants. $S = 7$ of the 10 are experienced. Write expressions for the following probabilities:
(a) Exactly three of those selected are experienced.
(b) Exactly four of those selected are experienced.

17-3 A company procurement analyst buys light bulbs by the gross (a dozen dozen, or 144). From each batch, $n = 5$ bulbs are selected and tested under excessive voltage (which destroys them for further use) in order to determine whether the entire batch of bulbs contains an excessive number of defectives. The analyst's rule for disposing of a batch is as follows: When the sample contains no defectives, the batch is accepted; otherwise the batch is rejected and returned to the supplier for full credit, with no charges made for the destroyed bulbs.
  (a) What is the probability that a "good" batch containing $S = 5$ defective bulbs will be rejected?
  (b) What is the probability that a "poor" batch containing $S = 30$ defective bulbs will be accepted?
  (c) Do you think that the rule provides protection against rejecting "good" batches and accepting "bad" batches? Explain.

17-4 The producer of the light bulbs in Exercise 17-3 occasionally takes a sample of $n = 50$ bulbs at random from the production line. Will the hypergeometric distribution help him to determine probabilities for the number of defectives to be found? Explain.

17-5 What is the probability for getting three or more kings when drawing five cards from a fully shuffled deck of cards? Using this answer, find the probability for getting three or more cards of the same denomination.

THE POISSON DISTRIBUTION   17-2

Many practical statistical problems involve events occurring over time. One of the most notable of these problems is how to design a facility to service customers arriving at unpredictable times. Knowledge of the probability distribution for the number of arrivals may help the designer to achieve an optimal balance between the amount of time the facility is idle and the time spent by customers waiting in line. For example, the number of tellers hired for a bank's new branch office will be affected by the pattern of customer arrivals over a period of time. During 5 minutes, the number of newly arrived customers can be expected to vary from 0, 1, 2, or perhaps 20 or more up to the limit of space available in the bank. There will be times when all tellers are busy, so that lines will form. Because customers do not enjoy waiting and may switch their business to a competing bank if they feel unduly delayed, the bank may decide to have ten tellers when five would be sufficient to handle any day's transactions. Many familiar events, such as airplanes landing at an airport with limited capacity, persons arriving at a bank of elevators in a busy office building, or telephone calls arriving at a switchboard, call for similar decisions.

Such decisions, which balance idle time against waiting time, are called waiting-line or *queuing* problems. In analyzing arrival patterns, we are concerned not with the nature of the event (an arrival) but rather with the number of event occurrences. Questions involving occurrences over time are also common in establishing inventory policies and in setting criteria for the reliability of a system. In some businesses, the distribution for customer orders over a period of time may be critical to finding the best inventory level. Enough stock must be available to fill orders, but too large an inventory could be quite costly. In designing systems to achieve a desired level of reliability, the pattern of failures over time is also an important consideration. For example, the probability distribution for the number of component malfunctions during the life of a communications satellite may be helpful in determining how much redundant capability ought to be provided. Likewise, a hospital would want a heart-lung machine to have a very small probability for malfunctioning during open-heart surgery.

The Poisson Process

There are many different patterns of unpredictable occurrences of events over time. A large class of situations in which events occur randomly can be characterized as a *Poisson process*, named after the eighteenth-century mathematician and physicist Siméon Poisson.

There are two very important probability distributions associated with a Poisson process. One, to be described here, provides probabilities for the *number* of events occurring in a time interval. This is the *discrete* Poisson distribution. The other, to be discussed in Section 17-4, is the *continuous* exponential distribution, used for finding the probabilities of *times between* the event occurrences.

The Poisson and Bernoulli processes are similar; however, it is important to note the differences between them. Recall that the binomial distribution provides probabilities for the number of events of a *particular kind* (successes) occurring in *n* independent trials of a random experiment. With a Poisson process, there is no fixed number of trials; the events occur randomly over time—which is continuous. There is only one kind of event, such as arrival of a customer or an equipment breakdown, instead of two complementary ones, success and failure, as in a Bernoulli process.

Expressing the Poisson Distribution

The probabilities for the number of events X that occur in a Poisson process in a period of duration t may be obtained from the following expression for the

POISSON DISTRIBUTION

$$\Pr[X = x] = \frac{e^{-\lambda t}(\lambda t)^x}{x!}$$

where $x = 0, 1, 2, \ldots$. The constant e is the base of natural logarithms and is equal to 2.7183. The parameter λ (the lowercase Greek *lambda*) *represents the* *mean rate at which events occur during the process.* That is, on the average, λ events per unit time will occur. Multiplying λ, a rate, by t, a duration of time, results in λt,* *the mean number of events occurring in a period of length t.* Note that we place no upper limit on the number of events that may occur; that is, X may assume any integral value from zero to infinity.

The Poisson distribution is completely defined by the process rate λ and the duration t. Its mean and variance are identical and are expressed as

$$\mu(X) = \lambda t$$

$$\sigma^2(X) = \lambda t$$

EXAMPLE: ARRIVALS AT A BARBERSHOP

Sammy Lee owns a one-man barbershop. One of his customers informs Sammy that during the lunch-time rush his customers arrive in a pattern that is approximately a Poisson process having a mean rate of 6 persons per hour. If Sammy takes 15 minutes (.25 hour) to cut one man's hair, what is the probability that exactly 2 new customers will arrive before he is finished?

We have $\lambda = 6$ customers per hour and $t = .25$ hour. Therefore,

$$\lambda t = 6(.25) = 1.50 \text{ customers}$$

We obtain

$$Pr[X = 2] = \frac{e^{-1.50}(1.50)^2}{2!}$$

From Appendix Table J, we find that $e^{-1.50} = .223130$, so that

$$Pr[X = 2] = \frac{.223130(1.50)^2}{2!} = .2510$$

If the duration t is increased, the probabilities for large numbers of arrivals will become greater. This seems logical, since there is more time for events to occur. For example, if $t = .4$ hour, then $\lambda t = 6(.4) = 2.4$, so that

$$Pr[X = 2] = \frac{e^{-2.4}(2.4)^2}{2!} = \frac{.090718(2.4)^2}{2!} = .2613$$

Analogously, the probabilities for few arrivals will be reduced when t is increased.

---

* In many books, $t = 1$, a standard unit of time, and λ represents the mean number of occurrences over a standard time period. There, $\lambda t = \lambda \times 1 = \lambda$.

Assumptions of a Poisson Process

As we have indicated, the events of a Poisson process occur randomly over time. Three additional assumptions are required to distinguish it from other types of occurrence patterns.

1. A Poisson process has *no memory*. That is, the number of events occurring in one interval of time is *independent* of what happened in previous time periods.
2. The process rate λ *must remain constant* for the entire duration considered.
3. It is extremely *rare* for more than one event to occur during a short interval of time; the shorter the duration, the rarer the occurrence of two or more events becomes. The probability for exactly one event occurring in such an interval is approximately λ times its duration.

If we divide a duration into small time segments, letting each represent a trial with just two possible outcomes—(a) exactly one event occurring or (b) a nonoccurrence—then the Poisson probabilities may be roughly represented by the binomial distribution. This is only approximately so, since more than one event may occur in each time segment (so that each "trial" actually has more than two possible outcomes, violating assumptions of the Bernoulli process). But our third assumption makes the probability of the occurrence of more than one event close to zero. When the time segments are very tiny, the approximation becomes quite close, and the Poisson distribution is obtained mathematically from the binomial formula.

Practical Limitations of the Poisson Distribution

A very serious mistake to make when applying the Poisson distribution is to assume that the mean event occurrence rate λ holds over an extended duration when it does not. Many queuing situations, for example, involve random arrivals whose rate changes with the time of day, day of week, season, or other circumstances. The mean rate of vehicle arrivals at a metropolitan toll plaza will be different at 9 A.M. than at 3 A.M.; it will be different on Fridays than on Mondays; and it will be greater in the fall than in the summer, when many drivers are on vacation. It is still proper to treat such situations as a Poisson process, but care must be taken to keep short the durations considered and to apply the appropriate value of λ. Thus a bank may keep only a third of its teller windows open at 10:30 A.M. on Tuesday, when λ is small, but it will probably keep them all open around 5 P.M. on a Friday that is also the first of the month, when transaction traffic proves heaviest.

TABLE 17-1
Poisson Probabilities
When $\lambda t = 1.50$

x	$\Pr[X = x]$
0	.2231
1	.3347
2	.2510
3	.1256
4	.0470
5	.0141
6	.0036
7	.0007
8	.0002
9	.0000

A second difficulty is that the Poisson distribution sets no limit on the number of customers who may arrive. In Table 17-1 the probabilities for several possible values of X have been calculated for $\lambda t = 1.50$. Although $\Pr[9] = .0000$, this is a nonzero value that has been rounded off. There is a nonzero probability that 9 or 20 or even 100 customers may arrive at the barbershop in any 15-minute interval, but the probability is quite tiny. As discussed in connection with the normal distribution, even absurd outcomes have nonzero probabilities. For instance, there is a nonzero probability that 10,000 men would try to crowd into Sammy's barbershop during a 15-minute period.

Finally, a random variable is only approximated by a Poisson distribution. However, the approximation is often empirically well justified. For example, a statistician has shown that the number of U.S. Supreme Court seats vacated in a given year very closely fits a Poisson distribution. Some studies have shown that the deaths of Prussian army recruits kicked by horses and the very rare freeze-ups of Lake Zurich are closely approximated by the Poisson process. Of far more practical significance is the fact that a vast number of waiting-line, reliability, and inventory situations are adequately characterized by the Poisson process.

Using Poisson Probability Tables

Computing Poisson probabilities by hand can be an onerous chore. Cumulative values of the Poisson probabilities are computed in Appendix Table K for levels of λt ranging from .1 to 20.

The table provides values of $\Pr[X \leq x]$. For example, to find the cumulative probability values for the number of cars arriving at a toll booth during an interval of $t = 10$ minutes when the arrival rate is $\lambda = 2$ per minute, we consult Appendix Table K where $\lambda t = 2(10) = 20$. The probability that the number of arriving cars is ≤ 15 is

$$\Pr[X \leq 15] = .1565$$

whereas the probability that ≤ 20 cars will arrive during the 10-minute interval is

$$\Pr[X \leq 20] = .5591$$

Like the cumulative binomial tables discussed in Chapter 6, it is possible to obtain from Appendix Table K Poisson probability values for other cases ($=$, $>$, $<$, or $\geq$ some number, or the interval between two numbers). For example, the probability that exactly 15 cars will arrive in 10 minutes is

$$\Pr[X = 15] = \Pr[X \leq 15] - \Pr[X \leq 14] = .1565 - .1049 = .0516$$

Similarly, we can obtain the probability that the number of cars arriving lies between two values. Thus, the probability that between 16 and 20 cars will arrive in 10 minutes is

$$\Pr[16 \leq X \leq 20] = \Pr[X \leq 20] - \Pr[X \leq 15] = .5591 - .1565 = .4026$$

The probability that > 20 cars will arrive is

$$\Pr[X > 20] = 1 - \Pr[X \leq 20] = 1 - .5591 = .4409$$

Applications Not Involving Time

Events occurring in space, as well as events occurring over time, may be characterized as Poisson processes. For instance, finding objects spread randomly over a space (such as misspelled names in a telephone directory) is viewed as an event, then encountering objects while the space (in this case, pages of the directory) is searched may be viewed as a Poisson process. Application of the Poisson process to events occurring in space has proven fruitful in quality control and in such esoteric areas as developing search techniques for radar, establishing tactics for ships sailing through mine fields, and hunting for submarines. Here λ would represent the mean number of events per unit distance (such as an inch), area (a square mile), or space (a cubic centimeter). More generally, λ may be the mean number of particular events per observation made of the phenomena in question. Thus λ may be 3 errors per page, or it may be 5 bad debts per 1,000 installment contracts. The "durations" would be analogous: the space searched, the number of pages scanned, the number of contracts written, and so on.

EXAMPLE: SEARCHING FOR TUNA

The commodore of a tuna-fishing fleet is searching the Pacific Ocean for a school of tuna. His plane can search for a distance of 600 miles, and he can spot any school of fish within a lateral distance of ten miles to either side of the aircraft. Thus the area searched in a day would be 2(10)600 = 12,000 square miles.

    The location of schools of tuna in this area is completely random. We will assume that historically the mean density of large tuna schools in the area has been 1 per 100,000 square miles.

    A storm is forecast to reach the fishing area in 6 days. The boats require at least one day to return to port. What is the probability that at least one school will be located no later than one day before the storm?

    Since there is 1 school per 100,000 square miles, $\lambda = .00001$ school per square mile. The value of t (here representing area) is 5 days times the area searched in a day, or $5 \times 12,000 = 60,000$ square miles. Thus, $\lambda t = .00001(60,000) = .60$, which represents the mean number of schools to be encountered in 5 days of searching. The probability that at least one school is encountered is 1.

$$Pr[X \geq 1] = 1 - Pr[X = 0]$$

$$= 1 - \frac{e^{-.60}(.60)^0}{0!}$$

$$= 1 - e^{-.60}$$

$$= 1 - .549 = .451$$

    Suppose that no tuna are found and that after docking for the storm, the search is resumed. We may find how many days must be committed to the search in order to ensure a probability of .95 that at least one school will be found. Since a Poisson process has no memory, we must start all over again. Otherwise the number of schools encountered in the first and second searches would be dependent events, violating the basic Poisson process assumptions. (The size of the remaining search area cannot even be reduced, since schools might enter the area previously scanned.) Thus, we want to find t such that

$$.95 = Pr[X \geq 1] = 1 - Pr[X = 0]$$

$$= 1 - \frac{e^{-\lambda t}(\lambda t)^0}{0!}$$

$$= 1 - e^{-\lambda t}$$

or

$$e^{-\lambda t} = 1 - .95 = .05$$

From Appendix Table J, we see that $e^{-y} = .05$ (rounded) when $y = 3.00$, so that using $\lambda t = 3.00$ we will have the smallest t for the above condition to hold. Thus, dividing 3.00 by the number of schools encountered per square mile, $\lambda = .00001$, we obtain the total search area t required:

$$t = \frac{\lambda t}{\lambda} = \frac{3.00}{.00001} = 300,000 \text{ square miles}$$

Dividing this area by the area searched per day, we obtain the days required to yield a .95 probability of finding at least one school of tuna:

$$\frac{300,000 \text{ square miles}}{12,000 \text{ square miles per day}} = 25 \text{ days}$$

These 25 days are *in addition to the 5 days already spent searching.*

EXERCISES

17-6 Between 9 and 10 A.M. on Saturday, the peak business period, customers arrive at Sammy Lee's barbershop at a mean rate of $\lambda = 5$ per hour. During the Thursday slump, between 2 and 3 P.M., customers arrive at the rate of $\lambda = 1$ per hour. In either case, the arrivals may be represented as a Poisson process.
   (a) Does this mean that there will always be more customers arriving during the peak than during the slump? Explain.
   (b) Compare the probability of exactly two customers arriving during Sammy's slump to that of the same number arriving during his peak.
   (c) If the mean rate of arrivals for the week as a whole is three customers per hour, can Sammy use the Poisson distribution with $\lambda = 3$ to find the probability that next week's arrivals will be between 100 and 150? Explain.

17-7 Suppose that a typesetter makes on the average $\lambda = .5$ error per page. What is the probability that he will make no errors in the first $t = 10$ pages? What is the probability that he will make exactly 5 errors in the first 10 pages?

17-8 During the late Friday rush at a bank, an average of $\lambda = 5$ customers per minute arrive. What is the probability that no customers arrive during a specified 1-minute interval?

17-9 A family receives an average of $\lambda = 2$ pieces of regular mail each delivery day. Assume that the Poisson distribution applies.
   (a) What is the probability that they will get no mail on the next two consecutive days?
   (b) What is the probability that, given two days without mail, they will get no mail for another two days?

17-10 Use Appendix Table K to find the probabilities that the stated number of events occur in a Poisson process at the rate of $\lambda = 8$/hour when the duration considered is
   (a) $t = 2$ hours, 10 or fewer events.

(b) $t = 1.5$ hours, 6 or more events.

(c) $t = 1$ hour, exactly 3 events.

(d) $t = .1$ hour, between 1 and 5 events.

17-11 A typist commits errors at the rate of .01 per word. Assuming that a Poisson process applies, use Appendix Table K to find the probabilities that the number of errors committed in a 500-word letter will be

(a) exactly 5; (b) zero; (c) more than 10; (d) between 3 and 7.

17-12 A California Central Valley tomato grower wishes to protect her crop against destruction by aphids. Two alternatives are open to her: Spray with a powerful insecticide that may harm the local ecology or wait for ladybugs to eat the aphids. Each spring, massive flights of ladybugs are wafted by wind currents across the California valleys from their Sierra Nevada hibernating points. Unfortunately, where and when a flight of ladybugs lands is purely a matter of chance. Wherever there are plenty of aphids, they lay eggs that rapidly hatch. If at least one flight of ladybugs lands on part of the grower's crop within five days, they and their larvae will devour all the aphids. Otherwise she will spray, killing all aphids and the unfortunate late-coming aphid-eaters as well, for by that time irreparable crop damage will have been suffered. On any given day, 10% of the Central Valley is covered by ladybugs. Assume that the arrivals of ladybugs may be considered a Poisson process. (The size of the farm may be ignored.)

(a) Assuming that the mean arrival rate is $\lambda = .1$ flight of ladybugs per day, what is the probability that no flights arrive during $t = 5$ days?

(b) What is the probability that the farmer won't have to spray?

17-13 Two tourists are enjoying the view from San Francisco's Golden Gate Bridge. They have decided to spend a half hour waiting for ships to pass under the bridge. Suppose that ship arrivals at the Golden Gate (leaving or entering San Francisco Bay) are a Poisson process at the mean rate of $\lambda = .2$ per hour. Find the probability that the tourists will see passing beneath them (a) no ships; (b) 1 ship; (c) 2 ships.

THE UNIFORM DISTRIBUTION 17-3

The uniform distribution provides probabilities for a random variable that has equal chances of assuming any value on a continuous scale. For example, it provides the probability that a wheel of fortune will stop so that its pointer lies within a particular segment. The uniform distribution has many interesting applications. One of the more notable is to generate random numbers, which are used for selecting a random sample.

The uniform distribution has the following probability density function:

$$f(x) = \begin{cases} \dfrac{1}{b-a} & \text{if } a \le x \le b \\ 0 & \text{otherwise} \end{cases}$$

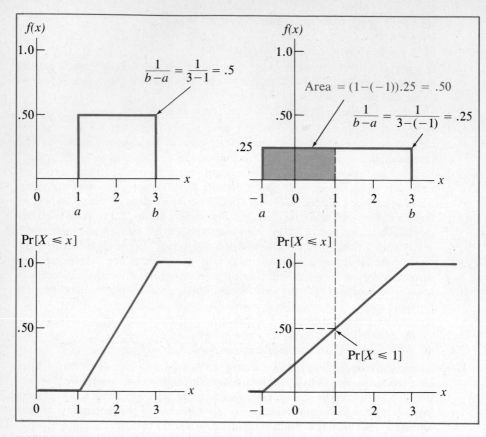

FIGURE 17-2

Top: Uniform density functions. Bottom: Cumulative distribution functions.

The top graphs in Figure 17-2 show the density functions for two different sets of values for a and b. Each function is a horizontal line segment of constant height $1/(b-a)$ over the interval from a to b. Outside the interval, $f(x) = 0$. This means that for a uniformly distributed random variable X, values below a or above b are impossible.

Recall from Chapter 6 that the probability that X will fall below a point is provided by the area under the density curve to the left of that point. The cumulative probability distribution function, $\Pr[X \leq x]$, provides this area. The cumulative function for values of x between a and b is the area of the rectangle of height $1/(b-a)$ and base $x - a$, found by multiplying these two values together. To the left of a, the cumulative probabilities must be zero, whereas the probability that X lies below points beyond b must be 1. The following expression thus provides

CUMULATIVE PROBABILITIES
FOR THE UNIFORM DISTRIBUTION

$$\Pr[X \le x] = \begin{cases} 0 & \text{if } x < a \\ \dfrac{x - a}{b - a} & \text{if } a \le x \le b \\ 1 & \text{if } b < x \end{cases}$$

The cumulative distribution functions for the two cases are shown in the bottom half of Figure 17-2.

EXAMPLE: PROBABILITY FOR BEING STUCK IN PEORIA

A traveling business woman must change planes in Peoria en route to New Haven. Her Peoria-bound plane is scheduled to arrive 20 minutes prior to the departure of the New Haven flight. If that flight is missed, she must spend 12 hours in Peoria waiting for the next plane. Assuming that her plane's arrival time is uniformly distributed over the interval from being 1 hour early to being 1 hour late, what is the probability that she will be stuck in Peoria (that is, that her Peoria-bound plane will be more than 20 minutes late)?

Here $a = -60$ minutes and $b = +60$ minutes. Letting X be the arrival time in Peoria, we have

$$\Pr[\text{stuck in Peoria}] = \Pr[X > 20]$$

$$= 1 - \Pr[X \le 20]$$

$$= 1 - \frac{20 - (-60)}{60 - (-60)}$$

$$= 1 - \frac{80}{120} = \frac{40}{120} = \frac{1}{3}$$

It is easy to obtain an expression for the expected value of a uniformly distributed random variable. The expected value of X is the long-run average result obtained from repeating the same random experiment. Since all values between a and b have the same chance of occurring, the average of a and b is the long-run average of the individual outcomes, or

$$\mu(X) = \frac{a + b}{2}$$

The variance of a uniform random variable must be obtained mathematically. It is given by

$$\sigma^2(X) = \frac{(b - a)^2}{12}$$

Often it is necessary to find the probability that a uniform random variable falls between two values. For instance, suppose that $a = 10$ and $b = 15$. Then the probability that X lies between 11 and 14 is equal to the rectangular area between 11 and 14. Therefore we subtract the smaller probability (area) from the larger one:

$$Pr[11 \leq X \leq 14] = Pr[X \leq 14] - Pr[X \leq 11]$$

$$= \frac{14 - 10}{15 - 10} - \frac{11 - 10}{15 - 10} = \frac{3}{5} = .6$$

(Like the normal and any other continuous distribution, $P[X \leq x] = P[X < x]$, so that the distinction between $\leq$ and $<$ is unimportant.)

EXERCISES

17-14 A lumber mill cuts logs into planks 10 feet long. After cutting, the trimmed plank ends are apt to be any length smaller than 10 feet. What proportion of the trimmings have lengths between (a) 4 and 6 feet; (b) 3 and 7 feet; (c) 2.5 and 6.4 feet; (d) 5.52 and 8.52 feet?

17-15 A pilot is navigating her airplane by "dead reckoning" over the ocean. She does not know her precise speed and actual heading, since she has no way to gauge the effects of wind. She is flying at night and intends to home in on a light ship. The minimum distance at which the plane will pass the light ship is uniformly distributed from 0 miles to 50 miles. Assuming clear visibility, what is the probability that the pilot will see the light ship when flying at an altitude providing a line-of-sight range of (a) 50 miles; (b) 40 miles; (c) 30 miles; (d) 10 miles?

17-16 A farmer is scheduled to begin reaping his wheat at 6 A.M. on July 1, the date on which the combines belonging to his cooperative will be available. Harvesting must begin within 4 days after ripening or the crop will be lost. It cannot be reaped before it is ripe. If his wheat is not ready for reaping on the scheduled date or if it ripens too early, the farmer must pay for private equipment. Assuming that the crop will ripen at a time uniformly distributed from 6 A.M. on June 23 through 6 A.M. on July 3, what is the probability that the farmer will not be forced to rent his combines from outside? (June has 30 days.)

THE EXPONENTIAL DISTRIBUTION   **17-4**

We have seen that the Poisson process can be applied to events occurring randomly over time (or space). The Poisson distribution provides the probabilities that a particular number of events will occur. Of great importance in business applications is the time duration (or space encountered) *between* the events. To a person waiting in a bank line, the number of persons arriving before him is of little importance if his line moves quickly; his prime concern is the length of his wait. The commodore whose fleet has enough capacity to capture just one school of tuna cares only about how long it will take to locate the first one.

In this section we will present the exponential distribution, which for a Poisson process provides the probabilities for the time (or space traversed) between events or until the first event occurrence.

Consider cars arriving at a toll-collection station. The cars are represented in Figure 17-3 by dots. The horizontal distance of each dot from the origin (at 9:00 A.M.) indicates when a car will arrive at the station. Such a graph could be constructed from an aerial photograph taken at 9:00 A.M. of the two miles of highway leading to the station. Assuming that all cars are traveling at the speed limit, we could directly translate each car's distance from the station into its arrival time. Note that the dots in Figure 17-3 are scattered across the page with no apparent pattern, as if placed there randomly. This pattern is typical of a Poisson process.

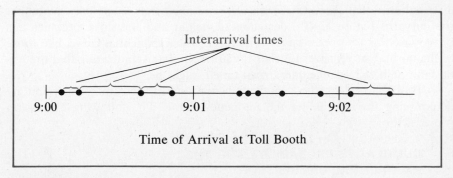

FIGURE 17-3
Times for random arrivals of cars at a toll booth.

Finding Interevent Probabilities

The gaps between the dots in Figure 17-3 represent the *interarrival times*, or times between successive arrivals of cars. The exponential distribution is concerned with the size of the gaps, measured in time units, separating successive cars. Although the dots are scattered randomly over time, the relative frequency of interarrival times of various sizes is predictable. Suppose that the cars arriving at the toll station are observed for several minutes, and that the time of each car's arrival is noted. Those data would provide a histogram similar to the one in Figure 17-4(a). This approximates the shape of the underlying frequency curve in Figure 17-4(b), the height of which may be determined for any interarrival time t (or, generally, for the time or space between any two events in a Poisson process) from

$$f(t) = \lambda e^{-\lambda t}$$

This is the probability density function for the exponential distribution.

The particular distribution applicable to a specific situation depends only on the level of λ. In our toll station illustration, the mean arrival rate is $\lambda = 4$ cars per minute. Note that the frequency curve intersects the vertical axis at height λ. The mean and the standard deviation of the exponential distribution are identical and may be expressed in terms of λ as

$$\mu(T) = 1/\lambda$$
$$\sigma(T) = 1/\lambda$$

where our random variable is T, the uncertain time (or space) between any two successive events. (Unlike the Poisson distribution, the *standard deviation* equals the mean, whereas the variance equals the mean squared.)

Note that the mean time between arrivals is the reciprocal of the mean rate of arrivals. Thus if $\lambda = 4$ *cars per minute*, then the mean time between arrivals is $1/\lambda = 1/4 = .25$ *minute per car*. Another feature of the exponential distribution is that *shorter durations are more likely than longer ones*. Note that the curve in Figure 17-4(b) decreases in height and the slope becomes less pronounced as t becomes larger. The tail of the exponential curve, like those of the normal curve, never touches the horizontal axis, indicating that there is no limit on how large the interarrival time t might be.

Probabilities for the exponential random variable may be found by determining the area under the frequency curve. The following expression provides the

CUMULATIVE PROBABILITY FOR THE EXPONENTIAL DISTRIBUTION

$$\Pr[T \leq t] = 1 - e^{-\lambda t}$$

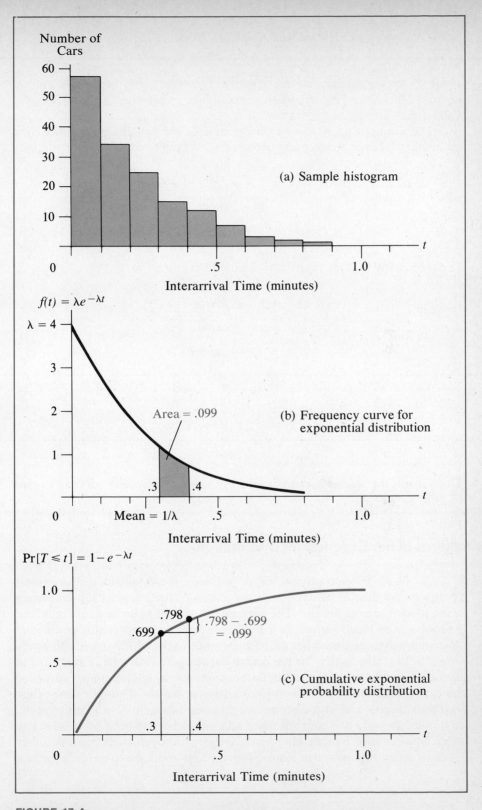

FIGURE 17-4

The exponential distribution for the interarrival times of cars at a toll station.

where T represents the time (or space) between any two successive events. Figure 17-4(c) shows the cumulative probability graph when $\lambda = 4$. Appendix Table J may be used to find values for $e^{-\lambda t}$.

For example, using $\lambda = 4$ cars per minute, we find the probability that the interarrival time between any two cars is less than or equal to $t = .4$ minute:

$$\Pr[T \le .4] = 1 - e^{-4(.4)} = 1 - e^{-1.6}$$

$$= 1 - .201897$$

$$= .798 \text{ (rounded)}$$

The probability that the time is $t = .3$ minute or less is

$$\Pr[T \le .3] = 1 - e^{-4(.3)} = 1 - e^{-1.2}$$

$$= 1 - .301194$$

$$= .699 \text{ (rounded)}$$

The probability that an interarrival time between .3 and .4 minute will be achieved is thus

$$P[.3 \le T \le .4] = .798 - .699$$

$$= .099$$

This value is the shaded area under the exponential frequency curve in Figure 17-4(b).

Applications of the Exponential Distribution

Because a Poisson process has no memory, the exponential distribution applies to the time T between any two events—whether near the beginning of the process or near the end. The burning out of light bulbs in a new building, a process often characterized as a Poisson process, illustrates this point. The probability that the time from the 99th burnout to the 100th exceeds 50 hours would be the same as that for the period between the 78th and 79th or for the period of initial connection to the 1st burnout. A very common application of the exponential distribution is to intermittent processes. Thus the time taken by a bank teller to service a customer might be an exponential random variable, so that the Poisson process ends when all customers have been processed. The teller is then "idle" (not making a transaction, although there may be other work to do during these occasions). The amount and pattern of idle time is a function of both the arrivals and the service. A new phase of the Poisson process begins with the arrival of the next customer.

One point should be made clear. There is no reason why the teller's service times must be exponentially distributed. Their historical frequency distribution may yield a histogram having a shape very different from that provided in Figure 17-4. In that case, the situation clearly would not fit the assumptions of a Poisson process. Whether the exponential distribution should be used is a matter of how closely it fits the situation. An exponential distribution, for instance, assigns low probabilities to long interevent times and high probabilities to short ones. But maybe both long and short are rare, which would make a normal distribution more appropriate.

Reliability Applications

The exponential distribution can be used to establish specifications for the reliability of equipment that will fail at some future time according to the probability pattern of this distribution. Operating requirements are often established in terms of mean time between failures (MTBF), which represents the mean number of failures per unit time: $1/\lambda$.

The MTBF figure itself may be established by a requirement that the equipment achieve some probability for lasting a minimum time before failing. For example, suppose that certain items must be designed so that 90% of them will last more than 100 hours. The requirements may be expressed in this way: Choose λ so that

$$.90 = \Pr[T > 100] = 1 - \Pr[T \le 100]$$
$$= 1 - (1 - e^{-\lambda(100)})$$
$$= e^{-\lambda(100)}$$

From Appendix Table J, we see that the closest value is $e^{-y} = .905$ (rounded) when $y = .10$. Thus, setting $\lambda(100) = .10$, we obtain

$$\lambda = \frac{.10}{100} = .001 \text{ failure per hour}$$

and

$$\text{MTBF} = \frac{1}{\lambda} = \frac{1}{.001} = 1,000 \text{ hours}$$

We may find the probability that a particular part built to these specifications will fail between 200 and 600 hours after operation by subtracting the probability that failure occurs by 200 hours from the probability that it occurs on or before 600 hours:

$$Pr[200 \le T \le 600] = Pr[T \le 600] - Pr[T \le 200]$$
$$= (1 - e^{-.001(600)}) - (1 - e^{-.001(200)})$$
$$= e^{-.2} - e^{-.6}$$
$$= .819 - .549 = .270$$

(As with the previously discussed continuous distributions, the distinction between $<$ and $\le$ is unimportant.)

EXAMPLE: EMERGENCY HOSPITAL POWER

A hospital is reviewing its contingency plans for coping with power failures. Certain hospital functions, such as surgery, require a continuous power supply. The hospital has one emergency generator connected to auxiliary electrical circuits in critical locations. Since the auxiliary power could fail during a blackout while essential services are being rendered, funds for a back-up generator are being sought.

Assuming that the present generator has a mean time between failures $(1/\lambda)$ rated as 100 hours, what is the probability that it will fail during the next 10-hour blackout?

Here $\lambda = .01$ failure per hour. Thus,

$$Pr[T \le 10] = 1 - e^{-.01(10)} = 1 - e^{-.10}$$
$$= 1 - .904837 = .095 \text{ (rounded)}$$

The probability is .095 that a failure will occur and .905 that it will not. The chance of failure is rather high, considering the stakes involved. Suppose that there are $n = 5$ such blackouts in one year. The probability that the generator works through all of them may be determined using $\pi = .905$ and $r = 5$ in the binomial formula:

$$\frac{5!}{5!0!}(.905)^5(.095)^0 = .61$$

an unsafe value, to be sure.

Consider how dramatically the probability changes with the purchase of a back-up generator. We will assume that the generators operate independently and simultaneously. The probability that both generators fail in a 10-hour period would be $(.095)^2 = .0090$. Thus the probability that at least one works throughout is

$$1 - .0090 = .9910$$

Considering five blackouts, the probability that at least one generator works during all the blackouts may be obtained from the binomial formula with $r = n = 5$ and $\pi = .991$:

$$C_5^5(.991)^5(.009)^0 = .956$$

If a third generator is purchased, the probability that all three fail in 10 hours becomes .00086, and the probability for no interruptions throughout five blackouts increases to .9957.

Queuing Applications

A very fruitful application of the exponential distribution has been its use in queuing problems. The following simple example illustrates one such possible application.

EXAMPLE: MACHINE-SHOP TOOL CAGE

Miller Machine Works has a single, centralized tool cage where machinists obtain parts for their equipment. The plant superintendent wants to replace the present clerk with an experienced machinist whose direct costs are $10.00 per hour. This runs counter to company policy, which dictates that the position should be filled by a clerk who is paid $5.00 per hour. The superintendent insists that the machinist, with his special knowledge, can provide service twice as quickly. And the superintendent has constructed an additional, very convincing argument.

He has assumed that the arrivals at the tool cage are a Poisson process with mean rate λ_A men per hour, and that the service time distribution for the clerk is exponential with mean rate λ_S men per hour (that is, counting just those times when someone is being serviced by the cage clerk). By a mathematical deduction, it may be established that the total time W that a machinist must wait at the tool cage—both in line and while being serviced—has an exponential distribution with expected value

$$\mu(W) = \frac{1}{\lambda_S - \lambda_A}$$

Thus, if $\lambda_S = 10$ per hour and $\lambda_A = 5$ per hour, then

$$\mu(W) = \frac{1}{10 - 5} = .2 \text{ hour}$$

Since a machinist is unproductive while waiting in line or being serviced, which takes an average of $1/\lambda_S$ hours, his direct costs of $10.00 per hour are lost during these times. Thus the total expected hourly cost of manning the tool cage with a clerk is the cost of the machinists' unproductive time plus the cost of the clerk. Since each machinist must be unproductive $\mu(W)$ hours, on the average, and there are λ_A arriving per hour, the total expected cost per hour is

$$\$\lambda_A[\mu(W)]10.00 + 5.00 = \$5(.2)10.00 + 5.00$$

$$= \$15.00$$

If an experienced machinist mans the cage, he can service at twice the clerk's rate, $\lambda_S = 2(10) = 20$ machinists per hour, so that

$$\mu(W) = \frac{1}{20 - 5} = \frac{1}{15} = .067 \text{ hour}$$

Thus using a cost of $10.00 per hour for the man in the tool room, we obtain a total expected cost per hour of

$$\$5(.067)10.00 + 10.00 = \$13.35$$

By using an experienced machinist, an average of $1.65 per hour can be saved.

EXERCISES

17-17 On Tuesday mornings, customers arrive at the Central Valley National Bank at a rate of $\lambda = 1$ per minute. What is the probability that the time between the next two successive arrivals will be (a) shorter than 1 minute; (b) longer than 5 minutes; (c) between 2 and 5 minutes?

17-18 The final settlement claims filed by policyholders of a casualty insurance company involving amounts of $100,000 or more are events from a Poisson process with a mean rate of $\lambda = 1$ per working day. Funds earmarked for large-claims settlement are invested in short-term government bonds at an interest rate of 5%, so that the timing of settlements affects company profits. What is the probability that the next large claim must be settled (a) within 5 days; (b) sometime after 2 days; (c) in between 2 and 5 days?

17-19 An automobile manufacturer wishes to determine the terminal mileage for the warranty coverage on the power train of its new compact car. A warranty is desired such that no more than 9.5% of the cars will have to use it. Assume that the time between power-train failures is exponentially distributed for each car. How many miles should warranty coverage last when the power trains fail at a mean rate of once every 100,000 miles driven, so that $\lambda = .00001$ failure per mile?

17-20 The tuna-fleet commodore described in the chapter has embarked from port and begins an air search for tuna. The airplane searches 20 square miles for every mile traveled. Schools of tuna are randomly distributed over the fishing area at a mean density of one school per 100,000 square miles.
   (a) Find λ, the rate at which tuna schools are located per mile flown.
   (b) What is the probability that the first school will be located before the search aircraft has flown 5,000 miles?
   (c) How many miles must the airplane be capable of traveling in order to guarantee a 90% probability of spotting at least one school of tuna?

17-21 C.A. Gopher & Sons is excavating a site requiring the removal of 100,000 cubic yards of material. Mr. Gopher has leased 10 trucks at $20 per hour. He has the choice of using a scoop loader or a shovel crane to load the dirt into the trucks. A scoop loader costs $40 per hour, whereas the cost of a shovel crane is $60 per hour.

Once work has been started, the trucks arrive according to a Poisson process with a mean rate of $\lambda_A = 7$ trucks per hour. The truck-filling times are exponentially distributed. A scoop loader can fill trucks at an average rate of $\lambda_S = 10$ per hour. The shovel crane is faster, filling on the average $\lambda_S = 15$ trucks per hour. (For simplicity, we assume the same truck arrival rate, regardless of equipment used.)

Since the number of truck arrivals required to excavate the site is fixed by the amount of dirt, the optimal choice of filling equipment will be one that will minimize the combined hourly costs of truck *unproductive time* plus the cost of the filling equipment. Adapting the procedures used in the machine-shop tool cage example, determine the optimal equipment choice.

17-22 One of the fundamental principles of designing a system to be reliable is that it have redundant (that is, duplicate) critical subsystems. Suppose that two power supplies are to be used simultaneously on a communications satellite. Assuming the lifetime of each to be exponentially distributed, operating independently, determine the probability that the satellite will function for 1,000 hours before total power failure when the mean time between failures $(1/\lambda)$ of each power supply is (a) 500 hours; (b) 1,000 hours; (c) 5,000 hours.

TESTING FOR GOODNESS OF FIT **17-5**

In this section we use a procedure called the *goodness-of-fit test* to make inferences regarding the type of distribution. The goodness-of-fit test determines whether or not sample data could have been generated according to a particular probability distribution, such as the Poisson or exponential. It can also indicate whether or not the frequency distribution for the sampled population has a particular shape, such as the normal curve. It achieves this by comparing the sample frequency distribution (histogram) to the assumed theoretical frequency distribution (frequency curve).

Importance of Knowing the Distribution

We have seen that many models used in statistical applications specify the probability distributions that must apply. For example, techniques developed for analyzing waiting lines in order to determine appropriate service facilities (such as the number of tellers for a bank) often assume that the number of arrivals has a Poisson distribution. Every rule developed in such queuing models depends on this assumption. (Other queuing models depend on other specified distributions.) As we have noted, the use of the Poisson distribution to characterize a random variable should be validated by collecting substantiating evidence. Recall that a Poisson process requires the following: (1) no memory (occurrences are independent over time); (2) constant mean occurrence rate; and (3) small likelihood that two or more events will occur during a short interval of time. The lack of any one of these requirements will make the Poisson distribution inappropriate. Only actual observation of the random process (or one believed to be identical in structure) can confirm that the above conditions are met.

When the Poisson distribution is wrongly used in place of the appropriate arrival distribution, the entire queuing solution is invalid. Consider, for example, the arrivals of baggage articles at the claims area of an airport. This may be a process with constant mean arrival rate and no memory, but it blatantly violates the requirement for rarity of simultaneous arrivals, as hundreds of pieces of luggage may arrive on a single flight. A facility designed to handle a fast-moving trickle of baggage (which is the case for Poisson arrivals) would be totally incapable of coping with an infrequent deluge of luggage.

Other statistical applications also rest on theoretical foundations that require particular population distributions. For example, the validity of the Student t distribution in making inferences about μ from small samples is limited to populations where the *normal curve* closely represents the underlying distribution. The random numbers list used in selecting the sample units is presumed to be *uniformly distributed* (that is, the number 21763 is just as likely to be the next number on a five-digit list as is any other value between 00000 and 99999). The *binomial distribution* provides probabilities for a variety of qualitative populations, but only if the underlying assumptions—constant π, independent and complementary trial outcomes—hold true.

There are practical reasons for wanting to know the population distribution rather than only the mean or variance. Many standardized tests used to measure aptitude and achievement are designed partly to provide scores that are normally distributed. This permits a common interpretation of resultant scores in terms of standard deviation units. Recall that *the distance of a person's score above or below the mean, in standard deviation units, can be directly translated into a percentile value if the normal curve applies to the population.* Thus, a personnel analyst or college admissions officer may readily resolve the conflicting scales on the various screening tests if their respective standard deviations are known. This is often done by converting data into *standard scores,* which are the same as normal deviates.

Conducting the Test: The Distribution of SAT Scores

In testing for goodness of fit, the following hypotheses apply:

H_0:   The sample is from a population having a designated distribution (for example, Poisson, normal).

H_A: The sample is from a population having some other distribution.

We must formulate a decision rule that can be expressed in terms of a critical value for some test statistic. If the computed value for this test statistic lies to one side of the critical value, we will accept the null hypothesis; otherwise, we will reject it. Again, this procedure will contain two possible kinds of error: Type I, or rejecting a true null hypothesis, and Type II, or accepting a false null hypothesis.

In testing for goodness of fit, the Type II error probability β is not well defined. This is because there are so many ways in which the null hypothesis can be false. And if the sample does come from some other distribution, this must be specified in order to calculate β. In general, the larger the sample, the better the protection against Type II errors will be.

The following illustration explains the goodness-of-fit test. The Scholastic Aptitude Test (SAT) has been designed to represent a population of high-school seniors who receive a mean score of 500, with a standard deviation of 100. Furthermore, this population is assumed to be normally distributed. A university president has asserted that her students—all of whom took the SAT before matriculation—are representative of that population. Based on a random sample of 200 student SAT scores, a statistics professor at the university constructed the histogram in Figure 17-5 and then superimposed on it the normal

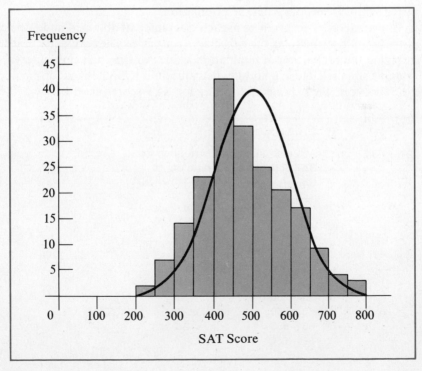

FIGURE 17-5
Sample histogram and assumed normal curve for student SAT scores.

curve assumed by the president. The professor disagrees with the president, since the histogram is positively skewed, with a lower mean and a greater variability than the supposed population frequency distribution indicates.

Not convinced by the graphical evidence, the president asks the statistician for a more conclusive argument. The professor begins by arranging the sample data in the frequency distribution shown in Table 17-2. (Note that the three lowest and the three highest class intervals from the histogram have been grouped into two broader categories; the reason for doing this will be explained later in this section.)

The professor must determine whether these results are typical for the null hypothesis that the university's SAT scores are normally distributed with a mean of $\mu = 500$ and a standard deviation of $\sigma = 100$. To do this he must compare them with the results expected under the null hypothesis.

In sampling from any population, there is no reason why the frequency distribution for sample results must precisely resemble that for the population. Because of sampling error, a perfect match is extremely unlikely. But a drastically different shape for the frequency distribution is also unlikely. For testing purposes, we will compare the actual sample result to the frequency distribution for the sample results expected under the assumption of the null hypothesis. This is the distribution obtainable "on the average" from a large number of repeated sample selections from the population.

The numbers of occurrences in each class interval obtained under the null hypothesis are referred to as the *expected frequencies*. They are determined by multiplying the proportion of population values within each class interval by the sample size used. Here, a normal distribution is hypothesized for the population. Therefore, the expected frequency for SAT scores at or below 350 can

TABLE 17-2
Actual Frequency Distribution for
SAT Scores of a Sample of
University Students

SAT Score	Number of Students f_a
≤ 350	23
351–400	23
401–450	42
451–500	33
501–550	25
551–600	21
601–650	17
> 650	16
	200

TABLE 17-3

Expected Frequencies for Student SAT Scores Under the Null Hypothesis

(1) SAT Score x = Upper Class Limit	(2) Normal Deviate $z = \dfrac{x - 500}{100}$	(3) Area to Left of x	(4) Area of Class Interval	(5) Expected Frequency f_e [(4) × 200]
350	−1.5	.0668	.0668	13.36
400	−1.0	.1587	.0919	18.38
450	−.5	.3085	.1498	29.96
500	0	.5000	.1915	38.30
550	.5	.6915	.1915	38.30
600	1.0	.8413	.1498	29.96
650	1.5	.9332	.0919	18.38
>650	∞	1.0000	.0668	13.36
			1.0000	200.00

be found by multiplying together (1) the area under the normal curve covering scores ≤ 350 and (2) the sample size $n = 200$.

To obtain the area, we determine the normal deviate:

$$z = \frac{x - \mu}{\sigma} = \frac{350 - 500}{100} = -1.5$$

Reading Appendix Table D, this corresponds to a lower tail area of .5000 − .4332 = .0668. Multiplying by the sample size, we obtain the expected frequency .0668 × 200 = 13.36 students. The calculations for the remaining expected frequencies appear in Table 17-3.

The Chi-Square Test Statistic

As the test statistic, we need a number that summarizes the amount of deviation between the frequencies of SAT scores actually obtained and the expected frequencies. The chi-square statistic serves this purpose.

We use the symbol f_e for the expected frequency of sample values in a particular class interval. Similarly, f_a represents the actual observed frequency of those values. To compute the test statistic, we use the following expression for the

CHI-SQUARE STATISTIC FOR A GOODNESS-OF-FIT TEST

$$\chi^2 = \sum \frac{(f_a - f_e)^2}{f_e}$$

TABLE 17-4

Calculation of the χ^2 Test Statistic for SAT Scores

SAT Score	Actual Frequency f_a	Expected Frequency f_e	$f_a - f_e$	$(f_a - f_e)^2$	$\dfrac{(f_a - f_e)^2}{f_e}$
≤ 350	23	13.36	9.64	92.9296	6.956
351–400	23	18.38	4.62	21.3444	1.161
401–450	42	29.96	12.04	144.9616	4.839
451–500	33	38.30	−5.30	28.0900	.733
501–550	25	38.30	−13.30	176.8900	4.619
551–600	21	29.96	−8.96	80.2816	2.680
601–650	17	18.38	−1.38	1.9044	.104
> 650	16	13.36	2.64	6.9696	.522
	200	200.00	0.00		$\chi^2 = 21.614$

In Table 17-4, these computations for the SAT scores are made using the actual and expected frequencies found in Table 17-3. We obtain the value $\chi^2 = 21.614$.

The degrees of freedom in a chi-square test correspond to the number of f_e values that we are "free" to set. Because the number of f_e's must sum to the sample size n, all but one f_e are free to vary. Thus, in testing for goodness of fit, *the number of degrees of freedom is equal to the number of categories minus one.* As we will see, this rule must be modified slightly in some cases.

The Hypothesis-Testing Steps

STEP 1

Formulate the null hypothesis. We are testing the null hypothesis that the SAT scores come from a particular normal population.

> H_0: University SAT scores are normally distributed with a mean $\mu = 500$ and a standard deviation $\sigma = 100$.

STEP 2

Select the test statistic. The chi-square statistic will be used.

STEP 3

Obtain the significance level α and identify the acceptance and rejection regions. The statistics professor desires an $\alpha = .01$ significance level. This means that he wants to have only a 1% chance of rejecting a true null hypothesis. From Appendix Table H, we use $8 - 1 = 7$ degrees of freedom and find the critical value $\chi^2_{.01} = 18.475$. Thus, if the university SAT score distribution is normal with a mean of 500 and a standard deviation of 100, there is a 1% chance that a sample will be obtained that yields a χ^2 value equal to or greater than 18.475. The acceptance and rejection regions are shown below.

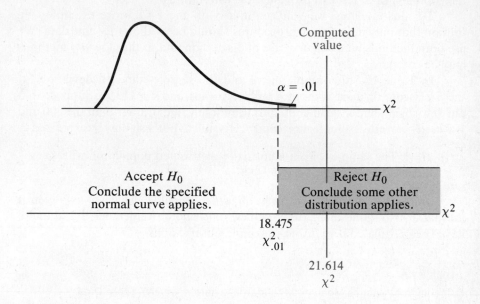

STEP 4
Collect the sample and compute the value of the test statistic. This step includes finding the values of f_a and f_e and computing χ^2. In Table 17-4 we found $\chi^2 = 21.614$.

STEP 5
Make the decision. The computed value of $\chi^2 = 21.614$ falls in the rejection region. The null hypothesis must be rejected. The professor must conclude that the university students' SAT scores are not normally distributed with the stated mean and standard deviation. The president now has some very convincing evidence that she is wrong.

Testing with Unknown Population Parameters

In the testing procedure just described, the population mean and standard deviation were specified in advance. Most often, we are not only concerned with identifying the family (such as normal, uniform, exponential, or Poisson) to which the underlying distribution belongs, but we may also need to find the appropriate particular distribution for that family. For example, in addition to not knowing whether a sample comes from a normal distribution, we may also lack knowledge of μ and σ. The unknown population parameters may themselves have to be estimated from the sample data. As an additional feature, the goodness-of-fit test can provide these estimations.

To illustrate this, we will continue with the SAT score example. We caution that only one of these procedures should be used on the same data. We use both methods here only for ease of discussion and so that the two methods may be compared.

In Table 17-5 the sample mean and the sample standard deviation for the SAT score data are computed to be $\bar{X} = 481$ and $s = 113.93$. Suppose that our statistician had instead chosen to use these values (rather than the 500 and 100 used before) as estimators of μ and σ. His null hypothesis may be expressed as

H_0:   The sample is from a normally distributed population where
$\mu = 481$ and $\sigma = 113.93$.

The alternative hypothesis includes all other possible distributions—even a normal distribution having other values for μ and σ. Remember that μ and σ have been estimated here from the sample observations.

TABLE 17-5

Calculation of Sample Mean and Standard Deviation for SAT Score Data

Class Interval	Frequency f	Midpoint X	fX	X^2	fX^2
201–250	2	225	450	50,625	101,250
251–300	7	275	1,925	75,625	529,375
301–350	14	325	4,550	105,625	1,478,750
351–400	23	375	8,625	140,625	3,234,375
401–450	42	425	17,850	180,625	7,586,250
451–500	33	475	15,675	225,625	7,445,625
501–550	25	525	13,125	275,625	6,890,625
551–600	21	575	12,075	330,625	6,943,125
601–650	17	625	10,625	390,625	6,640,625
651–700	9	675	6,075	455,625	4,100,625
701–750	4	725	2,900	525,625	2,102,500
751–800	3	775	2,325	600,625	1,801,875
	200		96,200		48,855,000

$$\bar{X} = \frac{\sum fX}{n} = \frac{96,200}{200} = 481$$

$$s = \sqrt{\frac{\sum fX^2 - n\bar{X}^2}{n-1}} = \sqrt{\frac{48,855,000 - 200(481)^2}{200-1}} = 113.93$$

TABLE 17-6

Expected Frequencies for Student SAT Scores When Parameters Are Estimated from Sample Data

(1) SAT Score x = Upper Class Limit	(2) Normal Deviate $z = \dfrac{x - 481}{113.93}$	(3) Area to Left of x	(4) Area of Class Interval	(5) Expected Frequency f_e [(4) × 200]
≤250	−2.03	.0212	.0212	4.24
300	−1.59	.0559	.0347	6.94
350	−1.15	.1251	.0692	13.84
400	−.71	.2388	.1137	22.74
450	−.27	.3936	.1548	30.96
500	.17	.5675	.1739	34.78
550	.61	.7291	.1616	32.32
600	1.04	.8508	.1217	24.34
650	1.48	.9306	.0798	15.96
700	1.92	.9726	.0420	8.40
750	2.36	.9909	.0183	3.66
>750		1.0000	.0091	1.82
			1.0000	200.00

The expected frequency calculations under the statistician's new hypothesis appear in Table 17-6, and the chi-square statistic is computed in Table 17-7. In Chapter 15 we indicated that when calculating χ^2, each expected frequency should round off to a value of at least 5. Thus, the first two categories are grouped, since the expected frequency 4.24 for ≤250 is smaller than 5. Likewise, the last two class intervals are grouped, yielding the expected frequency 5.48. These combinations reduce the number of final categories in Table 17-7 to ten.

When population parameters are estimated from the sample data, one additional degree of freedom is lost for each parameter. Thus, we have for the

RULE FOR THE NUMBER OF DEGREES OF FREEDOM

df = Number of categories − Number of parameters estimated − 1

In the present illustration we have

$$df = 10 - 2 - 1 = 7$$

Suppose that the statistician uses the same significance level as before. Because the degrees of freedom are unchanged, we have the same critical value of $\chi^2_{.01} = 18.475$. Since the computed value of $\chi^2 = 7.107$ (from Table 17-7) is

TABLE 17-7

Calculation of the χ^2 Test Statistic for SAT Scores When Parameters Are Estimated from Sample Data

SAT Score	Actual Frequency f_a	Expected Frequency f_e	$f_a - f_e$	$(f_a - f_e)^2$	$\dfrac{(f_a - f_e)^2}{f_e}$
≤ 250	2 } 9	4.24 } 11.18	-2.18	4.7524	.425
251–300	7	6.94			
301–350	14	13.84	.16	.0256	.002
351–400	23	22.74	.26	.0676	.003
401–450	42	30.96	11.04	121.8816	3.937
451–500	33	34.78	-1.78	3.1684	.091
501–550	25	32.32	-7.32	53.5824	1.658
551–600	21	24.34	-3.34	11.1556	.458
601–650	17	15.96	1.04	1.0816	.068
651–700	9	8.40	.60	.3600	.043
701–750	4 } 7	3.66 } 5.48	1.52	2.3104	.422
> 750	3	1.82			
	200	200.00	0.00		$\chi^2 = 7.107$

smaller than 18.475, H_0 must be *accepted*. This is the reverse of the conclusion that we reached using our previous null hypothesis. Although the data significantly refute a normally distributed population with a mean of $\mu = 500$ and a standard deviation of $\sigma = 100$ when these parameters are specified in advance, the results are quite consistent with the null hypothesis of normality when the parameters are estimated from the sample data.

Illustration: Poisson Probability Distribution for the Number of Typesetters' Errors

As we noted at the outset, whether or not a particular probability distribution applies can be crucial when using probability concepts for decision making. The following illustration shows how the goodness-of-fit test can be used to make a decision that depends on the data fitting a Poisson distribution.

A publisher wishes to insert a new clause in its contracts with typesetters composing its books that would allow it to send out notices of pending penalties or incentives. These notices would be based on the error rate in a sample of early typeset copy (galley proofs). By receiving such a notice *before books are completed*, the typesetters would be made aware of how their profits will be affected by continuation of poor performance or maintenance of good quality. From the publisher's point of view, such notification will result in a better quality end product.

Whether or not a sample may be used to determine errors will depend on the probability distribution for the number of errors per 100 lines of print. If the number of errors has a Poisson distribution, the occurrences of errors throughout the typesetting may be considered to be a Poisson process. This implies that it is extremely rare for several errors to be found adjacent to each other. Another property of a Poisson process is that the mean rate at which errors are committed would be constant, no matter which part of the book is sampled. Thus a sample taken from the first chapter may be used to infer the accuracy of the rest of the book.

On the other hand, if the number of errors closely fits some other distribution, such as the normal, the mean error rate need not be constant throughout the book. Also, the errors may tend to cluster in difficult portions, such as tables or mathematical expressions. Thus, galleys from all parts of the book may be required to make an assessment of its accuracy. Determination of penalties or incentives would have to wait for the composition of the entire book. If this is the case, there would be no need for sampling at all, since the ultimate settlement of accounts, adjusted for penalty or incentive, would take place after a time-consuming phase of thorough proofreading by publisher and author. There is nothing to be gained by haranguing the typesetter at such a late stage. Except for corrections, the composition work is largely done.

From files containing galley proofs of all recently published books, a random sample is taken of $n = 40$ segments of type 100 lines long and of comparable complexity. Table 17-8 shows the sample frequency distribution

TABLE 17-8
Frequency Distribution for Number of Typesetters' Errors

Number of Errors per 100 Lines	Total Number of Errors in Category	Observed Frequency f_a
0	0	5
1	9	9
2	10	5
3	21	7
4	16	4
5	10	2
6	18	3
7	14	2
8	8	1
9	0	0
10	20	2
	126	40

$$\bar{X} = 126/40 = 3.15$$

for the number of errors per 100 lines. The mean number of errors in the sample is $\bar{X} = 3.15$. Thus, the mean error rate per line times the number of lines is

$$\lambda t = 3.15$$

The following hypothesis steps apply.

STEP 1
Formulate the null hypothesis.

H_0: The typesetters' errors have a Poisson distribution with $\lambda t = 3.15$.

The alternative hypothesis includes all other possible distributions—possibly even a Poisson with λt at some other value. The value 3.15 has been estimated from the sample observations.

STEP 2
Select the test statistic. The chi-square statistic will be used.

TABLE 17-9
Chi-Square Calculations Under the Null Hypothesis of a Poisson Distribution

(1) Number of Errors x	(2) Actual Error Frequencies f_a	(3) Poisson Probabilities	(4) Expected Error Frequencies f_e	(5) $f_a - f_e$	(6) $(f_a - f_e)^2$	(7) $\dfrac{(f_a - f_e)^2}{f_e}$
0	5 ⎰14	.0428	1.71 ⎱7.11	6.89	47.472	6.68
1	9 ⎱	.1350	5.40 ⎰			
2	5	.2124	8.50	−3.50	12.250	1.44
3	7	.2233	8.93	−1.93	3.725	.42
4	4	.1759	7.04	−3.04	9.242	1.31
5	2	.1106	4.42			
6	3	.0581	2.32			
7	2	.0261	1.04			
8	1 ⎬10	.0106	.42 ⎬8.40	1.60	2.560	.31
9	0	.0036	.14			
10	2	.0011	.04			
11 or more	0	.0005	.02			
	40	1.0000	39.98*			10.16

$$f_e = 40Pr\,[X = x] = 40\,\frac{e^{-3.15}(3.15)^x}{x!}$$

$$\chi^2 = \sum \frac{(f_a - f_e)^2}{f_e} = 10.16$$

* Total would be 40, except for rounding errors.

STEP 3

Obtain the significance level α and identify the acceptance and rejection regions.
Suppose that the publisher chooses a significance level of $\alpha = .05$. Only 5
category groupings are used, because some categories have been combined to
achieve expected frequencies of 5 or more (see Table 17-9). The number of
degrees of freedom for this test is $5 - 1 - 1 = 3$, since only one parameter
(the mean λt of the Poisson distribution) is estimated from the sample data.
Appendix Table H provides the critical chi-square value of $\chi^2_{.05} = 7.815$. The
following acceptance and rejection regions apply.

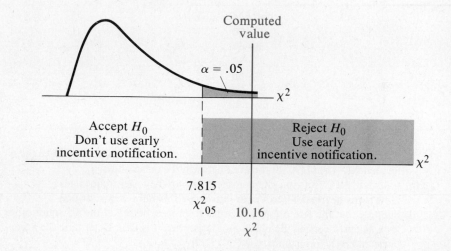

STEP 4

Collect the sample and compute the value of the test statistic. The expected
frequencies for each error category are calculated in column (4) of Table 17-9
by multiplying the Poisson probabilities in column (3) by $n = 40$. The chi-square
value, also computed in Table 17-9, is 10.16.

STEP 5

Make the decision. Because the computed value of $\chi^2 = 10.16$ falls in the rejection region, the null hypothesis that the error distribution is Poisson with
$\lambda t = 3.15$ must be *rejected*. The publisher therefore will not write an early
incentive-penalty notification clause into its typesetting contracts.

EXERCISES

17-23 The vice president for personnel of a large insurance company wishes to establish
hiring policies for the company's clerical personnel. A key element of the company's
screening procedures is a verbal aptitude test, which will be used to match applicants'

capabilities to job requirements. Personnel planning requires that the percentages of the applicant population having test scores at various levels be determined. This will involve specification of the frequency distribution, which can only be inferred from a limited sample. It will be significantly simpler to generate plans if the population is normally distributed. Thus the vice president wishes to conduct a goodness-of-fit test for a normal null hypothesis. The psychologists who designed the test maintain that an ordinary cross section of persons have achieved a mean of 72 and a standard deviation of 10. The observed results by a sample of 100 persons applying for positions at the company is provided below:

Score	Frequency
below 40	8
40–50	7
50–60	18
60–70	25
70–80	32
80–90	3
over 90	7
	100

(a) Using the values provided by the psychologist, construct a table of expected frequencies. (Use .4490 for the area between the mean and $z = 3.2$.)

(b) Calculate the χ^2 statistic. (Remember to group the lower frequencies.)

(c) How many degrees of freedom are associated with this test statistic?

(d) Suppose that the vice president wishes to protect herself at the .05 significance level against rejecting the null hypothesis when it is true. Determine the acceptance and rejection regions.

(e) Should the null hypothesis be accepted or rejected? What is the smallest significance level that you can determine at which the hypothesis can be rejected?

17-24 Suppose that the vice president in Exercise 17-23 does not believe that the psychologist's parameters are valid for her applicant population. Instead she wishes to use the sample results to estimate μ and σ. For the 100 test scores, the mean is 64.5 and the standard deviation is 16.8.

(a) Calculate the table of expected frequencies. Then calculate the test statistic. At a significance level of .05, should the null hypothesis be accepted or rejected?

(b) Is the lowest probability at which the null hypothesis can be rejected lower or higher than in 17-23(e) above?

17-25 The Seven-High Bottling Company has been using a refillable bottle that has proven to be relatively safe from exploding. To modernize its image, a sleeker Seven-High bottle is being sought. Because of the obvious dangers posed by bottles that explode, tests are being conducted on a new bottle for which a special seal has been devised. The tests will be carried out on a special device in which bottles filled with Seven-High are heated and shaken. For the old bottle, it has been established that the number of bottles tested between explosions on this device is approximately exponentially distributed, with a mean of 3 bottles between explosions. It is hoped that the new bottle will have the same distribution. A sample of 1,000 new bottles is tested. The frequency distributions obtained are provided below.

Number of Bottles Between Explosions	Class Interval	Actual Frequency	Expected Frequency
0	0– .999	97	86.75
1	1– 1.999	54	65.25
2	2– 2.999	58	49.50
3	3– 3.999	25	28.00
4	4– 4.999	27	21.00
5	5– 5.999	15	15.80
6	6– 6.999	10	12.70
7	7– 7.999	5	9.30
8	8– 8.999	11	6.20
9	9– 9.999	2⎫	3.10⎫
10	10–10.999	0⎭	3.10⎭
11	11–11.999	2⎫	3.10⎫
12	12–12.999	1⎬	3.10⎬
13 or more	≥ 13	3⎭	3.10⎭
		310	310.00

The null hypothesis is that the distribution for the number of new bottles tested between explosions is identical to that for the old bottles. The alternative is that some other distribution applies.

(a) Calculate χ^2.

(b) How many degrees of freedom are associated with this test statistic?

(c) State in a sentence the Type I error; the Type II error.

(d) If a significance level of .05 is desired, determine the acceptance and rejection regions. What does this number mean? Formulate the decision rule. Should the null hypothesis be rejected?

(e) If the significance level is instead .02, how do your answers to (d) change?

17-26 Before random number tables were widely available, some statisticians used telephone digits. One concern is whether such numbers occur with the proper frequency. Assuming that the last four digits in the Manhattan telephone directory are used to generate decimal values between 0 and 1, a random sample of 100 successive numbers was obtained, and the following frequency distribution was determined. (A 0000 suffix is counted as 1.0000.)

Class Interval	Number of Values Obtained
.0001– .1000	13
.1001– .2000	12
.2001– .3000	10
.3001– .4000	9
.4001– .5000	9
.5001– .6000	11
.6001– .7000	9
.7001– .8000	11
.8001– .9000	7
.9001–1.0000	9
	100

Use as your null hypothesis the assumption that the above numbers were obtained from a *uniformly distributed* population, so that all values between 0 and 1 are equally likely to occur.
(a) Construct a table of expected frequencies.
(b) Find the value of the χ^2 statistic.
(c) At the $\alpha = .10$ significance level, what conclusion can you make?

17-27 An automobile manufacturer wishes to determine the frequency distribution of warranty-financed repairs per car for its new minicar, the Colt. The number of such repairs per new automobile on its current line of cars has been established to have a Poisson distribution with parameter λ, the mean rate of repairs per car. The value of λ varies with the line of car. Management believes that the Colt will have nearly the same distribution for the number of repairs as the Pony, its next larger car. For the Pony, $\lambda = 3$ repairs per year. A random sample of 43 cars is selected for a year's study. Special records are kept on each year. The distribution for repairs per car is provided below.

Number of Repairs	Frequency of Cars
0	1
1	2
2	5
3	9
4	7
5	5
6	3
7	5
8 or more	6
	43

(a) Calculate the table of expected frequencies, using Poisson probabilities with $\lambda t = 3$.
(b) Determine the value of the test statistic.
(c) Should the null hypothesis be accepted or rejected at the .05 significance level?

REVIEW EXERCISES

17-28 A random sample of $n = 3$ persons is to be taken from a group of $N = 10$, $S = 4$ of whom are women. Find the probability that the sample contains exactly two women, assuming that
(a) the sampling is done with replacement.
(b) the sampling is done without replacement.

17-29 Suppose you deal yourself a poker hand consisting of 5 cards from a standard deck of playing cards. Find the probability that you will get three aces.

17-30 During noon hour on Fridays, customers arrive at a mean rate of 2 per minute to the tellers' windows of Chase-Haste Bank. The pattern of arrivals is considered to be purely random.

(a) Find the probability that exactly 7 customers arrive between 12:15 and 12:17.

(b) What is the probability that the first customer arrives before 12:01?

17-31 A computer program generates random decimal values in such a way that any value between 0 and 1 is equally likely to be obtained each time. Find the probability that the next number is (a) less than .5; (b) between .25 and .35; (c) greater than .99; (d) greater than or equal to .05.

17-32 A particular production process yields 10% defective items.

(a) Assuming that production is a Bernoulli process, find the probability that in the next 100 items produced there will be at least 5 defectives.

(b) Assuming that production is a Poisson process, with defectives occurring at the mean rate of 10 per hour, find the probability that at least 5 defectives will occur in any specific hour.

17-33 A communications satellite's power cells have an exponential time-to-failure distribution. The mean time between failures is 100 hours.

(a) What is the probability that any particular power cell will fail prior to 500 hours of operation?

(b) Suppose that there are four power cells altogether, that they operate independently, and that the satellite will function as long as there is at least one working cell. What is the probability that the satellite can still communicate at the end of 500 hours of operation?

17-34 A random number generator yields uniformly distributed random decimals between 0 and 1. Consider the sampling distribution for the mean $\bar{X}$ of 12 such decimals. $\bar{X}$ itself has an expected value (mean) of .5 and a standard deviation of 1/12.

(a) What probability distribution applies for $\bar{X}$?

(b) Find the probability that $\bar{X}$ falls between .4 and .6.

17-35 The actual frequency distribution for rush-hour automobile accidents reported in a certain city for a 100-day period is given in the table below. Assuming that these data represent a random sample from a population where the number of daily accidents has the Poisson distribution at a mean rate of 5 per day, we obtain the expected frequencies in the last column.

Number of Accidents per Day	Actual Frequency (days)	Expected Frequency (days)
≤2	13	13
3	8	14
4	22	18
5	24	18
6	15	15
7	8	10
8	5	6
≥9	5	6
	100	100

Are the sample data consistent with the indicated Poisson distribution? Use $\alpha = .05$.

17-36 The times between reported automobile accidents in a different city during rush hours are provided below for a random sample of 200 accidents. Also given are the expected frequencies that would apply if the population were exponentially distributed with a mean accident rate of one every 20 minutes

Time Between Accidents (minutes)	Actual Frequency	Expected Frequency
0–10	62	78.6
10–20	55	47.8
20–30	32	29.0
30–40	21	17.6
40–50	12	10.6
50–60	8	6.4
> 60	10	10.0
	200	200.0

Are the sample data consistent with the indicated exponential distribution? Use $\alpha = .10$.

Out of the mouths of babes and sucklings hath He perfected praise! In the last few evenings I have wrestled with a double humped curve and have overthrown it.

WALTER F.R. WELDON (1892)

18

Nonparametric Statistics

Most of the hypothesis-testing procedures discussed earlier in this book involved inferences concerning population parameters, such as the mean. These tests are referred to as *parametric tests*, and their test statistics are sometimes called *parametric statistics*. The sampling distributions of the statistics used in parametric tests are usually based on assumptions about the *populations* from which the samples were obtained. The assumptions of the Student t statistic (used to test means when sample sizes are small) and the F statistic (used in analysis of variance) are particularly stringent. An assumption of both the t and the F sampling distributions is that the sampled populations are normal.

Short of complete enumeration, there is no way we can be sure that the populations are actually normal. Even if the population distribution is known, the normal distribution serves only as a mathematically convenient representation, because we know that no real population is truly normal. Slight deviations from normality may be tolerated when the t and the F tests are used, but sometimes a population that does not meet the assumption of normality can invalidate the results of the statistical test.

18-1   THE NEED FOR NONPARAMETRIC STATISTICS

In this chapter, we will examine statistical tests that make no assumptions about the shape of the population distribution. For this reason, such tests are often said to be "distribution free." Because many of the test statistics require no assumptions about the population parameters, they are commonly referred to as *nonparametric statistics*. We have already used the nonparametric statistic χ^2 to test for independence. In doing this, we made no assumptions about the distribution for the populations, and our hypotheses did not involve suppositions about parameters.

Advantages of Nonparametric Statistics

One of the many convenient properties of nonparametric statistics is that they are easy to calculate. In addition, their sampling distributions can often be explained by the simplest laws of probability. But their major advantage is that they are burdened by very few restrictive assumptions, which is particularly important when, for various reasons, samples cannot be large. For instance, nonparametric statistics are vital to behavioral scientists, because large samples are rarely employed in long-term experiments involving human subjects.

In addition to explicit assumptions regarding the shape of a population, certain parametric tests are based on assumptions about the population parameters that are *not being tested*. For example, equal variances are assumed when the Student t statistic is used to test for differences in population means with independent samples (see Section 14-3). Another assumption often implicit in parametric tests involving means is that the population values are continuous. Many populations involve variates whose values are not continuous, and a large class of these values cannot be combined arithmetically.

Many practical research problems involve subjective ratings on a numerical scale. For example, a subject who rates the taste of brand A beets as 5, brand C beets as 4, and brand B as 3 is really expressing an order of preference: A is better than C, which is better than B. The numbers have no real meaning in themselves: Ratings of 999, 2, and -30, respectively, would express the same thing, and the mean of such numbers is not very meaningful.

Numbers that primarily express such things as preference, ranking, and hierarchy are convenient symbols that convey order, and their numerical values are said to belong to an *ordinal scale*. Because the distance between values has no inherent meaning, standard arithmetic operations cannot be consistently applied to ordinal numbers. Any statistical test requiring the calculations of means or variances would be largely invalid when applied to samples from populations with ordinal variates. *Usually, only nonparametric tests are valid for such populations.*

Another deficiency of parametric tests is a preoccupation with the mean and the variance, which is largely due to the convenient mathematical properties of the measures. Except in symmetrical distributions, the median is a more meaningful measure of central tendency than the mean. Various interfractile ranges depict dispersion more vividly than the standard deviation. For example, knowing that the incomes of 90% of all doctors in a particular region lie between $20,000 and $150,000 is more meaningful to most people than knowing that $\sigma = \$25,000$. Some nonparametric tests are best suited to testing hypotheses regarding medians, but the popular parametric tests are limited to the less meaningful mean. Furthermore, medians are equally valid measures whether the variates are continuous or ordinal.

Disadvantages of Nonparametric Statistics

The primary drawback of nonparametric statistics is a tendency to ignore much sample information that can be derived from parametric statistics. But although nonparametric statistics are less efficient, researchers often have more confidence in their methodology than in the unsubstantiable assumptions inherent in parametric tests like the Student t test. Another handicap of nonparametric tests is that they are so numerous that researchers must pay more attention to efficiency than when a parametric test is used.

COMPARING TWO POPULATIONS USING INDEPENDENT SAMPLES: THE WILCOXON RANK-SUM TEST   18-2

Two methods—independent sampling and matched-pairs sampling—for choosing the population with the greater mean were examined in Chapter 14, but these tests are adequate only when the populations are normally or nearly normal distributed. Another procedure—the *Wilcoxon rank-sum test*—that is free of the possibly invalid assumptions of normality can also be used to test independent samples. This test is named after the statistician Frank Wilcoxon, who first proposed it in 1945.

Description of the Test: Comparing Two Fertilizers

The Wilcoxon test compares two samples—a control and an experimental group—taken from two populations. All null hypotheses tested under this procedure share a common assumption that the samples were selected from identical populations, which is more stringent than assuming that they have identical means. The Wilcoxon test is therefore based on the principle that the

TABLE 18-1

Corn Yields Obtained in a Test of Two Fertilizers

	Yield (bushels)				
Fertilizer A	42.3	38.7	42.8	35.6	47.2
Fertilizer B	61.4	45.3	46.4	53.1	50.1

two samples may be treated as if they came from a common population. The data for the two samples may be combined under the various null hypotheses. The observed values in the pooled sample are then ranked from smallest to largest; the smallest value is assigned a rank of 1, the next smallest value is ranked 2, and so forth. The samples are then separated and the sums of the ranks are calculated for each sample. The rank sums obtained are used as test statistics.

As an illustration, we will compare the effectiveness of fertilizers A and B in increasing the yield of corn. In this experiment, ten Corn Belt farmers set aside one-acre plots for testing. Five acres chosen randomly will be treated with fertilizer A, and five will be treated with fertilizer B. It is assumed that the two fertilizers are equally effective, so that the hypotheses are

H_0:   Fertilizers A and B are equally effective.

H_A: Fertilizers A and B differ in effectiveness.

H_0 will be accepted or rejected on the basis of the yields of the sample plots. In effect, the *null hypothesis states that the population of yields for plots fertilized with A is identical to the corresponding population using B*. The alternative hypothesis is that the two populations differ somehow. Because the alternative may be true if A is better than B or if B is better than A, this experiment involves a *two-sided test*.

The yields obtained from the sample plots are provided in Table 18-1, and the pooled sample results are ranked in Table 18-2. If the fertilizers are equally effective, A should be ranked low and high about as many times as B. A convenient comparison can be made in terms of the sums of the ranks ob-

TABLE 18-2

Ranks of Fertilizer-Test Yields

Fertilizer	A	A	A	A	B	B	A	B	B	B
Yield	35.6	38.7	42.3	42.8	45.3	46.4	47.2	50.1	53.1	61.4
Rank	1	2	3	4	5	6	7	8	9	10

tained for the respective samples. *For this purpose, the sum of the ranks for A determines the test statistic.* Using the ranks in Table 18-2 and letting W represent this rank sum, we have

$$W = 1 + 2 + 3 + 4 + 7 = 17$$

Under the hypothesis of identical effectiveness, any five of the ten ranks for the pooled samples can belong to population A; that is, the first rank is no more likely to apply to A than to B. This holds for any rank. In other words, if the fertilizers are equally effective, then the smallest yield is just as likely to be obtained from a plot where fertilizer A has been applied as from a plot where fertilizer B has been used; this is also true of the highest yield or of any yield in between. Thus, assuming that the null hypothesis is true, we may view the sample results as only one possible outcome from an uncertain situation with as many equally likely outcomes as there are ways to select five ranks for sample A out of a total of ten ranks.

The sampling distribution of W can also be found by listing all of the rank combinations. Treating these elementary events as equally likely, we can determine exact probabilities for rank sums such as 16, 17, 18, 19, and so on. In most testing situations, however, this would be a tedious procedure.

Instead, we can approximate the distribution of W by using the normal curve, which applies to most sample sizes. For this purpose, the following expression is used to calculate the

NORMAL DEVIATE FOR THE WILCOXON RANK-SUM TEST

$$z = \frac{W - \dfrac{n_A(n_A + n_B + 1)}{2}}{\sqrt{\dfrac{n_A n_B (n_A + n_B + 1)}{12}}}$$

where n_A and n_B represent the number of sample observations made from populations A and B, respectively.

The usual decision rules apply to the rank-sum test. For one-sided tests, the critical normal deviate z_α that corresponds to the desired significance level is obtained from Appendix Table E. For two-sided tests, $z_{\alpha/2}$ is used. If a significance level of $\alpha = .05$ is desired in the fertilizer experiment, $z_{.025} = 1.96$. The decision rule is illustrated on the next page.

Small values of W, and therefore negative z's, indicate that smaller ranks have been assigned to sample A, which is consistent with a situation in which the values of population A are smaller than the values of B. For large W's and positive z's, the reverse holds.

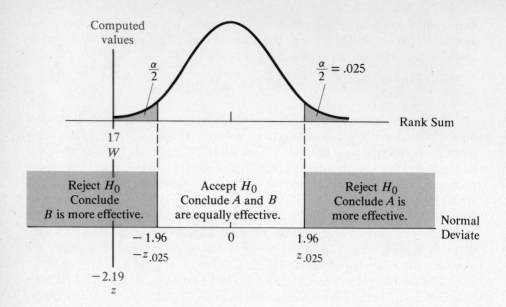

Substituting $n_A = n_B = 5$ and $W = 17$ into the expression for z, we find the normal deviate for the fertilizer-test results:

$$z = \frac{17 - \dfrac{5(5 + 5 + 1)}{2}}{\sqrt{\dfrac{5(5)(5 + 5 + 1)}{12}}} = \frac{17 - 27.5}{\sqrt{22.9167}} = -2.19$$

Because the computed value of the normal deviate $z = -2.19$ falls in the lower rejection region, H_0 must be *rejected*. The sample results indicate that fertilizer B is more effective at the $\alpha = .05$ significance level.

Application to One-Sided Tests: Evaluating Compensation Plans

To show how this procedure may be applied in a one-sided test, we will consider the case of a statistician for a toy manufacturer who wishes to compare two compensation plans. One would use an incentive wage; the other calls for a straight hourly wage. A sample of 10 employees are placed under the incentive plan (A). This group is to be compared with a sample of 15 employees who are paid by the hour (B). Table 18-3 shows the sample data that have been collected.

TABLE 18-3

Sample Data for Experiment Comparing Two Compensation Plans

Unit Output Under Incentive Compensation X_A	Unit Output Under Hourly Compensation X_B	X_A^2	X_B^2
148	143	21,904	20,449
146	137	21,316	18,769
152	149	23,104	22,201
157	151	24,649	22,801
144	146	20,736	21,316
148	155	21,904	24,025
160	147	25,600	21,609
154	150	23,716	22,500
154	147	23,716	21,609
156	138	24,336	19,044
1,519	145	230,981	21,025
	153		23,409
	155		24,025
	149		22,201
	138		19,044
	2,203		324,027

Sample	B	B	B	B	A	B	A	B	B	B	A	A	B	B
Output	137	138	138	143	144	145	146	146	147	147	148	148	149	149
Rank	1	2	3	4	5	6	7.5	7.5	9	10	11	12	13	14

	B	B	A	B	A	A	B	B	A	A	A
	150	151	152	153	154	154	155	155	156	157	160
	15	16	17	18	19	20	21	22	23	24	25

$$n_A = 10 \qquad\qquad n_B = 15$$

$$\bar{X}_A = \frac{1,519}{10} = 151.90 \qquad\qquad \bar{X}_B = \frac{2,203}{15} = 146.87$$

$$s_A^2 = \frac{230,981 - 10(151.90)^2}{9} \qquad\qquad s_B^2 = \frac{324,027 - 15(146.87)^2}{14}$$

$$= 27.21 \qquad\qquad\qquad = 33.22$$

STEP 1

Formulate the null hypothesis. As the null hypothesis, the statistician assumes that the output of the employees under the incentive plan is not greater than the productivity of the hourly employees:

$$H_0: \quad \text{Population } A \text{ values} \leq \text{Population } B \text{ values}$$

STEP 2
Select the test statistic. The test statistic is based on the output rank sum for the incentive-compensation group A. The normal deviate z is used.

STEP 3
Obtain the significance level α and identify the acceptance and rejection regions. The statistician selects a significance level of $\alpha = .05$. Thus, $z_{.05} = 1.64$. Because a larger number of high ranks for sample A should refute H_0, this test is upper-tailed. The acceptance and rejection regions are shown below.

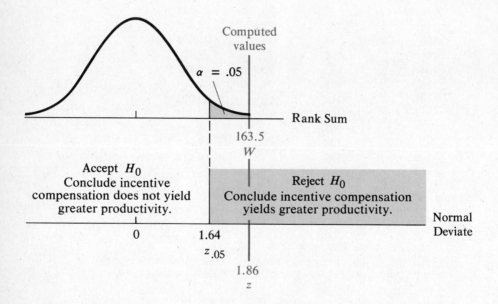

STEP 4
Collect the sample and compute the value of the test statistic. The combined sample data are ranked in Table 18-3. (The special handling of tying scores is explained on page 623.) The rank sum for the sample A outputs is

$$W = 5 + 7.5 + 11 + 12 + 17 + 19 + 20 + 23 + 24 + 25 = 163.5$$

With sample sizes $n_A = 10$ and $n_B = 15$, the normal deviate is

$$z = \frac{W - \dfrac{n_A(n_A + n_B + 1)}{2}}{\sqrt{\dfrac{n_A n_B(n_A + n_B + 1)}{12}}} = \frac{163.5 - \dfrac{10(10 + 15 + 1)}{2}}{\sqrt{\dfrac{10(15)(10 + 15 + 1)}{12}}} = \frac{33.5}{\sqrt{325}} = 1.86$$

STEP 5

Make the decision. The computed value of $z = 1.86$ falls in the rejection region, so the statistician must *reject* the null hypothesis that population A outputs are no larger than those for population B and conclude that incentive compensation yields greater productivity than does hourly compensation.

The Problem of Ties

Several ties were encountered in ranking the data in Table 18-3. For instance, two sample A workers shared the output level of 148, and two sample B workers achieved the same output of 147. *As long as ties occur within the same sample group, successive ranks may be assigned arbitrarily to the tying sample observations.* Thus, we ranked the two 147 outputs as 9 and 10 and the two 148 outputs as 11 and 12.

A difficulty arises, however, when ties occur between sample groups, because the choice of ranks affects W. *When ties occur between sample groups, each observation is assigned the average of the ranks for that value.* The ranks of 7 and 8 would have applied to the two outputs of 146 had they been in the same sample. However, since one worker in *each* sample achieved the same output level, the values receive equal ranks of $(7 + 8)/2 = 7.5$.

Comparison of the Wilcoxon Test and the Student *t* Test

The data from the compensation illustration may be used to compare the Wilcoxon rank-sum test to the Student t test for independent samples. Substituting the intermediate calculations from Table 18-3, we obtain

$$t = \frac{\bar{X}_A - \bar{X}_B}{\sqrt{\dfrac{(n_A - 1)s_A^2 + (n_B - 1)s_B^2}{n_A + n_B - 2}}\sqrt{\dfrac{1}{n_A} + \dfrac{1}{n_B}}}$$

$$t = \frac{151.90 - 146.87}{\sqrt{\dfrac{9(27.21) + 14(33.22)}{10 + 15 - 2}}\sqrt{\dfrac{1}{10} + \dfrac{1}{15}}}$$

$$= \frac{5.03}{\sqrt{30.8682}\sqrt{.1667}} = 2.218$$

Using $\alpha = .05$ with $n_A + n_B - 2 = 23$ degrees of freedom, Appendix Table F provides $t_{.05} = 1.714$. Since the computed value for t is larger than 1.714, the t test also indicates that H_0 must be *rejected*.

Similar results were obtained using both tests. (Once again, it should be emphasized that a particular test should be chosen in advance of sampling and that only one test would ordinarily be used on any given data. Both tests were applied here only to compare the two techniques.) Generally, the t test is more efficient or powerful than the Wilcoxon, so, using the same data, the t test should provide a lower probability α of incorrectly rejecting H_0. However, the Student t test requires that the *populations* be normally distributed, which is often not the case.* Also, the Wilcoxon test can be used for experiments where the observations are measured on an ordinal scale, whereas the t test cannot. *The Wilcoxon rank-sum test can therefore be more widely used.*

EXERCISES

18-1 A taxicab company intends to replace its brand A batteries with brand B batteries if testing shows that brand B is more effective. The null hypothesis is that brand A is at least as effective as B. Ten new brand A batteries were used in a random sample of taxicabs; 15 new brand B batteries were used in another random sample. The following results were obtained for the number of days until each battery required replacement:

Brand A   323, 178, 246, 195, 402, 603, 496, 328, 213, 187
Brand B   421, 327, 609, 433, 519, 504, 183, 455, 365, 615, 504, 312, 513, 497, 723

(a) Rank the sample results and calculate W.
(b) If management wishes to protect itself against incorrectly changing battery brands at a significance level of .05, should the battery brands be changed?

18-2 Suppose that the company in Exercise 18-1 installed 100 brand A batteries and 150 brand B batteries, and that $W = 11,500$. Should management reject the null hypothesis that brand A is at least as effective as B at the .05 significance level?

18-3 Suppose that the physician in Exercise 14-5 (page 477) applies the Wilcoxon rank-sum test to a sample of 8 patients treated with penicillin and a sample of 10 patients treated with sulfa. She obtains the following results for the number of days required to cure a virulent strain of venereal disease:

(A) *Penicillin*   15, 9, 12, 22, 14, 9, 10, 15,
(B) *Sulfa*   7, 8, 10, 6, 7, 7, 4, 13, 11, 5

(a) Rank the sample data and calculate W.
(b) The doctor's null hypothesis is that the treatments are equally effective. In the future, she will use the treatment that she finds most effective. Otherwise, she will test further. What action should the doctor take at the $\alpha = .05$ significance level?

* Both procedures require equal population variances.

18-4 A farm cooperative wishes to select the faster railroad to ship produce to Central City, which is served by both the CW&B and the Sun Belt lines. Random samples of less-than-carload shipments via each line were monitored, and the following results were obtained for the number of hours required to complete shipments:

$$(A) \ CW\&B \quad 10, 18, 12, 15, 27$$
$$(B) \ Sun \ Belt \quad 14, 19, 22, 23, 25, 19, 24, 31, 26$$

Shipments are currently made on the Sun Belt line. The null hypothesis is that the Sun Belt is at least as fast as the CW&B.
(a) Rank the sample results and calculate W.
(b) If protection against choosing the wrong railroad is desired at the $\alpha = .025$ significance level, should the null hypothesis be accepted or rejected?

18-5 Two security analysts for a mutual fund wished to compare their trading strategies, so each selected a random sample of 10 stocks and traded them for a year. Analyst A's procedure was somewhat unorthodox, so the null hypothesis was that analyst B's method would provide percentage rates of return at least as great as A's. The following results (in percentage rates of return) were obtained:

$$A \quad 10, -5, 15, 23, 113, 57, -51, 203, 33, 44$$
$$B \quad 11, 27, 9, 9, 18, -4, -8, 53, 112, 6$$

(a) Rank the sample results and calculate W.
(b) Should H_0 be accepted or rejected at the $\alpha = .10$ significance level?

THE MANN-WHITNEY *U* TEST   18-3

The *Mann-Whitney U test* is often used to compare two populations using independent samples. This procedure is equivalent to the Wilcoxon rank-sum test, and both tests lead to the same conclusions.

In the Mann-Whitney U test—as in the Wilcoxon test—the combined sample data are ranked, and then the sum W of the sample A ranks is calculated. The test statistic is

$$U = n_A n_B + \frac{n_A(n_A + 1)}{2} - W$$

The normal curve also serves as the approximate sampling distribution of U. The normal deviate for the sample results is provided by

$$z = \frac{U - \dfrac{n_A n_B}{2}}{\sqrt{\dfrac{n_A n_B(n_A + n_B + 1)}{12}}}$$

Because the Wilcoxon procedure is equivalent to the U test and involves a shorter computation, the use of the Wilcoxon rank-sum test is recommended.

18-4   COMPARING TWO POPULATIONS USING MATCHED PAIRS: THE SIGN TEST

In this section, we will discuss a nonparametric test for the differences between two populations that involves samples of *matched pairs*. This is the *sign test*, so named because it considers only the direction of difference in each sample pair, which may be expressed by either a plus or a minus sign. Like the Wilcoxon rank-sum test for independent samples, the sign test can be applied to a wider variety of situations than the parametric t test that we described in Chapter 14. However, the sign test makes no assumptions whatsoever about the shape or the parameters of the population frequency distributions. (The Wilcoxon test involves null hypotheses that consider the targeted experimental population and the control population to be identical.)

Description of the Test: Comparing Two Razor Blades

The Blue-Beard Razor Blade Company wishes to compare its prototype "Rapier" blade with a competitor's "Scimitar" brand to determine whether the Rapier is of superior quality. The Rapier blade will be marketed only if testing indicates that Rapier is superior to Scimitar; otherwise, more effort will be devoted to improving Rapier.

A random sample of a representative cross section of 20 men is chosen for the test. Each man is to shave one side of his face with the Rapier blade and the other side with Scimitar for five consecutive days. At the end of the test, each man will rate the two blades in each of five categories: closeness of shave, shaving comfort, durability, shaving ease, and residual facial discomfort during the day. The highest rating is 10 points, and the lowest is 0. The points in each category are added together, so that a blade that rates highest in all five categories will be given a score of 50 and a blade that rates lowest in all categories will be given a score of 0. Fractional scoring is allowed.

Blades are randomly assigned by tossing a coin to determine which side of each man's face will be shaved with the Rapier blade. The brands of the blades are not known to the men participating in the experiment. The same blade is to be used on the same side of a subject's face throughout the experiment.

The null and alternative hypotheses are expressed as

$$H_0: \quad \text{Rapier ratings} \leq \text{Scimitar ratings}$$

$$H_A: \quad \text{Rapier ratings} > \text{Scimitar ratings}$$

The data obtained from the experiment are provided in Table 18-4. A test statistic must now be determined that will enable the company to accept or reject the null hypothesis.

The null hypothesis includes the special case of the blades being of equal quality. If the blades are identical, any differences in the scores for Rapier and Scimitar will result from which side of the face each man finds easier to shave. For instance, the first subject in Table 18-4 has assigned a higher score to Rapier. This may be due to the blade, but it may also be due to the fact that he is right-handed and shaved the left side of his face with Scimitar, which would be more cumbersome. In the latter case, the coin flip—not the blade quality— would be responsible for difference in ratings. Under the presumption of equal

TABLE 18-4
Experimental Results of Razor Blade Test

Test Subject Number	Score Received		Difference	Sign
	Rapier	Scimitar		
1	48.0	37.0	+11.0	+
2	33.0	41.0	−8.0	−
3	37.5	23.4	+14.1	+
4	48.0	17.0	+31.0	+
5	42.5	31.5	+11.0	+
6	40.0	40.0	0.0	tie
7	42.0	31.0	+11.0	+
8	36.0	36.0	0.0	tie
9	11.3	5.7	+5.6	+
10	22.0	11.5	+10.5	+
11	36.0	21.0	+15.0	+
12	27.3	6.1	+21.2	+
13	14.2	26.5	−12.3	−
14	32.1	21.3	+10.8	+
15	52.0	44.5	+7.5	+
16	38.0	28.0	+10.0	+
17	17.3	22.6	−5.3	−
18	20.0	20.0	0.0	tie
19	21.0	11.0	+10.0	+
20	46.1	22.3	+23.8	+

quality, the opposite scores would be assigned if the coin toss resulted in a tail instead of a head: The score for Rapier would be 37.0, and the score for Scimitar would be 48.0.

The differences between the sample values of the matched pairs are then determined. In our illustration, the differences in Table 18-4 are obtained by subtracting the Scimitar score from the Rapier score.

The company can now determine whether the data support or refute the null hypothesis. For this purpose, only the signs of the difference for each pair are used. The rating scheme in the experiment resulted in three ties. In these cases, it is assumed that the subject liked the blades equally well. The tied outcomes are therefore eliminated from the analysis, leaving $20 - 3 = 17$ pairs with sign differences.

Because the larger number of positive signs in Table 18-4 indicates that Rapier is preferred over Scimitar, the decision can be based on the number of positive signs represented by R. In our illustration, $R = 14$. Although this value is larger than the number of negative signs because a majority of the subjects rated Rapier blades higher than Scimitar blades, the company must still determine whether R is large enough to warrant rejecting H_0. The first step will be to find the sampling distribution of R.

Presuming blade quality is identical, each sign in Table 18-4 could have been reversed; that is, the probability of a positive sign difference for each matched pair is .5, just as if the results had been determined by the toss of a coin. Thus, the probability of how untypical our results are under the null hypothesis can be determined by using the binomial distribution. The probability of obtaining a positive sign difference for any particular pair is $\pi = .5$, and $n = 17$ pairs.

Although the test may be performed directly using binomial probabilities, it is more convenient to use the normal approximation (which is satisfactory for most sample sizes encountered). The following expresssion applies for the

NORMAL DEVIATE FOR THE SIGN TEST

$$z = \frac{2R - n}{\sqrt{n}}$$

A significance level of $\alpha = .05$ is desired, which corresponds to a critical normal deviate of $z_{.05} = 1.64$. The test is upper-tailed, because a large value of R (of z) tends to refute the null hypothesis. The decision rule is given on the next page.

Because the computed value of $z = 2.67$ falls in the rejection region, the null hypothesis that Rapier is not superior to Scimitar must be *rejected* at the $\alpha = .05$ significance level. The Blue-Beard Razor Blade Company should market its new Rapier blade. There is less than a .05 chance that this action will prove incorrect.

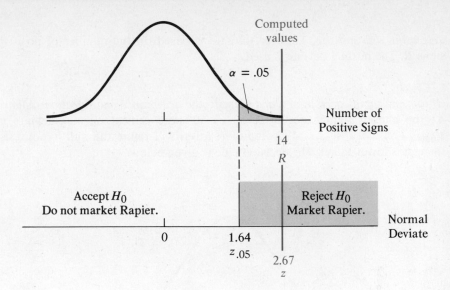

Illustration: Evaluating a New Reading Program

An educational researcher conducted an experiment to determine if elementary-school language programs would be improved by a new approach to vocabulary and spelling. The researcher believed that the same amount of time and effort would be more effectively devoted to additional reading than to the usual memorization methods. Fourth-grade students from various classes in several schools were divided into two groups. Children in the control group A took the standard language program in their homerooms. Students in the experimental group B went to special classes during their language periods, where they read, submitted book reports, and discussed special outside reading assignments.

Each child in group B was matched to a child in group A in the same classroom who had similar skills and aptitudes. Toward the end of the school year, both groups were tested. After subtracting the B child's score from the A child's score, the number of positive sign differences was determined.

STEP 1
Formulate the null hypothesis. As her null hypothesis, the researcher assumes that the new procedure does not improve language skills, as measured by scores on an achievement test. Thus

$$H_0: \quad A \text{ scores} \geq B \text{ scores} \quad \text{(new procedure is no improvement)}$$

STEP 2

Select the test statistic. The test statistic is based on the number of positive signs R. The normal deviate is used.

STEP 3

Obtain the significance level α and identify the acceptance and rejection regions. At the chosen significance level of $\alpha = .05$, the critical normal deviate is $z_{.05} = 1.64$. Because a small value of R (of z) will refute the null hypothesis, the test is lower-tailed. The decision rule is given below.

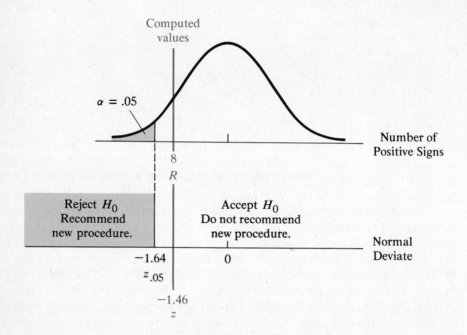

STEP 4

Collect the sample and compute the value of the test statistic. The data in Table 18-5 have been obtained. Eliminating the two ties leaves $n = 23$ pairs to consider. The number of positive signs is $R = 8$. Thus

$$z = \frac{2(8) - 23}{\sqrt{23}} = -1.46$$

STEP 5

Make the decision. Because $z = -1.46$ falls in the acceptance region, the researcher must *accept* H_0. The new procedure is no improvement and should not be adopted.

TABLE 18-5
Sample Data for Experiment Comparing Two Language-Skill Teaching Procedures

Pair i	Control Group A Score X_{A_i}	Experimental Group B Score X_{B_i}	Difference $d_i = X_{A_i} - X_{B_i}$	Sign	d_i^2
1	65	67	−2	−	4
2	72	66	+6	+	36
3	74	77	−3	−	9
4	81	90	−9	−	81
5	76	72	+4	+	16
6	95	95	0	tie	0
7	63	68	−5	−	25
8	85	95	−10	−	100
9	90	98	−8	−	64
10	64	60	+4	+	16
11	78	72	+6	+	36
12	86	94	−8	−	64
13	89	96	−7	−	49
14	75	73	+2	+	4
15	93	96	−3	−	9
16	78	90	−12	−	144
17	92	92	0	tie	0
18	67	63	+4	+	16
19	88	95	−7	−	49
20	79	88	−9	−	81
21	60	75	−15	−	225
22	90	95	−5	−	25
23	82	78	+4	+	16
24	73	85	−12	−	144
25	86	82	+4	+	16
			−81		1,229

$$\bar{d} = -\frac{81}{25} = -3.24$$

$$s_{d\text{-paired}} = \sqrt{\frac{1,229 - 25(-3.24)^2}{25 - 1}} = 6.35$$

A Comparison of the Sign Test and the Student t Test

The sign test can be compared to the analogous parametric procedure dexcribed in Chapter 14 for matched-pairs testing. There, we used the Student t test with small samples. In our current illustration, given $n = 25$ and substituting the mean of the matched-pairs differences and their estimated standard error (both calculated in Table 18-5), we find

$$t = \frac{\bar{d}}{s_{d\text{-paired}}/\sqrt{n}} = -\frac{3.24}{6.35/\sqrt{25}} = -2.551$$

From Appendix Table F for $25 - 1 = 24$ degrees of freedom at $\alpha = .05$, $t_{.05} = 1.711$. The acceptance and rejection regions are shown below.

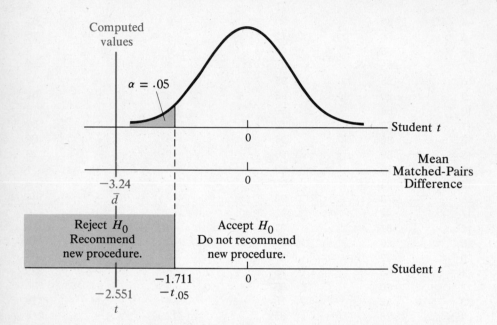

Because the computed value of $t = -2.551$ falls in the rejection region, the Student t test indicates that H_0 should be *rejected*. This conclusion is the reverse of the decision we reached using the sign test for the same data, reflecting the fact that *when applicable* the t test is more powerful and therefore more efficient than the sign test, as well as the Wilcoxon test. The t test is more discriminating because the relative sizes of the matched-pairs differences—as well as their signs—are used in calculating t. However, to use the t test, the *populations* of all potential matched-pairs differences must be normally distributed (or nearly so). *When this requirement is not met, the sign test is the valid test to use.*

EXERCISES

18-6   The sign test will be used to determine the appropriate action to be taken in each of the following situations. Find the acceptance and rejection regions, and indicate whether H_0 should be accepted or rejected at the stated significance level for each result provided.

(a)	(b)	(c)	(d)
$H_0:A$'s $\geq B$'s	$H_0:A$'s $\leq B$'s	$H_0:A$'s $\geq B$'s	$H_0:A$'s $\leq B$'s
$n = 18$	$n = 100$	$n = 50$	$n = 19$
$\alpha = .05$	$\alpha = .01$	$\alpha = .01$	$\alpha = .05$
$R = 4$	$R = 58$	$R = 13$	$R = 13$

18-7 A car-rental company must determine whether their cars should burn leaded (A) or unleaded (B) gasoline. A random sample of 15 pairs of cars are selected. One car from each pair uses leaded gasoline; the other uses unleaded. The cars in each pair are nearly identical and are driven over the same routes by drivers with similar driving styles. The null hypothesis is that unleaded gasoline provides at least the same mileage as leaded gasoline. The following sample results are obtained:

	Miles per Gallon			Miles per Gallon	
Pair	Leaded (A)	Unleaded (B)	Pair	Leaded (A)	Unleaded (B)
1	15.3	14.7	9	15.0	13.8
2	18.1	17.9	10	16.2	16.1
3	14.9	15.0	11	15.9	14.2
4	17.3	17.3	12	21.3	20.6
5	13.7	12.3	13	18.4	17.7
6	11.8	9.2	14	19.3	17.4
7	20.3	19.7	15	9.8	7.6
8	15.2	14.7			

(a) Determine the sign difference ($A - B$) for each pair.
(b) Assuming that leaded gasoline will be used if it provides better mileage, will A be used at a desired significance level of $\alpha = .001$?

18-8 A state highway patrol conducted a study similar to the one in Exercise 18-7, testing the same null hypothesis at the $\alpha = .05$ significance level for $n = 100$ automobile pairs. Assuming $R = 60$, which type of gasoline will be used?

18-9 Suppose that the researcher in the example on page 474 found the signs of differences in mean age for contracting angina pectoris by subtracting the onset age of the nonsmoker (B) from that of the smoker (A) and obtained $R = 30$ positive signs. As her null hypothesis, the researcher assumed that smoking does not affect the speed of the disease. At the $\alpha = .01$ significance level, what should she conclude about the effect of smoking on the age when angina occurs?

18-10 A medical researcher wishes to determine if a new appetite suppressant reduces weight more effectively than the one it now markets. The new suppressant will replace the current one if it proves more effective. A random sample of 40 persons are paired so that all factors believed to influence weight gain are nearly the same for each pair: Women are paired with women; smokers are paired with smokers; physically inactive people are paired; and so on. One pair member is given the new suppressant;

the other is given the old one. The null hypothesis is that the new suppressant is at most as effective as the old one. A suppressant is said to be more effective if it results in a greater weight reduction percentage after three months of use. The results are given below:

Pair	Reduction Percentage New Supplement	Reduction Percentage Old Supplement	Pair	Reduction Percentage New Supplement	Reduction Percentage Old Supplement
1	10	−2	11	15	8
2	5	3	12	13	12
3	7	1	13	14	10
4	8	10	14	13	5
5	4	2	15	7	8
6	15	11	16	11	3
7	12	13	17	−2	−3
8	18	5	18	0	−2
9	3	−2	19	16	9
10	8	12	20	9	8

(a) Determine the sign difference (new − old) for each pair.
(b) At the $\alpha = .05$ significance level, what course of action is indicated by the sample results?

18-11 A chemical supplier wishes to determine whether a new preservative will provide a longer shelf life for bread. A random sample of 10 bakeries have used the new preservative in some of their dough and have provided the supplier with two fresh loaves of standard bread—one baked with their regular preservative and one baked with the new preservative. The following shelf lives have been obtained:

Bakery	Shelf Life (days) Old Preservative	Shelf Life (days) New Preservative
1	5.7	6.3
2	4.2	3.9
3	6.5	6.7
4	3.4	3.6
5	6.1	6.3
6	5.3	5.7
7	4.9	5.2
8	3.7	3.7
9	2.8	3.7
10	4.3	4.5

(a) Use the sign test to determine whether to accept or reject at the $\alpha = .05$ level of significance the null hypothesis that the new preservation yields shelf lives that are at most as long as those provided by the old preservative.
(b) Repeat this test using the t statistic.

THE WILCOXON SIGNED-RANK **18-5**
TEST FOR MATCHED PAIRS

Thus far, we have described three nonparametric procedures. The first is based on rank sums and applies to independent samples; the second, the Mann-Whitney U test, is equivalent to the first; and the third, the sign test, is applicable when the sample data are matched into pairs. The first test deals only with the ranking of sample data and does not account for the relative magnitudes of the differences between sample groups. The third test considers the direction but not the size of the differences between sample pairs.

A test that considers both the direction and the magnitude of the differences between matched sample pairs is the *Wilcoxon signed-rank test*, proposed by Frank Wilcoxon. We will now describe this procedure, which is based on both rank sums and the signs of paired differences.

Description of the Test: Comparing Reading Programs

The Wilcoxon signed-rank test is based on matched-pairs differences. Ignoring signs, the *absolute values* of the differences are ranked from low to high, and the ranks corresponding to the original positive matched-pairs differences are summed. This procedure can be summarized in three steps:

1. Calculate the differences $d_i = X_{A_i} - X_{B_i}$ between all sample pairs.
2. Ignoring signs, rank the absolute values of the d_i's. Do *not* rank 0 differences.
3. Calculate V, the sum of ranks for positive d_i's.

The V statistic can be used to test the same hypotheses that are associated with the sign test. In every application H_0 includes the special case of the *A and B values being generated by identical populations*. If this is true, then each matched-pair difference has a 50–50 chance of having a positive or a negative sign, and positive or negative differences of the same absolute size should be equally likely. This indicates that under H_0, about half of the ranks should correspond to positive differences, and that the sum of these ranks V should be close to half of the total rank-sum value. To determine whether the value of V significantly refutes H_0, we must examine its sampling distribution.

As in the earlier Wilcoxon test, a detailed probability accounting will provide the exact sampling distribution of V. However, for most sample sizes, we can apply the normal distribution instead. The following expression is used to calculate the

NORMAL DEVIATE FOR THE WILCOXON SIGNED-RANK TEST

$$z = \frac{V - \dfrac{n(n+1)}{4}}{\sqrt{\dfrac{n(n+1)(2n+1)}{24}}}$$

The value of n is the number of nonzero differences found in the sample. As in the sign test, we ignore tied sample pairs, because they provide matched-pairs differences of zero.

To illustrate, we will reconsider our comparison of two methods of language-skill instruction. The data obtained by the researcher are shown in Table 18-6. As before, the A observations represent scores achieved by fourth-graders in the control group, who were exposed to a traditional memory-drill method. The B observations represent scores achieved by fourth-graders in the experimental group, who were enrolled in a special outside reading program instead of the regular reading class. The hypothesis to be tested is H_0: A scores

TABLE 18-6
Sample Data for Experiment Comparing Two Language-Skill Teaching Procedures with Assignment of Signed Ranks

Pair i	Control Group A Score X_{A_i}	Experimental Group B Score X_{B_i}	Difference $d_i = X_{A_i} - X_{B_i}$	Sign	Rank of Absolute Values of Matched-Pairs Differences
1	65	67	-2	$-$	1.5
2	72	66	$+6$	$+$	12
3	74	77	-3	$-$	3
4	81	90	-9	$-$	18
5	76	72	$+4$	$+$	5
6	95	95	0	tie	—
7	63	68	-5	$-$	10
8	85	95	-10	$-$	20
9	90	98	-8	$-$	16
10	64	60	$+4$	$+$	6
11	78	72	$+6$	$+$	13
12	86	94	-8	$-$	17
13	89	96	-7	$-$	14
14	75	73	$+2$	$+$	1.5
15	93	96	-3	$-$	4
16	78	90	-12	$-$	21
17	92	92	0	tie	—
18	67	63	$+4$	$+$	7
19	88	95	-7	$-$	15
20	79	88	-9	$-$	19
21	60	75	-15	$-$	23
22	90	95	-5	$-$	11
23	82	78	$+4$	$+$	8
24	73	85	-12	$-$	22
25	86	82	$+4$	$+$	9

are at least as great as B scores (the new procedure is no improvement over the old).

Because small values of V and the resulting negative z values indicate that the B scores are predominately larger than the A scores—the opposite of what the null hypothesis implies—this test is *lower-tailed*. At the $\alpha = .05$ significance level, $z_{.05} = 1.64$. This test is summarized below.

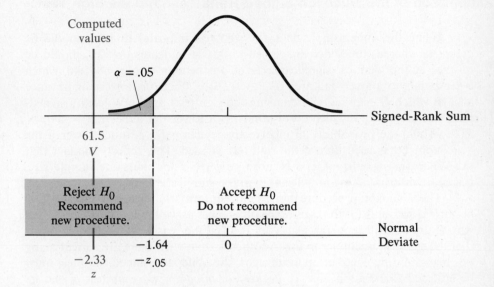

Ignoring the signs, the ranks for the nonzero matched-pairs differences in Table 18-6 were assigned to the absolute amounts, with 1 representing the lowest value. We assign absolute values to pairs of opposite signs in exactly the same way as we did in the Wilcoxon rank-sum text. Thus, the pairs with differences -2 and $+2$, which are tied for the smallest absolute value, both receive the average of the two lowest ranks $(1 + 2)/2 = 1.5$. As before, ties between positive or negative differences only are ignored.

The sum of the ranks for the eight positive differences is

$$V = 12 + 5 + 6 + 13 + 1.5 + 7 + 8 + 9 = 61.5$$

For $n = 23$ nonzero pair differences, the corresponding normal deviate is

$$z = \frac{61.5 - \dfrac{23(24)}{4}}{\sqrt{\dfrac{23(24)(46 + 1)}{24}}} = -\frac{76.5}{\sqrt{1,081}} = -2.33$$

Because the computed value of $z = -2.33$ falls in the rejection region, the Wilcoxon signed-rank test indicates that the null hypothesis should be *rejected*. The researcher should conclude that the experimental language program is an improvement over the traditional procedure.

A Comparison of the Wilcoxon Signed-Rank Test and the Sign Test

Using the same sample data, the sign test indicates that H_0 should be *accepted*, whereas the Wilcoxon signed-rank test indicates that H_0 should be *rejected* at the identical significance level. In the sign test, $z = -1.46$, which corresponds to a lower tail area of $.5 - .4279 = .0721$ (the lowest significance level at which H_0 can be rejected using the sign test). In the Wilcoxon signed-rank test, $z = -2.33$, which corresponds to a lower tail area of $.5 - .4901 = .0099$. Thus, the probability of incorrectly rejecting H_0 is much lower if the Wilcoxon test is used than if the sign test is used. This reflects the fact that the Wilcoxon signed-rank test is more powerful—for the same sample size, it is more discriminating or efficient than the sign test.

The Wilcoxon procedure derives more information from the sample data, because it accounts for the sizes of the differences as well as for their signs. Then why is the less efficient sign test used at all? One reason for its popularity is that the sign test is simpler to use. Another is that when there are a great many ties between differences of opposite signs, the Wilcoxon signed-rank test must be adjusted to be applicable. *A further advantage of the sign test is that its assumptions are less restrictive than the Wilcoxon signed-rank test.* The sign test does not assume that A and B values have identical population frequency distributions (and hence equal variances), as the Wilcoxon test does.

EXERCISES

18-12 The Wilcoxon signed-rank test will be used to determine the appropriate action to be taken in each of the following situations. Find the acceptance and rejection regions, and indicate whether H_0 should be accepted or rejected at the stated significance level for each result provided.

(a)	(b)	(c)	(d)
$H_0 : A\text{'s} \le B\text{'s}$	$H_0 : A\text{'s} \ge B\text{'s}$	$H_0 : A\text{'s} \le B\text{'s}$	$H_0 : A\text{'s} \ge B\text{'s}$
$n = 18$	$n = 100$	$n = 50$	$n = 19$
$\alpha = .05$	$\alpha = .01$	$\alpha = .01$	$\alpha = .05$
$V = 40$	$V = 2{,}065$	$V = 890$	$V = 44$

18-13 Consider the data in Table 14-3 (page 469) for the examination scores achieved by two groups of students, one using a new statistics book (A) and the other using an old statistics book (*B*). There, the professor chose the null hypothesis of identical scores for the two groups.

(a) Find the professor's acceptance and rejection regions at the $\alpha = .05$ significance level.

(b) Determine the value of the test statistic *V*. What conclusion should the professor reach, based on this value?

18-14 Using the data in Exercise 18-11, compute the test statistic *V*. Should the null hypothesis that the new preservative yields shelf lives that are at most as long as those provided by the old preservative be accepted or rejected at the $\alpha = .05$ significance level?

18-15 Use the sample data provided in Exercise 18-7 to apply the Wilcoxon signed-rank test. If leaded gasoline will be used only if it provides significantly better mileage than unleaded gasoline, which gasoline will be used at the $\alpha = .001$ level?

TESTING FOR RANDOMNESS: **18-6**
THE NUMBER-OF-RUNS TEST

The Need to Test for Randomness

We have seen that observations must be randomly obtained if probabilities are to be used to qualify inferences about populations based on sample results. Determining whether or not a sample is random is especially critical when observations about rare occurrences are collected over time. For example, data about production delays due to equipment malfunctions, the causes or costs of aircraft accidents, or the IQs of identical twins reared apart can be collected only historically—not by short-term random sampling. In such cases, the samples obtained are not random in the usual sense. But if the data appear to be random, they can be analyzed as if they were.

Sample data collected over time are not random if they exhibit some sort of serial dependency, so that the order in which particular attributes or variates of similar size occur is affected by previous events. To apply statistical methodology in such circumstances, we must verify that the order in which the observations are obtained is similar to the order that we would expect to find in a random sample.

The Number-of-Runs Test

A useful procedure for determining randomness is to separate the sample results into two opposite categories: defective, nondefective; above 80, 80 or below; fatal, nonfatal. Then the data can be represented chronologically as a string of categorical designations. For example, suppose a university department wishes to assess future admissions on the basis of test scores recorded for 20 graduate students admitted over the last two years. Denoting a score below the median by a and a score above the median by b, the results can be represented by a string of a's and b's, as in

Sequence 1: $a\ b\ b\ a\ a\ a\ b\ b\ a\ b\ b\ b\ b\ a\ a\ a\ b\ b\ a\ a$

The braces indicate *runs* of a particular category—in this sequence, 5 of a and 4 of b. *A run is a succession of one or more observations in the same category.* The number of runs is employed in hypothesis-testing procedures to determine randomness.

The rationale for using runs to test for randomness is that too few or too many runs are unlikely if the sample is truly random over time. For instance, consider the two-run result in

Sequence 2: $a\ a\ a\ a\ a\ a\ a\ a\ a\ a\ b\ b\ b\ b\ b\ b\ b\ b\ b\ b$

The first 10 students admitted all scored poorly, and the last 10 all scored above the median. Such an outcome is rarely the result of a random process. It might be explained by a change in screening procedure—the institution of a tougher admissions policy, for example. On the other hand, a random run of 20 results is also highly unlikely, as in

Sequence 3: $b\ a\ b\ a\ b\ a\ b\ a\ b\ a\ b\ a\ b\ a\ b\ a\ b\ a\ b\ a$

Such an outcome could result if the department attempted to maintain a balance in student abilities or a fixed ratio of disadvantaged students.

The number-of-runs test is based on the following null hypothesis:

H_0: The sampling is random. Therefore, each sequence position has the same prior chance of being assigned an a as any other.

The alternative is that the sample is not random.

The test statistic is based on the number of runs for category a, denoted by R_a. (We could just as easily use the number of runs for category b or the total number of runs; either of these would provide nearly the same results.) For the screening-test results of sequence 1, the number of a runs is $R_a = 5$; likewise, $R_a = 1$ for sequence 2, and $R_a = 10$ for sequence 3.

It will be easier to formulate an appropriate decision rule if we describe the sampling distribution of R_a first. Basic concepts of probability can be applied to show that R_a exhibits the *hypergeometric distribution*. The probabilities for this distribution—like those for the binomial distribution—are usually presented in tables, because they are tedious to compute. It is more convenient to approximate these probabilities from the normal distribution. The following expression can be used to calculate the

NORMAL DEVIATE FOR THE NUMBER-OF-RUNS TEST

$$z = \frac{R_a - \dfrac{n_a(n_b + 1)}{n_a + n_b}}{\sqrt{\dfrac{n_a(n_b + 1)(n_a - 1)}{(n_a + n_b)^2}\left(\dfrac{n_b}{n_a + n_b - 1}\right)}}$$

where n_a and n_b represent the number of a's and b's in the sample. For the three test-score sequences, $n_a = n_b = 10$, and the following normal deviates apply:

$$\text{Sequence 1:}\quad z = \frac{5 - \dfrac{10(10 + 1)}{10 + 10}}{\sqrt{\dfrac{10(10 + 1)(10 - 1)}{(10 + 10)^2}\left(\dfrac{10}{10 + 10 - 1}\right)}}$$

$$= \frac{5 - 5.5}{\sqrt{1.3026}} = \frac{5 - 5.5}{1.14} = -.44$$

$$\text{Sequence 2:}\quad z = \frac{1 - 5.5}{1.14} = -3.95$$

$$\text{Sequence 3:}\quad z = \frac{10 - 5.5}{1.14} = 3.95$$

To continue with our screening-test illustration, suppose we desire an $\alpha = .05$ significance level for the probability for incorrectly concluding that the test scores were randomly generated. *The testing procedure is two-sided*, because H_0 will be refuted by either too few or too many a runs. The critical normal deviate is $z_{.025} = 1.96$, and the appropriate decision rule is illustrated on the next page.

For sequence 1, the normal deviate $z = -.44$ lies within the acceptance region, so H_0 must be *accepted*. At the 5% significance level, we can conclude that the test scores in sequence 1 were randomly generated. Because the normal deviates for sequences 2 and 3 lie outside the acceptance region, we must *reject* H_0 for either sequence and conclude that in both cases the test scores were not randomly generated.

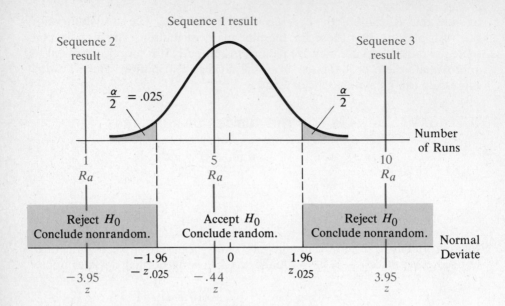

Testing the 1970 Draft Lottery for Randomness

We will now analyze further the 1970 draft lottery, which we mentioned in an earlier chapter as an example of a poor statistical procedure.

In December 1969, the U.S. Selective Service initiated a new procedure for determining the priorities for inducting young men into compulsory military service. Beginning with the 19-year-olds born in 1950, a revised policy was implemented to draft these younger men before the older ones. A 19-year-old man who was not drafted in 1970 would have a lower priority of being drafted in 1971 than a 19-year-old man born in 1951. Because the manpower requirements for draftees in 1970 were considerably less than the number of available men, a method was devised to determine the induction priorities for 1970 by lottery.

Capsules were to be drawn from a barrel. Those men whose birthdays were drawn first would almost certainly be drafted in 1970. Men who were born on dates chosen near the end of the lottery would almost certainly never have to enter military service.

Newspapers devoted wide coverage to the lottery. Some drew an imaginary line at 183 (half the number of days in a leap year), concluding that those with numbers 183 or below were "vulnerable" and those with numbers greater than 183 were "safe."

The calendar dates and the corresponding draft priority numbers obtained from the lottery in December 1969 are provided in Table 18-7. Note that a greater number of birthdays early in the year tend to be "safe" and that a greater number of birthdays late in the year tend to be "vulnerable." This leads us to

TABLE 18-7

Determination of Number of Runs of High-Priority (V) Numbers from the 1970 Draft Lottery

Draft Priority Numbers

Birthday	Jan.	Feb.	Mar.	Apr.	May	Jun.	Jul.	Aug.	Sep.	Oct.	Nov.	Dec.
1	S305	86V	108V	32V	S330	S249	93	111V	S225	S359	19V	129V
2	159V	144V	29V	S271	S298	S228	S350	45V	161V	125V	34V	S328
3	S251	S297	S267	83V	40V	S301	115V	S261	49V	S244	S348	157V
4	S215	S210	S275	81V	S276	20V	S279	145V	S322	S202	S266	165V
5	101V	S214	S293	S269	S364	28V	S188	54V	82V	24V	S310	56V
6	S224	S347	139V	S253	155V	110V	S327	114V	6V	87V	76V	10V
7	S306	91V	122V	147V	35V	85V	50V	168V	8V	S234	51V	12V
8	S199	181V	S213	S312	S321	S366	13V	48V	S184	S283	97V	105V
9	S194	S338	S317	S219	S197	S335	S277	106V	S263	S342	80V	43V
10	S325	S216	S323	S218	65V	S206	S284	21V	71V	S220	S282	41V
11	S329	150V	136V	14V	37V	134V	S248	S324	158V	S237	46V	39V
12	S221	68V	S300	S346	133V	S272	15V	142V	S242	72V	66V	S314
13	S318	152V	S259	124V	S295	69V	42V	S307	175V	138V	126V	163V
14	S238	4V	S354	S231	178V	S356	S331	S198	1V	S294	127V	26V
15	17V	89V	169V	S273	130V	180V	S322	102V	113V	171V	131V	S320
16	121V	S212	166V	148V	55V	S274	120V	44V	S207	S254	107V	96V
17	S235	S189	33V	S260	112V	73V	98V	154V	S255	S288	143V	S304
18	140V	S292	S332	90V	S278	S341	S190	141V	S246	5V	146V	128V
19	58V	25V	S200	S336	75V	104V	S227	S311	177V	S241	S203	S240
20	S280	S302	S239	S345	183V	S360	S187	S344	63V	S192	S185	135V
21	S186	S363	S334	62V	S250	60V	27V	S291	S204	S243	156V	70V
22	S337	S290	S265	S316	S326	S247	153V	S339	160V	117V	9V	53V
23	118V	57V	S256	S252	S319	109V	172V	116V	119V	S201	182V	162V
24	59V	S236	S258	2V	31V	358	23V	36V	S195	S196	S230	95V
25	52V	179V	S343	S351	S361	137V	67V	S286	149V	176V	132V	84V
26	92V	S365	170V	S340	S357	22V	S303	245V	18V	7V	47V	173V
27	S355	S205	S268	74V	S296	64V	S289	S352	S233	S264	S309	78V
28	77V	S299	S223	S262	S308	S222	88V	167V	S257	94V	78V	123V
29	S349	S285	S362	S191	S226	S353	S270	61V	151V	S229	99V	16V
30	164V		S217	S208	103V	S209	S287	S333	S315	38V	174V	3V
31	S211		30V		S313		S193	11V		79V		100V

question whether the lottery was truly random. To help identify the runs in priority numbers, safe values are labeled S and vulnerable values are labeled V.

We can use the number-of-runs test here to determine randomness, because the draft lottery fits all the necessary assumptions, with one slight interpretation. The capsules contained *dates*, so the sequence in which a date was drawn determined the draft priority number for that date. It will be simpler if we envision that capsules containing the numbers 1 through 366 are drawn one at a time—the first capsule corresponding to January 1, the second to January 2, and so forth throughout the year. The number inside each capsule represents the draft priority for that date. All possible sequences of priority numbers are equally likely, so that our underlying assumption regarding the sampling distribution of R_a is met.

STEP 1
Formulate the null hypothesis. We will test the null hypothesis that the lottery is random.

STEP 2
Select the test statistic. Based on the number of runs of vulnerable (high-priority) dates R_a, we use the normal deviate as the test statistic.

STEP 3
Obtain the significance level α and identify the acceptance and rejection regions. At a significance level of $\alpha = .01$, the critical normal deviate is $z_{.005} = 2.57$ from Appendix Table E. The acceptance and rejection regions are shown below.

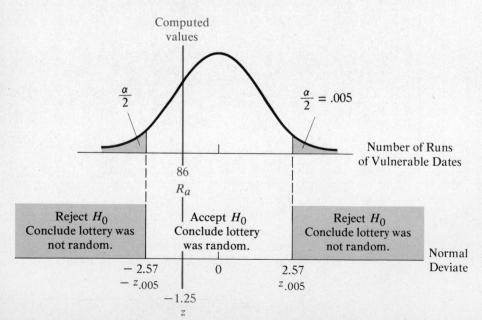

STEP 4

Collect the sample and compute the value of the test statistic. From Table 18-7, we observe that there are 86 runs of vulnerable (high-priority) a's and 86 runs of safe (low-priority) b's. Thus, $R_a = 86$. With $n_a = n_b = 183$, we calculate the corresponding normal deviate as

$$z = \frac{R_a - \dfrac{n_a(n_b + 1)}{n_a + n_b}}{\sqrt{\dfrac{n_a(n_b + 1)(n_a - 1)}{(n_a + n_b)^2}\left(\dfrac{n_b}{n_a + n_b - 1}\right)}}$$

$$= \frac{86 - \dfrac{183(183 + 1)}{183 + 183}}{\sqrt{\dfrac{183(183 + 1)(183 - 1)}{(183 + 183)^2}\left(\dfrac{183}{183 + 183 - 1}\right)}} = -1.25$$

STEP 5

Make the decision. Because the computed value of $z = -1.25$ falls in the acceptance region, we must conclude that *the draft lottery numbers were randomly selected.* Referring to Appendix Table D, $z = 1.25$ corresponds to a lower tail area of about .105. The null hypothesis can only be rejected at twice as great a level of significance, or approximately $\alpha = .21$.

The Meaning of the Draft Lottery Results

The draft lottery example has been included in this book partly because the U.S. Selective Service was vehemently criticized about the way in which the capsules were selected, due to the disparity in the draft priority numbers shown in Table 18-7. An investigation of the procedure revealed that the capsules were placed in the barrel on a monthly basis, beginning with January and ending with December. The capsules were superficially mixed, so that more December dates than January dates, for example, were on top. It has been convincingly argued that the 1970 draft lottery was unfair to young men whose birthdays were late in the year. Several lawsuits were initiated to invalidate the lottery for this reason.

But our number-of-runs test does not provide sufficient evidence to refute the hypothesis of random selection. The lowest possible significance level that can lead to rejection is very high—greater than .20. In many scientific applications, for example, a significance level as low as .01 or .001 may be required before the results are worthy of inclusion in the body of the theory. In general, a level of .20 is considered highly insignificant in establishing that the null hypothesis is untrue. At or above the .20 level, either judgment would be reserved or the null hypothesis would be accepted.

This example points out a major inadequacy of using hypothesis testing as a basis for drawing conclusions about how well data fit a particular assumption. This inadequacy is fundamental to the entire statistical approach of explaining outcomes in terms of probabilities. Our number-of-runs model is not refined enough to incorporate all information (for example, the manner in which the capsules were supposedly mixed and then selected). It focuses only on the results of the selection process; all other relevant information is ignored. The same statistician who would conclude that the *results* are not significant enough to warrant the rejection of H_0 would, on hearing a *description of the procedure* for capsule selection, be almost certain to dismiss the same null hypothesis with full conviction, *regardless of what results were obtained*.

One important question remains. Why was such a typical run statistic value calculated for the 1970 draft lottery, when it was so convincingly argued that the lottery was far from random? Although circumstances were far from ideal, there was *some mixing* of the capsules. Capsules were often drawn by plunging the fingers deeply into the barrel, so that all of the top capsules were not withdrawn early and all of the bottom capsules were not immune from an early grasp. These factors contributed some randomness to the lottery. As the numbers in Table 18-7 indicate, the number of vulnerable dates for February, May, June, July, September, and October did not deviate very much from half the number of days in these months respectively.

Undoubtedly embarrassed and goaded by public pressure, the Selective Service revised its procedures for the 1971 draft lottery. Two sets of capsules were used: One contained dates; the other, priority numbers. Mixing machines were employed, and the drawings consisted of simultaneously selecting a capsule from each barrel and matching the respective date and priority number. The randomness of this procedure was beyond reproach.

Other Tests for Randomness

We have only considered one type of test for randomness, based on the *number* of runs. A similar test that will not be described here is based on the *lengths* of runs. Number-of-runs tests consider only the *sequence* of numbers. In some situations, it is desirable to test for other features, such as the *frequency* of various values or their *serial correlation* (which measures the tendency of certain numbers to be followed by other numbers).

To illustrate the need to consider all these features, we will examine the problem of testing the suitability of using computer-generated random numbers (which only appear to be random and are properly called *pseudorandom numbers*) to select sample units. A number-of-runs test would accept as random the sequence of digits

1 9 9 9 1 1 1 9 9 1 1 1 1 9 9 9 1 1 1 9 9

Clearly, such a sequence would prove inadequate in selecting a random sample. The number-of-runs test does not detect the frequency with which each digit (in this case, 0 through 9) occurs. In a list of true random numbers, each digit is expected to occur 10% of the time, so that the uniform distribution represents the underlying population of values. A testing procedure (the *goodness-of-fit test*) can be used to determine whether or not the above sequence came from a randomly distributed population. Such a test would reject this particular sequence of computer-generated numbers as nonrandom.

A test for serial correlation could be applied to the sequence

$$2\ 7\ 6\ 9\ 5\ 3\ 2\ 1\ 0\ 7\ 6\ 9\ 5\ 4\ 3\ 8\ 7\ 6\ 4\ 0$$

to detect nonrandom patterns (that each occurrence of 7 is followed by a 6, for example). Such quirks are ignored by both number-of-runs and goodness-of-fit tests. Either of these tests would permit us to accept this sequence as consistent with its respective null hypothesis.

A detailed discussion of the various tests of randomness and how they may be combined in a series is beyond the scope of this book.*

EXERCISES

18-16 The following sequences represent the order in which 10 men and 10 women were admitted to four different graduate programs:

(1) *M W M W W W M M W M M W M W M W W M M W*
(2) *M W M M W M W W M W M M W M W W M W M W*
(3) *M M M M W W W W W W M M M M M M M W W W W W*
(4) *M W W M W M M W M W M W M W M W M W W M*

(a) At the $\alpha = .05$ significance level, identify the appropriate acceptance and rejection regions for testing the null hypothesis that the sexes of the graduates were randomly determined for each successive admission. Let R_a represent the number of runs of women.

(b) Applying your decision rule from (a), indicate whether an H_0 of randomness must be accepted or rejected for each sequence.

18-17 A fair coin tossed 16 times produced a total of 8 heads and 8 tails. Letting R_a represent the number of head runs, apply the normal approximation to determine whether the H_0 that the tosses were random events should be accepted or rejected for each of the following sequences at the $\alpha = .05$ significance level.

* For a comprehensive discussion about testing the randomness of number lists, see J.W. Schmidt and R.E. Taylor, *Simulation and Analysis of Industrial Systems* (Homewood, Ill.: Richard D. Irwin, 1970), Chapter 6.

(a) *H T T T H T T T H T T T H H T H H H*
(b) *H H H T T T T H H H H H T T T T*
(c) *H T H T H T T H H T H T H T H T*
(d) *H H T T T H H T T T H H H T H T T*

18-18 The results of the 1971 draft lottery are provided in Table 18-8.
(a) Determine the number of runs of vulnerable dates (those having a draft priority of 183 or less).

TABLE 18-8
1971 Draft Lottery Results

Birthday	Draft Priority Numbers											
	Jan.	Feb.	Mar.	Apr.	May	Jun.	Jul.	Aug.	Sep.	Oct.	Nov.	Dec.
1	133	335	14	224	179	65	104	326	283	306	243	347
2	195	354	77	216	96	304	322	102	161	191	205	321
3	336	186	207	297	171	135	30	279	183	134	294	110
4	99	94	117	37	240	42	59	300	231	266	39	305
5	33	97	299	124	301	233	287	64	295	166	286	27
6	285	16	296	312	268	153	164	251	21	78	245	198
7	159	25	141	142	29	169	365	263	265	131	72	162
8	116	127	79	267	105	7	106	49	108	45	119	323
9	53	187	278	223	357	352	1	125	313	302	176	114
10	101	46	150	165	146	76	158	359	130	160	63	204
11	144	227	317	178	293	355	174	230	288	84	123	73
12	152	262	24	89	210	51	257	320	314	70	255	19
13	330	13	241	143	353	342	349	58	238	92	272	151
14	71	260	12	202	40	363	156	103	247	115	11	348
15	75	201	157	182	344	276	273	270	291	310	362	87
16	136	334	258	31	175	229	284	329	139	34	197	41
17	54	345	220	264	212	289	341	343	200	290	6	315
18	185	337	319	138	180	214	90	109	333	340	280	208
19	188	331	189	62	155	163	316	83	228	74	252	249
20	211	20	170	118	242	43	120	69	261	196	98	218
21	129	213	246	8	225	113	356	50	68	5	35	181
22	132	271	269	256	199	307	282	250	88	36	253	194
23	48	351	281	292	222	44	172	10	206	339	193	219
24	177	226	203	244	22	236	360	274	237	149	81	2
25	57	325	298	328	26	327	3	364	107	17	23	361
26	140	86	121	137	148	308	47	91	93	184	52	80
27	173	66	254	235	122	55	85	232	338	318	168	239
28	346	234	95	82	9	215	190	248	309	28	324	128
29	277		147	111	61	154	4	32	303	259	100	145
30	112		56	358	209	217	15	167	18	332	67	192
31	60		38		350		221	275		311		126

(b) Remembering that there are only 365 dates, calculate the normal deviate corresponding to your result.

(c) What is the lowest significance level at which the null hypothesis of random selection can be rejected? (Recall that the number-of-runs test is two-sided.)

18-19 Toss a coin 30 times, recording whether a head or a tail occurs for each toss. Then test your results for randomness, using the number-of-runs test and applying the normal approximation. At a significance level of .10, can you reject the null hypothesis that you are a fair coin tosser?

18-20 The successive terms in the expansions of certain constants, such as $\pi = 3.1416\ldots$, have been proposed as substitutes for random numbers. A statistician testing for one property of these numbers obtained 45 four-digit numbers at or above 5,000 (a), and 55 numbers below 5,000 (b). Testing the null hypothesis of randomness (or, more correctly, the *appearance* of randomness) at the $\alpha = .05$ significance level, $R_a = 26$. What can the statistician conclude about the "randomness" of successive terms in the π expansion?

18-21 Before random number tables were constructed, a group of British statisticians used the London telephone directory to select random samples. Determine whether your local directory can be used for this purpose.

(a) Select a page from the telephone directory, and determine the number of runs of even last digits for the telephone numbers. Using the normal approximation, can you conclude that the sequence of odd and even last digits is not random at a significance level of .01?

(b) Regardless of your results in (a), do you think that telephone numbers would be suitable random numbers? Discuss.

THE RANK CORRELATION COEFFICIENT 18-7

In Chapter 10, the sample correlation coefficient r was introduced as an index used to measure the degree of association between two variables X and Y. Nonparametric statistics can be employed to provide alternative measures of correlation. One such statistic is the

SPEARMAN RANK CORRELATION COEFFICIENT

$$r_s = 1 - \frac{6\sum(X - Y)^2}{n(n^2 - 1)}$$

where X and Y are the *ranks* of the two variables being measured.

The rank correlation coefficient is derived directly from the conventional correlation coefficient, except that X and Y represent ranks instead of the observation values themselves. The fact that the rank means and the sums of the rank squares are automatically known for each sample size n (for example, when $n = 5$, $\sum X = 1 + 2 + 3 + 4 + 5 = 15$ and $\bar{X} = 15/5 = 3$, always) permits us to use the equivalent and simpler calculation for the rank correlation coefficient.

To illustrate the rank correlation coefficient, we will consider the data in Table 18-9. Here, observations have been made about the average number of weekly hours that a sample of $n = 10$ university students spent studying and their grade point averages (GPAs) for the term.

Beginning with 1 for the lowest value, the observations for each variable in Table 18-9 were ranked. *Tying observations were given the average of the successive ranks that would have been assigned if the values had been different.* Thus, both observations of 17 study hours were equally ranked as $(2 + 3)/2 = 2.5$, and the two GPAs of 3.6 each received a rank of $(7 + 8)/2 = 7.5$.

The differences in ranks were determined and their squares were calculated. The sum of the squared rank differences is $\sum(X - Y)^2 = 9.00$. The rank correlation coefficient between average weekly study hours and GPA is

$$r_s = 1 - \frac{6(9.00)}{10(10^2 - 1)} = 1 - \frac{54}{990} = .946$$

indicating a high correlation between study time and grade point average.

TABLE 18-9
Rank Correlation Calculations for Study Hours
and Grade Point Averages (GPA)

| Variables | | Ranks | | | |
Study Hours	GPA	For Study Hours X	For GPA Y	Rank Difference X − Y	(X − Y)²
24	3.6	6	7.5	−1.5	2.25
17	2.0	2.5	1	1.5	2.25
20	2.7	4	4	.0	.00
41	3.6	8	7.5	.5	.25
52	3.7	10	9	1.0	1.00
23	3.1	5	5	.0	.00
46	3.8	9	10	−1.0	1.00
17	2.5	2.5	3	−.5	.25
15	2.1	1	2	−1.0	1.00
29	3.3	7	6	1.0	1.00
				0.0	9.00

Usefulness of Rank Correlation

The rank correlation coefficient can be used in many situations for which the conventional correlation coefficient is unsuitable. In Chapter 10, we learned that the conventional r is based on the assumption that the underlying relationship is linear. The study time and GPA data in Table 18-9 are plotted in Figure 18-1. The graph shows a pronounced, curvilinear relationship between these variables, which is summarized by the colored curve. Here, GPA increases with study time, but each additional hour of study time produces a progressively smaller improvement in GPA. A least-squares regression would fit the black line to the sample data, and the conventional correlation coefficient (through the coefficient of determination r^2) would express the proportion of variation in GPA explained by this line. Obviously, the regression line poorly fits the data here, which is confirmed by the much lower conventional correlation value of $r = .58$ that we obtain.

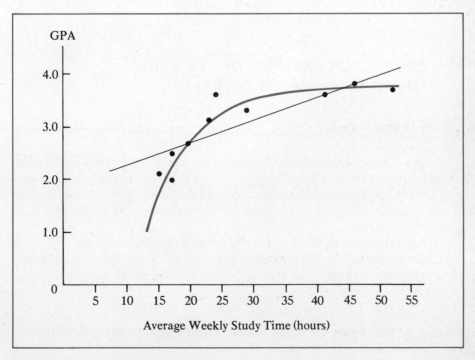

FIGURE 18-1
Graphical relationships between GPA and study time.

EXERCISES

18-22 The following daily data were obtained for a sample of $n = 10$ smokers who drink coffee:

Packs Smoked	Cups Drunk
.5	3
1.0	4
.5	5
1.5	6
2.0	4
2.5	6
1.0	3
.5	2
1.5	7
2.0	5

Calculate the rank correlation coefficient.

18-23 Referring to Exercise 10-18 on page 351, determine the rank correlation coefficient for prior and current productivity scores.

18-24 Referring to Exercise 10-23 on page 355, determine the rank correlation coefficient for mother's age and mean litter size of mice.

18-8 ONE-FACTOR ANALYSIS OF VARIANCE: THE KRUSKAL-WALLIS TEST

Alternative to the F Test

In Chapter 14, we examined a testing procedure for determining whether differences exist between several means—using the F test to perform analysis of variance. A very serious drawback to the F test is that it requires the populations to be normal, and, as we have already noted, this assumption may invalidate the testing procedure.

An alternative to the F test is the Kruskal-Wallis one-factor analysis of variance, introduced by W.H. Kruskal and W.A. Wallis in 1952. This procedure is actually an extension of the Wilcoxon rank-sum test to the analysis of several samples.

Description of the Test: College Major and Programming Aptitude

Its primary difference from the F test is that the Kruskal-Wallis test is based on a test statistic computed from ranks determined for pooled sample observations. Its null hypothesis is that the rank assigned to a particular observation has an equal chance of being any number between 1 and n, regardless of the sample group to which it belongs. We will describe this procedure with the following illustration.

A college computer-science instructor wishes to determine if there is really any difference in programming aptitudes between students in different majors. The students in his college can be classified into four major groups: science, liberal arts, business, and engineering. The instructor administered a computer-programming aptitude test to a random sample of students in each category. Their scores and score ranks appear in Table 18-10. (Note that there was no need to average ranks, since no ties between sample groups occurred.)

At a significance level of $\alpha = .05$, the instructor wishes to test the null hypothesis

H_0:   There are no programming aptitude differences between major groups.

The alternative hypothesis is that college major makes a difference in aptitude.

As in the Wilcoxon tests, the null hypothesis implies that the four sample groups are obtained from the same population. Under the null hypothesis, therefore, each score in Table 18-10 has the same prior probability for receiving any rank of 1 through 20.

TABLE 18-10
Test Scores and Ranks for Computer Programming by Major

(1) Science		(2) Liberal Arts		(3) Business		(4) Engineering	
Score	Rank	Score	Rank	Score	Rank	Score	Rank
85	12	95	19	67	3	90	15
73	7	54	1	74	8	65	2
96	20	72	6	84	11	92	17
91	16	81	10	68	4	94	18
18	14	69	5	87	13		
				77	9		

$$T_1 = 69 \qquad T_2 = 41 \qquad T_3 = 48 \qquad T_4 = 52$$

$$T_1^2 = 4{,}761 \qquad T_2^2 = 1{,}681 \qquad T_3^2 = 2{,}304 \qquad T_4^2 = 2{,}704$$

$$n_1 = 5 \qquad n_2 = 5 \qquad n_3 = 6 \qquad n_4 = 4$$

$$\frac{T_1^2}{n_1} = 952.2 \qquad \frac{T_2^2}{n_2} = 336.2 \qquad \frac{T_3^2}{n_3} = 384.0 \qquad \frac{T_4^2}{n_4} = 676.0$$

Score:	54	65	67	68	69	72	73	74	77	81	84	85	87	88	90	91	92	94	95	96
Rank:	1	2	3	4	5	6	7	8	9	10	11	12	13	14	15	16	17	18	19	20

For his test statistic, the instructor chooses to compare variabilities in the ranks within each column. The sum of ranks for each category T_j is computed. A sum of the squares of these sums—each term weighted by the reciprocal of the sample size n_j of the jth group—is then obtained. The following calculation provides the

KRUSKAL-WALLIS TEST STATISTIC

$$K = \frac{12}{n(n+1)} \left(\sum \frac{T_j^2}{n_j} \right) - 3(n+1)$$

where $n = \sum n_j$.

The sampling distribution of K is approximately a chi-square distribution with $m - 1$ degrees of freedom, where m is the number of categories. To find the critical value, we use Appendix Table H, which provides critical values for specified tail areas.

In our example, $m = 4$, so that there are $m - 1 = 4 - 1 = 3$ degrees of freedom. Letting $\alpha = .05$, the critical value from Appendix Table H is $\chi^2_{.05} = 7.815$. The decision rule is shown below.

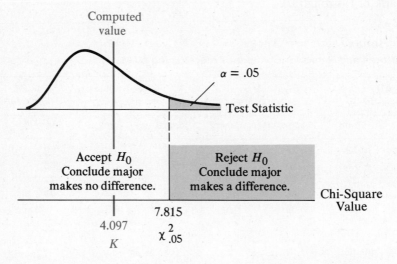

Using the values from Table 18-10, we calculate K, with $n = 20$:

$$K = \frac{12}{20(20+1)} (952.2 + 336.2 + 384.0 + 676.0) - 3(20+1)$$

$$= 4.097$$

Because K falls in the acceptance region, the instructor *must accept the null hypothesis that there are no aptitude differences between majors.*

EXERCISES

18-25 A chemical engineer wishes to know whether the mean time required to complete a chemical reaction is affected by the proportion of impurities present. The following reaction times have been obtained from sample reactions conducted at several different impurity levels:

	Impurity Level		
.001	.01	.05	.10
	Reaction Time (minutes)		
103	104	153	207
111	113	127	183
107	117	143	173
105	120	119	
	113	138	
		143	

(a) Rank the sample results and calculate the test statistic K. How many degrees of freedom are associated with this test statistic?

(b) At the $\alpha = .05$ significance level, can the null hypothesis that impurity level does not affect reaction time be rejected?

18-26 Referring to the sod-yield data in Table 16-1 (page 526), apply the Kruskal-Wallis test to the null hypothesis that mean yields are identical under all fertilizer treatments at the $\alpha = .05$ significance level. What conclusion should be made?

15-27 An insurance company wishes to determine if the type of profession makes any difference in the mean amount of whole life insurance held by its professional policyholders. The following results are obtained for a random sample of policyholders:

	Insurance Coverage	
Physicians	Lawyers	Dentists
$200,000	$ 50,000	$ 80,000
150,000	100,000	45,000
40,000	95,000	155,000
35,000	10,000	325,000
110,000	300,000	
	75,000	

Determine whether the null hypothesis of no difference must be accepted or rejected at the .05 significance level.

REVIEW EXERCISES

18-28 In testing the null hypothesis that drug A yields relief from sinus headaches for at least as long as drug B, a statistician selects two independent samples of 30 sinus sufferers and asks a physician to administer drug A to one group and drug B to the other. The number of pills taken over a fixed time period is used to compute each sufferer's mean relief duration score. In ranking the scores of the two groups, the statistician found that $W = 826$. Should H_0 be accepted or rejected at the $\alpha = .05$ significance level?

18-29 Suppose that the statistician in Exercise 18-28 grouped the patients into 30 pairs, matched on the basis of medical history, age, sex, and lifestyle. Also assume that the same individuals received the respective drugs. The difference in relief duration of each pair was found by subtracting the mean time of each drug *B* patient from the mean time of each corresponding drug *A* patient. A total of $R = 5$ positive differences resulted, and there were 5 ties. Should H_0 be accepted or rejected at the $\alpha = .05$ significance level? (*Note:* We would never use both the Wilcoxon rank-sum test and the sign test for matched pairs on the same data. This is done here only to illustrate how results may vary between the two tests.)

18-30 Repeat Exercise 18-29, applying the Wilcoxon signed-rank test when the sample data provide $V = 30$.

18-31 Referring to the data in Exercise 16-5 on page 540, apply the Kruskal-Wallis test to determine whether the dissolving time of aspirin tablets differs with the impurity level at the $\alpha = .01$ significance level.

18-32 Referring to the data in Exercise 10-5 on page 331, calculate the rank correlation coefficient.

18-33 A statistics professor asked her students to construct their own random number lists from scratch to demonstrate how faulty this procedure can be. One student obtained the following list. Proceeding down one column at a time, find the number of runs of small values (0 through 4), and then determine whether the null hypothesis of randomness should be accepted or rejected at the $\alpha = .05$ significance level.

4	6	9	9	0
8	1	2	3	6
3	9	8	8	2
7	2	3	2	3
9	4	7	7	4
0	8	1	5	8
2	7	7	1	5
1	6	3	7	3
8	0	4	0	8
5	3	6	5	0

18-9 OPTIONAL TOPIC:
TESTING FOR GOODNESS OF FIT—
THE KOLMOGOROV-SMIRNOV ONE-SAMPLE TEST

An Alternative to the Chi-Square Test

In Chapter 17, the chi-square test was introduced as a procedure for inferring whether a sample is obtained from a population having a particular distribution. As a test for goodness of fit, the chi-square test has some serious limitations. The most significant of these is the requirement for a large sample. Recall that a rule for using the chi-square distribution is that each class interval must have an expected frequency of at least 5. Unless n is quite large, only the most frequent class intervals will retain their data; the less frequent class intervals must be combined before computing χ^2 and, in doing this, information is lost. When only a very tiny sample is available, the chi-square test cannot be used at all.

An alternative goodness-of-fit test is the *Kolmogorov-Smirnov one-sample test*, named after A. Kolmogorov and N.V. Smirnov, two Russian mathematicians who provided the theoretical foundations for this test. The test compares the observed cumulative frequency distribution for the sample to that expected for the population specified by the null hypothesis. The test statistic obtained is the maximum deviation between the observed and the expected distributions.

Description of the Test: Aircraft Delay Times

The Kolmogorov-Smirnov test will be developed from the following illustration.

Before deciding on a new scheduling policy, Ace Airlines must determine whether the takeoff delay times at El Paso are normally distributed. Only 1% of the biweekly El Paso departures have resulted in unpredictable delays. Thus, over a ten-year period, only 11 delays have occurred. The durations are provided in Table 18-11. From studies of other airports, Ace has judged that

TABLE 18-11
Durations of Aircraft
Takeoff Delays at El Paso

Duration (hours) X	X^2
2.1	4.41
1.9	3.61
3.2	10.24
2.8	7.84
1.0	1.00
5.1	26.01
.9	.81
4.2	17.64
3.9	15.21
3.6	12.96
2.7	7.29
31.4	107.02

delays at El Paso should have a mean of 3 hours with a standard deviation of 1 hour. The null hypothesis is that the aircraft takeoff delays are normal with $\mu = 3$ hours and $\sigma = 1$ hour.

To find the value of the test statistic, we first establish the cumulative *relative* frequency distribution obtained for our sample results. This is shown in Table 18-12. Since the frequencies are relative, each sample value occurs with a frequency of $1/11 = .0909$. The observations are not grouped in the usual manner. Instead, the durations obtained are listed in column (1) of Table 18-12 in increasing order of size. The cumulative frequencies in column (2) are obtained by adding the relative frequencies. Thus, .1818 is the frequency of durations of 1.0 hour or less (obtained by adding together the .0909 frequencies for .9 and 1.0 hours), and .6363 is the frequency of delays of 3.2 hours or less. We denote the actual frequency of values less than or equal to x by $F_a(x)$.

The expected frequencies under the null hypothesis, denoted by $F_e(x)$, must coincide with the normal curve for $\mu = 3$ and $\sigma = 1$. In column (3) of Table 18-12, the normal deviates z are calculated for each value of x. The cumulative frequencies in column (4) are obtained from Appendix Table D. For negative values of z, the lower tail areas corresponding to the cumulative frequencies may be obtained by subtracting the area in Appendix Table D from .5. Thus, when $x = 1.9$, $z = -1.1$, and Table D provides the area .3643,

TABLE 18-12

Actual and Expected Cumulative Frequencies for Take-off Delay Durations, with Calculation of the Test Statistics

(1) Duration of Delays (in hours) from Sample x	(2) Actual Cumulative Frequency of Durations $\leq x$ $F_a(x)$	(3) $z = \dfrac{x - \mu}{\sigma}$	(4) Expected Cumulative Frequency of Durations $\leq x$ $F_e(x)$	(5) Deviation $F_a(x) - F_e(x)$
.9	.0909	−2.1	.0179	.0730
1.0	.1818	−2.0	.0228	.1590
1.9	.2727	−1.1	.1357	.1370
2.1	.3636	−.9	.1841	.1795
2.7	.4545	−.3	.3821	.0724
2.8	.5454	−.2	.4207	.1247
3.2	.6363	.2	.5793	.0570
3.6	.7272	.6	.7257	.0015
3.9	.8181	.9	.8159	.0022
4.2	.9090	1.2	.8849	.0241
5.1	.9999	2.1	.9821	.0178

$$D = \max|F_a(x) - F_e(x)| = .1795$$

so that the expected frequency is

$$F_e(1.9) = .5 - .3643 = .1357$$

For positive z values, Table D provides areas between the mean and z standard deviations. These must be added to .5 to find the cumulative frequencies. Thus, when $x = 4.2$, $z = 1.2$, so Table D provides the area .3849. The cumulative frequency is therefore

$$F_e(4.2) = .5 + .3849 = .8849$$

The deviation between the actual and the expected frequencies is calculated in column (5) of Table 18-12 for each duration. Large deviations indicate that the actual and the expected values have a poor fit. This type of comparison is illustrated in Figure 18-2, where the two cumulative frequency

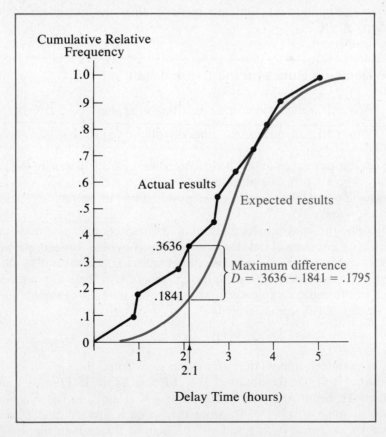

FIGURE 18-2

Actual and expected cumulative frequency distributions for El Paso takeoff delays.

distributions are plotted on the same graph. The actual frequencies lie above the expected ones for all values of x obtained from the sample. The vertical distance between the two curves indicates the magnitude of the deviations. The largest deviation occurs at $x = 2.1$, where

$$F_a(2.1) - F_e(2.1) = .1795$$

We will use this as our test statistic, denoting it by D. In general, the following expression provides the

KOLMOGOROV-SMIRNOV TEST STATISTIC

$$D = \max |F_a(x) - F_e(x)|$$

Note that we are interested only in absolute values; the sign of the difference is unimportant. Our test statistic, D, is the maximum frequency deviation.

Finding the Decision Rule and the Critical Value

Accept H_0 (conclude normal distribution applies)   if $D \leq D_\alpha$

Reject H_0 (conclude some other distribution applies) if $D > D_\alpha$

Large deviations tend to refute the null hypothesis. We will denote the critical value D_α, where α is the significance level at which a true null hypothesis will be rejected. To find the value of D_α, we must first find the sampling distribution of our test statistic D.

Because the sampling distribution of D is very complex and quite mathematical in nature, we will not describe its properties here. Instead, we will use Appendix Table L, which provides critical values D_α for several significance levels α and *sample sizes n*. (Unlike the chi-square test, these tabulated values depend on the entire sample size and not on the number of sample groups.) Thus, we can find D_α such that

$$\Pr[D \geq D_\alpha] = \alpha$$

For example, suppose that we wish to see whether the sample results are significant at $\alpha = .05$. We obtained $D = .1795$ in Table 18-12 with a sample size of $n = 11$. From Appendix Table L, for $n = 11$ and $\alpha = .05$, we obtain as our critical value $D_{.05} = .35242$. Since this value is greater than D, *the null hypothesis is accepted*. In fact, setting α as high as .10 leads to the same conclusion. The sample of takeoff delays appears to be quite consistent with the null hypothesis of normality.

TABLE 18-13

Calculation of the Test Statistic for
the Null Hypothesis of Uniform Distribution

(1) *Duration of Delays (in hours) from Sample* *x*	(2) $F_a(x)$	(3) $F_e(x)$	(4) $F_a(x) - F_e(x)$
.9	.0909	.0800	.0109
1.0	.1818	.1000	.0818
1.9	.2727	.2800	−.0073
2.1	.3636	.3200	.0436
2.7	.4545	.4400	.0145
2.8	.5454	.4600	.0854
3.2	.6363	.5400	.0963
3.6	.7272	.6200	.1072
3.9	.8181	.6800	.1381
4.2	.9090	.7400	.1690
5.1	.9999	.9200	−.0799

$$D = \max|F_a(x) - F_e(x)| = .1690$$

Investigating the Type II Error

It is interesting to note that in accepting the null hypothesis, there is a high likelihood of making a Type II error. Suppose that we choose instead a uniform distribution between .5 and 5.5 hours as our null hypothesis. Table 18-13 shows the calculation that we would obtain from this distribution. The $F_e(x)$ values are obtained by calculating the proportion of the distance between .5 and 5.5 that is represented by the corresponding value for *x*.

Although here $D = .1690$, our sample could reasonably be expected to have come from a uniform distribution. We have no basis for choosing between the normal and the uniform distributions. All we can say is that *there is no significant evidence to reject either*. Using a larger sample size is the only recourse we have for avoiding the Type II error.

Comparison with the Chi-Square Test

Comparisons have been made of the relative efficiencies of the χ^2 and Kolmogorov-Smirnov tests. The latter has been established to be more powerful for small samples; that is, for a fixed sample size, the Kolmogorov-Smirnov test provides a higher probability for rejecting a false null hypothesis.

A major disadvantage of the Kolmogorov-Smirnov test is that it does not allow us to estimate any of the population parameters from the sample data. *The population parameters must be specified in advance of testing.* Thus, in our example, we could not use $\bar{X}$ and s as estimates of μ and σ. (Had this been allowed, the fit would have been even closer.) The χ^2 test allows us to do this—at the price of a reduced number of degrees of freedom.

OPTIONAL EXERCISES

18-34 A postmaster wishes to establish whether the Friday lunchtime arrivals at a new branch post office follow a Poisson distribution. From previous tests, he has estimated that the mean arrival rate is 2 customers per minute. A sample of the arrivals in 100 random one-minute time segments has been collected over a two-month period. The results obtained and the expected sample frequencies appear below:

	Actual Observed Results			Expected Sample Results	
Number of Arrivals	Actual Frequency	Relative Frequency		Number of Arrivals	Relative Frequency
0	8	.08		0	.1353
1	19	.19		1	.2707
2	28	.28		2	.2707
3	24	.24		3	.1804
4	14	.14		4	.0902
5	2	.02		5	.0361
6	3	.03		6	.0120
7	1	.01		7	.0034
8	0	.00		8	.0009
9	1	.01		9	.0002
		1.00			.9999

(a) Determine the cumulative relative frequencies for the actual and the expected results.
(b) Calculate the maximum deviation D.
(c) What is the lowest significance level at which the null hypothesis of a Poisson distribution with $\lambda = 2$ can be rejected?
(d) Calculate the χ^2 statistic for this test. (Remember that χ^2 is based on whole frequencies instead of relative ones. Also recall that the low frequencies must be grouped.)
(e) What is the lowest significance level at which the null hypothesis may be rejected using the χ^2 test?
(f) In this example, which test do you believe is more discriminating—the χ^2 or the Kolmogorov-Smirnov? Explain.

18-35 Suppose that the postmaster in Exercise 18-34 estimated instead that 2.5 arrivals per minute apply. The following expected relative frequencies should be used:

Number of Arrivals	Relative Frequency
0	.0821
1	.2052
2	.2565
3	.2138
4	.1336
5	.0668
6	.0278
7	.0099
8	.0031
9	.0009
10	.0002
	.9999

(a) Determine the cumulative relative frequencies for the actual and the expected results.
(b) Calculate the maximum deviation D.
(c) At a significance level of $\alpha = .10$ can the postmaster reject the null hypothesis that the arrivals are a Poisson process?

18-36 Referring to the data in Exercise 17-23 (page 609), apply the Kolmogorov-Smirnov test at the $\alpha = .10$ significance level to determine whether or not the sample data are consistent with the indicated normal distribution.

There is no more miserable human being than one in whom nothing is habitual but indecision.

WILLIAM JAMES

Basic Concepts of Decision Making

Decision making under uncertainty has its roots in the area of study called *statistical decision theory*. Although most major developments in this field have occurred during the last 50 years, many contributions were made more than 200 years ago by the same pioneer mathematicians who formulated the theory of probability. In addition to probability, decision theory contains elements of statistics, economics, and psychology.

The central focus of this book is on the use of statistical procedures in decision making. This chapter considers the structure of decisions in general. We make a basic distinction between decision making under certainty, where no elements are left to chance, and decision making under uncertainty, where one or more random factors affect a choice's outcome.

We begin our discussion of decision theory by considering an underlying structural framework. This is followed by a presentation of the main decision-making criterion—the Bayes decision rule—which is based on expected values. In Chapter 20, our picture of decision theory will be enriched through consideration of several other criteria. In the remaining chapters we will consider special topics. Chapter 21 describes utility theory and provides the details for obtaining and using utility values. Chapter 22 is devoted to the practical problems of obtaining and using subjective probabilities. Finally, Chapters 23 and 24 discuss how samples and other types of experimental information may be systematically incorporated into decision making.

19-1 CERTAINTY AND UNCERTAINTY IN DECISION MAKING

The least complex applications of decision theory are encountered when we make decisions under certain conditions. Perhaps the simplest example of *decision making under certainty* is selecting what clothes to wear. Although the possibilities are numerous, we all manage to make this choice quickly and with little effort. But not all decisions are this easy to make. Remember how hard it was to choose from among an assortment of candy bars when you were a child? And not all decisions made under certainty are as trivial as choosing the day's apparel.

When the outcomes are only partly determined by choice, the decision-making process takes on an added complexity. So that we can see what is involved in structuring a decision under uncertainty, we will consider the choice of whether to carry an umbrella or some other rain protection. Here, we are faced with two alternatives: carrying an ungainly item that can, in the event of rain, help to defer a cleaning bill or a cough, or challenging the elements with hands free and hoping not to be caught in the rain. Because the weather prediction may be inaccurate, we are uncertain about whether it will rain. Yet faced with the needs of daily life, we must make a decision despite our uncertainty. This illustrates a common decision made under uncertainty: An action must be taken, even though its outcome is unknown and determined by chance.*

In this chapter, we will present a framework within which we can explain how and why particular choices are made. The decision that we must make in coping with the weather illustrates the essential features of making any decision under uncertainty. We choose to carry an umbrella when the chance of rain seems uncomfortably high, and we choose not to carry it when rain seems unlikely. But two people will make two different choices occasionally. Is there always one correct decision? If so, how can two persons make different choices? We can begin to answer these questions by identifying several key elements common to all decisions and then structuring them in a convenient form for analysis.

19-2 ELEMENTS OF DECISIONS

Every decision made under certainty exhibits two elements—*acts* and *outcomes*. The decision maker's choices are the acts. For example, when one

---

* This class of decisions is often divided into two categories—decision making under *risk*, where outcome probabilities are known, and decision making under *uncertainty*, where outcome probabilities are unknown. We will make no distinction here, but will assume that probabilities may always be found somehow—either objectively, through long-run frequency, or subjectively.

must choose between three television programs in the 9 P.M. time slot, each program represents a potential act. The outcomes can be characterized in terms of the enjoyment we may derive from each of the programs.

If the decision is made under uncertainty, a third element—*events*—exists. Continuing with our uncertainty about the rain, the acts are "carry an umbrella" and "leave the umbrella home." All decisions involve the selection of an act. But the outcomes resulting from each act are uncertain, because *an outcome is determined partly by choice and partly by chance.* For the act "carry an umbrella" there are two possible outcomes: (1) unnecessarily carting rain paraphernalia and (2) weathering a shower fully protected. For the other act, "leave the umbrella at home," the two outcomes are (1) getting wet unnecessarily and (2) remaining dry and unencumbered. Again, whether the first or second outcome occurs depends solely on the occurrence of rain. The outcome for any particular chosen act depends on which *event*, rain or no rain, occurs.

The Decision Table

To facilitate our analysis, we can summarize a decision problem by constructing a *decision table*, which indicates the relationship between pairs of decision elements. The decision table for the umbrella decision is provided in Table 19-1. Each row of the decision table corresponds to an event, and each column corresponds to an act. The outcomes appear as entries in the body of the table. There is a specific outcome for each act-event combination, reflecting the fact that the interplay between act and event determines the ultimate result.

Only the acts that the decision maker wishes to consider are included in this decision table. "Staying home" is another possible act, which we will exclude because it is not contemplated. The acts in Table 19-1 are mutually exclusive and collectively exhaustive, so that exactly one act will be chosen. The events in the table are also mutually exclusive and collectively exhaustive.

TABLE 19-1

Decision Table for the Umbrella Decision

Event	Act	
	Carry Umbrella	Leave Umbrella Home
Rain	Stay dry	Get wet
No Rain	Carry unnecessary burden	Be dry and free

The Decision Tree Diagram

A decision problem can also be conveniently illustrated with a *decision tree diagram* like the one shown in Figure 19-1. It is especially convenient to portray decision problems in the form of decision trees when choices must be made at different times over an extended period of time. The decision tree diagram is similar to the probability tree diagrams we used in Chapter 5. The choice of acts is shown as a fork with a separate branch for each act. The events are represented by separate branches in other forks. We distinguish between these two types of branching points by using squares for act-fork nodes and circles for event-fork nodes. A basic guideline for constructing a decision tree diagram is that the flow should be chronological from left to right. The acts are shown on the initial fork, because the decision must be made *before* the actual event is known. The events are therefore shown as branches in the second-stage forks. The outcome resulting from each act-event combination is shown as the end position of the corresponding path from the base of the tree.

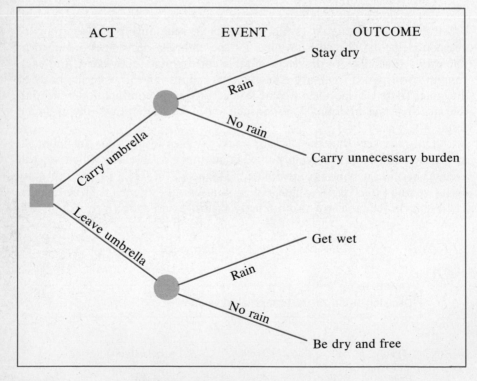

FIGURE 19-1
Decision tree diagram for the umbrella decision.

RANKING THE ALTERNATIVES   **19-3**
AND THE PAYOFF TABLE

In this section, we will consider how the choice of an act should be determined. We all cope with rain and manage to make umbrella decisions. But if we analyze the decision-making process, we will be able to make better decisions about more complex problems. Our analysis will focus on two measures: one for *uncertainty* and one for the comparative worth or *payoff* of the outcomes to the decision maker.

For now, we will consider examples with outcomes that have obvious payoffs, such as dollars. Every outcome—that is, every act-event combination—has a payoff value. Since the payoffs that the decision maker actually receives after choosing a particular act are conditional on whichever event occurs, these payoffs are sometimes referred to as *conditional values*. These values can be conveniently arranged in a *payoff table*, or a *conditional value table*, as shown in Table 19-2 for a gambling decision. The decision is to choose one of two acts: "gamble" or "don't gamble." Regardless of the choice made, a coin will be tossed, which will result in one of two possible events: "head" or "tail." The possible outcomes correspond to the four act-event combinations. A wager of $1 will be made if the decision maker chooses to gamble, and net winnings will be the payoff measure.

Objectives and Payoff Values

In determining appropriate payoffs, we will assume that decision makers will choose to take actions that will bring them closest to their objectives. Each outcome must somehow be ranked in terms of how close it is to the decision maker's goal. The payoff table should provide a meaningful basis for comparison and enhance the decision maker's ability to make a good choice.

For example, if a business decision maker's goal is to achieve a high level of profits, then a natural payoff would be the profit for each outcome. Profit

TABLE 19-2

Payoff Table for a Gambling Decision

Event	Act	
	Gamble	Don't Gamble
Head	$1	$0
Tail	− 1	0

is a valid measure for a limited set of objectives, but it is by no means the top concern of all business managers. The goal of the founder of a successful corporation may be to maintain personal control, and the founder may consciously keep profits low so that the firm will not be attractive to other merger-minded entrepreneurs.

In general, decision making involves different kinds of goals, each requiring a distinct payoff measure. Decision makers with different goals may even select dissimilar measures for payoffs when considering the same set of alternatives. The following example illustrates this.

EXAMPLE: FINDING THE WAY TO SAN JOSE

Charles Snyder, Herman Brown, and Sylvia Gold each wish to choose one of three routes from Los Angeles to San Jose: (1) Interstate 5, which is a freeway nearly all the way and has a high minimum speed limit; (2) State Highways 118 and 33, which are fairly direct and have no minimum speed limit; and (3) State Highway 1, which winds along the Pacific coast and is slow and long but very beautiful. Mr. Snyder is a salesman who travels to San Jose regularly; his goal is to reach his destination as quickly as possible. Mr. Brown is an economy "nut"; he wants to reach San Jose as cheaply as possible. Ms. Gold is on vacation and loves to drive on hilly, winding, scenic roads; she wishes to select the route that provides the greatest driving pleasure.

The payoffs that each person would assign to the three routes appear in Table 19-3. Time savings is Mr. Snyder's payoff measure, so he chooses Interstate 5, which yields the greatest payoff of 4 hours. Mr. Brown likes to drive at a moderate speed to obtain maximum gasoline mileage, while taking the shortest route possible. His payoff measure is fuel savings (based on the amount of gasoline required on the most expensive route), so he chooses the back roads, State Highways 118 and

TABLE 19-3

Payoffs for the Alternative Routes Relevant to the Goals of Three Decision Makers

	Payoff Measure		
	Mr. Snyder	*Mr. Brown*	*Ms. Gold*
Alternative Route	*Time Savings (hours)*	*Fuel Savings (gallons)*	*Enjoyment (subjective rating)*
Interstate 5	4	3	1
Highways 118 & 33	3	7	3
Highway 1	0	0	10

33, to save 7 gallons of gas. Ms. Gold has rated the routes in terms of points of interest, types of scenery, and number of hills and curves. This rating serves as her payoff measure, so her best route is scenic State Highway 1, which rates 10.

We can conclude that *a payoff measure should be selected so that the payoff will rank outcomes by the degree to which they attain the decision maker's goals*. The goal dictates which payoff measures are valid.

Tippi-Toes: A Case Illustration

As a detailed example of a business decision made under uncertainty, we will consider a hypothetical toy manufacturer who must choose between four prototype designs for Tippi-Toes, a dancing ballerina doll that does pirouettes and jetés. Each prototype represents a different technology for the moving parts, all powered by small, battery-operated motors. One prototype is a complete arrangement of gears and levers. The second is similar, with springs instead of levers. Another works on the principle of weights and pulleys. The movement of the fourth design is controlled pneumatically through a system of valves that open and close at the command of a small, solid-state computer housed in the head cavity. The dolls are identical in all functional aspects.

The choice of the movement design will be based solely on a comparison of the contributions to profits made by the four prototypes. The payoff table is provided in Table 19-4. The demand for Tippi-Toes is uncertain, but management feels that one of the following events will occur:

> Light demand (25,000 units)
> Moderate demand (100,000 units)
> Heavy demand (150,000 units)

TABLE 19-4

Payoff Table for the Tippi-Toes Decision

Event (Level of Demand)	Act (Choice of Movement)			
	Gears and Levers	Spring Action	Weights and Pulleys	Pneumatic
Light	$ 25,000	−$ 10,000	−$125,000	−$300,000
Moderate	400,000	440,000	400,000	300,000
Heavy	650,000	740,000	750,000	700,000

The toy manufacturer in this example is considering only three possible events—here, levels of demand—which greatly simplifies our analysis. Demand does not have to be precisely 25,000 or 100,000 units, however. The problem could be analyzed at several hundred thousand possible levels of demand—say, from 0 to 500,000 dolls. The techniques we will develop here can be applied to a more detailed situation. Due to the computational requirements, the demand probability distribution could be approximated to the nearest 100, 1,000, or 10,000 units.

Similarly, our example has only four alternatives or acts. In the practical, business decision-making environment, the number of possible acts can be quite large. For instance, deciding what mix of toys to sell could easily involve trillions of alternatives. *The decision analysis should include only the alternatives that the decision maker wishes to consider.* When there is no compelling reason for choosing one of the alternatives, "doing nothing" should be an alternative. The search for attractive alternatives is essential to sound decision making. However, decision analysis cannot tell us what factors should and should not be considered, although it can be used to guide our selection.

EXERCISES

19-1 A young bachelor is deciding whether to spend his Christmas vacation skiing at a resort in Colorado or surfing in Hawaii. He must commit himself to one of these alternatives in the early fall, since reservations have to be made months in advance. He really enjoys skiing more than surfing. Unfortunately, he cannot be certain about December snow conditions, and his ski trip will be ruined if there is poor snow. The trip to Hawaii would be a sure bet. But if he must go there when the snow is good elsewhere, his trip will be somewhat spoiled by regrets that he did not make arrangements to spend his vacation skiing.
(a) Construct the bachelor's decision table.
(b) Draw his decision tree diagram.

19-2 Peggy Jones, the founder of a computer-programming services firm, wishes to expand the firm's activities into the manufacture of peripheral equipment. Funds must be raised to build and operate the necessary facilities. Three financing alternatives are available: (1) Issue additional common stock, (2) sell bonds, and (3) issue nonvoting preferred stock. A common stock issue will provide a strong financial base for future expansion through borrowing, but it will considerably reduce Jones' percentage of ownership and control from its current 100%. New common stock will also divide future earnings into smaller amounts per share to existing shareholders. Bonds will allow existing shareholders to accrue all of the benefits of new earnings, but they will also increase the risk of forced liquidation if the new venture proves unsuccessful. Preferred stock will give its holders no claims on the firm's assets, but it will drastically reduce the rate of earnings participation on the part of existing common stockholders. The table on page 673 summarizes the forecast financial status of the firm if the manufacturing venture is successful.

Possible Payoff Measure	Financing Alternatives		
	Additional Common Stock	Bonds	Preferred Stock
1. Earnings after taxes and preferred dividends	$5,000,000	$3,500,000	$4,000,000
2. Common shares outstanding	1,000,000	500,000	500,000
3. Earnings per common share	$5.00	$7.00	$8.00
4. Jones' percentage of common ownership	50	100	100
5. Emergency line of credit	$1,000,000	$400,000	$500,000
6. Earnings available for common dividends	$5,000,000	$2,000,000	$4,000,000
7. Maximum possible dividends per share of common stock	$5.00	$4.00	$8.00

For each of the following goals, suggest an appropriate payoff measure. Then use this measure to identify the best and the worst alternative choices for financing in terms of the degree to which each *single* goal is met. Indicate any ties.

(a) Maintain a high percentage of control by Jones.
(b) Maximize the earnings of Jones' shares.
(c) Maximize the availability of short-term credit.
(d) Maximize the potential for cash dividends to Jones.

19-3 Shirley Smart has final examinations in accounting and finance on Monday and only 10 hours of study time left. The following anticipated grades apply, depending on the exam format.

Study Time (hours)	Multiple-Choice Format		Case Format	
	Accounting	Finance	Accounting	Finance
0	C	B	B	C
5	C	B	A	B
10	B	A	A	A

Shirley plans to allocate her study time in one of the following ways:

	Accounting	Finance
Plan 1	0 hours	10 hours
Plan 2	5	5
Plan 3	10	0

Although a single exam format will apply for any course, Shirley is uncertain which one each professor will pick, so any combination of multiple-choice and case formats is possible.

Using her combined accounting and finance grade point average (GPA) as her payoff measure (A = 4 points, B = 3 points, C = 2 points), construct Shirley's payoff table.

19-4   REDUCING THE NUMBER OF ALTERNATIVES: INADMISSIBLE ACTS

Regardless of the decision-making process that we ultimately employ to help us make a choice, an initial screening may be made to determine if there are any acts that will never be chosen. To illustrate this, consider the payoffs in Table 19-4. An interesting feature is exhibited by the payoffs of the acts "weights and pulleys" and "pneumatic": No matter which demand event occurs, the weights-and-pulleys act results in a greater payoff. If a light demand occurs, for instance, the payoff for weights and pulleys is −$125,000, which is more favorable than the −$300,000 payoff for the pneumatic movement. A similar finding results if we compare these two acts for the other possible demand events. Since the weights-and-pulleys movement will always be a superior choice to the pneumatic movement, we say that the first act *dominates* the second act. One act dominates another when it achieves a better or an equal payoff, no matter which events occur, and when it is strictly better for one or more events.

In general, whenever an act is dominated by another one, it is *inadmissible*. Thus, the pneumatic movement is an *inadmissible act*. The toy manufacturer's decision can be simplified by eliminating pneumatic movement from further consideration. Removing the pneumatic act leaves us with the modified payoff table in Table 19-5.

A simple way to determine if an act is inadmissible is to see if every entry in its column in the payoff table is less than or equal to the corresponding entry in some other column. It is easy to verify that this is not true for the entries in Table 19-5, so the remaining movement acts must be retained. The acts that remain are called *admissible acts*.

TABLE 19-5
Modified Payoff Table for the Tippi-Toes Decision

Event (Level of Demand)	Act (Choice of Movement)		
	Gears and Levers	Spring Action	Weights and Pulleys
Light	$ 25,000	−$ 10,000	−$125,000
Moderate	400,000	440,000	400,000
Heavy	650,000	740,000	750,000

EXERCISE

19-4 Identify any inadmissible acts in the following payoff table:

Event	Act				
	A_1	A_2	A_3	A_4	A_5
E_1	3	4	4	5	1
E_2	6	2	1	4	2
E_3	1	8	8	7	3

MAXIMIZING EXPECTED PAYOFF: 19-5
THE BAYES DECISION RULE

How does a decision maker choose an act? When there is no uncertainty, the answer is straightforward: Select the act that yields the highest payoff (although finding this particular optimal act can be very difficult when there are many alternatives). But when the events are uncertain, the act that yields the greatest payoff for one event may yield a lower payoff than a competing act for some other event.

Suppose that our toy manufacturer accepts the following probabilities for the demand for Tippi-Toes:

Light demand	.10
Moderate demand	.70
Heavy demand	.20
	1.00

We calculate the expected payoff for each act in Table 19-6 by multiplying each payoff by the respective event probability and summing the products from each column. We thus find that the spring-action movement results in the

TABLE 19-6
Calculation of Expected Payoffs for the Tippi-Toes Decision

Demand Event	Prob- ability	Gears and Levers		Spring Action		Weights and Pulleys	
		Payoff	Payoff × Probability	Payoff	Payoff × Probability	Payoff	Payoff × Probability
Light	.10	$ 25,000	$ 2,500	−$ 10,000	−$ 1,000	−$125,000	−$ 12,500
Moderate	.70	400,000	280,000	440,000	308,000	400,000	280,000
Heavy	.20	650,000	130,000	740,000	148,000	750,000	150,000
Expected payoff			$412,500		$455,000		$417,500

maximum expected payoff of $455,000. Thus, using maximum expected payoff as a decision-making criterion, our toy manufacturer would select the spring-action movement for the Tippi-Toes doll.

The criterion of selecting the act with the maximum expected payoff is sometimes referred to as the *Bayes decision rule*. This rule takes into account all the information about the chances for the various payoffs. But we will see that it is not a perfect device and can lead to a choice that is not actually the most desirable. However, we will also see that this criterion is a suitable basis for decision making under uncertainty when the payoff values are selected with great care.

EXERCISES

19-5 Consider the following payoff table:

		Act		
Event	Probability	A_1	A_2	A_3
E_1	.3	$10,000	$20,000	$ 5,000
E_2	.5	5,000	−10,000	10,000
E_3	.2	15,000	10,000	10,000

Compute the expected payoffs for each act. According to the Bayes decision rule, which act should be chosen?

19-6 Recompute the expected payoffs for the Tippi-Toes decision in Table 19-5, assuming that the following demand event probabilities now apply:

Light demand	.20
Moderate demand	.50
Heavy demand	.30

According to the Bayes decision rule, which act should the toy manufacturer choose?

19-7 A new product is to be evaluated. The main decision to be made is whether or not to market the product, in which case it will be a success (probability = .40) or a failure. The net payoff for a successful product is $10 million; a failure would result in a −$5 million payoff. Construct a payoff table for the decision, and then find the expected payoffs. Should the product be marketed?

19-6   DECISION TREE ANALYSIS

The decisions under uncertainty encountered thus far can be portrayed in terms of a payoff table. But some problems are too complex to be presented in a table. Difficulties arise when the same events do not apply for all acts. For example, a contractor might have to choose between bidding on a construction job for a dam or on one for an airport, not having sufficient resources to bid on both. Regardless of the job chosen, there is some probability (which may differ for the two projects) that the contractor will win the job bid on. Separate sets of events and probabilities are required for each act.

Decisions must often be made at two or more points in time, with uncertain events occurring between decisions. Sometimes these problems can be analyzed in terms of a payoff table, but usually the earlier choice of the act will have a bearing on the type, quantity, and probabilities of later events. At best, this makes it cumbersome to attempt to force the decision into the limited confines of the rectangular arrangement of a payoff table.

The decision tree diagram described earlier allows us to meaningfully arrange the elements of a complex decision problem without the restrictions of a tabular format. A further advantage of the decision tree is that it serves as an excellent management communication tool, because the tree clearly delineates every potential course of action and all possible outcomes.

Ponderosa Record Company: A Case Illustration

The president of Ponderosa Record Company, a small independent recording studio, has just signed a contract with a four-person rock group, called the Fluid Mechanics. A tape has been cut, and Ponderosa must decide whether or not to market the recording. If the record is to be test marketed, then a 5,000-record run will be made and promoted regionally; this may result in a later decision to distribute an additional 45,000 records nationally, for which a second pressing run will have to be made. If immediate national marketing is chosen, a pressing run of 50,000 records will be made. Regardless of the test-market results, the president may decide to enter the national market or decide not to enter it.

A Ponderosa record is either a complete success or a failure in its market. A recording is successful if all records that are pressed are sold; the sales of a failure are practically nil. Success in a regional market does not guarantee success nationally, but it is fairly reliable predictor.

The Decision Tree Diagram

The structure of the Ponderosa decision problem is presented in the decision tree diagram in Figure 19-2. Decisions are to be made at two different points in time, or stages. The immediate choice is to select one of two acts: "test market" or "don't test market." These acts are shown as branches on the initial fork at node *a*. If test marketing is chosen, then the result to be achieved in the test marketplace is uncertain. This is reflected by an event fork at node *b*, where the branches represent favorable and unfavorable outcomes. Regardless of which event occurs, a choice must be made between two new acts: "market nationally" or "abort." These acts occur at a later stage and are represented by a pair of act forks. Each fork corresponds to the two different conditions under

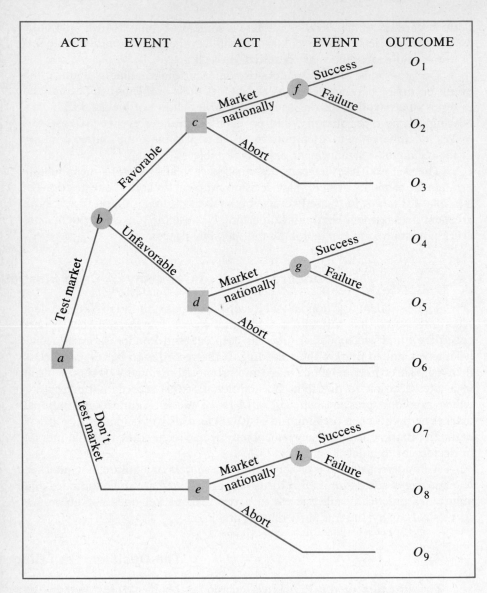

FIGURE 19-2
Decision tree diagram for the Ponderosa decision.

which this decision may be made: at node c, when the test marketing is favorable, and at node d, when it is unfavorable. If national marketing is chosen at either node c or node d, the success or failure of the recording still remains unknown, and the possible events are reflected on the decision tree as branches on the terminal event forks at nodes f and g.

If the initial choice at decision point *a* is "don't test market," then a further choice must be made at the act fork represented by decision point *e*: "market nationally" or "abort." As before, node *h* reflects the two uncertain events that will arise from the choice to market nationally. The "abort" path leading from node *e* contains a "dummy" branch—a diagrammatical convenience that allows event and act forks of similar form to appear at the same stage of the problem and permits all paths to terminate at a common stage. Thus, all "abort" acts are followed by a dummy branch.

Every path from the base of the decision tree leads to a terminal position corresponding to a decision outcome. Each possible combination of acts and events, or each path, has a distinct outcome. For instance, O_1 represents the following sequence of events and acts: "test market," "favorable," "market nationally," "success."

The first step in analyzing the decision problem is to obtain a payoff for each outcome.

Determining the Payoffs

The contract with the Fluid Mechanics calls for a $5,000 payment to the group if records are produced. Ponderosa arranges with a record manufacturer to make its pressings. For each pressing run, there is a $5,000 fixed cost plus a $.75 fee for each record. Record jackets, handling, and distribution cost an additional $.25 per record. The total variable cost per record is therefore $1.00. Using these figures, we can calculate the immediate cash effect of each act in the decision tree in Figure 19-2. Some of these cash effects, or *partial cash flows*, are computed in Table 19-7.

TABLE 19-7
Some Partial Cash Flows Used in Determining the Payoff Values
of Ponderosa's Outcomes

Act		Partial Cash Flow	
Test Market		−$ 5,000	(payment to group)
		−5,000	(fixed cost of pressing)
		−5,000	(variable costs of 5,000 records at $1.00)
	Total	−$15,000	
Don't Test Market		$0	
Market Nationally (without test)		−$ 5,000	(payment to group)
		−5,000	(fixed cost of pressing)
		−50,000	(variable costs of 50,000 records at $1.00)
	Total	−$60,000	
Abort		$0	

The negative cash flows indicate expenditures. The partial cash flows in Table 19-7 appear on the respective branches extending from decision points *a* and *e* of the decision tree in Figure 19-3. In a similar manner, we determine the partial cash flows for the acts at the forks at decision points *c* and *d*: $-\$50,000$ to market nationally ($5,000 fixed pressing cost plus $1.00 each in variable costs for 45,000 records), and $0 to abort.

Ponderosa receives $2 for each record it sells through retail outlets. Since the events "favorable" and "unfavorable" or "success" and "failure" represent sales of all and no records, respectively, the partial cash flows may be obtained by multiplying the number of records sold by $2. The partial cash flows for the events at the fork at node *b* are therefore $+\$10,000$ (for 5,000 records sold) and $0 (for no sales). The amounts for the events at nodes *f* and *g* are $+\$90,000$ (for 45,000 records sold) and $0, whereas the amounts for the events at node *h* are $+\$100,000$ and $0.

The payoff for each outcome may be obtained by adding the partial cash flows on the branches of the path leading to its terminal position. Thus, for O_1, we add the partial cash flows $-\$15,000$, $+\$10,000$, $-\$50,000$, and $+\$90,000$. The payoff for O_1 is therefore $+\$35,000$. The payoffs calculated for each outcome are shown at the respective terminal positions of the decision tree in Figure 19-3.

Assigning Event Probabilities

Ponderosa's management wishes to choose the act that will yield the maximum expected payoff. But before this choice can be made, probability values must be assigned to the events in the decision structure. Suppose that Ponderosa's president believes that the chance of favorably test marketing the recording is .50. The probability of unfavorably test marketing the recording is also .50. These probability values are placed in parentheses along the branches at node *b* in Figure 19-3. In assigning probability values to the success and failure events for national marketing, our decision maker is faced with three distinctly different situations. With no test marketing, the chance of national success is judged to be .50. A favorable test marketing indicates that a record appeals to the regional segment of the market, so the chance of national success in this case is judged to be a much higher .80; this is a *conditional probability*. Similarly, unfavorable test marketing is a likely indication of national appeal, so the conditional probability for success is judged to be .20 in this case. The following probability values are placed on the branches for the event at the remaining forks in the decision tree diagram: .80 for success and .20 for failure at node *f*; .20 for success and .80 for failure at node *g*; and .50 for success and .50 for failure at node *h*. (It is just coincidental that the probability for national marketing success after unfavorable test marketing is .20, which is also the probability for national marketing failure after favorable test marketing. Our decision maker could have selected another value, such as .10.)

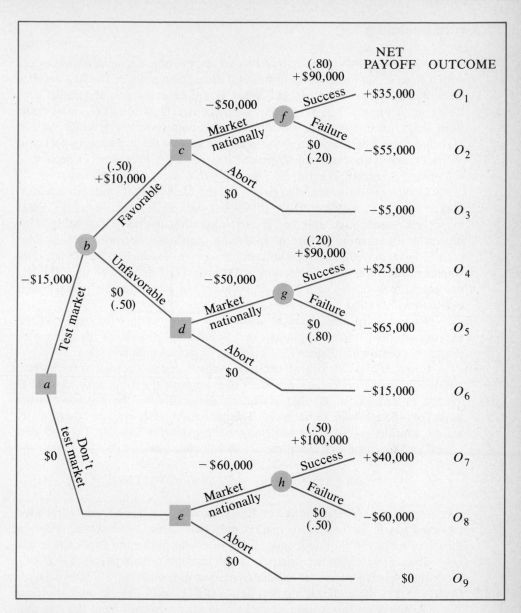

FIGURE 19-3

The Ponderosa decision tree diagram, showing partial cash flows and probabilities on the branches, with net cash-flow payoffs at the terminal positions.

Backward Induction

We are now ready to analyze Ponderosa's decision. Our decision maker wishes to select an initial or immediate act at decision point a. The first act that we will evaluate is "test market." What is the expected payoff for this act? Referring to Figure 19-3, we see that six outcomes, O_1 through O_6, may result from this choice. How can we translate the corresponding payoffs into an expected value? We cannot do this until we specify the intervening acts that will be chosen at nodes c and d. In general, *it is impossible to evaluate an immediate act without first considering all later decisions that result from this choice.*

Thus, to find the expected payoff for the "test market" act, our decision maker must first decide whether to market nationally or to abort if (1) test marketing proves favorable or (2) test marketing proves unfavorable. This illustrates an essential feature of analyzing multistage decisions: *Evaluations must be made in reverse of their natural chronological sequence.* Before deciding whether to test market, our decision maker must decide what to do if the test marketing is favorable or if it is unfavorable. The procedure for making such evaluations is called *backward induction.*

We can clarify this point by describing the procedure for our decision-making problem. For simplicity, the Ponderosa president's decision tree diagram is redrawn in Figure 19-4 without the partial cash flows.

Consider the act fork at decision point c. If the decision is to market nationally, then Ponderosa's president is faced with the event fork at node f. With a probability of .80 that marketing nationally will be a success, a net payoff of $+\$35,000$ will be achieved. The probability that marketing nationally will be a failure is .20, which leads to a net payoff of $-\$55,000$. The expected payoff for this event fork can therefore be calculated as

$$.80(+\$35,000) + .20(-\$55,000) = +\$17,000$$

The amount $+\$17,000$ is entered on the decision tree at node f, since this is the expected payoff for the act to market nationally. For convenience, we place the expected payoff for a sequence of acts or events above the applicable node.

The act to abort at decision point c will lead to a certain payoff of $-\$5,000$. Since the expected payoff for the act to market nationally ($+\$17,000$) is larger than $-\$5,000$, the choice to market nationally should be made over aborting. We may reflect on this future choice by *pruning* the branch from the tree that corresponds to the act "abort" at decision point c. This act merits no further consideration, since if decision point c is reached, the president will choose to market nationally. That is, if the decision maker initially decides to test market and the results turn out to be favorable, the president will choose to market nationally. Thus, $+\$17,000$ is the expected payoff resulting from making the best choice at decision point c. We bring back the amount $+\$17,000$ and enter it on the diagram above the node at c.

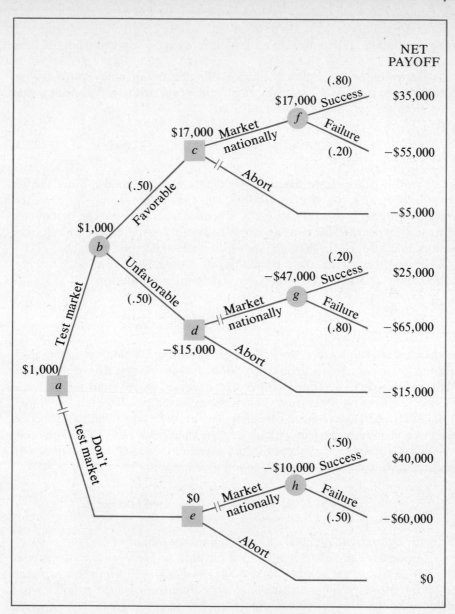

FIGURE 19-4

The Ponderosa decision tree diagram, showing backward induction analysis.

As a rule of thumb in performing backward induction, ultimately all but one act will be eliminated at each decision point (except in the case of ties), so that all branches except those leading to the greatest expected payoff will be pruned. Only the *best single payoff* from the later stage is brought backward to the preceding decision point (square). *Branch pruning takes place only in act forks—never in event forks.* Instead, event forks (circles) involve an expected value calculation, so that an average payoff is always computed from the later stage values.

The available choices when test marketing results are unfavorable may be handled in the same way. First, we calculate the expected payoff at node g that arises from the act "market nationally," or

$$.20(+\$25,000) + .80(-\$65,000) = -\$47,000$$

We place this figure on the decision tree diagram above node g. Since the act "abort" leads to a payoff of $-\$15,000$, which is larger than $-\$47,000$, the branch for the act to market nationally is pruned from the tree. The best choice when test marketing fails is to abort. We therefore bring back and enter the amount $-\$15,000$ on the diagram above the node at decision point d.

In a similar fashion, the expected payoff of the event fork at node h, when the initial choice is to market nationally, is determined as follows:

$$.50(+\$40,000) + .50(-\$60,000) = -\$10,000$$

At decision point e, the act "abort" is superior to the act "market nationally," and the latter branch is pruned from the tree. Our decision maker must now compare the acts at decision point a. The expected payoff from the act "test market" is still to be determined. This is the expected payoff for the event fork at node b, which has two event branches. The branch corresponding to success leads to a portion of the tree with an expected payoff of $+\$17,000$. The other branch leads to a choice with an expected payoff of $-\$15,000$. We use these two amounts to calculate the expected payoff at node b:

$$.50(+\$17,000) + .50(-\$15,000) = +\$1,000$$

We enter the amount $+\$1,000$ on the diagram above node b.

Ponderosa's president is now in a position to compare the two acts at decision point a: "test market" and "don't test market." Since the expected payoff for test marketing ($+\$1,000$) is higher than the expected payoff for not testing ($\$0$), our decision maker should choose to test market. The expected payoff of $\$1,000$ is brought back and placed above node a. The branch corresponding to "don't test market" is pruned, and our backward induction is complete.

A decision is indicated. Ponderosa's president should choose to test market the Fluid Mechanics' record. If the test marketing is favorable, then the president should market nationally; if the test marketing is unfavorable, then the president should abort the recording. This result is illustrated in Figure 19-4 by the unpruned branches that remain on the decision tree.

Additional Remarks

Choices that are made in later stages are not irrevocable, and this analysis does not preclude the fact that the decision maker's mind may change over time. Before the future decision must be made at node c, new information may be received that would, for instance, indicate a need to revise the probability for national success downward. If there is bad publicity about one of the Fluid Mechanics, for example, the expected payoff for national marketing might be smaller than the expected payoff for aborting. Possible changes in conditions do not invalidate the original backward induction analysis. In our example, *the choice to test market is the best decision that can be made based on the currently available information.*

The decision tree structure is suitable for analyzing decisions that extend over a long time period. It indicates the best course of action for the *current* decision. As time progresses, however, some uncertainties may be reduced and new ones may arise. Acts previously identified as optimal may turn out to be obviously poor choices, and brand new candidates may be determined. The relevant portion of the decision tree can be updated and revised prior to each new immediate decision. But each such decision is analyzed in the same general manner, using the best information available at the time a choice must be made.

Although a decision tree is analyzed by moving backward in time, the analysis is really forward-looking because it indicates the optimal course of action to take when future decision points are reached. The dollar amount brought backward to each node represents the best payoff the decision maker can expect to achieve if that position is reached at a later time. Regardless of what events have occurred, the optimal course of action for future choices is still indicated by the original analysis.

EXERCISES

19-8   Suppose that the president of the Ponderosa Record Company uses the following probability values to analyze the decision about the Fluid Mechanics' recording.

Pr[national marketing success|favorable test marketing] = .9

Pr[national marketing failure|unfavorable test marketing] = .6

Pr[national marketing success] = .7

Pr[test marketing success] = .75

Repeat the Ponderosa decision tree analysis to determine the company's optimal marketing strategy. Assume that all payoffs remain unchanged.

19-9   The manager of an oil company's data-processing operations personally interviews applicants for jobs as keypunchers. Employees who are hired with no previous experience are placed in a one-month training program on a trial basis. Satisfactory

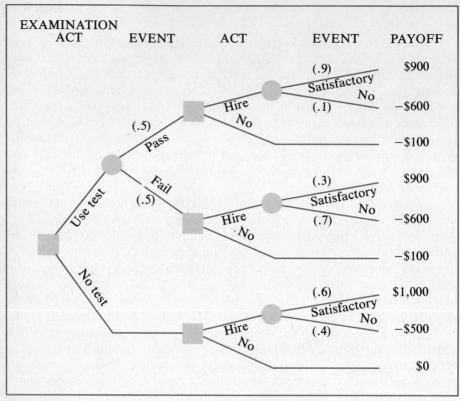

EXAMINATION
ACT · EVENT · ACT · EVENT · PAYOFF

(.9) Satisfactory — $900
No
Hire (.1) — −$600
No — −$100
(.5) Pass

Use test
(.5) Fail
(.3) Satisfactory — $900
No
Hire (.7) — −$600
No — −$100

No test

(.6) Satisfactory — $1,000
No
Hire (.4) — −$500
No — $0

FIGURE 19-5

employees are retained; all others are let go at the end of the month. Most of the people who have been let go in the past have been found to be lacking in aptitude. The manager is contemplating contracting the testing services of a personnel agency. For a fee the agency would administer a battery of aptitude tests. The manager has developed the decision tree in Figure 19-5 to help her make her hiring decisions. Perform backward induction analysis to determine the strategy or course of action that will maximize the manager's expected payoff.

19-10 Buzzy-B Toys must decide the course of action to follow in promoting a new whistling Yo-Yo. Initially, management must decide whether to market the Yo-Yo or to conduct a test marketing program. After test marketing the Yo-Yo, management must decide whether to abandon it or nationally distribute it.

A national success will increase profits by $500,000, and a failure will reduce profits by $100,000. Abandoning the product will not affect profits. The test marketing will cost Buzzy-B a further $10,000.

If no test marketing is conducted, the probability for a national success is judged to be .45. The assumed probability for a favorable test marketing result is .50. The conditional probability for national success given favorable test marketing is .80; for national success given unfavorable test results it is .10.

Construct the decision tree diagram and perform backward induction analysis to determine the optimal course of action if a net change in profits is the expected payoff.

19-11 Spillsberry Foods must determine whether or not to market a new cake mix. Management must also decide whether to conduct a consumer test marketing program that would cost $25,000. If the mix is successful, Spillsberry's profits will increase by $1,000,000; if the mix fails, the company will lose $250,000. Not marketing the product will not affect profits. The cake mix is considered to have a 60% chance of success without testing. The assumed probability for a favorable test marketing result is 50%. Given a favorable test result, the chance of product success is judged to be 85%. However, if the test results are unfavorable, the probability for the product's success is judged to be only 35%.

Construct a decision tree diagram that can be used to determine the optimal course of action that will provide the greatest expected payoff. Include the choice of whether or not to use the test. Perform backward induction analysis to determine which course of action maximizes the expected payoff.

REVIEW EXERCISES

19-12 Refer to the payoffs for the Tippi-Toes decision in Table 19-5. Suppose that the development cost of the spring-action movement is $40,000 higher than before (regardless of demand) and that revenues generated by the doll when demand is heavy are $50,000 lower than before (regardless of movement chosen).
(a) Construct a new payoff table reflecting these revisions.
(b) Using the payoff table, compute the expected payoff for each act. Which movement should be chosen under the Bayes decision rule?

19-13 Consider the following payoff table for a graphical design decision.

Texture Event	Prob-ability	Color Theme Act				
		Brown	Orange	Red	Green	Yellow
Full-tone	.2	100	110	50	120	90
Half-tone	.5	80	70	80	60	100
Mixed	.3	100	90	90	90	110

(a) Find any inadmissible color choices.
(b) Compute the expected payoff for each admissible act. Which color theme should be chosen?

19-14 A product manager for a soap manufacturer wishes to determine whether or not to market a new toothpaste. The manager can order a consumer testing program for $50,000. The present value of all future profits for a successful toothpaste is $1,000,000; the brand's failure would result in a net loss of $500,000. Not marketing the toothpaste will not affect profits. The manager judges that the toothpaste would have a 50–50 chance of success without testing. Customer testing will be either favorable (40% chance) or unfavorable. Given a favorable test result, the chance of product success is judged to be 80%. But for an unfavorable test result, the toothpaste's success probability is judged to be only 30%.

Construct the product manager's decision tree diagram. Perform backward induction analysis to determine which course of action provides the greatest expected profit.

19-15 The following two experiments with the product in Exercise 19-6 have been proposed. (Both experiments cannot be used.)

Test market at a cost of $1 million. Results will be "favorable" (probability = .48) or "unfavorable." Given a favorable result, the probability for product success is .75. Given an unfavorable result, the probability for product success is only .08.

Attitude survey at a cost of $.5 million. Results will be "warm" (probability = .40) or "cold." Given a warm response, the probability for product success is .70. Given a cold response, the probability for success is only .20.

In addition to these choices, the main decision to market or not to market may be made without obtaining any further information.

Construct a decision tree diagram for this problem, indicating all of the probabilities and payoffs. Perform backward induction analysis to determine which course of action maximizes expected payoff. (Specify all choices that may then have to be made.)

19-16 A government official wishes to determine the most effective way to control tree damage from the gypsy moth. There are three methods for attacking the pest: (1) spray with DDT; (2) use a scent to lure and trap males, so that the remaining males must compete for mating with a much larger number of males that have been sterilized in a laboratory and then released; and (3) spray with a juvenile hormone that prevents the larvae from developing into adult moths.

The net improvement in current and future tree losses using DDT is zero, because it is assumed that DDT will never completely eradicate the moth.

If the scent-lure program is instituted, the probability that it will leave a low number of native males is .5, with a .5 chance that it will leave a high number. Once the scent-lure results are known, a later choice must then be made either to spray with DDT or to release sterile males. The cost of the scent lures is $5 million and the cost of sterilization is an additional $5 million. But if this two-phase program is successful, the worth of present and future trees saved will be $30 million. If scent lures leave a small native male population, there is a 90% chance for success using sterile males; otherwise, there is only a 10% chance for success using sterile males. A failure results in no savings.

The juvenile hormone must be synthesized at a cost of $3 million. There is only a .20 probability that the resulting product will work. If it does, the worth of trees saved would be $50 million, because the gypsy moth would become extinct. If the hormone does not work, savings would be zero.

Construct a decision tree diagram for the official's decision. Using the worth of trees saved minus cost as the payoff measure (relative to using DDT, which has a payoff of zero), determine the course of action that will yield the maximum expected payoff.

Elements of Decision Theory

20

A primary focus of statistical decision theory is establishing systematic means for choosing an act, which is largely accomplished by using the payoff table introduced in Chapter 19. Various *decision-making criteria* may be employed in selecting the best act. The payoff measure itself is a key element in determining rules for decision making, and decision theory encompasses a variety of these measures.

We will begin our discussion of decision theory by examining some of the well-known criteria used in selecting a best act. Then we will describe *opportunity loss*—a payoff measure that enables us to assess the worth of the *information* that is obtained about uncertain events.

For various reasons, we will also see that the Bayes decision rule, which helps the decision maker select the act with the maximum expected payoff, is the favored criterion. The theoretical concepts of decision making largely expand on this rule, since this criterion makes use of all the information at the disposal of the decision maker. The Bayes decision rule always leads to the most desirable choice when the proper payoff measure, *utility payoff*, is used. *Utility theory*, a special field within the broader context of decision theory, is discussed in the next chapter.

20-1   DECISION CRITERIA

The Maximin Payoff Criterion

No decision-making theory could be complete without giving due consideration to various rules that might be used in selecting the most desirable act. We will begin with the simplest criterion, the *maximin payoff criterion*— a procedure that guarantees that the decision maker can do no worse than achieve the best of the poorest outcomes possible. As an illustration, we will use the payoff table given in Table 20-1, which presents the toy manufacturer's choices of movement for the Tippi-Toes doll.

Suppose that our toy manufacturer wishes to choose an act that will ensure a favorable outcome no matter what happens. This can be accomplished by taking a pessimistic viewpoint—that is, by determining the *worst* outcome for each act, regardless of the event. For the gears-and-levers movement, the lowest possible payoff is $25,000 when light demand occurs. The lowest payoff for the spring-action movement is a negative amount, −$10,000, also obtained when demand is light. For the weights-and-pulleys movement, the lowest payoff is −$125,000, again when demand is light. By choosing the act that yields the largest lowest payoff, our decision maker can guarantee a minimum return that is the best of the poorest outcomes possible. In this case, a gear-and-levers movement for the doll will guarantee the toy manufacturer a payoff of at least $25,000.

The gears-and-levers movement is the act with the maximum of the minimum payoffs. A more concise statement would be to say that gears and levers is the *maximin payoff act*. To show how the maximin payoff can be determined in general, we reconstruct the payoff table for the Tippi-Toes decision in Table 20-2.

In most decision-making illustrations so far, we have used profit as the measure of payoff. As we noted in Chapter 19, a variety of measures may be used to rank outcomes. In business applications, *cost* is often used for this

TABLE 20-1
Payoff Table for the Tippi-Toes Decision

Event (Level of Demand)	Act (Choice of Movement)		
	Gears and Levers	Spring Action	Weights and Pulleys
Light	$ 25,000	−$ 10,000	−$125,000
Moderate	400,000	440,000	400,000
Heavy	650,000	740,000	750,000

TABLE 20-2
Determining the Maximin Payoff Act for the Tippi-Toes Decision

Event (Level of Demand)	Act (Choice of Movement)		
	Gears and Levers	Spring Action	Weights and Pulleys
Light	$ 25,000	−$ 10,000	−$125,000
Moderate	400,000	440,000	400,000
Heavy	650,000	740,000	750,000
Column minimums	$ 25,000	−$ 10,000	−$125,000

Maximum of column minimums = $25,000
Maximin payoff act = Gears and levers

purpose when the goal is to minimize operational cost and when revenues are not subject to chance. We can apply the maximin payoff criterion to a situation involving costs by reversing our rule and selecting the act with the minimum of the maximum costs. The comparable terminology for this rule would be *minimax cost*. This criterion is identical to the maximin profit in the sense that cost can be viewed as negative profit, so that what is minimized and maximized must be reversed. Either criterion leads to choosing the best of the poorest outcomes. To avoid confusion, we will always use maximin and minimax as adjectives connected to a noun such as profit or cost.

The suitability of the maximin payoff criterion depends on the nature of the decision to be made. Consider the decision problem in Table 20-3. In this situation, the maximin decision maker chooses A_1 over A_2. A_2 may be a better choice if the probability of E_2 is high enough, but the maximin decision maker is giving up an opportunity to gain $10,000 in order to avoid a possible loss of $1. To avoid losing $1, the decision maker chooses an act that will guarantee the maintenance of the status quo. We can, however, envision circumstances in

TABLE 20-3
Determining the Maximin Payoff Act
for Hypothetical Decision A

Event	Act	
	A_1	A_2
E_1	$0	−$1
E_2	1	10,000
Column minimum	$0	−$1

Maximum of column minimums = $0
Maximin payoff act = A_1

TABLE 20-4

Payoff Table
for Hypothetical Decision *B*

Event	Act	
	B_1	B_2
E_1	$1	$10,000
E_2	−1	−10,000

which A_1 *would* be the better choice. If our decision maker had only $1 and had to use it to pay a debt to a loan shark or lose his life, the payoffs would not realistically represent the true values that the decision maker assigned to them.

Consider the situation represented by the payoffs given in Table 20-4. Here, the maximin payoff act is B_1. This would be the better choice for a decision maker who could not tolerate a loss of $10,000, no matter how unlikely it was. Few people would risk losing their businesses by choosing an act that could lead to bankruptcy unless the odds were extremely small. But an individual who could survive a loss of $10,000 would find B_2 a superior choice if the probability of E_2 were substantially lower than E_1.

Our examples illustrate a key deficiency of the maximin payoff: It is an extremely conservative decision criterion and can lead to some very bad decisions. Any alternative with a slightly larger risk is rejected in favor of a comparatively risk-free alternative, which may be far less attractive. Taken to a ludicrous extreme, a maximin payoff policy would force any firm out of business. No inventories would be stocked, because there would always be a possibility of unsold items. No new products would be introduced, because management could never be certain of their success. No credit would be granted, because there would always be some customer who would not pay.

Another major deficiency of the maximin payoff criterion exists if the probabilities of the various events are known. The maximin payoff is primarily suited to decision problems with unknown probabilities that cannot be reasonably assessed. As our illustrations indicate, it is usually in the extreme cases— the person hounded by loan sharks or the business that could go bankrupt— that the maximin payoff criterion leads to the best decision.

The Maximum Likelihood Criterion

Another rule that serves as a model for decision-making behavior is the *maximum likelihood criterion*, which focuses on the most likely event, to the exclusion of all others. Table 20-5 illustrates this criterion for the Tippi-Toes toy manufacturer's decision.

TABLE 20-5

Determining the Maximum Likelihood Act for the Tippi-Toes Decision

Event (Level of Demand)	Probability	Act (Choice of Movement)		
		Gears and Levers	Spring Action	Weights and Pulleys
Light	.10	$ 25,000	−$ 10,000	−$125,000
Moderate	.70	400,000	440,000	400,000
Heavy	.20	650,000	740,000	750,000

Most likely event = Moderate demand
Maximum row payoff = $440,000
Maximum likelihood act = Spring action

For this decision, we can see that the highest probability is .70 for a moderate demand. The maximum likelihood criterion tells us to ignore the light and heavy demand events completely—in effect, to assume that they will not occur. This rule then tells us to choose the best act assuming that a moderate demand will occur. In this example, the *maximum likelihood act* is to use the spring-action doll movement, which provides the greatest profit of $440,000 for a moderate demand.

How suitable is the maximum likelihood criterion for decision making? Using it in this example does not permit us to consider the range of outcomes for the spring-action movement, from a $10,000 loss if a light demand occurs to a $740,000 profit if demand is heavy. We also ignore most of the other possible outcomes, including the best (selecting weights and pulleys when demand is heavy, which yields a $750,000 profit) and the worst (selecting weights and pulleys when demand is light, which leads to a $125,000 loss). In a sense, the maximum likelihood criterion would have us "play ostrich," ignoring much that might happen. Then why is it discussed here?

We describe this criterion primarily because it seems to be so prevalent in the decision-making behavior of individuals and businesses. It can also be used to explain certain anomalies that would otherwise be hard to rationalize. These quirks are epitomized by the so-called "hog cycle" in the raising and marketing of pigs, which is related to the more or less predictable two-year-long pork price movement from higher to lower levels and back to higher levels again. Hog farmers have been blamed for this, since they expand their herds when prices are high, so that one year later the supply of mature hogs is excessive and prices are driven downward; then when prices are low, these same farmers reduce their herds, cutting the supply of marketable hogs, and next year's prices consequently rise.

Why don't the farmers break this cycle? It doesn't seem rational to be consistently wrong in timing hog production. One explanation is that the hog farmers use the maximum likelihood criterion. In their minds, the most likely future market price is the current one—and we know that this has proved to be a very poor judgment. Given such a premise, the maximum likelihood act is to increase herd sizes when current prices are high and to decrease them when prices are low.

The Criterion of Insufficient Reason

Another criterion employed in decision-making problems is the *criterion of insufficient reason*. This criterion may be used when a decision maker has no information about the event probabilities. In this case, no event may be regarded as more likely than any other event, and all events are assigned equal probability values. Since the events are collectively exhaustive and mutually exclusive, the probability of each event must be

$$\frac{1}{\text{Number of events}}$$

Using these event probabilities, the act with the maximum expected payoff is chosen.

A major criticism of the criterion of insufficient reason is that, except in a few situations, some knowledge of the relative chances that events will occur is always available. When more realistic probabilities can be obtained, employing the Bayes decision rule will provide more valid results.

The Preferred Criterion: The Bayes Decision Rule

The three decision-making criteria just discussed have obvious inadequacies. *None of them incorporates all of the information available to the decision maker.* Maximin payoff totally ignores event probabilities. Although it is argued that this is a strength when probabilities cannot be easily determined, judgment can be used to arrive at acceptable probability values in all but a few circumstances.

The maximum likelihood criterion ignores all events but the most likely one, even if that event happens to be a lot less likely than the rest combined. (Out of 20 events, for instance, the most likely event may have a probability of .10, leaving a .90 probability that one of the other 19 events will occur.)

The criterion of insufficient reason essentially asks us to ignore judgments and "willy nilly" assume that all events are equally likely. According to this criterion, even such events as "war" and "peace", and "prosperity" and "depression", have equal probabilities.

The *Bayes decision rule* has become the central focus of statistical decision theory. This makes the greatest use of all available information and is the

TABLE 20-6
Payoff Table for Hypothetical Decision C

Event	Probability	Act C_1		Act C_2	
		Payoff	Payoff × Probability	Payoff	Payoff × Probability
E_1	.5	−$1,000,000	−$ 500,000	$250,000	$125,000
E_2	.5	2,000,000	1,000,000	750,000	375,000
Expected payoff			$ 500,000		$500,000

only criterion that allows us to extend decision theory to incorporate sampling or experimental information. The major deficiency of the Bayes decision rule occurs when alternatives involve different magnitudes of risk. To illustrate this point, we will consider the decision structure in Table 20-6. Acts C_1 and C_2 are equally attractive according to the maximum expected payoff criterion (Bayes decision rule). Yet most decision makers would clearly prefer C_2, because it avoids the rather large risk of a $1,000,000 loss.

The paradox here may be resolved not by choosing another criterion, but by reconsidering the values chosen for the payoffs. The theory of utility presented in Chapter 21 will allow us to establish payoffs at values that express their true worth to the decision maker.

THE BAYES DECISION RULE AND UTILITY   20-2

We have presented a strong case for using expected values as the basis for decision making under uncertainty. As we have seen, a criterion based on expected values uses all the available probability data and assigns the proper weight to every outcome. The other decision-making criteria employ fewer structural elements from the decision. Expected values also provide us with a gauge for evaluating additional sources of information that can be used in decision making.

But when applied to *monetary* payoffs, the Bayes decision rule—maximizing expected profit and minimizing expected cost or loss—often leads to a less preferred choice.

Perhaps the best example of this occurs in casualty insurance decisions, where the choices are to buy or not to buy a policy. Most drivers have liability insurance for their cars, and most carry greater coverage than the legal minimum. We know that the policyholder's annual insurance costs exceed the expected loss from an accident. (This is because insurance companies must charge more than what they expect to pay in claims just to meet overhead costs and expenses.) But according to the Bayes decision rule, the best decision would be not to insure, because no insurance would have a greater expected monetary payoff (that is, the expected cost of no insurance is less than the cost of insurance). This course of action contradicts the true preference of most people.

Similar breakdowns of the Bayes decision rule occur whenever a person prefers a less risky alternative to one that involves considerable risk but actually has a greater expected monetary payoff. Since other decision-making criteria have serious defects, too, how should we objectively analyze decisions that involve great risk?

Fortunately, *decision theory accounts for attitudes toward risk by permitting an adjustment in the payoff values themselves.* This is accomplished by establishing a true-worth index called a *utility value* for every outcome. Thus, a decision may be analyzed using utilities instead of dollars or some other standard payoff measure.

In Chapter 21 we will describe the theory of utility and its application in great detail. One very important principle will be established there: *When the Bayes decision rule is applied to a decision-making problem with utility payoffs, it always indicates the most preferred course of action.* This makes that rule the theoretically perfect criterion for decision making, no matter how complex the decision happens to be.

EXERCISES

20-1 You have decided to participate in a gamble that offers the following monetary payoffs:

	Act	
Event	Choose Red	Choose Black
Red	$1	−$ 2
Black	−1	100

(a) Which act is the maximin payoff act?
(b) Supposing that the probability of red is .99, calculate the expected payoffs for each act. Which act is better according to the Bayes decision rule? Which act would you choose?
(c) Supposing that the probability of red is .5, calculate the expected payoffs for each act. Which act has the maximum expected payoff? Which act would you choose?
(d) In view of your answers to (b) and (c), what is your opinion of the maximin payoff decision criterion in this case?

20-2 A decision maker must choose one of three acts. The payoff table and the event probabilities are provided below:

Event	Probability	Act A_1	A_2	A_3
E_1	.3	$10	$15	$20
E_2	.4	15	20	15
E_3	.3	25	15	15

(a) Which act is the maximin payoff act?
(b) Which act is the maximum likelihood act?
(c) Calculate the expected payoffs. According to the Bayes decision rule, which act should be chosen?

20-3 A farmer intends to sign a contract to provide a cannery with her entire crop. She must choose to produce one of the following five vegetables: corn, tomatoes, beets, asparagus, or cauliflower. The farmer will plant her entire 1,000 acres with the selected crop. The yields of these vegetables will be affected by the weather to varying degrees. The following table indicates the approximate productivities for each vegetable in dry, moderate, and damp weather and also lists the price per bushel that the cannery has offered for each crop.

	Approximate Yield (bushels per acre)				
Weather	Corn	Tomatoes	Beets	Asparagus	Cauliflower
Dry	20	10	15	30	40
Moderate	35	20	20	25	40
Damp	40	10	30	20	40
Price per bushel	$1.00	$2.00	$1.50	$1.00	$.50

(a) Using as the payoff measure the approximate total cash receipts when the crop is sold, construct the payoff table for the farmer's decision.
(b) Identify any inadmissible acts and eliminate them from the payoff table.
(c) Which act is the maximin payoff act?
(d) Suppose that the following probabilities have been assigned to the types of weather. Calculate the expected payoff for each act; then identify the act that has the maximum expected payoff.

Weather	Probability
Dry	.3
Moderate	.5
Damp	.2

20-4 A newsdealer must decide how many copies of a particular magazine to stock in December. He will not stock less than the lowest possible demand or more than the highest possible demand. Each magazine costs him $.50 and sells for $1.00. At the end of the month the unsold magazines are thrown away. Three levels of monthly demand are equally likely: 10, 11, and 12. If demand exceeds stock, sales will equal stock.

(a) Using December profit as the payoff measure, construct the newsdealer's payoff table.

(b) According to the maximin criterion, how many copies should he stock?

(c) Which number of copies will provide the greatest expected payoff?

20-3   DECISION MAKING USING STRATEGIES

One important aspect of decision making is the use of information that might be helpful in making a choice. In establishing an employment policy based on a screening test, an applicant's score is the basis for hiring or rejecting that person. Regardless of the score achieved, a person who is hired will ultimately perform satisfactorily or not. When receiving components for assembly, manufacturers generally take a random sample to decide whether to accept or reject a shipment; the actual quality of the entire shipment will be known only after this decision has been made. As a further example, consider the choice between adding a new product to the line or abandoning it. This decision might be based on the results of a marketing research study; the success or failure of the new product will be known only after it has actually been marketed.

All of these situations are decisions with two points of uncertainty. The first uncertainty is the kind of information obtained—the screening-test score, the number of defective sample items, or the results of the marketing research study. The second uncertainty concerns the ultimate outcome—the new employee's performance, the quality of the shipment, the new product's performance. Between these points in time, a decision has to be made. The chosen act depends on which particular event has just occurred.

It is possible to determine the best acts to select for each informational event in advance. The resulting decision rule is called a *strategy*. The ultimate decision is what particular strategy to select. We will show how this is done by means of a case illustration that involves sampling and inspection.

The Cannery Inspector: A Case Illustration

A cannery inspector monitors tests for mercury-contamination levels before authorizing shipments of canned tuna. The procedure is to randomly select two crates of canned fish from a shipment and determine the parts per million of mercury. The number of these crates R exceeding government contamination guidelines is determined. The inspector may then approve (A) or disapprove (D) the shipment. If approved, the shipment is sent to distributors who perform more detailed testing to determine whether the average mercury levels of the entire shipment are excessive (*E*) or tolerable (*T*). An excessively contaminated shipment is returned to the cannery. If the company inspector originally disapproves a shipment, the production batch is sent to the rendering department to be converted into pet food. At this time, it is determined whether the entire shipment actually contains excessive average levels of mercury.

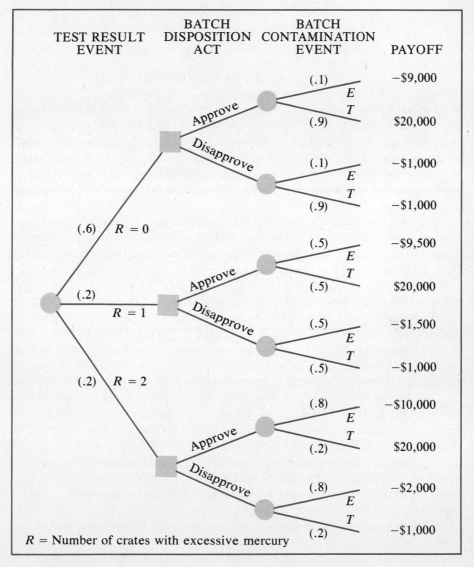

FIGURE 20-1
Decision tree diagram for the cannery inspector's problem.

TABLE 20-7

Strategies for the Cannery Inspector's Decision

Test Result Event	Strategy							
	S_1	S_2	S_3	S_4	S_5	S_6	S_7	S_8
$R = 0$	A	A	A	D	A	D	D	D
$R = 1$	A	A	D	A	D	A	D	D
$R = 2$	A	D	A	A	D	D	A	D

The decision tree diagram for the cannery inspector's decision is provided in Figure 20-1. Eight strategies, S_1 through S_8, are identified in Table 20-7. A strategy must specify which act—"approve" or "disapprove"—should be chosen for each possible test result. From Table 20-7, we can see that strategy S_1 is a decision rule specifying that the shipment must be approved no matter what the number of excessively contaminated crates R happens to be. Strategy S_2 specifies approval if $R = 0$ or $R = 1$, but disapproval if $R = 2$. Eight strategies are possible because there are 2 choices for each of the 3 events and therefore $2^3 = 8$ distinct decision rules.

As portrayed with the decision tree diagram, a strategy will be a particular pruned version of the tree. There are 8 distinct ways to prune the cannery inspector's tree, as shown in Figure 20-2. Only one of these versions will result from backward induction analysis.

A table can be constructed to indicate the payoff for each strategy-event combination. Such a payoff table, shown in Table 20-8, is identical in form to the payoff table for a single-stage decision, except that strategies are used in place of acts. Another difference is that there are uncertainties at two stages in the cannery decision: (1) how many excessively contaminated crates will be found in the sample and (2) whether the contamination level of the entire production batch will be found to be excessive or tolerable on the average. The six joint events are of the form $R = 0$ *and E, R = 2 and T.* The payoff values given in the table are the same as the payoff values on the decision tree in Figure 20-1 and correspond to the joint event that occurs for the specified strategy. Thus, if the joint event $R = 2$ *and E* occurs when S_1 is used, the payoff is -10 thousand dollars, since the inspector approves the shipment whenever $R = 2$, according to this particular strategy. If the inspector uses S_2, the same event indicates disapproval, and because the shipment contains excessive mercury, the payoff is -2 thousand dollars.

This strategy-selection decision can be analyzed by applying any of the various decision-making criteria we encountered earlier and then treating each strategy in the same way that an act is treated in a single-stage decision structure. However, we will continue to maximize expected payoff.

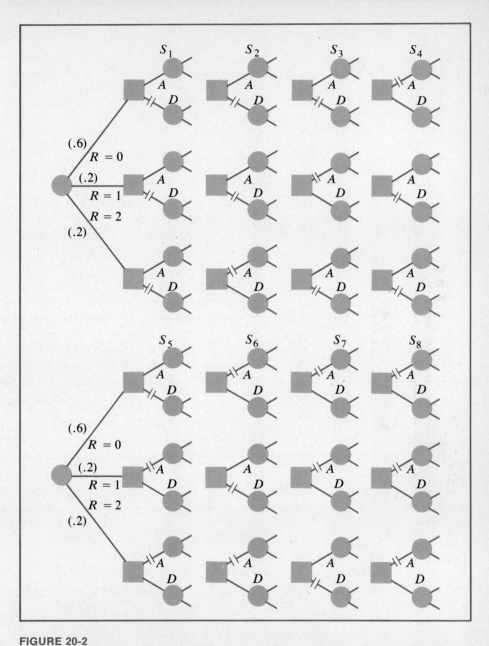

FIGURE 20-2

Pruned decision tree diagrams illustrating the eight strategies for the cannery inspector's problem given in Table 20-7.

TABLE 20-8

Payoff Table for the Cannery Inspector's Decision Using Strategies (payoffs in thousands of dollars)

Joint Event	Strategy							
	S_1	S_2	S_3	S_4	S_5	S_6	S_7	S_8
$R = 0$ and E	-9	-9	-9	-1	-9	-1	-1	-1
$R = 0$ and T	20	20	20	-1	20	-1	-1	-1
$R = 1$ and E	-9.5	-9.5	-1.5	-9.5	-1.5	-9.5	-1.5	-1.5
$R = 1$ and T	20	20	-1	20	-1	20	-1	-1
$R = 2$ and E	-10	-2	-10	-10	-2	-2	-10	-2
$R = 2$ and T	20	-1	20	20	-1	-1	20	-1

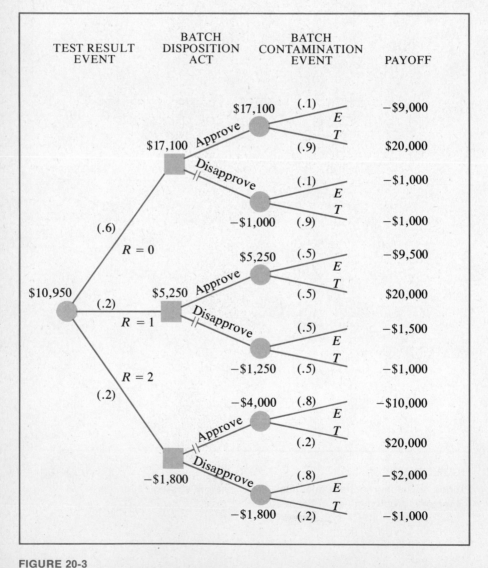

FIGURE 20-3

Extensive form analysis of the cannery inspector's problem, using the decision tree diagram.

Extensive and Normal Form Analysis

The cannery strategy-selection decision can be analyzed using the Bayes decision rule and maximizing expected payoff either by (1) backward induction on the decision tree or (2) direct computation from the values given in the payoff table to determine the strategy with the maximum expected payoff. *The two approaches will provide identical results.* When a decision tree is used, the procedure is called an *extensive form analysis.* When the analysis is based on the payoff table, it is referred to as *normal form analysis.*

Figure 20-3 illustrates extensive form analysis. The probability value determined for each event is shown on the corresponding branch of the decision tree. Backward induction indicates that the best procedure is to approve the shipment when $R = 0$ or $R = 1$ and to disapprove the shipment when $R = 2$. Referring to Table 20-7, we can see that this corresponds to strategy S_2.

The results of the normal form analysis for the cannery inspector's problem are shown in Table 20-9. Although similar calculations had to be made for each, there is room enough to illustrate only those for strategy S_2, which has the maximum expected payoff of 10.95 thousand dollars. This is the same result we obtained in extensive form analysis ($10,950). Fortunately, extensive form analysis requires that we prune the tree for just one strategy. This makes decision tree analysis superior to payoff table analysis in terms of computational efficiency. In backward induction, only the maximum expected payoffs need to be brought back to the earlier branching point. It is not even necessary to catalog the various strategies.

TABLE 20-9

Results of Normal Form Analysis for the Cannery Inspector's Decision Using Strategies, Showing the Expected Payoff Calculations for Strategy S_2 (payoffs in thousands of dollars)

(1) First-Stage Event Probability	(2) Second-Stage Event Probability	(3) Joint Probability (1) × (2)	(4) Payoff for S_2	(5) Payoff × Joint Probability (3) × (4)
$Pr[R = 0] = .6$	$Pr[E\|R = 0] = .1$	.06	−9	−.54
$Pr[R = 0] = .6$	$Pr[T\|R = 0] = .9$	.54	20	10.80
$Pr[R = 1] = .2$	$Pr[E\|R = 1] = .5$	.10	−9.5	−.95
$Pr[R = 1] = .2$	$Pr[T\|R = 1] = .5$	.10	20	2.00
$Pr[R = 2] = .2$	$Pr[E\|R = 2] = .8$	.16	−2	−.32
$Pr[R = 2] = .2$	$Pr[T\|R = 2] = .2$	.04	−1	−.04
			Expected payoff =	10.95

Strategy	S_1	S_2	S_3	S_4	S_5	S_6	S_7	S_8
Expected payoff	10.51	10.95	9.21	−.35	9.65	.09	−1.65	−1.21

Extensive form analysis using a decision tree is often the only possible approach, because the problem structure cannot be forced into the rectangular format of a payoff table. This is especially true of multistage problems that have two or more decision points, such as the Ponderosa Record Company problem diagrammed in Figure 19-4 (page 683). Only problems that result in a symmetrical decision tree like the one in Figure 20-2 can be analyzed either way in terms of expected payoff.

EXERCISES

20-5 Suppose that the following probabilities apply to the cannery illustration in the text.

Test Result	Probability	Conditional Probabilities	
$R = 0$	.4	.2 (E)	.8 (T)
$R = 1$	.3	.6 (E)	.4 (T)
$R = 2$	.3	.9 (E)	.1 (T)

(a) Conduct a new extensive form analysis (using a corrected decision tree) to determine the strategy that maximizes expected payoff.

(b) Conduct a new normal form analysis (using corrected joint probabilities) to select the strategy that maximizes expected payoff.

20-6 A cannery manager classifies each truckload of apricots purchased under contract from local orchards as underripe, ripe, or overripe. The manager must then decide whether a particular truckload will be used for dried apricots (D) or for apricot preserves (P). A truckload of apricots used for preserves yields a profit of \$6,000 if the fruit has a high sugar content, but only \$4,000 if the sugar contents is low (because costly extra sugar must be added). Regardless of sugar content, a truckload of dried apricots yields a profit of \$5,000. In either case, the actual sugar content can be determined only during final processing.

The probabilities are .3 for an underripe truckload, .5 for a ripe one, and .2 for an overripe one. The following probabilities for sugar content have been established for given levels of ripeness:

Sugar Content	Underripe	Ripe	Overripe
Low	.9	.4	.2
High	.1	.6	.8
	1.0	1.0	1.0

(a) Construct the manager's decision tree diagram, and perform an extensive form analysis to determine the maximum expected payoff strategy for disposing of a truckload of apricots.

(b) List the possible strategies for disposing of a truckload of apricots. Perform a normal form analysis to select the strategy that yields the greatest expected profit.

OPPORTUNITY LOSS AND THE 20-4
EXPECTED VALUE OF PERFECT INFORMATION

Is it worthwhile to buy information that may help us choose the best act? Information is usually not free. Resources, for example, are required to take a sample or to administer a test. In this section, we will attempt to place a value on such information. To do this, we will introduce the concept of *opportunity loss*.

Opportunity Loss

Suppose that we view each possible outcome in terms of a measure that expresses the difference between the payoff for the chosen act and the best payoff that could have been achieved. This measure is referred to as an *opportunity loss*.

DEFINITION
The opportunity loss for an outcome is the amount of payoff that is forgone by not selecting the act that has the greatest payoff for the event that actually occurs.

Table 20-10 shows how the opportunity losses are obtained for the payoffs for the toy manufacturer's doll-movement decision. To calculate the

TABLE 20-10

Determining the Opportunity Losses for the Tippi-Toes Decision

Event (Level of Demand)	Payoff			Row Maximum
	Gears and Levers	Spring Action	Weights and Pulleys	
Light	$ 25,000	−$ 10,000	−$125,000	$ 25,000
Moderate	400,000	440,000	400,000	440,000
Heavy	650,000	740,000	750,000	750,000

Row maximum − Payoff = Opportunity loss
(thousands of dollars)

Light	25 − 25 = 0	25 − (−10) = 35	25 − (−125) = 150
Moderate	440 − 400 = 40	440 − 440 = 0	440 − 400 = 40
Heavy	750 − 650 = 100	750 − 740 = 10	750 − 750 = 0

TABLE 20-11
Opportunity Loss Table for the Tippi-Toes Decision

Event (Level of Demand)	Act (Choice of Movement)		
	Gears and Levers	Spring Action	Weights and Pulleys
Light	$ 0	$35,000	$150,000
Moderate	40,000	0	40,000
Heavy	100,000	10,000	0

opportunity losses, the maximum payoff for each row is determined. Each payoff is then subtracted from its respective row maximum.

The *opportunity loss table* for the Tippi-Toes decision appears in Table 20-11. All opportunity loss values are non-negative, since they measure how much worse off the decision maker is made by choosing some act other than the best act for the event that occurs. Let us consider the meaning of the opportunity loss values. For example, suppose that the gears-and-levers movement is chosen and that a light demand occurs. The opportunity loss is zero, because we can see from Table 20-10 that no better payoff than $25,000 (the row maximum) could have been achieved if another act had been chosen. But if the gears-and-levers movement is chosen and a heavy demand occurs, the opportunity loss is $100,000, because the weights-and-pulleys movement has the greatest payoff for a heavy demand ($750,000). Since the gears-and-levers movement has a payoff of only $650,000, the payoff difference $750,000 − $650,000 = $100,000 represents the additional payoff forgone by not selecting the act with the greatest payoff. It should be emphasized that the $100,000 opportunity loss is not a loss in the accounting sense, because a net positive contribution of $650,000 to profits is obtained. Instead, the opportunity to achieve an additional $100,000 has been missed. We might say that should demand prove to be heavy, the decision maker would have $100,000 worth of *regret* by not choosing weights and pulleys instead of gears and levers.

The Bayes Decision Rule and Opportunity Loss

We can calculate the expected opportunity loss for each act and then select the act that has the minimum loss. This is done in Table 20-12 for the Tippi-Toes decision. The minimum expected opportunity loss is $5,500 for the spring-action movement, which is the *minimum expected opportunity loss act* in this example.

In Chapter 19, we saw that the spring-action movement was also the maximum expected payoff act and was therefore the best choice according to

TABLE 20-12

Calculation of Expected Opportunity Losses for the Tippi-Toes Decision

Event (Level of Demand)	Probability	Gears and Levers		Spring Action		Weights and Pulleys	
		Loss	Loss × Probability	Loss	Loss × Probability	Loss	Loss × Probability
Light	.10	$ 0	$ 0	$35,000	$3,500	$150,000	$15,000
Moderate	.70	40,000	28,000	0	0	40,000	28,000
Heavy	.20	100,000	20,000	10,000	2,000	0	0
Expected opportunity loss			$48,000		$5,500		$43,000

the Bayes decision rule. Our new criterion leads us to the same choice. It can be mathematically established that this will always be so. Since either criterion will always lead to the same choice, we can say that the *Bayes decision rule is to select the act that has the maximum expected payoff or the minimum expected opportunity loss.*

The Expected Value of Perfect Information

Up to this point, our toy manufacturer has selected an act without the benefit of any information except that acquired through experience with other toys. But it is possible to secure better information about next season's demand by test marketing, by taking opinion and attitude surveys, or by obtaining inside information concerning competitors' plans. How much should the decision maker be willing to pay for additional information?

It is helpful to know the payoff that can be expected from securing improved information about the events. We will consider the extreme case when the decision maker can acquire *perfect information.* Using this information, the decision maker can guarantee the selection of the act that yields the greatest payoff for whatever event actually occurs. Because we wish to investigate the worth of such information *before* it is obtained, we will determine the *expected payoff with perfect information.*

To calculate the expected payoff with perfect information, we determine the highest payoff for each event. This is illustrated for the Tippi-Toes decision in Table 20-13. The maximum payoff for each demand level is determined by finding the largest payoff in each row. Thus, for a light demand, we find that choosing the gears-and-levers movement yields the largest payoff ($25,000). If perfect information indicated that light demand was certain to occur, our

TABLE 20-13

Calculation of the Expected Payoff with Perfect Information for the Tippi-Toes Decision

Event (Level of Demand)	Prob- ability	Act			With Perfect Information		
		Gears and Levers	Spring Action	Weights and Pulleys	Maximum Payoff	Chosen Act	Payoff × Probability
Light	.10	$ 25,000	−$ 10,000	−$125,000	$ 25,000	G&L	$ 2,500
Moderate	.70	400,000	440,000	400,000	440,000	SA	308,000
Heavy	.20	650,000	740,000	750,000	750,000	W&P	150,000

Expected payoff with perfect information = $460,500

decision maker would choose this movement. Similarly, $440,000 is the maximum payoff possible for moderate demand, and this amount can be achieved only if the spring-action movement is chosen. Likewise, $750,000 is the maximum possible payoff when a heavy demand occurs, and this amount corresponds to a choice of the weights-and-pulleys movement. The last column of Table 20-13 shows the products of the maximum payoffs and their respective event probabilities. Summing these, we obtain $460,500 as the expected payoff with perfect information. This figure represents the average payoff if the toy manufacturer were faced with the same situation repeatedly and always selected the act that yielded the best payoff for the event indicated by the perfect information. Keep in mind that the $460,500 represents the expected payoff viewed from some point in time *before* the information becomes available. *After* the information has been obtained, exactly one of the payoffs, $25,000, $440,000, or $750,000, is bound to occur. When the information is actually obtained, the payoff is a certainty.

We can now answer the question regarding the worth of perfect information to the decision maker. As we have seen, the Bayes decision rule leads to the choice of the particular act that maximizes the expected payoff without regard to any additional information. Since this is the best act that our decision maker can select without any new information, and since the expected payoff with perfect information is the average payoff that can be anticipated with the best possible information, the worth of perfect information to the decision maker is expressed by the difference between these two amounts. We call the resulting number the *expected value of perfect information*, which is conveniently represented by the abbreviation EVPI.

EXPECTED VALUE OF PERFECT INFORMATION

EVPI = Expected payoff with perfect information
    − Maximum expected payoff (with no information)

For the toy manufacturer's decision, we obtain the EVPI by subtracting the maximum expected payoff of $455,000 (calculated in Table 19-6 on page 675) from the expected payoff with perfect information of $460,500:

$$\text{EVPI} = \$460,500 - \$455,000 = \$5,500$$

In this case, the EVPI represents the greatest amount of money that the decision maker would be willing to pay to obtain perfect information about what the demand will be. Stated differently, $5,500 is the increase in the decision maker's expected payoff that can be attributed to perfect knowledge of demand. Both $455,000 and $460,500 are meaningless values *after* the perfect information is obtained. Thus, the EVPI of $5,500 can be interpreted only *before* the perfect information has become known.

EVPI and Opportunity Loss

Note that $5,500 is the same amount as the minimum expected opportunity loss calculated in Table 20-12. Thus, we can see that *the expected value of perfect information is equal to the expected opportunity loss for the optimal act.*

Therefore, we can calculate the expected value of perfect information by calculating the expected opportunity losses. The minimum loss is then the EVPI. Table 20-14 summarizes the relationships between expected payoff, expected opportunity loss, and expected value of perfect information for the toy manufacturer's decision. Note that for any act, the sum of the expected payoff and the expected opportunity loss is equal to the expected payoff with perfect information.

Since perfect information is nonexistent in most real-world decision making, why are we interested in the EVPI? Our answer is that it helps to

TABLE 20-14

Relationships Between Expected Payoff, Expected Opportunity Loss, and EVPI for the Tippi-Toes Decision

	Gears and Levers	Spring Action	Weights and Pulleys
Expected payoff	$412,00	$455,000	$417,500
Expected opportunity loss	48,000	5,500	43,000
Expected payoff with perfect information	$460,500	$460,500	$460,500

Expected value of perfect information (EVPI) = $5,500   ↑—Optimal act establish a limit on the worth of less-than-perfect information. For example, if a marketing research study aimed at predicting demand costs $6,000, which exceeds the EVPI by $500, the study should not be conducted, regardless of its quality. We will investigate the concepts involved in decision making with experimental information further in Chapter 23.

EXERCISES

20-7 Use the payoff table below to construct an opportunity loss table.

		Act				
Event	Probability	A_1	A_2	A_3	A_4	A_5
E_1	.2	10	20	10	15	20
E_2	.2	−5	10	−5	10	−5
E_3	.6	15	5	10	10	10

Compute the expected opportunity loss for each act. Which act yields the lowest expected opportunity loss?

20-8 Answer the following questions based on the payoff table below.

		Act		
Event	Probability	A_1	A_2	A_3
E_1	.3	10	20	30
E_2	.5	40	−10	20
E_3	.2	20	50	20

(a) What is the maximum expected payoff? To which act does this payoff correspond?

(b) What is the expected payoff with perfect information?

(c) Use your answers from (a) and (b) to calculate the expected value of perfect information.

(d) What is the minimum expected opportunity loss?

(e) What do you notice about your answers to (c) and (d)?

20-9 B.F. Retread, a tire manufacturer, wishes to select one of three feasible prototype designs for a new longer-wearing radial tire. The costs of making the tires are given below:

Tire	Fixed Cost	Variable Cost per Unit
A	$ 60,000	$30
B	90,000	20
C	120,000	15

There are three levels of unit sales: 4,000 units, 7,000 units, and 10,000 units; the respective probabilities are .30, .50, and .20. The selling price will be $75 per tire.
(a) Construct the payoff table using total profit as the payoff measure.
(b) Determine the expected payoff for each act. According to the Bayes decision rule, which is the best act?
(c) Calculate the EVPI.
(d) Complete the opportunity loss table and compute the expected opportunity losses.

20-10 An oil wildcatter must decide whether to drill on a candidate drilling site. His judgment leads him to conclude that there is a 50–50 chance of oil. If the wildcatter drills and strikes oil, his profit will be $200,000. But if the well turns out to be dry, his net loss will be $100,000.
(a) According to the Bayes decision rule, should the wildcatter drill or abandon the site?
(b) What is the wildcatter's EVPI?
(c) A seismologist offers to conduct a highly reliable seismic survey. The results could help the wildcatter make his decision. What is the most that the wildcatter would consider paying for such seismic information?

REVIEW EXERCISES

20-11 Refer to the Tippi-Toes decision in Table 20-1. Suppose that reduction in the development cost of the gears-and-levers movement allows a $50,000 increase in all payoffs for that act. In addition, an upward revision of unit sales when demand is heavy results in a further $100,000 increase in all payoffs for that event. Finally, the following revised probabilities now apply to the demand levels: light, .50; moderate, .30; and heavy, .20.
(a) Construct the new Tippi-Toes payoff table.
(b) Determine the maximin payoff act.
(c) Determine the maximum likelihood act.
(d) Calculate the expected payoffs. According to the Bayes decision rule, which act should be chosen?

20-12 (*Continuation of Exercise* 20-11):
(a) Calculate the EVPI for the modified Tippi-Toes decision.
(b) Construct the opportunity loss table and compute the expected opportunity losses.

20-13 A product manager for a soap manufacturer wishes to determine whether or not to market a new toothpaste. The present value of all future profits from a successful toothpaste is $1,000,000, whereas failure of the brand would result in a net loss of $500,000. Not marketing the toothpaste would not affect profits. The manager has judged that the toothpaste would have a 50–50 chance of success.
(a) Construct the payoff table for this decision.
(b) Which act will maximize the expected payoff?
(c) Compute the decision maker's EVPI. What is the minimum expected opportunity loss?

20-14 Suppose that the manager in Exercise 20-13 wishes to implement a consumer testing program at a cost of $50,000. Consumer testing will be either favorable (a 40% chance) or unfavorable. Given a favorable test result, the chance of product success is judged to be 80%. For an unfavorable test result, the toothpaste's success probability is judged to be only 30%.

(a) Assuming that testing is used, construct a decision tree diagram. Then perform backward induction analysis to determine the optimal strategy to employ in using the test results.

(b) Identify the basic strategies involving the use of the results of the consumer testing program. Construct a payoff table with these strategies as the choices and the joint market outcomes and test results as the events. Then conduct a normal form analysis to determine which strategy maximizes the expected payoff.

Suppose that you are allowed to participate in a far-fetched lottery where the price is $100 if you win and death if you lose. Is there any probability for winning for which you would be indifferent between the status quo and playing?

Utility: Accounting for Attitude

T he goal of this chapter is to broaden the scope of decision theory through the introduction of a new payoff measure. We have seen that a good payoff measure should rank all possible outcomes in terms of how well they meet the decision maker's goals. This is often an easy task when there is no uncertainty. But the presence of uncertainty can severely complicate the issue when the possible outcomes to a decision are extreme. Such decisions contain elements of *risk*. Because people usually have different attitudes toward risk, two persons faced with an identical decision may actually prefer different courses of action.

The crucial role that attitude plays in any decision is illustrated by the divergent behavior of different persons faced with the same decision. The *umbrella situation* nicely demonstrates this point. *How can we explain why everyone does not carry an umbrella when we do*? To a certain extent, we can say that all individuals are not equally adept at selecting and exercising appropriate decision criteria. But this is only one possible explanation. With much justification we can conclude that the difference in behavior can also be explained by differing attitudes toward the consequences. Some people may enjoy getting wet, but others may view it as an invitation to pneumonia and possibly the first step toward a premature grave. Some people think it is chic to carry rain paraphernalia when it's not raining; others would rather lug around a ball and chain. Even if we can find two persons who have identical attitudes toward the decision consequences, they may still make opposite decisions because they may not have made identical *judgments* regarding the chance of rain. One person may rely on weather service prediction as a source of information, judging these subjective probabilities to be adequate. Another person may depend on lumbago pain as a fairly reliable measure of the probability for rain. (The role of judgment in establishing probabilities will be discussed further in Chapter 22.)

In this chapter, we will discuss utility as an alternative expression of payoff that reflects a person's attitudes. We will begin by examining the rationale for buying insurance. A brief historical discussion of utility and the underlying assumptions of a theory of utility will then be presented. Finally, a procedure will be introduced that can be used to determine utility values. The utility function so obtained provides a basis for our discussion of some basic attitudes toward risk.

21-1   ATTITUDES, PREFERENCES, AND UTILITY

In Chapter 20, we examined several procedures and criteria that help decision makers to make choices in the presence of uncertainty. In all cases, the payoff value of each outcome is required to analyze the decision. As we have seen, not all outcomes have an obvious numerical payoff. In this section, we will see how payoffs may be determined in such cases. Later in the chapter, we will develop methods of quantifying such consequences as reduced share of the market, loss of corporate control, and antitrust suits. Even when numerical payoffs can be naturally determined, we have seen that it may be unrealistic to select the act with the maximum expected payoff. In some cases, an extremely risky act fares better under the Bayes decision rule than an obviously preferred act. As noted in Chapter 20, this difficulty is not the fault of the Bayes criterion but is caused by payoff values that do not reflect their true worth to the decision maker.

The Decision to Buy Insurance

The inadequacy of using such obvious measures as dollar cost or profit to indicate payoffs can be vividly illustrated by evaluating an individual's decision of whether or not to buy fire insurance. Spiro Pyrophobis wishes to decide whether to buy a fire insurance policy for his home. Our decision maker's payoffs will be expressed in terms of his out-of-pocket costs, which we will represent by negative numbers. Our question is: Will the Bayes decision rule lead to the choice of the act that is actually preferred?

In answering this question, we will use the hypothetical payoff table provided in Table 21-1. Here, we have greatly simplified the decision. The acts

TABLE 21-1

Payoff Table for the Decision to Buy Fire Insurance

		Act			
		Buy Insurance		Don't Buy Insurance	
Event	Probability	Payoff	Payoff × Probability	Payoff	Payoff × Probability
Fire	.002	−$100	−$ .20	−$40,000	−$80.00
No fire	.998	−100	−99.80	0	0
Expected payoff			−$100.00		−$80.00

are to buy or not to buy an annual policy with a $100 premium charge. If there is a fire, we will assume that Spiro's home and all its contents, valued at $40,000, will be completely destroyed.

Insurance actuaries have established that historically 2 out of every 1,000 homes in the category of Spiro's home burn down each year. The probability that Spiro's home will burn down is therefore set at $2/1,000 = .002$. Thus, the complementary event—no fire—has a probability of $1 - .002 = .998$. We can use these probability values to calculate the expected payoffs for each act in Table 21-1. The maximum expected payoff is $- \$80$, which corresponds to the act "don't buy insurance" and is larger than the $- \$100$ payoff from buying fire insurance.

In this example, the *Bayes decision rule indicates that it is optimal to buy no insurance.* Insurance policy premiums are higher than the expected claim size, which is equivalent to the policyholder's expected dollar loss, so that the insurance company can pay wages and achieve profits. Thus, buying insurance can be considered an unfair gamble, where the payoff is not in the buyer's favor. Individuals can expect to pay more in insurance premiums than they will collect in claims,* and they feel fortunate if they never have to file a claim. Yet most persons faced with this decision choose to buy fire insurance. Loss of a home, which comprises the major portion of a lifetime's savings for many people, is a dreadful prospect. The expenditure of an annual premium, although not exactly appealing, buys a feeling of security that seems to outweigh the difference between the expected payoffs.

The Bayes decision rule selects the *less preferred act.* Does this mean that it is an invalid criterion? Rather than answer no immediately, let us consider the payoffs used. The true worth of the outcomes is not reflected by the dollar payoffs. A policyholder is willing to pay more than the expected dollar loss to achieve "peace of mind." We can say that the policyholder derives greater *utility* from having insurance. If dollar losses are valued on a scale of true worth

* This is not true of life insurance, which is ordinarily a form of savings.

or utility, then each additional dollar loss will make our decision maker feel disproportionately worse off. Thus, a 10% reduction in wealth may be more than twice as bad as a 5% reduction. The same is usually true for gains in dollar wealth; the second increase may not increase the decision maker's sense of well-being as much as the first. In the parlance of economics, *the policyholder's marginal utility for money is decreasing.* Each successive dollar gain buys a smaller increase in utility; each additional dollar loss reduces utility by a greater amount than before.

Thus, we may question the validity of using dollars as our payoff measure. Instead, it might be preferable to measure the payoff of an outcome in terms of its worth or utility.

21-2   NUMERICAL UTILITY VALUES

We wish to obtain *numerical utility values* that express the true worth of the payoffs that correspond to decision outcomes. We refer to such numbers as *utilities.* Much investigation has been made of the true worth of monetary payoffs. The early eighteenth-century mathematician Daniel Bernoulli—a pioneer in developing a measure of utility—proposed that *the true worth of an individual's wealth is the logarithm of the amount of money possessed.* Thus, a graphical relationship between utility and money would have the basic shape of the curve in Figure 21-1. Note that although the slope of this curve is always positive, it decreases as the amount of money increases, reflecting the assumption of decreasing marginal utility for money.

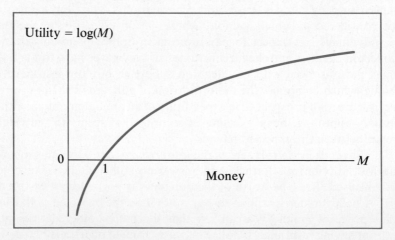

FIGURE 21-1
Bernoulli's utility function for money.

The Saint Petersburg Paradox

A gambling game called the *Saint Petersburg Paradox* led Bernoulli to his conclusion. In the game, a balanced coin is fairly tossed until the first head appears. The gambler's winnings are based on the number of tosses that are made before the game ends. If a head appears on the first toss, the player wins $2. If not, the "kitty" is doubled to $4—the reward if a head appears on the second toss. If a tail occurs on the second toss, the kitty is doubled again. The pot is doubled after every coin toss that results in a tail. The winnings are $2 raised to the power of the number of tosses until and including the first head. This procedure will be more interesting if you pause right now to think about what amount you would be willing to pay for the privilege of playing this game.

The probability that $n + 1$ tosses will occur before payment is the probability that there is a run of n tails and that the $(n + 1)$st toss is a head, or $(1/2)^{n+1}$. The payoff for $n + 1$ tosses is 2^{n+1}. We can therefore calculate the player's expected receipts from the sum

$$\$2(1/2) + \$2^2(1/2)^2 + \$2^3(1/2)^3 + \cdots = \$1 + \$1 + \$1 + \cdots = \$\infty$$

Since the number of $1s in this sum is unlimited, the *expected receipts from a play of this game are infinite*! Whatever amount you were willing to pay to play must have been a finite amount and therefore less than the expected receipts. Thus, the expected payoff for this gamble is also infinite, no matter what price is paid to play.

Few people are willing to pay more than $10 to play this game, and even at this price, a player would win only 1 out of 8 games on the average. A player paying $500 would show a profit in only 1 out of every 256 gambles on the average. The natural reticence of players to pay very much for this gamble led Bernoulli to his conclusion about the utility for money. In general, we say that a person who prefers not to participate in a gamble in which the expected receipts exceed the price to play has a *decreasing marginal utility for money*.

The Validity of Logarithmic Values

Via different paths of reasoning, other early mathematicians arrived at conclusions similar to Bernoulli's—that the marginal utility for money is decreasing—and proposed other utility curves with the same basic shape. A major fault of these early works is that they do not account for individual differences in the assignment of worth. A more modern treatment of utility in the abstract sense was advanced by John Von Neumann and Oskar Morgenstern in 1947 in their book *Theory of Games and Economic Behavior*. There, they proposed that a utility curve can be tailored for any individual, provided certain assumptions about the individual's preferences hold. These assumptions provide several valid, basic shapes for the utility curve, including curves similar to Bernoulli's. We will investigate some of these utility curves later in the chapter.

Outcomes Without A Natural Payoff Measure

Until now, the outcomes of our examples have had a *natural* numerical payoff measure, such as dollar profits, gallons of gasoline conserved, or time saved. But we have noted that some decisions have no numerical outcomes. As decision makers, we should be able to assess the relative worth of such an outcome.

In the case of most decisions, it is possible to determine preferences, although this is not always an easy task. Indeed, value judgments may be the most difficult step in analyzing a decision. Consider the student selecting a school from several top universities, the child choosing a candy bar, the single person contemplating getting married and forgoing the carefree life, the tired corporate founder pondering merger and retirement versus retaining control and delegating operating responsibility, or the innocent person choosing between pleading guilty to manslaughter or facing trial for murder. If we assume that we have the capability of ranking the consequences in order of preference, we can extend the notion of utility so that numerical payoffs can be made for the most intangible outcomes.

EXERCISES

21-1 Suppose that you are offered a gamble by Ms. I.M. Honest, a representative of a foundation studying human behavior. A fair coin is to be tossed. If a head occurs, you will receive $10,000 from Ms. Honest. But if a tail results, you must pay her foundation $5,000. If you do *not* have $5,000, a loan will be arranged, which must be repaid over a five-year period at $150 per month but can be deferred until you have graduated from school.
   (a) Calculate your expected profit from participating in this gamble.
   (b) Would you be willing to accept Ms. Honest's offer? Does your answer indicate that your marginal utility for money is decreasing?

21-2 Mr. Smith has offered you a gamble similar to that of Ms. Honest in Exercise 21-1. If a head occurs, he will hand you $1.00. But if a tail results you must pay Mr. Smith $.50.
   (a) Calculate your expected profit from participating in this gamble.
   (b) Would you be willing to accept Mr. Smith's offer?

21-3 A homeowner whose house is valued at $40,000 is offered tornado insurance at an annual premium of $500. Suppose that there are just two mutually exclusive outcomes—complete damage or no damage from a tornado—and that the probability for damage from a tornado is .0001.

(a) Construct the homeowner's payoff table for the decision of whether or not to buy tornado insurance.

(b) Calculate the decision maker's expected monetary payoff for each act. Which act has the maximum expected payoff?

(c) Suppose that the homeowner decides not to buy tornado insurance. Does this contradict the decreasing marginal utility for money? Explain.

THE ASSUMPTIONS OF UTILITY THEORY   21-3

The fundamental proposition of the modern treatment of utility is that it is possible to obtain a numerical expression for an individual's preferences. We can rank a set of outcomes by preference and then assign utility values that convey these preferences. The largest utility number is assigned to the most preferred outcome, the next largest number is assigned to the second most preferred outcome, and so forth. Suppose, for instance, that you are contemplating a menu. If you prefer New York steak to baked halibut and you wish to assign utility values to the entrees in accordance with your preferences, the utility for steak might be set at 5, so that u(steak) = 5, and u(halibut) will then be some number smaller than 5.

Before we describe how specific utility numbers can be obtained, we will discuss some of the assumptions underlying the theory of utility. Various assumptions have been made about the determination of utilities.* All of them have one feature in common—that the values obtained pertain only to a *single individual* who behaves *consistently* in accordance with his or her own tastes.

Preference Ranking

The first assumption of utility theory is that a person can determine for any pair of outcomes O_1 and O_2 whether he or she prefers O_1 to O_2, prefers O_2 to O_1, or regards both equally. This assumption is particularly advantageous when we consider monetary values, because then we can assume that more money is always better than less. But we have seen that it can be very difficult to rank preferences when qualitative alternatives are considered. Can a person always determine a preference for or establish an indifference toward outcomes? If not, then utilities cannot be found for these outcomes.

* Those discussed in this book are simplifications of the original axioms postulated by Von Neumann and Morgenstern.

Transitivity of Preference

The second assumption of utility theory is that if A is preferred to B and B is preferred to C, then A must be preferred to C. This property is called *transitivity of preference* and reflects an individual's consistency. Again, when we are dealing with monetary outcomes, we can usually assume transitivity.

The Assumption of Continuity

The third assumption of utility theory is that of *continuity*, which tells us that the individual considers some *gamble* having the best and worst outcomes as rewards to be equally preferable to some middle or in-between outcome. To illustrate continuity, we will consider the following example.

Homer Briant owns a small hardware store in a deteriorating neighborhood and is contemplating a move. Because Homer is still young and has no special skills, he will not consider leaving the hardware business. A move cannot be guaranteed to be successful, since relocating will involve the maximum extension of his credit and there will be no time for a gradual buildup of business. Therefore, moving will either improve Homer's present business or be disastrous. Thus, Homer is faced with one of the following outcomes:

> Most preferred O_3: increasing sales (if move is a success)
> O_2: decreasing sales (if Homer stays)
> Least preferred O_1: imminent bankruptcy (if move is a failure)

Whether a move will be a success depends largely on luck or chance. Our assumption of continuity presumes that there is some probability value for a successful move that will make Homer indifferent between staying and moving. Figure 21-2 presents the decision tree diagram for this decision. The fork at node b represents a gamble between O_3 and O_1 resulting from the act "move." Continuity may be justified by observing that if the value of Pr[success] is close to 1, so that a move will almost certainly be a success, Homer will prefer the gamble of moving to staying. But if Pr[success] is close to 0, making bankruptcy a near certainty, Homer will prefer to stay in his present location. Thus, there must be a success probability somewhere between 0 and 1 beyond which Homer's preference will pass from O_2 to the gamble. This value for Pr[success] makes the gamble as equally attractive as O_2.

Continuity is a crucial assumption of utility theory, but it may be hard to accept, especially if the outcomes include the ultimate one—death. Suppose that you are allowed to participate in a lottery that offers you $100 if you win and death if you lose. Is there any probability for winning that would make you indifferent between the status quo and playing? A natural response is that this is not a very meaningful gamble, so we will recast the situation. Suppose that you are informed by a reliable source that you can drive your car one mile down the road and someone will be passing out $100 bills, one to a person.

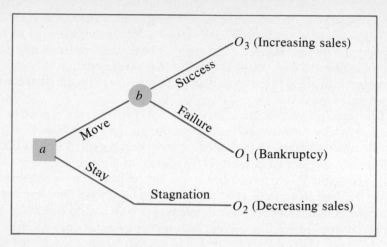

FIGURE 21-2
Homer's decision tree diagram for possible business relocation.

There are no gimmicks, and you will not be inconvenienced by a mob of people. Would you go? If your answer yes, then consider your chances of getting killed in an automobile accident on your journey. For the past several years, approximately 50,000 persons have been killed in such accidents in the United States annually. So although it is quite small, the probability that your rather untimely death will occur while you are collecting your $100 is not zero. Going to get your $100 is a gamble having death as a possible outcome, and you prefer the gamble to the status quo. Now suppose that we increase the chance of death. To reach your benefactor, you must cross a condemned bridge. Would you still go? Probably not, because the chance of death would be significantly higher. Somewhere in between these two extremes lies a probability for safely getting your $100 and a complementary probability for death that would make you indifferent between the status quo and the gamble.

The Assumption of Substitutability

A fourth assumption of utility theory allows us to revise a gamble by *substituting* one outcome for another outcome that is equally well regarded. The premise is that the individual will be indifferent between the original and the revised gambles. The substitutability assumption can be illustrated by means of an example.

A husband and wife cannot agree on how to spend Saturday night. In desperation, they decide to gamble by tossing a coin to determine the kind of entertainment they will select. If a head occurs, they will spend the evening at the opera (her preference), and if a tail occurs, they will go to a basketball game. Suppose that the wife changes her mind and wants to go to a dance instead. The husband dislikes dancing just as much as the opera, so he would be indifferent between tossing for the opera or basketball and a revised gamble between dancing and basketball. This will hold regardless of the odds, providing that the chance of going to the basketball game remains the same for the original and the revised gambles.

The principle of substitutability also holds if we treat a gamble as an outcome. For any outcome, we can substitute an *equivalent gamble* with two other outcomes as rewards that is equally as well regarded as the outcome that the gamble replaces. For example, suppose that the wife insists on a movie instead. She wants to see a romance story, but he feels that as compensation for being dragged to a movie, they should see an adventure film. Suppose that the husband is indifferent between an opera or a coin toss to determine which of the two movies to see. The second coin toss is an equivalent gamble to the opera outcome. Thus, the husband should be indifferent between the single- and the two-stage gambles in Figure 21-3.

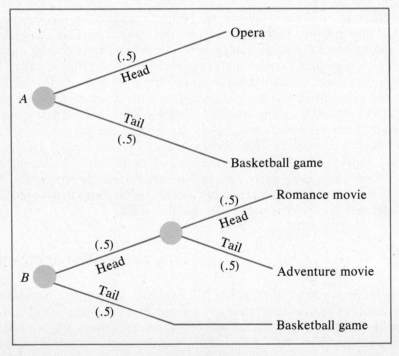

FIGURE 21-3

An illustration of the assumption of substitutability. The single-stage gamble at *A* and the two-stage gamble at *B* are equally well regarded.

Increasing Preference

The final assumption of utility theory concerns any pair of gambles with identical outcomes. The gamble that has the greater probability for the more desirable outcome must be preferred. Thus, the preference for gambles between the same two outcomes *increases* as the probability for attaining the better outcome increases. That this is plausible should be apparent. Suppose that when a coin is tossed, you are paid $100 if a head occurs and nothing if a tail occurs. The probability for winning $100 is 1/2. It should be obvious that this gamble would be decidedly inferior to a gamble with the same outcomes and a probability for winning greater than 1/2.

ASSIGNING UTILITY VALUES   21-4

Utility numbers must be assigned to outcomes in such a way that the outcome for which a person has a greater preference receives the greater value. The resulting values gauge that person's relative preferences. Any numbers satisfying these requirements will be suitable as utilities, and their absolute magnitudes may be arbitrarily set.

To lay the foundation for the methodology used in finding utility values, we again view the uncertain situation as a gamble.

Gambles and Expected Utility

We are presently concerned with making choices under uncertainty. Thus, the payoffs for decison acts are unknown, and each act can be viewed as a gamble with uncertain rewards. To evaluate such decisions, we must extend the concept of utility to gambles.

Recall that the Bayes decision rule involves comparisons between the expected payoffs of acts or strategies, so that the "optimal" choice has the maximum expected payoff. But the major difficulty with this criterion, as we have seen, is that the indicated course of action can be less attractive than some other action. For example, the Bayes decision rule tells us not to buy fire insurance when most people feel that insurance is desirable. We wish to overcome this obstacle by using utilities in place of dollar payoffs. We therefore require that the expected utility payoffs provide a valid means of comparing actions so that the action having the greatest expected utility is actually preferred to the alternative actions. Thus, buying fire insurance should have greater expected utility than not buying it.

But we can go one step further. Suppose that the most preferred action has the greatest expected utility, the next most preferred action has the next

(1) All outcomes are ranked. A convenient designation is to let a subscript denote the order of preference:

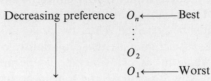

(2) Utilities for best and worst outcomes are arbitrarily assigned:

$u(O_n)$
$\vdots$ Utilities for intermediate outcomes are to be found.
$u(O_2)$
$u(O_1)$

(3) A reference lottery is formulated. This is a gamble having rewards O_n if won and O_1 if lost. The probability q for winning the reference lottery is treated as a variable, which can be changed at the will of the decision maker. The reference lottery is strictly hypothetical.

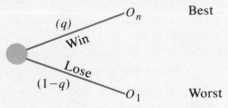

(4) For each intermediate outcome, the decision maker establishes the value of q that serves as a point of indifference between the outcome itself and the reference

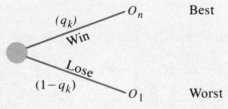

lottery. Thus, for intermediate outcome O_k, a win probability q_k is determined that results in a reference lottery that is equally as well regarded as O_k.

(5) The utility of O_k can now be determined. It is equal to the expected utility for the reference lottery with a win probability of q_k:

$$u(O_k) = q_k u(O_n) + (1 - q_k)u(O_1)$$

FIGURE 21-4

The procedure for assigning values to a set of outcomes.

greatest utility, and so forth. Then expected utility would express preference ranking, and the *expected utility values would themselves be utilities*. Each utility value would express the worth of a *gamble* between outcomes obtained by averaging the utility values of the outcomes, using their respective probabilities as weights. This may be stated more precisely as a property of utility theory:

In any gamble between outcome A and outcome B, with probabilities of q for A and $1 - q$ for B

$$u(\text{gamble}) = qu(A) + (1 - q)u(B)$$

Thus, the utility for a gamble between two outcomes is equal to the expected utility for the gamble. When acts having uncertain outcomes are viewed as gambles, the utility for an act is equal to the expected utility for its outcomes. *When payoffs are measured in terms of utilities, the Bayes decision rule will indicate that the act having the maximum expected utility is optimal,* so that this criterion can always be used to select the most preferred act or strategy.

We are now ready to assign utility values to outcomes. Figure 21-4 outlines this procedure. The numbers are obtained from a series of gambles between a pair of outcomes.

The Reference Lottery

The process begins with a preference ranking of all the outcomes to be considered. The most preferred and the least preferred outcomes are determined, and a gamble between these outcomes establishes the individual's utilities. We call this a *reference lottery.* It has two events: "win," which corresponds to achieving the best outcome, and "lose," which corresponds to attaining the worst outcome. Such a gamble is purely *hypothetical* and only provides a framework for assessing utility. The events "win" and "lose" do not relate to any events in the actual decision structure and are used to divorce the reference lottery from actual similar gambles. *The probability for winning the hypothetical reference lottery is a variable*, denoted by q, which changes according to the attitudes of the decision maker.

The initial assignment of utility values to the best and worst outcomes is *completely arbitrary.* It does not matter what values are chosen; assigning different values to these arbitrary utilities will result in different utility scales. This is similar to temperature measurement, in which two different and quite arbitrary values are used to define the Fahrenheit and Celsius scales. The choices of 32° Fahrenheit and 0° Celsius for the freezing point of water, and of 212°F and 100°C for its boiling point, result in quite different values on these two scales for any particular temperature.

Obtaining Utility Values

Once the extreme utility values are determined, the decision maker can use the reference lottery to obtain utilities for the intermediate outcomes. This is accomplished by varying the "win" probability q until the decision maker establishes a value of q that serves as a *point of indifference* between achieving that outcome for certain and letting the reward be determined by the reference lottery. That particular value of q makes the reference lottery a gamble that is equivalent to the intermediate outcome so evaluated. We have seen that the assumption of continuity makes this possible.

Again, we can add meaning to this procedure by considering its similarities to temperature measurement.

The decision maker's subjective evaluation is analogous to designing a thermometer. A thermometer is designed by determining a core diameter that will permit a substance such as mercury to rise to various levels within its tubular cavity. For each level of heat, there is a corresponding height to which the mercury must rise. On the Celsius scale, the 100° mark corresponds to the mercury's height when the thermometer is placed in boiling water. Various levels of heat between the freezing and boiling points of water correspond to marks at prescribed heights above the zero mark, allowing heat to be measured in relative degrees. Similarly, the values of q established to make the decision maker indifferent between respective intermediate outcomes and the reference lottery serve to measure his or her relative preferences. The indifference values of q are like the markings on a thermometer, and the different outcome preferences are analogous to different levels of heat. These values of q are established through introspection and have no more to do with the actual chance of winning than the design of a thermometer is related to tomorrow's temperature.

Once an indifference value of q has been established for an outcome, its utility value can be determined by calculating the expected value of the reference lottery using that value of q. Letting O_1 and O_n represent the least and the most preferred outcomes, we can then find the utility for an outcome O_k of intermediate preference from

$$u(O_k) = q_k u(O_n) + (1 - q_k)u(O_1)$$

Here, q_k is the value of q that makes the decision maker *indifferent* between the certain achievement of O_k and taking a chance with the reference lottery. The utility value $u(O_k)$ is analogous to a numerical degree value beside a marking on a thermometer.

To illustrate, we will continue with Homer Briant's contemplated business relocation. Homer has ranked his preferences for the outcomes of increasing

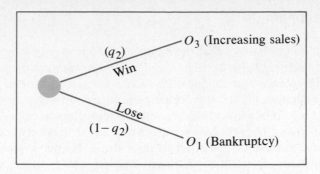

FIGURE 21-5
The reference lottery for Homer Briant's decision.

sales (O_3), decreasing sales (O_2), and bankruptcy (O_1), and the reference lottery is shown in Figure 21-5. Suppose that the utility values of the extreme outcomes are arbitrarily set at 10 and -5, so that

$$u(O_3) = 10 \qquad u(O_1) = -5$$

Now assume that Homer contemplates the reference lottery in terms of 100 marbles in a box, some labeled W for "win" and the rest labeled L for "lose." A marble is to be selected at random. If it is a W, then Homer will be guaranteed outcome O_3 (increasing sales), but if it is an L, he will go bankrupt for certain, achieving outcome O_1. Homer is then asked what number of W marbles would make him indifferent between facing declining sales (outcome O_2) or taking his chances with the lottery. After considerable thought, Homer replies that 75 W marbles would make him regard O_2 and the reference lottery equally well. This establishes a reference lottery win probability of q_2 that makes it an equivalent gamble to outcome O_2:

$$q_2 = 75/100 = .75$$

This probability can then be used to calculate the utility for declining sales:

$$
\begin{aligned}
u(O_2) &= q_2 u(O_3) + (1 - q_2)u(O_1) \\
&= .75(10) + .25(-5) \\
&= 6.25
\end{aligned}
$$

Attitude Versus Judgment

It must be emphasized that the value $q_2 = .75$ is merely a device used to establish indifference. *The selected probability for winning the lottery has nothing to do with the chance that the most favorable outcome will occur.* In setting $q_2 = .75$, the decision maker is expressing an *attitude* toward one outcome in terms of the rewards of a hypothetical gamble. This value was obtained through introspection in an attempt to balance tastes and aspirations between remaining in a declining business or gambling to improve it. Homer is assumed to be capable of switching from introspection to dispassionate *judgment* when asked later what he thinks the actual chance is that moving his business will be a success. To arrive at the probability for success, our decision maker must use his experience and knowledge of such factors as the history of failures by relocated businesses, prevailing economic conditions, and possible competitor reactions.

Suppose that Homer judges his chance of success after moving to be 1/2. We can now analyze his decision problem by applying the Bayes decision rule, using utilities as payoff values. The decision tree diagram is shown in Figure 21-6. The expected utility payoff for the event fork at b is 2.5, which is the utility achieved by moving. Since this value is smaller than the 6.25 utility achieved by remaining in his present location, Homer Briant should not move. Thus, we prune the branch corresponding to the act "move" and bring the 6.25 utility payoff back to node a.

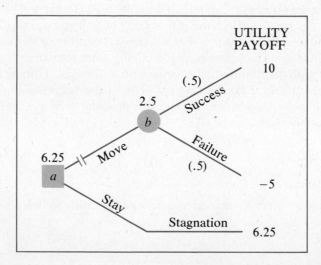

FIGURE 21-6

Homer Briant's decision tree diagram showing backward induction analysis with utility payoffs.

Utility and the Bayes Decision Rule

This example illustrates why the Bayes decision rule is valid when utility payoffs are used. The utility for a gamble is the expected value of the utilities assigned to its rewards. Since any act with uncertain outcomes may be viewed as a gamble, the act that provides the greatest utility—and therefore the one that must be preferred—is the act with the maximum expected utility payoff. The Bayes decision rule can therefore be viewed as an extension of utility theory, and the criterion serves only to translate the decision maker's preferences into a choice of act. Homer Briant decides to stay because this act provides the greater utility—which can only be the case, our theory states, if remaining in the present location is the preferred act. *In arriving at a choice, both the decision maker's attitudes toward the consequences and judgment regarding the chances of the events are considered and integrated.* The choice indicated by the Bayes criterion is optimal because it is preferred above all others.

If some other success probability (say, .90) had been determined, Homer would relocate, because doing so would have the higher utility: $.9(10) + .1(-5) = 8.5$. This might be the case, for example, if Homer learned that his major would-be competitor had just been taken over by his incompetent son. Changing event probabilities reflect only the decision maker's judgment regarding the factors that influence their occurrence. Only the *expected* utilities for uncertain *acts* can be affected by the revision of event probabilities. Regardless of the chance of the success of the relocation, the decision maker's utilities for the ultimate *outcomes* must remain unchanged. Only a change of taste or attitude, which might be caused by a death in the family or by a change in personal finances, can justify revising the ultimate outcome utilities.

EXERCISES

21-4 Actor Nathan Summers enjoys wearing costumes in front of audiences. Nathan likes dressing up like a little old lady the most and hates to dress up like an animal. Somewhere in between lies his preference for wearing a cowboy suit. Assigning a utility of 10 to being a lady and a utility of -5 to being an animal, what is Nathan's utility for playing a cowboy if he is indifferent between this outcome and a coin toss determining which of the other two roles he will play?

21-5 Ty Kuehn, a potential entrepreneur, is faced with the following outcomes:

O_3 Successfully established in his own business
O_2 Maintaining present employee status
O_1 Personal bankruptcy

Ty would be indifferent between remaining on his present job and opening a restaurant when the probability is q_2 for success and $1 - q_2$ for bankruptcy.

(a) At age 22 Ty finds that $q_2 = .50$ makes him indifferent. If he arbitrarily sets $u(O_3) - 200$ and $u(O_1) = -100$, calculate Ty's utility value for keeping his present job (and not going into business for himself).

(b) At age 30 Ty's outlook has changed drastically, and $q_2 = .90$ now applies. Find $u(O_2)$ again when $u(O_3) = 200$ and $u(O_1) = -100$.

(c) Undergoing a mid-life crises at age 40, Ty revises his indifference probability to .20. Preferring big numbers, he arbitrarily establishes $u(O_3) = 10,000$ and $u(O_1) = 0$. Find Ty's utility for remaining on somebody else's payroll.

21-6 You may achieve the following outcomes (no rights are transferable):

- 100 new record albums of your choice
- a grade of C on the next examination covering utility
- a year's assignment to Timbuktu, Mali
- confinement to an airport during a three-day storm
- a month of free telephone calls to anywhere

(a) Rank these outcomes in descending order of preference, designating them O_5, O_4, O_3, O_2, and O_1.

(b) Let the utilities be $u(O_5) = 100$ for the best outcome and $u(O_1) = 0$ for the worst outcome. Consider a box containing 1,000 marbles, some of which are labeled "win" and the remainder of which are labeled "lose." If a "win" marble is selected at random from the box, you will achieve O_5. If a "lose" marble is chosen, you will attain O_1. Determine how many marbles of each type would make you indifferent between gambling or achieving O_2. Determine the same for O_3 and O_4.

(c) The corresponding probabilities for winning q_k can be determined by dividing the respective number of "win" marbles by 1,000. Use these probabilities to calculate $u(O_4)$, $u(O_3)$, and $u(O_2)$.

21-5 THE UTILITY FOR MONEY AND ATTITUDES TOWARD RISK

Applying the Utility Function in Decision Analysis

The reference lottery can be used to construct a utility function for money. To do this, the best outcome is selected so that it is no smaller than the greatest possible payoff, and the worst outcome is selected so that it is no larger than the lowest possible payoff. Monetary outcomes offer some special advantages. A monetary amount can be measured on a continuous scale, so that the utility function itself will be continuous. This suggests that it may be determined by finding an appropriate smoothed curve relating money values to their utilities. To do this, only a few key dollar amounts and some knowledge of the curve's general shape are required. The curve obtained by connecting the points can then serve as an approximation of the utility function.

Such a curve is shown in Figure 21-7 for the Ponderosa Record Company decision in Chapter 19. This utility curve has been derived according to the

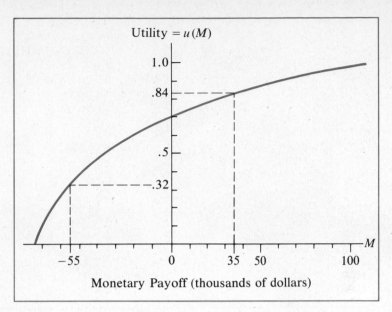

FIGURE 21-7
The utility function for the president of the Ponderosa Record Company.

procedures just described by applying a reference lottery and using a few key monetary amounts as the outcomes. The reference lottery that is used ranges from +$100,000 (for win) to −$75,000 (for lose), which for ease of evaluation are more extreme than any possible payoff. Arbitrary utility values of $u(+\$100,000) = 1$ and $u(-\$75,000) = 0$ have been set for simplicity.

In practice, a utility function is found empirically by personally interviewing the decision maker. Ordinarily, the function will be described graphically by reading the utilities directly from the curve rather than by using a mathematical equation.

We can use the utility curve in Figure 21-7 to analyze the Ponderosa president's decision problem. The original decision tree diagram is reconstructed in Figure 21-8. The utilities corresponding to each monetary payoff have been obtained from the utility curve and added to the tree. For instance, the utilities for the monetary payoffs $35,000 and −$55,000 are .84 and .32, respectively.

Backward induction is then performed using utilities instead of dollars. Here, we find that the optimal choice is "don't test market" and "abort." The two alternatives that involve marketing the record are too risky. Recall that we reached a different conclusion in Chapter 19 when expected monetary values were used. But since utility values express the true worth of monetary outcomes, our latest solution is the valid one.

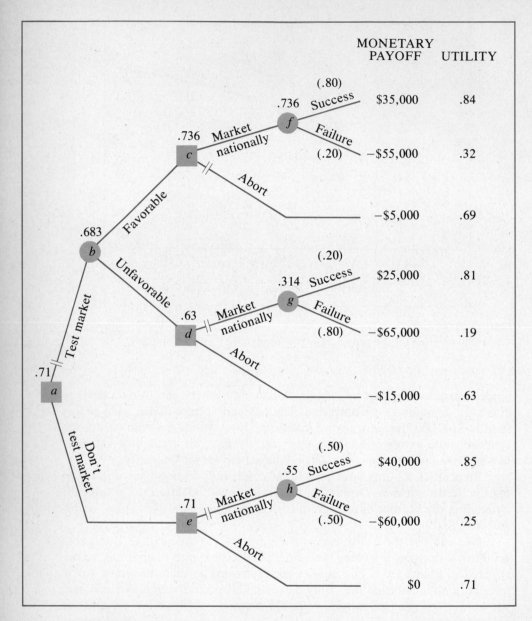

MONETARY PAYOFF   UTILITY

(.80) Success $35,000   .84
.736
f
Failure
(.20)   −$55,000   .32

.736 Market nationally
c
Abort   −$5,000   .69

Favorable
.683
b

(.20)
.314 Success $25,000   .81
g
Failure
(.80)   −$65,000   .19

Unfavorable
.63 Market nationally
d
Abort   −$15,000   .63

Test market
.71
a

Don't test market

(.50)
.55 Success $40,000   .85
h
Failure
(.50)   −$60,000   .25

.71 Market nationally
e
Abort $0   .71

FIGURE 21-8
The Ponderosa decision tree diagram showing backward induction analysis with utility payoffs.

Utility and the Decision to Buy Insurance

Earlier we showed how the Bayes decision rule will ordinarily indicate that the greatest expected *monetary* payoff is achieved by not buying casualty insurance. But when *utility* payoffs are employed, that same procedure reaches the opposite conclusion: *Utility is maximized by buying the insurance.* This confirms the choices most of us have made in determining our own insurance needs.

To illustrate, we will return to our fire insurance example. Suppose that Spiro Pyrophobis values the dollar changes in his assets according to the utility function

$$u(M) = \sqrt{M + 40{,}000} - 200$$

where M expresses the change in cash position associated with each outcome.

Table 21-2 provides the utility calculations and the expected utility calculations for the acts "buy insurance" and "don't buy insurance." For instance, if no insurance is bought and there is a fire, a loss results, so that $M = -\$40{,}000$ (a negative change in cash position). The utility for this outcome is obtained from the calculation

$$u(-\$40{,}000) = \sqrt{-40{,}000 + 40{,}000} - 200$$
$$= \sqrt{0} - 200 = -200$$

Calculating the utilities for the other outcomes in a similar manner, we find $u(0) = 0$, and $u(-\$100) = -.25$.

TABLE 21-2
Determination of Utilities for Outcomes of the Fire Insurance Decision
and Calculation of Expected Utilities

(1) Event	(2) Probability	(3) Cash Change M	(4) Utility $\sqrt{M + 40{,}000} - 200$	(5) Utility × Probability
		Buy Insurance		
Fire	.002	-$100	-.25	-.0005
No fire	.998	-100	-.25	-.2495
			Expected utility =	-.2500
		Don't Buy Insurance		
Fire	.002	-$40,000	-200	-.40
No fire	.998	0	0	0
			Expected utility =	-.40

Each act is a gamble. Buying insurance is a gamble having two identical outcomes in terms of dollar expenditure of −$100, since the same amount applies whether or not there is a fire. Buying no insurance is a gamble having cash outlays of −$40,000 if there is a fire and $0 if there is no fire. The utilities for the respective acts are therefore the expected utilities for the corresponding gambles. We can see from Table 21-2 that the expected utilities are −.25 for buying insurance and −.40 for not buying insurance. Since buying insurance has the higher utility, it must be preferred by the decision maker. Stated differently, the act "buy insurance" has the maximum expected utility payoff, so the Bayes decision rule indicates that this act is the optimal choice.

Thus, when we use utilities as payoff values, the Bayes decision rule indicates the "proper" result. However, this does not permit us to conclude that whenever utilities are used as payoffs, this criterion will lead to a decision to buy insurance. The choice depends on the relationship between the chance of fire and the price of the insurance policy. Suppose, for example, that the price of Spiro's policy is raised to $200, so that the utility for the dollar payoffs for buying insurance is

$$u(-\$200) = \sqrt{-200 + 40,000} - 200$$
$$= \sqrt{39,800} - 200 = -.50$$

The expected utility for buying insurance is −.50, so that if the probability for fire remains the same, the act "don't buy insurance" will have a utility of −.40, which is greater than −.50, making "don't buy insurance" the preferred act. This is the opposite outcome from our earlier decision. The insurance has become too expensive to be attractive.

Many people faced with the same circumstances would buy insurance even if the premium were raised to $1,000 or more. *Their tastes would be different, and this would be reflected by the different utility values that they would assign to each outcome.* Premium prices may partly explain the prevalence of fire insurance coverage and the paucity of protection against natural disasters, such as earthquakes, tornados, the floods. One reason why people do not generally buy insurance policies to cover natural disasters is the high premium required by insurance companies for such coverage (if it is offered at all) in relation to the probabilities for occurrence (which are difficult to obtain actuarily for such rare phenomena).

Attitudes Toward Risk and the Shape of the Utility Curve

The utility function for money can be used as the basis for describing an individual's attitudes toward risk. Three basic attitudes have been characterized. The polar cases are the *risk averter*, who will accept only favorable gambles, and the *risk seeker*, who will pay a premium for the privilege of participating in a gamble. Between these two extremes lies the *risk-neutral individual*, who considers the face value of money to be its true worth. The utility functions for each basic attitude appear in Figure 21-9. Each function has a particular shape, corresponding to the decision maker's fundamental outlook. All three utility functions show that utility increases with monetary gains. This reflects the underlying assumption of utility theory that utility increases with preference, which is combined with the additional assumption that more money will always increase an individual's well-being, so that the outcomes with greater payoffs are preferred. (This assumption may not be strictly true, but with the exception of eccentrics, most people behave in a manner that supports it.)

Throughout most of their lives, people are typically risk averters. These individuals buy plenty of casualty insurance. They avoid actions that involve high risks (chances of large monetary losses). Only gambles with high expected payoffs will be attractive to them. A risk averter's utility drops more and more severely as losses become larger, and the utilities for positive amounts do not grow as fast with monetary gains. The risk averter's marginal utility for money diminishes as the rewards increase, so that the risk averter's utility curve, shown in Figure 21-9(a), exhibits a decreasing positive slope as the level of monetary payoff becomes larger. Such a curve is *concave* when viewed from below.

The risk seeker's behavior is the opposite of the risk averter's behavior. Many of us are risk seekers at some stage of our lives. This attitude is epitomized by the "high roller," who may behave recklessly and who is motivated by the possibility of achieving the maximum reward in any gamble. This risk seeker will prefer *some* gambles with negative expected monetary payoffs to maintaining the status quo. The greater the maximum reward, the more the risk seeker's behavior will diverge from the risk averter's behavior. The risk seeker is typically self-insured, believing that the risk is superior to forgoing money spent on premiums. The risk seeker's marginal utility for money is increasing: Each additional dollar provides a disproportionately greater sense of well-being.

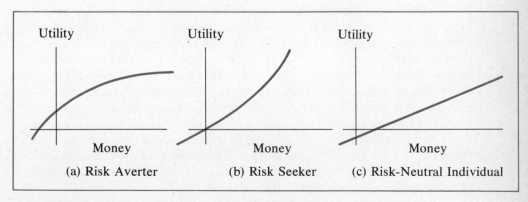

FIGURE 21-9
A graphical portrayal of utility functions for basic attitudes toward risk.

The loss of one more dollar is felt only slightly more severely for large absolute levels of loss than for small ones. Thus, the slope of the risk seeker's utility curve, shown in Figure 21-9(b), increases as the monetary change improves. This curve is *convex* when viewed from below.

Our third characterization of attitude toward risk is the risk-neutral individual, who prizes money at its face value. The utility function for such an individual is a straight line, as shown in Figure 21-9(c). His or her utility for a gamble is equal to the utility for the expected monetary payoff. Risk-neutral individuals buy no casualty insurance, since the premium charge is greater than the expected loss. Risk-neutral behavior is epitomized by individuals who are enormously wealthy. The decisions of large corporations are often based on the Bayes decision rule applied directly to monetary payoffs, reflecting that increments in dollar assets are valued at their face amount.

In general, risk neutrality holds only over a limited range of money values. For example, many large firms do not carry casualty insurance, but almost all giant corporations will insure against extremely large losses—airlines buy hijacking insurance, for example. The same holds for individuals. Many risk-averse persons are risk-neutral when the stakes are small. The player in the World Series office pool falls into this category; losses are hardly noticeable, and winnings permit the individual to indulge in some luxury. (Small gambles may add spice to a person's life—they are a form of entertainment. Thus, a person might play poker with more skillful players, where the expected payoff would be negative, just for the fun of it.) That people are risk-neutral for small risks is illustrated by their car insurance purchases. Many generally risk-averse people carry deductible comprehensive coverage when they first purchase an automobile, and they usually keep only the liability coverage when their car gets old. Again, this reflects risk neutrality over a limited range of monetary outcomes. This behavior does not contradict the curve shapes in Figure 21-9(a) and (b), because each curve can be approximated by a straight line segment throughout a narrow monetary interval.

Many people can be both risk averters and risk seekers, depending on the range of monetary values being considered. To an entrepreneur founding a business, the risks are very high—a lifetime's savings, plenty of hard work, burned career bridges, a heavy burden of debt, and a significant chance of bankruptcy. Those who embark on the hard road of self-employment may often be viewed as risk seekers. They are motivated primarily by the rewards—monetary and otherwise—of being their own boss. Once entrepreneurs become established and are viewed by peers as future pillars of the community, their attitudes toward risk will have evolved to a point where they can be characterized as risk averters. They are much more conservative (now there is something to conserve), and probably no venture imaginable could persuade them to risk everything they own to further their wealth.

We can conceive of an individual's attitudes varying between risk seeking and risk aversion over time. Usually a risk seeker has some definite goal or *aspiration level*, which can be achieved by obtaining a specific amount of money.

A young sports enthusiast might be willing to participate in an unfair gamble if winning would provide sufficient cash for a down payment on a first motor-cycle. The young professional may speculate in volatile stocks to try to earn enough money for a down payment on a fashionable home. To these risk seekers, losing is not much worse than maintaining the status quo. But once the goal is achieved, the risk seeker's outlook changes, and with a sated appetite, the risk seeker becomes a risk averter until some new goal enters the horizon.

A utility curve for such an individual appears in Figure 21-10. The hori-zontal axis measures total wealth in monetary units, rather than changes in

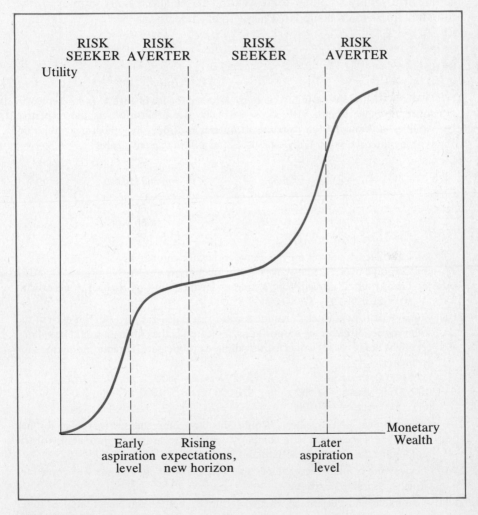

FIGURE 21-10

A Graphical Portrayal of the evolving utility function for an individual over a long period of time.

current cash position. Here, the utility function is convex—that is, the individual is a risk seeker—until an aspiration level is reached. Then, the individual becomes a conservative risk averter until more wealth permits the germination of a newer goal of a higher order. At this point, the cycle begins all over again with risk seeking, followed by another period of risk aversion.

Such a curve portrays behavior over a long period of time and is only an idealization of a long-run utility function. A great many factors can cause an individual's attitudes to change over time, and it may not be possible to obtain meaningful measures of the influence of remote goals. In general, the validity of a utility curve is very short-lived and is affected by changes in such factors as age, lifestyle, family size, and total wealth. To be successfully employed, the utility function should be updated prior to each decision.

EXERCISES

21-7 Willy B. Rich wants his utility curve constructed for the change M in his net worth over the range from $-\$10,000$ to $+\$20,000$. He arbitrarily sets the respective utilities at 0 and 100. In response to queries regarding hypothetical gambles involving these amounts, Willy establishes the following equivalences:

Equivalent Amount	Probability for Winning $20,000
$-\$5,000$	.60
0	.80
$+10,000$	.95

(a) Calculate his utilities for the above monetary amounts.
(b) On graph paper, sketch Willy's utility function.
(c) From your curve read Willy's utilities for the following changes in net worth: (1) $-\$2,000$; (2) $+\$2,000$; (3) $+\$5,000$.

21-8 Suppose that Alvin Black's attitude toward risk is generally averse. For each of the following 50–50 gambling propositions, indicate whether Alvin (1) would be willing, (2) might desire, or (3) would be unwilling to participate. Explain the reasons for your choices.
(a) $10,000 versus $0   (d) $500 versus $-$600
(b) $10,000 versus $-$1,000   (e) $20,000 versus $10,000
(c) $15,000 versus $-$10,000

21-9 Lucille Brown is risk-neutral. Would she buy comprehensive coverage for her automobile if she agreed with company actuaries regarding the probability distribution for future claim sizes? Explain.

21-10 Vicky White is a risk seeker. Does this necessarily imply that she will never buy casualty insurance? Explain.

21-11 A contractor must determine whether to buy or rent the equipment required to do a job up for bid. Because of lead-time requirements, she must decide whether to obtain the equipment before she knows if she has been awarded the contract. If she buys the equipment, the contract will result in $120,000 net profit after equip-

ment resale returns, but if she loses the job, the equipment will have to be sold at a $40,000 loss. By renting, her profit from the contract (if she wins it) will be only $50,000, but there will be no loss of money if the job is not won. The contractor's chances of winning are 50–50, and her utility function is $u(M) = \sqrt{M} + 40,000$.
(a) Construct the contractor's payoff table using profit as the payoff measure.
(b) Calculate the expected profit payoff for each act. According to the Bayes decision rule, which act should the contractor select?
(c) Construct the contractor's payoff table using utilities as the payoff measure.
(d) Calculate the expected utility payoff for each act. Which act provides the maximum expected utility?
(e) Which act should the decision maker choose? Explain.

21-12 Suppose that the Ponderosa Record Company's utility function for money is $u(M) = [(M + 65,000)/10,000]^2$.
(a) Redraw Figure 21-8, and calculate the utility for each end position.
(b) Perform backward induction analysis using the new utilities you have calculated. What strategy is optimal?
(c) On a piece of graph paper, plot the utilities you calculated in (a) as a function of monetary payoffs M. Sketch a curve through the points. What attitude toward risk is indicated by the shape of your curve?

REVIEW EXERCISES

21-13 An insurance policy would cost Hermie Hawks $1,000 per year to protect his home from tornado damage. Assume that any actual tornado damage to Hermie's house, valued at $100,000, would be totally destructive and that the probability that a tornado will hit his house during the year is .0025.
(a) If Hermie is risk-neutral, what would his optimal decision be regarding buying tornado insurance? Show your computations.
(b) How much above its expected claim size is the insurance company charging Hermie for its combined overhead and profit on the proposed policy?
(c) Hermie's utility function for any change in his monetary position for any amount M is

$$u(M) = 10,000 - (M/1,000)^2$$

What action should Hermie take to maximize his expected utility?
(d) What annual insurance premium charge would make Hermie indifferent between buying or not buying tornado insurance?

21-14 J.P. Tidewasser has just undergone the first traumatic phase of determining her utility function for a range of money values. By her response to a series of gambles, it has been established that she is indifferent between making the 50–50 gambles on the left and receiving the certain amounts of money shown on the right:

Rewards of Gamble		Equivalent Amount
+$30,000	−$10,000	$ 0
+30,000	0	+10,000
0	−10,000	−7,000
+10,000	−7,000	1,000

(a) If J.P. sets $u(\$30,000) = 1$ and $\mu(-\$10,000) = 0$, determine her utility for $0.
(b) Calculate J.P.'s utilities for $+\$10,000$ and $-\$7,000$.
(c) Calculate J.P.'s utility for $+\$1,000$. What, if any, inconsistencies do you notice between this and your previous answers?

21-15 Hoopla Hoops is a retail boutique catering to current crazes. The owner must decide whether or not to stock a batch of Water Wheelies. Each item cost $2 and sells for $4. Unsold items cannot be returned to the supplier, who sells them in batches of 500. The following probability distribution is assumed to apply for the

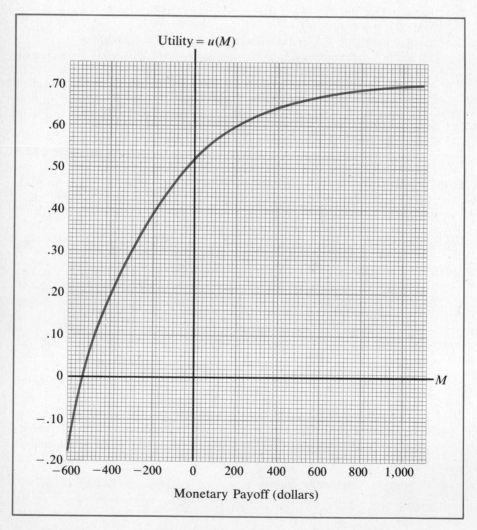

FIGURE 21-11

anticipated demand for Water Wheelies:

Demand	Probability
100	.05
200	.10
300	.15
400	.20
500	.20
600	.15
700	.10
800	.05
	1.00

Consider demand to mean the potential for sales. No more than what is demanded can be sold, but if demand exceeds on-hand inventory, not all of the demand can be fulfilled.

(a) Calculate the expected demand. If you assume that the expected demand will actually occur, what profit corresponds to this amount? Use the utility curve shown in Figure 21-11 to determine the corresponding utility value.

(b) Calculate the expected profit from stocking 500 Water Wheelies. Does this differ from the amount you found in (a)? Explain this. Then determine the utility for the expected profit.

(c) Calculate the expected utility for stocking 500 Water Wheelies. (First, calculate the profit for each level of possible demand; then find the utility for each level; finally, apply the probability weights.) Which act—stocking or not stocking Water Wheelies—provides the greatest expected utility?

21-16 Consider the plight of the decision maker in Exercise 19-9 (page 685). She must interview dozens of candidates annually for keypunching jobs. Losses of her recoverable training expenses can therefore be significant. Suppose that she has constructed the utility function shown in Figure 21-11.

(a) Redraw Figure 19-5 (page 686).

(b) For the monetary payoff value for each end position, determine the decision maker's approximate utility value from the curve in Figure 21-11.

(c) Perform backward induction analysis using utilities as payoffs. What strategy is optimal?

Almost every person can establish a subjective probability for any uncertain event. What really matters is whose numbers we use in making evaluations.

Subjective Probability: Accounting for Judgment

Although probability concepts have been extensively used in earlier chapters, little discussion has been devoted to *subjective probabilities*. Subjective probabilities are applicable to nonrepeatable circumstances, such as introducing a new product or drilling a wildcat oil well, and must be arrived at through *judgment*. This is in contrast to the long-run frequencies used to establish *objective probabilities*, which are valid only when elements of repeatability are present. Because so many business decisions involve one-shot situations that never recur exactly, there is a strong need for subjective probabilities when analyzing decision making under uncertainty. In this chapter, we will examine the procedures for translating judgment into the subjective probability values that are required when implementing Bayesian analysis to help solve decision problems in the real world.

PROBABILITIES OBTAINED FROM HISTORY   22-1

Historical experience can be a convenient starting point for assigning probabilities. To calculate the historical frequency of an event, we need to know only two things: the number of times the event has occurred in the past and the number of opportunities when it could have occurred. This is how fire insurance underwriters obtain probabilities for determining the expected claim sizes that are used to establish policy charges. With tens of thousands of buildings involved, *the event frequencies themselves define the probabilities*, because in the traditional sense, probability fundamentally expresses long-run frequency of occurrence.

There are inherent difficulties in using historical frequencies as probabilities. One is the limited extent of history—the available data may only provide a crude frequency estimate. Unless the number of similar past circumstances is large, statistical estimates of event frequencies can be unreliable. Past history may be suitable for setting fire insurance rates. But past frequencies cannot be wholly adequate—indeed, are unavailable—to determine the probability distributions for a great many variables encountered in business, such as the demand for a new product.

Another serious difficulty is that conditions change over time. The recent experience of automobile-casualty insurance firms serves as an example of how changing conditions can make historical frequencies unsuitable for obtaining probabilities. Car insurers, who have consistently complained about losing money on collision and comprehensive coverage, have found that past experience has proved to be a poor predictor of future levels of damage claims. This is due not to sampling error, which is virtually nonexistent because the data obtained constitute a census, but rather to changing circumstances. Cars are becoming less and less sturdy, so that minor impacts that would hardly have dented an older car can seriously damage a new one. Repair costs have also been rising in a pronounced inflationary spiral. More cars are sharing roads that are not increasing at the same rate, and driving habits are changing accordingly, affecting the accident rate. Automobile thefts have also been increasing as the result of new social pressures.

Using historical frequencies to estimate the probabilities for future automobile insurance claims may be compared to tossing a die, some side of which is shaved before each toss. We do not know which side has been shaved or by how much. Under these circumstances, we can never obtain a reasonable probability distribution for the respective sides from historical frequencies alone.

22-2   SUBJECTIVE PROBABILITIES

To apply basic decision-making models based on expected payoffs or utilities, we must employ probabilities. Past history can sometimes provide probability values that fit into the mold of long-run frequencies. But the applicability of such data is limited to events with a rich history, such as insurance claims. And even when they are available, these data can be misleading, owing to the forces of change.

In many business decisions, the only recourse is to use subjective probabilities, which are not tied to a long-run frequency of occurrence, because so many decisions involve one-shot situations that may be characterized by essentially nonrepeatable uncertainties. Good *judgment* may be the only method available to transform such uncertainties into a set of probabilities for the various events involved.

We have seen that decision making under uncertainty is analogous to gambling. Unlike card games, lotteries, dice, or roulette, however, most real-life gambles can be analyzed only with the help of subjective probabilities, which reflect the decision maker's judgment and experience. How do we obtain a subjective probability?

Betting Odds

Subjective probabilities can be considered betting odds; that is, they can be treated just like the probabilities that the decision maker would desire in a lottery situation of his or her own design in which the payoffs are identical in every respect to the possible payoffs from the actual decision being evaluated. For example, suppose that a contractor assigns a subjective probability of .5 to the event of winning a contract that will increase profits by $50,000 and that losing the contract will cost $10,000. This contractor ought to be indifferent between preparing a bid for the contract and gambling on a coin toss where a head provides a $50,000 win and a tail results in a $10,000 loss. The subjective probability for winning the contract can therefore be transformed directly into an "objective" .5 probability for obtaining a head from a coin toss. Assuming indifference between the real-life gamble and a hypothetical coin toss, we can then substitute the latter into the decision analysis.

One practical benefit of substituting a hypothetical gamble for an actual uncertainty is that subjective probabilities can be used in conjunction with the traditional long-run frequencies of occurrence. In effect, apples and oranges may be mixed, permitting wider acceptance of decision-theory analysis. More significantly, a hypothetical gamble or lottery can provide a convenient means of obtaining the subjective probability value itself. Consider the following example.

Substituting a Hypothetical Lottery for the Real Gamble

A project engineer must choose between two technologies in designing a prototype sonar system. She may use Doppler shift or acoustic ranging. If the Doppler shift is used, time becomes a crucial factor. To analyze this decision, the engineer must determine the probability for completing the project on time. If she is late, the project will be canceled and she will be out of a job. But if she is early or on time, her contract will be extended for two more years. The event fork of corncern appears in Figure 22-1(a).

Suppose that the engineer considers the hypothetical lottery shown in Figure 22-1(b), in which one marble is to be randomly selected from a box of 100. Some marbles are labeled E (for early); the rest are marked L (for late). In this hypothetical gamble, selecting an E marble will result in an extended contract, but drawing an L marble will result in a canceled contract. Our engineer can determine the mix of L and E marbles. She is asked what mixture will make her *indifferent* between letting her future be decided by trying the Doppler shift design or by selecting a marble from the box.

Suppose that the engineer determines that 70 E marbles and 30 L marbles would make her indifferent. This means that the probability for selecting an E marble is .70. This value can be considered the engineer's *judgmental assessment* that the project will be early or on time if the Doppler shift is used. Thus, our decision maker can let .70 represent the probability for being early or on time in analyzing her decision and place .70 on the early event branch of the fork in Figure 22-1(a).

Arriving at subjective probabilities by substituting the real-life gamble for an equally preferred lottery is a useful procedure when the number of possible events is small. But this method can be quite cumbersome when the situation involves more than a handful of events. In business applications, we are often faced with variables, such as product demand, which can be measured on many possible levels. It is best to use an entire *probability distribution* to represent such uncertain quantities.

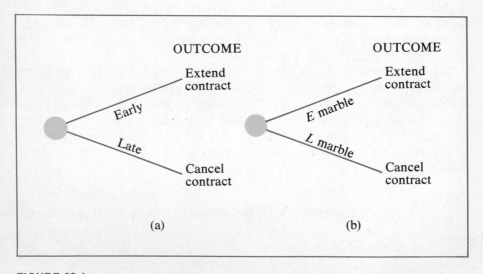

FIGURE 22-1

A project engineer's actual gamble (a) and a hypothetical lottery (b) yielding identical outcomes.

Subjective Probability Distributions

We have seen that probability distributions can be divided into two categories. *Discrete probability distributions* apply to variables, such as the demand for cars, that must be a whole number. *Continuous probability distributions* represent variables, such as time, that can be expressed on a continuous scale and measured to any degree of precision desired. When the number of possibilities is great, discrete variables are often treated as if they are approximately continuous. For this reason, we focus on finding continuous probability distributions.

Next we will consider using judgment to determine the normal distribution, which constitutes perhaps the most common distribution family encountered in business decision making with continuous variables. We will then discuss the more general problem of establishing a probability distribution for *any* uncertain quantity.

EXERCISES

22-1 Discuss whether historical frequencies can be meaningful in estimating the probabilities for each of the following cases:
(a) the first-year salary levels of business-school graduates
(b) the faces obtained by tossing an asymmetrical die
(c) the deaths during the next year of people in various age, health, sex, and occupational categories

22-2 Use your judgment to assess the probability that you will receive an A on your next examination. Imagine that your instructor will let you obtain your grade by lottery, so that 100 slips of paper (some labeled A; the rest, not A) will be put into a hat and mixed. You will draw one of these slips at random, and the letter obtained will be the grade that you receive. How many A slips must there be to make you indifferent between letting your grade be determined by lottery or by earning it?

DETERMINING A NORMAL CURVE JUDGMENTALLY   22-3

The normal distribution plays an important role in decision making. This is largely due to the fact that the probabilities for sample means can usually be characterized by a normal curve. But the normal curve can be applied in many situations other than sampling. The frequency patterns of physical measurements often approximate the normal curve. This feature makes it especially important in production applications, where natural fluctuations in size, density, concentration, and other factors cause individual units to vary according to the normal distribution. Test scores used to determine personal aptitude or achievement are also often characterized by the normal curve. Questions of facility design as it relates to waiting lines in manufacturing, retailing, or data-processing situations must take into account the time needed to produce units, service customers, or complete jobs; these service times are often normally distributed.

Since the normal distribution is prevalent in such a broad spectrum of decision-making situations, we will give it special emphasis in this chapter. We have seen that any normal distribution is uniquely defined in terms of two parameters—the mean μ and the standard deviation σ. Except in the rare circumstances when these parameters are known precisely, it is impossible to measure their values directly without expensive sampling procedures. It may be more convenient to exercise judgment in determining μ and σ. With these two quantities, we can specify the entire normal distribution. (In fact, it may be optimal not to take samples at all—a question that we will consider in Chapter 24. Often sampling itself is impossible because no population currently exists from which observations can be taken.)

Finding the Mean

The mean of the subjective normal curve for any quantity X believed to have a normal distribution can be established by selecting the *midpoint* of all possible values. *The subjective mean μ is that point judged to have a 50–50 chance for any value X to lie at or above it versus below it.* This value is actually the *median* level, since it is just as likely that X will fall below or above the identified point. Because the normal curve is symmetrical, this central value must also equal the mean.

For example, suppose that an engineer is evaluating a new teleprocessing terminal design to determine if a prototype should be fabricated. Based on the physical characteristics of the unit compared to existing equipment of similar scope, the engineer concludes that messages could be printed at a rate of between 30 and 70 lines per second, depending on the type of message. Because the unit has never been built, the true rate X for a typical message is uncertain. The engineer assumes that this quantity is normally distributed, because similar data in related applications have been found to fit well to the normal curve.

To determine the mean printing rate, the engineer is asked to establish a level such that it is equally likely for X to fall above or below it. This decision can be phrased in terms of a coin toss: "Suppose that you had to find a middle value such that it would be difficult for you to choose whether the actual printing rate X lies above or below it. If your professional reputation depended on being correct and you had to make a prediction for X, you would be willing to select one side or the other of that value by tossing a coin. Where would your midpoint lie?" After some thought the engineer might reply, "I think it is a coin-tossing proposition that the printing rate experienced in actual testing may fall above or below 50 lines per second." This establishes the desired midpoint and therefore the mean of the subjective probability distribution, so that $\mu = 50$ lines per second.

Relating the judgmental evaluation to coin tossing makes the problem easy for a person who is not used to dealing with probability. The 50–50 gamble is the easiest to envision. We may extend this concept to finding the standard deviation as well. *The subjective standard deviation σ can be found by establishing a middle range of values centered at μ such that X is judged to have an equal chance of lying inside or outside that interval.* To see how this works, let's review some properties of the normal curve.

Finding the Standard Deviation

In Chapter 7, we saw that the area under the normal curve between any two points is established by the distances separating each point from the mean, which are expressed in units of standard deviation. This standardized distance can be represented by a value of the normal deviate z, where for any particular point x, the corresponding normal deviate can be computed from

$$z = \frac{x - \mu}{\sigma}$$

We seek two possible values of X that are equally distant from μ such that the area between them is .50. This means that the area between μ and the upper limit must be one-half this value, or .25. From the normal curve areas in Appendix Table D, the normal deviate value of $z = .67$ provides the closest area, .2486. We can use $z = .67$ to find σ.

Figure 22-2 illustrates the underlying principles involved. The area between μ and $\mu + .67\sigma$ is about .25, so that the area in the interval $\mu \pm .67\sigma$ is

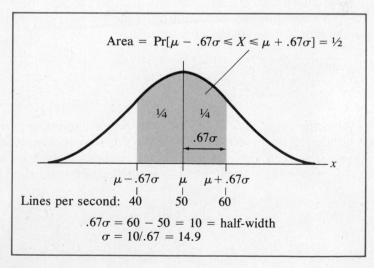

FIGURE 22-2
Finding the standard deviation of a subjective normal distribution.

about .50. If we know the distance separating the upper limit $\mu + 67\sigma$ and the mean μ, or the *half-width* of the interval, we can determine the corresponding value of σ by setting $.67\sigma$ equal to that distance:

$$.67\sigma = \text{Upper limit} - \mu$$

Dividing both sides by .67, we obtain the expression for the

JUDGMENTAL STANDARD DEVIATION

$$\sigma = \frac{\text{Upper limit} - \mu}{.67} = \frac{\text{Half-width}}{.67}$$

Thus, to find σ, we need to establish only the width of the middle range (covered by the shaded area in Figure 22-2). This quantity is sometimes called the *interquartile range*. This evaluation is tantamount to establishing the half-width such that there is a 50–50 chance that any particular value of X will fall within μ plus or minus this quantity.

The engineer in our example is now asked to establish the half-width of the central interval. This problem might be formulated: "Select the range of values centered at $\mu = 50$ lines per second so that the actual printing rate will be just as likely to fall inside or outside of it. To find the interval, determine an amount such that μ plus or minus that quantity establishes the range." The engineer might respond: "I would guess that ± 10 lines per second is suitable for this purpose. This means that it is a coin-tossing proposition that the actual printing rate will fall somewhere between $50 - 10 = 40$ and $50 + 10 = 60$ lines per second." This establishes the half-width of 10 lines per second for the middle range, and the standard deviation can be calculated:

$$\sigma = \frac{10}{67} = 14.9 \text{ lines per second}$$

The subjective probability distribution for the actual printing time of the proposed terminal is now specified. Combining this with economic data, the decision maker can then apply decision-theory concepts to evaluate various alternatives regarding the manufacturing or marketing of the proposed unit.

EXERCISES

22-3 An automobile production manager believes that the time taken to install a new car bumper is normally distributed. He has established that it is a 50–50 proposition that this task will take more than 60 seconds and that it is "even money" that the required time for any particular car will be between 45 and 75 seconds.

(a) Determine the mean and the standard deviation of the subjective probability distribution.
(b) Find the probabilities that the installation for a particular car will take
    (1) between 50 and 70 seconds.
    (2) less than 25 seconds.
    (3) more than 1 minute.
    (4) between 20 and 90 seconds.

22-4 Establish your own subjective probability distribution for the heights of adult males residing within 50 miles of your campus. Use the normal curve that you obtain to establish the probabilities that a randomly chosen man is (a) less than 6'2"; (b) taller than 5'6"; (c) taller than your father.

THE JUDGMENTAL PROBABILITY DISTRIBUTION: **22-4**
THE INTERVIEW METHOD

The procedure we have just examined is a limited one. Although the normal distribution is very common, it is the exception. We will now consider how judgment can be exercised to determine probability distributions. Our procedure applies to any variable with a large number of possible values, such as product demand.

A natural and fairly simple procedure for obtaining a probability distribution judgmentally is to use cumulative probabilities. By posing a series of 50–50 gambles, it is quite simple to make a judgmental determination of the cumulative probability distribution for a random variable. Each response provides a point that can be plotted on a graph; a smoothed curve can then be drawn through the points. This curve completely specifies the underlying probability distribution. The following illustration shows how this procedure may be carried out.

The president of a food-manufacturing concern wishes to obtain the probability distribution for the demand for a new snack product. This will be used to help the president decide whether or not to market the product. A statistical analyst asks the president a series of questions to obtain answers that will be used to formulate later questions. The interview follows.*

Q. What do you think the largest and smallest possible levels of demand are?
A. Certainly demand will exceed 500,000 units. But I would set an upper limit of 3,000,000 units. I don't think that under the most favorable circumstances we could sell more than this amount.

---

* This procedure was inspired by Howard Raiffa, *Decision Analysis: Introductory Lectures on Choices Under Uncertainty* (Reading, Mass.: Addison-Wesley, 1968).

Q. Okay, we have determined the range of possible demand. Now I want you to tell me what level of demand divides the possibilities into two equally likely ranges. For example, do you think demand will be just as likely to fall above 2,000,000 as below?

A. No. I'd rather pick 1,500,000 units as the 50–50 point.

Q. Very good. Now let's consider the demand levels below 1,500,000. If demand were to fall somewhere between 500,000 and 1,500,000 units, would you bet that it lies above or below 1,000,000?

A. Above. I would say that a demand of 1,250,000 units would be a realistic dividing point.

Q. We will use that amount as our 50–50 point. Let's do the same thing for the upper range of demand.

A. If I were to pick a number, I would choose 2,000,000 units. I feel that demand is just as likely to fall into the 1.5 to 2 million range as into the 2 to 3 million range.

Q. Excellent. We're making good progress. To get a finer fix on the points obtained so far, I now want you to tell me whether you think demand is just as likely to fall between 1.25 and 2 million units as it is to fall outside that range.

A. No. I think it is more likely to fall inside. I suppose this means I am being inconsistent.

Q. Yes, it does. Let's remedy this. Do you think that we ought to raise the 1,250,000 dividing point or lower the 2 million unit figure?

A. Lower the 2 million figure to 1.9 million.

Q. Let's check to see if this disturbs our other answers. Do you think that 1,500,000 splits demand over the range from 1,250,000 to 1,900,000 into two equally likely regions?

A. Yes, I am satisfied that it does.

Q. Just a few more questions. Suppose demand is above 1,900,000. What level splits this demand range into two equally likely regions?

A. I'd say 2,200,000.

Q. Good. Now if demand is between 2,200,000 and 3,000,000, where would you split?

A. I would guess that 2,450,000 units would be the 50–50 point.

Q. How about when demand is below 1,250,000?

A. Try 1,100,000 units.

Q. And when demand is between 500,000 and 1,100,000 units?

A. I think demand is far more likely to be close to the higher figure. I would bet on 950,000 units.

Table 22-1 shows the information obtained from this interview. The initial decision to divide demand at 1,500,000 units makes this level the 50% point or median. Since .5 has been judged the probability that demand will be below 1,500,000, we will refer to this as the .5 *fractile*. This means that the probability is .5 that the actual demand will be 1,500,000 units *or less*. Our decision maker has chosen to divide the range from 500,000 to 1,500,000 units at a demand level of 1,250,000. Believing that the chance of demand falling into this range is .5, the president has judged the chance that demand will be at or

TABLE 22-1

A Food Manufacturer's
Judgmental Assessment of
Fractiles for the Demand for
a New Snack Food

Fractile	Amount
0	500,000
.0625	950,000
.125	1,100,000
.25	1,250,000
.50	1,500,000
.75	1,900,000
.875	2,200,000
.9375	2,450,000
1.000	3,000,000

below 1,250,000 units to be $.5(.5) = .25$; thus, we refer to 1,250,000 units as the .25 fractile. This establishes a .25 probability that demand will be less than or equal to 1,250,000. The median of the range from 1,500,000 to 3,000,000 units is 1,900,000, which becomes the .75 fractile, since the probability is $.5 + .5(.5) = .75$ that demand will fall somewhere below 1,900,000 units. The analyst has proceeded to find the medians of the regions by working outward from previously determined 50% points. Thus, the .125 fractile of 1,100,000 units is the median demand for possible levels below the .25 fractile (1,250,000 units), which had to be determined first. The median of demands above 1,900,000 units is the .875 fractile of 2,200,000 units. Similarly, the median of the demands below 1,100,000 is the .0625 fractile of 950,000 units, whereas the median demand level above 2,200,000 is the .9375 fractile of 2,450,000.

The fractiles and the corresponding demands are plotted as points in Figure 22-3. The vertical axis represents the cumulative probability for demand. A curve has been smoothed through these points, which serves as an approximation of the cumulative probability distribution for the first-year demand for the snack product. The curve has an S shape; its slope increases initially and then decreases over higher levels of demand. The slope changes most rapidly for large and small demands, so that more points in these regions provide greater accuracy. This is why we work outward from the median in assessing the demand fractiles.

This example illustrates how we can obtain a very detailed measurement of judgment by posing a few 50–50 gambles. As a rule of thumb, the seven fractile values ranging from .0625 to .9375 in Table 22-1 are adequate for this purpose. Little can be gained from obtaining more fractiles, since further gambles might result in a lumpy curve and would probably not alter the basic shape anyway. Besides, there is no reason to "gild the lily" or to try the decision

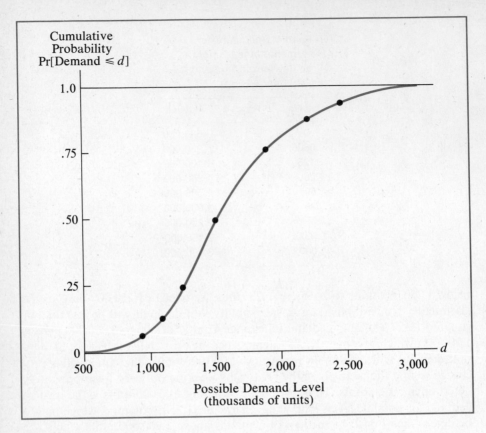

FIGURE 22-3
The cumulative probability distribution function, obtained by judgmental assessment, for a new snack product.

maker's patience. A curve obtained by following this procedure provides about as accurate a judgmental assessment as is humanly possible.

Common Shapes of Subjective Probability Curves

Ordinarily, subjective probability distributions obtained by judgmental assessment provide S-shaped graphs that are elongated at either the top or the bottom. Such graphs represent underlying probability distributions that are skewed to the left or to the right, as the corresponding frequency curves in Figure 22-4(a) and (b) show. Although skewed distributions are most common for business and economic variables, a symmetrical distribution, shown in Figure 22-4(c), is also possible.

Lumpy cumulative probability graphs with two stacked S-shaped curves, like the one in Figure 22-4(d), are to be avoided. The corresponding frequency

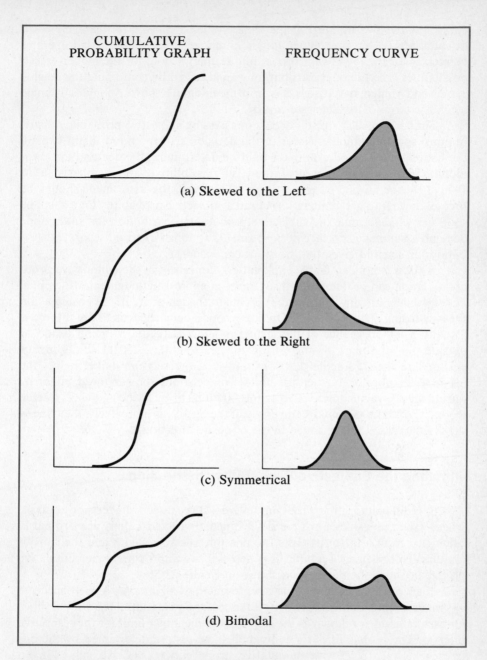

FIGURE 22-4
Possible shapes of subjective probability distributions.

curve has the two-humped shape that typifies a *bimodal distribution*. Such distributions reflect some underlying nonhomogeneous influence that operates differently for the lower-valued possibilities than it does for the higher-valued possibilities. The bimodal distribution is epitomized by combining the heights of men and women. In statistics, it is more meaningful to portray male and female heights in terms of two separate curves.

Similarly, if such a result occurs in assessing subjective probability distributions, some identifiable factor in the decision maker's mind might explain the bimodality. A good example would be determining the demand for automobiles in the next model year. If there is a possibility of an oil embargo, as in 1973–1974, or an energy crisis, as in 1978–1980, the assessment should be broken down into greater detail: (1) Find a subjective probability for a gasoline shortage; (2) determine the subjective probability distribution for automobile demand assuming a shortage occurs; and (3) establish a second separate distribution for demand given that no shortage occurs.

If there is no identifiable explanation, lumpiness in the cumulative probability graph can be due to inconsistencies in expressing judgment, which may be resolved by moving one or more points to the left or right and posing again the succeeding 50–50 gambles. One easy consistency check is to see if there is actually a 50–50 chance of the factor under consideration falling inside or outside the interquartile range, from the .25 fractile (1,250,000 units in our example) to the .75 fractile (1,900,000 units). If the decision maker judges the inside to be more likely, then the middle range should be narrowed, either by raising the .25 fractile (perhaps to 1,300,000 units) or by reducing the .75 fractile (perhaps to 1,800,000 units). Conversely, if the outside is more likely, then one of the fractiles should be changed in the opposite direction.

Approximating the Subjective Probability Distribution

It is difficult to employ the cumulative S curve directly in decision analysis, where expected values must be determined. Expected values are ordinarily calculated from a table that lists the possible variable values and their probabilities. To obtain such a table, it is necessary to approximate the cumulative probability curve by the method shown in Figure 22-5.

Each possible variable value is represented by an interval. A fairly accurate approximation is obtained with 10 intervals of equal width. The probabilities for each interval are shown as the step sizes at the upper limit for the respective interval. The probabilities for individual intervals are determined by the difference between successive cumulative probability values. All values in an interval are represented by a typical value. For this purpose, the midpoint is used.

To see how this is done, suppose that the decision maker now wishes to establish subjective probabilities for intervals of demand from 500,000 to 3,000,000 units in increments of 250,000. Table 22-2 shows how these demand

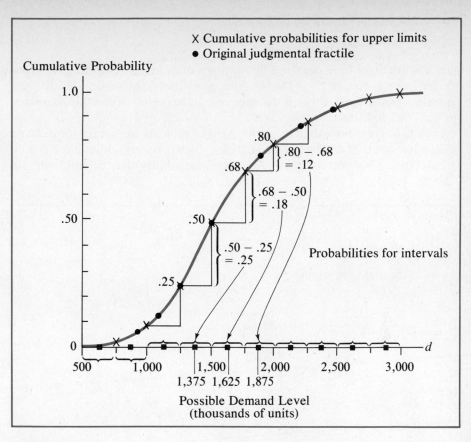

X Cumulative probabilities for upper limits
● Original judgmental fractile

Cumulative Probability

FIGURE 22-5

Approximating the new snack product's cumulative probability distribution using 10 intervals.

TABLE 22-2

Subjective Probabilities for Intervals of Demand
and an Approximate Calculation of the Expected Demand

(1) Demand Interval (thousands)	(2) Interval Midpoint (thousands)	(3) Probability for Demands at or Below Upper Limit (obtained from curve)	(4) Probability for Demand Interval	(5) Demand × Probability (2) × (4)
500–750	625	.02	.02	12.50
750–1,000	875	.08	.06	52.50
1,000–1,250	1,125	.25	.17	191.25
1,250–1,500	1,375	.50	.25	343.75
1,500–1,750	1,625	.68	.18	292.50
1,750–2,000	1,875	.80	.12	225.00
2,000–2,250	2,125	.89	.09	191.25
2,250–2,500	2,375	.95	.06	142.50
2,500–2,750	2,625	.98	.03	78.75
2,750–3,000	2,875	1.00	.02	57.50
			Approximate expected demand =	1,587.50

probabilities have been obtained by reading values from the cumulative probability curve in Figure 22-5. The resulting probability distribution can be used in further evaluations. From it the approximate expected demand is computed to be 1,587,500 units.

The interval probabilities can be used to plot the histogram for demand shown in Figure 22-6. The height of .25 for the bar covering the interval from 1,250,000 to 1,500,000 units represents the probability that demand will fall

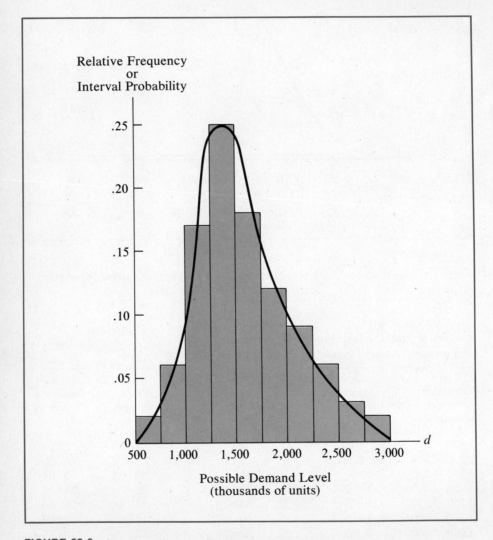

FIGURE 22-6

The frequency curve and the individual interval probabilities, obtained by judgmental assessment, for the demand for a new snack product.

somewhere between these amounts. Superimposed onto this histogram is a smoothed curve representing the judgmental frequency curve for demand. Note that the curve is positively skewed.

EXERCISES

22-5 Use the cumulative probability distribution in Figure 22-7 to determine the following probabilities:
(a) $\Pr[D > 500]$ (c) $\Pr[D \geq 300]$
(b) $\Pr[D \leq 150]$ (d) $\Pr[200 \leq D \leq 800]$

22-6 Use the cumulative probability distribution graph in Figure 22-7 to determine the following fractiles:
(a) .10 (b) .50 (c) .125 (d) .75 (e) .37

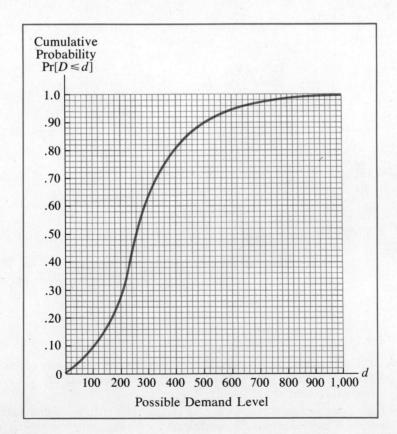

FIGURE 22-7

22-7 The following fractiles apply to the subjective probability distribution for the demand for a new product:

Fractile	Quantity
.0625	10,000
.125	25,000
.25	35,000
.50	40,000
.75	45,000
.875	55,000
.9375	75,000

Establish the probabilities that demand will fall within the following limits:
(a) 10,000 to 40,000   (c) 25,000 to 55,000
(b) 35,000 to 75,000   (d) 25,000 to 45,000

22-8 A real estate investor has established the following judgmental results regarding the rate of return on a proposed project. No value less than -20% or greater than 40% is possible.

Rate of Return	50–50 Point
all	15%
below 15%	7
above 15%	20
below 7%	3
above 20%	24
below 3%	−4
above 24%	28

(a) Complete the following table for the investor:

Fractile	Rate of Return
0	
.0625	
.125	
.25	
.50	
.75	
.875	
.9375	
1.000	

(b) Plot the cumulative probability distribution for the investor's rate of return.
(c) From your graph, find the investor's subjective probability that the rate of return is between 20% and 40%.

22-9 Envision your income during the first full calendar year after graduation. Establish your own subjective probability distribution for the adjusted gross income figure that you will report to the IRS. (If applicable, include your spouse's earnings, interest, dividends, and other income.) If your graph has an unusual shape, try to eliminate any inconsistencies or to identify the nonhomogeneous factors (such as pregnancy, unemployment, or divorce) that might explain the shape. Remember, you are the expert about yourself.

22-10 Use the cumulative probability distribution for demand in Figure 22-7 to construct the approximate probability distribution for demand, using five intervals in increments of 200. Select the midpoints of these intervals as representative values, and determine the approximate expected demand.

FINDING A PROBABILITY DISTRIBUTION **22-5** FROM ACTUAL/FORECAST RATIOS

The interview method is most suitable for use on a one-time basis when uncertain circumstances are involved that may never be encountered again. When judgmental forecasts of a single value are made more often, they provide a history that can be used to determine the underlying probability distribution.

We will illustrate this procedure with an example involving weekly sales forecasts for Blitz Beer made by the company's sales manager. The relevant data over a 10-month period are provided in Table 22-3. We will assume that the forecasts have been made solely from judgment. The actual sales values are divided by the respective forecast sales figures to provide *actual/forecast ratios*.

Actual sales of Blitz Beer for the first week were 1,133 barrels. The manager had forecast 1,200 barrels. Thus

$$\frac{\text{Actual sales}}{\text{Forecast sales}} = \frac{1,133}{1,200} = .94 \qquad \text{(actual/forecast ratio)}$$

TABLE 22-3
Blitz Beer Weekly Sales (in barrels)
Showing Judgment Forecasts
and Actual/Forecast Ratios

Actual	Fractile	Actual/Forecast
1,133	1,200	.94
1,422	1,150	1.24
1,288	1,300	.99
1,317	1,370	.96
1,080	1,410	.77
1,344	1,580	.85
1,506	1,650	.91
1,752	1,650	1.06
1,924	1,750	1.10
1,783	2,000	.89

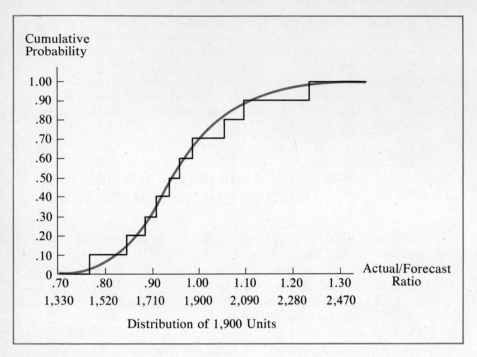

FIGURE 22-8
Subjective probability distribution for Blitz Beer sales based on actual/forecast ratios.

The sales manager's forecasting record is summarized in the cumulative probability graph in Figure 22-8. To plot this, the actual/forecast ratios are arranged in increasing value. There are 10 ratios, so each ratio is assigned a probability of 1/10. Thus, a .10 step in cumulative probability occurs at each value. A smoothed curve is then drawn freehand through the resulting cumulative probability stairway. This curve can be combined with the sales manager's next forecast to obtain the cumulative probability distribution for beer sales for that week. For instance, to find the probability distribution when the sales forecast is 1,900 barrels, the actual/forecast ratios on the horizontal axis are multiplied by 1,900.

EXERCISES

22-11 A sports writer has forecast Rod Carew's seasonal batting average for a 10-year period. The following values apply:

Actual	Forecast	Actual	Forecast
.273	.250	.350	.278
.332	.315	.364	.340
.366	.320	.359	.355
.307	.365	.331	.360
.318	.330	.388	.373

(a) Compute the actual/forecast ratios. Plot the cumulative probability graph, and then sketch a smoothed curve through the stairway.

(b) Suppose that the forecast batting average is .350 for the next baseball season. Read your graph to find the sports writer's subjective probability that Carew's batting average will be (1) $\leq .300$; (2) $\leq .360$; (3) $\leq .375$.

ADDITIONAL REMARKS   22-6

We have seen that probability values for decision making may sometimes be obtained from past history, but that past history is of limited use in business situations and may not exist at all for nonrepeatable circumstances. We have emphasized the direct assessment of a decision maker's judgment rather than traditional statistical techniques. But it should be noted that traditional techniques also rely heavily on judgmental inputs (usually of an indirect nature).

It is not necessary for a decision maker to obtain subjective probabilities personally. Such judgments can be delegated. For example, the chairman of General Motors might rely on various officers within the corporation to determine some or all of the probabilities to be used in analyzing a decision. After all, this is an area in which expert opinion should be relied on whenever possible. Although exercising judgment to find subjective probabilities is similar to assessing attitudes toward decision outcomes (constructing and then using a utility curve), it is dangerous for any decision maker to delegate the assessment of attitudes to others. Attitudes are highly personal and express unique tastes, whereas judgment can be shared. (The collective assessment of attitudes is prohibited by the axioms of utility theory. However, there is no reason why a committee cannot determine the subjective probabilities to be used.)

REVIEW EXERCISES

22-12 Use your judgment to assess probabilities for the following events:
(a) The New York Yankees will play in and win the next World Series.
(b) If you are presently single, you will marry within one year. If you are presently married, your spouse will change jobs within one year.
(c) You will replace one of your present automobiles in the coming year.

22-13 The dean of The Dover School of Business is assessing the probability distribution for next term's grade point average (GPA) for the entire school. It is "even money" that the GPA will fall at or below 2.75. The dean assigns a 50–50 probability that the low side will be at or below 2.70 and that the high side will be at or above 2.80. If the GPA is lower than 2.70, it is a coin-tossing proposition that it will fall at or below 2.60; similarly, the odds are even that the GPA will fall at or above 2.95, given that it lies above 2.80. Find the subjective probability that the GPA will lie within the following limits:

(a) 2.60 and 2.95   (c) 2.60 and 2.75
(b) 2.70 and 2.80   (d) 2.75 and 2.95

22-14 The yield of an active ingredient from a chemical process is assumed to be normally distributed. The 50–50 point is 30 grams per liter. The interquartile range (middle 50%) spans 4 grams per liter.

(a) Find the mean and the standard deviation of the subjective probability distribution.
(b) Find the probability that the yield falls (1) below 29 grams; (2) between 28.5 and 30.5 grams; (3) above 31.5 grams.

22-15 Consider the total size of the U.S. car market (passenger cars sold—both imported and domestic) from October of the current year to the following October. Establish your subjective probability distribution for this quantity, and plot the results on a cumulative probability graph.

22-16 The following sales data apply to Deuce Hardware (in thousands of dollars).

Actual	Forecast	Actual	Forecast
550	600	650	700
490	550	700	690
610	560	680	700
580	600	780	750
620	650	800	850

(a) Compute the actual/forecast ratios. Plot the cumulative probability graph, and then sketch a smoothed curve through the stairway.
(b) Suppose that the sales forecast for next year is 1,000 thousand dollars. Read from your curve the subjective probability that sales will fall at or below (1) 920; (2) 970; (3) 1,100.

23

Bayesian Analysis of Decisions Using Experimental Information

W e usually associate the term *experiment* with a test or an investigation. All experiments have one feature in common: *They provide information.* This information may serve to realign uncertainty. Information obtained by observing a solar eclipse can support hypotheses regarding the effect of the sun's gravity on stellar light rays. The way in which a person responds to your questions can help you decide whether you want him or her for a friend. *An experiment can help us make better decisions under uncertainty.*

However, most experiments are not conclusive. Any test can camouflage the truth. For instance, some potentially good employees will flunk well-designed employment screening tests, and some incompetents will pass them. Another good example is the seismic survey, which provides geological information about deep underground rock structures and is used to explore for oil deposits. Unfortunately, a seismic survey can deny the presence of oil in a field that is already producing oil and can confirm the presence of oil under a site that has already proved to be dry. Still, such imperfect experiments can be valuable. An unfavorable test result can increase the chance of rejecting a poor prospect—a job applicant or a drilling site—and a favorable test result can enhance the likelihood of selecting a good prospect.

In this chapter, we will incorporate experimentation into the framework of our decision-making analysis. The information we obtain will affect the probabilities of the events that determine the consequences of each act. We can revise the probabilities of these events upward or downward, depending on the evidence we obtain. Thus, a geologist will increase the subjective probability for oil if the seismic survey analysis is favorable and will lower this probability if the survey is unfavorable.

The seismic survey epitomizes the role of experimental information in decision making. In business situations, several other classic sources of such information are commonly employed. A marketing research study serves to realign uncertainty regarding the degree of success that a new product will achieve in the marketplace. An aptitude test is often used to help predict a job applicant's future success or failure if he or she is hired—a decision that involves considerable uncertainty. A sampling study is frequently employed to facilitate quality-control decisions related to how satisfactorily items are produced or how many defective items are arriving from a supplier.

First, we will investigate how to revise probabilities in accordance with experimental results by applying the probability concepts associated with Bayes' Theorem, which we discussed in Chapter 5. Then we will consider *posterior analysis*, which melds the probability information with all of the decision elements by means of a decision tree diagram. How much information is to be incorporated into the decision serves as a prelude to *preposterior analysis*, a more general analysis of the initial choice to experiment. For example, would an oil wildcatter use a seismic survey if it cost $50,000?

23-1 PROBABILITY ANALYSIS FOR DECISION MAKING

The decision maker can usually make some kind of judgment about the uncertain events, which may be expressed as a set of *prior probabilities* for the respective events. Occasionally, such a judgment must be quantified in terms of *subjective probabilities*, because the events in question frequently arise from nonrepeatable circumstances. At other times, the prior probabilities may be *objective* in nature. In accordance with the information obtained from the experiment, the event uncertainties are realigned to obtain *posterior probabilities*. Figure 23-1 presents the sequence of steps in this procedure—exactly the one originally proposed by Thomas Bayes.

Step 1

EXERCISE JUDGMENT

- Assign prior probabilities to main events. These may be either
  (1) objective probabilities
  or (2) subjective probabilities

For example: In drilling for oil on a wildcat site, there will be either oil *(O)* or a dry hole *(D)*.

$$\Pr[O] = .5$$
$$\Pr[D] = .5$$

Step 2

ASSESS THE EXPERIMENT

- Identify meaningful results
- Determine conditional result probabilities for the experiment given each main event using
  (1) reliability history
  or (2) probability logic

For example: Two results of a seismic survey are favorable *(F)* and unfavorable *(U)* predictions. The conditional result probabilities are

$$\Pr[F \mid O] = .9 \quad \Pr[U \mid O] = .1$$
$$\Pr[F \mid D] = .2 \quad \Pr[U \mid D] = .8$$

Step 3

REVISE THE PROBABILITIES

- Apply principles of Bayes' theorem to obtain
  (1) unconditional result probabilities for experimental outcomes
  (2) posterior probabilities for main events

For example: The unconditional result probabilities are

$$\Pr[F] = .55 \qquad \Pr[U] = .45$$

The posterior probabilities are

$$\Pr[O \mid F] = 9/11 \quad \Pr[D \mid F] = 2/11$$
$$\Pr[O \mid U] = 1/9 \quad \Pr[D \mid U] = 8/9$$

FIGURE 23-1

Steps in performing the probability portion of the decision-making analysis when experimental information is used.

The Oil Wildcatter: A Case Illustration

For the purpose of illustration, we will suppose that an oil wildcatter must decide whether to drill for oil on a leased site. The wildcatter is contemplating hiring a geologist to conduct a detailed seismic survey of the area. At present, we are concerned only with the probability portion of this problem.

As a first step, the wildcatter must *exercise judgment* regarding the likelihood of striking oil. Since no two unproved drilling sites are very similar, no historical frequency is available. The wildcatter must therefore rely on a subjective probability value. Suppose that he believes there is a 50–50 chance of striking oil. Letting O represent oil and D represent a dry hole, the prior probabilities for the basic events are

$$\Pr[O] = .5$$

$$\Pr[D] = .5$$

The next step the wildcatter takes is to *assess the experiment*—the seismic survey, in this case. He begins by contemplating what results would be meaningful to him. Although the seismic output might be highly complex and varied, we will assume, for simplicity, that the geologist's analysis can lead to only two meaningful results: a favorable (F) prediction for oil or an unfavorable one (U). It is necessary to obtain *conditional result probabilities* for the respective seismic outcomes given each possible basic event. Ordinarily, the conditional result probabilities for the experiment can be obtained objectively, either by estimation based on historical frequencies or by application of the underlying logic of probability. Thus, we may refer to these values as "logical-historical" probabilities to distinguish them from the several other types of probabilities that we will encounter.

In our present example, the geologist has recorded the "batting average" for the procedure. Historical records show that on 90% of all fields known to produce oil, the survey's prediction of oil has been favorable; that is, 90% of all similar seismic survey data have provided favorable oil predictions when oil did exist. Of course, such a percentage should not be biased by the fact that the seismic survey result may have affected the earlier decisions to drill on those sites. It is best to obtain a reliability figure by conducting a special test of the tester itself, which might be done by taking special seismic measurements on sites that are already producing oil. Similarly, by taking simulated readings on known dry holes, the geologist has determined that the survey is only 80% reliable in making an unfavorable prediction when no oil is present. The appropriate conditional result probabilities are

$$\Pr[F|O] = .90 \quad \text{and} \quad \Pr[U|O] = .10$$

$$\Pr[F|D] = .20 \quad \text{and} \quad \Pr[U|D] = .80$$

These are *historical probabilities* and can be regarded as statistical estimates of the underlying values, since they are based on limited samples of drilling sites. (Note that the conditional probability for a favorable result given oil is greater than the probability for an unfavorable prediction given a dry hole. There is no reason why a test must be equally discerning in both directions.)

In other situations, conditional result probabilities for an informational experiment can be obtained more directly without relying on historical frequencies. This would be true, for example, in assessing a quality-control sample. The precise probability distribution for the sample result can be determined through logical deduction, based only on the principles of probability and the type of events that characterize the sampled population. In Chapter 24, we will see how such *logical probabilities* can be determined using the binomial distribution.

The final step toward incorporating experimental information into the probability portion of decision analysis is to *revise the probabilities.* This revision ordinarily results in two kinds of probability values that are applicable at different stages of uncertainty: The *posterior probabilities* apply to the main events, and the *unconditional result probabilities* apply to the experimental outcomes themselves. Although the underlying concepts of Bayes' Theorem are used to arrive at these values, there is a more streamlined procedure that proves more convenient when using probability tree diagrams to analyze a decision.

Using Probability Trees

The probability tree diagram in Figure 23-2(a) depicts the *actual chronology* of events in our illustration. The first fork represents the events for the site status: oil or dry. The second forks represent the seismic survey results. This particular arrangement follows the sequence in which the events actually occur: first, nature determined (several million years ago) whether this site would cover an oil field; second, our geologist conducts a seismic test today. The actual chronology also adheres to the manner in which the probability data were initially obtained. The wildcatter has directly assessed the possibilities for the site-status events, and the geologist has indicated the reliabilities for the survey. Thus, the values given earlier for the prior probabilities for oil and dry and the conditional result probabilities are placed on the corresponding branches in the probability tree in (a).

The probability tree diagram in Figure 23-2(b) represents the *informational chronology.* This is the sequence in which the decision maker finds out what events occur. First, the wildcatter obtains the result for the seismic survey, which is portrayed by the initial event fork. Then, if he chooses to drill, he ultimately determines whether or not the site covers an oil field. This is the sequence of events as they would appear on a decision tree (which will be discussed later). But this particular chronology does not correspond directly to the initial probability data. Additional work is required to obtain the probability values shown on the tree diagram in (b).

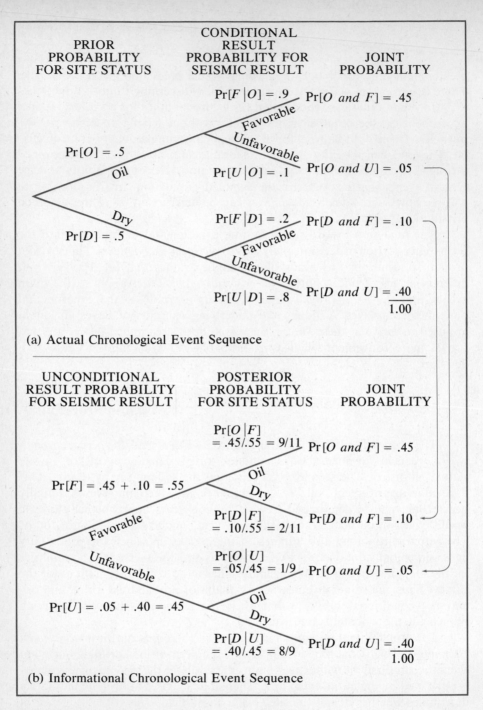

FIGURE 23-2

Probability tree diagrams showing the event chronologies for drilling a wildcat well and using a seismic survey.

We begin by multiplying the branch probabilities together on each path in the tree diagram in Figure 23-2(a) to obtain the corresponding joint probability values. The same numbers apply regardless of the chronology, so the joint probabilities can be transferred to tree diagram (b). This must be done with care, since the in-between joint outcomes are not listed in the same order in (b) as they are in (a), because the analogous paths (event sequences) differ between the diagrams. For example, in diagram (a) we obtain the joint probability for oil and an unfavorable seismic result

$$Pr[O \text{ and } U] = Pr[O] \times Pr[U|O] = .5 \times .1 = .05$$

This is the second joint probability in diagram (a) and corresponds to the third end position in diagram (b).

Next, we work entirely in diagram (b). First, we compute the unconditional result probabilities at the first stage. Here we use the addition law to obtain

$$Pr[F] = Pr[O \text{ and } F] + Pr[D \text{ and } F] = .45 + .10 = .55$$
$$Pr[U] = Pr[O \text{ and } U] + Pr[D \text{ and } U] = .05 + .40 = .45$$

These values are placed on the applicable branches at the first stage. Finally, the posterior probabilities for the second-stage events are computed using the basic property of conditional probability:

$$Pr[A|B] = \frac{Pr[A \text{ and } B]}{Pr[B]}$$

Thus, we determine the posterior probability for oil, given a favorable seismic survey result, to be

$$Pr[O|F] = \frac{Pr[O \text{ and } F]}{Pr[F]} = \frac{.45}{.55} = \frac{9}{11}$$

This value is placed on the second-stage branch for oil that is preceded by the earlier branch for a favorable result. Each of the other posterior probabilities shown in diagram (b) is found by dividing the respective end-position joint probability by the probability on the preceding branch.

Probabilities must be revised in this manner whenever experimental information is used in decision making. This happens because we ordinarily obtain our probabilities in the reverse chronology from the chronology required to analyze the problem.

EXERCISES

23-1 An oil wildcatter has assigned a .40 probability to striking oil on his property. He orders a seismic survey that has proved only 80% reliable in the past. Given oil, it predicts favorably 80% of the time; given no oil, it augurs unfavorably with a frequency of .8.

Construct probability trees for the actual and informational chronologies, and indicate the appropriate probability values for each branch and end position.

23-2 Your friend places two coins in a box. The coins are identical in all respects, except that one is two-headed. Without looking, you select one coin from the box and lay it on the table.

(a) What is the prior probability that you will select the two-headed coin?

(b) As a source of predictive information about the selected coin, you may examine the showing face. Construct the probability tree diagram for the actual chronology of events.

(c) After you have examined the showing face of the coin, you may then turn it over to see what is on the other side. Construct the probability tree diagram for the informational chronology.

(d) If a head shows before you turn the coin over, what is the posterior probability that you selected the two-headed coin?

23-3 Solve the weather-forecasting problem originally posed in Exercise 5-51 (page 150) by constructing probability trees for the actual and informational chronologies.

23-4 Solve the toothpaste-marketing problem posed in Exercise 5-52 (page 151) by constructing probability trees for the actual and informational chronologies.

23-2   POSTERIOR ANALYSIS

We have seen how to determine the probability values that incorporate the "main" events and experimental results. Decision makers can use that information in determining what actions to take. We will now show how such choices can be reached by employing decision tree analysis to maximize profit.

Decision Tree Analysis

Figure 23-3 is a decision tree diagram of the wildcatter's choices after the seismic results are known. Here, we have assumed that (1) the lease will be sold for $250,000 on striking oil; (2) the cost of drilling will be $100,000; and (3) the seismic survey will cost $25,000. The decision to drill or to abandon the lease follows the seismic survey result, since the wildcatter obviously would not reach a decision before finding out the geologist's prediction. The revised probabilities found earlier for the informational chronology are used. Since the posterior probabilities for the site-status events apply at the two decision

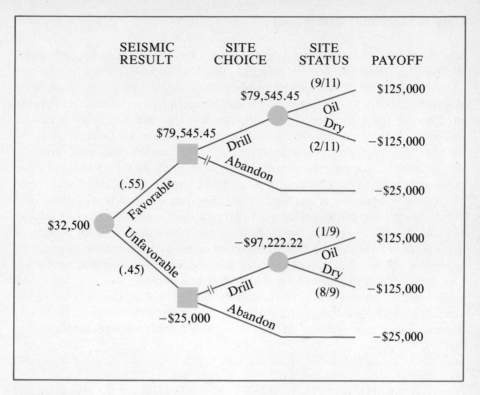

SEISMIC RESULT   SITE CHOICE   SITE STATUS   PAYOFF

FIGURE 23-3
The wildcatter's decision tree diagram after the seismic survey is taken.

points, using this decision tree as the basis for decision making is called *posterior analysis.*

Note that there is no event fork following the act to abandon the lease, because the oil wildcatter will never find out if there is oil unless he drills for it. (The uncertainty still exists, nevertheless; it would do no harm to have event forks for oil versus dry at those points, but the payoff would be −$25,000 in either case, and an identical conclusion would be reached.)

Performing backward induction, we see that the wildcatter would prune the abandon branch and drill if a favorable seismic result were obtained and would do the opposite in the case of an unfavorable prediction. Even though drilling will lead to identical payoffs for either seismic result, the posterior probabilities for oil and dry are different for the favorable and unfavorable predictions. The expected payoff from drilling is $79,545.45 for a favorable seismic result, but it is a negative value (−$97,222.22) for an unfavorable result. The wildcatter's expected payoff for the optimal strategy is $32,500. As we will see later, this amount will be helpful in determining whether the seismic survey should be used at all.

Obviously Nonoptimal Strategies

In the simpler decision structures, it may be convenient to streamline the decision tree diagram. We can conclude that the wildcatter would prune the same branches for almost any plausible payoffs that might apply. Since he is paying $25,000 for the seismic survey and this experiment provides fairly reliable predictions, the wildcatter should choose acts that are consistent with the information obtained. But the tree in Figure 23-3 allows for three other strategies: (1) drill regardless of the result (prune both abandon branches); (2) abandon in either case (prune the two drill branches); or (3) do the opposite of what is predicted (prune the drill branch if the seismic result is favorable, and prune the abandon branch if it is unfavorable). The last strategy is ridiculous and would never be considered. The other two strategies are inferior to two actions not shown on the present trees—drilling or abandoning without benefit of seismic results—since the $25,000 cost could be saved in either case by not even using the seismic. Such inferior strategies are *obviously nonoptimal strategies*.

Figure 23-4 illustrates how the wildcatter's decision tree diagram could have been drawn to exclude the obviously nonoptimal strategies. This representation can help to simplify an otherwise complex decision tree. However, for expository convenience we will always use the complete decision tree

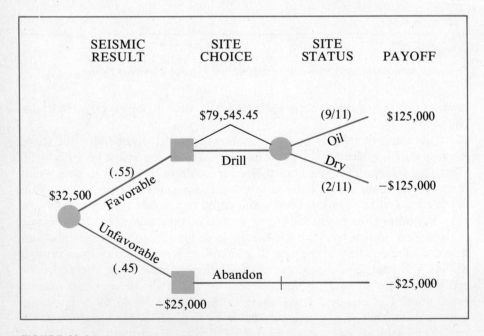

FIGURE 23-4

The simplified wildcatter's decision tree diagram with obviously nonoptimal strategies excluded.

diagram. When more than two acts or experimental results are involved, it is not easy to determine which strategies are obviously nonoptimal.

EXERCISES

23-5 The following payoff table of marketing choices for a new film has been determined by the management of a motion picture studio:

Box Office Result Events	Distribute as "A" Feature	Sell to TV Network	Distribute as "B" Feature
Success	$5,000,000	$1,000,000	$3,000,000
Failure	−2,000,000	1,000,000	−1,000,000

The prior probability for a box-office success has been judged to be .3. The studio plans a series of sneak previews. Historically; 70% of all the studio's successful films have received favorable previews, and 80% of all the studio's box-office failures have received unfavorable previews.
(a) Construct the probability tree diagrams for the actual and informational chronologies.
(b) Construct a table indicating all of the studio's possible strategies contingent on results of the sneak preview.
(c) Construct a decision tree diagram for the studio, assuming that the film will definitely be previewed.
(d) Perform backward induction analysis. What is the optimal course of action? To which strategy in (b) does this correspond?

23-6 The makers of Quicker Oats oatmeal have packaged this product in cylindrical containers for 50 years. Management believes that the cylindrical container is inseparable from the product's image. But consumer tastes change, and the new marketing vice president wonders if younger people will regard the round box as old-fashioned and unappealing. The vice president wishes to analyze whether or not to package Quicker Oats in a rectangular box that will save significantly on transportation costs by eliminating dead space in the packing cartons. It is also believed that the change can actually expand Quicker Oats' market by modernizing the product's image. But previous study has shown that a small segment of the existing market buys the oatmeal primarily for the round box; these customers would be lost if the package were changed. The following payoff table has been established for the present net worth of retaining the old box versus using the new box:

National Market Response to New Box Events	Act	
	Retain Old Box	Use New Box
Weak (W)	$0	−$2,000,000
Moderate (M)	0	0
Strong (S)	0	3,000,000

As prior probabilities for the new box response events, the marketing vice president arrived at the following estimates: $\Pr[W] = .20$; $\Pr[M] = .30$; $\Pr[S] = .50$. The new box is to be test marketed for six months in a "barometer" city. Three outcomes are possible: decreased sales (D), unchanged sales (U), and increased sales (I). Historical experience with other products has established the following conditional result probabilities:

$$\Pr[D|W] = .8 \qquad \Pr[D|M] = .2 \qquad \Pr[D|S] = 0$$
$$\Pr[U|W] = .2 \qquad \Pr[U|M] = .4 \qquad \Pr[U|S] = .1$$
$$\Pr[I|W] = 0 \qquad \Pr[I|M] = .4 \qquad \Pr[I|S] = .9$$

(a) Construct the probability tree diagrams for the actual and informational chronologies.
(b) Construct the Quicker Oats decision tree diagram, assuming that the new box will be test marketed.
(c) Perform backward induction. Then indicate the maximum expected payoff act for each test outcome. What is the optimal strategy?

23-3   THE DECISION TO EXPERIMENT: PREPOSTERIOR ANALYSIS

In the previous sections of this chapter, we illustrated the use of experimental information and described posterior analysis, which tells us what act we should select for each experimental outcome. Now we will incorporate into our analysis the additional choice of whether or not to obtain experimental information in the first place. The decision-making process is therefore expanded to include an initial stage involving the selection of acts concerning experimentation. The procedure employed to evaluate this expanded decision is sometimes referred to as *preposterior analysis*.

To illustrate how to incorporate the decision to use experimental information, we expand the oil wildcatter's decision. The expanded decision tree is presented in Figure 23-5. The additional decision of whether or not to take the seismic test is treated as an initial decision point and becomes the initial act fork with branches for taking and not taking the seismic test. If the seismic test is taken, an event fork follows that is related to the seismic result to be achieved; here, the unconditional result probabilities apply. These events are followed by the final decision to drill or to abandon the site. If the wildcatter drills, the last set of event forks represents the oil and dry events, and the posterior probabilities apply. If the wildcatter initially decides not to take the seismic survey, then the choice to drill or to abandon the site must be made without any information; this is represented as the act fork at the bottom of the tree. In this case, drilling

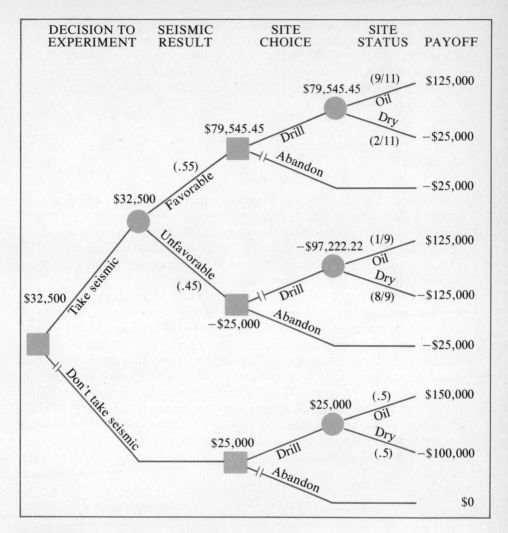

DECISION TO EXPERIMENT	SEISMIC RESULT	SITE CHOICE	SITE STATUS	PAYOFF

FIGURE 23-5

The wildcatter's decision tree diagram incorporating the initial decision regarding the seismic test.

leads to a final event fork for the site-status events. Here, the original prior probabilities apply for the oil and dry events. The payoffs in this bottom portion of the tree are $25,000 greater than their counterparts directly above, since the cost of the seismic survey is saved.

We now have a two-stage decision problem to analyze. Performing backward induction on the top portion of Figure 23-5, we obtain the result that we found earlier: Using the seismic survey yields an expected payoff of $32,500. We find that not using the survey leads to a smaller expected payoff of only $25,000. Thus, the branch "don't take seismic" is pruned at the first decision point. The course of action that will maximize the expected payoff is to take the seismic survey, and drill if it is favorable but abandon the site if it is unfavorable.

The Role of EVPI

In some decision-making situations, the forgoing procedure can be shortened considerably. Recall that the *expected value of perfect information*, or EVPI, indicates the worth of the best possible or ideal information about the events in the main decision. Information that is obtained through experiment is far from perfect in its predictive powers. If experimental evidence costs more than it can be worth at best, then it should obviously not be obtained.

In Table 23-1, the wildcatter's EVPI is calculated to be $50,000. (Here, the prior probabilities are used, since the seismic survey does not apply.) If the cost of the seismic survey were higher—say, $60,000—then it would be more profitable for the oil wildcatter not to bother to take the test, regardless of its reliability. In effect, the seismic branch would be pruned from the tree, and there would be no need to calculate posterior probabilities or to conduct any pre-posterior analysis. Of course, this shortcut applies only when the cost of the information exceeds the EVPI. Since the wildcatter has to spend only $25,000, which is a smaller amount than the $50,000 EVPI, the complete preposterior analysis is required in this case.

TABLE 23-1
Calculation of the Wildcatter's EVPI

Event	Probability	Payoff		Row Maximum	Row Maximum × Probability
		Drill	Abandon	Maximum	
Oil	.5	$150,000	$0	$150,000	$75,000
Dry	.5	−100,000	0	0	0
	1.0				$75,000

Expected payoff with perfect information = $75,000
Maximum expected payoff (with no information) = .5($150,000) + .5(−$100,000)
= $25,000
EVPI = $75,000 − $25,000 = $50,000

23-7 The exploration manager for a small oil company must decide whether to drill on a parcel of leased land or to abandon the lease. As an aid in making this choice, the manager must first decide whether to pay $30,000 for a seismic survey, which will confirm or deny the presence of the anticlinal structure necessary for oil. She has judged the prior probability for oil to be .30. For oil-producing fields of similar geology, her experience has shown that the chance of a confirming seismic is .9, but for dry holes with approximately the same characteristics, the probability that a seismic survey will deny oil has been established at only .7. Drilling costs have been firmly established at $200,000. If oil is struck, the manager's company plans to sell the lease for $500,000.
   (a) What is the manager's EVPI for the basic decision, using profit as the payoff measure? Comparing this value to the cost of the seismic survey, can you conclude definitely that no survey should be made?
   (b) Construct the manager's decision tree diagram, and determine the appropriate payoffs.
   (c) Find the revised probabilities for the informational chronology, and place these values on the corresponding branches of your decision tree diagram.
   (d) Perform backward induction analysis to determine the course of action that will provide the maximum expected profit.

23-8 A gas prospector must decide how to dispose of a particular lease that may be sold now for $20,000 or drilled on at a cost of $100,000. The drilling events and their prior probabilities are dry (D) at .6, low-pressure gas (L) at .3, and high-pressure gas (H) at .1. The lease will be abandoned for no receipts if D: it will be sold for $300,000 if L; and it will be sold for $500,000 if H.
   (a) Which act—sell or drill—will maximize expected profit?
   (b) What is the prospector's EVPI?
   (c) For a cost of $10,000, a 90% reliable seismic survey can predict gas favorably (F) with a probability of .90 if there is gas or unfavorably (U) with a probability of .90 if the site is dry, but it cannot measure pressure. Construct probability tree diagrams for the actual and informational chronologies, using the three gas events.
   (d) Perform a decision tree analysis to determine what course of action will maximize the prospector's expected profit.

23-9 The decision tree diagram in Figure 23-6 has been constructed by a marketing manager who wishes to determine how to introduce a new product. The manager judges that the prior probability for marketing success is .40.
   (a) From a consumer survey costing $30,000, the manager can obtain an 80% reliable indication of the product's impact in the marketplace. Thus, the probability for a favorable survey result given market success is .80, and the probability for an unfavorable result given market failure is .80. Find the posterior probabilities for the market events and the unconditional result probabilities for the survey events.
   (b) A sales program costing $50,000 might be conducted in a test region. The results are judged to be 95% reliable. Find the posterior and unconditional result probabilities.

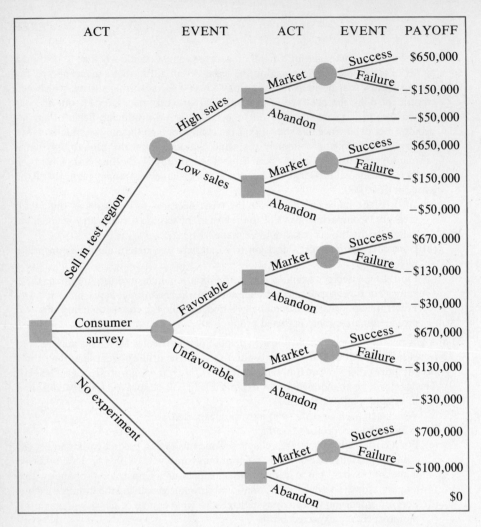

ACT	EVENT	ACT	EVENT	PAYOFF

Sell in test region

High sales
- Market
  - Success — $650,000
  - Failure — −$150,000
- Abandon — −$50,000

Low sales
- Market
  - Success — $650,000
  - Failure — −$150,000
- Abandon — −$50,000

Consumer survey

Favorable
- Market
  - Success — $670,000
  - Failure — −$130,000
- Abandon — −$30,000

Unfavorable
- Market
  - Success — $670,000
  - Failure — −$130,000
- Abandon — −$30,000

No experiment
- Market
  - Success — $700,000
  - Failure — −$100,000
- Abandon — $0

FIGURE 23-6

(c) Using the information given in the problem statement and your answers to (a) and (b), perform backward induction to find the manager's optimal course of action. (The payoffs in Figure 23-6 include the cost of experimenting.)

REVIEW EXERCISES

23-10 Solve the movie reviewer's prediction problem originally posed in Exercise 5-50 (page 150) by constructing the probability trees for the actual and informational chronologies.

23-11 Solve the employment screening-test problem originally posed in Exercise 5-54 (page 151) by constructing the probability trees for the actual and informational chronologies.

23-12 A box contains two pairs of dice. One pair is a fair one. The other pair consists of one die cube with a three on every side and one die cube with a four on every side. A pair is to be selected at random and tossed. You will only be able to see the top showing faces and not the sides. The main events of interest pertain to the crookedness or fairness of the tossed dice. (Both dice in the tossed pair will fall into the same category.) In each of the following cases, construct probability trees for the actual and informational chronologies.

(a) The experimental result is finding out whether or not a seven-sum (three and four, two and five, one and six) occurs.

(b) The experimental result is finding out whether or not a three-four combination occurs, which has a greater predictive worth than the result in (a).

23-13 Lucky Jones must decide whether to participate in a card game offered by Inscrutable Smith. For the price of $5, Jones will draw a card from an ordinary deck of playing cards. If the card is a king, Smith is to pay Jones $60 (so that Jones wins $55). But if the card is not a king, Jones will receive nothing for the $5. Smith, eager for action, offers Jones an additional enticement. For $3, Jones can draw a card without looking at it. Smith will then tell Jones whether or not the card is a face card. If Jones wishes to continue, an additional payment of $5 must be made for the game to proceed.

(a) Construct a decision tree diagram showing the structure of Jones' decision.

(b) Determine the probabilities for the events and the total profit for the end position.

(c) What course of action will provide Jones with the greatest expected payoff?

Decision theory does not burden the decision maker by requiring him or her to do everything at once by choosing a single number— "an α for all seasons."

Bayesian Analysis of Decisions Using Sample Information

In Chapter 23, we considered the general problem of using experimental information in decision making. We will now consider a decision, commonly encountered in business situations, that involves just *two acts*. The experiment is taking a *random sample* from a population whose characteristics will affect the ultimate payoffs. The decision maker's choice depends on the particular sample result obtained.

Two types of populations are encountered in these sampling experiments. The *qualitative population* is comprised of units that can be classified into categories. Examples are persons who can be categorized by occupation (blue-collar, professional), sex (male, female), or preferences (preferring a product, disliking a product); and production items that can be classified in terms of quality (satisfactory, unsatisfactory), weight (below the limit, above the limit), or color (light, medium, dark). The *quantitative population* associates a numerical value with each unit. For example, people can be measured in terms of income levels, aptitude test scores, or years of experience; and items can be assigned numerical values to indicate weight, volume, or quantity of an ingredient.

A sample from a qualitative population tells us how many sample units fall into a particular category, and this number reflects the prevalence of that attribute in the population, which in turn affects the payoff associated with the incidence of the attribute. For example, in deciding how to dispose of a supplier's shipment, a receiving inspector may discover that 12 of the items out of a random sample of 100 are defective, indicating that there is a high probability that the entire shipment is bad and that returning it might maximize expected payoff. The payoffs in such a problem are often expressed in terms of the proportion π of the population having the key attribute (for example, the proportion of defective items in the entire shipment). As in most decision making using experiments, the value of π itself and the number of defectives that will turn up in the sample are both uncertain. Probabilities for the number of defectives can be determined by using the binomial distribution.

A sample taken from a quantitative population provides a similar basis for action, such as accepting or rejecting a machine setting in a chemical process. Here, the mean quantity of a particular ingredient in each gallon may be the determining factor in establishing the payoff. The population mean μ measures the central quantity of the ingredient in all gallons made under that setting. The true value of μ is uncertain. A second area of uncertainty involves the quantities in the sample itself; here, the sample mean $\bar{X}$ may serve as the basis for decision making.

In this chapter, we will consider how to integrate sample information into the basic decision-making structure. As in Chapter 23, the procedure involves prior probabilities of π or μ, which must be revised to coincide with possible sample results to provide posterior probabilities for these population parameters. Backward induction with revised probabilities then provides the optimal decision rule, based on the sample statistic that is obtained. The analysis may be expanded to consider how many sample observations must be made.

24-1   DECISION MAKING WITH THE PROPORTION

Charles Stereo: An Acceptance-Sampling Illustration

Charles Stereo is a chain of retail outlets specializing in sound equipment. Because of its high volume, Charles stocks its inventory of stereo pickup cartridges by ordering lots of 100 units from various manufacturers. Bertrand Charles, the owner, wishes to decide whether to accept or reject a particular lot from one supplier, DICO. Each cartridge in a rejected lot is thoroughly inspected by Charles Stereo, and cartridges that are actually found to be defective are replaced by the supplier. An accepted lot is parceled without inspection to the retail stores for sale to customers, who are then the ones to find the defective cartridges, which Charles replaces without charge from its inventory. DICO will not give Charles credit for cartridges that have been used by retail customers, even if they were originally defective.

One of the Charles employees has suggested that the company sample incoming lots, using the information thereby obtained as a basis for deciding whether a lot should be accepted or rejected. Bertrand is skeptical about the advantages of sampling, since even after an inspection of randomly chosen items he cannot be certain of the value of the lot proportion defective π. An employee volunteers to analyze the decision and chooses the gross profit from a 100-item lot as the payoff measure.

Detailed records of all DICO cartridges received by Charles have been maintained, and the frequencies of lot proportions defective have been found. These frequencies serve as estimates of the following prior probabilities for values of π of .10, .20, and .30:

Possible Lot Proportion Defective π	Prior Probability
.10	.4
.20	.3
.30	.3
	1.0

To simplify our analysis, we will consider values of π only to the nearest whole 10%. (The same procedures would apply if we considered $\pi = .05, .06, \ldots, .35$).

Structuring the Decision

As an example, consider a sample of size $n = 2$ for the Charles decision problem diagrammed in Figure 24-1. An immediate choice, represented by the act fork at decision point a, is to determine whether or not to sample. If the choice is made to inspect the two randomly chosen items, the possible outcomes for the number of defectives are $R = 0$, $R = 1$, or $R = 2$, which are shown as events on the event fork at b. The choices to accept or reject the lot are represented by act branches in the forks at decision points c, d, e, and f. After the acts to accept or reject, the lot proportion defective π is determined either by a 100% inspection in the case of rejection or by a tally of returned cartridges if the lot is accepted. The various possible values of π are shown as events on forks g through n.

The payoffs for each end position are determined by adjusting the $1,000 markup for each lot downward to account for inspection costs and losses incurred when customers return used defective cartridges.

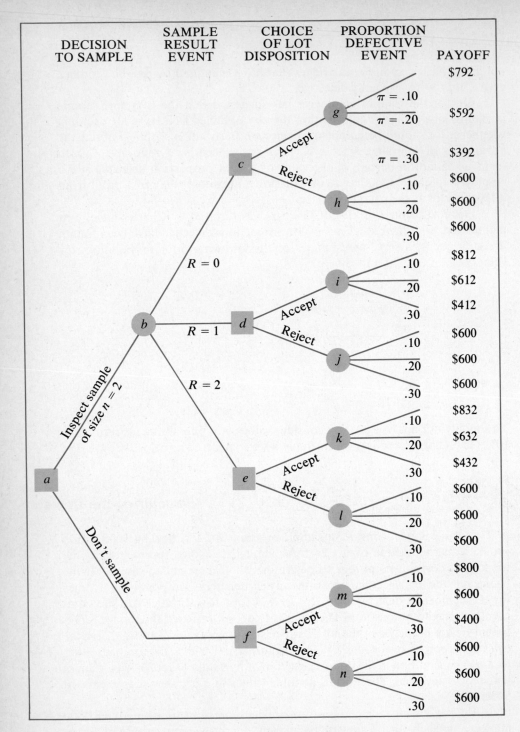

FIGURE 24-1
The decision tree for the Charles Stereo sampling decision.

Determining the Event Probabilities

The actual event chronology is provided in Figure 24-2. There, the first stage represents the prior probabilities for the proportion of defective cartridges in the shipment, and the second stage provides the conditional result probabilities for the number of defectives in the sample. The latter probabilities are

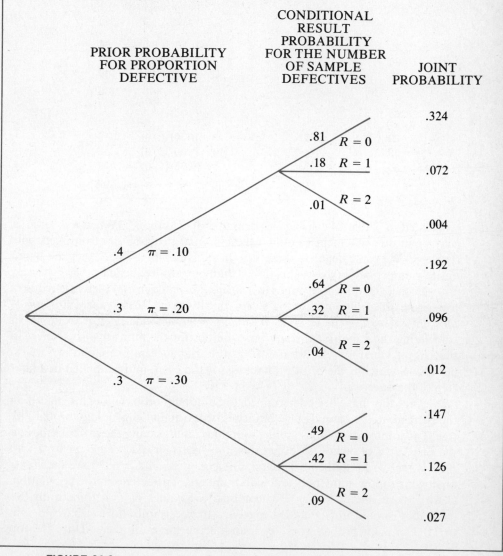

FIGURE 24-2
The actual chronology of events for the stereo-cartridge shipment decision.

TABLE 24-1

Binomial Conditional Result Probabilities
for the Number of Defectives in the Stereo-Cartridge Sample

π	r	$Pr[R = r] = \dfrac{n!}{r!(n-r)!}\, \pi^r(1-\pi)^{n-r}$
.10	0	$1(.10)^0(.90)^2 = .81$
.10	1	$2(.10)^1(.90)^1 = .18$
.10	2	$1(.10)^2(.90)^0 = .01$
		$\overline{1.00}$
.20	0	$1(.20)^0(.80)^2 = .64$
.20	1	$2(.20)^1(.80)^1 = .32$
.20	2	$1(.20)^2(.80)^0 = .04$
		$\overline{1.00}$
30	0	$1(.30)^0(.70)^2 = .49$
.30	1	$2(.30)^1(.70)^1 = .42$
.30	2	$1(.30)^2(.70)^0 = .09$
		$\overline{1.00}$

calculated in Table 24-1 using the binomial distribution.* (When n is large, it may be more convenient to obtain the binomial probabilities from Appendix Table C.) We can categorize these as logical probabilities, since they are based entirely on probability concepts rather than on past frequencies.

Each level of π corresponds to a *different population* and therefore requires a separate set of binomial probabilities. In calculating these values, *it is important not to confuse the level of π with its prior probability*. For example, in calculating the conditional result probabilities for the number of defectives in the sample when $\pi = .10$, we use .10 as the trial success probability π; we do not use .4, which is the prior probability for the event and is applied in a later step of the analysis.

After they have been determined, the binomial probabilities are placed on the corresponding second-stage branches of the actual chronology probability tree. The joint probabilities are then found by multiplying the probabilities on the respective branches. In doing this, *the events should not be confused with their probabilities*. Remember that $\pi = .10$, $\pi = .20$, and $\pi = .30$ are the *events* analogous to low, moderate, and high numbers of defectives in the population. (As events, it makes no sense to combine the values of π arithmetically. So since they should not be added together, they certainly do not have to sum to 1!) The prior probabilities .4, .3, and .3 are the multipliers. Thus, the top

---

* Here, the population of cartridges is fixed in size. Unless the sampling is done with replacement, some error results from using binomial probabilities. But when the population is large in relation to n, such errors are negligible.

joint probability in Figure 24-2 is

$$\Pr[\pi = .10 \; and \; R = 0] = .4 \times .81 = .324$$

and the other joint probabilities are calculated in the same way.

Figure 24-3 presents the informational chronology of events for the sampling experiment. The first stage provides the unconditional result probabilities for the number of sample items defective. This is in accordance with the

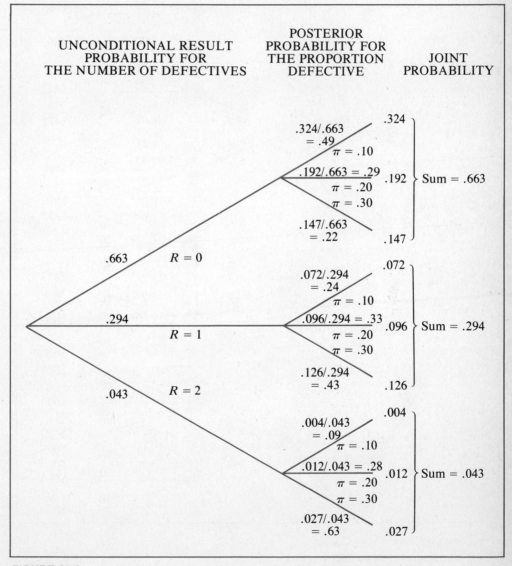

FIGURE 24-3
The informational chronology of events for the stereo-cartridge shipment decision.

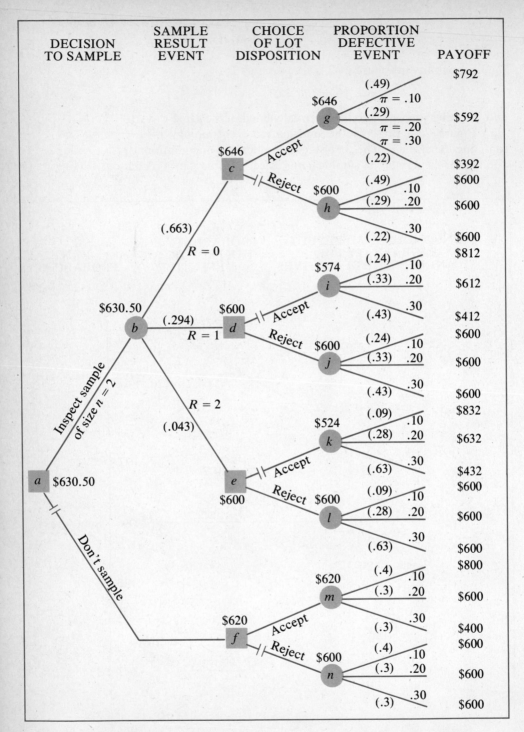

FIGURE 24-4
The Charles Stereo decision tree diagram.

sequence in which Charles Stereo discovers the applicable outcomes. The posterior probabilities for the population proportion of defective cartridges are next given. This informational chronology provides the revised probabilities that are used in the complete decision tree diagram in Figure 24-4.

The Decision Rule

The results of the backward induction analysis are shown in Figure 24-4. Observe that when the number of defectives is $R = 1$ or more, rejection yields a higher expected payoff. This demonstrates the principle that as the number of defective items in the sample becomes larger, the evidence favors larger π values. We refer to the largest value of R for which the lot will be accepted as the decision maker's *acceptance number*, which we denote by C. In this example, the acceptance number is $C = 0$. With a sample size of $n = 2$, our decision maker will therefore apply the following *decision rule*:

$$\text{Accept the lot if } R \leq C$$

$$\text{Reject the lot if } R > C$$

The objective in two-action decision problems involving sampling is to select a decision rule. The optimal strategy is therefore equivalent to finding the value of C that maximizes expected payoff.

In Section 24-3, we will consider whether or not a different sample size can provide an even greater expected payoff than the one that we obtained here for $n = 2$.

EXERCISES

24-1 The president of Admiral Mills believes that the proportion of children π who will like Crunchy Munchy has the following probability distribution:

Possible Proportion π	Probability
1/4	1/3
1/2	1/3
3/4	1/3
	1

The president wishes to revise these probabilities to devise a strategy for later test marketing. Three children have been chosen at random, given Crunchy Munchy, and then asked if they like it.

(a) Construct the actual chronological probability tree diagram for the outcomes of this experiment, letting the values of π represent the events of the branches in the first-stage fork and letting the number of children found to like Crunchy Munchy represent the events of the second-stage branches.

(b) Enter the above probabilities for the possible π values on the appropriate branches. Then use the binomial formula to find the probabilities for the branches in the remaining forks, and enter them on your diagram.

(c) Calculate the joint probabilities for each end position.

(d) Construct the probability tree diagram for the informational chronology, reversing the event sequences so that the branches of the first fork represent the number of children who like Crunchy Munchy, followed by forks for the values of π. Determine the probabilities for the events represented by each branch.

(e) If all three children like Crunchy Munchy, what are the posterior probabilities of π?

24-2 High Crock, the brewmaster and part owner of High & Higher Distilleries, has discovered a new fermentation process for malt ale that reclaims corn mash from whiskey vats to begin fermentation. Get, the marketing manager and joint owner of High & Higher, is excited about the revolutionary ale and has assigned the prior probabilities of $Pr[\pi = .10] = .20$, $Pr[\pi = .20] = .50$, and $Pr[\pi = .30] = .30$ to the proportion π of the market segment that he feels will buy the new beer. Get feels that the probabilities should be revised before a final product decision is made. A random sample of $n = 3$ connoisseurs has has been selected to test the new ale.

(a) Construct the actual chronological probability tree diagram, using the prior probabilities assigned by the marketing manager and the applicable binomial, conditional result probabilities obtained from Appendix Table C.

(b) From the probabilities generated in the actual chronology, construct the informational chronology indicating the unconditional result, posterior, and joint probabilities.

24-3 A presidential campaign manager asks two politicians what they believe to be the true proportion π of voters favoring their party's candidate. The respective replies are .40 and .50. The manager's judgment leads her to assign equal chances that either politician is correct. A random sample of $n = 100$ registered voters will be chosen, and the number R preferring the candidate will be found. The following results are of interest: $R < 40$; $40 \leq R \leq 50$; $R > 50$.

(a) Construct the actual chronological probability tree diagram, using the sample results in the second stage. (Appendix Table C provides the conditional result probabilities.)

(b) Determine the probability tree diagram for the informational chronology.

(c) Find the posterior probability that $\pi = .40$, given each of the following results:
    (1) $R < 40$ (2) $40 \leq R \leq 50$ (3) $R > 50$

24-2 DECISION MAKING WITH THE MEAN

A completely analogous procedure to decision making using binomial probabilities applies when samples are taken from quantitative populations.

A Computer Memory Device Decision

To illustrate decision making based on the sample mean, we will consider the decision of a computer-center manager regarding the kind of peripheral memory storage device to use in the computer system. The two proposed units are based on laser technology, and both units will operate more efficiently than the current memory storage device. One alternative is based on photographic principles and requires special film for storing the data. The other alternative employs holography—a process in which a three-dimensional image is retrieved from a special wafer. A photographic memory unit costs less to lease than a holographic unit, but it is slower and therefore more costly to operate. The storage capacities and reliabilities of the two units are identical.

The annual savings from using either alternative unit depends on the daily volume of peripheral memory access. Although the actual number of bits stored or retrieved varies daily, the mean daily access level can be used to establish an average annual access savings for each alternative. When this savings is added to the fixed lease cost, the resulting mean total annual savings serves as the payoff measure for this decision. This payoff depends on the mean daily

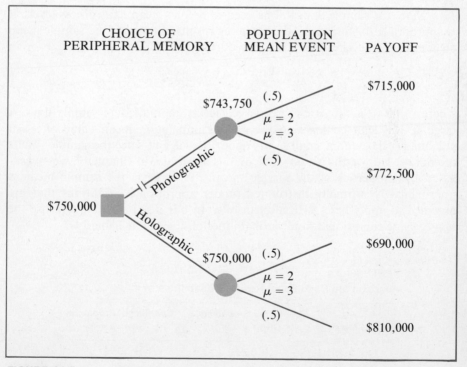

FIGURE 24-5
The computer-center manager's decision structure when no sample is used.

gigabits (billion bits) accessed μ, which represents the average volume over all days.

The computer-center manager is uncertain about the value of μ, since historical data on the density of peripheral memory traffic are incomplete.

Figure 24-5 presents the manager's decision structure when no sample information is available. Notice that different payoff values for mean annual savings are obtained for each type of unit and μ combination. Using prior probabilities of .5 for $\mu = 2$ and .5 for $\mu = 3$, we find that the holographic memory unit provides the greatest expected payoff of $750,000 in mean annual savings.

Decision Making with Sample Information

We will now consider the manager's analysis using sample data. The manager believes that for an extra few hundred dollars per day she can determine the precise level of peripheral memory access on sample days by adding a special accounting program to the software system. Any sampling cost arises from the slower processing that will result. Suppose that a sample of $n = 9$ days is to be used.

Since the sample data will be used to predict mean daily access levels, it is appropriate to summarize the sampling results in terms of the sample mean memory access level, which is computed from

$$\bar{X} = \frac{X_1 + X_2 + \cdots + X_n}{n}$$

where $X_1, X_2, \ldots, X_n$ are the observed levels for individual sample days. A large $\bar{X}$ will lend credence to the greater population mean value of $\mu = 3$ gigabits per day, and a small $\bar{X}$ will suport $\mu = 2$. But since the sample results are not yet known, the actual value of $\bar{X}$ is uncertain. In Chapter 7, we investigated the properties of the sample mean. For large n, the sample mean is approximately normally distributed (under appropriate conditions that are assumed to apply here) with a mean of μ. In our discussion, we will use the *approximate* conditional result probabilities of $\bar{X}$ given in Table 24-2.

TABLE 24-2
Approximate Conditional Result Probabilities of $\bar{X}$

Possible Mean	Conditional Probability Given $\mu = 2$	Conditional Probability Given $\mu = 3$
$\bar{X} = 1$	.35	.05
$\bar{X} = 2$	.35	.25
$\bar{X} = 3$	.25	.35
$\bar{X} = 4$	.05	.35

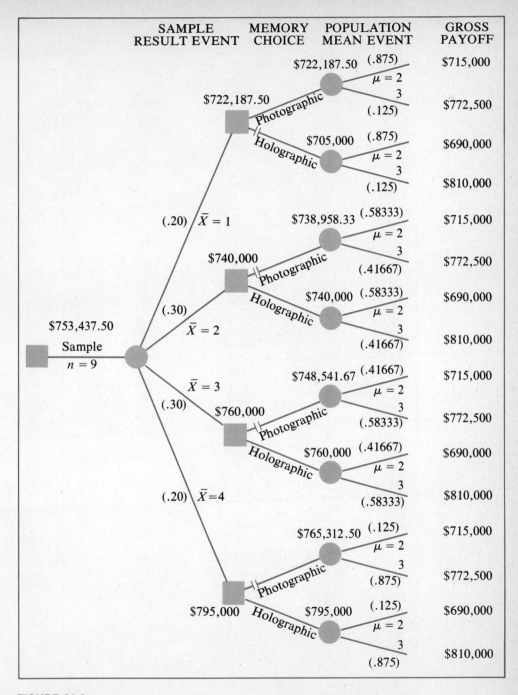

	SAMPLE RESULT EVENT	MEMORY CHOICE	POPULATION MEAN EVENT	GROSS PAYOFF
			$722,187.50 (.875)	$715,000
			$\mu = 2$	
			3	
	$722,187.50	Photographic	(.125)	$772,500
			$705,000 (.875)	$690,000
		Holographic	$\mu = 2$	
			3	
			(.125)	$810,000
	(.20) $\bar{X} = 1$		$738,958.33 (.58333)	$715,000
			$\mu = 2$	
			3	
	$740,000	Photographic	(.41667)	$772,500
			$740,000 (.58333)	$690,000
	(.30)	Holographic	$\mu = 2$	
$753,437.50 Sample $n = 9$	$\bar{X} = 2$		3 (.41667)	$810,000
	$\bar{X} = 3$		$748,541.67 (.41667)	$715,000
			$\mu = 2$	
	(.30)		3	
	$760,000	Photographic	(.58333)	$772,500
			$760,000 (.41667)	$690,000
		Holographic	$\mu = 2$	
			3	
	(.20) $\bar{X}=4$		(.58333)	$810,000
			$765,312.50 (.125)	$715,000
			$\mu = 2$	
			3	
		Photographic	(.875)	$772,500
	$795,000	Holographic	$795,000 (.125)	$690,000
			$\mu = 2$	
			3	
			(.875)	$810,000

FIGURE 24-6

The computer-center manager's decision structure using a sample of size $n = 9$.

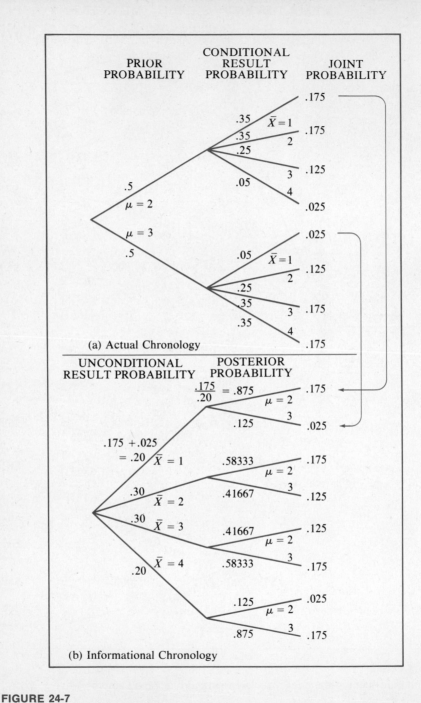

FIGURE 24-7

Probability trees for the actual and informational chronologies of events for the computer-center manager's decision.

The manager's decision tree diagram using sample information is provided in Figure 24-6. The revised probabilities used there are obtained from Figure 24-7. Backward induction in Figure 24-6 indicates that the maximum expected annual savings will be achieved by selecting the photographic memory unit for $\bar{X} = 1$ and the holographic peripheral storage unit for $\bar{X} = 2$, $\bar{X} = 3$, or $\bar{X} = 4$. We can express this result in terms of the decision rule

$$\text{Select photographic unit if } \bar{X} \leq C$$

$$\text{Select holographic unit if } \bar{X} > C$$

where $C = 1$ gigabit per day.

The expected payoff is $753,437.50 when the optimal decision rule is applied to a sample of size $n = 9$. We refer to this quantity as the *expected payoff with sample information*. It is based on gross payoffs and *does not reflect the cost of collecting the sample*.

Of course, sampling costs must ultimately be considered. Thus far, our analysis can be classified as *posterior*. We still have to consider the more basic decision of whether or not to sample in the first place and, if we sample, how large a sample to take. To answer these questions, we must move into the area of *preposterior analysis*.

EXERCISES

24-4 Consider a slight modification in the computer memory device decision. Suppose that the following prior probabilities for the mean memory access levels apply:

$$\mu = 2 \text{ gigabits per day:} \quad .6$$

$$\mu = 3 \text{ gigabits per day:} \quad .4$$

Compute the new probabilities for the actual and informational chronologies in Figure 24-7, assuming that the same conditional result probabilities apply.

24-5 The marketing manager of Blitz Beer must determine whether or not to sponsor Blitz Day with the Gotham City Hellcats. She is uncertain what the effect of the promotion will be in terms of the mean increase in daily sales volume that would result during the 100-day baseball season. The cost of sponsorship is $10,000, and each can of Blitz has a marginal cost of $.20 and sells for $.40. Two levels are judged equally likely for the mean increase in sales for the season: $\mu = 490$ and $\mu = 530$ cans per day.
(a) Construct the manager's payoff table.
(b) If the manager wishes to maximize expected payoff, what action should she take?

(c) Further experimental information may be desired before making a final decision. Calculate the EVPI. Can a sample from the underlying population—assuming it is cheap enough—be helpful in reaching this decision? Explain.

24-3   DECIDING ABOUT THE SAMPLE: PREPOSTERIOR ANALYSIS

In our first example of decision making using sample information, we found that rather than accepting each lot of stereo cartridges, it was preferable to take a sample of size $n = 2$ first and to accept only the lots in which no sample defectives were found. But we did not investigate the possibility that selecting a different sample size might have been an even greater improvement. Although more observations will add to the sampling costs, the increased sample reliability should provide a net gain in expected payoff. An analysis of various sample sizes helps determine exactly what kind of information to acquire.

Determining the Optimal Sample Size

How large a sample should be taken? We can investigate the choice of sample size in the same way that we determined whether or not to sample. By explicitly considering each possible level of n as an act in the decision structure, we can use the principles developed thus far to analyze the fully generalized decision. This decision involves not only the choice of whether or not to use experimental evidence but also the selection of the evidence to be obtained.

We can accomplish this for our acceptance-sampling illustration (Charles Stereo) by augmenting the decision structure as shown in Figure 24-8, where four possible sample sizes $n = 1$, $n = 2$, $n = 3$, and $n = 4$ are considered. Each sample size is a separate choice in an initial act fork. A sample of size n has $n + 1$ possible sample outcomes, since the number of defectives may be $R = 0$, $R = 1, \ldots$, or $R = n$. Thus, the number of events in the sample outcome forks varies with the size of n, and there are $n + 1$ branches for each sample size.

The posterior probabilities for the possible values of π will differ, depending on n and the sample outcome, and so they must be recalculated for each sample size being considered. The end-position payoffs for the paths of the decision tree depend on both the losses from returned defective cartridges and the sampling costs, and they must also be calculated separately. Once the payoffs have been determined, the sample size yielding the maximum expected payoff can then be arrived at by backward induction.

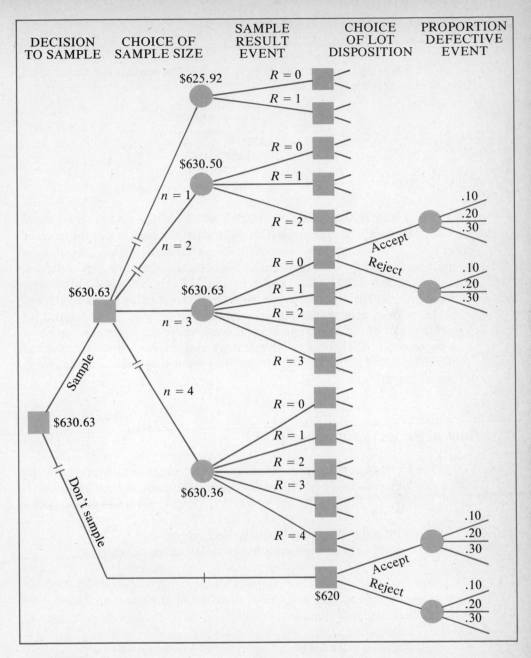

$625.92

$R = 0$
$R = 1$

$630.50

$R = 0$
$R = 1$
$R = 2$

$n = 1$

$n = 2$

Accept
Reject

.10
.20
.30

.10
.20
.30

$630.63

$630.63

$R = 0$
$R = 1$
$R = 2$
$R = 3$

$n = 3$

$n = 4$

$630.63

Sample

$R = 0$
$R = 1$
$R = 2$
$R = 3$

$630.36

$R = 4$

Don't sample

.10
.20
.30

Accept
Reject

.10
.20
.30

$620

FIGURE 24-8
The Charles Stereo decision tree diagram, incorporating the choice of sample size.

The following expected payoffs have been determined for the other values of n, as they were for $n = 2$:

Act	Expected Payoff
$n = 1$	$625.92
$n = 2$	630.50
$n = 3$	630.63 maximum
$n = 4$	630.36

Of the four sample sizes being considered, $n = 3$ has the maximum expected payoff of $630.63. The expected payoff rises when the sample size is increased from $n = 2$ to $n = 3$ and then declines for larger values of n. Thus , $n = 3$ is the *optimal sample size* for Charles Stereo to use in establishing an acceptance plan, and the branches of the decision tree in Figure 24-8 representing the other sizes of n are pruned. The expected payoff from the decision to sample is $630.63.

In a more general acceptance-sampling problem, a large number of possible values of n may have to be evaluated before the optimal sample size can be determined. This can involve a staggering amount of calculation and require the use of a digital computer to find the n that yields the maximum expected payoff.*

The Role of Perfect Information (EVPI)

The EVPI establishes the worth of perfect information and represents the difference in payoffs expected with perfect information and with no information at all, or

EVPI = Expected payoff with perfect information
  − Maximum expected payoff (with no information)

For the computer-center manager's decision discussed in Section 24-2, we have established that the best course of action when there is no information is to use the holographic unit, for which

Maximum expected payoff (with no information) = $750,000

If a perfect hypothetical predictor were available for the unknown value of the population mean daily access level, the manager would select the photographic

---

* In his book *Analysis of Decisions Under Uncertainty* (New York: McGraw-Hill, 1969), Robert Schlaifer describes a program developed at Harvard University to determine the optimal sample size for binomial or hypergeometric sampling.

memory, which has a payoff of $715,000, when $\mu = 2$, and the holographic memory, which has a payoff of $810,000, when $\mu = 3$. Thus, we have

$$\text{Expected payoff with perfect information} = .5(\$715,000) + .5(\$810,000)$$

$$= \$762,500$$

The expected value of perfect information is therefore

$$\text{EVPI} = \$762,500 - \$750,000 = \$12,500$$

The EVPI sets a limit on how much the manager is willing to pay for any kind of experimental information that will help predict the value of μ.

We have seen that the EVPI can be useful as a rough gauge for deciding whether or not it is worthwhile to pursue further information. If the EVPI is very small, then sample information will not be worth its cost, and an immediate decision should be made solely from prior knowledge.

The Expected Value of Sample Information

Another measure similar to the EVPI expresses the worth of the information contained in the sample. The *expected value of sample information* is calculated in a similar manner to the EVPI:

EXPECTED VALUE OF SAMPLE INFORMATION

$$\text{EVSI} = \text{Expected payoff with sample information}$$
$$- \text{Maximum expected payoff (with no information)}$$

To illustrate, we will continue with the computer-center manager's decision. In Figure 24-6, we obtained the expected payoff of $753,437.50 using a sample of size $n = 9$ and applying the indicated decision rule for $\bar{X}$. Thus, for the memory device decision,

$$\text{EVSI} = \$753,437.50 - \$750,000 = \$3,437.50$$

This amount represents how much better off the computer-center manager would be on the average if she had the $n = 9$ sample result instead of no information at all. The EVSI is totally analogous to the EVPI, but it applies to less reliable information gleaned from the sample. Like the EVPI, the EVSI establishes an upper limit on the amount a decision maker should pay to obtain the sample results.

The Expected Net Gain of Sampling

Preposterior analysis often begins with gross payoffs rather than net payoffs, and sampling costs must therefore be integrated at a later stage. (One reason for waiting to include the sampling costs is that it is sometimes easier to minimize expected opportunity loss than it is to maximize expected payoff. It is more convenient to include sampling costs at the end when such an approach is used.)

In terms of gross annual savings, the manager is better off by the EVSI of $3,437.50 if she obtains the sample results and then applies the optimal decision rule with $C = 1$. Thus, the manager should be willing to pay up to $3,437.50 for the sample but should not sample at a higher cost. *As long as the cost of sampling is less than the EVSI, the decision maker is better off with the sample than without it.*

To obtain the peripheral memory access levels for the sample days, a special program must be added to the operating system that will slow down the processing of each job. As a result, the computer must operate for a longer period of time, which creates an added expense. We will assume that this additional cost is $100 per day. The total cost of sampling for $n = 9$ days is therefore $900. Since this figure is smaller than the EVSI, the computer-center manager should prefer sampling to making a decision without the information that sampling provides.

How large should n be? As a practical matter, n can be no larger in this example than the number of days that remain before the peripheral unit must be ordered. However, this question is ordinarily a matter of economics. We could compute the EVSI for several levels of n, substracting the sampling costs in each case. Each resulting value would then represent the *expected net gain of sampling*, or ENGS. Treating n as a variable, the expected net gain of sampling is expressed as follows:

EXPECTED NET GAIN OF SAMPLING

$$ENGS(n) = EVSI(n) - Cost(n)$$

Thus, when $n = 9$,

$$ENGS(9) = EVSI(9) - Cost(9)$$
$$= \$3,437.50 - \$900$$
$$= \$2,537.50$$

Comparable figures could be obtained for other sizes of   and the optimal sample size would be the one with the greatest expected net ain.

The EVSI and ENGS results for the Charles Stereo ecision appear in Table 24-3. The optimal sample size of $n = 3$ (identified earlier in Figure 24-8)

TABLE 24-3
EVSI and ENGS Values for the Charles Stereo Decision

Sample Size n	(1) Expected Payoff (Net)	(2) Expected Payoff (Gross)	(3) Maximum Expected Payoff (with No Information)	(4) EVSI(n) [(2)–(3)]	(5) Cost (n)	(6) ENGS(n) [(4)–(5)]
1	$625.92	$629.92	$620	$ 9.92	$ 4	$ 5.92
2	630.50	638.50	620	18.50	8	10.50
3*	630.63	642.63	620	22.63	12	10.63
4	630.36	646.36	620	26.36	16	10.36

* Optimal n.

has the greatest ENGS at $10.63. Note that EVSI increases as n increases, reflecting the greater reliability of larger samples. But sampling costs increase by $4 per observation, and we can see in the table that ENGS peaks at $n = 3$ and drops thereafter. Similar results will be found for most sampling situations. Ordinarily, our search for the optimal n ends when we find an ENGS that is smaller than the preceding one.

EXERCISES

24-6 The decision tree diagram in Figure 24-9 has been constructed to determine whether or not to accept shipments from a particular supplier.
   (a) Perform backward induction. Which choice—not to sample, to sample with $n = 1$, or to sample with $n = 2$—should be made?
   (b) What is the optimal value of the acceptance number C if $n = 1$ is used? If $n = 2$ is used?

24-7 Suppose that the prior probabilities for the proportion π of CornChox buyers favoring a new package design are $\Pr[\pi = .4] = .5$ and $\Pr[\pi = .6] = .5$. Also suppose that if the new design is used, the present value of future profits will decrease by $10,000 when $\pi = .4$ and increase by $10,000 when $\pi = .6$. Sampling is very expensive, costing $500 per observation.
   (a) Which act provides the greatest expected payoff—not to sample or to sample with $n = 1$?
   (b) Does sampling with $n = 1$ or with $n = 2$ provide the greatest expected payoff? What can you conclude about the optimal sample size?

24-8 Suppose that the peripheral memory device decision is modified so that the following prior probabilities for the mean memory access levels apply:

$$\mu = 2 \text{ gigabits per day: } .3$$

$$\mu = 3 \text{ gigabits per day: } .7$$

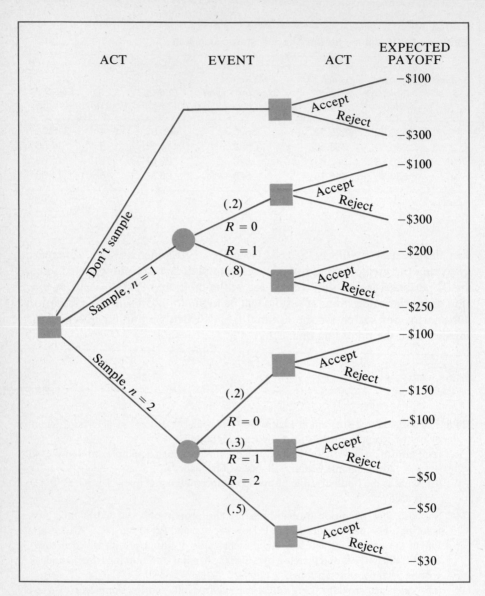

FIGURE 24-9

The decision tree diagram in Figure 24-10 now applies.
(a) Perform backward induction.
(b) What is the value of C that will maximize expected payoff? Formulate the optimal decision rule.
(c) Calculate the EVSI.
(d) Determine the expected net gain of sampling for a sample costing $900.

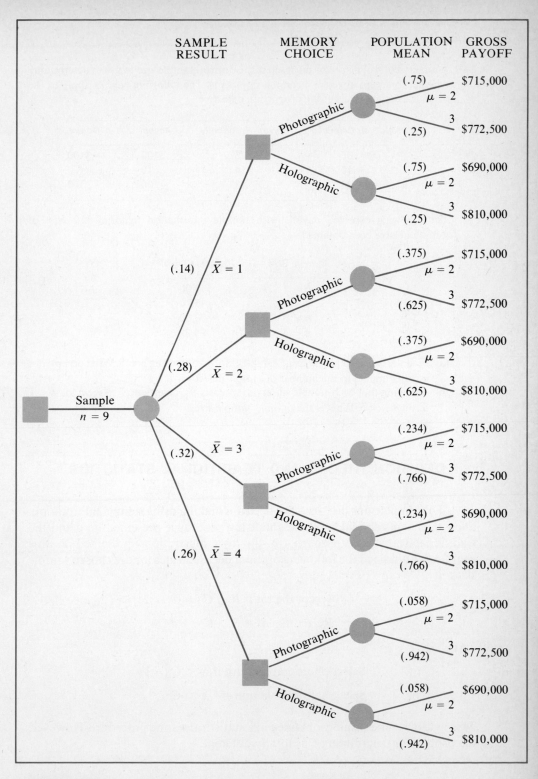

SAMPLE   MEMORY   POPULATION   GROSS
RESULT   CHOICE   MEAN   PAYOFF

Photographic
                                (.75) $715,000
                                μ = 2
                                3       $772,500
                                (.25)

Holographic
                                (.75) $690,000
                                μ = 2
                                3       $810,000
                                (.25)

(.14) X̄ = 1

Photographic
                                (.375) $715,000
                                μ = 2
                                3       $772,500
                                (.625)

Holographic
                                (.375) $690,000
                                μ = 2
                                3       $810,000
                                (.625)

(.28) X̄ = 2

Sample
n = 9

(.32) X̄ = 3

Photographic
                                (.234) $715,000
                                μ = 2
                                3       $772,500
                                (.766)

Holographic
                                (.234) $690,000
                                μ = 2
                                3       $810,000
                                (.766)

(.26) X̄ = 4

Photographic
                                (.058) $715,000
                                μ = 2
                                3       $772,500
                                (.942)

Holographic
                                (.058) $690,000
                                μ = 2
                                3       $810,000
                                (.942)

FIGURE 24-10

24-9 A quality-control inspector must select the optimal sample size to use in determining whether to accept or reject incoming shipments. The following payoffs apply to the various assumed levels of the proportion defective:

Proportion Defective	Prior Probability	Accept	Reject
$\pi = .05$	.3	$100	-$100
$\pi = .10$	.4	0	0
$\pi = .20$	.3	-100	100

The following expected payoffs with sample information (ignoring the cost of sampling) have been obtained:

Sample Size	Expected Payoff
$n = 1$	$ 9.00
$n = 2$	15.70
$n = 3$	20.50
$n = 4$	24.22

(a) Find the expected payoff using sample information when $n = 5$. What acceptance number applies to the number of sample defectives R?

(b) Assuming that each sample observation costs $3, determine ENGS($n$) for $n = 1$ through $n = 5$. What is the optimal sample size?

24-4 DECISION THEORY AND TRADITIONAL STATISTICS

To complete our discussion of decision making using sample information, a few comments should be made about the procedure presented here and the traditional statistical approach. In the two illustrations presented in this chapter, we obtained the following optimal decision rules for the chosen sample sizes:

$$\text{Accept the lot if } R \leq C \, (= 0)$$

$$\text{Reject the lot if } R > C$$

and

$$\text{Select photographic unit if } \bar{X} \leq C \, (= 1)$$

$$\text{Select holographic unit if } \bar{X} > C$$

This type of result is usually obtained in a statistical testing procedure. However, the process for determining C is totally different.

Hypothesis-Testing Concepts Reviewed

In traditional statistics, the basic uncertainty is couched in special terminology. The main events are expressed as *hypotheses*. Each decision has two kinds of hypotheses. One is referred to as the *null hypothesis* (originally used to represent "no change"), and the other is the complementary *alternative hypothesis*. In the language of classical statistics, the following hypotheses would apply to the quality-control inspection problem:

Null hypothesis: Lot is good $(\pi = .10)$

Alternative hypothesis: Lot is poor $(\pi > .10)$

The hypotheses for the computer memory decision would be

Null hypothesis: Photographic unit is best $(\mu = 2)$

Alternative hypothesis: Holographic unit is best $(\mu = 3)$

The decision rules that we formulated earlier can be expressed in terms of the respective hypotheses pairs. For the inspection decision, they would take the form

Accept the null hypothesis if $R \leq C$

Reject the null hypothesis if $R > C$

and for the computer memory device decision, they would take the form

Accept the null hypothesis if $\bar{X} \leq C$

Reject the null hypothesis if $\bar{X} > C$

(Rejecting the null hypothesis is the same as accepting the alternative hypothesis.)

Traditional statistics focuses on the worst outcomes of the decision. These are called *errors* and are of two types. The *Type I error* occurs when the null hypothesis is rejected when it is actually true; the *Type II error* occurs when the null hypothesis is accepted when it is actually false. In the quality-control inspection problem, these errors are

Type I error: Reject lot when it is good

Type II error: Accept lot when it is poor

In the computer memory device decision,

Type I error: Use holographic memory when photographic is better

Type II error: Select photographic memory when holographic is superior

The decision structure for traditional statistical analysis is presented in Table 24-4. The main events themselves are not assigned probabilities. Rather, the controlling factors in establishing the decision rule (the value of C) are the probabilities for the two kinds of errors:

$$\alpha = \Pr[\text{Type I error}] = \Pr[\text{rejecting the null hypothesis when it is true}]$$

$$\beta = \Pr[\text{Type II error}] = \Pr[\text{accepting the null hypothesis when it is false}]$$

These probabilities are usually established in advance of sampling and are actually conditional probabilities (for which the status of the null hypothesis—true or false—is the given event). Conventionally, the Greek letters α (alpha) and β (beta) are used to represent the error probabilities. In the quality-control inspection decision, α represents the probability for rejecting a good lot, which is sometimes referred to as the *producer's risk*, and β is the probability for accepting a poor lot, or the *consumer's risk*.

Both types of errors are undesirable, but obviously neither error can be avoided entirely. Traditional statistics is concerned with selecting the sample

TABLE 24-4
Decision Table for the Traditional Statistical Decision

	Act	
Event	*Accept Null Hypothesis.* (Accept lot; select photographic.)	*Reject Null Hypothesis.* (Reject lot; select holographic.)
Null hypothesis is true. (Lot good; photographic best.)	*Correct decision*	*Type I error* (Return good lot; select holographic when photographic is best.) $\alpha =$ Target probability
Null hypothesis is false. (Lot poor; holographic best.)	*Type II error* (Accept poor lot; select photographic when holographic is best.) $\beta =$ Target probability	*Correct decision*

size n and the decision rule (the value of C) that will achieve an acceptable balance between α and β. If n is fixed, one error probability can be reduced only by increasing the probability for the other error. Because the sample size itself is often dictated by economic or other considerations, it is usually possible to control only one error completely.

Generally, the null and alternative hypotheses are formulated in such a way that the Type I error is more serious. The tolerable probability for this error is specified at a level such as $\alpha = .01$, $\alpha = .05$, or $\alpha = .10$. A value of C is then chosen that guarantees that this level of α is not exceeded.

To illustrate, we will return to our quality-control inspection problem. Suppose that $\alpha = .05$, so that C must be the smallest value such that

$$\text{Pr[rejecting the null hypothesis when it is true]} \le \alpha$$

or

$$\text{Pr}[R > C | \pi = .10] \le .05$$

For $C = 0$, the probability for incorrectly rejecting a good lot (from Table 24-1) is

$$\text{Pr}[R > 0] = 1 - \text{Pr}[R = 0] = 1 - .81 = .19$$

which is greater than the targeted Type I error probability of $\alpha = .05$. Trying $C = 1$, we obtain

$$\text{Pr}[R > 1] = \text{Pr}[R = 2] = .01$$

which is smaller than $\alpha = .05$. The optimal decision rule using the traditional procedure is therefore $C = 1$.

This value of C differs from the one we found previously $(C = 0)$. *The optimal rule using decision-theory analysis will often differ from the optimal rule obtained using the traditional statistical hypothesis test.*

Contrasting the Two Approaches

Why do decision theory and traditional statistical analysis lead to different decision rules? This is because of the differences between these two procedures:

1. The decision-theory procedure considers the payoffs from every possible outcome. Payoffs are not explicitly considered in hypothesis testing, although they may influence the choice of α.
2. Prior probabilities are applied directly in the decision-theory approach. Like payoffs, prior probabilities ought to play some role in establishing α.

3. Hypothesis testing proceeds directly from the prescribed α to the decision rule. Decision-theory analysis arrives at the optimal decision rule by means of the Bayes decision rule, using the appropriate posterior probabilities and payoffs for each outcome. The resulting decision maximizes expected payoff.

Which procedure is preferable? Decision theory is plagued by subjective prior probabilities, which are considered by many to be the weakest link in its analytical chain. Many statisticians deny the existence of subjective probabilities, thereby relegating much of statistical decision theory to the ash heap. A major feature that makes traditional hypothesis-testing procedures more universally accepted is that no prior probabilities are required at all. *But the uncertainties regarding the population parameter still exist*, and traditional statistical procedures must also involve some sort of subjective assessment of these uncertainties when desired error probabilities are being established. In hypothesis testing, everything hinges on the prescribed significance level α (and also on β when there is the freedom or the capability to prescribe β). Unless α is carefully determined, inferior decisions are bound to occur. Decision theory permits the consistent and systematic treatment of chance and payoffs as well as attitude toward risk. In addition, decision theory does not burden the decision maker by requiring him or her to do everything at once by choosing a single number—"an α for all seasons."

When the outcomes do not have a natural numerical payoff measure or when the decision maker is risk-seeking or risk-averse, then the strengths of decision-theory analysis rest on a foundation of utility values. As we saw in Chapter 21, obtaining utilities requires a set of assumptions about attitudes, which are obtained by means of an elaborate procedure. Although any difficulties involved in employing utilities can be avoided by traditional statistics, α embodies everything, so that even more care must be exercised in choosing the appropriate target value.

The final and perhaps the most significant advantage of decision theory is that *it considers whether or not a sample should even be used*. We have seen that a greater expected payoff may be achieved by deciding what to do immediately, without incurring the expense of a sample. Traditional statistics never satisfactorily copes with this question. In addition, the choice of proper sample size is too often an *ad hoc* process in traditional statistics. (A sample size of 30 may be used, for example, because the Student t table stops at 30 degrees of freedom.) Decision theory explicitly considers the costs of the various sample sizes.

EXERCISES

24-10 A market researcher wishes to determine whether to accept the null hypothesis regarding the proportion π of Appleton smokers who will switch to a new mentholated version, Mapleton. Her null hypothesis is that $\pi = .1$. She takes a random sample of 100 current Appleton smokers, who will be contacted in six months to see if they have switched. The researcher wishes to protect against the Type I error with a probability of $\alpha = .005$. Using Appendix Table C, determine the smallest acceptance number C such that the probability that the number of switchers will exceed this number is less than or equal to the desired α. Then formulate the researcher's decision rule.

24-11 Referring to Problem 24-10, suppose that the following prior probabilities are obtained for π:

π	Probability
.05	1/3
.10	1/3
.15	1/3
	1

The following conditional result probabilities for the number of switchers R apply:

	$\pi = .05$	$\pi = .10$	$\pi = .15$
$R \le 17$	1.0000	.9900	.7633
$R = 18$	0	.0054	.0739
$R \ge 19$	0	.0046	.1628

The decision maker's payoffs, including the cost of sampling, are

	Accept Null Hypothesis	Reject Null Hypothesis
$\pi = .05$	$10,000	− $2,000
$\pi = .10$	5,000	0
$\pi = .15$	− 2,000	4,000

(a) Calculate the posterior probability distribution of π, given each of the sample result events above. Then find the unconditional result probabilities.

(b) Construct the market researcher's decision tree diagram. Enter on the branches the appropriate probabilities that you found in (a), and place the payoffs along the respective end positions.

(c) Should the decision maker accept or reject the null hypothesis if $R = 18$? Compare this result with your result in Problem 24-10.

REVIEW EXERCISES

24-12 Let π be the proportion of dogs that will like a new dog food, SuperPooch. Suppose that the following prior probabilities have been established for π: $\Pr[\pi = .20] = .1$, and $\Pr[\pi = .30] = .9$. A random sample of the responses of $n = 2$ dogs is obtained.

(a) Construct the actual chronological probability tree diagram.

(b) Construct the informational chronological probability tree diagram.

(c) Determine the posterior probability that $\pi = .20$ if

(1) no dogs are found to like the food.

(2) exactly one dog likes it.

(3) both dogs like it.

24-13 A chemical process yields a mean μ grams (g) of active ingredient for every liter of raw material processed. Due to variations in the raw material and in the control settings, the true population mean for any particular batch is unknown until processing is complete. From past history, the plant superintendent judges that the following prior probabilities for μ apply:

Possible Mean μ	Probability
25 g	.3
30	.4
35	.3

Early in the processing of the current batch, the superintendent plans to select a random sample of $n = 3$ liters and precisely determine the amount of active ingredient that each liter contains. The sample mean will then be calculated from these readings. Suppose that the following approximate conditional result probabilities apply:

	Probabilities for $\bar{X}$		
$\bar{X}$	$\mu = 25$ g	$\mu = 30$ g	$\mu = 35$ g
25 g	.7	.2	.2
30	.2	.6	.3
35	.1	.2	.5

Construct the actual chronological probability tree diagram. Then use this tree to find the tree for the informational chronology and to compute the unconditional result probabilities and the posterior probabilities for μ.

24-14 Referring to Exercise 24-13, the plant superintendent must decide whether or not to adjust the control settings in processing a chemical batch. Suppose that the following payoff table applies:

Population Mean	Acts	
μ	Adjust	Leave Alone
25 g	$ 500	− $500
30	0	0
35	− 1,000	500

(a) Assuming that no sample is taken, what course of action will maximize the expected payoff?

(b) Find the plant superintendent's EVPI. Would he use a sample to facilitate his decision if it cost $400? (Answer yes, no, or maybe.) If it cost $200? Explain.

24-15 Referring to Exercises 24-13 and 24-14 and to your answers to those exercises:

(a) Construct the plant superintendent's decision tree diagram, assuming that he is committed to taking a sample. Use gross payoff figures.

(b) Perform backward induction to determine the value of C (for accepting the need for adjustment) that will maximize expected gross payoff. Formulate the optimal decision rule that the superintendent should apply.

(c) Calculate the EVSI.

(d) Suppose that a sample of $n = 3$ liters costs $50. Calculate the superintendent's expected net gain of sampling.

Selected References

The Role of Statistics
Careers in Statistics. The American Statistical Association (no date available).
Huff, Darrell. *How to Lie with Statistics.* New York: W. W. Norton, 1954.
Moroney, M.J. *Facts from Figures.* Baltimore: Penguin Books, 1952.
Reichmann, W.J. *Use and Abuse of Statistics.* New York: Oxford University Press, 1962.
Wallis, W.A., and H.V. Roberts. *The Nature of Statistics.* New York: The Free Press, 1965.

Probability
Feller, William. *An Introduction to Probability Theory and Its Applications*, Vol. 1, 3rd ed. New York: John Wiley & Sons, 1968.
Hodges, J.L., Jr., and E.L. Lehmann. *Elements of Finite Probability.* San Francisco: Holden-Day, 1965.
Laplace, Pierre Simon, Marquis de. *A Philosophical Essay on Probabilities.* New York: Dover Publications, 1951.
Lindgren, Bernard W., and G.W. McElrath. *Introduction to Probability and Statistics*, 4th ed. New York: Macmillan, 1978.
Mosteller, F., R. Rourke, and G. Thomas, Jr. *Probability and Statistics.* Reading, Mass.: Addison-Wesley, 1961.
Parzen, Emmanuel. *Modern Probability Theory and Its Applications.* New York: John Wiley & Sons, 1960.

Regression and Correlation Analysis
Ezekiel, Mordecai, and Karl A. Fox. *Methods of Correlation and Regression Analysis*, 3rd ed. New York: John Wiley & Sons, 1959.

Johnston, John. *Econometric Methods*, 2nd ed. New York: McGraw-Hill, 1971.
Neter, John, and William Wasserman. *Applied Linear Statistical Models*. Homewood, Ill.: Richard D. Irwin, 1974.
Williams, E.J. *Regression Analysis*. New York: John Wiley & Sons, 1959.
Wonnacott, Ronald J., and Thomas H. Wonnacott. *Econometrics*, 2nd ed. New York: John Wiley & Sons, 1979.

Time Series Analysis and Index Numbers

Brown, Robert G. *Smoothing, Forecasting and Prediction of Discrete Time Series*. Englewood Cliffs, N.J.: Prentice-Hall, 1963.
McKinley, David H., Murray G. Lee, and Helene Duffy. *Forecasting Business Conditions*. The American Bankers Association, 1965.
Mudgett, Bruce D. *Index Numbers*. New York: John Wiley & Sons, 1951.
Spencer, Milton H., Colin G. Clark, and Peter W. Hoguet. *Business and Economic Forecasting: An Econometric Approach*. Homewood, Ill.: Richard D. Irwin, 1965.

Analysis of Variance and Design of Experiments

Cochran, William G., and Gertrude M. Cox. *Experimental Designs*, 2nd ed. New York: John Wiley & Sons, 1957.
Cox, David R. *Planning of Experiments*. New York: John Wiley & Sons, 1958.
Guenther, W.C. *Analysis of Variance*. Englewood Cliffs, N.J.: Prentice-Hall, 1964.
Mendenhall, William. *An Introduction to Linear Models and the Design and Analysis of Experiments*. Belmont, Calif.: Wadsworth, 1968.
Neter, John, and William Wasserman. *Applied Linear Statistical Models*. Homewood, Ill.: Richard D. Irwin, 1974.
Scheffé, Henry. *The Analysis of Variance*. New York: John Wiley & Sons, 1959.

Nonparametric Statistics

Bradley, James V. *Distribution-Free Statistical Tests*. Englewood Cliffs, N.J.: Prentice-Hall, 1968.
Conover, W.J. *Practical Nonparametric Statistics*. New York: John Wiley & Sons, 1971.
Gibbons, Jean D. *Nonparametric Statistical Inference*. New York: McGraw-Hill, 1970.
Hájek, Jaroslav. *Nonparametric Statistics*. San Francisco: Holden-Day, 1969.
Kraft, Charles H., and Constance van Eeden. *A Nonparametric Introduction to Statistics*. New York: Macmillan, 1968.
Noether, Gottfried E. *Introduction to Statistics: A Nonparametric Approach*, 2nd ed. Boston: Houghton Mifflin, 1976.
Siegel, Sidney. *Nonparametric Statistics for the Behavioral Sciences*. New York: McGraw-Hill, 1956.

Decision Theory and Utility

Aitchison, John. *Choice Against Chance*. Reading, Mass.: Addison-Wesley, 1970.
Brown, Rex V., A.S. Kahr, and C. Peterson. *Decision Analysis for the Manager*. New York: Holt, Rinehart and Winston, 1974.
Chernoff, Herman, and L.E. Moses. *Elementary Decision Theory*. New York: John Wiley & Sons, 1959.
Jones, J. Morgan. *Introduction to Decision Theory*. Homewood, Ill.: Richard D. Irwin, 1977.

Luce, R.D., and H. Raiffa. *Games and Decisions.* New York: John Wiley & Sons, 1957.

Miller, David W., and M.K. Starr. *Executive Decisions and Operations Research,* 2nd ed. Englewood Cliffs, N.J.: Prentice-Hall, 1969.

Morris, W.T. *Management Science: A Bayesian Introduction.* Englewood Cliffs, N.J.: Prentice-Hall, 1968.

Pratt, J.W., H. Raiffa, and R. Schlaifer. *Introduction to Statistical Decision Theory.* New York: McGraw-Hill, 1965.

Raiffa, H. *Decision Analysis: Introductory Lectures on Choices Under Uncertainty.* Reading, Mass.: Addison-Wesley, 1968.

Schlaifer, R. *Analysis of Decisions Under Uncertainty.* New York: McGraw-Hill, 1969.

———. *Introduction to Statistics for Business Decisions.* New York: McGraw-Hill, 1961.

Statistical Tables

Beyer, William H. (ed.). *Handbook of Tables for Probability and Statistics,* 2nd ed. Cleveland, Ohio: The Chemical Rubber Co., 1968.

Burington, Richard S., and Donald C. May. *Handbook of Probability and Statistics with Tables,* 2nd ed. New York: McGraw-Hill, 1970.

Fisher, Ronald A., and F. Yates. *Statistical Tables for Biological, Agricultural and Medical Research,* 6th ed. London: Longman Group, 1978.

National Bureau of Standards. *Tables of the Binomial Distribution.* Washington, D.C.: U.S. Government Printing Office, 1950.

Owen, D.B. *Handbook of Statistical Tables.* Reading, Mass.: Addison-Wesley, 1962.

Pearson, E.S., and H.O. Hartley. *Biometrika Tables for Statisticians,* 3rd ed. Cambridge, England: Cambridge University Press, 1966.

The Rand Corporation. *A Million Random Digits with* 100,000 *Normal Deviates.* New York: The Free Press, 1955.

Appendix

TABLE A   Squares, Square Roots, and Reciprocals, 820

TABLE B   Random Numbers, 821

TABLE C   Cumulative Values for the Binomial Probability Distribution, 822

TABLE D   Areas Under the Standard Normal Curve, 833

TABLE E   Normal Deviate Values, 834

TABLE F   Student t Distribution, 835

TABLE G   Four-Place Common Logarithms, 836

TABLE H   Chi-Square Distribution, 838

TABLE I F Distribution, 840

TABLE J   Exponential Functions, 842

TABLE K   Cumulative Probability Values for the Poisson Distribution, 843

TABLE L   Critical Values of D for Kolmogorov-Smirnov Maximum Deviation Test for Goodness of Fit, 848

Answers to Even-Numbered Exercises, 851

TABLE A

Squares, Square Roots, and Reciprocals

N	$\sqrt{N}$	N^2	$\sqrt{10N}$	$1000/N$	N	$\sqrt{N}$	N^2	$\sqrt{10N}$	$1000/N$
					50	7.07107	2500	22.36068	20.00000
1	1.00000	1	3.16228	1000.00000	51	7.14143	2601	22.58318	19.60784
2	1.41421	4	4.47214	500.00000	52	7.21110	2704	22.80351	19.23077
3	1.73205	9	5.47723	333.33333	53	7.28011	2809	23.02173	18.86792
4	2.00000	16	6.32456	250.00000	54	7.34847	2916	23.23790	18.51852
5	2.23607	25	7.07107	200.00000	55	7.41620	3025	23.45208	18.18182
6	2.44949	36	7.74597	166.66667	56	7.48331	3136	23.66432	17.85714
7	2.64575	49	8.36660	142.85714	57	7.54983	3249	23.87467	17.54386
8	2.82843	64	8.94427	125.00000	58	7.61577	3364	24.08319	17.24138
9	3.00000	81	9.48683	111.11111	59	7.68115	3481	24.28992	16.94915
10	3.16228	100	10.00000	100.00000	60	7.74597	3600	24.49490	16.66667
11	3.31662	121	10.48809	90.90909	61	7.81025	3721	24.69818	16.39344
12	3.46410	144	10.95445	83.33333	62	7.87401	3844	24.89980	16.12903
13	3.60555	169	11.40175	76.92308	63	7.93725	3969	25.09980	15.87302
14	3.74166	196	11.83216	71.42857	64	8.00000	4096	25.29822	15.62500
15	3.87298	225	12.24745	66.66667	65	8.06226	4225	25.49510	15.38462
16	4.00000	256	12.64911	62.50000	66	8.12404	4356	25.69047	15.15152
17	4.12311	289	13.03840	58.82353	67	8.18535	4489	25.88436	14.92537
18	4.24264	324	13.41641	55.55556	68	8.24621	4624	26.07681	14.70588
19	4.35890	361	13.78405	52.63158	69	8.30662	4761	26.26785	14.49275
20	4.47214	400	14.14214	50.00000	70	8.36660	4900	26.45751	14.28571
21	4.58258	441	14.49138	47.61905	71	8.42615	5041	26.64583	14.08451
22	4.69042	484	14.83240	45.45455	72	8.48528	5184	26.83282	13.88889
23	4.79583	529	15.16575	43.47826	73	8.54400	5329	27.01851	13.69863
24	4.89898	576	15.49193	41.66667	74	8.60233	5476	27.20294	13.51351
25	5.00000	625	15.81139	40.00000	75	8.66025	5625	27.38613	13.33333
26	5.09902	676	16.12452	38.46154	76	8.71780	5776	27.56810	13.15789
27	5.19615	729	16.43168	37.03704	77	8.77496	5929	27.74887	12.98701
28	5.29150	784	16.73320	35.71429	78	8.83176	6084	27.92848	12.82051
29	5.38516	841	17.02939	34.48276	79	8.88819	6241	28.10694	12.65823
30	5.47723	900	17.32051	33.33333	80	8.94427	6400	28.28427	12.50000
31	5.56776	961	17.60682	32.25806	81	9.00000	6561	28.46050	12.34568
32	5.65685	1024	17.88854	31.25000	82	9.05539	6724	28.63564	12.19512
33	5.74456	1089	18.16590	30.30303	83	9.11043	6889	28.80972	12.04819
34	5.83095	1156	18.43909	29.41176	84	9.16515	7056	28.98275	11.90476
35	5.91608	1225	18.70829	28.57143	85	9.21954	7225	29.15476	11.76471
36	6.00000	1296	18.97367	27.77778	86	9.27362	7396	29.32576	11.62791
37	6.08276	1369	19.23538	27.02703	87	9.32738	7569	29.49576	11.49425
38	6.16441	1444	19.49359	26.31579	88	9.38083	7744	29.66479	11.36364
39	6.24500	1521	19.74842	25.64103	89	9.43398	7921	29.83287	11.23596
40	6.32456	1600	20.00000	25.00000	90	9.48683	8100	30.00000	11.11111
41	6.40312	1681	20.24846	24.39024	91	9.53939	8281	30.16621	10.98901
42	6.48074	1764	20.49390	23.80952	92	9.59166	8464	30.33150	10.86957
43	6.55744	1849	20.73644	23.25581	93	9.64365	8649	30.49590	10.75269
44	6.63325	1936	20.97618	22.72727	94	9.69536	8836	30.65942	10.63830
45	6.70820	2025	21.21320	22.22222	95	9.74679	9025	30.82207	10.52632
46	6.78233	2116	21.44761	21.73913	96	9.79796	9216	30.98387	10.41667
47	6.85565	2209	21.67948	21.27660	97	9.84886	9409	31.14482	10.30928
48	6.92820	2304	21.90890	20.83333	98	9.89949	9604	31.30495	10.20408
49	7.00000	2401	22.13594	20.40816	99	9.94987	9801	31.46427	10.10101
50	7.07107	2500	22.36068	20.00000	100	10.00000	10000	31.62278	10.00000

TABLE B

Random Numbers

12651	61646	11769	75109	86996	97669	25757	32535	07122	76763
81769	74436	02630	72310	45049	18029	07469	42341	98173	79260
36737	98863	77240	76251	00654	64688	09343	70278	67331	98729
82861	54371	76610	94934	72748	44124	05610	53750	95938	01485
21325	15732	24127	37431	09723	63529	73977	95218	96074	42138
74146	47887	62463	23045	41490	07954	22597	60012	98866	90959
90759	64410	54179	66075	61051	75385	51378	08360	95946	95547
55683	98078	02238	91540	21219	17720	87817	41705	95785	12563
79686	17969	76061	83748	55920	83612	41540	86492	06447	60568
70333	00201	86201	69716	78185	62154	77930	67663	29529	75116
14042	53536	07779	04157	41172	36473	42123	43929	50533	33437
59911	08256	06596	48416	69770	68797	56080	14223	59199	30162
62368	62623	62742	14891	39247	52242	98832	69533	91174	57979
57529	97751	54976	48957	74599	08759	78494	52785	68526	64618
15469	90574	78033	66885	13936	42117	71831	22961	94225	31816
18625	23674	53850	32827	81647	80820	00420	63555	74489	80141
74626	68394	88562	70745	23701	45630	65891	58220	35442	60414
11119	16519	27384	90199	79210	76965	99546	30323	31664	22845
41101	17336	48951	53674	17880	45260	08575	49321	36191	17095
32123	91576	84221	78902	82010	30847	62329	63898	23268	74283
26091	68409	69704	82267	14751	13151	93115	01437	56945	89661
67680	79790	48462	59278	44185	29616	76531	19589	83139	28454
15184	19260	14073	07026	25264	08388	27182	22557	61501	67481
58010	45039	57181	10238	36874	28546	37444	80824	63981	39942
56425	53996	86245	32623	78858	08143	60377	42925	42815	11159
82630	84066	13592	60642	17904	99718	63432	88642	37858	25431
14927	40909	23900	48761	44860	92467	31742	87142	03607	32059
23740	22505	07489	85986	74420	21744	97711	36648	35620	97949
32990	97446	03711	63824	07953	85965	87089	11687	92414	67257
05310	24058	91946	78437	34365	82469	12430	84754	19354	72745
21839	39937	27534	88913	49055	19218	47712	67677	51889	70926
08833	42549	93981	94051	28382	83725	72643	64233	97252	17133
58336	11139	47479	00931	91560	95372	97642	33856	54825	55680
62032	91144	75478	47431	52726	30289	42411	91886	51818	78292
45171	30557	53116	04118	58301	24375	65609	85810	18620	49198
91611	62656	60128	35609	63698	78356	50682	22505	01692	36291
55472	63819	86314	49174	93582	73604	78614	78849	23096	72825
18573	09729	74091	53994	10970	86557	65661	41854	26037	53296
60866	02955	90288	82136	83644	94455	06560	78029	98768	71296
45043	55608	82767	60890	74646	79485	13619	98868	40857	19415
17831	09737	79473	75945	28394	79334	70577	38048	03607	06932
40137	03981	07585	18128	11178	32601	27994	05641	22600	86064
77776	31343	14576	97706	16039	47517	43300	59080	80392	63189
69605	44104	40103	95635	05635	81673	68657	09559	23510	95875
19916	52934	26499	09821	87331	80993	61299	36979	73599	35055
02606	58552	07678	56619	65325	30705	99582	53390	46357	13244
65183	73160	87131	35530	47946	09854	18080	02321	05809	04898
10740	98914	44916	11322	89717	88189	30143	52687	19420	60061
98642	89822	71691	51573	83666	61642	46683	33761	47542	23551
60139	25601	93663	25547	02654	94829	48672	28736	84994	13071

Source: The Rand Corporation. *A Million Random Digits with 100,000 Normal Deviates*. New York: The Free Press, 1955. Reproduced with permission of The Rand Corporation.

TABLE C
Cumulative Values for the Binomial Probability Distribution

$$\Pr[R \leq r] = \Pr[P \leq r/n]$$

$n = 1$

π r	.01	.05	.10	.20	.30	.40	.50
0	0.9900	0.9500	0.9000	0.8000	0.7000	0.6000	0.5000
1	1.0000	1.0000	1.0000	1.0000	1.0000	1.0000	1.0000

$n = 2$

π r	.01	.05	.10	.20	.30	.40	.50
0	0.9801	0.9025	0.8100	0.6400	0.4900	0.3600	0.2500
1	0.9999	0.9975	0.9900	0.9600	0.9100	0.8400	0.7500
2	1.0000	1.0000	1.0000	1.0000	1.0000	1.0000	1.0000

$n = 3$

π r	.01	.05	.10	.20	.30	.40	.50
0	0.9703	0.8574	0.7290	0.5120	0.3430	0.2160	0.1250
1	0.9997	0.9927	0.9720	0.8960	0.7840	0.6480	0.5000
2	1.0000	0.9999	0.9990	0.9920	0.9730	0.9360	0.8750
3	1.0000	1.0000	1.0000	1.0000	1.0000	1.0000	1.0000

$n = 4$

π r	.01	.05	.10	.20	.30	.40	.50
0	0.9606	0.8145	0.6561	0.4096	0.2401	0.1296	0.0625
1	0.9994	0.9860	0.9477	0.8192	0.6517	0.4752	0.3125
2	1.0000	0.9995	0.9963	0.9728	0.9163	0.8208	0.6875
3	1.0000	1.0000	0.9999	0.9984	0.9919	0.9744	0.9375
4	1.0000	1.0000	1.0000	1.0000	1.0000	1.0000	1.0000

$n = 5$

π r	.01	.05	.10	.20	.30	.40	.50
0	0.9510	0.7738	0.5905	0.3277	0.1681	0.0778	0.0313
1	0.9990	0.9774	0.9185	0.7373	0.5282	0.3370	0.1875
2	1.0000	0.9988	0.9914	0.9421	0.8369	0.6826	0.5000
3	1.0000	1.0000	0.9995	0.9933	0.9692	0.9130	0.8125
4	1.0000	1.0000	1.0000	0.9997	0.9976	0.9898	0.9688
5				1.0000	1.0000	1.0000	1.0000

TABLE C (*continued*)

n = 6

π r	.01	.05	.10	.20	.30	.40	.50
0	0.9415	0.7351	0.5314	0.2621	0.1176	0.0467	0.0156
1	0.9985	0.9672	0.8857	0.6554	0.4202	0.2333	0.1094
2	1.0000	0.9978	0.9841	0.9011	0.7443	0.5443	0.3438
3	1.0000	0.9999	0.9987	0.9830	0.9295	0.8208	0.6563
4	1.0000	1.0000	0.9999	0.9984	0.9891	0.9590	0.8906
5	1.0000	1.0000	1.0000	0.9999	0.9993	0.9959	0.9844
6				1.0000	1.0000	1.0000	1.0000

n = 7

π r	.01	.05	.10	.20	.30	.40	.50
0	0.9321	0.6983	0.4783	0.2097	0.0824	0.0280	0.0078
1	0.9980	0.9556	0.8503	0.5767	0.3294	0.1586	0.0625
2	1.0000	0.9962	0.9743	0.8520	0.6471	0.4199	0.2266
3	1.0000	0.9998	0.9973	0.9667	0.8740	0.7102	0.5000
4	1.0000	1.0000	0.9998	0.9953	0.9712	0.9037	0.7734
5	1.0000	1.0000	1.0000	0.9996	0.9962	0.9812	0.9375
6				1.0000	0.9998	0.9984	0.9922
7					1.0000	1.0000	1.0000

n = 8

π r	.01	.05	.10	.20	.30	.40	.50
0	0.9227	0.6634	0.4305	0.1678	0.0576	0.0168	0.0039
1	0.9973	0.9428	0.8131	0.5033	0.2553	0.1064	0.0352
2	0.9999	0.9942	0.9619	0.7969	0.5518	0.3154	0.1445
3	1.0000	0.9996	0.9950	0.9437	0.8059	0.5941	0.3633
4	1.0000	1.0000	0.9996	0.9896	0.9420	0.8263	0.6367
5	1.0000	1.0000	1.0000	0.9988	0.9887	0.9502	0.8555
6				0.9999	0.9987	0.9915	0.9648
7				1.0000	0.9999	0.9993	0.9961
8					1.0000	1.0000	1.0000

n = 9

π r	.01	.05	.10	.20	.30	.40	.50
0	0.9135	0.6302	0.3874	0.1342	0.0404	0.0101	0.0020
1	0.9966	0.9288	0.7748	0.4362	0.1960	0.0705	0.0195
2	0.9999	0.9916	0.9470	0.7382	0.4628	0.2318	0.0898
3	1.0000	0.9994	0.9917	0.9144	0.7297	0.4826	0.2539
4	1.0000	1.0000	0.9991	0.9804	0.9012	0.7334	0.5000
5	1.0000	1.0000	0.9999	0.9969	0.9747	0.9006	0.7461

TABLE C (*continued*)

π	.01	.05	.10	.20	.30	.40	.50

$n = 9$

π	.01	.05	.10	.20	.30	.40	.50
r							
6	1.0000	1.0000	1.0000	0.9997	0.9957	0.9750	0.9102
7				1.0000	0.9996	0.9962	0.9805
8					1.0000	0.9997	0.9980
9						1.0000	1.0000

$n = 10$

π	.01	.05	.10	.20	.30	.40	.50
r							
0	0.9044	0.5987	0.3487	0.1074	0.0282	0.0060	0.0010
1	0.9957	0.9139	0.7361	0.3758	0.1493	0.0464	0.0107
2	0.9999	0.9885	0.9298	0.6778	0.3828	0.1673	0.0547
3	1.0000	0.9990	0.9872	0.8791	0.6496	0.3823	0.1719
4	1.0000	0.9999	0.9984	0.9672	0.8497	0.6331	0.3770
5	1.0000	1.0000	0.9999	0.9936	0.9526	0.8338	0.6230
6	1.0000	1.0000	1.0000	0.9991	0.9894	0.9452	0.8281
7				0.9999	0.9999	0.9877	0.9453
8				1.0000	1.0000	0.9983	0.9893
9						0.9999	0.9990
10						1.0000	1.0000

$n = 11$

π	.01	.05	.10	.20	.30	.40	.50
r							
0	0.8953	0.5688	0.3138	0.0859	0.0198	0.0036	0.0005
1	0.9948	0.8981	0.6974	0.3221	0.1130	0.0302	0.0059
2	0.9998	0.9848	0.9104	0.6174	0.3127	0.1189	0.0327
3	1.0000	0.9984	0.9815	0.8369	0.5696	0.2963	0.1133
4	1.0000	0.9999	0.9972	0.9496	0.7897	0.5328	0.2744
5	1.0000	1.0000	0.9997	0.9883	0.9218	0.7535	0.5000
6	1.0000	1.0000	1.0000	0.9980	0.9784	0.9006	0.7256
7				0.9998	0.9957	0.9707	0.8867
8				1.0000	0.9994	0.9941	0.9673
9					1.0000	0.9993	0.9941
10						1.0000	0.9995
11							1.0000

TABLE C   *(continued)*

n = 12

π r	.01	.05	.10	.20	.30	.40	.50
0	0.8864	0.5404	0.2824	0.0687	0.0138	0.0022	0.0002
1	0.9938	0.8816	0.6590	0.2749	0.0850	0.0196	0.0032
2	0.9998	0.9804	0.8891	0.5583	0.2528	0.0834	0.0193
3	1.0000	0.9978	0.9744	0.7946	0.4925	0.2253	0.0730
4	1.0000	0.9998	0.9957	0.9274	0.7237	0.4382	0.1938
5	1.0000	1.0000	0.9995	0.9806	0.8821	0.6652	0.3872
6	1.0000	1.0000	0.9999	0.9961	0.9614	0.8418	0.6128
7	1.0000	1.0000	1.0000	0.9994	0.9905	0.9427	0.8062
8				0.9999	0.9983	0.9847	0.9270
9				1.0000	0.9998	0.9972	0.9807
10					1.0000	0.9997	0.9968
11						1.0000	0.9998
12							1.0000

n = 13

π r	.01	.05	.10	.20	.30	.40	.50
0	0.8775	0.5133	0.2542	0.0550	0.0097	0.0013	0.0001
1	0.9928	0.8646	0.6213	0.2336	0.0637	0.0126	0.0017
2	0.9997	0.9755	0.8661	0.5017	0.2025	0.0579	0.0112
3	1.0000	0.9969	0.9658	0.7473	0.4206	0.1686	0.0461
4	1.0000	0.9997	0.9935	0.9009	0.6543	0.3530	0.1334
5	1.0000	1.0000	0.9991	0.9700	0.8346	0.5744	0.2905
6	1.0000	1.0000	0.9999	0.9930	0.9376	0.7712	0.5000
7	1.0000	1.0000	1.0000	0.9988	0.9818	0.9023	0.7095
8				0.9998	0.9960	0.9679	0.8666
9				1.0000	0.9993	0.9922	0.9539
10					0.9999	0.9987	0.9888
11					1.0000	0.9999	0.9983
12						1.0000	0.9999
13							1.0000

TABLE C *(continued)*

$n = 14$

π	.01	.05	.10	.20	.30	.40	.50
r
0	0.8687	0.4877	0.2288	0.0440	0.0068	0.0008	0.0001
1	0.9916	0.8470	0.5846	0.1979	0.0475	0.0081	0.0009
2	0.9997	0.9699	0.8416	0.4481	0.1608	0.0398	0.0065
3	1.0000	0.9958	0.9559	0.6982	0.3552	0.1243	0.0287
4	1.0000	0.9996	0.9908	0.8702	0.5842	0.2793	0.0898
5	1.0000	1.0000	0.9985	0.9561	0.7805	0.4859	0.2120
6	1.0000	1.0000	0.9998	0.9884	0.9067	0.6925	0.3953
7	1.0000	1.0000	1.0000	0.9976	0.9685	0.8499	0.6047
8				0.9996	0.9917	0.9417	0.7880
9				1.0000	0.9983	0.9825	0.9102
10					0.9998	0.9961	0.9713
11					1.0000	0.9994	0.9935
12						0.9999	0.9991
13						1.0000	0.9999
14							1.0000

$n = 15$

π	.01	.05	.10	.20	.30	.40	.50
r
0	0.8601	0.4633	0.2059	0.0352	0.0047	0.0005	0.0000
1	0.9904	0.8290	0.5490	0.1671	0.0353	0.0052	0.0005
2	0.9996	0.9638	0.8159	0.3980	0.1268	0.0271	0.0037
3	1.0000	0.9945	0.9444	0.6482	0.2969	0.0905	0.0176
4	1.0000	0.9994	0.9873	0.8358	0.5155	0.2173	0.0592
5	1.0000	0.9999	0.9978	0.9389	0.7216	0.4032	0.1509
6	1.0000	1.0000	0.9997	0.9819	0.8689	0.6098	0.3036
7	1.0000	1.0000	1.0000	0.9958	0.9500	0.7869	0.5000
8				0.9992	0.9848	0.9050	0.6964
9				0.9999	0.9963	0.9662	0.8491
10				1.0000	0.9993	0.9907	0.9408
11					0.9999	0.9981	0.9824
12					1.0000	0.9997	0.9963
13						1.0000	0.9995
14							1.0000

TABLE C (*continued*)

$n = 16$

π r	.01	.05	.10	.20	.30	.40	.50
0	0.8515	0.4401	0.1853	0.0281	0.0033	0.0003	0.0000
1	0.9891	0.8108	0.5147	0.1407	0.0261	0.0033	0.0003
2	0.9995	0.9571	0.7892	0.3518	0.0994	0.0183	0.0021
3	1.0000	0.9930	0.9316	0.5981	0.2459	0.0651	0.0106
4	1.0000	0.9991	0.9830	0.7982	0.4499	0.1666	0.0384
5	1.0000	0.9999	0.9967	0.9183	0.6598	0.3288	0.1051
6	1.0000	1.0000	0.9995	0.9733	0.8247	0.5272	0.2272
7	1.0000	1.0000	0.9999	0.9930	0.9256	0.7161	0.4018
8	1.0000	1.0000	1.0000	0.9985	0.9743	0.8577	0.5982
9				0.9998	0.9929	0.9417	0.7728
10				1.0000	0.9984	0.9809	0.8949
11					0.9997	0.9951	0.9616
12					1.0000	0.9991	0.9894
13						0.9999	0.9979
14						1.0000	0.9997
15							1.0000

$n = 17$

π r	.01	.05	.10	.20	.30	.40	.50
0	0.8429	0.4181	0.1668	0.0225	0.0023	0.0002	0.0000
1	0.9877	0.7922	0.4818	0.1182	0.0193	0.0021	0.0001
2	0.9994	0.9497	0.7618	0.3096	0.0774	0.0123	0.0012
3	1.0000	0.9912	0.9174	0.5489	0.2019	0.0464	0.0064
4	1.0000	0.9988	0.9779	0.7582	0.3887	0.1260	0.0245
5	1.0000	0.9999	0.9953	0.8943	0.5968	0.2639	0.0717
6	1.0000	1.0000	0.9992	0.9623	0.7752	0.4478	0.1662
7	1.0000	1.0000	0.9999	0.9891	0.8954	0.6405	0.3145
8	1.0000	1.0000	1.0000	0.9974	0.9597	0.8011	0.5000
9				0.9995	0.9873	0.9081	0.6855
10				0.9999	0.9968	0.9652	0.8338
11				1.0000	0.9993	0.9894	0.9283
12					0.9999	0.9975	0.9755
13					1.0000	0.9995	0.9936
14						0.9999	0.9988
15						1.0000	0.9999
16							1.0000

TABLE C *(continued)*

$n = 18$

π r	.01	.05	.10	.20	.30	.40	.50
0	0.8345	0.3972	0.1501	0.0180	0.0016	0.0001	0.0000
1	0.9862	0.7735	0.4503	0.0991	0.0142	0.0013	0.0001
2	0.9993	0.9419	0.7338	0.2713	0.0600	0.0082	0.0007
3	1.0000	0.9891	0.9018	0.5010	0.1646	0.0328	0.0038
4	1.0000	0.9985	0.9718	0.7164	0.3327	0.0942	0.0154
5	1.0000	0.9998	0.9936	0.8671	0.5344	0.2088	0.0481
6	1.0000	1.0000	0.9988	0.9487	0.7217	0.3743	0.1189
7	1.0000	1.0000	0.9998	0.9837	0.8593	0.5634	0.2403
8	1.0000	1.0000	1.0000	0.9957	0.9404	0.7368	0.4073
9				0.9991	0.9790	0.8653	0.5927
10				0.9998	0.9939	0.9424	0.7597
11				1.0000	0.9986	0.9797	0.8811
12					0.9997	0.9942	0.9519
13					1.0000	0.9987	0.9846
14						0.9998	0.9962
15						1.0000	0.9993
16							0.9999
17							1.0000

$n = 19$

π r	.01	.05	.10	.20	.30	.40	.50
0	0.8262	0.3774	0.1351	0.0144	0.0011	0.0001	0.0000
1	0.9847	0.7547	0.4203	0.0829	0.0104	0.0008	0.0000
2	0.9991	0.9335	0.7054	0.2369	0.0462	0.0055	0.0004
3	1.0000	0.9868	0.8850	0.4551	0.1332	0.0230	0.0022
4	1.0000	0.9980	0.9648	0.6733	0.2822	0.0696	0.0096
5	1.0000	0.9998	0.9914	0.8369	0.4739	0.1629	0.0318
6	1.0000	1.0000	0.9983	0.9324	0.6655	0.3081	0.0835
7	1.0000	1.0000	0.9997	0.9767	0.8180	0.4878	0.1796
8	1.0000	1.0000	1.0000	0.9933	0.9161	0.6675	0.3238
9				0.9984	0.9674	0.8139	0.5000
10				0.9997	0.9895	0.9115	0.6762

TABLE C   (*continued*)

n = 19

π r	.01	.05	.10	.20	.30	.40	.50
11				0.9999	0.9972	0.9648	0.8204
12				1.0000	0.9994	0.9884	0.9165
13					0.9999	0.9969	0.9682
14					1.0000	0.9994	0.9904
15						0.9999	0.9978
16						1.0000	0.9996
17							1.0000

n = 20

π r	.01	.05	.10	.20	.30	.40	.50
0	0.8179	0.3585	0.1216	0.0115	0.0008	0.0000	0.0000
1	0.9831	0.7358	0.3917	0.0692	0.0076	0.0005	0.0000
2	0.9990	0.9245	0.6769	0.2061	0.0355	0.0036	0.0002
3	1.0000	0.9841	0.8670	0.4114	0.1071	0.0160	0.0013
4	1.0000	0.9974	0.9568	0.6296	0.2375	0.0510	0.0059
5	1.0000	0.9997	0.9887	0.8042	0.4164	0.1256	0.0207
6	1.0000	1.0000	0.9976	0.9133	0.6080	0.2500	0.0577
7	1.0000	1.0000	0.9996	0.9679	0.7723	0.4159	0.1316
8	1.0000	1.0000	0.9999	0.9900	0.8867	0.5956	0.2517
9	1.0000	1.0000	1.0000	0.9974	0.9520	0.7553	0.4119
10				0.9994	0.9829	0.8725	0.5881
11				0.9999	0.9949	0.9435	0.7483
12				1.0000	0.9987	0.9790	0.8684
13					0.9997	0.9935	0.9423
14					1.0000	0.9984	0.9793
15						0.9997	0.9941
16						1.0000	0.9987
17							0.9998
18							1.0000

TABLE C   *(continued)*

			$n = 50$				
π r	.01	.05	.10	.20	.30	.40	.50
0	0.6050	0.0769	0.0052	0.0000	0.0000	0.0000	0.0000
1	0.9106	0.2794	0.0338	0.0002	0.0000	0.0000	0.0000
2	0.9862	0.5405	0.1117	0.0013	0.0000	0.0000	0.0000
3	0.9984	0.7604	0.2503	0.0057	0.0000	0.0000	0.0000
4	0.9999	0.8964	0.4312	0.0185	0.0002	0.0000	0.0000
5	1.0000	0.9622	0.6161	0.0480	0.0007	0.0000	0.0000
6	1.0000	0.9882	0.7702	0.1034	0.0025	0.0000	0.0000
7	1.0000	0.9968	0.8779	0.1904	0.0073	0.0001	0.0000
8	1.0000	0.9992	0.9421	0.3073	0.0183	0.0002	0.0000
9	1.0000	0.9998	0.9755	0.4437	0.0402	0.0008	0.0000
10	1.0000	1.0000	0.9906	0.5836	0.0789	0.0022	0.0000
11	1.0000	1.0000	0.9968	0.7107	0.1390	0.0057	0.0000
12	1.0000	1.0000	0.9990	0.8139	0.2229	0.0133	0.0002
13	1.0000	1.0000	0.9997	0.8894	0.3279	0.0280	0.0005
14	1.0000	1.0000	0.9999	0.9393	0.4468	0.0540	0.0013
15	1.0000	1.0000	1.0000	0.9692	0.5692	0.0955	0.0033
16				0.9856	0.6839	0.1561	0.0077
17				0.9937	0.7822	0.2369	0.0164
18				0.9975	0.8594	0.3356	0.0325
19				0.9991	0.9152	0.4465	0.0595
20				0.9997	0.9522	0.5610	0.1013
21				0.9999	0.9749	0.6701	0.1611
22				1.0000	0.9877	0.7660	0.2399
23					0.9944	0.8438	0.3359
24					0.9976	0.9022	0.4439
25					0.9991	0.9427	0.5561
26					0.9997	0.9686	0.6641
27					0.9999	0.9840	0.7601
28					1.0000	0.9924	0.8389
29						0.9966	0.8987
30						0.9986	0.9405
31						0.9995	0.9675
32						0.9998	0.9836
33						0.9999	0.9923
34						1.0000	0.9967
35							0.9987
36							0.9995
37							0.9998
38							1.0000

TABLE C *(continued)*

$n = 100$

π r	.01	.05	.10	.20	.30	.40	.50
0	0.3660	0.0059	0.0000	0.0000	0.0000	0.0000	0.0000
1	0.7358	0.0371	0.0003	0.0000	0.0000	0.0000	0.0000
2	0.9206	0.1183	0.0019	0.0000	0.0000	0.0000	0.0000
3	0.9816	0.2578	0.0078	0.0000	0.0000	0.0000	0.0000
4	0.9966	0.4360	0.0237	0.0000	0.0000	0.0000	0.0000
5	0.9995	0.6160	0.0576	0.0000	0.0000	0.0000	0.0000
6	0.9999	0.7660	0.1172	0.0001	0.0000	0.0000	0.0000
7	1.0000	0.8720	0.2061	0.0003	0.0000	0.0000	0.0000
8	1.0000	0.9369	0.3209	0.0009	0.0000	0.0000	0.0000
9	1.0000	0.9718	0.4513	0.0023	0.0000	0.0000	0.0000
10	1.0000	0.9885	0.5832	0.0057	0.0000	0.0000	0.0000
11	1.0000	0.9957	0.7030	0.0126	0.0000	0.0000	0.0000
12	1.0000	0.9985	0.8018	0.0253	0.0000	0.0000	0.0000
13	1.0000	0.9995	0.8761	0.0469	0.0001	0.0000	0.0000
14	1.0000	0.9999	0.9274	0.0804	0.0002	0.0000	0.0000
15	1.0000	1.0000	0.9601	0.1285	0.0004	0.0000	0.0000
16	1.0000	1.0000	0.9794	0.1923	0.0010	0.0000	0.0000
17	1.0000	1.0000	0.9900	0.2712	0.0022	0.0000	0.0000
18	1.0000	1.0000	0.9954	0.3621	0.0045	0.0000	0.0000
19	1.0000	1.0000	0.9980	0.4602	0.0089	0.0000	0.0000
20	1.0000	1.0000	0.9992	0.5595	0.0165	0.0000	0.0000
21	1.0000	1.0000	0.9997	0.6540	0.0288	0.0000	0.0000
22	1.0000	1.0000	0.9999	0.7389	0.0479	0.0001	0.0000
23	1.0000	1.0000	1.0000	0.8109	0.0755	0.0003	0.0000
24				0.8686	0.1136	0.0006	0.0000
25				0.9125	0.1631	0.0012	0.0000
26				0.9442	0.2244	0.0024	0.0000
27				0.9658	0.2964	0.0046	0.0000
28				0.9800	0.3768	0.0084	0.0000
29				0.9888	0.4623	0.0148	0.0000
30				0.9939	0.5491	0.0248	0.0000
31				0.9969	0.6331	0.0398	0.0001
32				0.9984	0.7107	0.0615	0.0002
33				0.9993	0.7793	0.0913	0.0004
34				0.9997	0.8371	0.1303	0.0009
35				0.9999	0.8839	0.1795	0.0018

TABLE C   (*continued*)

	$n = 100$						
π	.01	.05	.10	.20	.30	.40	.50
r							
36				0.9999	0.9201	0.2386	0.0033
37				1.0000	0.9470	0.3068	0.0060
38					0.9660	0.3822	0.0105
39					0.9790	0.4621	0.0176
40					0.9875	0.5433	0.0284
41					0.9928	0.6225	0.0443
42					0.9960	0.6967	0.0666
43					0.9979	0.7635	0.0967
44					0.9989	0.8211	0.1356
45					0.9995	0.8689	0.1841
46					0.9997	0.9070	0.2421
47					0.9999	0.9362	0.3086
48					0.9999	0.9577	0.3822
49					1.0000	0.9729	0.4602
50						0.9832	0.5398
51						0.9900	0.6178
52						0.9942	0.6914
53						0.9968	0.7579
54						0.9983	0.8159
55						0.9991	0.8644
56						0.9996	0.9033
57						0.9998	0.9334
58						0.9999	0.9557
59						1.0000	0.9716
60							0.9824
61							0.9895
62							0.9940
63							0.9967
64							0.9982
65							0.9991
66							0.9996
67							0.9998
68							0.9999
69							1.0000

TABLE D
Areas Under the Standard Normal Curve

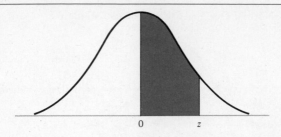

The following table provides the area between the mean and normal deviate value z.

Normal Deviate z	.00	.01	.02	.03	.04	.05	.06	.07	.08	.09
0.0	.0000	.0040	.0080	.0120	.0160	.0199	.0239	.0279	.0319	.0359
0.1	.0398	.0438	.0478	.0517	.0557	.0596	.0636	.0675	.0714	.0753
0.2	.0793	.0832	.0871	.0910	.0948	.0987	.1026	.1064	.1103	.1141
0.3	.1179	.1217	.1255	.1293	.1331	.1368	.1406	.1443	.1480	.1517
0.4	.1554	.1591	.1628	.1664	.1700	.1736	.1772	.1808	.1844	.1879
0.5	.1915	.1950	.1985	.2019	.2054	.2088	.2123	.2157	.2190	.2224
0.6	.2257	.2291	.2324	.2357	.2389	.2422	.2454	.2486	.2518	.2549
0.7	.2580	.2612	.2642	.2673	.2704	.2734	.2764	.2794	.2823	.2852
0.8	.2881	.2910	.2939	.2967	.2995	.3023	.3051	.3078	.3106	.3133
0.9	.3159	.3186	.3212	.3238	.3264	.3289	.3315	.3340	.3365	.3389
1.0	.3413	.3438	.3461	.3485	.3508	.3531	.3554	.3577	.3599	.3621
1.1	.3643	.3665	.3686	.3708	.3729	.3749	.3770	.3790	.3810	.3830
1.2	.3849	.3869	.3888	.3907	.3925	.3944	.3962	.3980	.3997	.4015
1.3	.4032	.4049	.4066	.4082	.4099	.4115	.4131	.4147	.4162	.4177
1.4	.4192	.4207	.4222	.4236	.4251	.4265	.4279	.4292	.4306	.4319
1.5	.4332	.4345	.4357	.4370	.4382	.4394	.4406	.4418	.4429	.4441
1.6	.4452	.4463	.4474	.4484	.4495	.4505	.4515	.4525	.4535	.4545
1.7	.4554	.4564	.4573	.4582	.4591	.4599	.4608	.4616	.4625	.4633
1.8	.4641	.4649	.4656	.4664	.4671	.4678	.4686	.4693	.4699	.4706
1.9	.4713	.4719	.4726	.4732	.4738	.4744	.4750	.4756	.4761	.4767
2.0	.4772	.4778	.4783	.4788	.4793	.4798	.4803	.4808	.4812	.4817
2.1	.4821	.4826	.4830	.4834	.4838	.4842	.4846	.4850	.4854	.4857
2.2	.4861	.4864	.4868	.4871	.4875	.4878	.4881	.4884	.4887	.4890
2.3	.4893	.4896	.4898	.4901	.4904	.4906	.4909	.4911	.4913	.4916
2.4	.4918	.4920	.4922	.4925	.4927	.4929	.4931	.4932	.4934	.4936
2.5	.4938	.4940	.4941	.4943	.4945	.4946	.4948	.4949	.4951	.4952
2.6	.4953	.4955	.4956	.4957	.4959	.4960	.4961	.4962	.4963	.4964
2.7	.4965	.4966	.4967	.4968	.4969	.4970	.4971	.4972	.4973	.4974
2.8	.4974	.4975	.4976	.4977	.4977	.4978	.4979	.4979	.4980	.4981
2.9	.4981	.4982	.4982	.4983	.4984	.4984	.4985	.4985	.4986	.4986
3.0	.49865	.4987	.4987	.4988	.4988	.4989	.4989	.4989	.4990	.4990
4.0	.49997									

TABLE E
Normal Deviate Values

Normal Deviates for Statistical Estimation

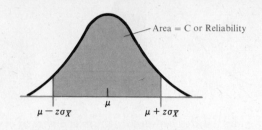

Reliability or Confidence Level C	Normal Deviate z
.80	1.28
.90	1.64
.95	1.96
.98	2.33
.99	2.57
.998	3.08
.999	3.27

Critical Normal Deviates for Hypothesis Testing

One–Sided Tests

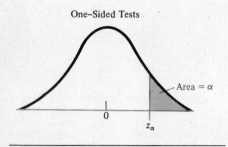

Two–Sided Tests

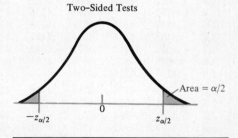

Significance Level α	Normal Deviate z_α	Significance Level α	Normal Deviate $z_{\alpha/2}$
.10	1.28	.10	1.64
.05	1.64	.05	1.96
.025	1.96	.025	2.24
.01	2.33	.01	2.57
.005	2.57	.005	2.81
.001	3.08	.001	3.27

TABLE F
Student *t* Distribution

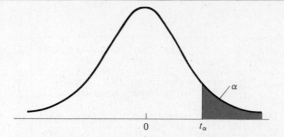

The following table provides the values of t_α that correspond to a given upper-tail area α and a specified number of degrees of freedom.

Degrees of Freedom	Upper-Tail Area α									
	.4	.25	.1	.05	.025	.01	.005	.0025	.001	.0005
1	0.325	1.000	3.078	6.314	12.706	31.821	63.657	127.32	318.31	636.62
2	.289	.816	1.886	2.920	4.303	6.965	9.925	14.089	22.327	31.598
3	.277	.765	1.638	2.353	3.182	4.541	5.841	7.453	10.214	12.924
4	.271	.741	1.533	2.132	2.776	3.747	4.604	5.598	7.173	8.610
5	0.267	0.727	1.476	2.015	2.571	3.365	4.032	4.773	5.893	6.869
6	.265	.718	1.440	1.943	2.447	3.143	3.707	4.317	5.208	5.959
7	.263	.711	1.415	1.895	2.365	2.998	3.499	4.029	4.785	5.408
8	.262	.706	1.397	1.860	2.306	2.896	3.355	3.833	4.501	5.041
9	.261	.703	1.383	1.833	2.262	2.821	3.250	3.690	4.297	4.781
10	0.260	0.700	1.372	1.812	2.228	2.764	3.169	3.581	4.144	4.587
11	.260	.697	1.363	1.796	2.201	2.718	3.106	3.497	4.025	4.437
12	.259	.695	1.356	1.782	2.179	2.681	3.055	3.428	3.930	4.318
13	.259	.694	1.350	1.771	2.160	2.650	3.012	3.372	3.852	4.221
14	.258	.692	1.345	1.761	2.145	2.624	2.977	3.326	3.787	4.140
15	0.258	0.691	1.341	1.753	2.131	2.602	2.947	3.286	3.733	4.073
16	.258	.690	1.337	1.746	2.120	2.583	2.921	3.252	3.686	4.015
17	.257	.689	1.333	1.740	2.110	2.567	2.898	3.222	3.646	3.965
18	.257	.688	1.330	1.734	2.101	2.552	2.878	3.197	3.610	3.922
19	.257	.688	1.328	1.729	2.093	2.539	2.861	3.174	3.579	3.883
20	0.257	0.687	1.325	1.725	2.086	2.528	2.845	3.153	3.552	3.850
21	.257	.686	1.323	1.721	2.080	2.518	2.831	3.135	3.527	3.819
22	.256	.686	1.321	1.717	2.074	2.508	2.819	3.119	3.505	3.792
23	.256	.685	1.319	1.714	2.069	2.500	2.807	3.104	3.485	3.767
24	.256	.685	1.318	1.711	2.064	2.492	2.797	3.091	3.467	3.745
25	0.256	0.684	1.316	1.708	2.060	2.485	2.787	3.078	3.450	3.725
26	.256	.684	1.315	1.706	2.056	2.479	2.779	3.067	3.435	3.707
27	.256	.684	1.314	1.703	2.052	2.473	2.771	3.057	3.421	3.690
28	.256	.683	1.313	1.701	2.048	2.467	2.763	3.047	3.408	3.674
29	.256	.683	1.311	1.699	2.045	2.462	2.756	3.038	3.396	3.659
30	0.256	0.683	1.310	1.697	2.042	2.457	2.750	3.030	3.385	3.646
40	.255	.681	1.303	1.684	2.021	2.423	2.704	2.971	3.307	3.551
60	.254	.679	1.296	1.671	2.000	2.390	2.660	2.915	3.232	3.460
120	.254	.677	1.289	1.658	1.980	2.358	2.617	2.860	3.160	3.373
∞	.253	.674	1.282	1.645	1.960	2.326	2.576	2.807	3.090	3.291

Source: E. S. Pearson and H. O. Hartley, *Biometrika Tables for Statisticians*, Vol. I. London: Cambridge University Press, 1966. Partly derived from Table III of Fisher and Yates, *Statistical Tables for Biological, Agricultural and Medical Research*, published by Longman Group Ltd., London (previously published by Oliver & Boyd, Edinburgh, 1963). Reproduced with permission of the authors and publishers.

TABLE G
Four-Place Common Logarithms

N	0	1	2	3	4	5	6	7	8	9	1	2	Proportional Parts 3	4	5	6	7	8	9
10	0000	0043	0086	0128	0170	0212	0253	0294	0334	0374	4	8	12	17	21	25	29	33	37
11	0414	0453	0492	0531	0569	0607	0645	0682	0719	0755	4	8	11	15	19	23	26	30	34
12	0792	0828	0864	0899	0934	0969	1004	1038	1072	1106	3	7	10	14	17	21	24	28	31
13	1139	1173	1206	1239	1271	1303	1335	1367	1399	1430	3	6	10	13	16	19	23	26	29
14	1461	1492	1523	1553	1584	1614	1644	1673	1703	1732	3	6	9	12	15	18	21	24	27
15	1761	1790	1818	1847	1875	1903	1931	1959	1987	2014	3	6	8	11	14	17	20	22	25
16	2041	2068	2095	2122	2148	2175	2201	2227	2253	2279	3	5	8	11	13	16	18	21	24
17	2304	2330	2355	2380	2405	2430	2455	2480	2504	2529	2	5	7	10	12	15	17	20	22
18	2553	2577	2601	2625	2648	2672	2695	2718	2742	2765	2	5	7	9	12	14	16	19	21
19	2788	2810	2833	2856	2878	2900	2923	2945	2967	2989	2	4	7	9	11	13	16	18	20
20	3010	3032	3054	3075	3096	3118	3139	3160	3181	3201	2	4	6	8	11	13	15	17	19
21	3222	3243	3263	3284	3304	3324	3345	3365	3385	3404	2	4	6	8	10	12	14	16	18
22	3424	3444	3464	3483	3502	3522	3541	3560	3579	3598	2	4	6	8	10	12	14	15	17
23	3617	3636	3655	3674	3692	3711	3729	3747	3766	3784	2	4	6	7	9	11	13	15	17
24	3802	3820	3838	3856	3874	3892	3909	3927	3945	3962	2	4	5	7	9	11	12	14	16
25	3979	3997	4014	4031	4048	4065	4082	4099	4116	4133	2	3	5	7	9	10	12	14	15
26	4150	4166	4183	4200	4216	4232	4249	4265	4281	4298	2	3	5	7	8	10	11	13	15
27	4314	4330	4346	4362	4378	4393	4409	4425	4440	4456	2	3	5	6	8	9	11	13	14
28	4472	4487	4502	4518	4533	4548	4564	4579	4594	4609	2	3	5	6	8	9	11	12	14
29	4624	4639	4654	4669	4683	4698	4713	4728	4742	4757	1	3	4	6	7	9	10	12	13
30	4771	4786	4800	4814	4829	4843	4857	4871	4886	4900	1	3	4	6	7	9	10	11	13
31	4914	4928	4942	4955	4969	4983	4997	5011	5024	5038	1	3	4	6	7	8	10	11	12
32	5051	5065	5079	5092	5105	5119	5132	5145	5159	5172	1	3	4	5	7	8	9	11	12
33	5185	5198	5211	5224	5237	5250	5263	5276	5289	5302	1	3	4	5	6	8	9	10	12
34	5315	5328	5340	5353	5366	5378	5391	5403	5416	5428	1	3	4	5	6	8	9	10	11
35	5441	5453	5465	5478	5490	5502	5514	5527	5539	5551	1	2	4	5	6	7	9	10	11
36	5563	5575	5587	5599	5611	5623	5635	5647	5658	5670	1	2	4	5	6	7	8	10	11
37	5682	5694	5705	5717	5729	5740	5752	5763	5775	5786	1	2	3	5	6	7	8	9	10
38	5798	5809	5821	5832	5843	5855	5866	5877	5888	5899	1	2	3	5	6	7	8	9	10
39	5911	5922	5933	5944	5955	5966	5977	5988	5999	6010	1	2	3	4	5	7	8	9	10
40	6021	6031	6042	6053	6064	6075	6085	6096	6107	6117	1	2	3	4	5	6	8	9	10
41	6128	6138	6149	6160	6170	6180	6191	6201	6212	6222	1	2	3	4	5	6	7	8	9
42	6232	6243	6253	6263	6274	6284	6294	6304	6314	6325	1	2	3	4	5	6	7	8	9
43	6335	6345	6355	6365	6375	6385	6395	6405	6415	6425	1	2	3	4	5	6	7	8	9
44	6435	6444	6454	6464	6474	6484	6493	6503	6513	6522	1	2	3	4	5	6	7	8	9
45	6532	6542	6551	6561	6571	6580	6590	6599	6609	6618	1	2	3	4	5	6	7	8	9
46	6628	6637	6646	6656	6665	6675	6684	6693	6702	6712	1	2	3	4	5	6	7	7	8
47	6721	6730	6739	6749	6758	6767	6776	6785	6794	6803	1	2	3	4	5	5	6	7	8
48	6812	6821	6830	6839	6848	6857	6866	6875	6884	6893	1	2	3	4	4	5	6	7	8
49	6902	6911	6920	6928	6937	6946	6955	6964	6972	6981	1	2	3	4	4	5	6	7	8
50	6990	6998	7007	7016	7024	7033	7042	7050	7059	7067	1	2	3	3	4	5	6	7	8
51	7076	7084	7093	7101	7110	7118	7126	7135	7143	7152	1	2	3	3	4	5	6	7	8
52	7160	7168	7177	7185	7193	7202	7210	7218	7226	7235	1	2	2	3	4	5	6	7	7
53	7243	7251	7259	7267	7275	7284	7292	7300	7308	7316	1	2	2	3	4	5	6	6	7
54	7324	7332	7340	7348	7356	7364	7372	7380	7388	7396	1	2	2	3	4	5	6	6	7
N	0	1	2	3	4	5	6	7	8	9	1	2	3	4	5	6	7	8	9

TABLE G (*continued*)

N	0	1	2	3	4	5	6	7	8	9	1	2	3	4	5	6	7	8	9
														Proportional Parts					
55	7404	7412	7419	7427	7435	7443	7451	7459	7466	7474	1	2	2	3	4	5	5	6	7
56	7482	7490	7497	7505	7513	7520	7528	7536	7543	7551	1	2	2	3	4	5	5	6	7
57	7559	7566	7574	7582	7589	7597	7604	7612	7619	7627	1	2	2	3	4	5	5	6	7
58	7634	7642	7649	7657	7664	7672	7679	7686	7694	7701	1	1	2	3	4	4	5	6	7
59	7709	7716	7723	7731	7738	7745	7752	7760	7767	7774	1	1	2	3	4	4	5	6	7
60	7782	7789	7796	7803	7810	7818	7825	7832	7839	7846	1	1	2	3	4	4	5	6	6
61	7853	7860	7868	7875	7882	7889	7896	7903	7910	7917	1	1	2	3	4	4	5	6	6
62	7924	7931	7938	7945	7952	7959	7966	7973	7980	7987	1	1	2	3	3	4	5	6	6
63	7993	8000	8007	8014	8021	8028	8035	8041	8048	8055	1	1	2	3	3	4	5	5	6
64	8062	8069	8075	8082	8089	8096	8102	8109	8116	8122	1	1	2	3	3	4	5	5	6
65	8129	8136	8142	8149	8156	8162	8169	8176	8182	8189	1	1	2	3	3	4	5	5	6
66	8195	8202	8209	8215	8222	8228	8235	8241	8248	8254	1	1	2	3	3	4	5	5	6
67	8261	8267	8274	8280	8287	8293	8299	8306	8312	8319	1	1	2	3	3	4	5	5	6
68	8325	8331	8338	8344	8351	8357	8363	8370	8376	8382	1	1	2	3	3	4	4	5	6
69	8388	8395	8401	8407	8414	8420	8426	8432	8439	8445	1	1	2	2	3	4	4	5	6
70	8451	8457	8463	8470	8476	8482	8488	8494	8500	8506	1	1	2	2	3	4	4	5	6
71	8513	8519	8525	8531	8537	8543	8549	8555	8561	8567	1	1	2	2	3	4	4	5	5
72	8573	8579	8585	8591	8597	8603	8609	8615	8621	8627	1	1	2	2	3	4	4	5	5
73	8633	8639	8645	8651	8657	8663	8669	8675	8681	8686	1	1	2	2	3	4	4	5	5
74	8692	8698	8704	8710	8716	8722	8727	8733	8739	8745	1	1	2	2	3	4	4	5	5
75	8751	8756	8762	8768	8774	8779	8785	8791	8797	8802	1	1	2	2	3	3	4	5	5
76	8808	8814	8820	8825	8831	8837	8842	8848	8854	8859	1	1	2	2	3	3	4	5	5
77	8865	8871	8876	8882	8887	8893	8899	8904	8910	8915	1	1	2	2	3	3	4	4	5
78	8921	8927	8932	8938	8943	8949	8954	8960	8965	8971	1	1	2	2	3	3	4	4	5
79	8976	8982	8987	8993	8998	9004	9009	9015	9020	9025	1	1	2	2	3	3	4	4	5
80	9031	9036	9042	9047	9053	9058	9063	9069	9074	9079	1	1	2	2	3	3	4	4	5
81	9085	9090	9096	9101	9106	9112	9117	9122	9128	9133	1	1	2	2	3	3	4	4	5
82	9138	9143	9149	9154	9159	9165	9170	9175	9180	9186	1	1	2	2	3	3	4	4	5
83	9191	9196	9201	9206	9212	9217	9222	9227	9232	9238	1	1	2	2	3	3	4	4	5
84	9243	9248	9253	9258	9263	9269	9274	9279	9284	9289	1	1	2	2	3	3	4	4	5
85	9294	9299	9304	9309	9315	9320	9325	9330	9335	9340	1	1	2	2	3	3	4	4	5
86	9345	9350	9355	9360	9365	9370	9375	9380	9385	9390	1	1	2	2	3	3	4	4	5
87	9395	9400	9405	9410	9415	9420	9425	9430	9435	9440	0	1	1	2	2	3	3	4	4
88	9445	9450	9455	9460	9465	9469	9474	9479	9484	9489	0	1	1	2	2	3	3	4	4
89	9494	9499	9504	9509	9513	9518	9523	9528	9533	9538	0	1	1	2	2	3	3	4	4
90	9542	9547	9552	9557	9562	9566	9571	9576	9581	9586	0	1	1	2	2	3	3	4	4
91	9590	9595	9600	9605	9609	9614	9619	9624	9628	9633	0	1	1	2	2	3	3	4	4
92	9638	9643	9647	9652	9657	9661	9666	9671	9675	9680	0	1	1	2	2	3	3	4	4
93	9685	9689	9694	9699	9703	9708	9713	9717	9722	9727	0	1	1	2	2	3	3	4	4
94	9731	9736	9741	9745	9750	9754	9759	9763	9768	9773	0	1	1	2	2	3	3	4	4
95	9777	9782	9786	9791	9795	9800	9805	9809	9814	9818	0	1	1	2	2	3	3	4	4
96	9823	9827	9832	9836	9841	9845	9850	9854	9859	9863	0	1	1	2	2	3	3	4	4
97	9868	9872	9877	9881	9886	9890	9894	9899	9903	9908	0	1	1	2	2	3	3	4	4
98	9912	9917	9921	9926	9930	9934	9939	9943	9948	9952	0	1	1	2	2	3	3	4	4
99	9956	9961	9965	9969	9974	9978	9983	9987	9991	9996	0	1	1	2	2	3	3	3	4
N	0	1	2	3	4	5	6	7	8	9	1	2	3	4	5	6	7	8	9

TABLE H
Chi-Square Distribution

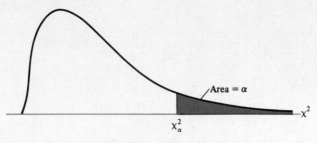

The following table provides the values of χ_α^2 that correspond to a given upper-tail area α and a specified number of degrees of freedom.

Degrees of Freedom	Upper-Tail Area α						
	.99	.98	.95	.90	.80	.70	.50
1	$.0^3157$	$.0^3628$	.00393	.0158	.0642	.148	.455
2	.0201	.0404	.103	.211	.446	.713	1.386
3	.115	.185	.352	.584	1.005	1.424	2.366
4	.297	.429	.711	1.064	1.649	2.195	3.357
5	.554	.752	1.145	1.610	2.343	3.000	4.351
6	.872	1.134	1.635	2.204	3.070	3.828	5.348
7	1.239	1.564	2.167	2.833	3.822	4.671	6.346
8	1.646	2.032	2.733	3.490	4.594	5.527	7.344
9	2.088	2.532	3.325	4.168	5.380	6.393	8.343
10	2.558	3.059	3.940	4.865	6.179	7.267	9.342
11	3.053	3.609	4.575	5.578	6.989	8.148	10.341
12	3.571	4.178	5.226	6.304	7.807	9.034	11.340
13	4.107	4.765	5.892	7.042	8.634	9.926	12.340
14	4.660	5.368	6.571	7.790	9.467	10.821	13.339
15	5.229	5.985	7.261	8.547	10.307	11.721	14.339
16	5.812	6.614	7.962	9.312	11.152	12.624	15.338
17	6.408	7.255	8.672	10.085	12.002	13.531	16.338
18	7.015	7.906	9.390	10.865	12.857	14.440	17.338
19	7.633	8.567	10.117	11.651	13.716	15.352	18.338
20	8.260	9.237	10.851	12.443	14.578	16.266	19.337
21	8.897	9.915	11.591	13.240	15.445	17.182	20.337
22	9.542	10.600	12.338	14.041	16.314	18.101	21.337
23	10.196	11.293	13.091	14.848	17.187	19.021	22.337
24	10.856	11.992	13.848	15.659	18.062	19.943	23.337
25	11.524	12.697	14.611	16.473	18.940	20.867	24.337
26	12.198	13.409	15.379	17.292	19.820	21.792	25.336
27	12.879	14.125	16.151	18.114	20.703	22.719	26.336
28	13.565	14.847	16.928	18.939	21.588	23.647	27.336
29	14.256	15.574	17.708	19.768	22.475	24.577	28.336
30	14.953	16.306	18.493	20.599	23.364	25.508	29.336

TABLE H   (*continued*)

Degrees of Freedom	Upper-Tail Area α						
	.30	.20	.10	.05	.02	.01	.001
1	1.074	1.642	2.706	3.841	5.412	6.635	10.827
2	2.408	3.219	4.605	5.991	7.824	9.210	13.815
3	3.665	4.642	6.251	7.815	9.837	11.345	16.268
4	4.878	5.989	7.779	9.488	11.668	13.277	18.465
5	6.064	7.289	9.236	11.070	13.388	15.086	20.517
6	7.231	8.558	10.645	12.592	15.033	16.812	22.457
7	8.383	9.803	12.017	14.067	16.622	18.475	24.322
8	9.524	11.030	13.362	15.507	18.168	20.090	26.125
9	10.656	12.242	14.684	16.919	19.679	21.666	27.877
10	11.781	13.442	15.987	18.307	21.161	23.209	29.588
11	12.899	14.631	17.275	19.675	22.618	24.725	31.264
12	14.011	15.812	18.549	21.026	24.054	26.217	32.909
13	15.119	16.985	19.812	22.362	25.472	27.688	34.528
14	16.222	18.151	21.064	23.685	26.873	29.141	36.123
15	17.322	19.311	22.307	24.996	28.259	30.578	37.697
16	18.418	20.465	23.542	26.296	29.633	32.000	39.252
17	19.511	21.615	24.769	27.587	30.995	33.409	40.790
18	20.601	22.760	25.989	28.869	32.346	34.805	42.312
19	21.689	23.900	27.204	30.144	33.687	36.191	43.820
20	22.775	25.038	28.412	31.410	35.020	37.566	45.315
21	23.858	26.171	29.615	32.671	36.343	38.932	46.797
22	24.939	27.301	30.813	33.924	37.659	40.289	48.268
23	26.018	28.429	32.007	35.172	38.968	41.638	49.728
24	27.096	29.553	33.196	36.415	40.270	42.980	51.179
25	28.172	30.675	34.382	37.652	41.566	44.314	52.620
26	29.246	31.795	35.563	38.885	42.856	45.642	54.052
27	30.319	32.912	36.741	40.113	44.140	46.963	55.476
28	31.391	34.027	37.916	41.337	45.419	48.278	56.893
29	32.461	35.139	39.087	42.557	46.693	49.588	58.302
30	33.530	36.250	40.256	43.773	47.962	50.892	59.703

Source: From Table IV of Fisher and Yates, *Statistical Tables for Biological, Agricultural and Medical Research*, published by Longman Group Ltd., London (previously published by Oliver & Boyd, Edinburgh, 1963). Reproduced with permission of the authors and publishers.

TABLE I
F Distribution

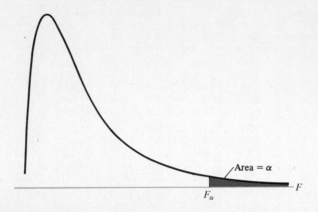

The following table provides the values of F_α that correspond to a given upper-tail area α and a specified degrees of freedom pair. The values of $F_{.05}$ are in lightface type, while those for $F_{.01}$ are given in boldface type. The number of degrees of freedom for the *numerator* mean square is indicated at the head of each *column*, while the number of degrees of freedom for the *denominator* mean square determines which *row* is applicable.

Degrees of Freedom in Denominator	Degrees of Freedom in Numerator											
	1	2	3	4	5	6	7	8	9	10	11	12
1	161	200	216	225	230	234	237	239	241	242	243	244
	4,052	**4,999**	**5,403**	**5,625**	**5,764**	**5,859**	**5,928**	**5,981**	**6,022**	**6,056**	**6,082**	**6,106**
2	18.51	19.00	19.16	19.25	19.30	19.33	19.36	19.37	19.38	19.39	19.40	19.41
	98.49	**99.00**	**99.17**	**99.25**	**99.30**	**99.33**	**99.36**	**99.37**	**99.39**	**99.40**	**99.41**	**99.42**
3	10.13	9.55	9.28	9.12	9.01	8.94	8.88	8.84	8.81	8.78	8.76	8.74
	34.12	**30.82**	**29.46**	**28.71**	**28.24**	**27.91**	**27.67**	**27.49**	**27.34**	**27.23**	**27.13**	**27.05**
4	7.71	6.94	6.59	6.39	6.26	6.16	6.09	6.04	6.00	5.96	5.93	5.91
	21.20	**18.00**	**16.69**	**15.98**	**15.52**	**15.21**	**14.98**	**14.80**	**14.66**	**14.54**	**14.45**	**14.37**
5	6.61	5.79	5.41	5.19	5.05	4.95	4.88	4.82	4.78	4.74	4.70	4.68
	16.26	**13.27**	**12.06**	**11.39**	**10.97**	**10.67**	**10.45**	**10.29**	**10.15**	**10.05**	**9.96**	**9.89**
6	5.99	5.14	4.76	4.53	4.39	4.28	4.21	4.15	4.10	4.06	4.03	4.00
	13.74	**10.92**	**9.78**	**9.15**	**8.75**	**8.47**	**8.26**	**8.10**	**7.98**	**7.87**	**7.79**	**7.72**
7	5.59	4.74	4.35	4.12	3.97	3.87	3.79	3.73	3.68	3.63	3.60	3.57
	12.25	**9.55**	**8.45**	**7.85**	**7.46**	**7.19**	**7.00**	**6.84**	**6.71**	**6.62**	**6.54**	**6.47**
8	5.32	4.46	4.07	3.84	3.69	3.58	3.50	3.44	3.39	3.34	3.31	3.28
	11.26	**8.65**	**7.59**	**7.01**	**6.63**	**6.37**	**6.19**	**6.03**	**5.91**	**5.82**	**5.74**	**5.67**
9	5.12	4.26	3.86	3.63	3.48	3.37	3.29	3.23	3.18	3.13	3.10	3.07
	10.56	**8.02**	**6.99**	**6.42**	**6.06**	**5.80**	**5.62**	**5.47**	**5.35**	**5.26**	**5.18**	**5.11**
10	4.96	4.10	3.71	3.48	3.33	3.22	3.14	3.07	3.02	2.97	2.94	2.91
	10.04	**7.56**	**6.55**	**5.99**	**5.64**	**5.39**	**5.21**	**5.06**	**4.95**	**4.85**	**4.78**	**4.71**
11	4.84	3.98	3.59	3.36	3.20	3.09	3.01	2.95	2.90	2.86	2.82	2.79
	9.65	**7.20**	**6.22**	**5.67**	**5.32**	**5.07**	**4.88**	**4.74**	**4.63**	**4.54**	**4.46**	**4.40**
12	4.75	3.88	3.49	3.26	3.11	3.00	2.92	2.85	2.80	2.76	2.72	2.69
	9.33	**6.93**	**5.95**	**5.41**	**5.06**	**4.82**	**4.65**	**4.50**	**4.39**	**4.30**	**4.22**	**4.16**
13	4.67	3.80	3.41	3.18	3.02	2.92	2.84	2.77	2.72	2.67	2.63	2.60
	9.07	**6.70**	**5.74**	**5.20**	**4.86**	**4.62**	**4.44**	**4.30**	**4.19**	**4.10**	**4.02**	**3.96**
14	4.60	3.74	3.34	3.11	2.96	2.85	2.77	2.70	2.65	2.60	2.56	2.53
	8.86	**6.51**	**5.56**	**5.03**	**4.69**	**4.46**	**4.28**	**4.14**	**4.03**	**3.94**	**3.86**	**3.80**

TABLE I *(continued)*

Degrees of Freedom in Denominator	Degrees of Freedom in Numerator											
	1	2	3	4	5	6	7	8	9	10	11	12
15	4.54	3.68	3.29	3.06	2.90	2.79	2.70	2.64	2.59	2.55	2.51	2.48
	8.68	**6.36**	**5.42**	**4.89**	**4.56**	**4.32**	**4.14**	**4.00**	**3.89**	**3.80**	**3.73**	**3.67**
16	4.49	3.63	3.24	3.01	2.85	2.74	2.66	2.59	2.54	2.49	2.45	2.42
	8.53	**6.23**	**5.29**	**4.77**	**4.44**	**4.20**	**4.03**	**3.89**	**3.78**	**3.69**	**3.61**	**3.55**
17	4.45	3.59	3.20	2.96	2.81	2.70	2.62	2.55	2.50	2.45	2.41	2.38
	8.40	**6.11**	**5.18**	**4.67**	**4.34**	**4.10**	**3.93**	**3.79**	**3.68**	**3.59**	**3.52**	**3.45**
18	4.41	3.55	3.16	2.93	2.77	2.66	2.58	2.51	2.46	2.41	2.37	2.34
	8.28	**6.01**	**5.09**	**4.58**	**4.25**	**4.01**	**3.85**	**3.71**	**3.60**	**3.51**	**3.44**	**3.37**
19	4.38	3.52	3.13	2.90	2.74	2.63	2.55	2.48	2.43	2.38	2.34	2.31
	8.18	**5.93**	**5.01**	**4.50**	**4.17**	**3.94**	**3.77**	**3.63**	**3.52**	**3.43**	**3.36**	**3.30**
20	4.35	3.49	3.10	2.87	2.71	2.60	2.52	2.45	2.40	2.35	2.31	2.28
	8.10	**5.85**	**4.94**	**4.43**	**4.10**	**3.87**	**3.71**	**3.56**	**3.45**	**3.37**	**3.30**	**3.23**
21	4.32	3.47	3.07	2.84	2.68	2.57	2.49	2.42	2.37	2.32	2.28	2.25
	8.02	**5.78**	**4.87**	**4.37**	**4.04**	**3.81**	**3.65**	**3.51**	**3.40**	**3.31**	**3.24**	**3.17**
22	4.30	3.44	3.05	2.82	2.66	2.55	2.47	2.40	2.35	2.30	2.26	2.23
	7.94	**5.72**	**4.82**	**4.31**	**3.99**	**3.76**	**3.59**	**3.45**	**3.35**	**3.26**	**3.18**	**3.12**
23	4.28	3.42	3.03	2.80	2.64	2.53	2.45	2.38	2.32	2.28	2.24	2.20
	7.88	**5.66**	**4.76**	**4.26**	**3.94**	**3.71**	**3.54**	**3.41**	**3.30**	**3.21**	**3.14**	**3.07**
24	4.26	3.40	3.01	2.78	2.62	2.51	2.43	2.36	2.30	2.26	2.22	2.18
	7.82	**5.61**	**4.72**	**4.22**	**3.90**	**3.67**	**3.50**	**3.36**	**3.25**	**3.17**	**3.09**	**3.03**
25	4.24	3.38	2.99	2.76	2.60	2.49	2.41	2.34	2.28	2.24	2.20	2.16
	7.77	**5.57**	**4.68**	**4.18**	**3.86**	**3.63**	**3.46**	**3.32**	**3.21**	**3.13**	**3.05**	**2.99**
26	4.22	3.37	2.98	2.74	2.59	2.47	2.39	2.32	2.27	2.22	2.18	2.15
	7.72	**5.53**	**4.64**	**4.14**	**3.82**	**3.59**	**3.42**	**3.29**	**3.17**	**3.09**	**3.02**	**2.96**

TABLE J

Exponential Functions

y	e^y	e^{-y}	y	e^y	e^{-y}
0.00	1.0000	1.000000	3.00	20.086	.049787
0.10	1.1052	.904837	3.10	22.198	.045049
0.20	1.2214	.818731	3.20	24.533	.040762
0.30	1.3499	.740818	3.30	27.113	.036883
0.40	1.4918	.670320	3.40	29.964	.033373
0.50	1.6487	.606531	3.50	33.115	.030197
0.60	1.8221	.548812	3.60	36.598	.027324
0.70	2.0138	.496585	3.70	40.447	.024724
0.80	2.2255	.449329	3.80	44.701	.022371
0.90	2.4596	.406570	3.90	49.402	.020242
1.00	2.7183	.367879	4.00	54.598	.018316
1.10	3.0042	.332871	4.10	60.340	.016573
1.20	3.3201	.301194	4.20	66.686	.014996
1.30	3.6693	.272532	4.30	73.700	.013569
1.40	4.0552	.246597	4.40	81.451	.012277
1.50	4.4817	.223130	4.50	90.017	.011109
1.60	4.9530	.201897	4.60	99.484	.010052
1.70	5.4739	.182684	4.70	109.95	.009095
1.80	6.0496	.165299	4.80	121.51	.008230
1.90	6.6859	.149569	4.90	134.29	.007447
2.00	7.3891	.135335	5.00	148.41	.006738
2.10	8.1662	.122456	5.10	164.02	.006097
2.20	9.0250	.110803	5.20	181.27	.005517
2.30	9.9742	.100259	5.30	200.34	.004992
2.40	11.023	.090718	5.40	221.41	.004517
2.50	12.182	.082085	5.50	244.69	.004087
2.60	13.464	.074274	5.60	270.43	.003698
2.70	14.880	.067206	5.70	298.87	.003346
2.80	16.445	.060810	5.80	330.30	.003028
2.90	18.174	.055023	5.90	365.04	.002739
3.00	20.086	.049787	6.00	403.43	.002479

TABLE K

Cumulative Probability Values for the Poisson Distribution

$$\Pr[X \leq x]$$

λt	0.1	0.2	0.3	0.4	0.5	0.6	0.7	0.8	0.9	1.0
x										
0	0.9048	0.8187	0.7408	0.6703	0.6065	0.5488	0.4966	0.4493	0.4066	0.3679
1	0.9953	0.9825	0.9631	0.9384	0.9098	0.8781	0.8442	0.8088	0.7725	0.7358
2	0.9998	0.9989	0.9964	0.9921	0.9856	0.9769	0.9659	0.9526	0.9371	0.9197
3	1.0000	0.9999	0.9997	0.9992	0.9982	0.9966	0.9942	0.9909	0.9865	0.9810
4	1.0000	1.0000	1.0000	0.9999	0.9998	0.9996	0.9992	0.9986	0.9977	0.9963
5	1.0000	1.0000	1.0000	1.0000	1.0000	1.0000	0.9999	0.9998	0.9997	0.9994
6	1.0000	1.0000	1.0000	1.0000	1.0000	1.0000	1.0000	1.0000	1.0000	0.9999
7	1.0000	1.0000	1.0000	1.0000	1.0000	1.0000	1.0000	1.0000	1.0000	1.0000

λt	1.1	1.2	1.3	1.4	1.5	1.6	1.7	1.8	1.9	2.0
x										
0	0.3329	0.3012	0.2725	0.2466	0.2231	0.2019	0.1827	0.1653	0.1496	0.1353
1	0.6990	0.6626	0.6268	0.5918	0.5578	0.5249	0.4932	0.4628	0.4338	0.4060
2	0.9004	0.8795	0.8571	0.8335	0.8088	0.7834	0.7572	0.7306	0.7037	0.6767
3	0.9743	0.9662	0.9569	0.9463	0.9344	0.9212	0.9068	0.8913	0.8747	0.8571
4	0.9946	0.9923	0.9893	0.9857	0.9814	0.9763	0.9704	0.9636	0.9559	0.9473
5	0.9990	0.9985	0.9978	0.9968	0.9955	0.9940	0.9920	0.9896	0.9868	0.9834
6	0.9999	0.9997	0.9996	0.9994	0.9991	0.9987	0.9981	0.9974	0.9966	0.9955
7	1.0000	1.0000	0.9999	0.9999	0.9998	0.9997	0.9996	0.9994	0.9992	0.9989
8	1.0000	1.0000	1.0000	1.0000	1.0000	1.0000	0.9999	0.9999	0.9998	0.9998
9	1.0000	1.0000	1.0000	1.0000	1.0000	1.0000	1.0000	1.0000	1.0000	1.0000

λt	2.1	2.2	2.3	2.4	2.5	2.6	2.7	2.8	2.9	3.0
x										
0	0.1225	0.1108	0.1003	0.0907	0.0821	0.0743	0.0672	0.0608	0.0550	0.0498
1	0.3796	0.3546	0.3309	0.3084	0.2873	0.2674	0.2487	0.2311	0.2146	0.1991
2	0.6496	0.6227	0.5960	0.5697	0.5438	0.5184	0.4936	0.4695	0.4460	0.4232
3	0.8386	0.8194	0.7993	0.7787	0.7576	0.7360	0.7141	0.6919	0.6696	0.6472
4	0.9379	0.9275	0.9162	0.9041	0.8912	0.8774	0.8629	0.8477	0.8318	0.8153
5	0.9796	0.9751	0.9700	0.9643	0.9580	0.9510	0.9433	0.9349	0.9258	0.9161
6	0.9941	0.9925	0.9906	0.9884	0.9858	0.9828	0.9794	0.9756	0.9713	0.9665
7	0.9985	0.9980	0.9974	0.9967	0.9958	0.9947	0.9934	0.9919	0.9901	0.9881
8	0.9997	0.9995	0.9994	0.9991	0.9989	0.9985	0.9981	0.9976	0.9969	0.9962
9	0.9999	0.9999	0.9999	0.9998	0.9997	0.9996	0.9995	0.9993	0.9991	0.9989
10	1.0000	1.0000	1.0000	1.0000	0.9999	0.9999	0.9999	0.9998	0.9998	0.9997
11	1.0000	1.0000	1.0000	1.0000	1.0000	1.0000	1.0000	1.0000	0.9999	0.9999
12	1.0000	1.0000	1.0000	1.0000	1.0000	1.0000	1.0000	1.0000	1.0000	1.0000

λt	3.1	3.2	3.3	3.4	3.5	3.6	3.7	3.8	3.9	4.0
x										
0	0.0450	0.0408	0.0369	0.0334	0.0302	0.0273	0.0247	0.0224	0.0202	0.0183
1	0.1847	0.1712	0.1586	0.1468	0.1359	0.1257	0.1162	0.1074	0.0992	0.0916
2	0.4012	0.3799	0.3594	0.3397	0.3208	0.3027	0.2854	0.2689	0.2531	0.2381
3	0.6248	0.6025	0.5803	0.5584	0.5366	0.5152	0.4942	0.4735	0.4533	0.4335
4	0.7982	0.7806	0.7626	0.7442	0.7254	0.7064	0.6872	0.6678	0.6484	0.6288
5	0.9057	0.8946	0.8829	0.8705	0.8576	0.8441	0.8301	0.8156	0.8006	0.7851
6	0.9612	0.9554	0.9490	0.9421	0.9347	0.9267	0.9182	0.9091	0.8995	0.8893
7	0.9858	0.9832	0.9802	0.9769	0.9733	0.9692	0.9648	0.9599	0.9546	0.9489
8	0.9953	0.9943	0.9931	0.9917	0.9901	0.9883	0.9863	0.9840	0.9815	0.9786
9	0.9986	0.9982	0.9978	0.9973	0.9967	0.9960	0.9952	0.9942	0.9931	0.9919

TABLE K   *(continued)*

λt x	3.1	3.2	3.3	3.4	3.5	3.6	3.7	3.8	3.9	4.0
10	0.9996	0.9995	0.9994	0.9992	0.9990	0.9987	0.9984	0.9981	0.9977	0.9972
11	0.9999	0.9999	0.9998	0.9998	0.9997	0.9996	0.9995	0.9994	0.9993	0.9991
12	1.0000	1.0000	1.0000	0.9999	0.9999	0.9999	0.9999	0.9998	0.9998	0.9997
13	1.0000	1.0000	1.0000	1.0000	1.0000	1.0000	1.0000	1.0000	0.9999	0.9999
14	1.0000	1.0000	1.0000	1.0000	1.0000	1.0000	1.0000	1.0000	1.0000	1.0000

λt x	4.1	4.2	4.3	4.4	4.5	4.6	4.7	4.8	4.9	5.0
0	0.0166	0.0150	0.0136	0.0123	0.0111	0.0101	0.0091	0.0082	0.0074	0.0067
1	0.0845	0.0780	0.0719	0.0663	0.0611	0.0563	0.0518	0.0477	0.0439	0.0404
2	0.2238	0.2102	0.1974	0.1851	0.1736	0.1626	0.1523	0.1425	0.1333	0.1247
3	0.4142	0.3954	0.3772	0.3595	0.3423	0.3257	0.3097	0.2942	0.2793	0.2650
4	0.6093	0.5898	0.5704	0.5512	0.5321	0.5132	0.4946	0.4763	0.4582	0.4405
5	0.7693	0.7531	0.7367	0.7199	0.7029	0.6858	0.6684	0.6510	0.6335	0.6160
6	0.8735	0.8675	0.8558	0.8436	0.8311	0.8180	0.8046	0.7908	0.7767	0.7622
7	0.9427	0.9361	0.9290	0.9214	0.9134	0.9049	0.8960	0.8867	0.8769	0.8666
8	0.9755	0.9721	0.9683	0.9642	0.9597	0.9549	0.9497	0.9442	0.9382	0.9319
9	0.9905	0.9889	0.9871	0.9851	0.9829	0.9805	0.9778	0.9749	0.9717	0.9682
10	0.9966	0.9959	0.9952	0.9943	0.9933	0.9922	0.9910	0.9896	0.9880	0.9863
11	0.9989	0.9986	0.9983	0.9980	0.9976	0.9971	0.9966	0.9960	0.9953	0.9945
12	0.9997	0.9996	0.9995	0.9993	0.9992	0.9990	0.9988	0.9986	0.9983	0.9980
13	0.9999	0.9999	0.9998	0.9998	0.9997	0.9997	0.9996	0.9995	0.9994	0.9993
14	1.0000	1.0000	1.0000	0.9999	0.9999	0.9999	0.9999	0.9999	0.9998	0.9998
15	1.0000	1.0000	1.0000	1.0000	1.0000	1.0000	1.0000	1.0000	0.9999	0.9999
16	1.0000	1.0000	1.0000	1.0000	1.0000	1.0000	1.0000	1.0000	1.0000	1.0000

λt x	5.1	5.2	5.3	5.4	5.5	5.6	5.7	5.8	5.9	6.0
0	0.0061	0.0055	0.0050	0.0045	0.0041	0.0037	0.0033	0.0030	0.0027	0.0025
1	0.0372	0.0342	0.0314	0.0289	0.0266	0.0244	0.0224	0.0206	0.0189	0.0174
2	0.1165	0.1088	0.1016	0.0948	0.0884	0.0824	0.0768	0.0715	0.0666	0.0620
3	0.2513	0.2381	0.2254	0.2133	0.2017	0.1906	0.1801	0.1700	0.1604	0.1512
4	0.4231	0.4061	0.3895	0.3733	0.3575	0.3422	0.3272	0.3127	0.2987	0.2851
5	0.5984	0.5809	0.5635	0.5461	0.5289	0.5119	0.4950	0.4783	0.4619	0.4457
6	0.7474	0.7324	0.7171	0.7017	0.6860	0.6703	0.6544	0.6384	0.6224	0.6063
7	0.8560	0.8449	0.8335	0.8217	0.8095	0.7970	0.7842	0.7710	0.7576	0.7440
8	0.9252	0.9181	0.9106	0.9026	0.8944	0.8857	0.8766	0.8672	0.8574	0.8472
9	0.9644	0.9603	0.9559	0.9512	0.9462	0.9409	0.9352	0.9292	0.9228	0.9161
10	0.9844	0.9823	0.9800	0.9775	0.9747	0.9718	0.9686	0.9651	0.9614	0.9574
11	0.9937	0.9927	0.9916	0.9904	0.9890	0.9875	0.9859	0.9840	0.9821	0.9799
12	0.9976	0.9972	0.9967	0.9962	0.9955	0.9949	0.9941	0.9932	0.9922	0.9912
13	0.9992	0.9990	0.9988	0.9986	0.9983	0.9980	0.9977	0.9973	0.9969	0.9964
14	0.9997	0.9997	0.9996	0.9995	0.9994	0.9993	0.9991	0.9990	0.9988	0.9986
15	0.9999	0.9999	0.9999	0.9998	0.9998	0.9998	0.9997	0.9996	0.9996	0.9995
16	1.0000	1.0000	1.0000	0.9999	0.9999	0.9999	0.9999	0.9999	0.9999	0.9998
17	1.0000	1.0000	1.0000	1.0000	1.0000	1.0000	1.0000	1.0000	1.0000	0.9999
18	1.0000	1.0000	1.0000	1.0000	1.0000	1.0000	1.0000	1.0000	1.0000	1.0000

TABLE K (continued)

λt	6.1	6.2	6.3	6.4	6.5	6.6	6.7	6.8	6.9	7.0
x										
0	0.0022	0.0020	0.0018	0.0017	0.0015	0.0014	0.0012	0.0011	0.0010	0.0009
1	0.0159	0.0146	0.0134	0.0123	0.0113	0.0103	0.0095	0.0087	0.0080	0.0073
2	0.0577	0.0536	0.0498	0.0463	0.0430	0.0400	0.0371	0.0344	0.0320	0.0296
3	0.1425	0.1342	0.1264	0.1189	0.1119	0.1052	0.0988	0.0928	0.0871	0.0818
4	0.2719	0.2592	0.2469	0.2351	0.2237	0.2127	0.2022	0.1920	0.1823	0.1730
5	0.4298	0.4141	0.3988	0.3837	0.3690	0.3547	0.3407	0.3270	0.3137	0.3007
6	0.5902	0.5742	0.5582	0.5423	0.5265	0.5108	0.4953	0.4799	0.4647	0.4497
7	0.7301	0.7160	0.7018	0.6873	0.6728	0.6581	0.6433	0.6285	0.6136	0.5987
8	0.8367	0.8259	0.8148	0.8033	0.7916	0.7796	0.7673	0.7548	0.7420	0.7291
9	0.9090	0.9016	0.8939	0.8858	0.8774	0.8686	0.8596	0.8502	0.8405	0.8305
10	0.9531	0.9486	0.9437	0.9386	0.9332	0.9274	0.9214	0.9151	0.9084	0.9015
11	0.9776	0.9750	0.9723	0.9693	0.9661	0.9627	0.9591	0.9552	0.9510	0.9466
12	0.9900	0.9887	0.9873	0.9857	0.9840	0.9821	0.9801	0.9779	0.9755	0.9730
13	0.9958	0.9952	0.9945	0.9937	0.9929	0.9920	0.9909	0.9898	0.9885	0.9872
14	0.9984	0.9981	0.9978	0.9974	0.9970	0.9966	0.9961	0.9956	0.9950	0.9943
15	0.9994	0.9993	0.9992	0.9990	0.9988	0.9986	0.9984	0.9982	0.9979	0.9976
16	0.9998	0.9997	0.9997	0.9996	0.9996	0.9995	0.9994	0.9993	0.9992	0.9990
17	0.9999	0.9999	0.9999	0.9999	0.9998	0.9998	0.9998	0.9997	0.9997	0.9996
18	1.0000	1.0000	1.0000	1.0000	0.9999	0.9999	0.9999	0.9999	0.9999	0.9999
19	1.0000	1.0000	1.0000	1.0000	1.0000	1.0000	1.0000	1.0000	1.0000	0.9999
20	1.0000	1.0000	1.0000	1.0000	1.0000	1.0000	1.0000	1.0000	1.0000	1.0000

λt	7.1	7.2	7.3	7.4	7.5	7.6	7.7	7.8	7.9	8.0
x										
0	0.0008	0.0007	0.0007	0.0006	0.0006	0.0005	0.0005	0.0004	0.0004	0.0003
1	0.0067	0.0061	0.0056	0.0051	0.0047	0.0043	0.0039	0.0036	0.0033	0.0030
2	0.0275	0.0255	0.0236	0.0219	0.0203	0.0188	0.0174	0.0161	0.0149	0.0138
3	0.0767	0.0719	0.0674	0.0632	0.0591	0.0554	0.0518	0.0485	0.0453	0.0424
4	0.1641	0.1555	0.1473	0.1395	0.1321	0.1249	0.1181	0.1117	0.1055	0.0996
5	0.2881	0.2759	0.2640	0.2526	0.2414	0.2307	0.2203	0.2103	0.2006	0.1912
6	0.4349	0.4204	0.4060	0.3920	0.3782	0.3646	0.3514	0.3384	0.3257	0.3134
7	0.5838	0.5689	0.5541	0.5393	0.5246	0.5100	0.4956	0.4812	0.4670	0.4530
8	0.7160	0.7027	0.6892	0.6757	0.6620	0.6482	0.6343	0.6204	0.6065	0.5926
9	0.8202	0.8096	0.7988	0.7877	0.7764	0.7649	0.7531	0.7411	0.7290	0.7166
10	0.8942	0.8867	0.8788	0.8707	0.8622	0.8535	0.8445	0.8352	0.8257	0.8159
11	0.9420	0.9371	0.9319	0.9265	0.9208	0.9148	0.9085	0.9020	0.8952	0.8881
12	0.9703	0.9673	0.9642	0.9609	0.9573	0.9536	0.9496	0.9453	0.9409	0.9362
13	0.9857	0.9841	0.9824	0.9805	0.9784	0.9762	0.9739	0.9714	0.9687	0.9658
14	0.9935	0.9927	0.9918	0.9908	0.9897	0.9886	0.9873	0.9859	0.9844	0.9827
15	0.9972	0.9968	0.9964	0.9959	0.9954	0.9948	0.9941	0.9934	0.9926	0.9918
16	0.9989	0.9987	0.9985	0.9983	0.9980	0.9978	0.9974	0.9971	0.9967	0.9963
17	0.9996	0.9995	0.9994	0.9993	0.9992	0.9991	0.9989	0.9988	0.9986	0.9984
18	0.9998	0.9998	0.9998	0.9997	0.9997	0.9996	0.9996	0.9995	0.9994	0.9993
19	0.9999	0.9999	0.9999	0.9999	0.9999	0.9999	0.9998	0.9998	0.9998	0.9997
20	1.0000	1.0000	1.0000	1.0000	1.0000	0.9999	0.9999	0.9999	0.9999	0.9999
21	1.0000	1.0000	1.0000	1.0000	1.0000	1.0000	1.0000	1.0000	1.0000	1.0000

TABLE K

λt	8.1	8.2	8.3	8.4	8.5	8.6	8.7	8.8	8.9	9.0
x
0	0.0003	0.0003	0.0002	0.0002	0.0002	0.0002	0.0002	0.0002	0.0001	0.0001
1	0.0028	0.0025	0.0023	0.0021	0.0019	0.0018	0.0016	0.0015	0.0014	0.0012
2	0.0127	0.0118	0.0109	0.0100	0.0093	0.0086	0.0079	0.0073	0.0068	0.0062
3	0.0396	0.0370	0.0346	0.0323	0.0301	0.0281	0.0262	0.0244	0.0228	0.0212
4	0.0941	0.0887	0.0837	0.0789	0.0744	0.0701	0.0660	0.0621	0.0584	0.0550
5	0.1822	0.1736	0.1653	0.1573	0.1496	0.1422	0.1352	0.1284	0.1219	0.1157
6	0.3013	0.2896	0.2781	0.2670	0.2562	0.2457	0.2355	0.2256	0.2160	0.2068
7	0.4391	0.4254	0.4119	0.3987	0.3856	0.3728	0.3602	0.3478	0.3357	0.3239
8	0.5786	0.5647	0.5508	0.5369	0.5231	0.5094	0.4958	0.4823	0.4689	0.4557
9	0.7041	0.6915	0.6788	0.6659	0.6530	0.6400	0.6269	0.6137	0.6006	0.5874
10	0.8058	0.7955	0.7850	0.7743	0.7634	0.7522	0.7409	0.7294	0.7178	0.7060
11	0.8807	0.8731	0.8652	0.8571	0.8487	0.8400	0.8311	0.8220	0.8126	0.8030
12	0.9313	0.9261	0.9207	0.9150	0.9091	0.9029	0.8965	0.8898	0.8829	0.8758
13	0.9628	0.9595	0.9561	0.9524	0.9486	0.9445	0.9403	0.9358	0.9311	0.9262
14	0.9810	0.9791	0.9771	0.9749	0.9726	0.9701	0.9675	0.9647	0.9617	0.9585
15	0.9908	0.9898	0.9887	0.9875	0.9862	0.9847	0.9832	0.9816	0.9798	0.9780
16	0.9958	0.9953	0.9947	0.9941	0.9934	0.9926	0.9918	0.9909	0.9899	0.9889
17	0.9982	0.9979	0.9976	0.9973	0.9970	0.9966	0.9962	0.9957	0.9952	0.9947
18	0.9992	0.9991	0.9990	0.9989	0.9987	0.9985	0.9983	0.9981	0.9978	0.9976
19	0.9997	0.9996	0.9996	0.9995	0.9995	0.9994	0.9993	0.9992	0.9991	0.9989
20	0.9999	0.9999	0.9998	0.9998	0.9998	0.9997	0.9997	0.9997	0.9996	0.9996
21	1.0000	0.9999	0.9999	0.9999	0.9999	0.9999	0.9999	0.9999	0.9998	0.9998
22	1.0000	1.0000	1.0000	1.0000	1.0000	1.0000	1.0000	0.9999	0.9999	0.9999
23	1.0000	1.0000	1.0000	1.0000	1.0000	1.0000	1.0000	1.0000	1.0000	1.0000

λt	9.1	9.2	9.3	9.4	9.5	9.6	9.7	9.8	9.9	10.0
x
0	0.0001	0.0001	0.0001	0.0001	0.0001	0.0001	0.0001	0.0001	0.0001	0.0000
1	0.0011	0.0010	0.0009	0.0009	0.0008	0.0007	0.0007	0.0006	0.0005	0.0005
2	0.0058	0.0053	0.0049	0.0045	0.0042	0.0038	0.0035	0.0033	0.0030	0.0028
3	0.0198	0.0184	0.0172	0.0160	0.0149	0.0138	0.0129	0.0120	0.0111	0.0103
4	0.0517	0.0486	0.0456	0.0429	0.0403	0.0378	0.0355	0.0333	0.0312	0.0293
5	0.1098	0.1041	0.0987	0.0935	0.0885	0.0838	0.0793	0.0750	0.0710	0.0671
6	0.1973	0.1892	0.1808	0.1727	0.1650	0.1575	0.1502	0.1433	0.1366	0.1301
7	0.3123	0.3010	0.2900	0.2792	0.2687	0.2584	0.2485	0.2388	0.2294	0.2202
8	0.4426	0.4296	0.4168	0.4042	0.3918	0.3796	0.3676	0.3558	0.3442	0.3328
9	0.5742	0.5611	0.5479	0.5349	0.5218	0.5089	0.4960	0.4832	0.4705	0.4579
10	0.6941	0.6820	0.6699	0.6576	0.6453	0.6330	0.6205	0.6080	0.5955	0.5830
11	0.7932	0.7832	0.7730	0.7626	0.7520	0.7412	0.7303	0.7193	0.7081	0.6968
12	0.8684	0.8607	0.8529	0.8448	0.8364	0.8279	0.8191	0.8101	0.8009	0.7916
13	0.9210	0.9156	0.9100	0.9042	0.8981	0.8919	0.8853	0.8786	0.8716	0.8645
14	0.9552	0.9517	0.9480	0.9441	0.9400	0.9357	0.9312	0.9265	0.9216	0.9165
15	0.9760	0.9738	0.9715	0.9691	0.9665	0.9638	0.9609	0.9579	0.9546	0.9513
16	0.9878	0.9865	0.9852	0.9838	0.9823	0.9806	0.9789	0.9770	0.9751	0.9730
17	0.9941	0.9934	0.9927	0.9919	0.9911	0.9902	0.9892	0.9881	0.9869	0.9857
18	0.9973	0.9969	0.9966	0.9962	0.9957	0.9952	0.9947	0.9941	0.9935	0.9928
19	0.9988	0.9986	0.9985	0.9983	0.9980	0.9978	0.9975	0.9972	0.9969	0.9965
20	0.9995	0.9994	0.9993	0.9992	0.9991	0.9990	0.9989	0.9987	0.9986	0.9984
21	0.9998	0.9998	0.9997	0.9997	0.9996	0.9996	0.9995	0.9995	0.9994	0.9993
22	0.9999	0.9999	0.9999	0.9999	0.9998	0.9998	0.9998	0.9998	0.9997	0.9997
23	1.0000	1.0000	1.0000	0.9999	0.9999	0.9999	0.9999	0.9999	0.9999	0.9999
24	1.0000	1.0000	1.0000	1.0000	1.0000	1.0000	1.0000	1.0000	0.9999	0.9999
25	1.0000	1.0000	1.0000	1.0000	1.0000	1.0000	1.0000	1.0000	1.0000	1.0000

TABLE K (*continued*)

λt / x	11.0	12.0	13.0	14.0	15.0	16.0	17.0	18.0	19.0	20.0
0	0.0000	0.0000	0.0000	0.0000	0.0000	0.0000	0.0	0.0	0.0	0.0
1	0.0002	0.0001	0.0000	0.0000	0.0000	0.0000	0.0000	0.0000	0.0000	0.0
2	0.0012	0.0005	0.0002	0.0001	0.0000	0.0000	0.0000	0.0000	0.0000	0.0000
3	0.0049	0.0023	0.0011	0.0005	0.0002	0.0001	0.0000	0.0000	0.0000	0.0000
4	0.0151	0.0076	0.0037	0.0018	0.0009	0.0004	0.0002	0.0001	0.0000	0.0000
5	0.0375	0.0203	0.0107	0.0055	0.0028	0.0014	0.0007	0.0003	0.0002	0.0001
6	0.0786	0.0458	0.0259	0.0142	0.0076	0.0040	0.0021	0.0010	0.0005	0.0003
7	0.1432	0.0895	0.0540	0.0316	0.0180	0.0100	0.0054	0.0029	0.0015	0.0008
8	0.2320	0.1550	0.0998	0.0621	0.0374	0.0220	0.0126	0.0071	0.0039	0.0021
9	0.3405	0.2424	0.1658	0.1094	0.0699	0.0433	0.0261	0.0154	0.0089	0.0050
10	0.4599	0.3472	0.2517	0.1757	0.1185	0.0774	0.0491	0.0304	0.0183	0.0108
11	0.5793	0.4616	0.3532	0.2600	0.1847	0.1270	0.0847	0.0549	0.0347	0.0214
12	0.6887	0.5760	0.4631	0.3585	0.2676	0.1931	0.1350	0.0917	0.0606	0.0390
13	0.7813	0.6815	0.5730	0.4644	0.3632	0.2745	0.2009	0.1426	0.0984	0.0661
14	0.8540	0.7720	0.6751	0.5704	0.4656	0.3675	0.2808	0.2081	0.1497	0.1049
15	0.9074	0.8444	0.7636	0.6694	0.5681	0.4667	0.3714	0.2866	0.2148	0.1565
16	0.9441	0.8987	0.8355	0.7559	0.6641	0.5660	0.4677	0.3750	0.2920	0.2211
17	0.9678	0.9370	0.8905	0.8272	0.7489	0.6593	0.5640	0.4686	0.3784	0.2970
18	0.9823	0.9626	0.9302	0.8826	0.8195	0.7423	0.6549	0.5622	0.4695	0.3814
19	0.9907	0.9787	0.9573	0.9235	0.8752	0.8122	0.7363	0.6509	0.5606	0.4703
20	0.9953	0.9884	0.9750	0.9521	0.9170	0.8682	0.8055	0.7307	0.6472	0.5591
21	0.9977	0.9939	0.9859	0.9711	0.9469	0.9108	0.8615	0.7991	0.7255	0.6437
22	0.9989	0.9969	0.9924	0.9833	0.9672	0.9418	0.9047	0.8551	0.7931	0.7206
23	0.9995	0.9985	0.9960	0.9907	0.9805	0.9633	0.9367	0.8989	0.8490	0.7875
24	0.9998	0.9993	0.9980	0.9950	0.9888	0.9777	0.9593	0.9317	0.8933	0.8432
25	0.9999	0.9997	0.9990	0.9974	0.9938	0.9869	0.9747	0.9554	0.9269	0.8878
26	1.0000	0.9999	0.9995	0.9987	0.9967	0.9925	0.9848	0.9718	0.9514	0.9221
27	1.0000	0.9999	0.9998	0.9994	0.9983	0.9959	0.9912	0.9827	0.9687	0.9475
28	1.0000	1.0000	0.9999	0.9997	0.9991	0.9978	0.9950	0.9897	0.9805	0.9657
29	1.0000	1.0000	1.0000	0.9999	0.9996	0.9989	0.9973	0.9940	0.9881	0.9782
30	1.0000	1.0000	1.0000	0.9999	0.9998	0.9994	0.9985	0.9967	0.9930	0.9865
31	1.0000	1.0000	1.0000	1.0000	0.9999	0.9997	0.9992	0.9982	0.9960	0.9919
32	1.0000	1.0000	1.0000	1.0000	0.9999	0.9999	0.9996	0.9990	0.9978	0.9953
33	1.0000	1.0000	1.0000	1.0000	1.0000	0.9999	0.9998	0.9995	0.9988	0.9973
34	1.0000	1.0000	1.0000	1.0000	1.0000	1.0000	0.9999	0.9997	0.9994	0.9985
35	1.0000	1.0000	1.0000	1.0000	1.0000	1.0000	0.9999	0.9999	0.9997	0.9992
36	1.0000	1.0000	1.0000	1.0000	1.0000	1.0000	1.0000	0.9999	0.9998	0.9996
37	1.0000	1.0000	1.0000	1.0000	1.0000	1.0000	1.0000	1.0000	0.9999	0.9998
38	1.0000	1.0000	1.0000	1.0000	1.0000	1.0000	1.0000	1.0000	1.0000	0.9999
39	1.0000	1.0000	1.0000	1.0000	1.0000	1.0000	1.0000	1.0000	1.0000	0.9999
40	1.0000	1.0000	1.0000	1.0000	1.0000	1.0000	1.0000	1.0000	1.0000	1.0000

TABLE L
Critical Values of D for Kolmogorov-Smirnov
Maximum Deviation Test for Goodness of Fit

The following table provides the critical values D_α corresponding to an upper-tail probability α of the test statistic D. The following relationship holds

$$\Pr[D_\alpha \leqslant D] = \alpha$$

n	$\alpha = .10$	$\alpha = .05$	$\alpha = .025$	$\alpha = .01$	$\alpha = .005$
1	.90000	.95000	.97500	.99000	.99500
2	.68377	.77639	.84189	.90000	.92929
3	.56481	.63604	.70760	.78456	.82900
4	.49265	.56522	.62394	.68887	.73424
5	.44698	.50945	.56328	.62718	.66853
6	.41037	.46799	.51926	.57741	.61661
7	.38148	.43607	.48342	.53844	.57581
8	.35831	.40962	.45427	.50654	.54179
9	.33910	.38746	.43001	.47960	.51332
10	.32260	.36866	.40925	.45662	.48893
11	.30829	.35242	.39122	.43670	.46770
12	.29577	.33815	.37543	.41918	.44905
13	.28470	.32549	.36143	.40362	.43247
14	.27481	.31417	.34890	.38970	.41762
15	.26588	.30397	.33760	.37713	.40420
16	.25778	.29472	.32733	.36571	.39201
17	.25039	.28627	.31796	.35528	.38086
18	.24360	.27851	.30936	.34569	.37062
19	.23735	.27136	.30143	.33685	.36117
20	.23156	.26473	.29408	.32866	.35241
21	.22617	.25858	.28724	.32104	.34427
22	.22115	.25283	.28087	.31394	.33666
23	.21645	.24746	.27490	.30728	.32954
24	.21205	.24242	.26931	.30104	.32286
25	.20790	.23768	.26404	.29516	.31657
26	.20399	.23320	.25907	.28962	.31064
27	.20030	.22898	.25438	.28438	.30502
28	.19680	.22497	.24993	.27942	.29971
29	.19348	.22117	.24571	.27471	.29466
30	.19032	.21756	.24170	.27023	.28987
31	.18732	.21412	.23788	.26596	.28530
30	.18445	.21085	.23424	.26189	.28094
33	.18171	.20771	.23076	.25801	.27677
34	.17909	.20472	.22743	.25429	.27279
35	.17659	.20185	.22425	.25073	.26897
36	.17418	.19910	.22119	.24732	.26532
37	.17188	.19646	.21826	.24404	.26180
38	.16966	.19392	.21544	.24089	.25843
39	.16753	.19148	.21273	.23786	.25518
40	.16547	.18913	.21012	.23494	.25205

TABLE L   (*continued*)

n	α = .10	α = .05	α = .025	α = .01	α = .005
41	.16349	.18687	.20760	.23213	.24904
42	.16158	.18468	.20517	.22941	.24613
43	.15974	.18257	.20283	.22679	.24332
44	.15796	.18053	.20056	.22426	.24060
45	.15623	.17856	.19837	.22181	.23798
46	.15457	.17665	.19625	.21944	.23544
47	.15295	.17481	.19420	.21715	.23298
48	.15139	.17302	.19221	.21493	.23059
49	.14987	.17128	.19028	.21277	.22828
50	.14840	.16959	.18841	.21068	.22604
51	.14697	.16796	.18659	.20864	.22386
52	.14558	.16637	.18482	.20667	.22174
53	.14423	.16483	.18311	.20475	.21968
54	.14292	.16332	.18144	.20289	.21768
55	.14164	.16186	.17981	.20107	.21574
56	.14040	.16044	.17823	.19930	.21384
57	.13919	.15906	.17669	.19758	.21199
58	.13801	.15771	.17519	.19590	.21019
59	.13686	.15639	.17373	.19427	.20844
60	.13573	.15511	.17231	.19267	.20673
61	.13464	.15385	.17091	.19112	.20506
62	.13357	.15263	.16956	.18960	.20343
63	.13253	.15144	.16823	.18812	.20184
64	.13151	.15027	.16693	.18667	.20029
65	.13052	.14913	.16567	.18525	.19877
66	.12954	.14802	.16443	.18387	.19729
67	.12859	.14693	.16322	.18252	.19584
68	.12766	.14587	.16204	.18119	.19442
69	.12675	.14483	.16088	.17990	.19303
70	.12586	.14381	.15975	.17863	.19167
71	.12499	.14281	.15864	.17739	.19034
72	.12413	.14183	.15755	.17618	.18903
73	.12329	.14087	.15649	.17498	.18776
74	.12247	.13993	.15544	.17382	.18650
75	.12167	.13901	.15442	.17268	.18528
76	.12088	.13811	.15342	.17155	.18408
77	.12011	.13723	.15244	.17045	.18290
78	.11935	.13636	.15147	.16938	.18174
79	.11860	.13551	.15052	.16832	.18060
80	.11787	.13467	.14960	.16728	.17949
81	.11716	.13385	.14868	.16626	.17840
82	.11645	.13305	.14779	.16526	.17732
83	.11576	.13226	.14691	.16428	.17627
84	.11508	.13148	.14605	.16331	.17523
85	.11442	.13072	.14520	.16236	.17421

TABLE L *(continued)*

n	α = .10	α = .05	α = .025	α = .01	α = .005
86	.11376	.12997	.14437	.16143	.17321
87	.11311	.12923	.14355	.16051	.17223
88	.11248	.12850	.14274	.15961	.17126
89	.11186	.12779	.14195	.15873	.17031
90	.11125	.12709	.14117	.15786	.16938
91	.11064	.12640	.14040	.15700	.16846
92	.11005	.12572	.13965	.15616	.16755
93	.10947	.12506	.13891	.15533	.16666
94	.10889	.12440	.13818	.15451	.16579
95	.10833	.12375	.13746	.15371	.16493
96	.10777	.12312	.13675	.15291	.16408
97	.10722	.12249	.13606	.15214	.16324
98	.10668	.12187	.13537	.15137	.16242
99	.10615	.12126	.13469	.15061	.16161
100	.10563	.12067	.13403	.14987	.16081

Source: Reprinted by permission from L. H. Miller, "Table of Percentage Points of Kolmogorov Statistics," *Journal of the American Statistical Association*, 51 (1956), pages 111–121.

Answers to Even-Numbered Exercises

2-2 Answers will vary.

2-4 4.63, rounded up to 5.00

2-6 (a)

Class Interval	Frequency	Class Interval	Frequency
$ 3,000–under 5,000	4	15,000–under 17,000	2
5,000–under 7,000	6	17,000–under 19,000	1
7,000–under 9,000	13	19,000–under 21,000	0
9,000–under 11,000	13	21,000–under 23,000	1
11,000–under 13,000	7	23,000–under 25,000	1
13,000–under 15,000	1	25,000–under 27,000	1
			50

2-8

Letter	Frequency	Letter	Frequency	Letter	Frequency
a	20	j	0	s	7
b	2	k	0	t	22
c	7	l	5	u	4
d	8	m	1	v	3
e	26	n	20	w	5
f	3	o	17	x	0
g	6	p	2	y	2
h	9	q	1	z	0
i	14	r	16		200

2-10

Class Interval (years)	Frequency
10–under 20	18
20–under 30	57
30–under 40	60
40–under 50	53
50–under 60	67
60–under 70	96
70–under 80	12

2-12 (a) ambiguous designations
(b) overlapping intervals
(c) some values not included

2-14

Consecutive Days Absent	(a) Relative Frequency	(b) Cumulative Relative Frequency
0–under 5	.620	.620
5–under 10	.162	.782
10–under 15	.102	.884
15–under 20	.064	.948
20–under 25	.024	.972
25–under 30	.006	.978
30–under 60	.022	1.000
	1.000	

2-16

Number of Shares	(a) Relative Frequency	(b) Frequency [(a) × 1,000)]	(c) Cumulative Frequency
0–under 5,000	.36	360	360
5,000–under 10,000	.27	270	630
10,000–under 15,000	.18	180	810
15,000–under 20,000	.09	90	900
20,000–under 25,000	.07	70	970
25,000–under 30,000	.03	30	1,000
	1.00	1,000	

2-18 Answers may vary.

2-20

Age (years)	Relative Frequency	Cumulative Frequency
20–under 25	.18	18
25–under 30	.23	41
30–under 35	.15	56
35–under 40	.13	69
40–under 45	.07	76
45–under 50	.06	82
50–under 55	.05	87
55–under 60	.05	92
60–under 65	.08	100
	1.00	

2-22 (a)

Sex	Frequency
Male	220
Female	130
	350

(b)

Marital Status	Frequency
Married	171
Single	179
	350

(c)

Occupation	Frequency
Blue collar	155
White collar	173
Professional	22
	350

3-2 Choose CompuQuick.

3-4 26.15

3-6 (a) 2.75 (b) 2 (c) 2

3-8 $2,129

3-10 Answers will vary.

3-12 $s^2 = 184.30$
$s = 13.58$

3-14 21.11

3-16 $\bar{X} = 2.75$ $s = .63$

3-18

A	B	C	D
.0624	.178	.0629	.113

E
.125

Firm A

3-20

	(1)	(2)	(3)	(4)
(a) P	.07	.05	.016	.039
(b)	reject	accept	accept	accept

3-22 (a) 4 (c) 1.33
(b) 2.0 (d) 1.15

3-24 (a) 1,400 to 1,600; 68%
(b) 1,300 to 1,700; 95.5%
(c) 1,200 to 1,800; 99.7%

3-26 $\bar{X} = 126.2$ $s^2 = 385.41$

3-28 (a) $\bar{X} = 77.615$ $s^2 = 208.55$
(b) $\bar{X} = 78.077$ $s^2 = 176.14$
(c) $-.462$

4-2 Answers will vary.

4-4 Answers will vary.

4-6 Answers will vary.

4-8 Answers will vary.

4-10 (a) convenience
(b) judgment
(c) random
(d) judgment or convenience

4-12 Answers will vary.

4-14 A judgment-convenience sample applies.

4-16 Solti, Maag, Abbado, Rowicki, Bloomfield, Krips, Prêtre, Frühbeck de Burgos, Newman, Schippers

4-18 Beecham, Dragon, Golschmann, Karajan, Krips, Pedrotti, Rignold, Scherchen, Stein, Svetlanov

5-2 (a) (1) .20   (3) .40
(2) .35   (4) .05
(b) (1) .15   (3) .20
(2) .65
(c) (1) .06   (3) .01
(2) .04

5-4 (a) 15/36   (d) 1/6
(b) 15/36   (e) 15/36
(c) 1/4   (f) 1/9

5-6 Only the events in (c) are collectively exhaustive.

5-8 (a) .70   (b) .80
(c) no, not independent

5-10 (a) (1) .55   (2) .40   (3) .80
(b) (1) .13   (2) .07

5-12 (a) .12   (c) .08
(b) .48   (d) .32

5-14 (a) 1/4   (d) 1/5
(b) 1/2   (e) 9/20
(c) 3/10   (f) 3/10

5-16 (a) (1) 11/20   (3) 4/5
(2) 3/4
(b) (1) 3/5   (3) 3/5
(2) 9/20

5-18 (a) .20   (d) .15
(b) .80   (e) 0
(c) .57

5-20 (a) .60   (d) .67
(b) .43   (e) .20
(c) .40   (f) .33

5-24 (a) .18   (c) .054
(b) .16

5-26 (a) (1) .80   (3) .10
(2) .16
(b)

Score	Performance S	U	Marginal Probability
L	.08	.08	.16
H	.72	.12	.84
Marginal Probability	.80	.20	1.00

(c) (1) .50   (2) .40

5-28 (a) .729   (b) .001   (c) .243

5-30 (b) (1) .6   (4) .7
(2) .4   (5) .6
(3) .4   (6) .3
(c) (1) .24   (3) .28
(2) .36   (4) .12
(d) (1) .64   (2) .36

5-32 (a) 1/5   (c) 7/10
(b) 2/5   (d) 4/5

5-34 (a) (1) 1/15   (6) 1/6
(2) 1/12   (7) 1/8
(3) 1/24   (8) 17/24
(4) 1/4   (9) 31/60
(5) 4/15   (10) 2/3

(b) (1) no   (3) all
    (2) no

5-36 (a) .000005   (b) five

5-38 (a) .75   (c) .66
    (b) .15

5-40 (c) (1) 1/4   (3) 1/2
       (2) 1/4   (4) 1/12

5-42 (a) 720   (d) 138,600
    (b) 364   (e) 66,045
    (c) 40,320

5-44 (a) 6   (c) 6
    (b) 3   (d) 1,944

5-46 C_{13}^{52} P_{13}^{52}

5-48 (a) 20!   (c) $2 \times 10!$
    (b) $2 \times 10! \times 10!$

5-50 (a) .467   (b) .086

5-52 (a) .809   (b) .005

5-54 (a) .50   (c) .11
    (b) .82

6-2   *Point*
      Spread

x	$\Pr[X = x]$
0	6/36
1	10/36
2	8/36
3	6/36
4	4/36
5	2/36
	36/36

6-4 (a)

w	$\Pr[W = w]$
$-\$1$	20/36
$+1$	14/36
$+2$	1/36
$+3$	1/36
	1

    (b)

w	$\Pr[W = w]$
$-\$2$	20/36
$+2$	14/36
$+4$	1/36
$+6$	1/36
	1

6-6

w	$\Pr[W = w]$
$-\$1$	20/38
$+1$	18/38
	1

$$\mu(W) = -\$2/38 = -\$0.053$$

6-8 $\mu(X) = 2.1$ $\sigma^2(X) = 1.29$

6-10 (a) $155.25 for High-Volatility
          Engineering
          $103.50 for Stability Power

6-12 (a)

$\bar{x}$	$\Pr[\bar{X} = \bar{x}]$
900	1/25
950	4/25
1,000	4/25
1,050	4/25
1,100	8/25
1,200	4/25
	1

(b)
$\mu(\bar{X}) = 1,060$
$\sigma^2(\bar{X}) = 7,200$
$\sigma(\bar{X}) = 84.90$

6-14 (a) *Mean*

$\bar{x}$	$\Pr[X = \bar{x}]$
1.0	1/25
1.5	2/25
2.0	3/25
2.5	4/25
3.0	5/25
3.5	4/25
4.0	3/25
4.5	2/25
5.0	1/25
	1

(b) $\mu(\bar{X}) = 3.0$ $\sigma^2(\bar{X}) = 1.0$

6-16 (a) 6   (c) 21
    (b) 20   (d) 15

6-18 (a) (1) .1641   (3) .0078
       (2) .2734   (4) .2734
    (b) The answers to (2) and (4) must
        be the same.

6-20 (a) .1390   (d) .0848
    (b) .1224   (e) .0009
    (c) .7822   (f) .3105

6-22 $\mu(P) = .10$

6-24 (a) .0000 (d) .0815
(b) .5905 (e) .0815
(c) .4095

6-26 0.0000 (rounded)

6-28 (a) .1347 (d) .3743
(b) .1285 (e) .0942
(c) .9672 (f) .9096

6-30

x	$Pr[X = x]$
5	36/108
10	48/108
20	16/108
50	4/108
100	4/108
	1

6-32 $\mu(X) = 1.0$ $\sigma^2(X) = 1.0$

6-34 (a) .04 (b) .10 (c) .03

6-36 (a)

$\bar{x}$	$Pr[\bar{X} = \bar{x}]$
19.5	1/10
20.0	1/10
20.5	1/10
21.0	1/10
21.5	2/10
22.0	2/10
22.5	1/10
23.5	1/10
	1

$$\mu(\bar{X}) = 21.40$$
$$\sigma^2(\bar{X}) = 1.29$$
$$\sigma_{\bar{x}} = 1.14$$

(b) $\sigma = 1.85$ $\sigma_{\bar{x}} = 1.31$

7-2 (a) .4332 (e) .0968
(b) .1915 (f) .9861
(c) .2420 (g) .97585
(d) .0062 (h) .0606

7-4 (a) 5′4.175″ (d) 5′10.625″
(b) 5′5.9″ (e) 6′2.1″
(c) 5′8.7″ (f) 6′3.825″

7-6 (a) .0456 (c) .50003
(b) .0228

7-8 (a) .01 (c) .50 (e) 10.00
(b) .10 (d) 2.00

7-10 (a) 4 (d) .0401
(b) .4938 (e) .0994
(c) .99865 (f) .3085

7-12 .9544

7-14 (a) .9876 (b) .7888
(c) Greater variability reduces reliability.
(d) 1 (approximately)
(e) Larger sample sizes increase reliability.

7-16 (a) .9544 (c) .49865
(b) .99730 (d) .5000

7-18 (a) .0062 (b) .0116

7-20 (a) .0228 (c) .99865
(b) .9544 (d) .1587

7-22 (a) .6826 (b) .9876

7-24 .00135

7-26 (a) .0207 (b) .0036

7-28 (a) .4525 (b) .0475
(c) 0 (approximately)
(d) 0 (approximately) (e) .9734

7-30 (a) .0228 (b) .0062

7-32 (a) .3164 (d) .3000
(b) 1.3791 (e) .4359
(c) .1000

7-34 (a) .5000 (b) .9429
(c) 1 (approximately)
Choose rule (c).

8-2 (a) $14,800 (c) .6
(b) $11,708 (d) .1

8-4 (a) interval (c) point
(b) either (d) interval

8-6 (a) 100.53 ± 4.96
(b) $69.2 \pm .15$ (c) $12.00 \pm .73$

8-8 30.3 ± 2.11

8-10 (a) $52,346 \pm 1,058$

8-12 (a) 1.812 (d) 2.358
(b) 2.650 (e) 2.750
(c) 2.080

8-14 (a) 8.00 ± 4.12
(b) $15.03 \pm .40$
(c) 27.30 ± 1.43

8-16 (a) $\bar{X} = 12.33$ $s = 6.86$
$\mu = 12.33 \pm 3.80$

8-18 $73,249 \pm 14,817$

8-20 (a) 3 ± 1.43 (b) negligible

8-22 (a) $.25 \pm .050$ (c) $.05 \pm .032$
(b) $.01 \pm .0081$

8-24 (a) $.2 \pm .025$ (c) $.01 \pm .0087$
(b) $.04 \pm .0086$

8-26 (a) 661   (b) 385
(c) 166; one-fourth as large as 661
(d) 2,642; sample size is four times as large.

8-28 (a) .8414   (b) 1.39

8-30 (a) 385   (b) 350; 35

8-32 $1,246 \pm 22.5$

8-34 (a) 8,068   (c) 9,604
(b) 9,220   (d) 3,458

8-36 100 ± 12.7

8-38 (a) 1,537   (c) 97
(b) 385   (d) 16

8-40 (a) 151   (b) 2,401
(c) $151 for stopwatch
$120.05 for work sampling

9-2 (a) (1) correct
(2) Type II error
(3) correct
(4) Type I error
(b) (1) Type II error
(2) correct
(3) Type I error
(4) correct
(c) (1) Type I error
(2) correct
(3) correct
(4) Type II error

9-4 (a) Type I error:
Adopt the new process when it

is no improvement.
Type II error:
Don't adopt the new process when it is an improvement.
(b) Type I error:
Sign the contract when output will not increase.
Type II error:
Don't sign the contract when output will increase.
(c) Type I error:
Lobby for the new law when it does not meet greater approval.
Type II error:
Don't lobby for the new law when it does meet greater approval.

9-6 $\alpha = .0918$ $\beta = .0228$

9-8 $\alpha = .0294$ $\beta = .0023$
Both α and β decrease (are smaller).

9-10 (c) $z = -5.00$ *Reject H_0.*

9-12 (a) $z_\alpha = 2.33$ (upper-tailed)
(b) $z = 6.25$ *Reject H_0.* Yes.

9-14 The officials should *not* revise the test. *Accept $H_0 : \mu \leq 80$.*

9-16 accepted

9-18 $\mu = 1.8 \pm .26$
Reject H_0 and conclude marijuana *lengthens* dream time.

9-20 $t_{.05} = 1.711$ $t = 1.67$
Accept H_0.

9-22 accepted

9-24 yes

9-26 $z_\alpha = 1.28$ $z = 1.00$
Accept H_0.

9-28 accepted

9-30 (a) .1335   (b) .0668

9-32 (a) *Reject* and don't campaign.
(b) *Accept* and campaign.

(c) *Accept* and campaign.

(d) *Reject* and don't campaign.

9-34   *Accept H_0 and don't take corrective action.*

10-2   Answers will vary. $a = 20$; $b = .2727$; $\hat{Y}_X = 20 + .2727X$

10-4

Dependent Variable	Independent Variable
(a) Time deposits	Employee income
(b) Income	Stock purchases
(c) Savings	Wages

10-6   (a) $\hat{Y}_X = 10.0 + 1.0X$

(b) $\hat{Y}_X = 30.0 - 1.0X$

(c) $\hat{Y}_X = -15.0 + .5X$

10-8 $\hat{Y}_X = -4 + .20X$

10-10   (c) $\hat{Y}_1 = 1,100$

$\hat{Y}_2 = 900$

$\hat{Y}_3 = 700$

10-12   (a) $1,600 \pm 41.38$ and ± 211.00

(b) $2,275 \pm 139.82$ and ± 249.71

(c) $2,950 \pm 270.29$ and ± 340.39

10-14   (a) $.05 \pm .017$ (b) $.05 \pm .038$

10-16   Answers will vary.

10-18   (a) $\hat{Y}_X = 18.5725 + .7443X$

(b) $.649$ (c) 64.9%

10-20 $.561$

10-22 $.760$

10-24   (a) $\hat{Y}_X = -.40 + .1X$

10-26   (a) $12,500 \pm 1,209.3$

(b) $10,000 \pm 1,263.0$

(c) $15,000 \pm 1,263.0$

10-28   (a) $\$80,000 \pm 78.40$

(b) $\$80,000 \pm 1,028$

10-30   (a) $-.002 + .00185$

(b) rejected

(c) accepted

11-2   (a) 3.189   (c) 2.975

(b) 3.254   (d) 3.708

11-4   (a) $\hat{Y} = -26 + 4X_1 + 6X_2$

11-6   (a) $\hat{Y}_{X_1} = 4 + X_1$

(b) $s_{Y \cdot X_1} = 9.325$

$S_{Y \cdot 12} = 3.02$

(c) $Y_I = 12 \pm 6.39$

$\mu_{Y \cdot 12} = 12 \pm 1.25$

11-8   (a) $\hat{Y} = 4.2 + .7668X_1 + 192.9X_2 - .634X_3$

(b) 422.7 (c) yes

11-10   (a) $\hat{Y} = 36.6584 - .0670X_1 - 1.6525X_2 - 1.6013X_3$

(b) 1.79 (c) yes

11-12   (b) $\hat{Y} = -.488 + .0594X_1 + 23.3509X_2$

11-14   (b) $\hat{Y}_{X_1} = 375.0 - .1778X_1$

(c) $\hat{Y} = 121.55 + .2606X_1 + 166.7X_2$

11-16   (a) $.465$ (b) $.960$

11-18   (a) $r^2_{Y1 \cdot 2} = .5$ $r^2_{Y2 \cdot 1} = .5$

11-20   (a) $\hat{Y} = -.412 + .631X_1 + .00223X_2$

(b) (1) 2.96   (3) 3.06

(2) 2.25   (4) 2.93

11-22 $\hat{Y} = 2.900 + .637X_1 + .002X_2 - .007X_3$

12-2

Fall 19X1	$1,000,000.00
Winter 19X2	1,090,000.00
Spring	1,188,100.00
Summer	1,295,029.00
Fall	1,411,581.61

12-4   For Winter, $Y_t = \$540,350$; for Spring, $I_t = 101.42$; for Summer, $S_t = 80.00$; for Fall, $C_t = 104.00$.

12-6   (b) $\hat{Y}_X = 200.67 + 9.673X$ ($X = 0$ in 1972)

(c) 374.78

12-8   (a) (1) $\hat{Y}_X = 945,000 + 10,000X$

(2) $\hat{Y}_X = 955,000 + 30,000X$

(b) (1) $\hat{Y}_X = 8,525 + 50X$

(2) $\hat{Y}_X = 8,575 + 150X$

12-10 (b,c,d)

Quarter	Moving Average	Percentage of Moving Average	Seasonal Index	Deseasonalized Data
1977 W			93.4	4.2
S			112.0	5.4
S	6.8	63.2	69.2	6.2
F	7.8	138.5	125.4	8.6
1978 W	8.7	89.7	93.4	8.4
S	9.4	112.8	112.0	9.5
S	10.3	67.0	69.2	10.0
F	11.6	116.4	125.4	10.8
1979 W	12.6	102.4	93.4	13.8
S	13.6	111.8	112.0	13.6
S	14.4	71.5	69.2	14.9
F	14.4	129.8	125.4	14.9
1980 W	14.3	97.2	93.4	14.9
S	14.1	102.1	112.0	12.9
S	13.9	73.4	69.2	14.7
F	14.3	121.0	125.4	13.8
1981 W	15.3	88.2	93.4	14.5
S	16.2	112.3	112.0	16.3
S			69.2	20.5
F			125.4	16.5

12-12 (a)

	Deseasonalized Data				
Month	1977	1978	1979	1980	1981
J	6.8	7.3	8.8	6.3	10.5
F	5.4	6.4	7.3	8.1	9.6
M	6.6	7.2	8.1	5.6	6.3
A	4.0	6.8	6.3	8.7	10.3
M	5.5	7.1	7.4	8.4	7.9
J	8.1	8.2	7.7	9.0	9.2
J	7.2	6.1	7.2	8.4	7.3
A	8.1	8.0	7.4	7.8	8.2
S	7.4	7.5	7.9	8.8	9.1
O	7.6	7.6	8.2	9.0	9.6
N	6.4	6.0	8.1	9.0	8.6
D	6.8	5.6	7.8	8.2	12.2

12-14

Period t	Forecast Sales F_t ($\alpha = .50$)	Period t	Forecast Sales F_t ($\alpha = .50$)
1	—	11	5,193.0
2	4,890	12	5,206.5
3	4,900.0	13	5,243.3
4	4,935.0	14	5,286.6
5	4,972.5	15	5,333.3
6	5,016.3	16	5,386.7
7	5,058.1	17	5,423.3
8	5,054.1	18	5,471.7
9	5,112.0	19	5,480.8
10	5,146.0	20	5,515.4

12-16 (a) $\hat{Y}_X = 142.4 + 64.0X$ ($X = 0$ in 1977)
(b) 398.4

12-18

Quarter	Deseasonalized Data	Trend $\hat{Y}_X$
1977 W	35.0	35
S	35.0	38
S	33.3	41
F	30.0	44
1978 W	58.3	47
S	45.0	50
S	50.6	53
F	35.0	56
1979 W	65.0	59
S	68.3	62
S	65.0	65
F	95.0	68
1980 W	130.0	71
S	121.7	74
S	75.6	77
F	75.0	80
1981 W	90.0	83
S	95.0	86
S	88.9	89
F	72.5	92

12-20 W 1981 19.58
S 1981 21.21
S 1981 16.38
F 1981 19.61

13-2 7.4%

13-4 (a) $I = 101.3$, a 1.3% increase.
(b) $I = 98.7$, a 1.3% increase from 1980 to 1981.

13-6 1978 100.0
1979 106.6
1980 130.4
1981 142.2
42.2%

13-8 (a) $I = 108.41$; 8.41%
(b) $I = 92.24$; 8.41%

13-10 1978 100
1979 110.0
1980 141.9
1981 161.8

13-12 1963 $108.65   1970 $114.99
1964 110.84   1971 117.43
1965 113.79   1972 123.46
1966 115.58   1973 124.76
1967 114.90   1974 119.43
1968 117.57   1975 117.56
1969 117.95

13-14 1978 100
1979 104.7
1980 109.3
1981 120.8
20.8%

13-16 1967 55.1   1974 81.4
1968 57.4   1975 88.8
1969 60.5   1976 94.9
1970 64.1   1977 100.0
1971 66.8   1978 107.7
1972 69.0   1979 119.8
1973 73.3   1980 136.5

14-2 $.19 \pm .412$

14-4 $2 \pm .392$

14-6 *Reject H_0 and use ladybugs.*

14-8 *Accept H_0.*

14-10 *Reject H_0 and conclude that the mean gasoline consumption is better using radial tires.*

14-12 $1.5 \pm .86$

14-14 (a) $2,000 \pm 2,184.9$
(b) *Accept H_0.*

14-16 no

14-18 commission only

14-20 Support for the candidate is equally strong in both areas.

14-22 (a) Batting averages are higher for home games.
(b) $.028 \pm .023$

14-24 (a) $\bar{d} = 5.4$; $s_{d\text{-paired}} = 6.096$
(b) rejected

14-26 (a) $H_0: \mu_A = \mu_B$
$z_{\alpha/2} = 1.96$, $z = 2.14$
rejected

(b) $H_0: \mu_A \geq \mu_B$ (c) $H_0: \mu_A \leq \mu_B$
   $z_\alpha = 1.64$ $z_\alpha = 1.64$
   accepted   rejected

14-28 rejected

15-2 (a) rejected   (c) rejected
     (b) rejected   (d) accepted

15-4 (a) (1) $df = 1$
        (2) $\chi^2_{.01} = 6.635$

     (b) (1)

	(1) Single	(2) Married	Total
(1) A	20 / 12	10 / 18	30
(2) B	20 / 28	50 / 42	70
Total	40	60	100

       (2) 12.699
     (c) *Reject H_0.*

15-6 (a)

	(1) Male	(2) Female	Total
(1)	27.65	22.35	50
(2)	24.34	19.66	44
(3)	32.08	25.92	58
(4)	18.81	15.19	34
(5)	22.12	17.88	40
Total	125	101	226

     (b) 10.683
     (c) 4
     (d) 13.277   yes
     (e) 9.488   *Reject H_0.*

15-8 Quality is affected by temperature.

15-10 $\chi^2 = 1.15$ *Accept H_0.*
      $\alpha = .70$

15-12 $\chi^2 = 12.256$ *Reject H_0.*

15-14 *Accept H_0* and conclude equal regional preference.

15-16 no

15-18 (a) (1) lower-tailed
        (2) 13.848
        (3) $\chi^2 = 28.5$; accepted
     (b) (1) upper-tailed
        (2) 21.666
        (3) $\chi^2 = 9.45$; accepted
     (c) (1) lower-tailed
        (2) 10.865
        (3) $\chi^2 = 16.594$; accepted
     (d) (1) upper-tailed
        (2) 26.296
        (3) $\chi^2 = 26.492$; rejected

15-20 (a) $56.20 \leq \sigma^2 \leq 83.68$
     (b) *Accept H_0.*

16-2 (a) 6.93; accepted
     (b) 3.88; rejected
     (c) 3.88; rejected

16-4 (a) (1) numerator $df = 2$;
          denominator $df = 12$
        (2) $F_{.05} = 3.88$
     (b) (1) $\bar{X}_1 = 87.8$; $\bar{X}_2 = 64.2$;
          $\bar{X}_3 = 76.0$; $\bar{\bar{X}} = 76.0$
        (2) $SST = 1,392.4$;
          $SSE = 2,001.6$
        (3) $MST = 696.2$;
          $MSE = 166.8$;
          $F = 4.17$
     (c) rejected

16-6 (a) no   (b) Answers will vary.

16-8 (a) 87.8 ± 12.59
     (b) 23.6 ± 17.80 yes

16-10 (a) $SSE = 72$; $MST = 21.0$;
        $MSE = 6.0$; $F = 3.5$
     (b) accepted

16-12 (a) $MST = 200,000$;
        $SSE = 1,200,000$;
        $MSE = 80,000$; $F = 2.5$
        no

(b) $MST = 300,000;$
$SSE = 1,100,000;$
$MSE = 68,750; F = 4.36$
yes

(c) $MSA = 200,000;$
$MSB = 300,000;$
$SSE = 300,000;$
$MSE = 25,000;$
$F = 8.0$ for A;
$F = 12.0$ for B
Reject both.

(d) Two-factor analysis is more discriminating.

16-14 (a) yes (c) no (e) yes
(b) no (d) yes

16-16 (a) $SSE = 75; MST = 39.25;$
$MSCOL = 19.25;$
$MSROW = 35.75;$
$F = 6.28$ for X;
$F = 3.08$ for Y;
$F = 5.72$ for Z

(b) The population means are significantly different only for X and Z.

16-18 (a) $\bar{X}_1 = 26; \bar{X}_2 = 34; \bar{X}_3 = 50;$
$\bar{X}_4 = 58; SST = 3,200;$
$SSE = 1,368;$
Total $SS = 4,568$

Variation	df	Sum of Squares	Mean Square	F
Treatments	3	3,200	1,066.67	12.48
Error	16	1,368	85.5	
Total	19	4,568		

(b) yes
(c) 32 ± 12.4 yes

16-20 (a) $\bar{X}_A = 58; \bar{X}_B = 82; \bar{X}_C = 103;$
$\bar{X}_1. = 78; \bar{X}_2. = 81; \bar{X}_3. = 84;$
$\bar{X}_{.1} = 74; \bar{X}_{.2} = 82; \bar{X}_{.3} = 87;$
$SST = 3,042; SSE = 18;$
$SSCOL = 258;$
Total $SS = 3,372;$
$SSROW = 54$

(b)

Variation	df	Sum of Squares	Mean Square	F
Treatments	2	3,042	1,521	169.00
Columns	2	258	129	
Rows	2	54	27	
Error	2	18	9	
Total	8	3,372		

(c) *Reject H_0 and conclude that skill affects performance.*

16-22 (a)

Variation	df	Sum of Squares	Mean Square	F
Treatments	3	25,271.5	8,423.83	.88
Error	12	114,504.5	9,542.04	
Total	15	139,776		

no

(b)

Variation	df	Sum of Squares	Mean Square	F
Treatments	3	25,271.5	8,423.83	4.17
Blocks	3	96,326	32,108.67	
Error	9	18,178.5	2,019.83	
Total	15	139,776		

yes

(c) Their inclusion increases the discrimination of the testing procedure.

17-2 (a) 5/12 (b) 5/12

17-4 no

17-6 (a) no
(b) .183930 for slump
.084225 for peak
(c) no

17-8 .006738

17-10 (a) .1755 (c) .0137
(b) .0067 (d) .7419

17-12 (a) .606531   (b) .393469

17-14 (a) .20   (c) .39
(b) .40   (d) .30

17-16 .40

17-18 (a) .993   (b) .135   (c) .128

17-20 (a) .0002 school per mile
(b) .632   (c) 11,500 miles

17-22 (a) .2524   (c) .9671
(b) .6004

17-24 (a)

Score	f_e
below 40	7.21
40–50	12.28
50–60	19.87
60–70	23.57
70–80	19.19
80–90	11.45
over 90	6.43

$\chi^2 = 17.47$; rejected
(b) between .01 and .001

17-26 (b) $\chi^2 = 2.8$ (c) *Accept H_0.*

17-28 (a) .2880   (b) .3000

17-30 (a) .0596   (b) .8647

17-32 (a) .9763   (b) .9707

17-34 (a) normal   (b) .7698

17-36 yes

18-2 yes

18-4 (a) $W = 25$ (b) accepted

18-6 (a) rejected   (c) rejected
(b) accepted   (d) accepted

18-8 leaded

18-10 (b) *Reject H_0 and conclude that
the new supplement is more
effective.*

18-12 (a) accepted   (c) rejected
(b) accepted   (d) rejected

17-14 $V = 5.5$ rejected

18-16 (b) (1) $R_a = 7$ accepted
(2) $R_a = 8$ rejected
(3) $R_a = 2$ rejected
(4) $R_a = 8$ rejected

18-18 (a) 95   (b) .68   (c) .4964

18-20 *Accept H_0 and conclude that
there is the appearance of
randomness.*

18-22 .61

18-24 $-.50$

18-26 $K = 5.18$
*Accept H_0 and conclude that
identical treatment means exist.*

18-28 accepted

18-30 rejected

18-32 .88

18-34 (a)

x	$F_a(x)$	$F_e(x)$
0	.08	.1353
1	.27	.4060
2	.55	.6767
3	.79	.8571
4	.93	.9473
5	.95	.9834
6	.98	.9954
7	.99	.9988
8	.99	.9997
9	1.00	.9999

(b) $D = .1360$ (d) 10.56
(c) .025   (e) .10
(f) Kolmogorov-Smirnov test

18-36 *Reject H_0 and conclude non-
normality.*

19-2

	Payoff Measure	Best Act	Worst Act
(a)	4	(tie) bonds, preferred	common stock
(b)	3	preferred stock	common stock
(c)	5	common stock	bonds
(d)	7	preferred stock	bonds

19-4 A_3 and A_5

19-6

G & L	S.A.	W & P
$400,000	$440,000	$400,000

spring action

19-8 Test market; then market if that is favorable, and abort otherwise.

19-10 Test market; then market if that is favorable, and abandon otherwise.

19-12 (a)

Demand Event	Probability	Act		
		Gears and Levers	Spring Action	Weights and Pulleys
Light	.10	$ 25,000	−$ 50,000	−$125,000
Moderate	.70	400,000	400,000	400,000
Heavy	.20	600,000	650,000	700,000
Expected Payoff:		$402,500	$405,000	$407,500

(b) weights and pulleys

19-14 Market without benefit of the test.

19-16 Use the juvenile hormone.

20-2 (a) A_2 and A_3 (tie)
(b) A_2 (c) Choose A_2.

20-4 (a)

Event	Act		
	Stock 10	Stock 11	Stock 12
Demand 10	$5.00	$4.50	$4.00
Demand 11	5.00	5.50	5.00
Demand 12	5.00	5.50	6.00

(b) 10   (c) 11

20-6 (a) Dry underripe loads and preserve the rest.
(b)

Ripeness	Strategy							
	S_1	S_2	S_3	S_4	S_5	S_6	S_7	S_8
Underripe	D	D	D	P	D	P	P	P
Ripe	D	D	P	D	P	D	P	P
Overripe	D	P	D	D	P	P	D	P

Choose S_5.

20-8 (a) 27 for A_1 (b) 39
(c) 12   (d) 12   (e) same

20-10 (a) drill   (c) $50,000
(b) $50,000

20-12 (a) $22,000
(b)

Demand Event	Probability	Act		
		Gears and Levers	Spring Action	Weights and Pulleys
Light	.50	$ 0	$85,000	$200,000
Moderate	.30	40,000	0	40,000
Heavy	.20	50,000	10,000	0
Expected Opportunity Loss:		$22,000	$44,500	$112,000

20-14 (a) Market only if test is favorable.
(b) Same as in (a)

21-2 (a) $.25
(b) Answers will vary.

21-4 2.5

21-6 Answers will be subjective.

21-8 (a) willing
(b) might be willing
(c) might be willing
(d) unwilling
(e) willing

21-10 no

21-12 (b) Market nationally without testing.
(c) risk seeking

21-14 (a) 1/2   (b) 3/4 and 1/4
(c) 1/2. Two different dollar amounts have identical utilities.

21-16 (c) Use the test, hiring only those who pass.

22-2 Answer will be subjective.

22-4 Answer will be subjective.

22-6 (a) 110   (c) 130   (e) 230
(b) 260   (d) 360

22-8 (a)

Fractile	Rate of Return
0	−20%
.0625	−4
.125	3
.25	7
.5	15
.75	20
.875	24
.9375	28
1.000	40

(c) .25

22-10

Demand	Probability
100	.26
300	.54
500	.14
700	.05
900	.01

expected demand = 302

22-12 Answers will vary.

22-14 (a) $\mu = 30$ $\sigma = 2.985$
(b) (1) .3669   (3) .3085
    (2) .2590

22-16 (b) (1) .20   (2) .57   (3) .95

23-2 (a) 1/2   (d) 2/3

23-4 (a) .809   (b) .005

23-6 (c) Retain old box if sales decrease; use new box otherwise.

23-8 (a) drill   (b) $72,000
(d) Take seismic; drill only if favorable.

23-10 (a) .467   (b) .086

24-6 (a) Sample with $n = 2$.
(b) (1) $C = 1$ (2) $C = 0$

24-8 (b) $C = 1$. Select photographic unit if $\bar{X} \leq 1$. Select holographic unit if $\bar{X} > 1$.
(c) $1,317.30   (d) $417.30

24-10 $C = 18$. *Accept* H_0 if $R \leq 18$; *reject* H_0 if $R > 18$.

24-12 (c) (1) .127   (2) .078   (3) .047

24-14 (a) leave alone
(b) $300   no

Index

A

Absolute value, 56
Acceptance number, 230, 278, 791. *See also*
    Critical value
Acceptance sampling, 230, 285–87, 784–85
Accounting applications of statistics, 101,
    260, 272, 331, 374, 385–86, 455
Act, 666–67
Act fork, 668
Actual chronology, 769
Actual/forecast ratio, 761
Actual frequency, 499
Addition law, 108–11
  complementary events and, 111–12
  contrasted with multiplication law,
    132–33
  for mutually exclusive events, 109
  for mutually exclusive and collectively
    exhaustive events, 111
Admissible acts, 674
Aggregate consumption function, 386
Aggregate price indexes, 441–47. *See also*
    Price relatives indexes
  Laspeyres index, 443–45, 449, 452
  Paasche index, 444–45, 449

Agricultural applications of statistics, 320,
    342, 356, 389, 456, 477, 485–86,
    492–93, 495–96, 526–28, 544–46,
    583–84, 585, 588, 617–19, 686, 697,
    698–700, 704
Alternative hypothesis, 277. *See also*
    Hypotheses
Analysis of variance, 525–67, 652–56
  completely randomized design for, 552,
    557
  confidence intervals with, 541–43
  design considerations in, fixed effects, 557
    random effects, 557
    randomization, 558
    replication, 557–58
  factorial design for, 557
  interactions and, 557
  Latin square design for, 558–62
  multiple comparisons and, 543–44
  and multiple regression, 565–67
  nonparametric procedure. *See*
    Kruskal-Wallis test
  one-factor, 526–38, 652–55
  randomized block design for, 545–51
  test statistic for, 534
  three-factor, 556–62

Analysis of variance (*cont.*)
  two-factor, 545–56
ANOVA table, 533
Arithmetic mean. *See* Mean
Arithmetic scale, 420
Association, degree of. *See* Correlation
    analysis
Attitude, accounting for. *See* Utility
Attribute, 11. *See also* Qualitative
    population
Average, arithmetic, 41–42
  moving, 422
  weighted, 43
  See also Mean

B

Backward induction, 682–84
Base period, for index numbers, 441
  for time series analysis, 411
Bayes, Thomas, 146
Bayes decision rule, 675–76, 694–95
  opportunity loss and, 706–707
  utility and, 694–95, 729
Bayes' theorem, 146–48
Bayesian analysis. *See* Decision making
Bernoulli, Daniel, 716
Bernoulli, Jacob, 162n
Betting odds, 745
Between-groups variation. *See* Treatments
    variation
Bias, 5. *See also* Bias of nonresponse;
    Induced bias; Prestige bias; Sampling
    bias; Unbiased estimator
Bias of nonresponse, 79
Bimodal distribution, 49–50
Binomial distribution 170–86, 224
  applicability of, 598
  in Bayesian decision making, 785–91
  compared to hypergeometric distribution,
    576
  family of, 178–81
  mean of, 180
  normal approximation to, 224–32
  normal distribution and, 179
  for proportion, 177–78
  shape of, 178–81
  table of probabilities, 822–32
  tables, using, 183–85
  as used with sign test, 628
  variance of, 180
Binomial formula, 175–76
Birthday problem. *See* Matching birthdays
    problem

Block, 545
Blocking factor, 545
Blocks mean square, 549
Blocks sum of squares, 547, 560
Blocks variation, 545
Branching point. *See* Act fork; Event fork
Buffa, Elwood S., 267n
Bureau of Labor Statistics, U.S., 440

C

Carli, G.R., 440
Causal relation, 350, 354
Census, 71
  limitations on conducting, 73
  versus sampling, 71–74
Census of Retail Trade, 452
Central limit theorem, 219–23
Central tendency, measures of, 39–40,
    41–44. *See also* Mean; Median;
    Mode
  limitations of, 53–54
Certain event, 100
Chances. *See* Probability
Chance variation. *See* Unexplained
    variation
Chebyshev's Theorem, 62
Chi-square distribution, 503–505
  as approximation for Kruskal-Wallis test,
    654
  limitations on use of, 507
  normal approximation to, 522–23
  table of values, 838–39
Chi-square statistic, 501–503
  degree of freedom for, 505, 517, 602, 605
Chi-square test, compared to
    Kolmogorov-Smirnov test, 661–62
  for equality of proportions, 510–12
  for goodness of fit, 597–609
  for independence, 496–507
  for value of variance, 516–23
Chronology of probability tree diagram, 769
Class frequency, 12
Class interval, 12
  midpoint of, 16
  number of, 18–20
  width of, 12, 18–19
Classical statistics, decision theory and,
    806–10
  deficiencies of, 809–10
  nonparametric statistics and, 616–17
Cluster sample, 83
Coale, Ansley J., 74n
Coefficient of determination, 344–50

interpretation of, 345–46
in multiple regression. *See* Coefficient of
multiple determination
partial. *See* Coefficient of partial
determination
for population, 346
for sample, 346–47
Coefficient of multiple determination,
390–91
Coefficient of partial determination, 391–94
Collectively exhaustive events, 105
Combinations, 141–43
number of combinations, 141
Complementary events, 111–12
Component events, 102
Composite events, 97
Computers,
use of in analysis of variance, 532
use of in regression analysis, 375–80
Conditional mean, 333
Conditional probability, 116–20
computed from joint probability table,
118–19
life insurance and, 117
posterior probability and, 148–49
Conditional result probability, 768
Confidence, 246
distinguished from reliability, 261
Confidence interval, 244–49. *See also*
Prediction interval
Confidence interval estimate, with analysis
of variance, 541–44
desired features of, 248–49
for difference in means, 462–70, 478–81,
541–44
and hypothesis testing, 295
of mean, 244–57
of proportion, 257–59
of slope of regression line, 357–58
of variance, 518–19
Confidence level, 246
Consistent estimator, 240
Consumer applications of statistics, 242,
283, 524
Consumer Price Index, 84, 422, 440, 445,
450–53
Consumer's risk, 231. *See also* Type I error
Contingency table, 499
Continuity, utility theory assumption of,
720–21
Continuity correction, 228
Continuous probability distribution, 188–93
Continuous random variable, 188–93
Control group, 92
Convenience sample, 83–84

Correlation, 312, 319
causality and, 350, 354
explained without regression, 349
multiple, 390–94
partial, 391–94
perfect, 319
rank, 649–51
serial, 646
spurious, 350
Correlation analysis, 312, 343–52, 390–94
See also Multiple correlation analysis
limitations of, 352–54
Correlation coefficient, 346–49
multiple, 391
partial, 392
sign of, 347
Counting techniques, 136–43
Critical normal deviate, 286
Critical value, 278
Cross-sectional study, 381
Cryptanalysis, 21–24
Cumulative frequency, 26, 29–31
Cumulative frequency distribution, 29–31
Cumulative probability, 181–83, 206–207
Cumulative probability distribution, 181–83,
193, 581–82
Curvilinear regression, 354–55
Curvilinear relationship, 320
Cyclical time series movement, 403–407, 432

D

Data, raw, 11–12
Data processing applications of statistics,
43, 74, 272, 556, 653, 748–49, 793–94
Decision to buy insurance, 714–16, 733–34
Decision criteria, 690–95, 706–707
Bayes decision rule, 675–76, 694–96,
706–707
insufficient reason, 694
maximin payoff, 690–92
maximum likelihood, 692–94
Decision-making, Bayesian, 765–813
classical. *See* Hypothesis testing
with decision tree diagram, 676–85
objectives and, 669–71
with sample information, 783–813
theory of, 665–712
with utility function, 730–37
Decision-making using experimental
information, 772–81, 783–813
contrasted to classical hypothesis testing,
806–10
decision to experiment, 776–813

Decision-making using experimental
  information (*cont.*)
  role of EVPI in, 778
Decision point. *See* Act fork
Decision rule, 230, 278
  in Bayesian decision-making, 791. *See
    also* Strategy
  selection of, 281
Decision to sample. *See* Preposterior
  analysis
Decision table, 667. *See also* Payoff table
Decision theory, 665–712
  classical statistics and, 806–10
Decision tree analysis, 676–85
  backward induction in, 682–84
Decision tree diagram, 677–79
Decisions under risk, 666n
Deductive statistics, 4
Degrees of freedom, 251
  number of, in analysis of variance, 534,
    547, 561
  for contingency table, 505
  in goodness of fit test, 605
Dependent variable, 314
Descriptive analysis, 17
Descriptive statistics, 3
Deseasonalized data, 437
Design of experiments. *See* Experimental
  designs
Directly related variables, 317
Discrete probability distributions, 156–57
Discrete random variables, 153–57
Dispersion. *See* Variability
Distribution-free statistics. *See*
  Nonparametric statistics
Distributions. *See* Probability distribution
  and also under individual listings
Dominated acts. *See* Inadmissible acts
Draft lottery, 85, 642–46
Dummy variable, 156–57, 383
  techniques in multiple regression, 381–90

E

Economics, applications of statistics to, 323,
    332, 351, 356, 373, 380, 386–88,
    388–89, 395, 400, 409–10, 555. *See
    also* Consumer Price Index; Gross
    National Product
Education, applications of statistics to, 25,
    50–51, 63, 68, 92, 113–15, 122,
    165–68, 208, 233, 282, 373–74, 396,
    463–69, 470, 539, 556, 564–65,

599–600, 629–31, 636, 640, 650, 653
  See also Personnel management
    applications of statistics
Efficiency, of estimator, 239–40
  of test. *See* Power
Elementary event, 96
Elementary unit, 8–9
Environmental and safety applications of
    statistics, 93–94, 124–25, 135, 259,
    472–74, 477, 585, 614, 688
Equally likely events, 99
Error. *See also* Nonsampling error;
    Sampling error
  in estimation, 259–60
  in testing, Type I, 278–79
    Type II, 278–79
  tolerable, 261
Error mean square, 533, 549
Error probabilities, 279. *See also*
    Significance level; Type I error; Type
    II error
Error sum of squares, 530, 548, 561
Estimation, 237–73
  interval estimate, 234. *See also*
    Confidence interval estimate
  point estimate, 243
Estimator, 243
  criteria for choosing, 239–42
    consistency, 240
    efficiency, 239–40
    unbiasedness, 239
  of population mean, 240–41
  of population proportion, 242
  of population variance, 241
Event, 99, 667
  certain, 100
  component, 102
  composite, 97
  in decision structure, 667
  elementary, 96
  impossible, 100
  joint, 103, 116–17
Event chronology, 769
Event fork, 119, 668
Event relationships, 102–103
  collective exhaustiveness, 105
  complementarity, 111–12
  equal likelihood, 98–99
  independence, 105–106, 112–13
  intersection of, 103
  mutual exclusiveness, 109–10
  union of, 103
Event space. *See* Sample space
Expected frequency, 500

Expected net gain of sampling (ENGS), 802
Expected opportunity loss, 706–707
Expected payoff, 675–76
Expected payoff with perfect information, 707–708
Expected utility, 723–25
Expected value, 158–62
  of continuous random variable, 183
  of a function of a random variable, 159–60
  meaning of, 154
  properties of, 154–55
  of sample mean, 164–65
  of sample proportion, 185–86
  of sample variance, 241
  sampling and, 157–58
Expected value of perfect information (EVPI), 707–10
  role in decision-making, 778, 800
Expected value of sample information (EVSI), 801
Experiment, decision to, 776–78, 798–803
Experimental designs, 556–58
  factorial design, 557
  fixed effects with, 557
  random effects with, 557
  randomization with, 558
  replication in, 557–58
  types of, completely randomized, 552, 557
    Latin square, 558–62
    randomized block, 545–51
Experimental group, 92
Experimental information, 765–66
  decision-making analysis with,
    pre-posterior, 776–78, 798–803
    posterior, 772–75, 784–97
Explained variation, in analysis of variance, 534
  in regression analysis, 345, 390, 566
Exponential distribution, 35, 569, 589–97
  curves for, 591
  mean of, 590
  variance of, 590
Exponential function, table of values, 842
Exponential growth curve, 410
Exponential smoothing, 432–36
Extensive form analysis, 703
Extrapolation, 340
  in time series, 408

F

F distribution, 535–37
  applicability of, 538
  degrees of freedom for, 534, 547, 561
  density curves for, 535
  table of values for, 840
F statistic, in analysis of variance, 534, 550, 561
F test. *See* Analysis of variance
Factor, 528
Factorial, 138–39
Factorial design, 557
  complete, 552, 557
  fractional, 557
  incomplete, 557
  Latin square, 558–62
Financial applications of statistics, 26–27, 32, 118, 162–63, 247–48, 260, 321, 342, 355, 431, 446, 450, 625, 672, 710
Finite population correction factor, 234
Fixed-effects experiment, 557
Forecasting. *See* Time series analysis
Fractile, 54–55
Fractional design, 557
Frame, 9
Frequency, 2, 11–12
  actual, 499
  cumulative, 26–30
  cumulative relative, 33
  expected, 500
  relative, 26–29
Frequency analysis, 22–23
Frequency curve, 15–17, 32–36
  bell-shaped, 33
  bimodal, 49–51
  shape of and the central limit theorem, 217–18
  skewed, 33, 48–49
  symmetrical, 33, 48
Frequency distribution, 11–24
  common forms of, 32–35
    bimodal, 49–50
    exponential, 35
    skewed, 33, 48–49
    symmetrical, 33, 48
    uniform, 35
  construction of, 11–12, 20–24
  cumulative, 29–30
  limitations of, 39
  for qualitative populations, 21–22
  relative, 26–29
  smoothed-curve approximation to, 189–190
Frequency polygon, 15–16
Function, probability density, 190–92
  of random variable, 160

G

Galton, Sir Francis, 312
Gambles, equivalent, 722
    hypothetical, 751–56
    reference lottery, 724
    substitutability of, 721–22
    utility and, 723–24
Gompertz curve, 410
Goodness-of-fit test, 570, 597–612, 656–62
Gosset, W.S., 251
Grand mean, 529, 547
Graphs, frequency polygon, 15–16
    histogram, 13–15
    ogive, 30
Gross national product (GNP), U.S.,
        399–400, 408–409, 453–54
    GNP deflator, 453
Grouped approximation, accuracy of, 43
    for mean, 42–44
    for standard deviation, 61
    for variance, 59

H

Hawthorne studies, 267n
Health applications of statistics. *See*
        Scientific and health applications of
        statistics
Histogram, 13–15
Human engineering applications of
        statistics, 206–207
Hypergeometric distribution, 569–76, 641
    approximated by normal distribution, 641
    compared to binomial distribution, 575–76
    mean of, 575
    variance of, 575
Hyperplane. *See* Regression plane
Hypotheses, 276–77
    formulation of, 281–82
    two-sided, 282
Hypothesis testing, 275–309. *See also*
        Hypothesis tests
    basic concepts of, 275–84
    choosing the test, 304–307
    contrasted with Bayesian
        decision-making, 806–10
    decision rules in. *See* Decision rule
    efficiency and, 307. *See also* Power
    factors influencing decision, 281–82
    inadequacies and limitations of, 307–308
    lower- and upper-tailed procedures in, 289
    two-sided procedures in, 293–96

    using confidence intervals, 295
    using *t* statistic, 296–301
Hypothesis tests, for comparing two means,
        472–478, 481–84, 543–44
    for comparing two proportions, 486–90
    for equality of several means. *See*
        Analysis of variance
    for equality of several proportions,
        510–14
    for goodness of fit, 597–612, 656–63
    for independence, 496–507
    nonparametric. *See* Nonparametric
        statistics
    for population mean, 275–301
        using small samples, 296–301
    for population proportion, 301–304
    for population variance, 520–23
    for randomness, 639–49
    for slope of regression line, 358–59
Hypothesis tests using two samples, 472–78,
        481–84, 486–90. *See also* Two-
        sample inferences for means, for
        proportions
    with independent samples, 472–74,
        481–83, 486–90, 617–25
    with matched pairs, 474–78, 483–84,
        626–39

I

Impossible event, 100
Inadmissible act, 674
Increasing preference, 723
Independence, 105, 120–21
    multiplication law and, 123
    between qualitative variables, 495–98
    random sampling and, 127
    testing for, 496–509
Independent random sample, 168
Independent sample observations, 168
Independent sampling, 92, 462–65
    compared to matched-pairs sampling, 470
Independent variables in regression, 314,
        361
    predictive worth of, 368–71
Index numbers, 439–60. *See also* Aggregate
        price indexes; Consumer Price Index;
        Price relatives indexes; Quantity
        indexes
    shifting the base of, 457–58
    splicing of, 458–59
Induced bias, 81
Inferential statistics, 4. *See also* Hypothesis

testing; Confidence interval estimate;
    Correlation analysis; Regression
    analysis
Information from experiments. *See*
    Experimental information
Informational chronology, 769
Insufficient reason, criterion of, 694
Insurance applications of statistics, 272,
    596, 655, 714–16, 718, 739
Interfractile range, 55
Interquartile range, 56
Intersection, 103
Interval estimate. *See* Confidence interval
    estimate
Inventory control applications of statistics,
    697–98, 740
Inverse relationship between variables, 319
Irregular time series variation, 401, 405, 432

J

Joint event, 103, 116–17
Joint probability, 116–19
    multiplication law and, 123–25
Joint probability table, 116–18
Judgment sample, 84–85
Judgmental probability. *See* Subjective
    probability

K

Kahn, David, 22n
Kolmogorov, A., 657
Kolmogorov-Smirnov one-sample test,
    656–63
    contrasted with chi-square test, 656–57,
    661–62
    table of critical values for, 848–50
Kruskal, W.H., 652
Kruskal-Wallis test, 652–56
    as alternative to F test, 652
    chi-square approximation with, 654

L

Latin square, 559
Latin square design, 558–62
Laspeyres index, 443–45, 447–49, 452, 456
Law of large numbers, 106
Least squares method, 323–32

desirable features of, 334–35
for determining trend of time series,
    410–14
in multiple regression, 362–75
rationale for, 324–25
Legal applications of statistics, 133, 149,
    194, 606–609
Length-of-runs test, 646
Likelihood. *See* Maximum likelihood
    estimator; Probability
Line, equation for, 315–16
Linear regression. *See* Least squares
    method; Regression analysis
Literary Digest, 79
Location, measures of. *See* Central
    tendency, measures of
Logarithmic curve, 322
Logarithmic trend line, 417
Logarithms, table of, 836–37
Logistic curve, 410
Lot. *See* Population
Lot proportion defective, 785
Lower class limit, 12
Lower-tailed test, 289

M

Management applications of statistics, 249,
    283, 342, 374, 381, 471, 485, 486,
    492–93, 596, 612, 624, 633–34, 687,
    704, 720, 729, 739, 745–46, 775–76
Mann-Whitney U test, 625–26
Marginal probability, 117–18
Marginal propensity to consume, 386
Marginal utility for money, 716
Marketing applications of statistics, 91, 365,
    377, 515, 677–79, 686–87, 711, 764
    in advertising and pricing, 40–41, 83, 88,
    138, 488–90, 508–509, 510, 675–76,
    797
    in distribution, 83, 313, 331, 498–507, 624
    in forecasting sales or demand, 395,
    401–405, 407–408, 414–21, 422–30,
    432–36, 751–54, 764
    in marketing research, 102, 114–15, 122,
    150–51, 187, 258, 280, 492, 508–509,
    515–16, 626–30, 791–92, 803, 810–11
    in product decisions, 276, 539–40, 575–76,
    633–34, 775–76, 779
Matched-pairs sampling, 93, 465–70
    compared to independent sampling, 470
Matching birthdays problem, 125–26
Maximin payoff criterion, 690–92

Maximum expected payoff criterion. *See* Bayes decision rule
Maximum frequency deviation, 660
Maximum likelihood criterion, 692–94
Maximum likelihood estimator, 335n
Mean, 41–44. *See also* Sample mean; Population mean
  calculation of, 42–43
    from grouped data, 43–44
  compared to median, 46
  estimation of, 240
  expected value of, 167
  normal distribution and, 60–61
  properties of, 46, 48–49
Mean deviation, 56
Mean squares, 532–33, 549, 561, 566–67
Mean time between failures (MTBF), 292, 593
Median, 45–46
  compared to mean, 46
  properties and difficulties of, 46–49
  used to calculate seasonal indexes, 427
Midpoint of class interval, 16
Minimax criterion. *See* Maximin payoff criterion
Modal class, 47
Mode, 45–49
  found from frequency curve, 47
Monte Carlo simulation, 570
Morgenstern, Oskar, 717, 719n
Mortality table, 117
Moving averages, 422
Mudgett, Bruce D., 440n
Multiple comparisons, 543–44
Multiple correlation analysis, 390–94
Multiple correlation coefficient, 391
Multiple regression analysis, 361–90
  and analysis of variance, 565–67
  dummy variable techniques in, 381–90
  stepwise, 379
  using computer, 375–81
Multiplication, principle of, 137–38
Multiplication law, 123–25
  contrasted with addition law, 132–33
  for independent events, 112
  misapplication of, 132–33
Multivariate analysis. *See* Analysis of variance; Multiple regression
Mutually exclusive events, 104–105

N

Nonparametric statistics, 615–64
  advantages of, 616–17

  disadvantages of, 617
  Kolmogorov-Smirnov test, 656–63
  Kruskal-Wallis test, 652–55
  Mann-Whitney U test, 625–26
  number-of-runs test, 639–49
  sign test, 626–34
  Spearman rank correlation coefficient, 649–52
  Wilcoxon rank-sum test, 617–26
  Wilcoxon signed-rank test, 635–39
Nonsampling error, 80–81
Normal approximation, to binomial distribution, 224–32
  to chi-square distribution, 522–23
  to hypergeometric distribution, 641
  with number-of-runs test, 641
  with sign test, 628
  with Wilcoxon rank-sum test, 619
  with Wilcoxon signed-rank test, 635
Normal curve, 33, 61, 200–210
  area under, 202–10
  table of areas, 833
Normal deviate, 201
Normal distribution, 195–231
  binomial distribution and, 168
  characteristics of, 195–97
  compared to Student t, 253
  implications and pitfalls in interpretation of, 209
  importance of, 195, 218
  subjective, 747–50
  table of areas, 833
Normal equations, 325, 365, 411
  for curvilinear regression, 354
  for logarithmic trend line, 417
Normal form analysis, 703
Null hypothesis, 277. *See also* Hypotheses
Number-of-runs test, 639–49
  applied to 1970 draft lottery, 642–46
  normal approximation with, 641

O

Objectives in decision-making, 669–71
Observation, 7
Odds. *See* Probability
Ogive, 31
Operating characteristic curve. *See* Power curve.
Opportunity loss, 705
Ordinal scale, 616
Outcome, in decision structure, 666–67

P

Paasche index, 444–45, 449
Parameters. *See* Population parameters
Parametric statistics, 615
Partial correlation, 391–94
Partial correlation coefficient, 392
Partial regression coefficients, 363
Payoff, 669–70
   in decision tree diagrams, 679–81
   objectives and, 669–71
   utilities as measures of, 716
Payoff table, 669
Percentile, 56, 206–207
Permutations, 140–41
   number of, 140
Personal probability. *See* Subjective
      probability
Personnel management applications of
      statistics, 24–25, 31–32, 44–45, 66,
      130, 144, 151, 250, 256–57, 272, 283,
      292, 309, 332, 351–52, 486, 544,
      558–62, 563–64, 576, 609–10, 620–23,
      781
Point estimate, 243
Poisson, Siméon, 578
Poisson distribution, 569–70, 577–85
   applications of, not involving time,
      582–84
   as approximated by binomial, 580
   limitations of, 580–81
   mean of, 579
   table of values for, 843–47
   using tables with, 581–82
   variance of, 579
Poisson process, 576–80
   assumptions of, 580
Political forecasting applications of
      statistics, 73, 79, 89, 229, 258, 309,
      490, 513, 792
Population, 3, 7–11
   description of, 8
   elementary units of, 9
   nonhomogeneous, 50–51
   qualitative, 10–11
   quantitative, 10–11
   joint, 117–19. *See also* Joint probability
   laws, of addition, 108–11
      of multiplication, 123–27
   marginal, 117
   posterior, 148–49. *See also* Bayes'
      theorem
   prior, 147. *See also* Bayes' theorem
   subjective, 100. *See also* Subjective
      probability

Probability density function, 190
Probability distribution, 153–56
   continuous, 188–92
   cumulative, 181–83
   discrete, 158–59
   judgmental assessment of, 747–59
   mean of, 163–66,
   of sample mean. *See* Sampling
      distribution of mean
   of sample proportion. *See* Sampling
      distribution of proportion
   subjective, 747–59
   variance of, 158–160
   See also under individual listings:
      Binomial; Chi-square; Exponential;
      F; Hypergeometric; Normal;
      Poisson; Student *t;* Uniform
Probability sample. *See* Random sample
Probability tree diagram, 128–29
Producer's risk, 231. *See also* Type II error
Product development applications of
      statistics, 66, 144, 187, 276, 332, 338,
      539–40, 575–76, 671–72, 710, 711.
      See also Marketing applications of
      statistics, in product decisions

Production management applications of
      statistics, 52, 62, 64, 130–31, 144,
      201–205, 216, 254–55, 266–69, 296,
      301, 308–309, 323, 374, 388, 392, 513,
      526–28, 539, 544–46, 588, 610–11,
      750–51, 764

Proportion, 40, 64–65. *See also* Sample
      proportion
   estimation of, 242
Pseudorandom numbers, 646

Q

Qualitative population, 10–11
Quality control applications of statistics,
      127–30, 131, 233, 250, 259, 270,
      285–86, 293–95, 296, 301, 304, 309,
      514, 577, 606–609, 698–703, 784–91,
      806
Quantitative population, 10–11
Quantity indexes, 455–56
Quartiles, 56
Questionnaire design, 81–82
Queuing applications of statistics, 33–34,
      111, 162, 194, 223, 520–22, 577, 579,
      584, 589–92, 595–96, 662
Quota sampling, 452

R

Raiffa, Howard, 751n
Random-effects experiment, 557
Random numbers, 86–87, 611, 646–47
    computer generated, 646–47
    table of, 821
Random sample, 4, 85–87
    cluster, 89–90
    independent, 168
    simple, 87
    statistical independence and, 127–30
    stratified, 88–89
    systematic, 88–90
    testing for randomness of, 639–49
Random variable, 154–56
    continuous, 188–92
    discrete, 156–57
    expected value of, 158–61
    function of, 155
    standard deviation of, 157
Randomized design, complete, 522
    for blocks, 545–51
    fractional, 557
Randomness, need for, 639
    testing for, 639–49
Range, 39, 56, 243
    interquartile, 56
Rank correlation, 649–52
Rank sum, 619
Reciprocals, table of, 820
Reference lottery, 724
Regression analysis, 311–42. See also
        Multiple regression analysis
    assumptions of, 332–35
    curvilinear, 354–58
    with logarithmic trend line, 414–20
    multiple, 361–90
Regression coefficients, estimated, 316, 326,
        363
    interpretation of, 327–28, 367
    partial, 363
    true, 334
Regression curves, 320–22
Regression equation, 316, 362
Regression line, estimated, 316
    estimation of, 334–35
    meaning and use of, 327–28
    true, 334
Regression plane, 362
Regression toward mean, 312
Regret. See Opportunity loss
Relative frequency, 26–29
Relative frequency distribution, 27–29
Reliability, of sample results, 259–62
    distinguished from confidence, 269
    and sample size, 265
Reliability applications of statistics, 45, 63,
        210, 270, 292, 322, 491, 578, 593–95,
        597, 613
Replication, in analysis of variance, 557–58
Residual variation. See Unexplained
        variation
Revised probability. See Posterior
        probabilities
Robustness, 299
Rothwell, Doris P., 452n

S

Saint Petersburg Paradox, 717
Sample, 3, 7, 71, 75. See also Random
        sample
    cluster sample, 90
    convenience sample, 84
    as distinguished from population, 8
    judgment sample, 84–85
    stratified, 88–89
    systematic random, 88–90
Sample mean, 43. See also Mean
    as consistent estimator of population
        mean, 240
    as efficient estimator of population mean,
        239–40
    grouped approximation for, 42–44
    sampling distribution of, 163–70, 211–24
    as unbiased estimator of population mean,
        239
Sample median, 45–46. See also Median
Sample proportion, 64. See also Proportion
    as consistent estimator of population
        proportion, 242
    sampling distribution of, 185–86, 224–32
    as unbiased estimator of population
        proportion, 242
Sample size, 259–71
    determination of, 263–66
    to estimate μ by $\overline{X}$, 263
    to estimate π by P, 265–66
    influences upon, 264–65
    in matched-pairs sampling, 470
    optimal, 798–800
    using a computer to find, 800
Sample selection, 83–90
Sample space, 97
Sample standard deviation, 58. See also
        Standard deviation
Sampling. See also Statistical sampling
        study

accuracy and, 74
expected value and, 161–62
independent observations and, 168
random. *See* Random sample
reasons for, 71–74
   destructive observations, 73
   economic advantages of, 72
   inaccessible population, 73
   large population, 73
   timeliness, 72
with replacement, 129–30, 167–68, 234
without replacement, 129–30, 167–68, 234
Sampling bias, 79–80
control of, 85
Sampling distribution, 163–70
Sampling distribution of mean, effect of
     sample size upon, 217–18
mean of, 166–67
for normal population, 211–17
when population is not normal, 217–24
for small populations, 234–36
for small samples. *See* Student *t*
   distribution
standard deviation of, 169. *See also*
   Standard error of mean
Sampling distribution of proportion, 185–86,
   224–32. *See also* Binomial
   distribution; Hypergeometric
   distribution
mean of, 186
normal aproximation to, 224–32
standard deviation of, 186. *See also*
   Standard error of proportion
Sampling error, 78
Scales, arithmetic, 420
ordinal, 616
semilogarithmic, 420
Scatter diagram, 320–22
Schlaifer, Robert, 800n
Schmidt, J.W., 647n
Scientific and health applications of
     statistics, 78, 91, 93, 210, 216, 222,
     232, 250, 256, 276–77, 289, 292, 297,
     301–302, 309, 380, 471, 474–77, 479,
     481–83, 485–86, 492–93, 514, 519,
     523, 524, 551–54, 563, 624, 633, 652,
     656
Seasonal index, 403–407
calculation of, 422–32
forecasting with, 429–30, 436
Seasonal time series fluctuation, *See*
   Seasonal index
Secular trend. *See* Trend
Semilogarithmic scale, 420

Sequences, number of. *See* Factorial;
   Permutations
Set. *See* Event set
Shifting base of index numbers, 457–58
Sign test, 626–34
compared to Student *t* test, 631–32
compared to Wilcoxon signed-rank test,
   638
normal approximation with, 628
Significance, 278
Significance level, 279. *See also* Error
   probabilities
Significance test. *See* Hypothesis testing
Skew, 33–34, 48–49
Slope of line, 315
Slope of regression line, estimation of,
   357–58
testing for value of, 348–49
Smirnov, N.V., 657
Smoothing constant, 433
Spearman rank correlation coefficient,
   649–52
Splicing index numbers, 458–59
Spurious correlation, 362
Squares and square roots, table of, 820
Standard deviation, 58–62
calculated from grouped data, 60
estimation of, 518–19
meaning of, 61–62
of population. *See* Population standard
   deviation
of a random variable, 160
of sample mean, 169
of sample proportion, 186
Standard error of estimate, 329–30
in multiple regression, 370–71
Standard error of the mean, 169, 211, 213–16
distinguished from *s*, 216
role of, 213–16
for small populations, 234
Standard error of proportion, 186, 225
Standard normal curve, 201
Standard normal distribution, 207–209
Standard normal random variable, 207–209
Standard score, 208
Statistic, 40
Statistical Abstract of the United States, 3
Statistical applications. *See separate*
   listings under Accounting;
   Agricultural; Consumer; Data
   processing; Economics; Education;
   Environmental and safety; Financial;
   Human engineering; Insurance;
   Inventory control; Legal;
   Management; Marketing; Personnel

Statistical applications (*cont.*)
    management; Political forecasting;
    Product development; Production
    management; Quality control;
    Queuing; Reliability; Scientific and
    health; Work measurement
Statistical decisions. *See* Decision-making;
    Hypothesis testing
Statistical decision theory. *See* Decision
    theory; Decision-making
Statistical error, 5. *See also* Error
Statistical estimation. *See* Estimation
Statistical independence. *See* Independence
Statistical inference, 4. *See also* Correlation
    analysis; Estimation; Hypothesis
    testing; Regression analysis
Statistical quality control. *See* Quality
    control applications of statistics
Statistical sampling study, 71, 75–78
    procedures of, 91–94
    stages of, 75–78
        data analysis and conclusions, 77–78
        data collection, 77
        planning, 75–77
Statistical significance, 278
Statistical test. *See* Hypothesis testing
Statistics, deductive, 4
    defined, 2
    descriptive, 3
    inductive, 5
    inferential, 3–4. *See also* Statistical
        inference
    vital, 2–3
Stephen, Frederick F., 74n
Stepwise multiple regression, 379
Strata, 88
Strategy, 698. *See also* Decision rule
    obviously non-optimal, 774–75
Stratified sample, 88–89
Student *t* distribution, 251–53
    applicability of, 299
    compared to normal, 252
    density curve for, 251
    table of values, 835
Student *t* test, compared to sign test, 631–32
    compared to Wilcoxon rank-sum test,
        623–24
    used in regression analysis, 357–59
Subjective probability, 100, 743–47
    actual/forecast ratios and, 761–63
    betting odds, 745
    determination of, 743–47
    interview method for finding, 751–54
    with normal curve, 747–51
Subjective probability distribution, 751–63

Subscripts, 41
Substitutability of gambles, in finding
        subjective probabilities, 745–46
    in utility theory, 721–22
Sum of squares, 530
    blocks, 547, 560
    error, 530, 548, 561
    total, 531
    treatments, 530, 560
Summation sign, 42
Systematic random sample, 88

T

t distribution. *See* Student *t* distribution
t statistic. *See* Student *t* statistic
Target population, 9
Taylor, R.E., 647n
Test. *See* Hypothesis tests
Test statistic, 277
Thorp, Edward, 159n
Time-and-motion study. *See* Work sampling
Time series, 381, 399–437
    classical model for, 401–407
    exponential smoothing with, 432–36
    deflating using index numbers, 453–55
    relevancy of past data of, in regression
        analysis, 353
    using dummy variable techniques with,
        386–88
Time-series analysis, 399–437
Time-series components, 399–401
    cyclical movement, 401–407, 432
    irregular variation, 401, 405–407, 408, 437
    seasonal fluctuation, 401, 403–405,
        422–32
    secular trend, 401–402, 408–422
Tolerable-error level, 261
Total sum of squares, 531
Total variation, 566
Transitivity of preference, 720
Treatments, 526
Treatments mean square, 532
Treatments sum of squares, 530, 560
Tree diagram. *See* Decision tree diagram;
    Probability tree diagram
Trend, 401–402, 408–22
    analysis of, 408–22
    exponential, 414–21
    modifying for shorter periods, 413–14
    nonlinear, 414–21
Trend curve. *See* Trend line
Trend line, 401
    logarithmic, 417
    projecting with, 411–12

Two-sample inferences for means, 462–86.
    See also Confidence interval
    estimate; Hypothesis tests using two
    samples
  using independent samples, 88, 462–65,
    472–74, 478–80, 481–83
  using large samples, 462–78
  using matched pairs, 93, 465–69, 474–77,
    480–81, 483–84
  using small samples, 478–86
Two-sample inferences for proportions,
    486–90
Two-sided hypotheses, 282
Type I error, 278
  probability of. *See* Error probabilities
Type II error, 278
  in goodness-of-fit test, 599
  in Kolmogorov-Smirnov test, 661
  probability of. *See* Error probabilities

U

Unbiased estimator, 239
Unconditional probability, 121
Unconditional result probability, 769
Unexplained variation, in analysis of
    variance, 531
  in regression, 345, 392, 566
Uniform distribution, 34, 569, 585–89
  mean of, 588
  variance of, 588
Union of events, 103
Universe. *See* Population
Upper class limit, 12
Upper-tailed test, 289
Utility, 695–96, 713–41
  and Bayes decision rule, 695–96, 729
  expected, 723–27
  gambles and, 723–24
  obtaining values for, 726–27
  scales for, 724
Utility curve, 716, 730–31, 734–38
  with aspiration levels, 737
  for risk averter, 735
  for risk-neutral individual, 735
  for risk seeker, 735
  shapes of, 734–38
Utility for money, 730–38
Utility function, 716, 730–32
Utlity theory,
  assumptions of, 719–23
    continuity, 720–21
    increasing preference, 723
    preference ranking, 719
    substitutability, 721–22

transitivity of preference, 720
U.S. Bureau of Labor Statistics, 440

V

Value weights for index numbers, 448
Variability, 53–64
  importance of, 53, 54
  measures of, 54–59
    average measures of, 54–56
    distance measures of, 56
Variable, dependent, 314
  dummy, 156–57, 383
  independent, 314, 361
  random, 154–56
Variance, 56–58. *See also* Standard
    deviation
  calculated from grouped data, 60
  inferences regarding, 516–24
  of population. *See* Population variance
Variate, 11
Variation, error, 530
  explained. *See* Explained variation
  total. *See* Total variation
  treatments, 530
  unexplained. *See* Unexplained variation
Von Neumann, John, 717, 719n

W

Waiting line applications. *See* Queuing
    applications of statistics
Wallis, W.A., 652
Wilcoxon, Frank, 617
Wilcoxon rank-rum test, 617–25
  advantages of, 623–24
  compared to Student *t* test, 623–24
  extended to several sample case. *See*
    Kruskal-Wallis test
  normal approximation for, 619
Wilcoxon signed-rank test, 635–39
  compared to sign test, 638
  normal approximation with, 635
Within-groups variation. *See* Unexplained
    variation
Work measurement, 74, 266–69
Work sampling, 74, 266–69
Working population, 10

Y

Y intercept of line, 315